ELECTRODYNAMICS OF CONTINUOUS MEDIA

by

L. D. LANDAU and E. M. LIFSHITZ

Institute of Physical Problems, USSR Academy of Sciences

Volume 8 *of Course of Theoretical Physics*

Translated from the Russian by

J. B. SYKES, J. S. BELL and M. J. KEARSLEY

Second Edition revised and enlarged

by E. M. LIFSHITZ and L. P. PITAEVSKII

PERGAMON PRESS

OXFORD · NEW YORK · TORONTO · SYDNEY · PARIS · FRANKFURT

U.K.	Pergamon Press Ltd., Headington Hill Hall, Oxford OX3 0BW, England
U.S.A.	Pergamon Press Inc., Maxwell House, Fairview Park, Elmsford, New York 10523, U.S.A.
CANADA	Pergamon Press Canada Ltd., Suite 104, 150 Consumers Road, Willowdale, Ontario M2J 1P9, Canada
AUSTRALIA	Pergamon Press (Aust.) Pty. Ltd., P.O. Box 544, Potts Point, N.S.W. 2011, Australia
FRANCE	Pergamon Press SARL, 24 rue des Ecoles, 75240 Paris, Cedex 05, France
FEDERAL REPUBLIC OF GERMANY	Pergamon Press GmbH, Hammerweg 6, D-6242 Kronberg-Taunus, Federal Republic of Germany

First English edition 1960

Reprinted 1963, 1975, 1981, 1982

Second Revised edition 1984

Translated from the second edition of *Elektrodinamika sploshnykh sred*, Izdatel'stvo "Nauka", Moscow, 1982

Library of Congress Cataloging in Publication Data

Landau, L. D. (Lev Davidovich), 1908–1968.
Electrodynamics of continuous media.
(Course of theoretical physics; v. 8)
Translation of: Elektrodinamika sploshnykh sred.
"Translated from the second edition . . . Izdatel'stvo "Nauka", Moscow 1982."—T.p. verso.
1. Electromagnetic waves. 2. Electrodynamics.
3. Continuum mechanics. I. Lifshits, E. M. (Evgeniĭ Mikhaĭlovich) II. Pitaevskiĭ, L. P. (Lev Petrovich)
III. Title. IV. Series: Landau, L. D. (Lev Davidovich), 1908–1968. Course of theoretical physics; v. 8.
QC661.L2413 1984 537 83–24997

British Library Cataloguing in Publication Data

Landau, L. D.

Electrodynamics of continuous media.—2nd ed.
—(Course of theoretical physics; V. 8)
1. Electrodynamics
I. Title II. Lifshitz, E. M. III. Series
537.6 QC631

ISBN 0-08-030276-9 (Hardcover)
ISBN 0-08-030275-0 (Flexicover)

Printed in Great Britain by A Wheaton & Co. Exeter

CONTENTS

PREFACE TO THE SECOND EDITION

TWENTY-FIVE years have passed since the writing of this volume in its first edition. Such a long interval has inevitably made necessary a fairly thorough revision and expansion of the book for its second edition.

The original choice of material was such that, with some very slight exceptions, it has not become obsolete. In this part, only some relatively minor additions and improvements have been made.

It has, however, been necessary to incorporate a considerable amount of new material. This relates in particular to the theory of the magnetic properties of matter and the theory of optical phenomena, with new chapters on spatial dispersion and non-linear optics.

The chapter on electromagnetic fluctuations has been deleted, since this topic is now dealt with, in a different way, in Volume 9 of the *Course*.

As with the other volumes, invaluable help in the revision has been derived from the comments of scientific colleagues, who are too numerous to be named here in their entirety, but to whom we offer our sincere thanks. Particularly many comments came from V. L. Ginzburg, B. Ya. Zel'dovich and V. P. Kraĭnov. It was most useful to be able to hold regular discussions of questions arising, with A. F. Andreev, I. E. Dzyaloshinskiĭ and I. M. Lifshitz. We are particularly grateful to S. I. Vaĭnshteĭn and R. V. Polovin for much assistance in revising the chapter on magnetohydrodynamics. Lastly, our thanks are due to A. S. Borovik-Romanov, V. I. Grigor'ev and M. I. Kaganov for reading the manuscript and for a number of useful remarks.

Moscow
July, 1981

E. M. LIFSHITZ
L. P. PITAEVSKIĬ

PREFACE TO THE FIRST ENGLISH EDITION

THE present volume in the *Course of Theoretical Physics* deals with the theory of electromagnetic fields in matter and with the theory of the macroscopic electric and magnetic properties of matter. These theories include a very wide range of topics, as may be seen from the Contents.

In writing this book we have experienced considerable difficulties, partly because of the need to make a selection from the extensive existing material, and partly because the customary exposition of many topics to be included does not possess the necessary physical clarity, and sometimes is actually wrong. We realize that our own treatment still has many defects, which we hope to correct in future editions.

We are grateful to Professor V. L. Ginzburg, who read the book in manuscript and made some useful comments. I. E. Dzyaloshinskiĭ and L. P. Pitaevskiĭ gave great help in reading the proofs of the Russian edition. Thanks are due also to Dr Sykes and Dr Bell, who not only carried out excellently the arduous task of translating the book, but also made some useful comments concerning its contents.

Moscow
June, 1959

L. D. LANDAU
E. M. LIFSHITZ

NOTATION

Electric field **E**

Electric induction **D**

Magnetic field **H**

Magnetic induction **B**

External electric field $\mathfrak{E}$, magnitude $\mathfrak{E}$

External magnetic field $\mathfrak{H}$, magnitude $\mathfrak{H}$

Dielectric polarization **P**

Magnetization **M**

Total electric moment of a body $\mathscr{P}$

Total magnetic moment of a body $\mathscr{M}$

Permittivity ε

Dielectric susceptibility κ

Magnetic permeability μ

Magnetic susceptibility χ

Current density **j**

Conductivity σ

Absolute temperature (in energy units) T

Pressure P

Volume V

Thermodynamic quantities:	per unit volume	for a body
entropy	S	$\mathscr{S}$
internal energy	U	$\mathscr{U}$
free energy	F	$\mathscr{F}$
thermodynamic potential (Gibbs free energy)	Φ	$\mathscr{G}$

Chemical potential ζ

A complex periodic time factor is always taken as $e^{-i\omega t}$.

Volume element dV or d^3x; surface element **df**.

The summation convention always applies to three-dimensional (Latin) and two-dimensional (Greek) suffixes occurring twice in vector and tensor expressions.

References to other volumes in the *Course of Theoretical Physics*:

Mechanics = Vol. 1 (*Mechanics*, third English edition, 1976).

Fields = Vol. 2 (*The Classical Theory of Fields*, fourth English edition, 1975).

QM = Vol. 3 (*Quantum Mechanics—Non-relativistic theory*, third English edition, 1977).

QED = Vol. 4 (*Quantum Electrodynamics*, second English edition, 1982).

SP 1 = Vol. 5 (*Statistical Physics*, Part 1, third English edition, 1980).

FM = Vol. 6 (*Fluid Mechanics*, English edition, 1959).

TE = Vol. 7 (*Theory of Elasticity*, second English edition, 1970).

SP 2 = Vol. 9 (*Statistical Physics*, Part 2, English edition, 1980).

PK = Vol. 10 (*Physical Kinetics*, English edition, 1981).

All are published by Pergamon Press.

CHAPTER I

ELECTROSTATICS OF CONDUCTORS

§1. The electrostatic field of conductors

MACROSCOPIC electrodynamics is concerned with the study of electromagnetic fields in space that is occupied by matter. Like all macroscopic theories, electrodynamics deals with physical quantities averaged over elements of volume which are "physically infinitesimal", ignoring the microscopic variations of the quantities which result from the molecular structure of matter. For example, instead of the actual "microscopic" value of the electric field $\mathbf{e}$, we discuss its averaged value, denoted by $\mathbf{E}$:

$$\bar{\mathbf{e}} = \mathbf{E}. \tag{1.1}$$

The fundamental equations of the electrodynamics of continuous media are obtained by averaging the equations for the electromagnetic field in a vacuum. This method of obtaining the macroscopic equations from the microscopic was first used by H. A. Lorentz (1902).

The form of the equations of macroscopic electrodynamics and the significance of the quantities appearing in them depend essentially on the physical nature of the medium, and on the way in which the field varies with time. It is therefore reasonable to derive and investigate these equations separately for each type of physical object.

It is well known that all bodies can be divided, as regards their electric properties, into two classes, *conductors* and *dielectrics*, differing in that any electric field causes in a conductor, but not in a dielectric, the motion of charges, i.e. an *electric current*.†

Let us begin by studying the static electric fields produced by charged conductors, that is, the *electrostatics of conductors*. First of all, it follows from the fundamental property of conductors that, in the electrostatic case, the electric field inside a conductor must be zero. For a field $\mathbf{E}$ which was not zero would cause a current; the propagation of a current in a conductor involves a dissipation of energy, and hence cannot occur in a stationary state (with no external sources of energy).

Hence it follows, in turn, that any charges in a conductor must be located on its surface. The presence of charges inside a conductor would necessarily cause an electric field in it;‡ they can be distributed on its surface, however, in such a way that the fields which they produce in its interior are mutually balanced.

Thus the problem of the electrostatics of conductors amounts to determining the electric field in the vacuum outside the conductors and the distribution of charges on their surfaces.

At any point far from the surface of the body, the mean field $\mathbf{E}$ in the vacuum is almost

† The conductor is here assumed to be homogeneous (in composition, temperature, etc.). In an inhomogeneous conductor, as we shall see later, there may be fields which cause no motion of charges.

‡ This is clearly seen from equation (1.8) below.

the same as the actual field **e**. The two fields differ only in the immediate neighbourhood of the body, where the effect of the irregular molecular fields is noticeable, and this difference does not affect the averaged field equations. The exact microscopic Maxwell's equations in the vacuum are

$$\operatorname{div} \mathbf{e} = 0. \tag{1.2}$$

$$\mathbf{curl}\, \mathbf{e} = -(1/c)\partial \mathbf{h}/\partial t, \tag{1.3}$$

where **h** is the microscopic magnetic field. Since the mean magnetic field is assumed to be zero, the derivative $\partial \mathbf{h}/\partial t$ also vanishes on averaging, and we find that the static electric field in the vacuum satisfies the usual equations

$$\operatorname{div} \mathbf{E} = 0, \qquad \mathbf{curl}\, \mathbf{E} = 0, \tag{1.4}$$

i.e. it is a potential field with a potential ϕ such that

$$\mathbf{E} = -\mathbf{grad}\, \phi, \tag{1.5}$$

and ϕ satisfies Laplace's equation

$$\triangle \phi = 0. \tag{1.6}$$

The boundary conditions on the field **E** at the surface of a conductor follow from the equation **curl E** = 0, which, like the original equation (1.3), is valid both outside and inside the body. Let us take the z-axis in the direction of the normal **n** to the surface at some point on the conductor. The component E_z of the field takes very large values in the immediate neighbourhood of the surface (because there is a finite potential difference over a very small distance). This large field pertains to the surface itself and depends on the physical properties of the surface, but is not involved in our electrostatic problem, because it falls off over distances comparable with the distances between atoms. It is important to note, however, that, if the surface is homogeneous, the derivatives $\partial E_z/\partial x$, $\partial E_z/\partial y$ along the surface remain finite, even though E_z itself becomes very large. Hence, since $(\mathbf{curl}\, \mathbf{E})_x = \partial E_z/\partial y - \partial E_y/\partial z = 0$, we find that $\partial E_y/\partial z$ is finite. This means that E_y is continuous at the surface, since a discontinuity in E_y would mean an infinity of the derivative $\partial E_y/\partial z$. The same applies to E_x, and since $\mathbf{E} = 0$ inside the conductor, we reach the conclusion that the tangential components of the external field at the surface must be zero:

$$\mathbf{E}_t = 0. \tag{1.7}$$

Thus the electrostatic field must be normal to the surface of the conductor at every point. Since $\mathbf{E} = -\mathbf{grad}\, \phi$, this means that the field potential must be constant on the surface of any particular conductor. In other words, the surface of a homogeneous conductor is an equipotential surface of the electrostatic field.

The component of the field normal to the surface is very simply related to the charge density on the surface. The relation is obtained from the general electrostatic equation $\operatorname{div} \mathbf{e} = 4\pi\rho$, which on averaging gives

$$\operatorname{div} \mathbf{E} = 4\pi\bar{\rho}, \tag{1.8}$$

$\bar{\rho}$ being the mean charge density. The meaning of the integrated form of this equation is well known: the flux of the electric field through a closed surface is equal to the total charge inside that surface, multiplied by 4π. Applying this theorem to a volume element lying between two infinitesimally close unit areas, one on each side of the surface of the conductor, and using the fact that $\mathbf{E} = 0$ on the inner area, we find that $E_n = 4\pi\sigma$, where σ is the surface charge density, i.e. the charge per unit area of the surface of the conductor. Thus the distribution of charges over the surface of the conductor is given by the formula

$$4\pi\sigma = E_n = -\partial\phi/\partial n, \tag{1.9}$$

the derivative of the potential being taken along the outward normal to the surface. The total charge on the conductor is

$$e = -\frac{1}{4\pi}\oint\frac{\partial\phi}{\partial n}\,\mathrm{d}f, \tag{1.10}$$

the integral being taken over the whole surface.

The potential distribution in the electrostatic field has the following remarkable property: the function $\phi(x, y, z)$ can take maximum and minimum values only at boundaries of regions where there is a field. This theorem can also be formulated thus: a test charge e introduced into the field cannot be in stable equilibrium, since there is no point at which its potential energy $e\phi$ would have a minimum.

The proof of the theorem is very simple. Let us suppose, for example, that the potential has a maximum at some point A not on the boundary of a region where there is a field. Then the point A can be surrounded by a small closed surface on which the normal derivative $\partial\phi/\partial n < 0$ everywhere. Consequently, the integral over this surface $\oint(\partial\phi/\partial n)\,\mathrm{d}f < 0$. But by Laplace's equation $\oint(\partial\phi/\partial n)\mathrm{d}f = \int\triangle\phi\,\mathrm{d}V = 0$, giving a contradiction.

§2. The energy of the electrostatic field of conductors

Let us calculate the total energy $\mathscr{U}$ of the electrostatic field of charged conductors,†

$$\mathscr{U} = \frac{1}{8\pi}\int\mathbf{E}^2\,\mathrm{d}V, \tag{2.1}$$

where the integral is taken over all space outside the conductors. We transform this integral as follows:

$$\mathscr{U} = -\frac{1}{8\pi}\int\mathbf{E}\cdot\mathbf{grad}\,\phi\,\mathrm{d}V = -\frac{1}{8\pi}\int\mathrm{div}\,(\phi\mathbf{E})\,\mathrm{d}V + \frac{1}{8\pi}\int\phi\,\mathrm{div}\,\mathbf{E}\,\mathrm{d}V.$$

The second integral vanishes by (1.4), and the first can be transformed into integrals over the surfaces of the conductors which bound the field and an integral over an infinitely remote surface. The latter vanishes, because the field diminishes sufficiently rapidly at infinity (the arbitrary constant in ϕ is assumed to be chosen so that $\phi = 0$ at infinity). Denoting by ϕ_a the constant value of the potential on the ath conductor, we have‡

$$\mathscr{U} = \frac{1}{8\pi}\sum_a\oint\phi E_n\,\mathrm{d}f = \frac{1}{8\pi}\sum_a\phi_a\oint E_n\,\mathrm{d}f.$$

† The square $\mathbf{E}^2$ is not the same as the mean square $\overline{\mathbf{e}^2}$ of the actual field near the surface of a conductor or inside it (where $\mathbf{E} = 0$ but, of course, $\overline{\mathbf{e}^2} \neq 0$). By calculating the integral (2.1) we ignore the internal energy of the conductor as such, which is here of no interest, and the affinity of the charges for the surface.

‡ In transforming volume integrals into surface integrals, both here and later, it must be borne in mind that E_n is the component of the field along the outward normal to the conductor. This direction is opposite to that of the outward normal to the region of the volume integration, namely the space outside the conductors. The sign of the integral is therefore changed in the transformation.

Finally, since the total charges e_a on the conductors are given by (1.10) we obtain

$$\mathscr{U} = \tfrac{1}{2}\sum_a e_a \phi_a, \tag{2.2}$$

which is analogous to the expression for the energy of a system of point charges.

The charges and potentials of the conductors cannot both be arbitrarily prescribed; there are certain relations between them. Since the field equations in a vacuum are linear and homogeneous, these relations must also be linear, i.e. they must be given by equations of the form

$$e_a = \sum_b C_{ab}\phi_b, \tag{2.3}$$

where the quantities C_{aa}, C_{ab} have the dimensions of length and depend on the shape and relative position of the conductors. The quantities C_{aa} are called *coefficients of capacity*, and the quantities $C_{ab}\,(a \neq b)$ are called *coefficients of electrostatic induction.* In particular, if there is only one conductor, we have $e = C\phi$, where C is the *capacitance*, which in order of magnitude is equal to the linear dimension of the body. The converse relations, giving the potentials in terms of the charges, are

$$\phi_a = \sum_b C^{-1}{}_{ab} e_b, \tag{2.4}$$

where the coefficients $C^{-1}{}_{ab}$ form a matrix which is the inverse of the matrix C_{ab}.

Let us calculate the change in the energy of a system of conductors caused by an infinitesimal change in their charges or potentials. Varying the original expression (2.1), we have $\delta\mathscr{U} = (1/4\pi)\int \mathbf{E}\cdot\delta\mathbf{E}\,\mathrm{d}V$. This can be further transformed by two equivalent methods. Putting $\mathbf{E} = -\,\mathbf{grad}\,\phi$ and using the fact that the varied field, like the original field, satisfies equations (1.4) (so that $\operatorname{div}\delta\mathbf{E} = 0$), we can write

$$\begin{aligned}\delta\mathscr{U} &= -\frac{1}{4\pi}\int \mathbf{grad}\,\phi\cdot\delta\mathbf{E}\,\mathrm{d}V = -\frac{1}{4\pi}\int \operatorname{div}(\phi\,\delta\mathbf{E})\,\mathrm{d}V \\ &= \frac{1}{4\pi}\sum_a \phi_a \oint \delta E_n\,\mathrm{d}f,\end{aligned}$$

that is

$$\delta\mathscr{U} = \sum_a \phi_a\,\delta e_a, \tag{2.5}$$

which gives the change in energy due to a change in the charges. This result is obvious; it is the work required to bring infinitesimal charges δe_a to the various conductors from infinity, where the field potential is zero.

On the other hand, we can write

$$\begin{aligned}\delta\mathscr{U} &= -\frac{1}{4\pi}\int \mathbf{E}\cdot\mathbf{grad}\,\delta\phi\,\mathrm{d}V = -\frac{1}{4\pi}\int \operatorname{div}(\mathbf{E}\,\delta\phi)\,\mathrm{d}V \\ &= \frac{1}{4\pi}\sum_a \delta\phi_a \oint E_n\,\mathrm{d}f,\end{aligned}$$

that is

$$\delta \mathscr{U} = \sum_a e_a \delta\phi_a, \tag{2.6}$$

which expresses the change in energy in terms of the change in the potentials of the conductors.

Formulae (2.5) and (2.6) show that, by differentiating the energy $\mathscr{U}$ with respect to the charges, we obtain the potentials of the conductors, and the derivatives of $\mathscr{U}$ with respect to the potentials are the charges:

$$\partial \mathscr{U}/\partial e_a = \phi_a, \quad \partial \mathscr{U}/\partial \phi_a = e_a. \tag{2.7}$$

But the potentials and charges are linear functions of each other. Using (2.3) we have $\partial^2 \mathscr{U}/\partial\phi_a\partial\phi_b = \partial e_b/\partial\phi_a = C_{ba}$, and by reversing the order of differentiation we get C_{ab}. Hence it follows that

$$C_{ab} = C_{ba}, \tag{2.8}$$

and similarly $C^{-1}{}_{ab} = C^{-1}{}_{ba}$. The energy $\mathscr{U}$ can be written as a quadratic form in either the potentials or the charges:

$$\mathscr{U} = \tfrac{1}{2}\sum_{a,b} C_{ab}\phi_a\phi_b = \tfrac{1}{2}\sum_{a,b} C^{-1}{}_{ab}e_a e_b. \tag{2.9}$$

This quadratic form must be positive definite, like the original expression (2.1). From this condition we can derive various inequalities which the coefficients C_{ab} must satisfy. In particular, all the coefficients of capacity are positive:

$$C_{aa} > 0 \tag{2.10}$$

(and also $C^{-1}{}_{aa} > 0$).†

All the coefficients of electrostatic induction, on the other hand, are negative:

$$C_{ab} < 0 \quad (a \neq b). \tag{2.11}$$

That this must be so is seen from the following simple arguments. Let us suppose that every conductor except the ath is earthed, i.e. their potentials are zero. Then the charge induced by the charged ath conductor on another (the bth, say) is $e_b = C_{ba}\phi_a$. It is obvious that the sign of the induced charge must be opposite to that of the inducing potential, and therefore $C_{ab} < 0$. This can be more rigorously shown from the fact that the potential of the electrostatic field cannot reach a maximum or minimum outside the conductors. For example, let the potential ϕ_a of the only conductor not earthed be positive. Then the potential is positive in all space, its least value (zero) being attained only on the earthed conductors. Hence it follows that the normal derivative $\partial\phi/\partial n$ of the potential on the surfaces of these conductors is positive, and their charges are therefore negative, by (1.10). Similar arguments show that $C^{-1}{}_{ab} > 0$.

The energy of the electrostatic field of conductors has a certain extremum property, though this property is more formal than physical. To derive it, let us suppose that the

---

† We may also mention that another inequality which must be satisfied if the form (2.9) is positive is $C_{aa}C_{bb} > C_{ab}{}^2$.

charge distribution on the conductors undergoes an infinitesimal change (the total charge on each conductor remaining unaltered), in which the charges may penetrate into the conductors; we ignore the fact that such a charge distribution cannot in reality be stationary. We consider the change in the integral $\mathscr{U} = (1/8\pi)\int E^2 \mathrm{d}V$, which must now be extended over all space, including the volumes of the conductors themselves (since after the displacement of the charges the field $\mathbf{E}$ may not be zero inside the conductors). We write

$$\begin{aligned}\delta\mathscr{U} &= -\frac{1}{4\pi}\int \mathbf{grad}\,\phi\cdot\delta\mathbf{E}\,\mathrm{d}V \\ &= -\frac{1}{4\pi}\int \operatorname{div}(\phi\delta\mathbf{E})\,\mathrm{d}V + \frac{1}{4\pi}\int \phi\,\operatorname{div}\delta\mathbf{E}\,\mathrm{d}V.\end{aligned}$$

The first integral vanishes, being equivalent to one over an infinitely remote surface. In the second integral, we have by (1.8) $\operatorname{div}\delta\mathbf{E} = 4\pi\delta\bar{\rho}$, so that $\delta\mathscr{U} = \int\phi\delta\bar{\rho}\,\mathrm{d}V$. This integral vanishes if ϕ is the potential of the true electrostatic field, since then ϕ is constant inside each conductor, and the integral $\int\delta\bar{\rho}\,\mathrm{d}V$ over the volume of each conductor is zero, since its total charge remains unaltered.

Thus the energy of the actual electrostatic field is a minimum† relative to the energies of fields which could be produced by any other distribution of the charges on or in the conductors (*Thomson's theorem*).

From this theorem it follows, in particular, that the introduction of an uncharged conductor into the field of given charges (charged conductors) reduces the total energy of the field. To prove this, it is sufficient to compare the energy of the actual field resulting from the introduction of the uncharged conductor with the energy of the fictitious field in which there are no induced charges on that conductor. The former energy, since it has the least possible value, is less than the latter energy, which is also the energy of the original field (since, in the absence of induced charges, the field would penetrate into the conductor, and remain unaltered). This result can also be formulated thus: an uncharged conductor remote from a system of given charges is attracted towards the system.

Finally, it can be shown that a conductor (charged or not) brought into an electrostatic field cannot be in stable equilibrium under electric forces alone. This assertion generalizes the theorem for a point charge proved at the end of §1, and can be derived by combining the latter theorem with Thomson's theorem. We shall not pause to give the derivation in detail.

Formulae (2.9) are useful for calculating the energy of a system of conductors at finite distances apart. The energy of an uncharged conductor in a uniform external field $\mathfrak{E}$, which may be imagined as due to charges at infinity, requires special consideration. According to (2.2), this energy is $\mathscr{U} = \frac{1}{2}e\phi$, where e is the remote charge which causes the field, and ϕ is the potential at this charge due to the conductor. $\mathscr{U}$ does not include the energy of the charge e in its own field, since we are interested only in the energy of the conductor. The charge on the conductor is zero, but the external field causes it to acquire an electric dipole moment, which we denote by $\mathscr{P}$. The potential of the electric dipole field at a large distance $\mathbf{r}$ from it is $\phi = \mathscr{P}\cdot\mathbf{r}/r^3$. Hence $\mathscr{U} = e\mathscr{P}\cdot\mathbf{r}/2r^3$. But $-e\mathbf{r}/r^3$ is just the field $\mathfrak{E}$ due to the charge e. Thus

$$\mathscr{U} = -\tfrac{1}{2}\mathscr{P}\cdot\mathfrak{E}. \tag{2.12}$$

† We shall not give here the simple arguments which demonstrate that the extremum is a minimum.

Since all the field equations are linear, it is evident that the components of the dipole moment $\mathscr{P}$ are linear functions of the components of the field $\mathfrak{E}$. The coefficients of proportionality between $\mathscr{P}$ and $\mathfrak{E}$ have the dimensions of length cubed, and are therefore proportional to the volume of the conductor:

$$\mathscr{P}_i = V\alpha_{ik}\mathfrak{E}_k, \tag{2.13}$$

where the coefficients α_{ik} depend only on the shape of the body. The quantities $V\alpha_{ik}$ form a tensor, which may be called the *polarizability tensor* of the body. This tensor is symmetrical: $\alpha_{ik} = \alpha_{ki}$, a statement which will be proved in §11. Accordingly, the energy (2.12) is

$$\mathscr{U} = -\tfrac{1}{2}V\alpha_{ik}\mathfrak{E}_i\mathfrak{E}_k. \tag{2.14}$$

PROBLEMS

PROBLEM 1. Express the mutual capacitance C of two conductors (with charges $\pm e$) in terms of the coefficients C_{ab}.

SOLUTION. The *mutual capacitance* of two conductors is defined as the coefficient C in the relation $e = C(\phi_2 - \phi_1)$, and the energy of the system is given in terms of C by $\mathscr{U} = \frac{1}{2}e^2/C$. Comparing with (2.9), we obtain

$$\begin{aligned} 1/C &= C^{-1}{}_{11} - 2C^{-1}{}_{12} + C^{-1}{}_{22} \\ &= (C_{11} + 2C_{12} + C_{22})/(C_{11}C_{22} - C_{12}{}^2). \end{aligned}$$

PROBLEM 2. A point charge e is situated at O, near a system of earthed conductors, and induces on them charges e_a. If the charge e were absent, and the ath conductor were at potential ϕ'_a, the remainder being earthed, the field potential at O would be ϕ'_0. Express the charges e_a in terms of ϕ'_a and ϕ'_0.

SOLUTION. If charges e_a on the conductors give them potentials ϕ_a, and similarly for e'_a and ϕ'_a, it follows from (2.3) that

$$\sum_a \phi_a e'_a = \sum_{a,b} \phi_a C_{ab}\phi'_b = \sum_a \phi'_a e_a.$$

We apply this relation to two states of the system formed by all the conductors and the charge e (regarding the latter as a very small conductor). In one state the charge e is present, the charges on the conductors are e_a, and their potentials are zero. In the other state the charge e is zero, and one of the conductors has a potential $\phi'_a \neq 0$. Then we have $e\phi'_0 + e_a\phi'_a = 0$, whence $e_a = -e\phi'_0/\phi'_a$.

For example, if a charge e is at a distance r from the centre of an earthed conducting sphere with radius a ($< r$), then $\phi'_0 = \phi'_a a/r$, and the charge induced on the sphere is $e_a = -ea/r$.

As a second example, let us consider a charge e placed between two concentric conducting spheres with radii a and b, at a distance r from the centre such that $a < r < b$. If the outer sphere is earthed and the inner one is charged to potential ϕ'_a, the potential at distance r is

$$\phi'_0 = \phi'_a \frac{1/r - 1/b}{1/a - 1/b}.$$

Hence the charge induced on the inner sphere by the charge e is $e_a = -ea(b-r)/r(b-a)$. Similarly the charge induced on the outer sphere is $e_b = -eb(r-a)/r(b-a)$.

PROBLEM 3. Two conductors, with capacitances C_1 and C_2, are placed at a distance r apart which is large compared with their dimensions. Determine the coefficients C_{ab}.

SOLUTION. If conductor 1 has a charge e_1, and conductor 2 is uncharged, then in the first approximation $\phi_1 = e_1/C_1$, $\phi_2 = e_1/r$; here we neglect the variation of the field over conductor 2 and its polarization. Thus $C^{-1}{}_{11} = 1/C_1$, $C^{-1}{}_{12} = 1/r$, and similarly $C^{-1}{}_{22} = 1/C_2$. Hence we find†

$$C_{11} = C_1\left(1 + \frac{C_1C_2}{r^2}\right), \quad C_{12} = -\frac{C_1C_2}{r}, \quad C_{22} = C_2\left(1 + \frac{C_1C_2}{r^2}\right).$$

† The subsequent terms in the expansion are in general of order (in $1/r$) one higher than those given. If, however, r is taken as the distance between the "centres of charge" of the two bodies (for spheres, between the geometrical centres), then the order of the subsequent terms is two higher.

PROBLEM 4. Determine the capacitance of a ring (radius b) of thin conducting wire of circular cross-section (radius $a \ll b$).

SOLUTION. Since the wire is thin, the field at the surface of the ring is almost the same as that of charges distributed along the axis of the wire (for a right cylinder, it would be exactly the same). Hence the potential of the ring is

$$\phi_a = \frac{e}{2\pi b}\oint \frac{dl}{r},$$

where r is the distance from a point on the surface of the ring to an element dl of the axis of the wire, the integration being over all such elements. We divide the integral into two parts corresponding to $r < \Delta$ and $r > \Delta$, Δ being a distance such that $a \ll \Delta \ll b$. Then for $r < \Delta$ the segment of the ring concerned may be regarded as straight, and therefore

$$\oint_{\Delta > r} \frac{dl}{r} = \int_{-\Delta}^{\Delta} \frac{dl}{\sqrt{(l^2 + a^2)}} \cong 2\log(2\Delta/a).$$

In the range $r > \Delta$ the thickness of the wire may be neglected, i.e. r may be taken as the distance between two points on its axis. Then

$$\oint_{r > \Delta} \frac{dl}{r} = 2\int_{\phi_0}^{\pi} \frac{b\,d\phi}{2b\sin\frac{1}{2}\phi} = -2\log\tan\tfrac{1}{4}\phi_0,$$

where ϕ is the angle subtended at the centre of the ring by the chord r, and the lower limit of integration is such that $2b\sin\frac{1}{2}\phi_0 = \Delta$, whence $\phi_0 \cong \Delta/b$. When the two parts of the integral are added, Δ cancels, and the capacitance of the ring is

$$C = \frac{e}{\phi_a} = \frac{\pi b}{\log(8b/a)}.$$

§3. Methods of solving problems in electrostatics

The general methods of solving Laplace's equation for given boundary conditions on certin surfaces are studied in mathematical physics, and we shall not give a detailed description of them here. We shall merely mention some of the more elementary procedures and solve various problems of intrinsic interest.†

(1) *The method of images.* The simplest example of the use of this method is to determine the field due to a point charge e outside a conducting medium which occupies a half-space. The principle of the method is to find fictitious point charges which, together with the given charge or charges, produce a field such that the surface of the conductor is an equipotential surface. In the case just mentioned, this is achieved by placing a fictitious charge $e' = -e$ at a point which is the image of e in the plane which bounds the conducting medium. The potential of the field due to the charge e and its *image* e' is

$$\phi = e\left(\frac{1}{r} - \frac{1}{r'}\right), \tag{3.1}$$

† The solutions of many more complex problems are given by W. R. Smythe, *Static and Dynamic Electricity*, 3rd ed., McGraw-Hill, New York, 1968; G. A. Grinberg, *Selected Problems in the Mathematical Theory of Electric and Magnetic Phenomena* (*Izbrannye voprosy matematicheskoĭ teorii élektricheskikh i magnitnykh yavleniĭ*), Moscow, 1948.

where r and r' are the distances of a point from the charges e and e'. On the bounding plane, $r = r'$ and the potential has the constant value zero, so that the necessary boundary condition is satisfied and (3.1) gives the solution of the problem. It may be noted that the charge e is attracted to the conductor by a force $e^2/(2a)^2$ (the *image force*; a is the distance of the charge from the conductor), and the energy of their interaction is $-e^2/4a$.

The distribution of surface charge induced on the bounding plane by the point charge e is given by

$$\sigma = -\frac{1}{4\pi}\left[\frac{\partial\phi}{\partial n}\right]_{r=r'} = -\frac{e}{2\pi}\frac{a}{r^3}. \tag{3.2}$$

It is easy to see that the total charge on the plane is $\int \sigma \mathrm{d}f = -e$, as it should be.

The total charge induced on an originally uncharged insulated conductor by other charges is, of course, zero. Hence, if in the present case the conducting medium (in reality a large conductor) is insulated, we must suppose that, besides the charge $-e$, a charge $+e$ is also induced, which, however, has a vanishingly small density, being distributed over the large surface of the conductor.

Next, let us consider a more difficult problem, that of the field due to a point charge e near a spherical conductor. To solve this problem, we use the following result, which can easily be proved by direct calculation. The potential of the field due to two point charges e and $-e'$, namely $\phi = e/r - e'/r'$, vanishes on the surface of a sphere whose centre is on the line joining the charges (but not between them). If the radius of the sphere is R and its centre is distant l and l' from the two charges, then $l/l' = (e/e')^2$, $R^2 = ll'$.

Let us first suppose that the spherical conductor is maintained at a constant potential $\phi = 0$, i.e. it is earthed. Then the field outside the sphere due to the point charge e at A (Fig. 1), at a distance l from the centre of the sphere, is the same as the field due to two charges, namely the given charge e and a fictitious charge $-e'$ at A' inside the sphere, at a distance l' from its centre, where

$$l' = R^2/l, \qquad e' = eR/l. \tag{3.3}$$

The potential of this field is

$$\phi = \frac{e}{r} - \frac{eR}{lr'}, \tag{3.4}$$

r and r' being as shown in Fig. 1. A non-zero total charge $-e'$ is induced on the surface of the sphere. The energy of the interaction between the charge and the sphere is

$$\mathscr{U} = -\tfrac{1}{2}ee'/(l-l') = -\tfrac{1}{2}e^2R/(l^2-R^2), \tag{3.5}$$

and the charge is attracted to the sphere by a force $F = -\partial\mathscr{U}/\partial l = -e^2lR/(l^2-R^2)^2$.

If the total charge on the spherical conductor is kept equal to zero (an insulated uncharged sphere), a further fictitious charge must be introduced, such that the total charge induced on the surface of the sphere is zero, and the potential on that surface is still constant. This is done by placing a charge $+e'$ at the centre of the sphere. The potential of the required field is then given by the formula

$$\phi = \frac{e}{r} - \frac{e'}{r'} + \frac{e'}{r_0}. \tag{3.6}$$

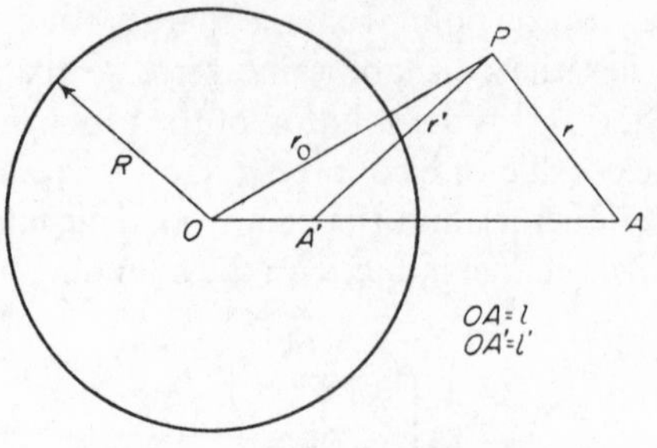

FIG. 1

The energy of interaction in this case is

$$\mathscr{U} = \tfrac{1}{2}ee'\left(\frac{1}{l} - \frac{1}{l-l'}\right) = -\frac{e^2R^3}{2l^2(l^2-R^2)}. \tag{3.7}$$

Finally, if the charge e is at A' (Fig. 1) in a spherical cavity in a conducting medium, the field inside the cavity must be the same as the field due to the charge e at A' and its image at A outside the sphere, regardless of whether the conductor is earthed or insulated:

$$\phi = \frac{e}{r'} - \frac{eR}{l'r}. \tag{3.8}$$

(2) *The method of inversion.* There is a simple method whereby in some cases a known solution of one electrostatic problem gives the solution of another problem. This method is based on the invariance of Laplace's equation with respect to a certain transformation of the variables.

In spherical polar coordinates Laplace's equation has the form

$$\frac{1}{r^2}\frac{\partial}{\partial r}\left(r^2\frac{\partial\phi}{\partial r}\right) + \frac{1}{r^2}\triangle_\Omega\phi = 0,$$

where $\triangle_\Omega$ denotes the angular part of the Laplacian operator. It is easy to see that this equation is unaltered in form if the variable r is replaced by a new variable r' such that

$$r = R^2/r' \tag{3.9}$$

(the *inversion* transformation) and at the same time the unknown function ϕ is replaced by ϕ' such that

$$\phi = r'\phi'/R. \tag{3.10}$$

Here R is some constant having the dimensions of length (the *radius of inversion*). Thus, if the function $\phi(\mathbf{r})$ satisfies Laplace's equation, then so does the function

$$\phi'(\mathbf{r}') = R\phi(R^2\mathbf{r}'/r'^2)/r'. \tag{3.11}$$

Let us assume that we know the electrostatic field due to some system of conductors, all at the same potential ϕ_0, and point charges. The potential $\phi(\mathbf{r})$ is usually defined so as to vanish at infinity. Here, however, we shall define $\phi(\mathbf{r})$ so that it tends to $-\phi_0$ at infinity. Then $\phi = 0$ on the conductors.

We may now ascertain what problem of electrostatics will be solved by the transformed function (3.11). First of all, the shapes and relative positions of all the conductors of finite size will be changed. The boundary condition of constant potential on their surfaces will be automatically satisfied, since $\phi' = 0$ if $\phi = 0$. Furthermore, the positions and magnitudes of all the point charges will be changed. A charge e at a point $\mathbf{r}_0$ moves to $\mathbf{r}'_0 = R^2\mathbf{r}_0/r_0{}^2$ and takes a value e' which can be determined as follows. As $\mathbf{r} \to \mathbf{r}_0$ the potential $\phi(\mathbf{r})$ tends to infinity as $e/|\delta\mathbf{r}|$, where $\delta\mathbf{r} = \mathbf{r} - \mathbf{r}_0$. Differentiating the relation $\mathbf{r} = R^2\mathbf{r}'/r'^2$, we find that the magnitudes of the small differences $\delta\mathbf{r}$ and $\delta\mathbf{r}' = \mathbf{r}' - \mathbf{r}'_0$ are related by $(\delta\mathbf{r})^2 = R^4(\delta\mathbf{r}')^2/r'_0{}^4$. Hence, as $\mathbf{r}' \to \mathbf{r}'_0$, the function ϕ' tends to infinity as $eR/r'_0|\delta\mathbf{r}| = er'_0/R|\delta\mathbf{r}'|$, corresponding to a charge

$$e' = er'_0/R = eR/r_0. \tag{3.12}$$

Finally, let us examine the behaviour of the function $\phi'(\mathbf{r}')$ near the origin. For $\mathbf{r}' = 0$ we have $\mathbf{r} \to \infty$ and $\phi(\mathbf{r}) \to -\phi_0$. Hence, as $\mathbf{r}' \to 0$, the function ϕ' tends to infinity as $-R\phi_0/r'$. This means that there is a charge $e_0 = -R\phi_0$ at the point $\mathbf{r}' = 0$.

We shall give, for reference, the way in which certain geometrical figures are transformed by inversion. A spherical surface with radius a and centre $\mathbf{r}_0$ is given by the equation $(\mathbf{r} - \mathbf{r}_0)^2 = a^2$. On inversion, this becomes $([R^2\mathbf{r}'/r'^2] - \mathbf{r}_0)^2 = a^2$, which, on multiplying by r'^2 and rearranging, can be written $(\mathbf{r}' - \mathbf{r}'_0)^2 = a'^2$, where

$$\mathbf{r}'_0 = -R^2\mathbf{r}_0/(a^2 - r_0{}^2), \qquad a' = aR^2/|a^2 - r_0{}^2|. \tag{3.13}$$

Thus we have another sphere, with radius a' and centre $\mathbf{r}'_0$. If the original sphere passes through the origin ($a = r_0$), then $a' = \infty$. In this case the sphere is transformed into a plane perpendicular to the vector $\mathbf{r}_0$ and distant $r'_0 - a' = R^2/(a + r_0) = R^2/2a$ from the origin.

(3) *The method of conformal mapping.* A field which depends on only two Cartesian co-ordinates (x and y, say) is said to be *two-dimensional*. The theory of functions of a complex variable is a powerful means of solving two-dimensional problems of electrostatics. The theoretical basis of the method is as follows.

An electrostatic field in a vacuum satisfies two equations: $\mathbf{curl}\,\mathbf{E} = 0$, $\operatorname{div}\mathbf{E} = 0$. The first of these makes it possible to introduce the field potential, defined by $\mathbf{E} = -\mathbf{grad}\,\phi$. The second equation shows that we can also define a *vector potential* $\mathbf{A}$ of the field, such that $\mathbf{E} = \mathbf{curl}\,\mathbf{A}$. In the two-dimensional case, the vector $\mathbf{E}$ lies in the xy-plane, and depends only on x and y. Accordingly, the vector $\mathbf{A}$ can be chosen so that it is perpendicular to the xy-plane. Then the field components are given in terms of the derivatives of ϕ and $\mathbf{A}$ by

$$E_x = -\partial\phi/\partial x = \partial A/\partial y, \qquad E_y = -\partial\phi/\partial y = -\partial A/\partial x. \tag{3.14}$$

These relations between the derivatives of ϕ and A are, mathematically, just the well-known Cauchy–Riemann conditions, which express the fact that the complex quantity

$$w = \phi - iA \tag{3.15}$$

is an analytic function of the complex argument $z = x + iy$. This means that the function $w(z)$ has a definite derivative at every point, independent of the direction in which the derivative is taken. For example, differentiating along the x-axis, we find $\mathrm{d}w/\mathrm{d}z = \partial\phi/\partial x - i\partial A/\partial x$, or

$$\mathrm{d}w/\mathrm{d}z = -E_x + iE_y. \tag{3.16}$$

The function w is called the *complex potential*.

The lines of force are defined by the equation $dx/E_x = dy/E_y$. Expressing E_x and E_y as derivatives of A, we can write this as $(\partial A/\partial x)dx + (\partial A/\partial y)dy = dA = 0$, whence $A(x, y) = \text{constant}$. Thus the lines on which the imaginary part of the function $w(z)$ is constant are the lines of force. The lines on which its real part is constant are the equipotential lines. The orthogonality of these families of lines is ensured by the relations (3.14), according to which

$$\frac{\partial\phi}{\partial x}\frac{\partial A}{\partial x} + \frac{\partial\phi}{\partial y}\frac{\partial A}{\partial y} = 0.$$

Both the real and the imaginary part of an analytic function $w(z)$ satisfy Laplace's equation. We could therefore equally well take im w as the field potential. The lines of force would then be given by re w = constant. Instead of (3.15) we should have $w = A + i\phi$.

The flux of the electric field through any section of an equipotential line is given by the integral $\oint E_n dl = -\oint(\partial\phi/\partial n)dl$, where dl is an element of length of the equipotential line and **n** the direction of the normal to it. According to (3.14) we have $\partial\phi/\partial n = -\partial A/\partial l$, the choice of sign denoting that l is measured to the left when one looks along **n**. Thus $\oint E_n dl = \oint(\partial A/\partial l)dl = A_2 - A_1$, where A_2 and A_1 are the values of A at the ends of the section. In particular, since the flux of the electric field through a closed contour is $4\pi e$, where e is the total charge enclosed by the contour (per unit length of conductors perpendicular to the plane), it follows that

$$e = (1/4\pi)\Delta A, \tag{3.17}$$

where ΔA is the change in A on passing counterclockwise round the closed equipotential line.

The simplest example of the complex potential is that of the field of a charged straight wire passing through the origin and perpendicular to the plane. The field is given by $E_r = 2e/r$, $E_\theta = 0$, where r, θ are polar coordinates in the xy-plane, and e is the charge per unit length of the wire. The corresponding complex potential is

$$w = -2e \log z = -2e \log r - 2ie\theta. \tag{3.18}$$

If the charged wire passes through the point (x_0, y_0) instead of the origin, the complex potential is

$$w = -2e \log(z - z_0), \tag{3.19}$$

where $z_0 = x_0 + iy_0$.

Mathematically, the functional relation $w = w(z)$ constitutes a *conformal mapping* of the plane of the complex variable z on the plane of the complex variable w. Let C be the cross-sectional contour of a conductor in the xy-plane, and ϕ_0 its potential. It is clear from the above discussion that the problem of determining the field due to this conductor amounts to finding a function $w(z)$ which maps the contour C in the z-plane on the line $w = \phi_0$, parallel to the axis of ordinates, in the w-plane. Then re w gives the potential of the field. (If the function $w(z)$ maps the contour C on a line parallel to the axis of abscissae, then the potential is im w.)

(4) *The wedge problem.* We shall give here, for reference, formulae for the field due to a point charge e placed between two intersecting conducting half-planes. Let the z-axis of a system of cylindrical polar coordinates (r, θ, z) be along the apex of the wedge, the angle θ

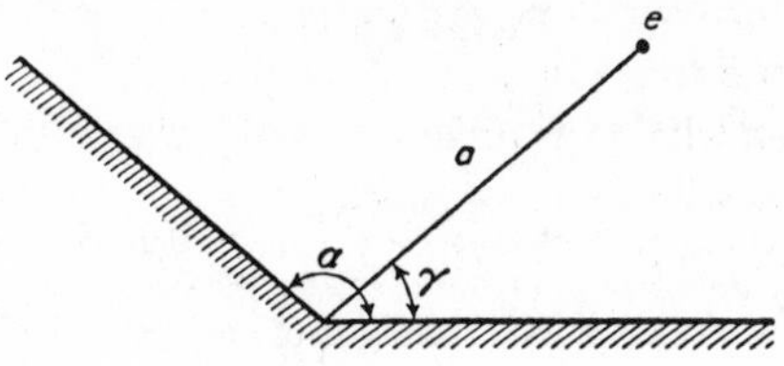

FIG. 2

being measured from one of the planes, and let the position of the charge e be $(a, \gamma, 0)$ (Fig. 2). The angle α between the planes may be either less or greater than π; in the latter case we have a charge outside a conducting wedge.

The field potential is given by

$$\phi = \frac{e}{\alpha\sqrt{(2ar)}} \int_\eta^\infty \left\{ \frac{\sinh(\pi\zeta/\alpha)}{\cosh(\pi\zeta/\alpha) - \cos[\pi(\theta-\gamma)/\alpha]} - \frac{\sinh(\pi\zeta/\alpha)}{\cosh(\pi\zeta/\alpha) - \cos[\pi(\theta+\gamma)/\alpha]} \right\}$$

$$\times \frac{d\zeta}{\sqrt{(\cosh\zeta - \cosh\eta)}}, \quad \cosh\eta = (a^2 + r^2 + z^2)/2ar, \quad \eta > 0. \tag{3.20}$$

The potential $\phi = 0$ on the surface of the conductors, i.e. for $\theta = 0$ or α. This formula was first given by H. M. Macdonald (1895)†.

In particular, for $\alpha = 2\pi$ we have a conducting half-plane in the field of a point charge. In this case the integral in (3.20) can be evaluated explicitly, giving

$$\left.\begin{aligned} \phi &= \frac{e}{\pi}\left\{\frac{1}{R}\cos^{-1}\left(\frac{-\cos\frac{1}{2}(\theta-\gamma)}{\cosh\frac{1}{2}\eta}\right) - \frac{1}{R'}\cos^{-1}\left(\frac{-\cos\frac{1}{2}(\theta+\gamma)}{\cosh\frac{1}{2}\eta}\right)\right\}, \\ R^2 &= a^2 + r^2 + z^2 - 2ar\cos(\gamma - \theta), \\ R'^2 &= a^2 + r^2 + z^2 - 2ar\cos(\gamma + \theta). \end{aligned}\right\} \tag{3.21}$$

In the limit as the point (r, θ, z) tends to the position of the charge e, the potential (3.21) becomes

$$\phi = \phi' + e/R, \quad \text{where } \phi' = -\frac{e}{2\pi a}\left[1 + \frac{\pi - \gamma}{\sin\gamma}\right]. \tag{3.22}$$

The second term is just the Coulomb potential, which becomes infinite as $R \to 0$, while ϕ' is the change caused by the conductor in the potential at the position of the charge. The energy of the interaction between the charge and the conducting half-plane is

$$\mathscr{U} = \tfrac{1}{2}e\phi' = -\frac{e^2}{4\pi a}\left[1 + \frac{\pi - \gamma}{\sin\gamma}\right]. \tag{3.23}$$

† Its derivation is given by him in *Electromagnetism*, Bell, London, 1934, p. 79, and by V. V. Batygin and I. N. Toptygin, *Problems in Electrodynamics*, 2nd ed., Academic Press, London, 1978, p. 47.

PROBLEMS

PROBLEM 1. Determine the field near an uncharged conducting sphere with radius R placed in a uniform external electric field $\mathfrak{E}$.

SOLUTION. We write the potential in the form $\phi = \phi_0 + \phi_1$, where $\phi_0 = -\mathfrak{E} \cdot \mathbf{r}$ is the potential of the external field and ϕ_1 is the required change in potential due to the sphere. By symmetry, the function ϕ_1 can depend only on the constant vector $\mathfrak{E}$. The only such solution of Laplace's equation which vanishes at infinity is

$$\phi_1 = -\text{constant} \times \mathfrak{E} \cdot \mathbf{grad}(1/r) = \text{constant} \times \mathfrak{E} \cdot \mathbf{r}/r^3,$$

the origin being taken at the centre of the sphere. On the surface of the sphere ϕ must be constant, and so the constant in ϕ_1 is R^3, whence

$$\phi = -\mathfrak{E}\, r \cos\theta \left(1 - \frac{R^3}{r^3}\right),$$

where θ is the angle between $\mathfrak{E}$ and $\mathbf{r}$. The distribution of charge on the surface of the sphere is given by

$$\sigma = -(1/4\pi)[\partial\phi/\partial r]_{r=R} = (3\mathfrak{E}/4\pi)\cos\theta.$$

The total charge $e = 0$. The dipole moment of the sphere is most easily found by comparing ϕ_1 with the potential $\mathscr{P} \cdot \mathbf{r}/r^3$ of an electric dipole field, whence $\mathscr{P} = R^3\mathfrak{E}$.

PROBLEM 2. The same as Problem 1, but for an infinite cylinder in a uniform transverse field.

SOLUTION. We use polar coordinates in a plane perpendicular to the axis of the cylinder. The solution of the two-dimensional Laplace's equation which depends only on a constant vector is

$$\phi_1 = \text{constant} \times \mathfrak{E} \cdot \mathbf{grad}(\log r) = \text{constant} \times \mathfrak{E} \cdot \mathbf{r}/r^2.$$

Adding $\phi_0 = -\mathfrak{E} \cdot \mathbf{r}$ and putting the constant equal to R^2, we have

$$\phi = -\mathfrak{E}\, r \cos\theta \left(1 - \frac{R^2}{r^2}\right).$$

The surface charge density is $\sigma = (\mathfrak{E}/2\pi)\cos\theta$. The dipole moment per unit length of the cylinder can be found by comparing ϕ with the potential of a two-dimensional dipole field, namely $2\mathscr{P} \cdot \mathbf{grad}(\log r) = 2\mathscr{P} \cdot \mathbf{r}/r^2$, so that $\mathscr{P} = \frac{1}{2}R^2\mathfrak{E}$.

PROBLEM 3. Determine the field near a wedge-shaped projection on a conductor.

SOLUTION. We take polar coordinates r, θ in a plane perpendicular to the apex of the wedge, the origin being at the vertex of the angle θ_0 of the wedge (Fig. 3). The angle θ is measured from one face of the wedge, the region outside the conductor being $0 \leqslant \theta \leqslant 2\pi - \theta_0$. Near the apex of the wedge the potential can be expanded in powers of r, and we shall be interested in the first term of the expansion (after the constant term), which contains the lowest power of r. The solutions of the two-dimensional Laplace's equation which are proportional to r^n are $r^n \cos n\theta$ and $r^n \sin n\theta$. The solution having the smallest n which satisfies the condition ϕ = constant for $\theta = 0$ and $\theta = 2\pi - \theta_0$ (i.e. on the surface of the conductor) is

$$\phi = \text{constant} \times r^n \sin n\theta, \qquad n = \pi/(2\pi - \theta_0).$$

The field varies as r^{n-1}. For $\theta_0 < \pi\,(n < 1)$, therefore, the field becomes infinite at the apex of the wedge. In particular, for a very sharp wedge ($\theta_0 \ll 1, n \cong \frac{1}{2}$) E increases as $r^{-\frac{1}{2}}$ as $r \to 0$. Near a wedge-shaped concavity in a conductor ($\theta_0 > \pi$, $n > 1$) the field tends to zero.

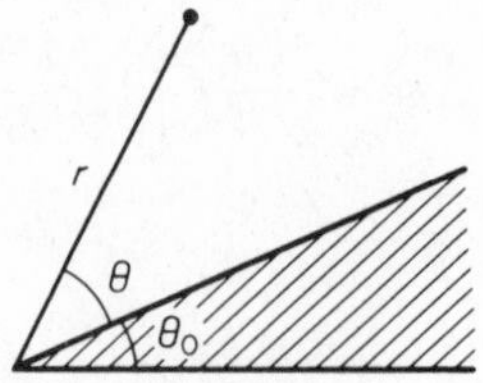

FIG. 3

The value of the constant can be determined only by solving the problem for the whole field. For example, for a very sharp wedge in the field of a point charge e, the passage to the limit of small r in (3.21) confirms that

$$\phi = \text{constant} \times \sqrt{r} \sin \tfrac{1}{2}\theta,$$

the constant being $[4e\sqrt{a}/\pi(a^2+z^2)]\sin\frac{1}{2}\gamma$. In this case, "near the wedge" means that $r \ll a$, under which condition the $\partial^2\phi/\partial z^2$ term in Laplace's equation may be neglected.

PROBLEM 4. Determine the field near the end of a sharp conical point on the surface of a conductor.

SOLUTION. We take spherical polar coordinates, with the origin at the vertex of the cone and the polar axis along the axis of the cone. Let the angle of the cone be $2\theta_0 \ll 1$, so that the region outside the conductor corresponds to polar angles in the range $\theta_0 \leqslant \theta \leqslant \pi$. As in Problem 3, we seek a solution for the variable part of the potential, which is symmetrical about the axis, in the form

$$\phi = r^n f(\theta), \tag{1}$$

with the smallest possible value of n. Laplace's equation

$$\frac{1}{r^2}\frac{\partial}{\partial r}\left(r^2\frac{\partial\phi}{\partial r}\right) + \frac{1}{r^2 \sin\theta}\frac{\partial}{\partial\theta}\left(\sin\theta\frac{\partial\phi}{\partial\theta}\right) = 0,$$

after substitution of (1), gives

$$\frac{1}{\sin\theta}\frac{\mathrm{d}}{\mathrm{d}\theta}\left(\sin\theta\frac{\mathrm{d}f}{\mathrm{d}\theta}\right) + n(n+1)f = 0. \tag{2}$$

The condition of constant potential on the surface of the cone means that we must have $f(\theta_0) = 0$.

For small θ_0 we seek a solution by assuming that $n \ll 1$ and $f(\theta)$ is of the form constant $\times$ $[1+\psi(\theta)]$, where $\psi \ll 1$. (For $\theta_0 \to 0$, i.e. an infinitely sharp point, we should expect that ϕ tends to a constant almost everywhere near the cone.) The equation for ψ is

$$\frac{1}{\sin\theta}\frac{\mathrm{d}}{\mathrm{d}\theta}\left(\sin\theta\frac{\mathrm{d}\psi}{\mathrm{d}\theta}\right) = -n. \tag{3}$$

The solution having no singularities outside the cone (in particular, at $\theta = \pi$) is $\psi(\theta) = 2n \log \sin\frac{1}{2}\theta$.

For $\theta \sim \theta_0 \ll 1$, ψ is no longer small. Nevertheless, this expression remains valid, since the second term in equation (2) may be neglected because θ is small. To determine the constant n in the first approximation we must require that the function $f = 1+\psi$ vanish for $\theta = \theta_0$. Thus† $n = -1/2\log\theta_0$. The field increases to infinity as $r^{-(1-n)}$ in the neighbourhood of the vertex, i.e. essentially as $1/r$.

PROBLEM 5. The same as Problem 4, but for a sharp conical depression on the surface of a conductor.

SOLUTION. The region outside the conductor now corresponds to the range $0 \leqslant \theta \leqslant \theta_0$. As in Problem 4, we seek ϕ in the form (1), but now $n \gg 1$. Since $\theta \ll 1$ for all points in the field, equation (2) becomes

$$\frac{1}{\theta}\frac{\mathrm{d}}{\mathrm{d}\theta}\left(\theta\frac{\mathrm{d}f}{\mathrm{d}\theta}\right) + n^2 f = 0.$$

This is Bessel's equation, and the solution having no singularities in the field is $J_0(n\theta)$. The value of n is determined as the smallest root of the equation $J_0(n\theta_0) = 0$, whence $n = 2.4/\theta_0$.

PROBLEM 6. Determine the energy of the attraction between an electric dipole and a plane conducting surface.

SOLUTION. We take the x-axis perpendicular to the surface of the conductor, and passing through the dipole; let the dipole moment vector $\mathscr{P}$ lie in the xy-plane. The image of the dipole is at the point $-x$ and has a moment $\mathscr{P}'_x = \mathscr{P}_x$, $\mathscr{P}'_y = -\mathscr{P}_y$. The required energy of attraction is the energy of the interaction between the dipole and its image, and is $\mathscr{U} = -(2\mathscr{P}_x^2 + \mathscr{P}_y^2)/8x^3$.

PROBLEM 7. Determine the mutual capacitance per unit length of two parallel infinite conducting cylinders with radii a and b, their axes being at a distance c apart.‡

† A more exact formula $n = 1/2\log(2/\theta_0)$, containing a coefficient in the (large) logarithm, cannot really be obtained by the simple method given here. A more rigorous calculation, however, leads, as it happens, to this same formula.

‡ The corresponding problem for two spheres cannot be solved in closed form. The difference arises because, in the field of two parallel wires bearing equal and opposite charges, all the equipotential surfaces are circular cylinders, whereas in the field of two equal and opposite point charges the equipotential surfaces are not spheres.

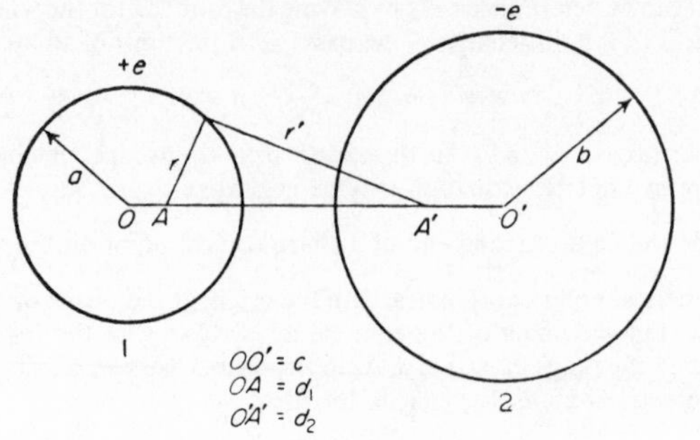

FIG. 4

SOLUTION. The field due to the two cylinders is the same as that which would be produced (in the region outside the cylinders) by two charged wires passing through certain points A and A' (Fig. 4). The wires have charges $\pm e$ per unit length, equal to the charges on the cylinders, and the points A and A' lie on OO' in such a way that the surfaces of the cylinders are equipotential surfaces. For this to be so, the distances OA and $O'A'$ must be such that $OA \cdot OA' = a^2$, $O'A' \cdot O'A = b^2$, i.e. $d_1(c-d_2) = a^2$, $d_2(c-d_1) = b^2$. Then, for each cylinder, the ratio r/r' of the distances from A and A' is constant. On cylinder 1, $r/r' = a/OA' = a/(c-d_2) = d_1/a$, and on cylinder 2, $r'/r = d_2/b$. Accordingly, the potentials of the cylinders are $\phi_1 = -2e\log(r/r') = -2e\log(d_1/a)$, $\phi_2 = 2e\log(d_2/b)$, $\phi_2 - \phi_1 = 2e\log(d_1d_2/ab)$. Hence we find the required mutual capacitance $C = e/(\phi_2 - \phi_1)$:

$$1/C = 2\log(d_1d_2/ab) = 2\cosh^{-1}[(c^2-a^2-b^2)/2ab].$$

In particular, for a cylinder with radius a at a distance $h(>a)$ from a conducting plane, we put $c = b+h$ and take the limit as $b \to \infty$, obtaining $1/C = 2\cosh^{-1}(h/a)$.

If two hollow cylinders are placed one inside the other ($c < b-a$), there is no field outside, while the field between the cylinders is the same as that due to two wires with charges $\pm e$ passing through A and A' (Fig. 5). The same method gives

$$1/C = 2\cosh^{-1}[(a^2+b^2-c^2)/2ab].$$

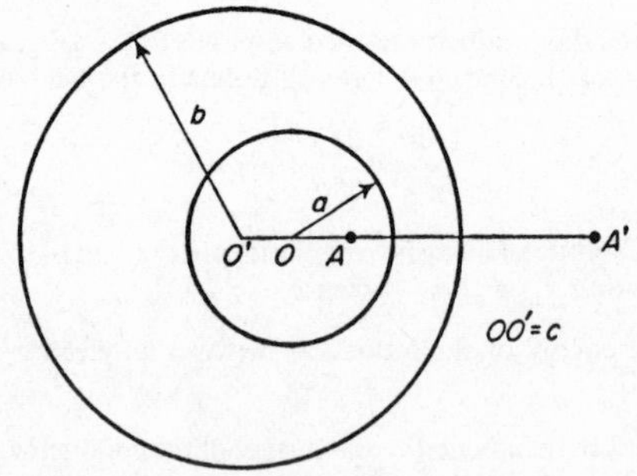

FIG. 5

PROBLEM 8. The boundary of a conductor is an infinite plane with a hemispherical projection. Determine the charge distribution on the surface.

SOLUTION. In the field determined in Problem 1, whose potential is

$$\phi = \text{constant} \times z\left(1 - \frac{R^3}{r^3}\right),$$

the plane $z = 0$ with a projection $r = R$ is an equipotential surface, on which $\phi = 0$. Hence it can be the surface of a conductor, and the above formula gives the field outside the conductor. The charge distribution on the plane part of the surface is given by

$$\sigma = -\frac{1}{4\pi}\left[\frac{\partial\phi}{\partial z}\right]_{z=0} = \sigma_0\left(1-\frac{R^3}{r^3}\right);$$

we have taken the constant in ϕ as $-4\pi\sigma_0$, so that σ_0 is the charge density far from the projection. On the surface of the projection we have

$$\sigma = -\frac{1}{4\pi}\left[\frac{\partial\phi}{\partial r}\right]_{r=R} = 3\sigma_0\frac{z}{R}.$$

PROBLEM 9. Determine the dipole moment of a thin conducting cylindrical rod, with length $2l$ and radius $a \ll l$, in an electric field $\mathfrak{E}$ parallel to its axis.

SOLUTION. Let $\tau(z)$ be the charge per unit length induced on the surface of the rod, and z the coordinate along the axis of the rod, measured from its midpoint. The condition of constant potential on the surface of the conductor is

$$-\mathfrak{E}z + \frac{1}{2\pi}\int_0^{2\pi}\int_{-l}^{l}\frac{\tau(z')\mathrm{d}z'\mathrm{d}\phi}{R} = 0,$$

$$R^2 = (z'-z)^2 + 4a^2\sin^2\tfrac{1}{2}\phi,$$

where ϕ is the angle between planes passing through the axis of the cylinder and through two points on its surface at a distance R apart. We divide the integral into two parts, putting $\tau(z') \equiv \tau(z) + [\tau(z') - \tau(z)]$. Since $l \gg a$, we have for points not too near the ends of the rod

$$\frac{\tau(z)}{2\pi}\int\int\frac{\mathrm{d}z'\mathrm{d}\phi}{R} \cong \frac{\tau(z)}{2\pi}\int_0^{2\pi}\log\frac{l^2-z^2}{a^2\sin^2\frac{1}{2}\phi}\mathrm{d}\phi = \tau(z)\log\frac{4(l^2-z^2)}{a^2},$$

using the result that $\int_0^\pi \log\sin\phi\,\mathrm{d}\phi = -\pi\log 2$. In the integral which contains the difference $\tau(z') - \tau(z)$, we can neglect the a^2 term in R, since it no longer causes the integral to diverge. Thus

$$\mathfrak{E}z = \tau(z)\log 4(l^2-z^2)/a^2 + \int_{-l}^{l}\frac{\tau(z')-\tau(z)}{|z'-z|}\mathrm{d}z'.$$

The quantity τ is almost proportional to z, and in this approximation the integral gives $-2\tau(z)$, the result being

$$\tau(z) = \frac{\mathfrak{E}z}{\log[4(l^2-z^2)/a^2]-2}.$$

This expression is invalid near the ends of the rod, but in calculating the dipole moment that region is unimportant. In the above approximation we have

$$\mathscr{P} = \int_{-l}^{l}\tau(z)z\,\mathrm{d}z = \frac{\mathfrak{E}}{L}\int_0^l\left\{z^2 - \frac{z^2}{2L}\log\left(1-\frac{z^2}{l^2}\right)\right\}\mathrm{d}z$$

$$= \frac{\mathfrak{E}l^3}{3L}\left\{1+\frac{1}{L}\left(\frac{4}{3}-\log 2\right)\right\},$$

where $L = \log(2l/a) - 1$ is large, or (with the same accuracy)

$$\mathscr{P} = \frac{\mathfrak{E}l^3}{3\log(4l/a)-7}.$$

PROBLEM 10. Determine the capacitance of a hollow conducting cap of a sphere.

SOLUTION. We take the origin O at a point on the rim of the cap (Fig. 6), and carry out the inversion transformation $r = l^2/r'$, where l is the diameter of the cap. The cap then becomes the half-plane shown by the dashed line in Fig. 6, which is perpendicular to the radius AO of the cap and passes through the point B on its rim.

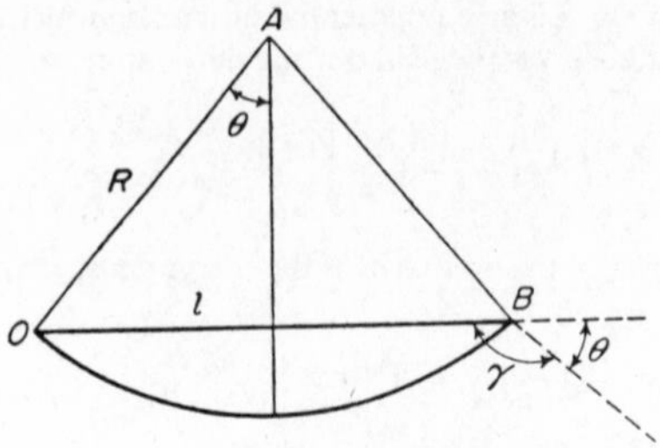

FIG. 6

The angle $\gamma = \pi - \theta$, where 2θ is the angle subtended by the diameter of the cap at the centre of the sphere.

If the charge on the cap is e and its potential is taken as zero, then as $r \to \infty$ the potential $\phi \to -\phi_0 + e/r$. Accordingly, in the transformed problem, as $r' \to 0$ the potential is $\phi' \to l\phi/r' \cong -l\phi_0/r' + e/l$, where the first term corresponds to a charge $e' = -l\phi_0$ at the origin.

According to formula (3.22) we have

$$\phi' = \frac{e'}{r'} - \frac{e'}{2\pi l}\left(1 + \frac{\theta}{\sin\theta}\right)$$

(the potential near a charge e' at a distance l from the edge of a conducting half-plane at zero potential). Comparing the two expressions, we have for the required capacitance $C = e/\phi_0$

$$C = \frac{l}{2\pi}\left(1 + \frac{\theta}{\sin\theta}\right) = \frac{R}{\pi}(\sin\theta + \theta),$$

where R is the radius of the cap.

PROBLEM 11. Determine the correction due to edge effects on the value $C = S/4\pi d$ for the capacitance of a plane capacitor (S being the area of the plates, and $d \ll \sqrt{S}$ the distance between them).

SOLUTION. Since the plates have free edges, the distribution of charge over them is not uniform. To determine the required correction in a first approximation, we consider points which are at distances x from the edge such that $d \ll x \ll \sqrt{S}$. For example, taking the upper layer (at potential $\phi = \frac{1}{2}\phi_0$, Fig. 7a) and neglecting its distance $\frac{1}{2}d$ from the midplane (the equipotential surface $\phi = 0$), we have the problem of the field near the boundary between two parts of a plane having different potentials (Fig. 7b). The solution is elementary†, and the excess charge density (relative to the value of σ far from the edge) is $\Delta\sigma = E_n/4\pi = \phi_0/8\pi^2 x$, so that the total excess charge is $L\int \Delta\sigma\, dx = (\phi_0 L/8\pi^2)\log(\sqrt{S}/d)$, where L is the perimeter of the plate. In calculating the logarithmically divergent integral, we have taken the limits as those of the region $d \ll x \ll \sqrt{S}$. Hence we find the capacitance

$$C = \frac{S}{4\pi d} + \frac{L}{8\pi^2}\log\frac{\sqrt{S}}{d}.$$

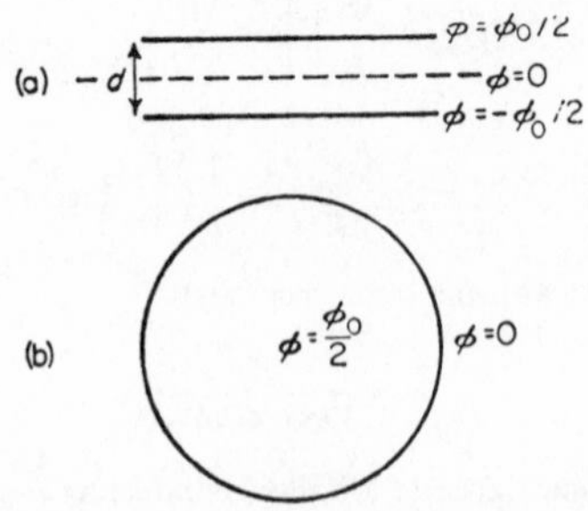

FIG. 7

† See §23. In formula (23.2) for the potential we must here put $\phi_{ab} = \frac{1}{2}\phi_0$, $\alpha = \pi$.

A more exact calculation (determining the coefficient in the argument of the logarithm) demands considerably more elaborate methods, and the result depends the shape of the plates. If these are circular, with radius R, we obtain *Kirchhoff's formula*

$$C = \frac{R^2}{4d} + \frac{R}{4\pi}\left(\log\frac{16\pi R}{d} - 1\right).$$

§4. A conducting ellipsoid

The problem of the field of a charged conducting ellipsoid and that of an ellipsoid in a uniform external field are solved by the use of *ellipsoidal coordinates*. These are related to Cartesian coordinates by the equation

$$\frac{x^2}{a^2+u} + \frac{y^2}{b^2+u} + \frac{z^2}{c^2+u} = 1 \quad (a > b > c). \tag{4.1}$$

This equation, a cubic in u, has three different real roots ξ, η, ζ, which lie in the following ranges:

$$\xi \geqslant -c^2, \quad -c^2 \geqslant \eta \geqslant -b^2, \quad -b^2 \geqslant \zeta \geqslant -a^2. \tag{4.2}$$

These three roots are the ellipsoidal coordinates of the point x, y, z. Their geometrical significance is seen from the fact that the surfaces of constant ξ, η and ζ are respectively ellipsoids and hyperboloids of one and two sheets, all confocal with the ellipsoid

$$x^2/a^2 + y^2/b^2 + z^2/c^2 = 1. \tag{4.3}$$

One surface of each of the three families passes through each point in space, and the three surfaces are orthogonal. The formulae for transformation from ellipsoidal to Cartesian coordinates are given by solving three simultaneous equations of the type (4.1), and are

$$\left.\begin{aligned} x &= \pm\sqrt{\left[\frac{(\xi+a^2)(\eta+a^2)(\zeta+a^2)}{(b^2-a^2)(c^2-a^2)}\right]}, \\ y &= \pm\sqrt{\left[\frac{(\xi+b^2)(\eta+b^2)(\zeta+b^2)}{(c^2-b^2)(a^2-b^2)}\right]}, \\ z &= \pm\sqrt{\left[\frac{(\xi+c^2)(\eta+c^2)(\zeta+c^2)}{(a^2-c^2)(b^2-c^2)}\right]}. \end{aligned}\right\} \tag{4.4}$$

The element of length in ellipsoidal coordinates is

$$\left.\begin{aligned} &\mathrm{d}l^2 = {h_1}^2\mathrm{d}\xi^2 + {h_2}^2\mathrm{d}\eta^2 + {h_3}^2\mathrm{d}\zeta^2, \\ &h_1 = \sqrt{[(\xi-\eta)(\xi-\zeta)]}/2R_\xi, \quad h_2 = \sqrt{[(\eta-\zeta)(\eta-\xi)]}/2R_\eta, \\ &h_3 = \sqrt{[(\zeta-\xi)(\zeta-\eta)]}/2R_\zeta, \quad {R_u}^2 = (u+a^2)(u+b^2)(u+c^2), \\ &u = \xi, \eta, \zeta. \end{aligned}\right\} \tag{4.5}$$

Accordingly, Laplace's equation in these coordinates is

$$\triangle\phi = \frac{4}{(\xi-\eta)(\zeta-\xi)(\eta-\zeta)} \times\left[(\eta-\zeta)R_\xi\frac{\partial}{\partial\xi}\left(R_\xi\frac{\partial\phi}{\partial\xi}\right) + (\zeta-\xi)R_\eta\frac{\partial}{\partial\eta}\left(R_\eta\frac{\partial\phi}{\partial\eta}\right) + (\xi-\eta)R_\zeta\frac{\partial}{\partial\zeta}\left(R_\zeta\frac{\partial\phi}{\partial\zeta}\right)\right] = 0. \tag{4.6}$$

If two of the semiaxes a, b, c become equal, the system of ellipsoidal coordinates degenerates. Let $a = b > c$. Then the cubic equation (4.1) becomes a quadratic,

$$\frac{\rho^2}{a^2+u} + \frac{z^2}{c^2+u} = 1, \qquad \rho^2 = x^2+y^2, \tag{4.7}$$

with two roots whose values lie in the ranges $\xi \geqslant -c^2$, $-c^2 \geqslant \eta \geqslant -a^2$. The coordinate surfaces of constant ξ and η become respectively confocal oblate spheroids and confocal hyperboloids of revolution of one sheet (Fig. 8). As the third coordinate we can take the polar angle ϕ in the xy-plane ($x = \rho\cos\phi$, $y = \rho\sin\phi$). For $a = b$ the ellipsoidal coordinate ζ degenerates to a constant, $-a^2$. Its relation to the angle ϕ is given by the way in which it tends to $-a^2$ as b tends to a, namely

$$\cos\phi = \sqrt{[(a^2+\zeta)/(a^2-b^2)]} \quad \text{as} \quad b \to a. \tag{4.8}$$

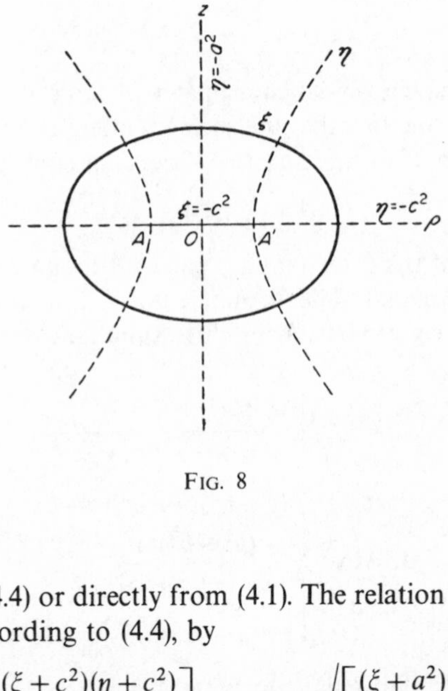

FIG. 8

This is easily seen from (4.4) or directly from (4.1). The relation between the coordinates z, ρ and ξ, η is given, according to (4.4), by

$$z = \pm\sqrt{\left[\frac{(\xi+c^2)(\eta+c^2)}{c^2-a^2}\right]}, \qquad \rho = \sqrt{\left[\frac{(\xi+a^2)(\eta+a^2)}{a^2-c^2}\right]}. \tag{4.9}$$

The coordinates ξ, η, ϕ are called *oblate spheroidal coordinates*.†

Similarly, for $a > b = c$ ellipsoidal coordinates become *prolate spheroidal coordinates*. Two coordinates ξ and ζ are roots of the equation

$$\frac{x^2}{a^2+u} + \frac{\rho^2}{b^2+u} = 1, \qquad \rho^2 = y^2+z^2, \tag{4.10}$$

where $\xi \geqslant -b^2$, $-b^2 \geqslant \zeta \geqslant -a^2$. The surfaces of constant ξ and ζ are prolate spheroids

† We here define spheroidal coordinates to be the limit of ellipsoidal coordinates. Other definitions are used in the literature, but are easily related to ours.

and hyperboloids of revolution of two sheets (Fig. 9). The coordinate η degenerates to a constant, $-b^2$, for $c \to b$, and we have

$$\cos\phi = \sqrt{[(b^2+\eta)/(b^2-c^2)]}, \tag{4.11}$$

where ϕ is the polar angle in the yz-plane. The relation between the coordinates x, ρ and ξ, ζ is given by

$$x = \pm\sqrt{\left[\frac{(\xi+a^2)(\zeta+a^2)}{a^2-b^2}\right]}, \quad \rho = \sqrt{\left[\frac{(\xi+b^2)(\zeta+b^2)}{b^2-a^2}\right]}. \tag{4.12}$$

In a system of oblate spheroidal coordinates the foci of the spheroids and hyperboloids lie on a circle of radius $\sqrt{(a^2-c^2)}$ in the xy-plane; in Fig. 8 AA' is a diameter of this circle. Let us draw a plane passing through the z-axis and some point P. It intersects the focal circle at two points; let their distances from P be r_1, r_2. If the coordinates of P are ρ, z, then

$$r_1^{\,2} = [\rho-\sqrt{(a^2-c^2)}]^2+z^2, \qquad r_2^{\,2} = [\rho+\sqrt{(a^2-c^2)}]^2+z^2.$$

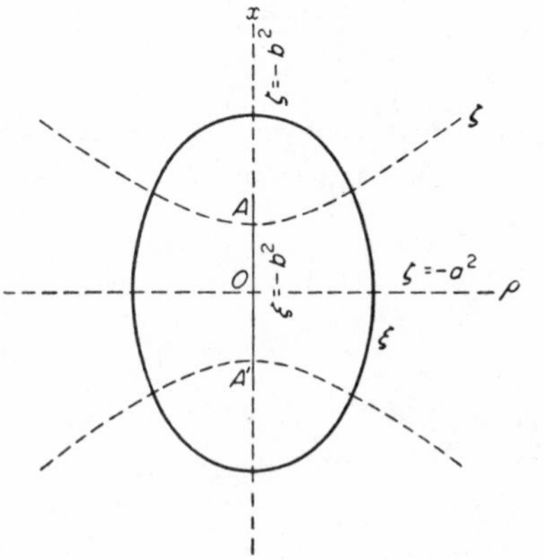

FIG. 9

The spheroidal coordinates ξ, η are given in terms of r_1, r_2 by

$$\xi = \tfrac{1}{4}(r_1+r_2)^2-a^2, \qquad \eta = \tfrac{1}{4}(r_2-r_1)^2-a^2. \tag{4.13}$$

In a system of prolate spheroidal coordinates the foci are the points $x = \pm\sqrt{(a^2-b^2)}$ on the x-axis (the points A, A' in Fig. 9). If r_1 and r_2 are the distances of these foci from P, then

$$r_1^{\,2} = \rho^2+[x-\sqrt{(a^2-b^2)}]^2, \quad r_2^{\,2} = \rho^2+[x+\sqrt{(a^2-b^2)}]^2,$$

and the spheroidal coordinates ξ, ζ are given in terms of r_1, r_2 by the same formulae (4.13), with ζ in place of η.

Let us now turn to the problem of the field of a charged ellipsoid whose surface is given by the equation (4.3). In ellipsoidal coordinates this is the surface $\xi = 0$. It is therefore clear that, if we seek the field potential as a function of ξ only, all the ellipsoidal surfaces ξ = constant, and in particular the surface of the conductor, will be equipotential surfaces.

Laplace's equation (4.6) then becomes

$$\frac{d}{d\xi}\left(R_\xi \frac{d\phi}{d\xi}\right) = 0,$$

whence

$$\phi(\xi) = A \int_\xi^\infty \frac{d\xi}{R_\xi}.$$

The upper limit of integration is taken so that the field is zero at infinity. The constant A is most simply determined from the condition that at large distances r the field must become a Coulomb field and $\phi \cong e/r$, where e is the total charge on the conductor. When $r \to \infty$, $\xi \to \infty$, and $\xi \cong r^2$, as we see from equation (4.1) with $u = \xi$. For large ξ we have $R_\xi \cong \xi^{3/2}$, and $\phi \cong 2A/\sqrt{\xi} = 2A/r$. Hence $2A = e$, and therefore

$$\phi(\xi) = \tfrac{1}{2}e \int_\xi^\infty \frac{d\xi}{R_\xi}. \tag{4.14}$$

The integral is an elliptic integral of the first kind. The surface of the conductor corresponds to $\xi = 0$, and so the capacitance of the conductor is given by

$$\frac{1}{C} = \tfrac{1}{2} \int_0^\infty \frac{d\xi}{R_\xi} \tag{4.15}$$

The distribution of charge on the surface of the ellipsoid is determined by the normal derivative of the potential:

$$\sigma = -\frac{1}{4\pi}\left[\frac{\partial\phi}{\partial n}\right]_{\xi=0} = -\frac{1}{4\pi}\left[\frac{1}{h_1}\frac{d\phi}{d\xi}\right]_{\xi=0} = \frac{e}{4\pi}\frac{1}{\sqrt{(\eta\zeta)}}.$$

From equations (4.4) we easily see that for $\xi = 0$

$$\frac{x^2}{a^4} + \frac{y^2}{b^4} + \frac{z^2}{c^4} = \frac{\eta\zeta}{a^2b^2c^2}.$$

Hence

$$\sigma = \frac{e}{4\pi abc}\left(\frac{x^2}{a^4} + \frac{y^2}{b^4} + \frac{z^2}{c^4}\right)^{-\frac{1}{2}} \tag{4.16}$$

For a spheroid the integrals (4.14), (4.15) degenerate and can be expressed in terms of elementary functions. For a prolate spheroid ($a > b = c$) the field potential is

$$\phi = \frac{e}{\sqrt{(a^2 - b^2)}} \tanh^{-1} \sqrt{\frac{a^2 - b^2}{\xi + a^2}}, \tag{4.17}$$

and the capacitance is

$$C = \frac{\sqrt{(a^2 - b^2)}}{\cosh^{-1}(a/b)}. \tag{4.18}$$

For an oblate spheroid ($a = b > c$) we have

$$\phi = \frac{e}{\sqrt{(a^2 - c^2)}} \tan^{-1} \sqrt{\frac{a^2 - c^2}{\xi + c^2}}, \qquad C = \frac{\sqrt{(a^2 - c^2)}}{\cos^{-1}(c/a)}. \tag{4.19}$$

In particular, for a circular disc ($a = b$, $c = 0$)

$$C = 2a/\pi. \tag{4.20}$$

Let us now consider the problem of an uncharged conducting ellipsoid in a uniform external electric field $\mathfrak{E}$. Without loss of generality we may take the field $\mathfrak{E}$ to be along one of the axes of the ellipsoid. In any other case this field may be resolved into components along the three axes, and the resultant field is a superposition of those arising from each component separately.

The potential of a uniform field $\mathfrak{E}$ along the x-axis (the a-axis of the ellipsoid) is, in ellipsoidal coordinates,

$$\phi_0 = -\mathfrak{E}x = -\mathfrak{E}\sqrt{[(\xi + a^2)(\eta + a^2)(\zeta + a^2)/(b^2 - a^2)(c^2 - a^2)]}. \tag{4.21}$$

We write the field potential outside the ellipsoid as $\phi = \phi_0 + \phi'$, where ϕ' gives the required perturbation of the external field by the ellipsoid, and seek ϕ' in the form

$$\phi' = \phi_0 F(\xi). \tag{4.22}$$

In this function the factors depending on η and ζ are the same as in ϕ_0; this enables us to satisfy the boundary condition at $\xi = 0$ for arbitrary η, ζ (i.e. on the surface of the ellipsoid). Substituting (4.22) in Laplace's equation (4.6), we obtain for $F(\xi)$ the equation

$$\frac{d^2F}{d\xi^2} + \frac{dF}{d\xi}\frac{d}{d\xi}\log[R_\xi(\xi + a^2)] = 0.$$

One solution of this equation is F = constant, and the other is

$$F(\xi) = A\int_\xi^\infty \frac{d\xi}{(\xi + a^2)R_\xi}. \tag{4.23}$$

The upper limit of integration is taken so that $\phi' \to 0$ for $\xi \to \infty$. The integral is an elliptic integral of the second kind.

We must have ϕ = constant on the surface of the ellipsoid. For this condition to be satisfied with $\xi = 0$ and arbitrary η, ζ, the constant value of ϕ must be zero. Determining the coefficient A in $F(\xi)$ so that $F(0) = -1$, we obtain the following final expression for the field potential near the ellipsoid:

$$\phi = \phi_0\left\{1 - \int_\xi^\infty \frac{ds}{(s + a^2)R_s} \bigg/ \int_0^\infty \frac{ds}{(s + a^2)R_s}\right\}. \tag{4.24}$$

Let us find the form of the potential ϕ' at large distances r from the ellipsoid. For large r, the coordinate ξ is large, and $\xi \cong r^2$, as follows at once from equation (4.1). Hence

$$\int_\xi^\infty \frac{ds}{(s + a^2)R_s} \cong \int_{r^2}^\infty \frac{ds}{s^{5/2}} = \frac{2}{3r^3},$$

and the potential $\phi' = \mathfrak{E}xV/4\pi n^{(x)}r^3$, where $V = \frac{4}{3}\pi abc$ is the volume of the ellipsoid and $n^{(x)}$, $n^{(y)}$, $n^{(z)}$ are defined by

$$\left.\begin{aligned} n^{(x)} &= \tfrac{1}{2}abc\int_0^\infty \frac{ds}{(s+a^2)R_s}, \quad n^{(y)} = \tfrac{1}{2}abc\int_0^\infty \frac{ds}{(s+b^2)R_s}, \\ n^{(z)} &= \tfrac{1}{2}abc\int_0^\infty \frac{ds}{(s+c^2)R_s}. \end{aligned}\right\} \tag{4.25}$$

The expression for ϕ' is, as we should expect, the potential of an electric dipole: $\phi' = x\mathscr{P}_x/r^3$, where the dipole moment of the ellipsoid is

$$\mathscr{P}_x = \mathfrak{E}_x V/4\pi n^{(x)}. \tag{4.26}$$

Analogous expressions give the dipole moment when the field $\mathfrak{E}$ is along the y or z axis.

The positive constants $n^{(x)}$, $n^{(y)}$, $n^{(z)}$ depend only on the shape of the ellipsoid, and not on its volume; they are called the *depolarizing factors*.† If the coordinate axes do not necessarily coincide with those of the ellipsoid, formula (4.26) must be written in the tensor form

$$(4\pi/V)n_{ik}\mathscr{P}_k = \mathfrak{E}_i. \tag{4.27}$$

The quantities $n^{(x)}$, $n^{(y)}$, $n^{(z)}$ are the principal values of the symmetrical tensor n_{ik} of rank two. Comparison with the definition (2.13) shows that $\alpha_{ik} = n^{-1}{}_{ik}/4\pi$ is the polarizability tensor of the conducting ellipsoid.

In the general case of arbitrary a, b, c, it follows from the definitions of $n^{(x)}$, $n^{(y)}$, $n^{(z)}$ that

$$n^{(x)} < n^{(y)} < n^{(z)} \quad \text{if} \quad a > b > c. \tag{4.28}$$

Further, by adding the integrals for $n^{(x)}$, $n^{(y)}$, $n^{(z)}$ and using as the variable of integration $u = R_s^2$, we find

$$n^{(x)} + n^{(y)} + n^{(z)} = \tfrac{1}{2}abc \int_{(abc)^2}^\infty \frac{du}{u^{3/2}},$$

whence

$$n^{(x)} + n^{(y)} + n^{(z)} = 1. \tag{4.29}$$

The sum of the three depolarizing factors is thus unity; in tensor notation, $n_{ii} = 1$. Since these coefficients are positive, none can exceed unity.

For a sphere ($a = b = c$) it is evident from symmetry that

$$n^{(x)} = n^{(y)} = n^{(z)} = \tfrac{1}{3}. \tag{4.30}$$

For a cylinder with its axis in the x-direction ($a \to \infty$), we have‡

$$n^{(x)} = 0, \quad n^{(y)} = n^{(z)} = \tfrac{1}{2}. \tag{4.31}$$

† The same coefficients occur in problems concerning a dielectric ellipsoid in an external electric field, or a magnetic ellipsoid in a magnetic field (§8). Tables and graphs of the coefficients for spheroids and ellipsoids have been given by E. C. Stoner (*Philosophical Magazine* [7] **36**, 803, 1945) and J. A. Osborn (*Physical Review* **67**, 351, 1945).

‡ These values for a sphere and a cylinder agree, of course, with those found in §3, Problems 1 and 2.

The limiting case $a, b \to \infty$ (a flat plate) corresponds to the obvious values

$$n^{(x)} = n^{(y)} = 0, \qquad n^{(z)} = 1.$$

The elliptic integrals (4.25) can be expressed in terms of elementary functions if the ellipsoid is a spheroid. For a prolate spheroid ($a > b = c$) of eccentricity $e = \sqrt{(1-b^2/a^2)}$,

$$n^{(x)} = \frac{1-e^2}{2e^3}\left(\log\frac{1+e}{1-e} - 2e\right), \quad n^{(y)} = n^{(z)} = \tfrac{1}{2}(1-n^{(x)}). \tag{4.32}$$

If the spheroid is nearly spherical ($e \ll 1$) we have approximately

$$n^{(x)} = \tfrac{1}{3} - \tfrac{2}{15}e^2, \quad n^{(y)} = n^{(z)} = \tfrac{1}{3} + \tfrac{1}{15}e^2. \tag{4.33}$$

For an oblate spheroid ($a = b > c$)

$$n^{(z)} = \frac{1+e^2}{e^3}(e - \tan^{-1}e), \quad n^{(x)} = n^{(y)} = \tfrac{1}{2}(1-n^{(z)}), \tag{4.34}$$

where $e = \sqrt{(a^2/c^2 - 1)}$. If $e \ll 1$, then

$$n^{(z)} = \tfrac{1}{3} + \tfrac{2}{15}e^2, \quad n^{(x)} = n^{(y)} = \tfrac{1}{3} - \tfrac{1}{15}e^2. \tag{4.35}$$

PROBLEMS

PROBLEM 1. Find the field of a charged conducting circular disc with radius a, expressing it in cylindrical coordinates. Find the distribution of charge on the disc.

SOLUTION. The charge distribution is obtained by taking the limit of formula (4.16) as $c \to 0$, $z \to 0$, with $z/c = \sqrt{(1-r^2/a^2)}$ (where $r^2 = x^2 + y^2$), in accordance with (4.3). This gives

$$\sigma = \frac{e}{4\pi a^2}\left(1 - \frac{r^2}{a^2}\right)^{-\frac{1}{2}}.$$

The field potential is given in all space by formula (4.19), where we put $c = 0$ and express ξ in terms of r and z by means of equation (4.1) with $c = 0$, $u = \xi$, $a = b$:

$$\phi = \frac{e}{a}\tan^{-1}\left[\frac{2a^2}{r^2+z^2-a^2+\sqrt{[(r^2+z^2-a^2)^2+4a^2z^2]}}\right]^{\frac{1}{2}}.$$

Near the edge of the disc, we replace r and z by coordinates ρ and θ such that $z = \rho\sin\theta$, $r = a - \rho\cos\theta$ (Fig. 10, p. 26; $\rho \ll a$), obtaining

$$\phi \cong \frac{e}{a}\left(\tfrac{1}{2}\pi - \sqrt{\frac{2\rho}{a}}\sin\tfrac{1}{2}\theta\right),$$

in agreement with the general result derived in §3, Problem 3.

PROBLEM 2. Determine the electric quadrupole moment of a charged ellipsoid.

SOLUTION. The *quadrupole moment tensor* of a charged conductor is defined as $D_{ik} = e(3\overline{x_i x_k} - \overline{r^2}\delta_{ik})$, where e is the total charge, and the bar denotes an average such as

$$\overline{x_i x_k} = \frac{1}{e}\oint x_i x_k \sigma \, df.$$

It is evident that the axes of the ellipsoid are also the principal axes of the tensor D_{ik}. Using formula (4.16) for σ, and for the element of surface of the ellipsoid the expression

$$df = \frac{dx\,dy}{\nu_z} = \frac{dx\,dy}{z/c^2}\sqrt{\left[\frac{x^2}{a^4}+\frac{y^2}{b^4}+\frac{z^2}{c^4}\right]},$$

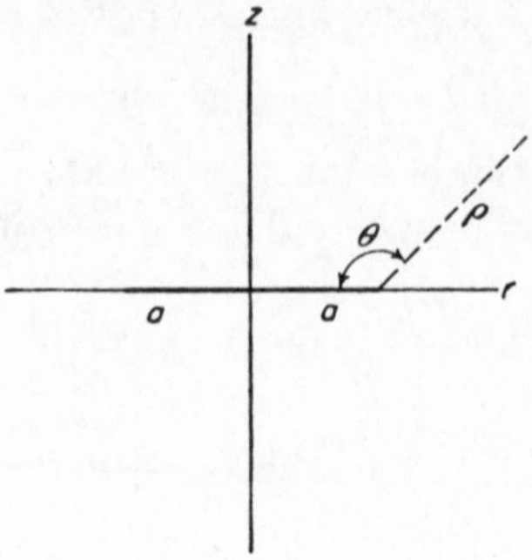

FIG. 10

where $\mathbf{v}$ is a unit vector normal to the surface, we obtain

$$\overline{z^2} = \frac{c}{4\pi ab}\int z\,dx\,dy = \tfrac{1}{3}c^2;$$

the integration over x and y covers twice the area of the cross-section of the ellipsoid by the xy-plane. Thus

$$D_{xx} = \tfrac{1}{5}e(2a^2-b^2-c^2),\quad D_{yy} = \tfrac{1}{5}e(2b^2-c^2-a^2),\quad D_{zz} = \tfrac{1}{5}e(2c^2-a^2-b^2).$$

PROBLEM 3. Determine the distribution of charge on the surface of an uncharged conducting ellipsoid placed in a uniform external field.

SOLUTION. According to formula (1.9) we have

$$\sigma = -\frac{1}{4\pi}\left[\frac{\partial\phi}{\partial n}\right]_{\xi=0} = -\left[\frac{1}{4\pi h_1}\frac{\partial\phi}{\partial\xi}\right]_{\xi=0};$$

by (4.5) the element of length along the normal to the surface of the ellipsoid is $h_1\,d\xi$. Substituting (4.24) and using the fact that

$$\nu_x = \left[\frac{1}{h_1}\frac{\partial x}{\partial\xi}\right]_{\xi=0} = \left[\frac{x}{2a^2h_1}\right]_{\xi=0},$$

we have $\sigma = \mathfrak{E}\nu_x/4\pi n^{(x)}$ when the external field is in the x-direction. When the direction of the external field is arbitrary this becomes

$$\sigma = \frac{1}{4\pi}\nu_i n^{-1}{}_{ik}\mathfrak{E}_k = \frac{1}{4\pi}\left[\frac{\nu_x}{n^{(x)}}\mathfrak{E}_x + \frac{\nu_y}{n^{(y)}}\mathfrak{E}_y + \frac{\nu_z}{n^{(z)}}\mathfrak{E}_z\right].$$

PROBLEM 4. The same as Problem 3, but for a plane circular disc with radius a lying parallel to the field.† Determine also the dipole moment of the disc.

SOLUTION. Let us regard the disc as the limit of a spheroid when the semiaxis c tends to zero. The depolarizing factor along this axis (the z-axis) tends to 1, and those along the x and y axes tend to zero: $n^{(z)} = 1-\pi c/2a$, $n^{(x)} = n^{(y)} = \pi c/4a$, by (4.34). The component ν_x of the unit vector along the normal to the surface of the spheroid tends to zero:

$$\nu_x = \frac{x}{a^2}\left(\frac{x^2+y^2}{a^4}+\frac{z^2}{c^4}\right)^{-\frac{1}{2}} \to \frac{x}{a^2}\frac{c^2}{z} = \frac{xc}{a^2}\left(1-\frac{x^2+y^2}{a^2}\right)^{-\frac{1}{2}}.$$

Hence the charge density is

$$\sigma = \frac{\mathfrak{E}}{4\pi}\frac{\nu_x}{n^{(x)}} = \frac{\mathfrak{E}\rho\cos\phi}{\pi^2\sqrt{(a^2-\rho^2)}},$$

where ρ and ϕ are polar coordinates in the plane of the disc.

† The problem for a disc lying perpendicular to the field is trivial: the field remains uniform in all space, and charges $\sigma = \pm\mathfrak{E}/4\pi$ are induced on the two sides of the disc.

The dipole moment of the disc is obtained from formula (4.26), and is $\mathscr{P} = 4a^3\mathfrak{E}/3\pi$. Thus it is proportional to a^3, and not to the "volume" a^2c of the disc.

PROBLEM 5. Determine the field potential outside an uncharged conducting spheroid with its axis of symmetry parallel to a uniform external field.

SOLUTION. For a prolate spheroid ($a > b = c$, with the field $\mathfrak{E}$ in the x-direction) we find, on calculating the integral in formula (4.24),

$$\phi = -\mathfrak{E}x\left\{1 - \frac{\tanh^{-1}\sqrt{[(a^2-b^2)/(\xi+a^2)]} - \sqrt{[(a^2-b^2)/(\xi+a^2)]}}{\tanh^{-1}\sqrt{(1-b^2/a^2)} - \sqrt{(1-b^2/a^2)}}\right\}.$$

The coordinate ξ is related to x and $\rho = \sqrt{(y^2+z^2)}$ by

$$\frac{\rho^2}{b^2+\xi} + \frac{x^2}{a^2+\xi} = 1,$$

with $0 \leqslant \xi \leqslant \infty$ in the space outside the ellipsoid.

For an oblate spheroid ($a = b > c$) the field $\mathfrak{E}$ is along the z-axis. We must therefore replace $s + a^2$ by $s + c^2$ and put $\phi_0 = -\mathfrak{E}z$ in the integrals in (4.24). Then

$$\phi = -\mathfrak{E}z\left\{1 - \frac{\sqrt{[(a^2-c^2)/(\xi+c^2)]} - \tan^{-1}\sqrt{[(a^2-c^2)/(\xi+c^2)]}}{\sqrt{(a^2/c^2-1)} - \tan^{-1}\sqrt{(a^2/c^2-1)}}\right\},$$

where the coordinate ξ is related to z and $\rho = \sqrt{(x^2+y^2)}$ by

$$\frac{\rho^2}{a^2+\xi} + \frac{z^2}{c^2+\xi} = 1.$$

PROBLEM 6. The same as Problem 5, but with the axis of symmetry perpendicular to the external field.

SOLUTION. For a prolate spheroid (with the field along the z-axis)

$$\phi = -\mathfrak{E}z\left\{1 - \frac{\sqrt{(\xi+a^2)/(\xi+b^2)} - (a^2-b^2)^{-\frac{1}{2}}\tanh^{-1}\sqrt{[(a^2-b^2)/(\xi+a^2)]}}{a/b^2 - (a^2-b^2)^{-\frac{1}{2}}\tanh^{-1}\sqrt{(1-b^2/a^2)}}\right\}.$$

For an oblate spheroid (with the field along the x-axis)

$$\phi = -\mathfrak{E}x\left\{1 - \frac{(a^2-c^2)^{-\frac{1}{2}}\tan^{-1}\sqrt{[(a^2-c^2)/(\xi+c^2)]} - \sqrt{(\xi+c^2)/(\xi+a^2)}}{(a^2-c^2)^{-\frac{1}{2}}\tan^{-1}\sqrt{(a^2/c^2-1)} - c/a^2}\right\}.$$

PROBLEM 7. A uniform field $\mathfrak{E}$ in the z-direction (in the half-space $z < 0$) is bounded by an earthed conducting plane at $z = 0$, containing a circular aperture. Determine the field and charge distribution on the plane.

SOLUTION. The xy-plane with a circular aperture of radius a and centre at the origin may be regarded as the limit of the hyperboloids of revolution of one sheet

$$\frac{\rho^2}{a^2-|\eta|} - \frac{z^2}{|\eta|} = 1, \qquad \rho^2 = x^2+y^2,$$

as $|\eta| \to 0$. These hyperboloids are one of the families of coordinate surfaces in a system of oblate spheroidal coordinates with $c = 0$. The Cartesian coordinate z, according to (4.9), is given in terms of ξ and η by $z = \sqrt{(\xi|\eta|)}/a$, and $\sqrt{\xi}$ must be taken with the positive and negative sign in the upper and lower half-space respectively.

Let us seek a solution in the form $\phi = -\mathfrak{E}z\,F(\xi)$. For the function $F(\xi)$ we obtain

$$F(\xi) = \text{constant} \times \int \frac{d\xi}{\xi^{\frac{3}{2}}(\xi+a^2)} = \text{constant} \times \left[\frac{a}{\sqrt{\xi}} - \tan^{-1}\frac{a}{\sqrt{\xi}}\right];$$

the constant of integration is put equal to zero in accordance with the condition $\phi = 0$ for $z \to +\infty$, i.e. $\sqrt{\xi} \to +\infty$. The inverse tangent of a negative quantity must be taken as $\tan^{-1}(a/-\sqrt{\xi}) = \pi - \tan^{-1}(a/\sqrt{\xi})$, and not as $-\tan^{-1}(a/\sqrt{\xi})$ since the potential would then be discontinuous at the aperture ($\xi = 0$). The constant coefficient is chosen so that, for $z \to -\infty$ (i.e. for $\sqrt{\xi} \to -\infty$ and $\tan^{-1}(a/\sqrt{\xi}) \to \pi$), $\phi \to -\mathfrak{E}z$, and so we finally have

$$\phi = -\frac{\mathfrak{E}z}{\pi}\left[\tan^{-1}\frac{a}{\sqrt{\xi}} - \frac{a}{\sqrt{\xi}}\right] = -\frac{\mathfrak{E}}{\pi}\sqrt{|\eta|}\left[\frac{\sqrt{\xi}}{a}\tan^{-1}\frac{a}{\sqrt{\xi}} - 1\right].$$

On the conducting plane $\eta = 0$ and the potential is zero, as it should be.

At large distances $r = \sqrt{(z^2+\rho^2)}$ from the aperture we have $\xi \cong r^2$, and the potential (in the upper half-space) is

$$\phi \cong \frac{\mathfrak{E}a^2}{3\pi}\frac{\sqrt{-\eta}}{\xi} = \mathfrak{E}a^3 z/3\pi r^3,$$

i.e. we have a dipole field, the moment of the dipole being $\mathscr{P} = \mathfrak{E}a^3/3\pi$.

The field decreases as $1/r^3$, and therefore the flux of the field through an infinitely remote surface (in the half-space $z > 0$) is zero. This means that all the lines of force passing through the aperture reach the upper side of the conducting plane.

The distribution of charge on the conducting plane is given by

$$\sigma = \mp\frac{1}{4\pi}\left[\frac{\partial\phi}{\partial z}\right]_{z=0} = \mp\frac{a}{4\pi\sqrt{\xi}}\frac{\partial\phi}{\partial\sqrt{-\eta}} = \pm\frac{\mathfrak{E}}{4\pi^2}\left[\tan^{-1}\frac{a}{\sqrt{\xi}} - \frac{a}{\sqrt{\xi}}\right],$$

where the upper and lower signs refer to the upper and lower sides of the plane respectively. According to the formula

$$\frac{\rho^2}{a^2+\xi} + \frac{z^2}{\xi} = 1,$$

which relates ξ to ρ, z, we have $\sqrt{\xi} = \pm\sqrt{(\rho^2-a^2)}$ on the plane $z = 0$. Thus the charge distribution on the lower side of the conducting plane is given by the formula

$$\sigma = -\frac{\mathfrak{E}}{4\pi^2}\left(\pi - \sin^{-1}\frac{a}{\rho} + \frac{a}{\sqrt{(\rho^2-a^2)}}\right).$$

As $\rho \to \infty$ we have $\sigma = -\mathfrak{E}/4\pi$, as we should expect. On the upper side

$$\sigma = -\frac{\mathfrak{E}}{4\pi^2}\left(\frac{a}{\sqrt{(\rho^2-a^2)}} - \sin^{-1}\frac{a}{\rho}\right).$$

The total induced charge on the upper side of the plane is finite:

$$e' = \int_a^\infty \sigma\cdot 2\pi\rho\cdot d\rho = -\tfrac{1}{8}a^2\mathfrak{E}.$$

PROBLEM 8. The same as Problem 7, but for a plane with a slit of width $2b$.

SOLUTION. The xy-plane with a slit along the x-axis may be regarded as the limit of the hyperbolic cylinders

$$\frac{y^2}{b^2-|\eta|} - \frac{z^2}{|\eta|} = 1$$

as $|\eta| \to 0$. These hyperbolic cylinders are one of the families of coordinate surfaces in a system of ellipsoidal coordinates with $a \to \infty$, $c \to 0$. The Cartesian coordinate $z = \sqrt{(\xi|\eta|)}/b$.

As in Problem 7, we seek a solution in the form $\phi = -\mathfrak{E}zF(\xi)$, obtaining for the function F

$$F = \text{constant} \times \int \frac{d\xi}{\xi^{\frac{3}{2}}\sqrt{(\xi+b^2)}}.$$

Here the coefficient and the constant of integration are determined by the conditions that $F = 0$ and 1 for $z \to +\infty$ and $-\infty$ respectively (i.e. for $\sqrt{\xi} \to +\infty$ and $-\infty$), and the final result is

$$\phi = \frac{\mathfrak{E}}{2b}[\sqrt{(\xi+b^2)} \mp \sqrt{\xi}]\sqrt{|\eta|},$$

where we now take $\sqrt{\xi}$ positive and the two signs $\mp$ correspond to the regions $z > 0$ and $z < 0$.

At large distances from the slit we have in the upper half-space $\xi \cong y^2 + z^2 = r^2$, and the potential is

$\phi \cong \frac{1}{4}b\mathfrak{E}\sqrt{(|\eta|/\xi)} = \frac{1}{4}\mathfrak{E}b^2z/r^2$, i.e. the field of a two-dimensional dipole of moment $\frac{1}{8}\mathfrak{E}b^2$ per unit length of the slit (see the formula in §3, Problem 2).

The distribution of charge on the conducting plane is given by

$$\sigma = -\frac{\mathfrak{E}}{8\pi}\left(\frac{|y|}{\sqrt{(y^2-b^2)}} \mp 1\right).$$

The total induced charge on the upper side of the plane, per unit length of the slit, is

$$e' = 2\int_b^\infty \sigma\, dy = -\mathfrak{E}b/4\pi.$$

Near the edge of the slit we can take $\xi \to 0$ in the expression for $\phi(\xi, \eta)$, obtaining

$$\eta \cong -2b\rho \sin^2\tfrac{1}{2}\theta,$$

where ρ and θ are polar coordinates in the yz-plane, measured from the slit edge ($y = b + \rho\cos\theta$, $z = \rho\sin\theta$). Then

$$\phi \cong \mathfrak{E}\sqrt{(\tfrac{1}{2}b\rho)}\sin\tfrac{1}{2}\theta,$$

in agreement with the result in §3, Problem 3, for the case $\theta_0 \ll 1$.

§5. The forces on a conductor

In an electric field certain forces act on the surface of a conductor. These forces are easily calculated as follows.

The momentum flux density in an electric field in a vacuum is given by the *Maxwell stress tensor*:†

$$-\sigma_{ik} = \frac{1}{4\pi}(\tfrac{1}{2}E^2\delta_{ik} - E_iE_k).$$

The force on an element $d\mathbf{f}$ of the surface of the body is just the flux of momentum through it from outside, and is therefore $\sigma_{ik}df_k = \sigma_{ik}n_k df$ (the sign is changed because the normal vector $\mathbf{n}$ is outwards and not inwards). The quantity $\sigma_{ik}n_k$ is thus the force $\mathbf{F}_s$ per unit area of the surface. Since, at the surface of a conductor, the field $\mathbf{E}$ has no tangential component, we obtain

$$\mathbf{F}_s = \mathbf{n}E^2/8\pi, \tag{5.1}$$

or, introducing the surface charge density σ,

$$\mathbf{F}_s = 2\pi\sigma^2\mathbf{n} = \tfrac{1}{2}\sigma\mathbf{E}.$$

We therefore conclude that a "negative pressure" acts on the surface of a conductor; it is directed along the outward normal to the surface, and its magnitude is equal to the energy density in the field.

The total force $\mathbf{F}$ on the conductor is obtained by integrating the force (5.1) over the whole surface‡:

$$\mathbf{F} = \oint (E^2/8\pi)\, d\mathbf{f}. \tag{5.2}$$

† See *Fields*, §33. The stress tensor σ_{ik} is usually defined with the opposite sign to the momentum flux density tensor. This definition has been used elsewhere in the *Course of Theoretical Physics*, but by an oversight the definition of σ_{ik} in *Fields*, §33, had the other sign.

‡ In the present case we are applying this formula to a surface which does not precisely coincide with that of the body, but is some distance away, in order to exclude the effect of the field structure near the surface (see §1).

Usually, however, it is more convenient to calculate this quantity from the general laws of mechanics, by differentiating the energy $\mathscr{U}$. The force, in the direction of a coordinate q, acting on a conductor is $-\partial\mathscr{U}/\partial q$, where the derivative signifies the rate of change of energy when the body is translated in the q-direction. The energy must be expressed in terms of the charges on the conductors (which give rise to the field), and the differentiation is performed with the charges constant. Denoting this by the suffix e, we write

$$F_q = -(\partial\mathscr{U}/\partial q)_e. \tag{5.3}$$

Similarly, the projection, on any axis, of the total moment of the forces on the conductor is

$$K = -(\partial\mathscr{U}/\partial\psi)_e, \tag{5.4}$$

where ψ is the angle of rotation of the body about that axis.

If, however, the energy is expressed as a function of the potentials of the conductors, and not of their charges, the calculation of the forces from the energy requires special consideration. The reason is that, to maintain constant the potential of a moving conductor, it is necessary to use other bodies. For example, the potential of a conductor can be kept constant by connecting it to another conductor of very large capacitance, a "charge reservoir". On receiving a charge e_a, the conductor takes it from the reservoir, whose potential ϕ_a is unchanged on account of its large capacitance, although its energy is reduced by $e_a\phi_a$. When the whole system of conductors receives charges e_a, the energy of the reservoirs connected to them changes by a total of $-\Sigma e_a\phi_a$. Only the energy of the conductors, and not that of the reservoirs, appears in $\mathscr{U}$. In this sense we can say that $\mathscr{U}$ pertains to a system which is not energetically closed. Thus, for a system of conductors whose potentials are kept constant, the part of the mechanical energy is played not by $\mathscr{U}$, but by

$$\tilde{\mathscr{U}} = \mathscr{U} - \sum_a e_a\phi_a. \tag{5.5}$$

Substituting (2.2), we find that $\mathscr{U}$ and $\tilde{\mathscr{U}}$ differ only in sign:

$$\tilde{\mathscr{U}} = -\mathscr{U}. \tag{5.6}$$

The force F_q is obtained by differentiating $\tilde{\mathscr{U}}$ with respect to q for constant potentials, i.e.

$$F_q = -(\partial\tilde{\mathscr{U}}/\partial q)_\phi = (\partial\mathscr{U}/\partial q)_\phi. \tag{5.7}$$

Thus the forces acting on a conductor can be obtained by differentiating $\mathscr{U}$ either for constant charges or for constant potentials, the only difference being that the derivative must be taken with the minus sign in the first case and with the plus sign in the second.

The same result could be obtained more formally by starting from the differential identity

$$\mathrm{d}\mathscr{U} = \sum_a \phi_a \mathrm{d}e_a - F_q\mathrm{d}q, \tag{5.8}$$

in which $\mathscr{U}$ is regarded as a function of the charges on the conductors and the coordinate q. This identity states that $\partial\mathscr{U}/\partial e_a = \phi_a$ and $\partial\mathscr{U}/\partial q = -F_q$. Using the variables ϕ_a instead of e_a, we have

$$\mathrm{d}\tilde{\mathscr{U}} = -\sum_a e_a\mathrm{d}\phi_a - F_q\mathrm{d}q, \tag{5.9}$$

which gives (5.7).

At the end of §2 we have discussed the energy of a conductor in a uniform external electric field. The total force on an uncharged conductor in a uniform field is, of course, zero. The expression for the energy (2.14) can, however, be used to determine the force acting on a conductor in a quasi-uniform field $\mathfrak{E}$, i.e. a field which varies only slightly over the dimensions of the conductor. In such a field the energy can still be calculated, to a first approximation, from formula (2.14), and the force **F** is the gradient of this energy:

$$\mathbf{F} = -\mathbf{grad}\ \mathscr{U} = \tfrac{1}{2}\alpha_{ik} V\ \mathbf{grad}\ (\mathfrak{E}_i \mathfrak{E}_k). \tag{5.10}$$

The total torque **K** is in general non-zero even in a uniform external field. By the general laws of mechanics **K** can be determined by considering an infinitesimal virtual rotation of the body. The change in energy in such a rotation is related to **K** by $\delta\mathscr{U} = -\mathbf{K}\cdot\delta\boldsymbol{\psi}$, $\delta\boldsymbol{\psi}$ being the angle of the rotation. A rotation through an angle $\delta\boldsymbol{\psi}$ in a uniform field is equivalent to a rotation of the field through an angle $-\delta\boldsymbol{\psi}$ relative to the body. The change in the field is $\delta\mathfrak{E} = -\delta\boldsymbol{\psi}\times\mathfrak{E}$, and the change in energy is

$$\delta\mathscr{U} = (\partial\mathscr{U}/\partial\mathfrak{E})\cdot\delta\mathfrak{E} = -\delta\boldsymbol{\psi}\cdot\mathfrak{E}\times\partial\mathscr{U}/\partial\mathfrak{E}.$$

But $\partial\mathscr{U}/\partial\mathfrak{E} = -\mathscr{P}$, as we see from a comparison of formulae (2.13) and (2.14). Hence $\delta\mathscr{U} = -\mathscr{P}\times\mathfrak{E}\cdot\delta\boldsymbol{\psi}$, whence

$$\mathbf{K} = \mathscr{P}\times\mathfrak{E}, \tag{5.11}$$

in accordance with the usual expression given by the theory of fields in a vacuum.

If the total force and torque on a conductor are zero, the conductor remains at rest in the field, and effects involving the deformation of the body (called *electrostriction*) become important. The forces (5.1) on the surface of the conductor result in changes in its shape and volume. Because the force is an extending one, the volume of the body increases. A complete determination of the deformation requires a solution of the equations of the theory of elasticity, with the given distribution of forces (5.1) on the surface of the body. If, however, we are interested only in the change in volume, the problem can be solved very simply.

To do so, we must bear in mind that, if the deformation is slight (as in fact is true for electrostriction), the effect of the change of shape on the change of volume is of the second order of smallness. In the first approximation, therefore, the change in volume can be regarded as the result of deformation without change in shape, i.e. as a volume expansion under the action of some effective excess pressure ΔP which is uniformly distributed over the surface of the body and replaces the exact distribution given by (5.1). The relative change in volume is obtained by multiplying ΔP by the coefficient of uniform expansion of the substance. The pressure ΔP is given, according to a well-known formula, by the derivative of the electric energy $\mathscr{U}$ of the body with respect to its volume: $\Delta P = -\partial\mathscr{U}/\partial V$.†

Let the deforming field be due to the charged conductor itself. Then the energy $\mathscr{U} = \frac{1}{2}e^2/C$, and the pressure is $\Delta P = -\frac{1}{2}e^2\partial C^{-1}/\partial V$. For a given shape, the capacitance of the body (having the dimensions of length) is proportional to the linear dimension, i.e. to $V^{1/3}$. Hence

$$\Delta P = e^2/6CV = e\phi/6V. \tag{5.12}$$

† The quantity thus determined is the pressure exerted on the surface by the body itself; the pressure acting on the surface from outside is obtained by changing the sign.

If an uncharged conductor is situated in a uniform external field $\mathfrak{E}$, its energy is given by formula (2.14). The extending pressure is therefore

$$\Delta P = \tfrac{1}{2}\alpha_{ik}\mathfrak{E}_i\mathfrak{E}_k. \tag{5.13}$$

PROBLEMS

PROBLEM 1. A small conductor with capacitance c (equal in order of magnitude to its dimension) is at a distance r from the centre of a spherical conductor with large radius a ($\gg c$). The distance $r-a$ from the conductor to the surface of the sphere is supposed large compared with c, but not large compared with a. The two conductors are joined by a thin wire, so that they are at the same potential ϕ. Determine the force of their mutual repulsion.

SOLUTION. Since the conductor c is small, we can suppose that its potential is the sum of the potential $\phi a/r$ at a distance r from the centre of the large sphere and the potential e/c due to the charge e on the conductor itself. Hence $\phi = \phi a/r + e/c$, or $e = c\phi\,(1-a/r)$. The required force of interaction F is the Coulomb repulsion between the charge e on the conductor and the charge $a\phi$ on the sphere:

$$F = \frac{ac\phi^2}{r^2}\left(1-\frac{a}{r}\right).$$

This expression is correct to within terms of higher order in c. The force is greatest when $r = 3a/2$, and its value there is $F_{\max} = 4c\phi^2/27a$, decreasing on either side of this distance.

PROBLEM 2. A charged conducting sphere is cut in half. Determine the force of repulsion between the hemispheres.†

SOLUTION. We imagine the hemispheres separated by an infinitely narrow slit, and determine the force F on each of them by integrating over the surface the force $(E^2/8\pi)\cos\theta$, which is the component of (5.1) in a direction perpendicular to the plane of separation of the hemispheres. In the slit $E = 0$, and on the outer surface $E = e/a^2$, where a is the radius of the sphere and e the total charge on it. The result is $F = e^2/8a^2$.

PROBLEM 3. The same as Problem 2, but for an uncharged sphere in a uniform external field $\mathfrak{E}$ perpendicular to the plane of separation.

SOLUTION. As in Problem 2, except that the field on the surface of the sphere is $E = 3\,\mathfrak{E}\cos\theta$ (§3, Problem 1). The required force is $F = 9a^2\mathfrak{E}^2/16$.

PROBLEM 4. Determine the change in volume and in shape of a conducting sphere in a uniform external electric field.

SOLUTION. The change in volume $\Delta V/V = \Delta P/K$, where K is the modulus of volume expansion of the material, and ΔP is given by formula (5.13). For a sphere, $\alpha_{ik} = \delta_{ik}\alpha = 3\delta_{ik}/4\pi$ (§3, Problem 1), so that $\Delta V/V = 3\mathfrak{E}^2/8\pi K$.

As a result of the deformation, the sphere is changed into a prolate spheroid. To determine the eccentricity, we may regard the deformation as a uniform pure shear in the volume of the body, just as, to determine the change in the total volume, we regarded it as a uniform volume expansion.

The condition of equilibrium for a deformed body may be formulated as requiring that the sum of the electrostatic and elastic energies should be a minimum. The former is, by (2.12) and (4.26),

$$\mathcal{U}_{\text{es}} = -\frac{V}{8\pi n}\mathfrak{E}^2 \cong -\frac{3V\mathfrak{E}^2}{8\pi} - \frac{3V}{10\pi}\frac{a-b}{R}\mathfrak{E}^2,$$

where R is the original radius of the sphere, a and b the semiaxes of the spheroid, and $n \cong \frac{1}{3} - 4(a-b)/15R$ is the depolarizing factor (see (4.33).)

Since the deformation is axially symmetrical about the direction of the field (the x-axis), only the components u_{xx} and $u_{yy} = u_{zz}$ of the strain tensor are non-zero. Since we are considering equilibrium with respect to a change in shape, we can regard the volume as unchanged, i.e. $u_{ii} = 0$. Hence the elastic energy may be written

$$\mathcal{U}_{\text{el}} = \tfrac{1}{2}u_{ik}\sigma_{ik}V = \tfrac{1}{3}(\sigma_{xx}-\sigma_{yy})(u_{xx}-u_{yy})V,$$

† In Problems 2 and 3 we assume that the hemispheres are at the same potential.

where σ_{ik} is the elastic stress tensor (*TE*, §4). We have $\sigma_{xx} - \sigma_{yy} = 2(u_{xx} - u_{yy})$, where μ is the modulus of rigidity of the material, and $u_{xx} - u_{yy} = (a-b)/R$. Hence

$$\mathscr{U}_{\mathrm{el}} = \tfrac{2}{3}\mu(a-b)^2 V/R^2.$$

Making the sum $\mathscr{U}_{\mathrm{es}} + \mathscr{U}_{\mathrm{el}}$ a minimum, we have $(a-b)/R = 9\mathfrak{E}^2/40\pi\mu$.

PROBLEM 5. Find the relation between frequency and wavelength for waves propagated on a charged plane surface of a liquid conductor (in a gravitational field). Obtain the condition for this surface to be stable (Ya. I. Frenkel', 1935).

SOLUTION. Let the wave be propagated along the x-axis, with the z-axis vertically upwards. The vertical displacement of points on the surface of the liquid is $\zeta = ae^{i(kx-\omega t)}$. When the surface is at rest, the field above it is $E_z = E = 4\pi\sigma_0$, and its potential $\phi = -4\pi\sigma_0 z$, where σ_0 is the surface charge density. The potential of the field above the oscillating surface can be written as $\phi = -4\pi\sigma_0 z + \phi_1$, with $\phi_1 = \text{constant} \times e^{i(kx-\omega t)}e^{-kz}$, ϕ_1 being a small correction which satisfies the equation $\triangle\phi_1 = 0$ and vanishes for $z \to \infty$. On the surface itself, the potential must have a constant value, which we take to be zero, and so $\phi_1 = 4\pi\sigma_0\zeta$ for $z = 0$.

According to (5.1), an additional negative pressure acts on the charged surface of the liquid; this pressure is, as far as terms of the first order in ϕ_1, $E^2/8\pi \cong E_z{}^2/8\pi \cong 2\pi\sigma_0{}^2 + [k\sigma_0\phi_1]_{z=0} = 2\pi\sigma_0{}^2 + 4\pi\sigma_0{}^2 k\zeta$. The constant term $2\pi\sigma_0{}^2$ is of no importance, since it can be included in the constant external pressure.

The consideration of the hydrodynamical motion in the wave is entirely analogous to the theory of capillary waves (*FM*, §61), differing only by the presence of the additional pressure mentioned above. At the surface of the liquid we have the boundary condition $\rho g\zeta + \rho[\partial\Phi/\partial t]_{z=0} - \alpha\partial^2\zeta/\partial x^2 - 4\pi\sigma_0{}^2 k\zeta = 0$, where α is the surface-tension coefficient, ρ the density of the liquid, and Φ its velocity potential. Φ and ζ are also related by $\partial\zeta/\partial t = [\partial\Phi/\partial z]_{z=0}$. Substituting in these two relations $\zeta = ae^{i(kx-\omega t)}$ and $\Phi = Ae^{i(kx-\omega t)}e^{-kz}$ (Φ satisfies the equation $\triangle\Phi = 0$) and eliminating a and A, we find the required relation between k and ω:

$$\omega^2 = k(g\rho - 4\pi\sigma_0{}^2 k + \alpha k^2)/\rho. \tag{1}$$

If the surface of the liquid is to be stable, the frequency ω must be real for all values of k (since otherwise there would be complex ω with a positive imaginary part, and the factor $e^{-i\omega t}$ would increase indefinitely). The condition for the right-hand side of (1) to be positive is $(4\pi\sigma_0{}^2)^2 - 4g\rho\alpha < 0$, or $\sigma_0{}^4 < g\rho\alpha/4\pi^2$. This is the condition for stability.

PROBLEM 6. Find the condition of stability for a charged spherical drop (Rayleigh, 1882).

SOLUTION. The sum of the electrostatic and surface energies of the drop is $\mathscr{U} = e^2/2C + \alpha S$, where α is the surface-tension coefficient of the liquid, C the capacitance of the drop and S its surface area. Instability occurs (with increasing e) with respect to deformation of the sphere into a spheroid, and does so when $\mathscr{U}$ becomes a decreasing function of the eccentricity (for a given volume). The spherical shape always corresponds to an extremum of $\mathscr{U}$; the stability condition is therefore $[\partial^2\mathscr{U}/\partial(a-b)^2]_{a=b} > 0$, where a and b are the semiaxes of the spheroid, and the differentiation is carried out with $ab^2 = \text{constant}$. Using the formula for the surface of a spheroid and (4.18) for its capacitance, we find after a somewhat lengthy calculation $e^2 < 16\pi a^3\alpha$.

This condition ensures stability of the drop with respect to small deformations. It is found to be weaker than the condition for stability with respect to large deformations that divide the drop into two equal drops with charge $\frac{1}{2}e$ and radius $a/2^{1/3}$:

$$e^2 < 16\pi a^3\alpha\,(2^{1/3}-1)/(2-2^{1/3}) = 0.35 \times 16\pi a^3\alpha.$$

CHAPTER II

ELECTROSTATICS OF DIELECTRICS

§6. The electric field in dielectrics

WE SHALL now go on to consider a static electric field in another class of substances, namely dielectrics. The fundamental property of dielectrics is that a steady current cannot flow in them. Hence the static electric field need not be zero, as in conductors, and we have to derive the equations which describe this field. One equation is obtained by averaging equation (1.3), and is again

$$\mathbf{curl\,E} = 0. \tag{6.1}$$

A second equation is obtained by averaging the equation div $\mathbf{e} = 4\pi\rho$:

$$\operatorname{div}\mathbf{E} = 4\pi\bar{\rho}. \tag{6.2}$$

Let us suppose that no charges are brought into the dielectric from outside, which is the most usual and important case. Then the total charge in the volume of the dielectric is zero; even if it is placed in an electric field we have $\int\bar{\rho}\,\mathrm{d}V = 0$. This integral equation, which must be valid for a body of any shape, means that the average charge density can be written as the divergence of a certain vector, which is usually denoted by $-\mathbf{P}$:

$$\bar{\rho} = -\operatorname{div}\mathbf{P}, \tag{6.3}$$

while outside the body $\mathbf{P} = 0$. For, on integrating over the volume bounded by a surface which encloses the body but nowhere enters it, we find $\int\bar{\rho}\,\mathrm{d}V = -\int\operatorname{div}\mathbf{P}\,\mathrm{d}V = -\oint\mathbf{P}\cdot\mathbf{df} = 0$. $\mathbf{P}$ is called the *dielectric polarization*, or simply the *polarization*, of the body. A dielectric in which $\mathbf{P}$ differs from zero is said to be *polarized*. The vector $\mathbf{P}$ determines not only the volume charge density (6.3), but also the density σ of the charges on the surface of the polarized dielectric. If we integrate formula (6.3) over an element of volume lying between two neighbouring unit areas, one on each side of the dielectric surface, we have, since $\mathbf{P} = 0$ on the outer area (cf. the derivation of formula (1.9)),

$$\sigma = P_n, \tag{6.4}$$

where P_n is the component of the vector $\mathbf{P}$ along the outward normal to the surface.

To see the physical significance of the quantity $\mathbf{P}$ itself, let us consider the total dipole moment of all the charges within the dielectric; unlike the total charge, the total dipole moment need not be zero. By definition, it is the integral $\int\mathbf{r}\bar{\rho}\,\mathrm{d}V$. Substituting $\bar{\rho}$ from (6.3) and again integrating over a volume which includes the whole body we have

$$\int\mathbf{r}\bar{\rho}\,\mathrm{d}V = -\int\mathbf{r}\operatorname{div}\mathbf{P}\,\mathrm{d}V = -\oint\mathbf{r}(\mathbf{df}\cdot\mathbf{P}) + \int(\mathbf{P}\cdot\mathbf{grad})\mathbf{r}\,\mathrm{d}V.$$

The integral over the surface is zero, and in the second term we have $(\mathbf{P}\cdot\mathbf{grad})\mathbf{r} = \mathbf{P}$, so that

$$\int\mathbf{r}\,\bar{\rho}\,\mathrm{d}V = \int\mathbf{P}\,\mathrm{d}V. \tag{6.5}$$

Thus the polarization vector is the dipole moment (or *electric moment*) per unit volume of the dielectric.†

Substituting (6.3) in (6.2), we obtain the second equation of the electrostatic field in the form

$$\operatorname{div} \mathbf{D} = 0, \tag{6.6}$$

where we have introduced a quantity **D** defined by

$$\mathbf{D} = \mathbf{E} + 4\pi\mathbf{P}, \tag{6.7}$$

called the *electric induction*. The equation (6.6) has been derived by averaging the density of charges in the dielectric. If, however, charges not belonging to the dielectric are brought in from outside (we shall call these *extraneous charges*), then their density must be added to the right-hand side of equation (6.6):

$$\operatorname{div} \mathbf{D} = 4\pi\rho_{\text{ex}}. \tag{6.8}$$

On the surface of separation between two different dielectrics, certain boundary conditions must be satisfied. One of these follows from the equation **curl E** = 0. If the surface of separation is uniform as regards physical properties,‡ this condition requires the continuity of the tangential component of the field:

$$\mathbf{E}_{t1} = \mathbf{E}_{t2}; \tag{6.9}$$

cf. the derivation of the condition (1.7). The second condition follows from the equation div **D** = 0, and requires the continuity of the normal component of the induction:

$$D_{n1} = D_{n2}. \tag{6.10}$$

For a discontinuity in the normal component $D_n = D_z$ would involve an infinity of the derivative $\partial D_z/\partial z$, and therefore of div **D**.

At a boundary between a dielectric and a conductor, $\mathbf{E}_t = 0$, and the condition on the normal component is obtained from (6.8):

$$\mathbf{E}_t = 0, \quad D_n = 4\pi\sigma_{\text{ex}}, \tag{6.11}$$

where σ_{ex} is the charge density on the surface of the conductor; cf. (1.8), (1.9).

§7. The permittivity

In order that equations (6.1) and (6.6) should form a complete set of equations determining the electrostatic field, they must be supplemented by a relation between the induction **D** and the field **E**. In the great majority of cases this relation may be supposed linear. It corresponds to the first terms in an expansion of **D** in powers of **E**, and its correctness is due to the smallness of the external electric fields in comparison with the internal molecular fields.

The linear relation between **D** and **E** is especially simple in the most important case, that

---

† It should be noticed that the relation (6.3) inside the dielectric and the condition **P** = 0 outside do not in themselves determine **P** uniquely; inside the dielectric we could add to **P** any vector of the form **curl f.** The exact form of **P** can be completely determined only by establishing its connection with the dipole moment.

‡ That is, as regards composition of the adjoining media, temperature, etc. If the dielectric is a crystal, the surface must be a crystallographic plane.

of an isotropic dielectric. It is evident that, in an isotropic dielectric, the vectors **D** and **E** must be in the same direction. The linear relation between them is therefore a simple proportionality:†

$$\mathbf{D} = \varepsilon \mathbf{E}. \tag{7.1}$$

The coefficient ε is the *permittivity* or *dielectric permeability* or *dielectric constant* of the substance and is a function of its thermodynamic state.

As well as the induction, the polarization also is proportional to the field:

$$\mathbf{P} = \kappa \mathbf{E} \equiv (\varepsilon - 1)\mathbf{E}/4\pi. \tag{7.2}$$

The quantity κ is called the *polarization coefficient* of the substance, or its *dielectric susceptibility*. Later (§14) we shall show that the permittivity always exceeds unity; the polarizability, accordingly, is always positive. The polarizability of a rarefied medium (a gas) may be regarded as proportional to its density.

The boundary conditions (6.9) and (6.10) on the surface separating two isotropic dielectrics become

$$\mathbf{E}_{t1} = \mathbf{E}_{t2}, \quad \varepsilon_1 E_{n1} = \varepsilon_2 E_{n2}. \tag{7.3}$$

Thus the normal component of the field is discontinuous, changing in inverse proportion to the permittivity of the medium.

In a homogeneous dielectric, $\varepsilon =$ constant, and then it follows from $\operatorname{div} \mathbf{D} = 0$ that $\operatorname{div} \mathbf{P} = 0$. By the definition (6.3) this means that the volume charge density in such a body is zero (but the surface density (6.4) is in general not zero). On the other hand, in an inhomogeneous dielectric we have a non-zero volume charge density

$$\bar{\rho} = -\operatorname{div} \mathbf{P} = -\operatorname{div} \frac{\varepsilon - 1}{4\pi\varepsilon} \mathbf{D} = -\frac{1}{4\pi} \mathbf{D} \cdot \mathbf{grad} \frac{\varepsilon - 1}{\varepsilon} = -\frac{1}{4\pi\varepsilon} \mathbf{E} \cdot \mathbf{grad}\, \varepsilon.$$

If we introduce the electric field potential by $\mathbf{E} = -\mathbf{grad}\, \phi$, then equation (6.1) is automatically satisfied, and the equation $\operatorname{div} \mathbf{D} = \operatorname{div} \varepsilon \mathbf{E} = 0$ gives

$$\operatorname{div} (\varepsilon\, \mathbf{grad}\, \phi) = 0. \tag{7.4}$$

This equation becomes the ordinary Laplace's equation only in a homogeneous dielectric medium. The boundary conditions (7.3) can be rewritten as the following conditions on the potential:

$$\left.\begin{aligned} \phi_1 &= \phi_2, \\ \varepsilon_1 \partial\phi_1/\partial n &= \varepsilon_2 \partial\phi_2/\partial n; \end{aligned}\right\} \tag{7.5}$$

the continuity of the tangential derivatives of the potential is equivalent to the continuity of ϕ itself.

In a dielectric medium which is piecewise homogeneous, equation (7.4) reduces in each homogeneous region to Laplace's equation $\triangle \phi = 0$, so that the permittivity appears in the solution of the problem only through the conditions (7.5). These conditions, however,

† This relation, which assumes that **D** and **E** vanish simultaneously, is, strictly speaking, valid only in dielectrics which are homogeneous as regards physical properties (composition, temperature, etc.). In inhomogeneous bodies **D** may be non-zero even when $\mathbf{E} = 0$, and is determined by the gradients of thermodynamic quantities which vary through the body. The corresponding terms, however, are very small, and we shall use the relation (7.1) in what follows, even for inhomogeneous bodies.

involve only the ratio of the permittivities of two adjoining media. In particular, the solution of an electrostatic problem for a dielectric body with permittivity ε_2, surrounded by a medium with permittivity ε_1, is the same as for a body with permittivity $\varepsilon_2/\varepsilon_1$, surrounded by a vacuum.

Let us consider how the results obtained in Chapter I for the electrostatic field of conductors will be modified if these conductors are not in a vacuum but in a homogeneous and isotropic dielectric medium. In both cases the potential distribution satisfies the equation $\triangle\phi = 0$, with the boundary condition that ϕ is constant on the surface of the conductor, and the only difference is that, instead of $E_n = -\partial\phi/\partial n = 4\pi\sigma$, we have

$$D_n = -\varepsilon\,\partial\phi/\partial n = 4\pi\sigma, \tag{7.6}$$

giving the relation between the potential and the surface charge. Hence it is clear that the solution of the problem of the field of a charged conductor in a vacuum gives the solution of the same problem with a dielectric in place of the vacuum if we make the formal substitution $\phi \to \varepsilon\phi$, $e \to e$ or $\phi \to \phi$, $e \to e/\varepsilon$. For given charges on the conductors, the potential and the field are reduced by a factor ε in comparison with their values in a vacuum. This reduction in the field can be explained as the result of a partial "screening" of the charge on the conductor by the surface charges on the adjoining polarized dielectric. If, on the other hand, the potentials of the conductors are maintained, then the field is unchanged but the charges are increased by a factor ε.†

Finally, it may be noted that in electrostatics we may formally regard a conductor (uncharged) as a body of infinite permittivity, in the sense that its effect on an external electric field is the same as that of a dielectric (of the same form) as $\varepsilon \to \infty$. For, since the boundary condition on the induction **D** is finite, **D** must remain finite in the body even for $\varepsilon \to \infty$. This means that $\mathbf{E} \to 0$, in accordance with the properties of conductors.

PROBLEMS

PROBLEM 1. Determine the field due to a point charge e at a distance h from a plane boundary separating two different dielectric media.

SOLUTION. Let O be the position of the charge e in medium 1, and O' its image in the plane of separation, situated in medium 2 (Fig. 11, p. 38). We shall seek the field in medium 1 in the form of the field of two point charges, e and a fictitious charge e' at O' (cf. the method of images, §3): $\phi_1 = e/\varepsilon_1 r + e'/\varepsilon_1 r'$, where r and r' are the distances from O and O' respectively. In medium 2 we seek the field as that of a fictitious charge e'' at O: $\phi_2 = e''/\varepsilon_2 r$. On the boundary plane ($r = r'$) the conditions (7.5) must hold, leading to the equations $e - e' = e''$, $(e+e')/\varepsilon_1 = e''/\varepsilon_2$, whence

$$e' = e(\varepsilon_1 - \varepsilon_2)/(\varepsilon_1 + \varepsilon_2), \qquad e'' = 2\varepsilon_2 e/(\varepsilon_1 + \varepsilon_2). \tag{1}$$

For $\varepsilon_2 \to \infty$ we have $e' = -e$, $\phi_2 = 0$, i.e. the result obtained in §3 for the field of a point charge near a conducting plane.

The force acting on the charge e (the *image force*) is

$$F = \frac{ee'}{(2h)^2\varepsilon_1} = \left(\frac{e}{2h}\right)^2 \frac{\varepsilon_1 - \varepsilon_2}{\varepsilon_1(\varepsilon_1 + \varepsilon_2)};$$

$F > 0$ corresponds to repulsion.

PROBLEM 2. The same as Problem 1, but for an infinite charged straight wire parallel to a plane boundary surface at a distance h.

---

† From this it follows, in particular, that when a capacitor is filled with a dielectic its capacitance increases by a factor ε.

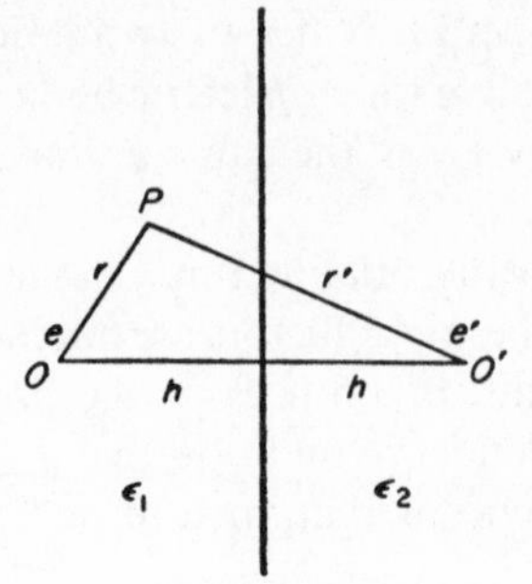

FIG. 11

SOLUTION. As in Problem 1, except that the field potentials in the two media are $\phi_1 = -(2e/\varepsilon_1)\log r - (2e'/\varepsilon_1)\log r'$, $\phi_2 = -(2e''/\varepsilon_2)\log r$, where e, e', e'' are the charges per unit length of the wire and of its images, and r, r' are the distances in a plane perpendicular to the wire. The same expressions (1) are obtained for e', e'', and the force on unit length of the wire is $F = 2ee'/2h\varepsilon_1 = e^2(\varepsilon_1-\varepsilon_2)/h\varepsilon_1(\varepsilon_1+\varepsilon_2)$.

PROBLEM 3. Determine the field due to an infinite charged straight wire in a medium with permittivity ε_1, lying parallel to a cylinder with radius a and permittivity ε_2, at a distance $b\,(> a)$ from its axis.†

SOLUTION. We seek the field in medium 1 as that produced in a homogeneous dielectric (with ε_1) by the actual wire (passing through O in Fig. 12), with charge e per unit length, and two fictitious wires with charges e' and $-e'$ per unit length, passing through A and O' respectively. The point A is at a distance a^2/b from the axis of the cylinder. Then, for all points on the circumference, the distances r and r' from O and A are in a constant ratio $r'/r = a/b$, and so it is possible to satisfy the boundary conditions on this circumference. In medium 2 we seek the field as that produced in a homogeneous medium (with ε_2) by a fictitious charge e'' on the wire passing through O.

The boundary conditions on the surface of separation are conveniently formulated in terms of the potential ϕ ($\mathbf{E} = -\mathbf{grad}\,\phi$) and the vector potential $\mathbf{A}$ (cf. §3), defined by $\mathbf{D} = \mathbf{curl}\,\mathbf{A}$ (in accordance with the equation $\operatorname{div}\mathbf{D} = 0$). In a two-dimensional problem, $\mathbf{A}$ is in the z-direction (perpendicular to the plane of the figure). The conditions of continuity for the tangential components of $\mathbf{E}$ and the normal component of $\mathbf{D}$ are equivalent to $\phi_1 = \phi_2$, $A_1 = A_2$.

For the field of a charged wire we have in polar coordinates r, θ the equation $\phi = -(2e/\varepsilon)\log r + \text{constant}$, $A = 2e\theta + \text{constant}$; cf. (3.18). Hence the boundary conditions are

$$\frac{2}{\varepsilon_1}(-e\log r - e'\log r' + e'\log a) = -\frac{2e''}{\varepsilon_2}\log r + \text{constant},$$

$$2[e\theta + e'\theta' - e'(\theta+\theta')] = 2e''\theta,$$

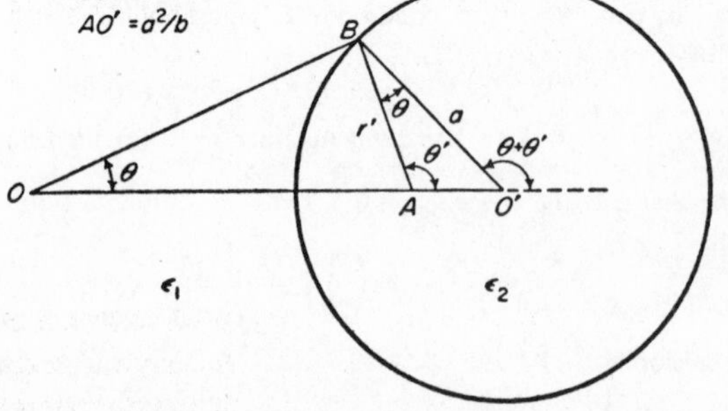

FIG. 12

† The corresponding problem of a point charge near a dielectric sphere cannot be solved in closed form.

where the angles are as shown in Fig. 12, and we have used the fact that $OO'B$ and $BO'A$ are similar triangles. Hence $\varepsilon_2(e+e') = \varepsilon_1 e''$, $e-e' = e''$, and the expressions for e' and e'' are again formulae (1) of Problem 1.

The force acting on unit length of the charged wire is parallel to OO', and is

$$F = eE = \frac{2ee'}{\varepsilon_1}\left(\frac{1}{OA} - \frac{1}{OO'}\right) = \frac{2e^2(\varepsilon_1-\varepsilon_2)a^2}{\varepsilon_1(\varepsilon_1+\varepsilon_2)b(b^2-a^2)};$$

$F > 0$ corresponds to repulsion. In the limit $a, b \to \infty$, $b-a \to h$, this gives the result in Problem 1.

PROBLEM 4. The same as Problem 3, but for the case where the wire is inside a cylinder with permittivity ε_2 $(b < a)$.

SOLUTION. We seek the field in medium 2 as that due to the actual wire, with charge e per unit length (O in Fig. 13), and a fictitious wire with charge e' per unit length passing through A, which is now outside the cylinder. In medium 1 we seek the field as that of wires with charges e'' and $e-e''$ passing through O and O' respectively. By the same method as in the preceding problem we find $e' = -e(\varepsilon_1-\varepsilon_2)/(\varepsilon_1+\varepsilon_2)$, $e'' = 2\varepsilon_1 e/(\varepsilon_1+\varepsilon_2)$. For $\varepsilon_2 > \varepsilon_1$ the wire is repelled from the surface of the cylinder by a force

$$F = \frac{2ee'}{\varepsilon_2}\frac{1}{OA} = \frac{2e^2(\varepsilon_2-\varepsilon_1)b}{\varepsilon_2(\varepsilon_1+\varepsilon_2)(a^2-b^2)}.$$

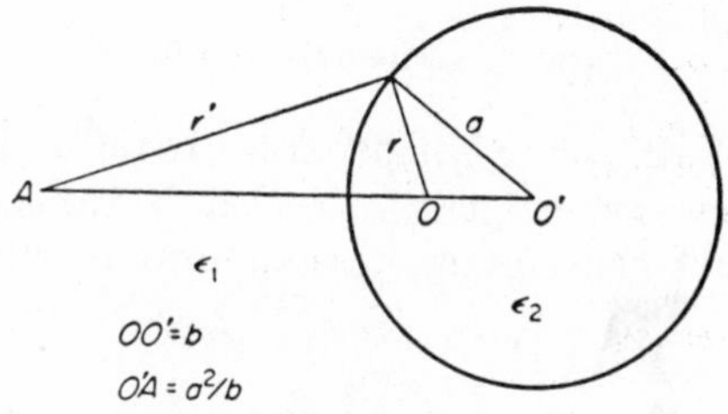

FIG. 13

PROBLEM 5. Show that the field potential $\phi_A(\mathbf{r}_B)$ at a point $\mathbf{r}_B$ in an arbitrary inhomogeneous dielectric medium, due to a point charge e at $\mathbf{r}_A$, is equal to the potential $\phi_B(\mathbf{r}_A)$ at $\mathbf{r}_A$ due to the same charge at $\mathbf{r}_B$.

SOLUTION. The potentials $\phi_A(\mathbf{r})$ and $\phi_B(\mathbf{r})$ satisfy the equations

$$\operatorname{div}(\varepsilon\, \mathbf{grad}\, \phi_A) = -4\pi e\delta(\mathbf{r}-\mathbf{r}_A), \quad \operatorname{div}(\varepsilon\, \mathbf{grad}\, \phi_B) = -4\pi e\delta(\mathbf{r}-\mathbf{r}_B).$$

Multiplying the first by ϕ_B and the second by ϕ_A and subtracting, we have

$$\operatorname{div}(\phi_B\varepsilon\,\mathbf{grad}\,\phi_A) - \operatorname{div}(\phi_A\varepsilon\,\mathbf{grad}\,\phi_B) = -4\pi e\delta(\mathbf{r}-\mathbf{r}_A)\phi_B(\mathbf{r}) + 4\pi e\delta(\mathbf{r}-\mathbf{r}_B)\phi_A(\mathbf{r}).$$

Integration of this equation over all space gives the required relation:

$$\phi_A(\mathbf{r}_B) = \phi_B(\mathbf{r}_A).$$

§8. A dielectric ellipsoid

The polarization of a dielectric ellipsoid in a uniform external electric field has some unusual properties which render this example particularly interesting.

Let us consider first a simple special case, that of a dielectric sphere in an external field $\mathfrak{E}$. We denote its permittivity by $\varepsilon^{(i)}$, and that of the medium surrounding it by $\varepsilon^{(e)}$. We take the origin of spherical polar coordinates at the centre of the sphere, and the direction of $\mathfrak{E}$ as the axis from which the polar angle θ is measured, and seek the field potential outside the sphere in the form $\phi^{(e)} = \mathfrak{E}\cdot\mathbf{r} + A\mathfrak{E}\cdot\mathbf{r}/r^3$; the first term is the potential of the external field imposed, and the second, which vanishes at infinity, gives the required change in potential due to the sphere (cf. §3, Problem 1, solution). Inside the sphere, we seek the field potential in the form $\phi^{(i)} = -B\mathfrak{E}\cdot\mathbf{r}$, the only function which satisfies Laplace's equation, remains finite at the centre of the sphere, and depends only on the constant vector $\mathfrak{E}$ (which is the only parameter of the problem).

The constants A and B are determined by the boundary conditions on the surface of the sphere. It may be seen at once, however, that the field in the sphere $\mathbf{E}^{(i)} = B\mathfrak{E}$ is uniform and differs only in magnitude from the applied field $\mathfrak{E}$.

The boundary condition of continuity of the potential gives $\mathbf{E}^{(i)} = \mathfrak{E}(1 - A/R^3)$, where R is the radius of the sphere, and the condition of continuity of the normal component of the induction gives

$$\mathbf{D}^{(i)} = \varepsilon^{(e)}\mathfrak{E}(1 + 2A/R^3).$$

Eliminating A from these two equations, we obtain

$$\tfrac{1}{3}(\mathbf{D}^{(i)} + 2\varepsilon^{(e)}\mathbf{E}^{(i)}) = \varepsilon^{(e)}\mathfrak{E} \tag{8.1}$$

or, substituting $\mathbf{D}^{(i)} = \varepsilon^{(i)}\mathbf{E}^{(i)}$,

$$\mathbf{E}^{(i)} = 3\varepsilon^{(e)}\mathfrak{E}/(2\varepsilon^{(e)} + \varepsilon^{(i)}). \tag{8.2}$$

The problem of an infinite dielectric cylinder in an external field perpendicular to its axis is solved in an entirely similar manner (cf. §3, Problem 2). The field inside the cylinder, like that inside the sphere in the above example, is uniform. It satisfies the relation

$$\tfrac{1}{2}(\mathbf{D}^{(i)} + \varepsilon^{(e)}\mathbf{E}^{(i)}) = \varepsilon^{(e)}\mathfrak{E}, \tag{8.3}$$

or

$$\mathbf{E}^{(i)} = 2\varepsilon^{(e)}\mathfrak{E}/(\varepsilon^{(e)} + \varepsilon^{(i)}). \tag{8.4}$$

The relations (8.1) and (8.3), in which the permittivity $\varepsilon^{(i)}$ of the sphere or cylinder does not appear explicitly, are particularly important because their validity does not depend on a linear relation between $\mathbf{E}$ and $\mathbf{D}$ within the body; they hold whatever the form of this relation (e.g. for anisotropic bodies). The analogous relations

$$\mathbf{E}^{(i)} = \mathfrak{E} \tag{8.5}$$

for a cylinder in a longitudinal field and

$$\mathbf{D}^{(i)} = \varepsilon^{(e)}\mathfrak{E} \tag{8.6}$$

for a flat plate in a field perpendicular to it are similarly valid; these relations are evident at once from the boundary conditions.

The property of causing a uniform field within itself on being placed in a uniform external field is found to pertain to any ellipsoid, whatever the ratio of the semiaxes a, b, c. The problem of the polarization of a dielectric ellipsoid is solved by the use of ellipsoidal coordinates, in the same way as the corresponding problem for a conducting ellipsoid in §4.

Let the external field be again in the x-direction. The field potential outside the ellipsoid may again be sought in the form (4.22): $\phi'_e = \phi_0 F(\xi)$, with the function $F(\xi)$ given by (4.23). Such a function cannot, however, appear in the field potential ϕ_i inside the ellipsoid, since it does not satisfy the condition that the field must be finite everywhere inside the ellipsoid.

For let us consider the surface $\xi = -c^2$, which is an ellipse in the xy-plane, with semiaxes $\sqrt{(a^2-c^2)}$ and $\sqrt{(b^2-c^2)}$, lying within the ellipsoid. For $\xi \to -c^2$, the integral (4.23) behaves as $\sqrt{(\xi+c^2)}$. The field, i.e. the potential gradient, therefore behaves as $1/\sqrt{(\xi+c^2)}$, and becomes infinite at $\xi = -c^2$. Thus the only solution suitable for the field inside the ellipsoid is $F(\xi) =$ constant, so that ϕ_i must be sought in the form $\phi_i = B\phi_0$. We see that the potential ϕ_i differs only by a constant factor from the potential ϕ_0 of the uniform field. In other words, the field inside the ellipsoid is also uniform.

We shall not pause to write out the formulae for the field outside the ellipsoid. The uniform field inside the ellipsoid can be found without actually writing out the boundary conditions, by using some results already known.

Let us first suppose that the ellipsoid is in a vacuum ($\varepsilon^{(e)} = 1$). Then there must be a linear relation between the vectors $\mathbf{E}^{(i)}$, $\mathbf{D}^{(i)}$ and $\mathfrak{E}$ (which are all in the x-direction), of the form $aE^{(i)}{}_x + bD^{(i)}{}_x = \mathfrak{E}_x$, where the coefficients a, b depend only on the shape of the ellipsoid, and not on its permittivity $\varepsilon^{(i)}$. The existence of such a relation follows from the form of the boundary conditions, as we saw above in the examples of the sphere and the cylinder.

To determine a and b we notice that, in the trivial particular case $\varepsilon^{(i)} = 1$, we have simply $\mathbf{E} = \mathbf{D} = \mathfrak{E}$, and so $a+b = 1$. Another particular case for which the solution is known is that of a conducting ellipsoid. In a conductor $\mathbf{E}^{(i)} = 0$, and the induction $\mathbf{D}^{(i)}$, though it has no direct physical significance, may be regarded formally as being related to the total dipole moment of the ellipsoid by $\mathbf{D}^{(i)} = 4\pi\mathbf{P} = 4\pi\,\mathscr{P}/V$. According to (4.26) we then have $D^{(i)}{}_x = \mathfrak{E}_x/n^{(x)}$, i.e. $b = n^{(x)}$, and so $a = 1-n^{(x)}$. Thus we conclude that

$$(1-n^{(x)})\,E^{(i)}{}_x + n^{(x)}\,D^{(i)}{}_x = \mathfrak{E}_x, \tag{8.7}$$

or

$$E^{(i)}{}_x = \mathfrak{E}_x - 4\pi n^{(x)} P_x. \tag{8.8}$$

The quantity $4\pi n^{(x)}P_x$ is called the *depolarizing field*.† Similar relations, but with coefficients $n^{(y)}$, $n^{(z)}$, hold for the fields in the y and z directions. Like the particular formulae (8.1) and (8.3), they are valid whatever the relation between $\mathbf{E}$ and $\mathbf{D}$ inside the ellipsoid.

The field inside the ellipsoid is found from (8.7) by putting $D^{(i)}{}_x = \varepsilon^{(i)}E^{(i)}{}_x$:

$$E^{(i)}{}_x = \mathfrak{E}_x/[1+(\varepsilon^{(i)}-1)n^{(x)}], \tag{8.9}$$

and the total dipole moment of the ellipsoid is

$$\mathscr{P}_x = VP_x = (\varepsilon^{(i)}-1)VE^{(i)}{}_x/4\pi = \tfrac{1}{3}abc(\varepsilon^{(i)}-1)\mathfrak{E}_x/[1+(\varepsilon^{(i)}-1)n^{(x)}]. \tag{8.10}$$

If the field $\mathfrak{E}$ has components along all three axes, then the field inside the ellipsoid is still uniform, but in general not parallel to $\mathfrak{E}$. For an arbitrary choice of coordinate axes we can write the relation (8.7) in the general form

$$E^{(i)}{}_i + n_{ik}(D^{(i)}{}_k - E^{(i)}{}_k) = \mathfrak{E}_i. \tag{8.11}$$

The transition to the case where the permittivity of the medium differs from unity is effected by simply replacing $\varepsilon^{(i)}$ by $\varepsilon^{(i)}/\varepsilon^{(e)}$. Then formula (8.7) becomes

$$(1-n^{(x)})\varepsilon^{(e)}E^{(i)}{}_x + n^{(x)}D^{(i)}{}_x = \varepsilon^{(e)}\mathfrak{E}_x. \tag{8.12}$$

† Similar formulae hold for a magnetized ellipsoid in a uniform external magnetic field (see §29). In this case $n^{(x)}$, $n^{(y)}$, $n^{(z)}$ are called *demagnetizing factors*.

This formula can be applied, in particular, to the field inside an ellipsoidal cavity in an infinite dielectric medium. In this case $\varepsilon^{(i)} = 1$.

PROBLEMS†

PROBLEM 1. Determine the torque on a spheroid in a uniform electric field.

SOLUTION. According to the general formula (16.13), the torque on an ellipsoid is $\mathbf{K} = \mathscr{P} \times \mathfrak{E}$, where $\mathscr{P}$ is the dipole moment of the ellipsoid. In a spheroid, the vector $\mathscr{P}$ is in a plane passing through the axis of symmetry and the direction of $\mathfrak{E}$. The torque is perpendicular to this plane, and a calculation of its magnitude from formulae (8.10) gives

$$K = \frac{(\varepsilon-1)^2|1-3n|\mathfrak{E}^2 V \sin 2\alpha}{8\pi[n\varepsilon+1-n][(1-n)\varepsilon+1+n]},$$

where α is the angle between the direction of $\mathfrak{E}$ and the axis of symmetry of the spheroid, and n is the depolarizing factor along the axis (so that the depolarizing factors in the directions perpendicular to the axis are $\frac{1}{2}(1-n)$). The torque is directed so that it tends to turn the axis of symmetry of a prolate ($n < \frac{1}{3}$) or oblate ($n > \frac{1}{3}$) ellipsoid parallel or perpendicular to the field respectively.

For a conducting ellipsoid ($\varepsilon \to \infty$) we have

$$K = \frac{|1-3n|}{8\pi n(1-n)} V\mathfrak{E}^2 \sin 2\alpha.$$

PROBLEM 2. A hollow dielectric sphere (with permittivity ε and internal and external radii b and a) is in a uniform external field $\mathfrak{E}$. Determine the field in the cavity.

SOLUTION. As above in the problem of a solid sphere, we seek the field potentials in the vacuum outside the sphere (region 1) and in the cavity (region 3) in the forms $\phi_1 = -\mathfrak{E}\cos\theta\,(r - A/r^2)$, $\phi_3 = -B\mathfrak{E}r\cos\theta$, and that in the dielectric (region 2) as $\phi_2 = -C\mathfrak{E}\cos\theta\,(r - D/r^2)$, where A, B, C, D are constants determined from the conditions of continuity of ϕ and $\varepsilon\,\partial\phi/\partial r$ at the boundaries $1-2$ and $2-3$. Thus the field $\mathbf{E}_3 = B\mathfrak{E}$ in the cavity is uniform, but the field $\mathbf{E}_2$ in the sphere is not. A calculation of the constants gives the result

$$\mathbf{E}_3 = 9\varepsilon\mathfrak{E}/[(\varepsilon+2)(2\varepsilon+1) - 2(\varepsilon-1)^2(b/a)^3].$$

PROBLEM 3. The same as Problem 2, but for a hollow cylinder in a uniform transverse field.‡

SOLUTION. As in Problem 2, with the result

$$\mathbf{E}_3 = 4\varepsilon\mathfrak{E}/[(\varepsilon+1)^2 - (\varepsilon-1)^2(b/a)^2].$$

§9. The permittivity of a mixture

If a subtance is a finely dispersed mixture (an emulsion, powder mixture, etc.), we can consider the electric field averaged over volumes which are large compared with the scale of the inhomogeneities. The mixture is a homogeneous and isotropic medium with respect to such an average field, and so may be characterized by an effective permittivity, which we denote by ε_{mix}. If $\overline{\mathbf{E}}$ and $\overline{\mathbf{D}}$ are the field and induction averaged in this way, then, by the definition of ε_{mix},

$$\overline{\mathbf{D}} = \varepsilon_{\text{mix}}\overline{\mathbf{E}}. \tag{9.1}$$

If all the particles in the mixture are isotropic, and the differences in their permittivities are small in comparison with ε itself, it is possible to calculate ε_{mix} in a general form which is correct as far as terms of the second order in these differences.

---

† In these three Problems the body is assumed to be in a vacuum.

‡ In a longitudinal field the solution is clearly $\mathbf{E}_3 = \mathfrak{E}$.

We write the local field as $\mathbf{E} = \overline{\mathbf{E}} + \delta\mathbf{E}$, and the local permittivity as $\bar{\varepsilon} + \delta\varepsilon$, where

$$\bar{\varepsilon} = (1/V)\int \varepsilon \mathrm{d}V \tag{9.2}$$

is obtained by averaging over the volume. Then the mean induction is

$$\overline{\mathbf{D}} = \overline{(\bar{\varepsilon} + \delta\varepsilon)(\overline{\mathbf{E}} + \delta\mathbf{E})} = \bar{\varepsilon}\overline{\mathbf{E}} + \overline{\delta\varepsilon\delta\mathbf{E}}, \tag{9.3}$$

since the mean values of $\delta\varepsilon$ and $\delta\mathbf{E}$ are zero by definition. In the zero-order approximation $\varepsilon_{\text{mix}} = \bar{\varepsilon}$; the first non-zero correction term will, of course, be of the second order in $\delta\varepsilon$, as we see from (9.3).

From the non-averaged equation div $\mathbf{D} = 0$ we have, as far as small terms of the first order,

$$\operatorname{div}[(\bar{\varepsilon} + \delta\varepsilon)(\overline{\mathbf{E}} + \delta\mathbf{E})] = \bar{\varepsilon}\operatorname{div}\delta\mathbf{E} + \overline{\mathbf{E}}\cdot\mathbf{grad}\,\delta\varepsilon = 0. \tag{9.4}$$

The averaging of the product $\delta\varepsilon\delta\mathbf{E}$ in (9.3) is done in two stages. We first average over the volume of particles of a given kind, i.e. for a given $\delta\varepsilon$. The value of $\delta\mathbf{E}$ thus averaged is easily obtained from equation (9.4): on account of the isotropy of the mixture as a whole.

$$\frac{\partial}{\partial x}\overline{\delta E_x} = \frac{\partial}{\partial y}\overline{\delta E_y} = \frac{\partial}{\partial z}\overline{\delta E_z} = \tfrac{1}{3}\operatorname{div}\overline{\delta\mathbf{E}}.$$

If $\overline{\mathbf{E}}$ is in the x-direction, say, we have from (9.4)

$$3\bar{\varepsilon}\frac{\partial}{\partial x}\overline{\delta E_x} = -\overline{E}_x\frac{\partial\delta\varepsilon}{\partial x},$$

whence

$$\overline{\delta E_x} = -(\overline{E}_x/3\bar{\varepsilon})\delta\varepsilon.$$

Since the direction of the x-axis is chosen arbitrarily, this equation may be written in the vector form $\overline{\delta\mathbf{E}} = -(1/3\bar{\varepsilon})\overline{\mathbf{E}}\delta\varepsilon$. Multiplying by $\delta\varepsilon$ and effecting the final averaging over all components of the mixture, we obtain $\overline{\delta\varepsilon\delta\mathbf{E}} = -(1/3\bar{\varepsilon})\overline{\mathbf{E}}\overline{(\delta\varepsilon)^2}$. Finally, substituting this expression in (9.3) and comparing with (9.1), we have the required result:

$$\varepsilon_{\text{mix}} = \bar{\varepsilon} - (1/3\bar{\varepsilon})\overline{(\delta\varepsilon)^2}. \tag{9.5}$$

This formula can be written in another manner if we put

$$\overline{\varepsilon^{\frac{1}{3}}} = \overline{(\bar{\varepsilon} + \delta\varepsilon)^{\frac{1}{3}}} = \bar{\varepsilon}^{\frac{1}{3}}\left(1 - \frac{\overline{(\delta\varepsilon)^2}}{9\bar{\varepsilon}^2}\right);$$

this is accurate to terms of the second order. Then

$$\varepsilon_{\text{mix}}{}^{\frac{1}{3}} = \overline{\varepsilon^{\frac{1}{3}}}. \tag{9.6}$$

Thus we can say that, in this approximation, the cube root of ε is additive.

Another limiting case for which an exact treatment is possible concerns the permittivity of an emulsion having an arbitrary difference between the permittivities ε_1 of the medium and ε_2 of the disperse phase but only a small concentration of the latter, whose particles are assumed spherical.

In the integral

$$\frac{1}{V}\int(\mathbf{D} - \varepsilon_1\mathbf{E})\mathrm{d}V \equiv \overline{\mathbf{D}} - \varepsilon_1\overline{\mathbf{E}}$$

the integrand is zero except within particles of the emulsion. It is therefore proportional to the volume concentration c of the emulsion, and in calculating it we can assume that the particles are in an external field which equals the mean field $\overline{\mathbf{E}}$. Assuming the particles spherical and using formula (8.2), we obtain for the proportionality coefficient between $\overline{\mathbf{D}}$ and $\overline{\mathbf{E}}$

$$\varepsilon_{\text{mix}} = \varepsilon_1 + 3c\varepsilon_1(\varepsilon_2 - \varepsilon_1)/(\varepsilon_2 + 2\varepsilon_1). \tag{9.7}$$

This formula is correct to terms of the first order in c. When ε_1 and ε_2 are nearly equal it is the same (to the first order in c and the second in $\varepsilon_2 - \varepsilon_1$) as the result given by formula (9.5) for small c.

§10. Thermodynamic relations for dielectrics in an electric field

The question of the change in thermodynamic properties owing to the presence of an electric field does not arise for conductors. Since there is no electric field inside a conductor, any change in its thermodynamic properties amounts simply to an increase in its total energy by the energy of the field which it produces in the surrounding space.† This quantity is quite independent of the thermodynamic state (and, in particular, of the temperature) of the body, and so does not affect the entropy, for example.

On the other hand, an electric field penetrates into a dielectric and so has a great effect on its thermodynamic properties. To investigate this effect, let us first determine the work done on a thermally insulated dielectric when the field in it undergoes an infinitesimal change.

The electric field in which the dielectric is placed must be imagined as due to various external charged conductors, and the change in the field can then be regarded as resulting from changes in the charges on these conductors.‡ Let us suppose for simplicity that there is only one conductor, with charge e and potential ϕ. The work which must be done to increase its charge by an infinitesimal amount δe is

$$\delta R = \phi \delta e; \tag{10.1}$$

this is the mechanical work done by the given field on a charge δe brought from infinity (where the field potential is zero) to the surface of the conductor, i.e. through a potential difference of ϕ. We shall put δR in a form which is expressed in terms of the field in the space filled with dielectric which surrounds the conductor.

If D_n is the component of the electric induction vector in the direction of the normal to the surface of the conductor (out of the dielectric and into the conductor), then the surface charge density on the conductor is $-D_n/4\pi$, so that

$$e = -\frac{1}{4\pi}\oint D_n \mathrm{d}f = -\frac{1}{4\pi}\oint \mathbf{D}\cdot\mathbf{df}.$$

† We here neglect the energy of the attachment of the charge to the substance of the conductor; this will be discussed in §23.

‡ The final results which we shall obtain involve only the values of the field inside the dielectric, and therefore are independent of the origin of the field. For this reason there is no need for special discussion of the case where the field is produced, not by charged conductors, but (for instance) by extraneous charges placed in the dielectric itself or by pyroelectric polarization of it (§13).

Since the potential ϕ is constant on the surface of the conductor, we can write

$$\delta R = \phi \delta e = -\frac{1}{4\pi}\oint \phi \delta \mathbf{D}\cdot d\mathbf{f} = -\frac{1}{4\pi}\int \operatorname{div}(\phi\delta\mathbf{D})dV.$$

The last integral is taken over the whole volume outside the conductor. Since the varied field, like the original field, must satisfy the field equations, we have $\operatorname{div}\delta\mathbf{D} = 0$, and so $\operatorname{div}(\phi\delta\mathbf{D}) = \phi \operatorname{div}\delta\mathbf{D} + \delta\mathbf{D}\cdot\mathbf{grad}\,\phi = -\mathbf{E}\cdot\delta\mathbf{D}$. Thus the following important formula is obtained:

$$\delta R = \int (\mathbf{E}\cdot\delta\mathbf{D}/4\pi)dV. \tag{10.2}$$

It should be emphasized that the integration in (10.2) is over the whole field, including the vacuum if the dielectric does not occupy all space outside the conductor.

The work done on a thermally insulated body is just the change in its energy at constant entropy. Hence the expression (10.2) must be included in the thermodynamic relation which gives the infinitesimal change in the total energy of the body; the latter contains also the energy of the electric field. Denoting the total energy by $\mathscr{U}$, we therefore have

$$\delta\mathscr{U} = T\delta\mathscr{S} + \frac{1}{4\pi}\int \mathbf{E}\cdot\delta\mathbf{D}\,dV, \tag{10.3}$$

where T is the temperature of the body and $\mathscr{S}$ its entropy.†

Accordingly we have for the total free energy‡ $\mathscr{F} = \mathscr{U} - T\mathscr{S}$

$$\delta\mathscr{F} = -\mathscr{S}\delta T + \frac{1}{4\pi}\int \mathbf{E}\cdot\delta\mathbf{D}\,dV. \tag{10.4}$$

Similar thermodynamic relations can be obtained for the quantities pertaining to unit volume of the body. Let U, S and ρ be the internal energy, entropy and mass of unit volume. It is well known that the ordinary thermodynamic relation (in the absence of a field) for the internal energy of unit volume is $dU = TdS + \zeta d\rho$, where ζ is the chemical potential of the substance.§ In the presence of a field in a dielectric, there must be added the integrand in (10.3):

$$dU = TdS + \zeta d\rho + \mathbf{E}\cdot d\mathbf{D}/4\pi. \tag{10.5}$$

For the free energy per unit volume of the dielectric, $F = U - TS$, we therefore have

$$dF = -SdT + \zeta d\rho + \mathbf{E}\cdot d\mathbf{D}/4\pi. \tag{10.6}$$

These relations are the basis of the thermodynamics of dielectrics.

We see that U and F are the thermodynamic potentials with respect to S, ρ, $\mathbf{D}$ and T, ρ, $\mathbf{D}$ respectively. In particular, we can obtain the field by differentiating these potentials with respect to the components of the vector $\mathbf{D}$:

$$\mathbf{E} = 4\pi(\partial U/\partial\mathbf{D})_{S,\rho} = 4\pi(\partial F/\partial\mathbf{D})_{T,\rho}. \tag{10.7}$$

---

† In (10.3) and (10.4) the volume of the body is assumed constant, but in general it becomes inhomogeneous in an electric field, and so the volume no longer characterizes the state of the body.

‡ This quantity is meaningful only when the temperature is constant throughout the body.

§ See *SP* 1, §24. Instead of the mass density we there use the number of particles N per unit volume; $\rho = Nm$, where m is the mass of one particle. The chemical potentials ζ per unit mass and μ per particle are related by $\zeta = \mu/m$.

The use of the letter ρ for the mass density as well as the charge density cannot lead to any misunderstanding, because the two quantities never appear together.

The free energy is more convenient in this respect, since it is to be differentiated at constant temperature, whereas the internal energy must be expressed in terms of the entropy, which is less easy.

Together with U and F, it is convenient to introduce thermodynamic potentials in which the components of the vector **E**, instead of **D**, are the independent variables. Such are

$$\tilde{U} = U - \mathbf{E}\cdot\mathbf{D}/4\pi, \quad \tilde{F} = F - \mathbf{E}\cdot\mathbf{D}/4\pi. \tag{10.8}$$

On differentiating these we have

$$\left.\begin{aligned} d\tilde{U} &= T dS + \zeta d\rho - \mathbf{D}\cdot d\mathbf{E}/4\pi, \\ d\tilde{F} &= -S dT + \zeta d\rho - \mathbf{D}\cdot d\mathbf{E}/4\pi. \end{aligned}\right\} \tag{10.9}$$

Hence, in particular,

$$\mathbf{D} = -4\pi(\partial\tilde{U}/\partial\mathbf{E})_{S,\rho} = -4\pi(\partial\tilde{F}/\partial\mathbf{E})_{T,\rho}. \tag{10.10}$$

It should be noticed that the relation between the thermodynamic quantities with and without the tilde is exactly that which occurs in §5 for the energy of the electrostatic field of conductors in a vacuum. For the integral $\int \mathbf{E}\cdot\mathbf{D}\,dV$ can be transformed in an exactly similar manner to the one at the beginning of §2, with the equation $\operatorname{div}\mathbf{D} = 0$ inside the dielectric and the boundary condition $D_n = 4\pi\sigma$ on the surfaces of conductors:

$$\begin{aligned} \frac{1}{4\pi}\int \mathbf{E}\cdot\mathbf{D}\,dV &= -\frac{1}{4\pi}\int \mathbf{grad}\,\phi\cdot\mathbf{D}\,dV \\ &= \frac{1}{4\pi}\sum_a \int \phi_a D_n\,df = \sum_a \phi_a e_a. \end{aligned} \tag{10.11}$$

Hence we have for the internal energy, for example,

$$\tilde{\mathscr{U}} = \mathscr{U} - \int \frac{\mathbf{E}\cdot\mathbf{D}}{4\pi}\,dV = \mathscr{U} - \sum_a \phi_a e_a, \tag{10.12}$$

in agreement with the definition (5.5).

It is useful to derive also the formulae for infinitesimal changes in these quantities, expressed in terms of the charges and potentials of the conductors (the sources of the field). For example, the variation in the free energy (for a given temperature) is

$$(\delta\mathscr{F})_T = \delta R = \sum_a \phi_a \delta e_a. \tag{10.13}$$

For the variation of $\tilde{\mathscr{F}}$ we have

$$(\delta\tilde{\mathscr{F}})_T = (\delta\mathscr{F})_T - \delta\sum_a \phi_a e_a = -\sum_a e_a \delta\phi_a. \tag{10.14}$$

We can say that the quantities without the tilde are the thermodynamic potentials with respect to the charges on the conductors, while those with it are thermodynamic potentials with respect to their potentials.

It is known from thermodynamics that the various thermodynamic potentials have the property of being minima in a state of thermodynamic equilibrium, relative to various changes in the state of the body. In formulating these conditions of equilibrium in an electric field, it is necessary to say whether changes of state with constant charges on the conductors (the field sources) or those with constant potentials are being considered. For example, in equilibrium $\mathscr{F}$ and $\tilde{\mathscr{F}}$ are minima with respect to changes in state occurring at constant temperature and (respectively) constant charges and potentials of the conductors (the same is true for $\mathscr{U}$ and $\tilde{\mathscr{U}}$ at constant entropy).

If any processes (such as chemical reactions) which are not directly related to the electric field can occur in the body, the condition of equilibrium with respect to these processes is that F be a minimum for given density, temperature and induction $\mathbf{D}$, or that $\tilde{F}$ be a minimum for constant density, temperature and field $\mathbf{E}$.

Hitherto we have made no assumptions concerning the dependence of $\mathbf{D}$ on $\mathbf{E}$, so that all the thermodynamic relations derived above are valid whatever the nature of this dependence. Let us now apply them to an isotropic dielectric, where a linear relation $\mathbf{D} = \varepsilon\mathbf{E}$ holds. In this case integration of (10.5) and (10.6) gives

$$\left.\begin{aligned} U &= U_0(S, \rho) + D^2/8\pi\varepsilon, \\ F &= F_0(T, \rho) + D^2/8\pi\varepsilon, \end{aligned}\right\} \tag{10.15}$$

where U_0 and F_0 pertain to the dielectric in the absence of the field. Thus in this case the quantity

$$D^2/8\pi\varepsilon = \varepsilon E^2/8\pi = ED/8\pi \tag{10.16}$$

is the change in the internal energy (for given entropy and density) or in the free energy (for given temperature and density), per unit volume of the dielectric medium, resulting from the presence of the field.

The expressions for the potentials $\tilde{U}$ and $\tilde{F}$ are similarly

$$\left.\begin{aligned} \tilde{U} &= U_0(S, \rho) - \varepsilon E^2/8\pi, \\ \tilde{F} &= F_0(T, \rho) - \varepsilon E^2/8\pi. \end{aligned}\right\} \tag{10.17}$$

We see that the differences $U - U_0$ and $\tilde{U} - U_0$ in this case differ only in sign, as they did for an electric field in a vacuum (§5). In a dielectric medium, however, this simple result holds good only when there is a linear relation between $\mathbf{D}$ and $\mathbf{E}$.

We shall write out also, for future reference, formulae for the entropy density S and the chemical potential ζ, which follow from (10.15):

$$\begin{aligned} S = -\left(\frac{\partial F}{\partial T}\right)_{\rho,\mathbf{D}} &= S_0(T, \rho) + \frac{D^2}{8\pi\varepsilon^2}\left(\frac{\partial \varepsilon}{\partial T}\right)_\rho \\ &= S_0(T, \rho) + \frac{E^2}{8\pi}\left(\frac{\partial \varepsilon}{\partial T}\right)_\rho, \end{aligned} \tag{10.18}$$

$$\zeta = \left(\frac{\partial F}{\partial \rho}\right)_{T,\mathbf{D}} = \zeta_0(T, \rho) - \frac{E^2}{8\pi}\left(\frac{\partial \varepsilon}{\partial \rho}\right)_T. \tag{10.19}$$

These quantities, of course, differ from zero only inside the dielectric.

The total free energy is obtained by integrating (10.15) over all space. By (10.11) we have

$$\mathscr{F} - \mathscr{F}_0 = \int \mathbf{E}\cdot\mathbf{D}\,\mathrm{d}V/8\pi = \tfrac{1}{2}\sum e_a\phi_a. \tag{10.20}$$

This last expression is formally identical with the energy of the electrostatic field of conductors in a vacuum. The same result can be obtained directly by starting from the variation $\delta\mathscr{F}$ (10.13) for an infinitesimal change in the charges on the conductors. In the present case, when **D** and **E** are linearly related, all the field equations and their boundary conditions are also linear. Hence the potentials of the conductors must (as for the field in a vacuum) be linear functions of their charges, and integration of equation (10.13) gives (10.20).

It should be emphasized that these arguments do not presuppose the dielectric to fill all space outside the conductors. If, however, this is so, we can go further and use the results at the end of §7 to draw the following conclusion. For given charges on the conductors, the presence of the dielectric medium reduces by a factor ε both the potentials of the conductors and the field energy, as compared with the values for a field in a vacuum. If, on the other hand, the potentials of the conductors are maintained constant, then their charges and the field energy are increased by a factor ε.

PROBLEM

Determine the height h to which a liquid rises in a vertical plane capacitor.

SOLUTION. For given potentials on the capacitor plates, $\mathscr{F}$ must be a minimum. $\mathscr{F}$ includes the energy $\frac{1}{2}\rho g h^2$ of the liquid under gravity. From this condition we easily obtain $h = (\varepsilon - 1)E^2/8\pi\rho g$.

§11. The total free energy of a dielectric

The total free energy $\mathscr{F}$ (or the total internal energy $\mathscr{U}$), as defined in §10, includes the energy of the external electric field which polarizes the dielectric; this field may be imagined as being produced by a particular assembly of conductors with specified total charges. It is also meaningful to consider the total free energy less the energy of the field which would be present in all space if the body were absent. We denote this field by $\mathfrak{E}$. Then the total free energy in this sense is

$$\int (F - \mathfrak{E}^2/8\pi)\,\mathrm{d}V, \tag{11.1}$$

where F is the free energy density. Here we shall denote this quantity by the letter $\mathscr{F}$, which in §10 signified $\int F\,\mathrm{d}V$. It should be emphasized that the difference between the two definitions of $\mathscr{F}$ is a quantity independent of the thermodynamic state and properties of the dielectric, and hence it has no effect on the fundamental differential relations of thermodynamics pertaining to this quantity.†

Let us calculate the change in $\mathscr{F}$ resulting from an infinitesimal change in the field which occurs at constant temperature and does not destroy the thermodynamic equilibrium of the medium. Since $\delta F = \mathbf{E}\cdot\delta\mathbf{D}/4\pi$, we have $\delta\mathscr{F} = \int(\mathbf{E}\cdot\delta\mathbf{D} - \mathfrak{E}\cdot\delta\mathfrak{E})\mathrm{d}V/4\pi$. This expression is identically equal to

$$\delta\mathscr{F} = \int(\mathbf{D} - \mathfrak{E})\cdot\delta\mathfrak{E}\,\mathrm{d}V/4\pi + \int\mathbf{E}\cdot(\delta\mathbf{D} - \delta\mathfrak{E})\mathrm{d}V/4\pi - \int(\mathbf{D} - \mathbf{E})\cdot\delta\mathfrak{E}\,\mathrm{d}V/4\pi. \tag{11.2}$$

---

† It may be noted that there would be no sense in subtracting $E^2/8\pi$ from F, because **E** is the field as modified by the presence of the dielectric, and so the difference $F - E^2/8\pi$ could not be regarded as the free energy density of the dielectric as such.

In the first integral we write $\delta\mathfrak{E} = -\mathbf{grad}\,\delta\phi_0$ (where ϕ_0 is the potential of the field $\mathfrak{E}$) and integrate by parts:

$$\int \mathbf{grad}\,\delta\phi_0 \cdot (\mathbf{D}-\mathfrak{E})\,dV = \oint \delta\phi_0(\mathbf{D}-\mathfrak{E})\cdot d\mathbf{f} - \int \delta\phi_0 \operatorname{div}(\mathbf{D}-\mathfrak{E})\,dV.$$

It is easy to see that both the integrals on the right-hand side are zero. For the volume integral this follows at once from the equations $\operatorname{div}\mathbf{D} = 0$ and $\operatorname{div}\mathfrak{E} = 0$ which the induction in the dielectric and the field in the vacuum must respectively satisfy. The surface integral is taken over the surfaces of the conductors which produce the field and over an infinitely distant surface. The latter of these is, as usual, zero, and for each of the conductors $\delta\phi_0 =$ constant, so that $\oint\delta\phi_0(\mathbf{D}-\mathfrak{E})\cdot d\mathbf{f} = \delta\phi_0\oint(\mathbf{D}-\mathfrak{E})\cdot d\mathbf{f}$. The field $\mathfrak{E}$, by definition, is produced by the same sources as the field **E** and induction **D** (i.e. by the same conductors with given total charges e). Hence the two integrals $\oint D_n df$ and $\oint \mathfrak{E}_n df$ are both equal to $4\pi e$, and their difference is zero.

Similarly, we can see that the second term in (11.2) is also zero, by putting $\mathbf{E} = -\mathbf{grad}\,\phi$ and using the same transformation. Finally, we have

$$\delta\mathscr{F} = -\int(\mathbf{D}-\mathbf{E})\cdot\delta\mathfrak{E}\,dV/4\pi = -\int\mathbf{P}\cdot\delta\mathfrak{E}\,dV. \tag{11.3}$$

It should be noticed that the integral in this expression need be taken only over the volume of the dielectric medium, since outside it $\mathbf{P} = 0$.

However, we must emphasize that the integrand $\mathbf{P}\cdot\delta\mathfrak{E}$ cannot be interpreted as the variation of the free energy density in the same way as was done with formulae (10.3), (10.4). First of all, this density must exist outside the body, which modifies the field in the surrounding space also. It is clear, moreover, that the energy density at any point in the body can depend only on the field actually present there, and not on the field which would be present if the body were removed.

If the external field $\mathfrak{E}$ is uniform, then

$$\delta\mathscr{F} = -\delta\mathfrak{E}\cdot\int\mathbf{P}\,dV = -\mathscr{P}\cdot\delta\mathfrak{E}, \tag{11.4}$$

where $\mathscr{P}$ is the total electric dipole moment of the body. Hence the thermodynamic identity for the free energy can be written in this case as

$$d\mathscr{F} = -\mathscr{S}\,dT - \mathscr{P}\cdot d\mathfrak{E}. \tag{11.5}$$

The total electric moment of the body can therefore be obtained by differentiating the total free energy:

$$\mathscr{P} = -(\partial\mathscr{F}/\partial\mathfrak{E})_T. \tag{11.6}$$

The latter formula can also be obtained directly from the general statistical formula

$$\overline{\partial\hat{\mathscr{H}}/\partial\lambda} = (\partial\mathscr{F}/\partial\lambda)_T,$$

where $\hat{\mathscr{H}}$ is the Hamiltonian of the body as the system of its component particles, and λ is any parameter characterizing the external conditions in which the body is placed; see *SP* 1, (11.4), (15.11). For a body in a uniform external field $\mathfrak{E}$, the Hamiltonian contains a term $-\mathfrak{E}\cdot\hat{\mathscr{P}}$, where $\hat{\mathscr{P}}$ is the dipole moment operator. Taking $\mathfrak{E}$ as the parameter λ, we obtain the required formula.

If **D** and **E** are connected by the linear relation $\mathbf{D} = \varepsilon\mathbf{E}$, we can similarly calculate explicitly not only the variation $\delta\mathscr{F}$ but $\mathscr{F}$ itself. We have

$$\mathscr{F} - \mathscr{F}_0 = \int(\mathbf{E}\cdot\mathbf{D} - \mathfrak{E}^2)\,dV/8\pi.$$

This can be identically transformed into

$$\mathscr{F} - \mathscr{F}_0 = \int (\mathbf{E} + \mathfrak{E}) \cdot (\mathbf{D} - \mathfrak{E}) \mathrm{d}V/8\pi - \int \mathfrak{E} \cdot (\mathbf{D} - \mathbf{E}) \mathrm{d}V/8\pi.$$

The first term on the right is zero, as we see by putting

$$\mathbf{E} + \mathfrak{E} = -\mathbf{grad}\,(\phi + \phi_0)$$

and again using the same transformation. Hence we have

$$\mathscr{F} - \mathscr{F}_0(V, T) = -\tfrac{1}{2} \int \mathfrak{E} \cdot \mathbf{P}\, \mathrm{d}V. \tag{11.7}$$

In particular, in a uniform external field

$$\mathscr{F} - \mathscr{F}_0(V, T) = -\tfrac{1}{2} \mathfrak{E} \cdot \mathscr{P}. \tag{11.8}$$

This last equation can also be obtained by direct integration of the relation (11.3) if we notice that, since all the field equations are linear when $\mathbf{D} = \varepsilon \mathbf{E}$, the electric moment $\mathscr{P}$ must be a linear function of $\mathfrak{E}$.

The linear relation between the components of $\mathscr{P}$ and $\mathfrak{E}$ can be written

$$\mathscr{P}_i = V\alpha_{ik} \mathfrak{E}_k, \tag{11.9}$$

as for conductors (§2). For a dielectric, however, the polarizability depends not only on the shape but also on the permittivity. The symmetry of the tensor α_{ik}, mentioned in §2, follows at once from the relation (11.6); it is sufficient to notice that the second derivative $\partial^2 \mathscr{F}/\partial \mathfrak{E}_k \partial \mathfrak{E}_i = -\partial \mathscr{P}_i / \partial \mathfrak{E}_k = -V\alpha_{ik}$ is independent of the order of differentiation.

Formula (11.7) becomes still simpler in the important case where ε is close to 1, i.e. the dielectric susceptibility $\kappa = (\varepsilon - 1)/4\pi$ is small. In this case, in calculating the energy, we can neglect the modification of the field due to the presence of the body, putting $\mathbf{P} = \kappa \mathbf{E} \cong \kappa \mathfrak{E}$. Then

$$\mathscr{F} - \mathscr{F}_0 = -\tfrac{1}{2}\kappa \int \mathfrak{E}^2 \,\mathrm{d}V, \tag{11.10}$$

the integral being taken over the volume of the body. In a uniform field, the dipole moment $\mathscr{P} = V\kappa \mathfrak{E}$, and the free energy is

$$\mathscr{F} - \mathscr{F}_0 = -\tfrac{1}{2}\kappa V \mathfrak{E}^2. \tag{11.11}$$

In the general case of an arbitrary relation between **D** and **E**, the simple formulae (11.7) and (11.8) do not hold. Here the formula

$$\mathscr{F} = \int \left(F - \frac{\mathfrak{E}^2}{8\pi} \right) \mathrm{d}V = \int \left[F - \frac{\mathbf{E} \cdot \mathbf{D}}{8\pi} - \tfrac{1}{2} \mathbf{P} \cdot \mathfrak{E} \right] \mathrm{d}V \tag{11.12}$$

may be useful in calculating $\mathscr{F}$; its derivation is obvious after the above discussion. The two integrands differ by

$$-\frac{\mathbf{E} \cdot \mathbf{D}}{8\pi} - \tfrac{1}{2} \mathfrak{E} \cdot \mathbf{P} + \frac{\mathfrak{E}^2}{8\pi} = -\frac{1}{8\pi} (\mathbf{D} - \mathfrak{E}) \cdot (\mathbf{E} + \mathfrak{E});$$

after substitution of $\mathbf{E} = -\mathbf{grad}\,\phi$, $\mathfrak{E} = -\mathbf{grad}\,\phi_0$ and integration over all space, the result is zero. In (11.12), as in (11.7), the second integrand is zero outside the body, where $\mathbf{P} = 0$ and $F = E^2/8\pi$, so that the integration is taken only over the volume of the body.

PROBLEM

Derive the formula which replaces (11.7) when the body is not in a vacuum but in a medium of permittivity $\varepsilon^{(e)}$.

SOLUTION. Using the same transformations as before, we find

$$\mathcal{F} - \mathcal{F}_0 = -\frac{1}{8\pi}\int \mathfrak{E}\cdot(\mathbf{D} - \varepsilon^{(e)}\mathbf{E})\,dV.$$

§12. Electrostriction of isotropic dielectrics

For a solid dielectric in an electric field the concept of pressure cannot be defined as for an isotropic body in the absence of a field, because the forces acting on a dielectric (which we shall determine in §§15, 16) vary over the body, and are anisotropic even if the body itself is isotropic. An exact determination of the deformation (*electrostriction*) of such a body involves the solution of a complex problem in the theory of elasticity.

However, matters are much simpler if we are interested only in the change in the total volume of the body. As we saw in §5, the shape of the body may then be regarded as unchanged, i.e. the deformation may be regarded as a uniform volume compression or expansion.

We shall neglect the dielectric properties of the external medium (the atmosphere, for instance) in which the body is situated, i.e. we suppose that $\varepsilon = 1$. This medium thus serves merely to exert a uniform pressure on the surface of the body, which we shall denote by P. If $\mathcal{F}$ is the total free energy of the body, then we have the thermodynamic relation $P = -(\partial\mathcal{F}/\partial V)_T$, and accordingly the expression for the differential $d\mathcal{F}$ contains a term $-PdV$. For example, in a uniform external field, (11.5) becomes

$$d\mathcal{F} = -\mathcal{S}dT - PdV - \mathcal{P}\cdot d\mathfrak{E}.$$

We introduce the total thermodynamic potential (Gibbs function) of the body in accordance with the usual thermodynamic relation

$$\Phi = \mathcal{F} + PV. \tag{12.1}$$

The differential of this quantity in a uniform external field is

$$d\Phi = -\mathcal{S}dT + VdP - \mathcal{P}\cdot d\mathfrak{E}. \tag{12.2}$$

The change in the thermodynamic quantities in an external electric field is usually a relatively small quantity. It is known from the theorem of small increments, *SP* 1 (15.12), that a small change in the free energy (for given T and V) is equal to the small change in the thermodynamic potential (for given T and P). Hence, besides (11.8), we can write analogously

$$\Phi = \Phi_0 - \tfrac{1}{2}\mathfrak{E}\cdot\mathcal{P} \tag{12.3}$$

for the thermodynamic potential of a body in a uniform external field. Here Φ_0 is the value for the body in the absence of the field and for given values of P and T, while $\mathcal{F}_0$ in (11.8) is the free energy in the absence of the field and for given values of V and T.

Making explicit the dependence of the dipole moment on V and $\mathfrak{E}$ according to (11.9), we can rewrite (12.3) as

$$\Phi = \Phi_0(P,T) - \tfrac{1}{2}V\alpha_{ik}\mathfrak{E}_i\mathfrak{E}_k, \tag{12.4}$$

where the correction term must be expressed as a function of temperature and pressure by means of the equation of state for the body in the absence of the field. In particular, for a substance with small dielectric susceptibility this formula becomes simply

$$\mathscr{G} = \mathscr{G}_0(P,T) - \tfrac{1}{2}\kappa V \mathfrak{E}^2; \tag{12.5}$$

cf. (11.11).

The required change in volume $V - V_0$ in the external field can now be obtained immediately by differentiating $\mathscr{G}$ with respect to pressure for constant T and $\mathfrak{E}$. For example, from (12.5) we have

$$V - V_0 = -\tfrac{1}{2}\mathfrak{E}^2 [\partial(\kappa V)/\partial P]_T. \tag{12.6}$$

This quantity may be either positive or negative (whereas, in electrostriction of conductors, the volume is always greater in the presence of the field).

Similarly, we can calculate the amount of heat Q absorbed in a dielectric when an external electric field is isothermally applied (the external pressure being constant).† Differentiation of $\mathscr{G} - \mathscr{G}_0$ with respect to temperature gives the change in the entropy of the body, and by multiplying this by T we obtain the required quantity of heat. For example, from (12.5) we obtain

$$Q = \tfrac{1}{2}\mathfrak{E}^2 T [\partial(\kappa V)/\partial T]_P. \tag{12.7}$$

Positive values of Q correspond to absorption of heat.

PROBLEMS

PROBLEM 1. Determine the change in volume and the electrocaloric effect for a dielectric ellipsoid in a uniform electric field parallel to one of its axes.

SOLUTION. From formulae (12.3) and (8.10) we have

$$\mathscr{G} = \mathscr{G}_0 - \frac{V}{8\pi}\frac{\varepsilon - 1}{n\varepsilon + 1 - n}\mathfrak{E}^2.$$

The relative change in volume is found to be‡

$$\frac{V - V_0}{V} = \frac{\mathfrak{E}^2}{8\pi}\left[\frac{\varepsilon - 1}{n\varepsilon + 1 - n}\frac{1}{K} - \frac{1}{(n\varepsilon + 1 - n)^2}\left(\frac{\partial \varepsilon}{\partial P}\right)_T\right],$$

and the electrocaloric effect

$$Q = \frac{TV\mathfrak{E}^2}{8\pi}\left[\frac{\varepsilon - 1}{n\varepsilon + 1 - n}\alpha + \frac{1}{(n\varepsilon + 1 - n)^2}\left(\frac{\partial \varepsilon}{\partial T}\right)_P\right],$$

where $1/K = -(1/V)(\partial V/\partial P)_T$ is the compressibility of the body, and $\alpha = (1/V)(\partial V/\partial T)_P$ the thermal expansion coefficient.

In particular, for a plane disc in a field perpendicular to it, $n = 1$, so that

$$\frac{V - V_0}{V} = \frac{\mathfrak{E}^2}{8\pi}\left[\frac{\varepsilon - 1}{\varepsilon}\frac{1}{K} - \frac{1}{\varepsilon^2}\left(\frac{\partial \varepsilon}{\partial P}\right)_T\right],$$

$$Q = \frac{TV\mathfrak{E}^2}{8\pi}\left[\frac{\varepsilon - 1}{\varepsilon}\alpha + \frac{1}{\varepsilon^2}\left(\frac{\partial \varepsilon}{\partial T}\right)_P\right].$$

---

† If the body is thermally insulated, the application of the field results in a change of temperature $\Delta T = -Q/\mathscr{C}_P$, where $\mathscr{C}_P$ is the heat capacity at constant pressure.

‡ In the limit $\varepsilon \to \infty$, we find as the change in volume of a conducting ellipsoid $(V - V_0)/V = \mathfrak{E}^2/8\pi K n$. For a sphere, $n = \frac{1}{3}$, and we recover the result in §5, Problem 4.

For a similar disc (or any cylinder) in a longitudinal field, $n = 0$, and

$$\frac{V-V_0}{V} = \frac{\mathfrak{E}^2}{8\pi}\left[\frac{\varepsilon-1}{K} - \left(\frac{\partial\varepsilon}{\partial P}\right)_T\right], \quad Q = \frac{TV\mathfrak{E}^2}{8\pi}\left[(\varepsilon-1)\alpha + \left(\frac{\partial\varepsilon}{\partial T}\right)_P\right].$$

PROBLEM 2. Determine the difference between the heat capacity $\mathscr{C}_\phi$ of a plane disc in a field perpendicular to it, with a constant potential difference between its faces, and the heat capacity $\mathscr{C}_D$ at constant induction, the external pressure being maintained constant in each case.†

SOLUTION. According to the results of Problem 1, the entropy of the disc is

$$\mathscr{S} = -\left(\frac{\partial \tilde{\mathscr{g}}}{\partial T}\right)_{P,\mathfrak{E}} = \mathscr{S}_0(P,T) + \frac{V\mathfrak{E}^2}{8\pi}\left[\frac{\varepsilon-1}{\varepsilon}\alpha + \frac{1}{\varepsilon^2}\left(\frac{\partial\varepsilon}{\partial T}\right)_P\right].$$

The induction inside the disc is the same as the external field: $D = \mathfrak{E}$. Hence, to calculate the heat capacity $\mathscr{C}_D$, we must differentiate $\mathscr{S}$ for constant $\mathfrak{E}$. The potential difference between the faces of the disc is $\phi = El = \mathfrak{E}l/\varepsilon$, where l is its thickness. For a uniform compression or expansion of a body, l is proportional to $V^{\frac{1}{3}}$. Hence, to calculate the heat capacity $\mathscr{C}_\phi$, we must differentiate $\mathscr{S}$ for constant $\mathfrak{E}V^{\frac{1}{3}}/\varepsilon$. The required difference is found to be

$$\mathscr{C}_\phi - \mathscr{C}_D = \frac{TV\mathfrak{E}^2}{4\pi\varepsilon}\left[(\varepsilon-1)\alpha + \frac{1}{\varepsilon}\left(\frac{\partial\varepsilon}{\partial T}\right)_P\right]\left[\frac{1}{\varepsilon}\left(\frac{\partial\varepsilon}{\partial T}\right)_P - \tfrac{1}{3}\alpha\right].$$

PROBLEM 3. Determine the electrocaloric effect in a homogeneous dielectric whose total volume is kept constant.

SOLUTION. Strictly speaking, when an external field is applied the density of the body changes (and ceases to be uniform), even if the total volume is kept constant. In calculating the change in the total entropy, however, we can ignore this and assume the density ρ constant at every point.‡

According to (10.18) the total entropy of the body is

$$\mathscr{S} = \mathscr{S}_0(\rho,T) + \frac{1}{8\pi}\left(\frac{\partial\varepsilon}{\partial T}\right)_\rho \int E^2\,\mathrm{d}V,$$

where the integration is over the volume of the body. The amount of heat absorbed is

$$Q = \frac{T}{8\pi}\left(\frac{\partial\varepsilon}{\partial T}\right)_\rho \int E^2\,\mathrm{d}V.$$

PROBLEM 4. Determine the difference $\mathscr{C}_\phi - \mathscr{C}_D$ (see Problem 2) when the total volume of the disc is kept constant.

SOLUTION. When the volume, and therefore the thickness, of the disc are constant, differentiation for constant potential difference is equivalent to differentiation for constant field E. Using the formula of Problem 3 for the entropy we have

$$\mathscr{C}_E - \mathscr{C}_D = \frac{TVE^2}{4\pi\varepsilon}\left(\frac{\partial\varepsilon}{\partial T}\right)_\rho^2 = \frac{TV\mathfrak{E}^2}{4\pi\varepsilon^3}\left(\frac{\partial\varepsilon}{\partial T}\right)_\rho^2.$$

PROBLEM 5. A capacitor consists of two conducting surfaces at a distance h apart which is small compared with their dimensions; the space between them is filled with a substance of permittivity ε_1. A sphere of radius $a \ll h$ and permittivity ε_2 is placed in the capacitor. Determine the change in capacitance.

SOLUTION. Let the sphere be placed in the capacitor in such a way that the potential difference ϕ between the plates remains unchanged. The free energy for constant potentials of the conductors is $\tilde{\mathscr{F}}$. In the absence of the sphere, $\tilde{\mathscr{F}} = -\frac{1}{2}C_0\phi^2$, where C_0 is the original capacitance. Since the sphere is small, we may imagine it to be brought into a uniform field $\mathfrak{E} = \phi/h$, and the change in $\tilde{\mathscr{F}}$ is small. The small change in $\tilde{\mathscr{F}}$ at constant potentials is equal to the small change in $\mathscr{F}$ at constant charges on the sources of the field. Using the formula derived in §11, Problem, and (8.2), we have

$$\tilde{\mathscr{F}} = -\tfrac{1}{2}C_0\phi^2 - \tfrac{1}{2}a^3\varepsilon_1(\varepsilon_2-\varepsilon_1)\phi^2/(2\varepsilon_1+\varepsilon_2)h^2,$$

---

† $\mathscr{C}_\phi$ is the heat capacity of a disc between the plates of a plane capacitor in circuit with a constant e.m.f. In an unconnected capacitor with constant charges on the plates, the heat capacity of the disc is $\mathscr{C}_D$.

‡ The change in density $\delta\rho$ is of the second order with respect to the field ($\propto E^2$), and the consequent change in the total entropy is of the fourth order: the term in the change of total entropy which is linear in $\delta\rho$ is $(\partial S_0/\partial\rho)\int\delta\rho\,\mathrm{d}V$, and the integral is zero because the total mass of the body is unaltered.

whence the required capacitance is

$$C = C_0 + a^3\varepsilon_1(\varepsilon_2-\varepsilon_1)/(2\varepsilon_1+\varepsilon_2)h^2.$$

§13. Dielectric properties of crystals

In an anisotropic dielectric medium (a single crystal) the linear relation between the electric induction and the electric field is less simple, and does not reduce to a simple proportionality.

The most general form of such a relation is

$$D_i = D_{0i} + \varepsilon_{ik}E_k, \tag{13.1}$$

where $\mathbf{D}_0$ is a constant vector, and the quantities ε_{ik} form a tensor of rank two, called the *permittivity tensor* (or the *dielectric tensor*). The inhomogeneous term $\mathbf{D}_0$ in (13.1) does not, however, appear for all crystals. The majority of the types of crystal symmetry do not admit this constant vector (see below), and we then have simply

$$D_i = \varepsilon_{ik}E_k. \tag{13.2}$$

The tensor ε_{ik} is symmetrical:

$$\varepsilon_{ik} = \varepsilon_{ki}. \tag{13.3}$$

In order to prove this, it is sufficient to use the thermodynamic relation (10.10) and to observe that the second derivative $-4\pi\partial^2\tilde{F}/\partial E_k\partial E_i = \partial D_i/\partial E_k = \varepsilon_{ik}$ is independent of the order of differentiation.

For $\tilde{F}$ itself we have (when (13.2) holds) the expression

$$\tilde{F} = F_0 - \varepsilon_{ik}E_iE_k/8\pi. \tag{13.4}$$

The free energy F is

$$F = \tilde{F} + E_iD_i/4\pi = F_0 + \varepsilon^{-1}{}_{ik}D_iD_k/8\pi. \tag{13.5}$$

Like every symmetrical tensor of rank two, the tensor ε_{ik} can be brought to diagonal form by a suitable choice of the coordinate axes. In general, therefore, the tensor ε_{ik} is determined by three independent quantities, namely the three principal values $\varepsilon^{(1)}, \varepsilon^{(2)}, \varepsilon^{(3)}$. All these are necessarily greater than unity, just as $\varepsilon > 1$ for an isotropic body (see §14).

The number of different principal values of the tensor ε_{ik} may be less than three for certain symmetries of the crystal.

In crystals of the triclinic, monoclinic and orthorhombic systems, all three principal values are different; such crystals are said to be *biaxial*.† In crystals of the triclinic system, the directions of the principal axes of the tensor ε_{ik} are not uniquely related to any directions in the crystal. In those of the monoclinic system, one of the principal axes must coincide with the twofold axis of symmetry or be perpendicular to the plane of symmetry of the crystal. In crystals of the orthorhombic system, all three principal axes of the tensor ε_{ik} are crystallographically fixed.

Next, in crystals of the tetragonal, rhombohedral and hexagonal systems, two of the

† This name refers to the optical properties of the crystals; see §§98, 99.

three principal values are equal, so that there are only two independent quantities; such crystals are said to be *uniaxial*. One of the principal axes coincides with the fourfold, threefold or sixfold axis of crystal symmetry, but the directions of the other two principal axes can be chosen arbitrarily.

Finally, in crystals of the cubic system all three principal values of the tensor ε_{ik} are the same, and the directions of the principal axes are entirely arbitrary. This means that the tensor ε_{ik} is of the form $\varepsilon\delta_{ik}$, i.e. it is determined by a single scalar ε. In other words, as regards their dielectric properties, crystals of the cubic system are no different from isotropic bodies.

All these fairly obvious symmetry properties of the tensor ε_{ik} become particularly clear if we use a concept from tensor algebra, the *tensor ellipsoid*, the lengths of whose semiaxes are proportional to the principal values of a symmetrical tensor of rank two. The symmetry of the ellipsoid corresponds to that of the crystal. For instance, in a uniaxial crystal the tensor ellipsoid degenerates into a spheroid completely symmetrical about the longitudinal axis; it should be emphasized that, as regards the physical properties of the crystal which are determined by a symmetrical tensor of rank two, the presence of a threefold or higher axis of symmetry is equivalent to complete isotropy in the plane perpendicular to this axis. In cubic crystals, the tensor ellipsoid degenerates into a sphere.

Let us now examine the dielectric properties of crystals for which the constant term $\mathbf{D}_0$ appears in (13.1). The presence of this term signifies that the dielectric is spontaneously polarized even in the absence of an external electric field. Such bodies are said to be *pyroelectric*. The magnitude of this spontaneous polarization is, however, in practice always very small (in comparison with the molecular fields). This is because large values of $\mathbf{D}_0$ would lead to strong fields within the body, which is energetically very unfavourable and therefore could not correspond to thermodynamic equilibrium. The smallness of $\mathbf{D}_0$ also ensures the legitimacy of an expansion of $\mathbf{D}$ in powers of $\mathbf{E}$, of which (13.1) represents the first two terms.

The thermodynamic quantities for a pyroelectric body are found by integrating the relation $-4\pi\partial\tilde{F}/\partial E_i = D_i = D_{0i} + \varepsilon_{ik}E_k$, whence

$$\tilde{F} = F_0 - \varepsilon_{ik}E_iE_k/8\pi - E_iD_{0i}/4\pi. \tag{13.6}$$

The free energy is

$$\begin{aligned} F = \tilde{F} + E_iD_i/4\pi &= F_0 + \varepsilon_{ik}E_iE_k/8\pi \\ &= F_0 + \varepsilon^{-1}{}_{ik}(D_i - D_{0i})(D_k - D_{0k})/8\pi. \end{aligned} \tag{13.7}$$

It should be noted that the term in $\tilde{F}$ linear in E_i does not appear in F.†

The total free energy of a pyroelectric can be calculated from formula (11.12) by substituting (13.7) and (13.1). If there is no external field, $\mathfrak{E} = 0$, and we have simply

$$\mathscr{F} = \int [F_0 - (\mathbf{E}\cdot\mathbf{D}_0/8\pi)]\,\mathrm{d}V. \tag{13.8}$$

It is remarkable that the free energy of a pyroelectric in the absence of an external field depends, like the field $\mathbf{E}$, not only on the volume of the body but also on its shape.

† It should also be noted that in these formulae we neglect the *piezoelectric effect*, i.e. the effect of internal stresses on the electric properties of a body; see §17. The formulae given here are therefore, strictly speaking, applicable only when the fields are uniform throughout the body, and internal stresses do not arise.

As has already been pointed out, the phenomenon of pyroelectricity is not possible for every crystal symmetry. Since, in any symmetry transformation, all the properties of the crystal must remain unchanged, it is clear that the only crystals which can be pyroelectric are those in which there is a direction which is unchanged (and, in particular, not reversed) in all symmetry transformations, and that this will be the direction of the constant vector $\mathbf{D}_0$.

This condition is satisfied only by those symmetry groups which consist of a single axis together with planes of symmetry which pass through the axis. In particular, crystals having a centre of symmetry certainly cannot be pyroelectric. We may enumerate those out of the 32 crystal classes in which pyroelectricity occurs:

triclinic system: C_1
monoclinic system: C_s, C_2
orthorhombic system: C_{2v}
tetragonal system: C_4, C_{4v}
rhombohedral system: C_3, C_{3v}
hexagonal system: C_6, C_{6v}.

There are, of course, no pyroelectric cubic crystals. In a crystal of class C_1 the direction of the pyroelectric vector $\mathbf{D}_0$ is not related to any direction fixed in the crystal; in one of class C_s, it must lie in the plane of symmetry. In all the remaining classes listed above the direction of $\mathbf{D}_0$ is that of the axis of symmetry.†

It should be mentioned that, under ordinary conditions, pyroelectric crystals have zero total electric dipole moment, although their polarization is not zero. The reason is that there is a non-zero field $\mathbf{E}$ inside a spontaneously polarized dielectric. Since a body usually has a small but non-zero conductivity, the presence of a field gives rise to a current, which flows until the free charges formed on the surface of the body annihilate the field inside it. The same effect is produced by ions deposited on the surface from the air. Experimentally, pyroelectric properties are observed when a body is heated and a change in its spontaneous polarization is detected.

PROBLEMS

PROBLEM 1. Determine the field of a pyroelectric sphere in a vacuum.

SOLUTION. The field inside the sphere is uniform, and the field and induction are related by $2\mathbf{E} = -\mathbf{D}$ (as follows from (8.1) when $\mathfrak{E} = 0$, i.e. when there is no applied external field). Substituting in (13.1), we obtain the equation $2E_i + \varepsilon_{ik}E_k = -D_{0i}$. We take the coordinate axes to be the principal axes of the tensor ε_{ik}. Then this equation gives $E_i = -D_{0i}/(2+\varepsilon^{(i)})$. The polarization of the sphere is $P_i = (D_i - E_i)/4\pi = 3D_{0i}/4\pi(2+\varepsilon^{(i)})$. The field outside the sphere is that of an electric dipole with moment $\mathscr{P} = \mathbf{P}V$.

PROBLEM 2. Determine the field of a point charge in a homogeneous anisotropic medium.‡

SOLUTION. The field of a point charge is given by the equation div $\mathbf{D} = 4\pi e\delta(\mathbf{r})$ (the charge being at the origin). In an anisotropic medium $D_i = \varepsilon_{ik}E_k = -\varepsilon_{ik}\partial\phi/\partial x_k$; taking the coordinate axes x, y, z along the principal axes of the tensor ε_{ik}, we obtain for the potential the equation

$$\varepsilon^{(x)}\partial^2\phi/\partial x^2 + \varepsilon^{(y)}\partial^2\phi/\partial y^2 + \varepsilon^{(z)}\partial^2\phi/\partial z^2 = -4\pi e\,\partial(\mathbf{r}).$$

† In referring to symmetry conditions, we are regarding the crystal as an infinite medium. For a finite crystal, the exact value of the total dipole moment may depend (in an ionic crystal) on which crystal planes form its faces and whether these planes contain ions of only one sign or are electrically neutral. However, in macroscopic electrodynamics, which implies averaging over physically infinitesimal volumes, it is reasonable to consider that the position of the faces relative to the crystal lattice is averaged also. In consequence of this averaging, $\mathbf{D}_0$ vanishes in any non-pyroelectric finite crystal, and in a pyroelectric one is independent of the face configuration.

‡ In Problems 2–6 the anisotropic dielectric is assumed to be non-pyroelectric.

By the introduction of new variables

$$x' = x/\sqrt{\varepsilon^{(x)}}, \qquad y' = y/\sqrt{\varepsilon^{(y)}}, \qquad z' = z/\sqrt{\varepsilon^{(z)}}, \tag{1}$$

this becomes

$$\triangle'\phi = -\frac{4\pi e}{\sqrt{(\varepsilon^{(x)}\varepsilon^{(y)}\varepsilon^{(z)})}}\,\delta(\mathbf{r}')$$

which formally differs from the equation for the field in a vacuum only in that e is replaced by $e' = e/\sqrt{(\varepsilon^{(x)}\varepsilon^{(y)}\varepsilon^{(z)})}$. Hence

$$\phi = \frac{e'}{r'} = \frac{e}{\sqrt{(\varepsilon^{(x)}\varepsilon^{(y)}\varepsilon^{(z)})}}\left[\frac{x^2}{\varepsilon^{(x)}} + \frac{y^2}{\varepsilon^{(y)}} + \frac{z^2}{\varepsilon^{(z)}}\right]^{-\frac{1}{2}}.$$

In tensor notation, independent of the system of coordinates chosen, we have

$$\phi = e/\sqrt{(|\varepsilon|\varepsilon^{-1}{}_{ik}x_i x_k)},$$

where $|\varepsilon|$ is the determinant of the tensor ε_{ik}.

PROBLEM 3. Determine the capacitance of a conducting sphere, with radius a, in an anisotropic dielectric medium.

SOLUTION. By the transformation shown in Problem 2, the determination of the field of a sphere with charge e in an anisotropic medium reduces to the determination of the field in a vacuum due to a charge e' distributed over the surface of the ellipsoid $\varepsilon_{ik}x'_i x'_k = \varepsilon^{(x)}x'^2 + \varepsilon^{(y)}y'^2 + \varepsilon^{(z)}z'^2 = a^2$. Using formula (4.14) for the potential due to an ellipsoid, we find the required capacitance to be given by

$$\frac{1}{C} = \frac{1}{2\sqrt{(\varepsilon^{(x)}\varepsilon^{(y)}\varepsilon^{(z)})}}\int_0^\infty \left[\left(\xi + \frac{a^2}{\varepsilon^{(x)}}\right)\left(\xi + \frac{a^2}{\varepsilon^{(y)}}\right)\left(\xi + \frac{a^2}{\varepsilon^{(z)}}\right)\right]^{-\frac{1}{2}} d\xi.$$

PROBLEM 4. Determine the field in a flat anisotropic plate in a uniform external field $\mathfrak{E}$.

SOLUTION. From the condition of continuity of the tangential component of the field it follows that $\mathbf{E} = \mathfrak{E} + A\mathbf{n}$, where $\mathbf{E}$ is the uniform field inside the plate, $\mathbf{n}$ a unit vector normal to its surface, and A a constant. The constant is determined from the condition of continuity of the normal component of the induction, $\mathbf{n}\cdot\mathbf{D} = \mathbf{n}\cdot\mathfrak{E}$, or $n_i\varepsilon_{ik}E_k = n_i\varepsilon_{ik}\mathfrak{E}_k + A\varepsilon_{ik}n_i n_k = \mathfrak{E}_i n_i$. Hence $A = -(\varepsilon_{ik} - \delta_{ik})n_i\mathfrak{E}_k/\varepsilon_{lm}n_l n_m$.

In particular, if the external field is along the normal to the plate (the z-direction), we have

$$A = \mathfrak{E}(1 - \varepsilon_{zz})/\varepsilon_{zz}.$$

If it is parallel to the plate and in the x-direction, then

$$A = -\mathfrak{E}\varepsilon_{zx}/\varepsilon_{zz}.$$

PROBLEM 5. Determine the torque on an anisotropic dielectric sphere, with radius a, in a uniform external field $\mathfrak{E}$ in a vacuum.

SOLUTION. According to (8.2) we have for the field inside the sphere $E_x = 3\mathfrak{E}_x/(\varepsilon^{(x)} + 2)$, and similarly for E_y, E_z. Here the axes of x, y, z are taken to be the principal axes of the tensor ε_{ik}. Hence the components of the dipole moment of the sphere are

$$\mathscr{P}_x = \frac{4}{3}\pi a^3 P_x = \frac{\varepsilon^{(x)} - 1}{\varepsilon^{(x)} + 2}a^3\mathfrak{E}_x, \text{ etc.}$$

The components of the torque on the sphere are

$$K_z = (\mathscr{P}\times\mathfrak{E})_z = 3a^3\mathfrak{E}_x\mathfrak{E}_y(\varepsilon^{(x)} - \varepsilon^{(y)})/(\varepsilon^{(x)} + 2)(\varepsilon^{(y)} + 2),$$

and similarly for K_x, K_y.

PROBLEM 6. An infinite anisotropic medium contains a spherical cavity with radius a. Express the field in the cavity in terms of the uniform field $E^{(e)}$ far from the cavity.

SOLUTION. The transformation (1) of Problem 2 reduces the equation for the field potential in the medium to Laplace's equation for the field in a vacuum. The equation for the field potential in the cavity is transformed into that for the potential in a medium with permittivities $1/\varepsilon^{(x)}, 1/\varepsilon^{(y)}, 1/\varepsilon^{(z)}$. Moreover, the sphere is transformed into an ellipsoid with semiaxes $a/\sqrt{\varepsilon^{(x)}}$, $a/\sqrt{\varepsilon^{(y)}}$, $a/\sqrt{\varepsilon^{(z)}}$. Let $n^{(x)}$, $n^{(y)}$, $n^{(z)}$ be the depolarizing factors of such an ellipsoid (given by formulae (4.25)). Applying formula (8.7) to the field of this ellipsoid, we obtain the relation

$$(1-n^{(x)})\frac{\partial\phi^{(i)}}{\partial x'}+\frac{n^{(x)}}{\varepsilon^{(x)}}\frac{\partial\phi^{(i)}}{\partial x'}=\frac{\partial\phi^{(e)}}{\partial x'},$$

and similarly for the y' and z' directions. Returning to the original coordinates, we have $\partial\phi/\partial x' = \sqrt{\varepsilon^{(x)}}\partial\phi/\partial x = -\sqrt{\varepsilon^{(x)}}E_x$, so that the field in the cavity is

$$E^{(i)}{}_x = \frac{\varepsilon^{(x)}}{\varepsilon^{(x)}-n^{(x)}(\varepsilon^{(x)}-1)}E^{(e)}{}_x.$$

§14. The sign of the dielectric susceptibility

To elucidate the way in which the thermodynamic quantities for a dielectric in a field depend on its permittivity, let us consider the formal problem of the change in the electric component of the total free energy of the body when ε undergoes an infinitesimal change.

For an isotropic (not necessarily homogeneous) body we have by (10.20) $\mathscr{F}-\mathscr{F}_0 = \int (D^2/8\pi\varepsilon)\mathrm{d}V$. When ε changes, so does the induction, and the variation in the free energy is therefore

$$\delta\mathscr{F} = \int\frac{\mathbf{D}\cdot\delta\mathbf{D}}{8\pi\varepsilon}\mathrm{d}V - \int\frac{D^2}{8\pi\varepsilon^2}\delta\varepsilon\,\mathrm{d}V.$$

The first term on the right is the same as (10.2), which gives the work done in an infinitesimal change in the field sources (i.e. charges on conductors). In the present case, however, we are considering a change in the field but no change in the sources. This term therefore vanishes, leaving

$$\delta\mathscr{F} = -\int(\delta\varepsilon/\varepsilon^2)(D^2/8\pi)\mathrm{d}V = -\int\delta\varepsilon(E^2/8\pi)\mathrm{d}V. \quad (14.1)$$

From this formula it follows that any increase in the permittivity of the medium, even if in only a part of it (the sources of the field remaining unchanged), reduces the total free energy. In particular, we can say that the free energy is always reduced when uncharged conductors are brought into a dielectric medium, since these conductors may (in electrostatics) be regarded as bodies whose permittivity is infinite. This conclusion generalizes the theorem (§2) that the energy of the electrostatic field in a vacuum is diminished when an uncharged conductor is placed in it.

The total free energy is diminished also when any charge is brought up to a dielectric body from infinity (a process which may be regarded as an increase of ε in a certain volume of the field round the charge). In order to conclude from this that any charge is attracted to a dielectric, we should, strictly speaking, prove also that $\mathscr{F}$ cannot attain a minimum for any finite distance between the charge and the body. We shall not pause here to prove this statement, especially as the presence of an attractive force between a charge and a dielectric may be regarded as a fairly evident consequence of the interaction between the charge and the dipole moment of the dielectric, which it polarizes.

We can deduce immediately from formula (14.1) the direction of motion of a dielectric body in an almost uniform electric field, i.e. one which may be regarded as uniform over the dimensions of the body. In this case E^2 is taken outside the integral, and the difference $\mathscr{F}-\mathscr{F}_0$ is a negative quantity, proportional to E^2. In order to take a position in which its free energy is a minimum, the body will therefore move in the direction of E increasing.

It can be shown independently of (14.1) that the total change in the free energy of a dielectric when it is placed in an electric field is negative.† This can be done by the use of thermodynamic perturbation theory, the change in the free energy of the body being regarded as the result of a perturbation of its quantum energy levels by the external electric field. According to this theory we have

$$\mathscr{F} - \mathscr{F}_0 = \overline{V}_{nn} - \frac{1}{2}\sum_n \sum_m{}' \frac{|V_{nm}|^2 (w_m - w_n)}{E_n^{(0)} - E_m^{(0)}} - \frac{1}{2T}\overline{(V_{nn} - \overline{V}_{nn})^2}; \tag{14.2}$$

see *SP* 1, (32.6). Here $E_n^{(0)}$ are the unperturbed levels, V_{mn} the matrix elements of the perturbing energy, and the bar denotes a statistical averaging with respect to the Gibbs distribution $w_n = \exp\{(\mathscr{F}_0 - E_n^{(0)})/T\}$.

The term $\overline{V}_{nn}$ in formula (14.2), which is linear in the field, is zero except in pyroelectric bodies. The quadratic change in the free energy, which is of interest here, is given by the remaining terms. It is evident that they are negative.

On the other hand, it is clear from the derivation of (14.2) that the total free energy $\mathscr{F}$ must be taken in this formula as described in §11, omitting the energy of the field which would exist in the absence of the body. The difference $\mathscr{F} - \mathscr{F}_0$ is therefore given by the thermodynamic formula (11.7). Let us consider a long narrow cylinder placed parallel to a uniform external field $\mathfrak{E}$. The field within the cylinder is then $\mathfrak{E}$ also, and its polarization $\mathbf{P} = (\varepsilon - 1)\mathfrak{E}/4\pi$, so that

$$\mathscr{F} - \mathscr{F}_0 = -(\varepsilon - 1)V\mathfrak{E}^2/8\pi.$$

Thus $\mathscr{F} - \mathscr{F}_0$ is negative only if $\varepsilon > 1$. This leads to the conclusion mentioned in §7 and already made use of, namely that the permittivity of all bodies exceeds unity, and the dielectric susceptibility $\kappa = (\varepsilon - 1)/4\pi$ is therefore positive.

In the same way we can prove the inequalities $\varepsilon^{(i)} > 1$ for the principal values of the tensor ε_{ik} in an anisotropic dielectric medium. To do so, it is sufficient to consider the energy of a field parallel to each of the three principal axes in turn.

§15. Electric forces in a fluid dielectric

The problem of calculating the forces (called *ponderomotive* forces) which act on a dielectric in an arbitrary non-uniform electric field is fairly complicated and requires separate consideration for fluids (liquids or gases) and for solids. We shall take first the simpler case, that of fluid dielectrics. We denote by $\mathbf{f}\,dV$ the force on a volume element dV, and call the vector $\mathbf{f}$ the *force density*.

It is well known that the forces acting on any finite volume in a body can be reduced to forces applied to the surface of that volume (see *TE*, §2). This is a consequence of the law of conservation of momentum. The force acting on the matter in a volume dV is the change in its momentum per unit time. This change must be equal to the amount of momentum entering the volume through its surface per unit time. If we denote the momentum flux

† The change proportional to the square of the field is meant. It may be recalled that, in pyroelectric bodies, the change in the free energy contains also a term linear in the field, which is of no interest here.

tensor by $-\sigma_{ik}$, then†

$$\int f_i \mathrm{d}V = \oint \sigma_{ik} \mathrm{d}f_k, \tag{15.1}$$

where the integration on the right is over the surface of the volume V. The tensor σ_{ik} is called the *stress tensor*. It is evident that $\sigma_{ik}\mathrm{d}f_k = \sigma_{ik}n_k\mathrm{d}f$ is the ith component of the force on a surface element $\mathrm{d}f$ ($\mathbf{n}$ being a unit vector along the normal to the surface outwards from the volume under consideration).

Similarly, the total torque acting on a given volume also reduces to a surface integral, by virtue of the law of conservation of angular momentum. This reduction is possible because of the symmetry of the stress tensor ($\sigma_{ik} = \sigma_{ki}$), which thus expresses the conservation law mentioned.

On transforming the surface integral in (15.1) into a volume integral, we obtain $\int f_i \mathrm{d}V = \int (\partial\sigma_{ik}/\partial x_k)\mathrm{d}V$, whence, since the volume of integration is arbitrary,

$$f_i = \partial\sigma_{ik}/\partial x_k. \tag{15.2}$$

This is a well-known formula giving the body forces in terms of the stress tensor.

Let us now calculate the stress tensor. Any small region of the surface may be regarded as plane, and the properties of the body and the electric field near it as uniform. Hence, to simplify the derivation, we can with no loss of generality consider a plane-parallel layer of material (with thickness h and uniform composition, density and temperature) in an electric field which is uniform.‡ This field may be imagined to be due to conducting planes, such as the plates of a capacitor, applied to the surfaces of the layer.

Following the general method for determining forces, we subject one of the plates (the upper one, say) to a virtual translation over an infinitesimal distance $\boldsymbol{\xi}$, whose direction is arbitrary and need not be that of the normal $\mathbf{n}$. We shall suppose that the potential of the conductor remains unchanged at every point, and that the homogeneous deformation of the dielectric layer, resulting from the translation, is isothermal.

A force $-\sigma_{ik}n_k$ is exerted by the layer on unit area of the surface. In the virtual displacement this force does work $-\sigma_{ik}n_k\xi_i$. The work done in an isothermal deformation at constant potential is equal to the decrease in $\int \tilde{F}\mathrm{d}V$, i.e. in $h\tilde{F}$ per unit surface area. Thus

$$\sigma_{ik}\xi_i n_k = \delta(h\tilde{F}) = h\delta\tilde{F} + \tilde{F}\delta h. \tag{15.3}$$

The thermodynamic quantities for the fluid depend (for given temperature and field) only on its density; deformations which do not change the density (i.e. pure shears) do not affect the thermodynamic state. We can therefore write for an isothermal variation $\delta\tilde{F}$ in a fluid

$$\begin{aligned}\delta\tilde{F} &= \left(\frac{\partial\tilde{F}}{\partial\mathbf{E}}\right)_{T,\rho}\cdot\delta\mathbf{E} + \left(\frac{\partial\tilde{F}}{\partial\rho}\right)_{\mathbf{E},T}\delta\rho \\ &= -\frac{\mathbf{D}\cdot\delta\mathbf{E}}{4\pi} + \left(\frac{\partial\tilde{F}}{\partial\rho}\right)_{\mathbf{E},T}\delta\rho.\end{aligned} \tag{15.4}$$

† The force component f_i is not to be confused with the surface area component $\mathrm{d}f_k$.

‡ We thus ignore any terms in the stress tensor depending on the gradients of temperature, field, etc. These terms, however, are vanishingly small in comparison with terms which do not contain derivatives, in the same way as any terms containing derivatives which might appear in the relation between $\mathbf{D}$ and $\mathbf{E}$.

The change in the density of the layer is related to the change in its thickness by $\delta\rho = -\rho\delta h/h$. The variation of the field is calculated as follows. At a given point in space (with position vector **r**) there appears matter which was originally at $\mathbf{r}-\mathbf{u}$, where **u** is the particle displacement vector in the layer. Since, under the conditions stated (homogeneous deformation, and constant potential on the plates), each particle carries its potential with it, the change in the potential at a given point in space is $\delta\phi = \phi(\mathbf{r}-\mathbf{u})-\phi(\mathbf{r}) = -\mathbf{u}\cdot\mathbf{grad}\,\phi = \mathbf{u}\cdot\mathbf{E}$, where **E** is the uniform field in the undeformed layer. Since the deformation is homogeneous, however, we have

$$\mathbf{u} = z\boldsymbol{\xi}/h, \tag{15.5}$$

where z is the distance from the lower surface. Hence the variation of the field is

$$\delta\mathbf{E} = -\mathbf{n}(\mathbf{E}\cdot\boldsymbol{\xi})/h. \tag{15.6}$$

Substituting the above expressions in (15.4) and using also the fact that $\delta h = \xi_z = \boldsymbol{\xi}\cdot\mathbf{n}$, we obtain

$$\sigma_{ik}\xi_i n_k = \frac{1}{4\pi}(\mathbf{n}\cdot\mathbf{D})(\boldsymbol{\xi}\cdot\mathbf{E}) - \boldsymbol{\xi}\cdot\mathbf{n}\rho\frac{\partial\tilde{F}}{\partial\rho} + \boldsymbol{\xi}\cdot\mathbf{n}\tilde{F}$$

$$= \left\{\frac{E_i D_k}{4\pi} - \rho\frac{\partial\tilde{F}}{\partial\rho}\delta_{ik} + \tilde{F}\delta_{ik}\right\}\xi_i n_k.$$

Hence we have finally the following expression for the stress tensor:

$$\sigma_{ik} = [\tilde{F} - \rho(\partial\tilde{F}/\partial\rho)_{\mathbf{E},T}]\delta_{ik} + E_i D_k/4\pi. \tag{15.7}$$

In isotropic media, which are those here considered, **E** and **D** are parallel. Hence $E_i D_k = E_k D_i$, and the tensor (15.7) is symmetrical, as it should be.†

If the linear relation $\mathbf{D} = \varepsilon\mathbf{E}$ holds, then

$$\tilde{F} = F_0(\rho, T) - \varepsilon E^2/8\pi; \tag{15.8}$$

see (10.17). F_0 is the free energy per unit volume in the absence of the field. According to a well-known thermodynamic relation, the derivative of the free energy per unit mass with respect to the specific volume is the pressure:

$$\left[\frac{\partial}{\partial(1/\rho)}\left(\frac{F_0}{\rho}\right)\right]_T = F_0 - \rho\left(\frac{\partial F_0}{\partial\rho}\right)_T = -P_0;$$

$P_0 = P_0(\rho, T)$ is the pressure which would be found in the medium in the absence of a field and for given values of ρ and T. Hence, substituting (15.8) in (15.7), we have

$$\sigma_{ik} = -P_0(\rho, T)\delta_{ik} - \frac{E^2}{8\pi}\left[\varepsilon - \rho\left(\frac{\partial\varepsilon}{\partial\rho}\right)_T\right]\delta_{ik} + \frac{\varepsilon E_i E_k}{4\pi}. \tag{15.9}$$

In a vacuum, this expression becomes the familiar Maxwell stress tensor of the electric field.‡

---

† It is not important that in this derivation **E** is parallel to **n**, since σ_{ik} can obviously depend only on the direction of **E**, not on that of **n**.

‡ See the first footnote to §5.

The forces exerted on the surface of separation by two adjoining media must be equal and opposite: $\sigma_{ik}n_k = -\sigma'_{ik}n'_k$, where the quantities with and without the prime refer to the two media. The normal vectors **n** and **n**$'$ are in opposite directions, so that

$$\sigma_{ik}n_k = \sigma'_{ik}n_k. \tag{15.10}$$

At the boundary of two isotropic media the condition of equality of the tangential forces is satisfied identically. For, substituting (15.7) in (15.10) and taking the tangential component, we obtain $\mathbf{E}_t D_n = \mathbf{E}'_t D'_n$. This equation is satisfied by virtue of the boundary conditions of continuity on $\mathbf{E}_t$ and D_n. The condition of equality of the normal forces is, however, a non-trivial condition on the pressure difference between the two media.

For example, let us consider a boundary between a liquid and the atmosphere (for which we can put $\varepsilon = 1$). Denoting by a prime quantities pertaining to the atmosphere, and using formula (15.9) for σ_{ik}, we have

$$-P_0(\rho, T) + \frac{E^2}{8\pi}\rho\left(\frac{\partial\varepsilon}{\partial\rho}\right)_T + \frac{\varepsilon}{8\pi}(E_n{}^2 - E_t{}^2) = -P_{\text{atm}} + \frac{1}{8\pi}(E'_n{}^2 - E'_t{}^2).$$

Using the boundary conditions $E_t = E'_t$, $D_n = \varepsilon E_n = D'_n = E'_n$, we can rewrite this equation as

$$P_0(\rho, T) - P_{\text{atm}} = \frac{\rho E^2}{8\pi}\left(\frac{\partial\varepsilon}{\partial\rho}\right)_T - \frac{\varepsilon - 1}{8\pi}(\varepsilon E_n{}^2 + E_t{}^2). \tag{15.11}$$

This relation is to be taken as determining the density ρ of the liquid near its surface from the electric field in it.

Let us now determine the body forces acting in a dielectric medium. Differentiating (15.9) in accordance with (15.2) gives

$$f_i = \frac{\partial}{\partial x_i}\left[-P_0 + \frac{E^2}{8\pi}\rho\left(\frac{\partial\varepsilon}{\partial\rho}\right)_T\right] - \frac{E^2}{8\pi}\frac{\partial\varepsilon}{\partial x_i} + \frac{1}{4\pi}\left[-\tfrac{1}{2}\varepsilon\frac{\partial}{\partial x_i}E^2 + \frac{\partial}{\partial x_k}(E_i D_k)\right].$$

On using the equation $\operatorname{div}\mathbf{D} \equiv \partial D_k/\partial x_k = 0$, the expression in the brackets in the last term can be reduced to

$$-\varepsilon E_k\frac{\partial E_k}{\partial x_i} + D_k\frac{\partial E_i}{\partial x_k} = -D_k\left(\frac{\partial E_k}{\partial x_i} - \frac{\partial E_i}{\partial x_k}\right),$$

which is zero, since $\mathbf{curl}\,\mathbf{E} = 0$. Thus we have

$$\mathbf{f} = -\mathbf{grad}\,P_0(\rho, T) + \frac{1}{8\pi}\mathbf{grad}\left[E^2\rho\left(\frac{\partial\varepsilon}{\partial\rho}\right)_T\right] - \frac{E^2}{8\pi}\mathbf{grad}\,\varepsilon \tag{15.12}$$

(H. Helmholtz, 1881).

If the dielectric contains extraneous charges with density ρ_{ex}, the force **f** contains a further term $\mathbf{E}\operatorname{div}\mathbf{D}/4\pi$, or, since $\operatorname{div}\mathbf{D} = 4\pi\rho_{\text{ex}}$,

$$\rho_{\text{ex}}\mathbf{E}; \tag{15.13}$$

however, it should not be supposed that this result is obvious (cf. §16, Problem 3).

In a gas, as already mentioned in §7, we can assume the difference $\varepsilon - 1$ to be proportional to the density. Then $\rho\partial\varepsilon/\partial\rho = \varepsilon - 1$, and formula (15.12) takes the simpler form

$$\mathbf{f} = -\mathbf{grad}\,P_0 + \frac{\varepsilon - 1}{8\pi}\mathbf{grad}\,E^2. \tag{15.14}$$

Formula (15.12) is valid for media of both uniform and non-uniform composition. In the latter case ε is a function not only of ρ and T but also of the concentration of the mixture, which varies through the medium. In a body of uniform composition, on the other hand, ε is a function only of ρ and T, and **grad** ε can be written as

$$\mathbf{grad}\,\varepsilon = (\partial\varepsilon/\partial T)_\rho\,\mathbf{grad}\,T + (\partial\varepsilon/\partial\rho)_T\,\mathbf{grad}\,\rho.$$

Then (15.12) becomes

$$\mathbf{f} = -\mathbf{grad}\,P_0(\rho, T) + \frac{\rho}{8\pi}\,\mathbf{grad}\left[E^2\left(\frac{\partial\varepsilon}{\partial\rho}\right)_T\right] - \frac{E^2}{8\pi}\left(\frac{\partial\varepsilon}{\partial T}\right)_\rho\mathbf{grad}\,T. \tag{15.15}$$

If the temperature also is constant throughout the body, the third term on the right is zero, and in the first term **grad** P_0 can be replaced by ρ **grad** ζ_0, in accordance with the thermodynamic identity for the chemical potential in the absence of a field, $\rho d\zeta_0 = dP_0 - S_0 dT$. Thus

$$\begin{aligned}\mathbf{f} &= -\rho\,\mathbf{grad}\left[\zeta_0 - \frac{E^2}{8\pi}\left(\frac{\partial\varepsilon}{\partial\rho}\right)_T\right]\\ &= -\rho\,\mathbf{grad}\,\zeta,\end{aligned} \tag{15.16}$$

where ζ is the chemical potential in an electric field (see (10.19)).

In particular, the condition of mechanical equilibrium $\mathbf{f} = 0$ is, for constant temperature,

$$\zeta = \zeta_0 - (E^2/8\pi)(\partial\varepsilon/\partial\rho)_T = \text{constant}, \tag{15.17}$$

in accordance with the thermodynamic condition of equilibrium. This condition can usually be written still more simply. The change in density of the medium due to the field is proportional to E^2. Hence, if the medium is of uniform density in the absence of the field, we can put ρ = constant in the last two terms in (15.15) when the field is present; an allowance for the change in ρ is beyond the accuracy of formulae which assume the linear relation $\mathbf{D} = \varepsilon\mathbf{E}$. Then, equating to zero $\mathbf{f}$ from (15.15), we obtain the equilibrium condition at constant temperature in the form

$$P_0(\rho, T) - (\rho E^2/8\pi)(\partial\varepsilon/\partial\rho)_T = \text{constant}, \tag{15.18}$$

which differs from (15.17) in that ζ_0 is replaced by P_0/ρ.

To close this section, we shall show how the expression (15.12) for the force may be derived directly from (14.1) if the calculation of the stress tensor is not required.

Let us consider an infinite inhomogeneous dielectric medium subjected to a small isothermal deformation that is zero at infinity. The variation $\delta\varepsilon$ is made up of two parts: (1) the change

$$\varepsilon(\mathbf{r} - \mathbf{u}) - \varepsilon(\mathbf{r}) = -\mathbf{u}\cdot\mathbf{grad}\,\varepsilon$$

due to the fact that a particle is brought by the deformation from $\mathbf{r} - \mathbf{u}$ to $\mathbf{r}$, and (2) the change $-(\partial\varepsilon/\partial\rho)_T\rho\,\mathrm{div}\,\mathbf{u}$, due to the change in the density of the substance at the point $\mathbf{r}$; it is known (see *TE*, §1) that div $\mathbf{u}$ is the relative change in the volume element, so that the change in the density is $\delta\rho = -\rho\,\mathrm{div}\,\mathbf{u}$. The variation in the free energy is therefore

$$\begin{aligned}\delta\mathscr{F} &= \delta\mathscr{F}_0 - \int\delta\varepsilon(E^2/8\pi)dV\\ &= -\int P_0\,\mathrm{div}\,\mathbf{u}\,dV + \int(E^2/8\pi)\left[\mathbf{u}\cdot\mathbf{grad}\,\varepsilon + (\partial\varepsilon/\partial\rho)_T\rho\,\mathrm{div}\,\mathbf{u}\right]dV;\end{aligned} \tag{15.19}$$

the first term is the variation in the free energy when the field is absent. Integrating by parts the div **u** terms in (15.19) and comparing the result with the expression $\delta\mathscr{F} = -\int \mathbf{u}\cdot\mathbf{f}\,dV$ for the free energy variation in terms of the work done by the forces **f**, we arrive at (15.12).

§16. Electric forces in solids

The dielectric properties of a solid body change not only when its density changes (as with liquids) but also under deformations (pure shears) which do not affect the density. Let us first consider bodies which are isotropic in the absence of the field. In general, the deformed body is no longer isotropic; in consequence, its dielectric properties also become anisotropic, and the scalar permittivity ε is replaced by the dielectric tensor ε_{ik}.

The state of a slightly deformed body is described by the strain tensor

$$u_{ik} = \frac{1}{2}\left(\frac{\partial u_i}{\partial x_k} + \frac{\partial u_k}{\partial x_i}\right),$$

where $\mathbf{u}(x, y, z)$ is the displacement vector for points in the body. Since these quantities are small, only the first-order terms in u_{ik} need be retained in the variation of the components ε_{ik}. Accordingly, we represent the dielectric tensor of the deformed body as

$$\varepsilon_{ik} = \varepsilon_0\delta_{ik} + a_1 u_{ik} + a_2 u_{ll}\delta_{ik}. \tag{16.1}$$

Here ε_0 is the permittivity of the undeformed body, and the other two terms, which contain the scalar constants a_1, a_2, form the most general tensor of rank two which can be constructed linearly from the components u_{ik}.

Let us now see where the derivation given in §15 must be modified. Since, in a solid body, $\tilde{F}$ depends on all the components of the strain tensor, we must replace (15.4) by $\delta\tilde{F} = -\mathbf{D}\cdot\delta\mathbf{E}/4\pi + (\partial\tilde{F}/\partial u_{ik})\delta u_{ik}$. For the virtual displacement considered, the vector **u** is given by formula (15.5), so that the strain tensor is $u_{ik} = (\xi_i n_k + \xi_k n_i)/2h$. Substituting this in $\delta\tilde{F}$ and using the symmetry of the tensor u_{ik}, and therefore of the derivatives $\partial\tilde{F}/\partial u_{ik}$, we obtain

$$\delta\tilde{F} = -\mathbf{D}\cdot\delta\mathbf{E}/4\pi + (\xi_i n_k/h)\partial\tilde{F}/\partial u_{ik}. \tag{16.2}$$

It is now evident that we find, instead of (15.7), the following expression for the stress tensor:†

$$\sigma_{ik} = \tilde{F}\delta_{ik} = (\partial\tilde{F}/\partial u_{ik})_{T,\mathbf{E}} + E_i D_k/4\pi. \tag{16.3}$$

Formula (16.3) is valid whatever the relation between **D** and **E**. For a body which is neither pyroelectric nor piezoelectric, so that $D_i = \varepsilon_{ik}E_k$, $\tilde{F}$ is given by formula (13.4) and the required derivatives are $\partial\tilde{F}/\partial u_{ik} = \partial F_0/\partial u_{ik} - (a_1 E_i E_k + a_2 E^2\delta_{ik})/8\pi$. We then put $\varepsilon_{ik} = \varepsilon_0\delta_{ik}$ everywhere in (16.3) and obtain the following formula for the stress tensor:

$$\sigma_{ik} = \sigma^{(0)}{}_{ik} + (2\varepsilon_0 - a_1)E_i E_k/8\pi - (\varepsilon_0 + a_2)E^2\delta_{ik}/8\pi. \tag{16.4}$$

† The quantity $\tilde{F}$ in this formula, and in all preceding formulae, is the free energy per unit volume. In the theory of elasticity, however, a somewhat different definition is usual: the thermodynamic quantities are referred to the amount of matter contained in unit volume of the undeformed body, which may after deformation occupy some other volume. It is easy to go from one definition to the other by expressing the relative volume change in the deformation in terms of the tensor u_{ik}; on account of the presence of the derivative with respect to u_{ik} in (16.3), this must be done with allowance for second-order terms. As a result, the first two terms on the right of (16.3) combine into one of the form $\partial\tilde{F}/\partial u_{ik}$, in accordance with the usual formula of elasticity theory.

$\sigma^{(0)}{}_{ik}$ is the stress tensor in the absence of an electric field, determined by the moduli of rigidity and compression according to the ordinary formulae of the theory of elasticity.

Let us now make similar calculations for anisotropic solids.† The necessary modification of the above argument is as follows. When the layer undergoes a virtual deformation, its crystallographic axes are rotated, and their orientation relative to the electric field is therefore changed. On account of the anisotropy of the dielectric properties of the crystal, this leads to an additional change in $\tilde{F}$ not shown in (16.2). To calculate this change we can equally well suppose that the crystal axes rotate through some angle $\delta\boldsymbol{\phi}$ relative to the field **E**, or that the field rotates through an angle $-\delta\boldsymbol{\phi}$ relative to the axes, and the latter approach is the more convenient.

Thus the variation of the field (15.6) considered above must be augmented by the change in **E** on rotation through an angle $-\delta\boldsymbol{\phi}$:

$$\delta\mathbf{E} = -\mathbf{n}(\mathbf{E}\cdot\boldsymbol{\xi})/h - \delta\boldsymbol{\phi}\times\mathbf{E}.$$

The angle $\delta\boldsymbol{\phi}$ is related to the displacement vector **u** in the deformation by $\delta\boldsymbol{\phi} = \frac{1}{2}\mathbf{curl}\,\mathbf{u}$; this equation is easily obtained by noticing that, when the body rotates through an angle $\delta\boldsymbol{\phi}$, its points are displaced by $\mathbf{u} = \delta\boldsymbol{\phi}\times\mathbf{r}$. Substituting **u** from (15.5), we find $\delta\boldsymbol{\phi} = \mathbf{curl}\,z\boldsymbol{\xi}/2h = \mathbf{n}\times\boldsymbol{\xi}/2h$, and $\delta\mathbf{E} = -\mathbf{n}(\mathbf{E}\cdot\boldsymbol{\xi})/h + \mathbf{E}\times(\mathbf{n}\times\boldsymbol{\xi})/2h = -[\mathbf{n}(\mathbf{E}\cdot\boldsymbol{\xi}) + \boldsymbol{\xi}(\mathbf{n}\cdot\mathbf{E})]/2h$. The first term in (16.2) becomes

$$-\frac{1}{4\pi}\mathbf{D}\cdot\delta\mathbf{E} = \frac{1}{8\pi h}[(\mathbf{n}\cdot\mathbf{D})(\boldsymbol{\xi}\cdot\mathbf{E}) + (\boldsymbol{\xi}\cdot\mathbf{D})(\mathbf{n}\cdot\mathbf{E})] = \frac{1}{4\pi h}\xi_i n_k\cdot\tfrac{1}{2}(E_i D_k + E_k D_i).$$

Hence we see that the product $E_i D_k$ in (16.3) must be replaced by the second factor in the last expression:

$$\sigma_{ik} = \tilde{F}\delta_{ik} + \frac{\partial\tilde{F}}{\partial u_{ik}} + \frac{1}{8\pi}(E_i D_k + E_k D_i). \tag{16.5}$$

This expression is symmetrical in the suffixes i and k, as it should be.

The expression (16.1) for the dielectric tensor, involving two scalar constants, must be replaced in the case of a deformed crystal by

$$\varepsilon_{ik} = \varepsilon^{(0)}{}_{ik} + a_{iklm}u_{lm}, \tag{16.6}$$

where a_{iklm} is a constant tensor of rank four, symmetrical with respect to the pairs of suffixes i, k and l, m (but not with respect to an interchange of these pairs). The number of independent non-zero components of this tensor depends on the crystal class.

We shall not pause to write out here the formula for the stress tensor (analogous to (16.4)) which is obtained by using (16.6).

The formulae which we have obtained give the stresses inside a solid dielectric. They are not needed, however, if we wish to determine the total force **F** or the total torque **K** exerted on the body by the external field. Let us consider a body immersed in a fluid medium and kept at rest there. The total force on it is equal to the integral $\oint \sigma_{ik}n_k\,\mathrm{d}f$, taken over the surface. Since the force $\sigma_{ik}n_k$ is continuous, it does not matter whether this integral is calculated from the values of σ_{ik} given by (16.4) or from formula (15.9), which relates to the

† We shall see in §17 that the phenomenon of electrostriction in crystals may, for some types of symmetry, differ markedly from that in isotropic bodies. Such crystals are said to be *piezoelectric*. Here, however, we discuss only electrostriction in non-piezoelectric crystals.

medium surrounding the body. Let us suppose that this medium is in mechanical and thermal equilibrium. Then the calculation is further simplified if we use the condition of equilibrium (15.18). From this condition, part of the stress tensor (15.9) is constant through the body, being a uniform compressing or expanding pressure and making no contribution to the total force **F** and torque **K** acting on the body. These can therefore be calculated by writing σ_{ik} as

$$\sigma_{ik} = (\varepsilon/4\pi)(E_i E_k - \tfrac{1}{2}E^2\delta_{ik}) \tag{16.7}$$

simply, where **E** is the field in the fluid and ε its permittivity; this expression differs only by a factor ε from the Maxwell stress tensor of the electric field in a vacuum. Thus

$$\mathbf{F} = (\varepsilon/4\pi)\oint [\mathbf{E}(\mathbf{n}\cdot\mathbf{E}) - \tfrac{1}{2}E^2\mathbf{n}]\,df, \tag{16.8}$$

$$\mathbf{K} = (\varepsilon/4\pi)\oint [\mathbf{r}\times\mathbf{E}(\mathbf{n}\cdot\mathbf{E}) - \tfrac{1}{2}E^2\mathbf{r}\times\mathbf{n}]\,df. \tag{16.9}$$

It may also be noted that, since the fluid is in equilibrium, we can take these integrals over any closed surface which surrounds the body in question (but, of course, does not enclose any of the charged bodies which are sources of the field).

The calculation of the total force on a dielectric in an electric field in a vacuum can also be approached in another way by expressing this force, not in terms of the actual field, but in terms of the field $\mathfrak{E}$ which would be produced by the given sources in the absence of the dielectric; this is the "external field" in which the body is placed. Here it is assumed that the distribution of charges producing the field is unchanged when the body is brought in. This condition may not be fulfilled in practice—for example, if the charges are distributed over the surface of an extended conductor and the dielectric is brought to a finite distance from it.

In a virtual translation of the body over an infinitesimal distance **u**, the total free energy of the body varies, according to (11.3), by $\delta\mathscr{F} = -\int \mathbf{P}\cdot\delta\mathfrak{E}\,dV$, where $\delta\mathfrak{E} = \mathfrak{E}(\mathbf{r}+\mathbf{u}) - \mathfrak{E}(\mathbf{r}) = (\mathbf{u}\cdot\mathbf{grad})\mathfrak{E}$ is the change in the field at any given point in the body. Since $\mathbf{u}$ = constant and $\mathbf{curl}\,\mathfrak{E} = 0$, we have $\mathbf{P}\cdot(\mathbf{u}\cdot\mathbf{grad})\mathfrak{E} = \mathbf{P}\cdot\mathbf{grad}(\mathbf{u}\cdot\mathfrak{E}) = \mathbf{u}\cdot(\mathbf{P}\cdot\mathbf{grad})\mathfrak{E}$, so that

$$\delta\mathscr{F} = -\mathbf{u}\cdot\int(\mathbf{P}\cdot\mathbf{grad})\mathfrak{E}\,dV.$$

But $\delta\mathscr{F} = -\mathbf{u}\cdot\mathbf{F}$, and we therefore have for the required force†

$$\mathbf{F} = \int(\mathbf{P}\cdot\mathbf{grad})\mathfrak{E}\,dV. \tag{16.10}$$

Similarly, the total torque on the body can be determined. We shall not go through the calculation, but merely give the result:

$$\mathbf{K} = \int \mathbf{P}\times\mathfrak{E}\,dV + \int \mathbf{r}\times(\mathbf{P}\cdot\mathbf{grad})\mathfrak{E}\,dV. \tag{16.11}$$

In an almost uniform field, which may be regarded as constant over the dimensions of the body, formula (16.10) gives to a first approximation

$$\mathbf{F} = \left(\int \mathbf{P}\,dV\cdot\mathbf{grad}\right)\mathfrak{E} = (\mathscr{P}\cdot\mathbf{grad})\mathfrak{E}, \tag{16.12}$$

† It should be emphasized, however, that the integrand in (16.10) cannot be interpreted as the force density. The reason is that the local forces in the dielectric arise not only from the field $\mathfrak{E}$ but also from the internal fields which, by Newton's third law, contribute nothing to the total force, though they modify the distribution of forces over the volume of the body.

where $\mathscr{P}$ is the total dipole moment of the polarized dielectric; this result, of course, could have been obtained by direct differentiation of $\mathscr{F}$ from (11.8). In formula (16.11) we neglect the second term in the first approximation and reach the natural conclusion that

$$\mathbf{K} = \mathscr{P} \times \mathfrak{E}. \tag{16.13}$$

PROBLEMS

PROBLEM 1. A dielectric sphere with radius a in a uniform external field $\mathfrak{E}$ is cut in half by a plane perpendicular to the field. Determine the force of attraction between the hemispheres.

SOLUTION. We imagine the hemispheres separated by an infinitely narrow slit and determine the force from formula (16.8) with $\varepsilon = 1$, integrating over the surface of a hemisphere; $\mathbf{E}$ is the field in the vacuum near the surface. According to (8.2) the field $\mathbf{E}^{(i)}$ inside the sphere is uniform and equal to $3\mathfrak{E}/(2+\varepsilon)$, where ε is the permittivity of the sphere. The field in the slit is perpendicular to the surface and is $\mathbf{E} = \mathbf{D}^{(i)} = 3\varepsilon\mathfrak{E}/(2+\varepsilon)$. On the outer surface of the sphere we have

$$E_r = D^{(i)}{}_r = \frac{3\varepsilon}{2+\varepsilon}\mathfrak{E}\cos\theta, \qquad E_\theta = E^{(i)}{}_\theta = -\frac{3}{2+\varepsilon}\mathfrak{E}\sin\theta,$$

where θ is the angle between the position vector and the direction of $\mathfrak{E}$. A calculation of the integral gives an attractive force†

$$F = 9(\varepsilon-1)^2 a^2 \mathfrak{E}^2/16(\varepsilon+2)^2.$$

PROBLEM 2. Determine the change in shape of a dielectric sphere in a uniform external electric field.

SOLUTION. As in §5, Problem 4. In determining the change in shape, we assume the volume of the sphere to be unchanged.‡ The elastic part of the free energy is given by the same expression as in §5, Problem 4. The electric part is given by (8.9):

$$-\tfrac{1}{2}\mathscr{P}\cdot\mathfrak{E} = -\frac{V}{8\pi}\frac{\varepsilon^{(x)}-1}{1+n(\varepsilon^{(x)}-1)}\mathfrak{E}^2,$$

and the permittivity in the x-direction is, by (16.1), $\varepsilon^{(x)} = \varepsilon_0 + a_1 u_{xx} = \varepsilon_0 + \frac{2}{3}a_1(u_{xx}-u_{yy}) = \varepsilon_0 + \frac{2}{3}a_1(a-b)/R$. From the condition that the total free energy be a minimum we find (since the quantity concerned is small)

$$\frac{a-b}{R} = \frac{9\mathfrak{E}^2}{40\pi\mu}\frac{(\varepsilon_0-1)^2+5a_1/2}{(\varepsilon_0+2)^2}.$$

For $\varepsilon_0 \to \infty$ this tends to the value for a conducting sphere.

PROBLEM 3. Determine the body forces in an isotropic solid dielectric, assumed homogeneous, when extraneous charges are present in it.

SOLUTION. Assuming ε_0, a_1, a_2 constant and using the equations $\mathbf{curl}\,\mathbf{E} = 0$, $\operatorname{div}\mathbf{D} \cong \varepsilon_0 \operatorname{div}\mathbf{E} = 4\pi\rho_{\text{ex}}$, we have from (16.4)

$$f_i = \frac{\partial\sigma_{ik}}{\partial x_k} = \frac{\partial\sigma^{(0)}{}_{ik}}{\partial x_k} - \frac{1}{8\pi}(\tfrac{1}{2}a_1+a_2)\frac{\partial E^2}{\partial x_i} + \left(1-\frac{a_1}{2\varepsilon_0}\right)\rho_{\text{ex}}E_i.$$

§17. Piezoelectrics

The internal stresses which occur in an isotropic dielectric in an electric field are proportional to the square of the field. The effect is similar in crystals belonging to some of

† It is by chance that, in the limit $\varepsilon \to \infty$, this expression tends to the result obtained in §5, Problem 3, for a conducting sphere (indeed, the forces are in opposite directions). The two cases are evidently not physically equivalent, because there is no field in the slit between two conducting hemispheres at the same potential, whereas in this problem there is a field in the slit.

‡ The change in volume is determined in §12, Problem 1.

the crystal classes. For certain types of symmetry, however, the electrostriction properties of the crystals are quite different. The internal stresses in these *piezoelectric* bodies resulting from an electric field are proportional to the field itself. The converse effect also occurs: the deformation of a piezoelectric is accompanied by the appearance in it of a field proportional to the deformation.

Since in a piezoelectric only the principal (linear) effect is of interest, we can neglect the terms quadratic in the field in the general formula (16.5). Then $\sigma_{ik} = \tilde{F}\delta_{ik} + (\partial \tilde{F}/\partial u_{ik})_{T,\mathbf{E}}$. In this section we shall use the thermodynamic quantities referred to the matter in unit volume of the undeformed body (see the first footnote to §16). Taking $\tilde{F}$ in this sense, we have simply

$$\sigma_{ik} = (\partial \tilde{F}/\partial u_{ik})_{T,\mathbf{E}}. \tag{17.1}$$

Accordingly, the thermodynamic relation for the differential $d\tilde{F}$ is

$$d\tilde{F} = -S\,dT + \sigma_{ik}\,du_{ik} - \mathbf{D}\cdot d\mathbf{E}/4\pi. \tag{17.2}$$

The following remark should be made concerning the last term. In the form given here, this term (taken from (10.9)) pertains, strictly speaking, to unit volume of the deformed body. By ignoring this fact, we commit an error which, in the case of a piezoelectric, is of a higher order of smallness than the remaining terms in (17.2).

The independent variables in (17.2) include the components of the tensor u_{ik}. It is sometimes convenient to use instead the components σ_{ik}. To do so, we must introduce the thermodynamic potential, defined as

$$\tilde{\Phi} = \tilde{F} - u_{ik}\sigma_{ik}. \tag{17.3}$$

For the differential of this quantity we have

$$d\tilde{\Phi} = -S\,dT - u_{ik}\,d\sigma_{ik} - \mathbf{D}\cdot d\mathbf{E}/4\pi. \tag{17.4}$$

It must be emphasized that the use of the thermodynamic potential $\tilde{\Phi}$ in electrodynamics in accordance with formulae (17.3) and (17.4) rests on the validity of (17.1) and so is possible only for piezoelectric bodies.

Having thus defined the necessary thermodynamic quantities, let us now ascertain the piezoelectric properties of crystals. If σ_{ik} and E_k are taken as independent variables, the induction $\mathbf{D}$ must be regarded as a function of them, and an expansion of this function must retain the terms linear in them. The linear terms in the expansion of the components of a vector in powers of the components of a tensor of rank two can be written, in the most general case, as $4\pi\gamma_{i,kl}\,\sigma_{kl}$, where the constants $\gamma_{i,kl}$ form a tensor of rank three, and the factor 4π is introduced for convenience. Since the tensor σ_{kl} is symmetrical, it is clear that the tensor $\gamma_{i,kl}$ may also be supposed to have the symmetry property

$$\gamma_{i,kl} = \gamma_{i,lk}. \tag{17.5}$$

For clarity we separate the symmetrical suffixes from the remaining one by a comma. We call $\gamma_{i,kl}$ the *piezoelectric tensor*. If it is known, the piezoelectric properties of the crystal are entirely determined.

Adding the piezoelectric terms to the expression (13.1) for the electric induction in the crystal, we have

$$D_i = D_{0i} + \varepsilon_{ik}E_k + 4\pi\gamma_{i,kl}\sigma_{kl}. \tag{17.6}$$

Corresponding additional terms appear in the thermodynamic quantities. The thermodynamic potential of a non-piezoelectric crystal in the absence of a field is $\tilde{\Phi} = \Phi = \Phi_0 - \frac{1}{2}\mu_{iklm}\sigma_{ik}\sigma_{lm}$, where Φ_0 pertains to the undeformed body, and the second term is the ordinary elastic energy, determined by the *elastic constant tensor* μ_{iklm}.† For a piezoelectric we have

$$\tilde{\Phi} = \Phi_0 - \tfrac{1}{2}\mu_{iklm}\sigma_{ik}\sigma_{lm} - \varepsilon_{ik}E_iE_k/8\pi - E_iD_{0i}/4\pi - \gamma_{i,kl}E_i\sigma_{kl}. \tag{17.7}$$

The form of the last three terms is given by the fact that the derivatives of $\tilde{\Phi}$ with respect to E_i (for given temperature and internal stresses), found from the relation $D_i = -4\pi\partial\tilde{\Phi}/\partial E_i$, must accord with (17.6).

Knowing $\tilde{\Phi}$. we can obtain from (17.4) a formula giving the strain tensor in terms of the stresses σ_{ik} and the field **E**:

$$u_{ik} = -(\partial\tilde{\Phi}/\partial\sigma_{ik})_{T,\mathbf{E}} = \mu_{iklm}\sigma_{lm} + \gamma_{l,ik}E_l. \tag{17.8}$$

It should be mentioned that to regard the quantities μ_{iklm} and ε_{ik} for a piezoelectric as elastic constants and permittivity is to some extent conventional. With the definitions used here, they give respectively the strains as functions of the elastic stresses for a given field, and the induction as a function of the field for given stresses. If, however, the deformation occurs with a given value of the induction, or we consider the induction as a function of the field for given strains, the elastic constants and the permittivity will be represented by other quantities, which can be expressed as somewhat complex functions of the components of the tensors μ, ε and γ.

The field in a piezoelectric body must be determined together with its deformation, leading to a problem in both electrostatics and elasticity theory. We must seek a simultaneous solution of the electrostatic equations

$$\operatorname{div}\mathbf{D} = 0, \qquad \mathbf{curl}\,\mathbf{E} = 0, \tag{17.9}$$

with **D** given by (17.6), and the equations of elastic equilibrium

$$\partial\sigma_{ik}/\partial x_k = 0, \tag{17.10}$$

with the appropriate boundary conditions at the surface of the body and use of the relation (17.8) between σ_{ik} and the strains. In general this problem is very complex.

The problem is much simplified for a body of ellipsoidal form with a free surface (i.e. one subject to no external mechanical forces). In this case (§8), the field inside the body is uniform; the deformation is therefore homogeneous, and the elastic stresses $\sigma_{ik} = 0$.

Finally, let us consider which types of crystal symmetry allow the existence of piezoelectricity; in other words, what are the restrictions imposed on the components of the tensor $\gamma_{i,kl}$ by the symmetry conditions. In general, this tensor (which is symmetrical in

---

† The tensor μ_{iklm} determines the relation between stress and strain:

$$u_{ik} = -\partial\Phi/\partial\sigma_{ik} = \mu_{iklm}\sigma_{lm}.$$

In *TE*, §10, the converse relation $\sigma_{ik} = \lambda_{iklm}u_{lm}$ is used. It is evident that the symmetry properties of the tensor μ_{iklm} are exactly the same as those of λ_{iklm}.

The free energy F contains the elastic energy with the plus sign:

$$F_{\mathrm{el}} = \tfrac{1}{2}\lambda_{iklm}u_{ik}u_{lm}.$$

The thermodynamic potential is obtained from F by subtracting $\sigma_{ik}u_{ik}$, and so

$$\Phi_{\mathrm{el}} = F_{\mathrm{el}} - \sigma_{ik}u_{ik} = -\tfrac{1}{2}\lambda_{iklm}u_{ik}u_{lm} = -\tfrac{1}{2}\mu_{iklm}\sigma_{ik}\sigma_{lm}.$$

the suffixes k and l) has 18 independent non-zero components, but in reality the number of independent components is usually much smaller.

In all symmetry transformations of a given crystal, the components of the tensor $\gamma_{i,kl}$ must remain unaltered in value. Hence it follows at once that no piezoelectric body can have a centre of symmetry or, in particular, be isotropic. For, on reflection in the centre (i.e. change of sign of all three coordinates), the components of a tensor of rank three change sign.

Of the 32 crystal classes, only 20 allow piezoelectricity. These comprise the ten enumerated in §13 as allowing pyroelectricity (all pyroelectrics are also piezoelectrics) and the ten following classes:

orthorhombic system: $\boldsymbol{D}_2$
tetragonal system: $\boldsymbol{D}_4$, $\boldsymbol{D}_{2d}$, $\boldsymbol{S}_4$
rhombohedral system: $\boldsymbol{D}_3$
hexagonal system: $\boldsymbol{D}_6$, $\boldsymbol{C}_{3h}$, $\boldsymbol{D}_{3h}$
cubic system: T, T_d.

The non-zero components of the piezoelectric tensor for each class are given in the Problems below.

Mention may also be made here of a phenomenon akin to piezoelectricity, which results from the "deformation" of a liquid crystal. We shall consider what are called *nematic* crystals (*SP* 1, §140), liquids in which there is a distinctive direction of preferred orientation of the molecules. At each point in the medium, this direction is specified by a unit vector **d**, the *director* of the crystal. In an undeformed liquid crystal, **d** has the same direction everywhere, but in a deformed one this direction is a function of the coordinates. The expansion (17.6) corresponds to an expression for the induction in a liquid crystal in the form

$$D_i = \varepsilon_{ik} E_k + 4\pi e_1 d_i \operatorname{div} \mathbf{d} + 4\pi e_2 (\mathbf{curl\, d} \times \mathbf{d})_i, \tag{17.11}$$

where e_1 and e_2 are scalar coefficients (R. B. Meyer, 1969).† The last two terms, which describe the effect in question, constitute the most general polar vector that can be formed from **d** and its first derivatives with respect to the coordinates. The expression (17.11) is automatically invariant under a change in the sign of **d**.

The permittivity tensor of a nematic crystal has the same symmetry as for uniaxial crystals, the axis of symmetry being represented by the local (at each point) direction of the director. The tensor ε_{ik} may be expressed as

$$\varepsilon_{ik} = \varepsilon_0 \delta_{ik} + \varepsilon_a d_i d_k, \tag{17.12}$$

with two independent constants ε_0 and ε_a.

PROBLEMS

PROBLEM 1. Determine the non-zero components of the tensor $\gamma_{i,kl}$ for non-pyroelectric crystal classes which allow piezoelectricity.

SOLUTION. The class $\boldsymbol{D}_2$ has three mutually perpendicular twofold axes of symmetry, which we take as the axes of x, y and z. Rotations through 180° about these axes change the sign of two out of the three coordinates.

† Pyroelectricity in nematic crystals is in practice unknown, and we therefore put $\mathbf{D}_0 = 0$.

Since the components $\gamma_{i,kl}$ are transformed as the products $x_i x_k x_l$, the only non-zero components are those with three different suffixes: $\gamma_{x,yz}$, $\gamma_{z,xy}$, $\gamma_{y,zx}$. (The other non-zero components are equal to these, since $\gamma_{i,kl} = \gamma_{i,lk}$.) Accordingly, the piezoelectric part of the thermodynamic potential is†

$$\tilde{\Phi}_{\text{pie}} = -2(\gamma_{x,yz}E_x\sigma_{yz} + \gamma_{y,xz}E_y\sigma_{xz} + \gamma_{z,xy}E_z\sigma_{xy}). \tag{1}$$

The class $\boldsymbol{D}_{2d}$ is obtained by adding to the axes of class $\boldsymbol{D}_2$ two planes of symmetry passing through one axis (the z-axis, say) and bisecting the angles between the other two. Reflection in one of these planes gives the transformation $x \to y$, $y \to x$, $z \to z$. Hence the components $\gamma_{i,kl}$ which differ by interchange of x and y must be equal, so that only two out of the three coefficients in (1) are now independent: $\gamma_{z,xy}$, $\gamma_{x,yz} = \gamma_{y,xz}$.

The class $\boldsymbol{T}$ is obtained from the class $\boldsymbol{D}_2$ by adding four diagonal threefold axes of symmetry, rotations about which effect a cyclic permutation of x, y, z, e.g. $x \to z$, $y \to x$, $z \to y$. Hence all three coefficients in (1) are equal: $\gamma_{x,yz} = \gamma_{y,zx} = \gamma_{z,xy}$. The same result is obtained for the cubic class $\boldsymbol{T}_d$.

The class $\boldsymbol{D}_4$ has one fourfold axis of symmetry (the z-axis, say) and four twofold axes lying in the xy-plane. Here the symmetry elements of the class $\boldsymbol{D}_2$ are supplemented by a rotation through 90° about the z-axis, i.e. the transformation $x \to y$, $y \to -x$, $z \to z$. Consequently, one of the coefficients in (1) must be zero ($\gamma_{z,xy} = -\gamma_{z,yx} = -\gamma_{z,xy} = 0$), and the other two are equal, but opposite in sign: $\gamma_{x,yz} = -\gamma_{y,xz}$. The same result is obtained for the class $\boldsymbol{D}_6$.

The class $\boldsymbol{S}_4$ includes the transformations $x \to y$, $y \to -x$, $z \to -z$ and $x \to -x$, $y \to -y$, $z \to z$. The non-zero components are $\gamma_{z,xy}$, $\gamma_{x,yz} = \gamma_{y,xz}$, $\gamma_{z,xx} = -\gamma_{z,yy}$, $\gamma_{x,zx} = -\gamma_{y,zy}$. One of these can be made to vanish by a suitable choice of the x and y axes.

The class $\boldsymbol{D}_3$ has one threefold axis of symmetry (the z-axis, say), and three twofold axes lying in the xy-plane; let one of these be the x-axis. To find the restrictions imposed by the presence of a threefold axis, we make a formal transformation by introducing the complex "coordinates" $\xi = x + iy$, $\eta = x - iy$; the coordinate z remains unchanged. We must also transform the tensor $\gamma_{i,kl}$ to these new coordinates, in which the suffixes take the values ξ, η, z. In a rotation through 120° about the z-axis these coordinates undergo the transformation $\xi \to \xi e^{2\pi i/3}$, $\eta \to \eta e^{-2\pi i/3}$, $z \to z$. The only components of the tensor $\gamma_{i,ki}$ which remain unchanged and so may be different from zero are $\gamma_{z,\eta\xi}$, $\gamma_{\eta,z\xi}$, $\gamma_{\xi,z\eta}$, $\gamma_{\xi,\xi\xi}$, $\gamma_{\eta,\eta\eta}$ and $\gamma_{z,zz}$. A rotation through 180° about the x-axis gives the transformation $x \to x$, $y \to -y$, $z \to -z$, or $\xi \to \eta$, $\eta \to \xi$, $z \to -z$; $\gamma_{z,\eta\xi}$ and $\gamma_{z,zz}$ change sign and so must be zero, while the remaining components listed above are mutually transformed in pairs, giving $\gamma_{\eta,z\xi} = -\gamma_{\xi,z\eta}$, $\gamma_{\xi,\xi\xi} = \gamma_{\eta,\eta\eta}$. In order to write an expression of $\tilde{\Phi}_{\text{pie}}$, we must form the sum $-\gamma_{i,kl}E_i\sigma_{kl}$, in which the suffixes take the values ξ, η, z:

$$\tilde{\Phi}_{\text{pie}} = -2\gamma_{\eta,z\xi}(E_\eta\sigma_{z\xi} - E_\xi\sigma_{z\eta}) - \gamma_{\xi,\xi\xi}(E_\xi\sigma_{\xi\xi} + E_\eta\sigma_{\eta\eta}).$$

Here the components E_i and σ_{ik} in the coordinates ξ, η, z must also be expressed in terms of those in the original coordinates x, y, z. This is easily done by using the fact that the components of a tensor are transformed as the products of the corresponding coordinates. Hence, for example, from $\xi\xi = xx - yy + 2ixy$, we have $\sigma_{\xi\xi} = \sigma_{xx} - \sigma_{yy} + 2i\sigma_{xy}$. The result is

$$\tilde{\Phi}_{\text{pie}} = 2a(E_y\sigma_{zx} - E_x\sigma_{zy}) + b[2E_y\sigma_{xy} - E_x(\sigma_{xx} - \sigma_{yy})], \tag{2}$$

where $a = 2i\gamma_{\eta,z\xi}$ and $b = 2\gamma_{\xi,\xi\xi}$ are real constants. The relations between the components $\gamma_{i,kl}$ in the coordinates x, y, z are, as we see from (2),‡

$$\gamma_{y,zx} = -\gamma_{x,zy} \equiv -a, \qquad \gamma_{y,xy} = -\gamma_{x,xx} = \gamma_{x,yy} \equiv -b.$$

The class $\boldsymbol{D}_{3h}$ is obtained from the class $\boldsymbol{D}_3$ by adding a plane of symmetry (the xy-plane) perpendicular to the threefold axis. Reflection in this plane changes the sign of z, and so $\gamma_{\eta,z\xi} = 0$, so that only the term with the coefficient b remains in (2).

The class $\boldsymbol{C}_{3h}$ has a threefold axis and a plane of symmetry perpendicular to it. Reflection in this plane changes the sign of z, and so all components $\gamma_{i,kl}$ whose suffixes contain z an odd number of times must be zero. Taking into account also the restrictions derived above which are imposed by the threefold axis of symmetry, we find that only the two components $\gamma_{\eta,\eta\eta}$ and $\gamma_{\xi,\xi\xi}$ are not zero. These quantities must be complex conjugates in order that $\tilde{\Phi}$

---

† To avoid misunderstanding it should be recalled that, if we calculate the components of the strain tensor u_{ik} by direct differentiation of the actual expression for $\tilde{\Phi}$ with respect to σ_{ik}, the derivatives with respect to components σ_{ik} with $i \neq k$ give twice the corresponding components u_{ik}, because the expressions $u_{ik} = -\partial\tilde{\Phi}/\partial\sigma_{ik}$ are essentially meaningful only as representing the fact that $d\tilde{\Phi} = -u_{ik}d\sigma_{ik}$, and the terms containing the differentials of non-diagonal components of the symmetrical tensor σ_{ik} appear twice in the sum $u_{ik}d\sigma_{ik}$.

‡ In non-orthogonal coordinates such as ξ, η, z the covariant and contravariant components of tensors must be distinguished. This should have been done in returning to the original coordinates x, y, z: if the components E_i and σ_{kl} transform contravariantly, then those of the tensor $\gamma_{i,kl}$ transform covariantly. We avoid this necessity, however, by obtaining the required relations between the components $\gamma_{i,kl}$ in the coordinates x, y, z directly from the form of the scalar combination (2).

should be real. Putting $2\gamma_{\eta,\eta\eta} = a - ib$, $2\gamma_{\xi,\xi\xi} = a + ib$, we find

$$\tilde{\Phi}_{\text{pie}} = a[2E_y\sigma_{xy} - E_x(\sigma_{xx} - \sigma_{yy})] + b[2E_x\sigma_{xy} + E_y(\sigma_{xx} - \sigma_{yy})]. \tag{3}$$

Either a or b can be made to vanish by a suitable choice of the x and y axes.

PROBLEM 2. The same as Problem 1, but for the crystal classes which allow pyroelectricity.

SOLUTION. Let the z-axis be the twofold, threefold, fourfold or sixfold axis of symmetry, or in the class C_s be perpendicular to the plane of symmetry. In the classes C_{nv} the xz-plane is a plane of symmetry.

We give below for each class all the components $\gamma_{i,kl}$ which are not zero.

Class C_1: all $\gamma_{i,kl}$.
C_s: all those in which the suffix z appears twice or not at all.
C_{2v}: $\gamma_{z,xx}, \gamma_{z,yy}, \gamma_{z,zz}, \gamma_{x,xz}, \gamma_{y,yz}$.
C_2: the same, together with $\gamma_{x,yz}, \gamma_{y,xz}, \gamma_{z,xy}$.
C_{4v}: $\gamma_{z,xx} = \gamma_{z,yy}, \gamma_{z,zz}, \gamma_{x,xz} = \gamma_{y,yz}$.
C_4: the same, together with $\gamma_{x,yz} = -\gamma_{y,xz}$.
C_{3v}: $\gamma_{z,zz}, \gamma_{x,xz} = \gamma_{y,yz}, \gamma_{x,xx} = -\gamma_{x,yy} = -\gamma_{y,xy}, \gamma_{z,xx} = \gamma_{z,yy}$.
C_3: the same, together with $\gamma_{x,yz} = -\gamma_{y,xz}, \gamma_{y,xx} = -\gamma_{y,yy} = \gamma_{x,xy}$.
C_{6v}: $\gamma_{z,zz}, \gamma_{x,xz} = \gamma_{y,yz}, \gamma_{z,xx} = \gamma_{z,yy}$.
C_6: the same, together with $\gamma_{x,yz} = -\gamma_{y,xz}$.

By a suitable choice of directions of the x, y, z axes three more components can be made zero in the class C_1, and by a choice of the x and y axes one more component can be made zero in the classes C_s, C_2, C_3; in C_4 and C_6, the expression $\gamma_{i,kl}E_i\sigma_{kl}$ is invariant under rotation through any angle about the z axis, and therefore no further reduction in the number of non-zero components $\gamma_{i,kl}$ is possible.

PROBLEM 3. Determine Young's modulus (the coefficient of proportionality between the extending stress and the relative extension) for a flat slab of a non-pyroelectric piezoelectric in the following cases: (a) where the slab is stretched by the plates of a short-circuited capacitor, (b) where it is stretched by those of an uncharged capacitor, (c) where it is stretched parallel to its plane with no external field.

SOLUTION. (a) In this case the field **E** inside the slab is zero. The only non-zero component of the tensor σ_{ik} is the extending stress σ_{zz} (the z-axis being perpendicular to the slab).† From (17.8) we have $u_{zz} = \mu_{zzzz}\sigma_{zz}$, whence Young's modulus is $E = 1/\mu_{zzzz}$.

(b) In this case we have in the slab $E_x = E_y = 0$, $D_z = 0$. From (17.6) and (17.8) we have $D_z = \varepsilon_{zz}E_z + 4\pi\gamma_{z,zz}\sigma_{zz} = 0$, $u_{zz} = \mu_{zzzz}\sigma_{zz} + \gamma_{z,zz}E_z$. Eliminating E_z, we obtain $1/E = \mu_{zzzz} - 4\pi\gamma_{z,zz}{}^2/\varepsilon_{zz}$.

(c) In this case also, $E_x = E_y = 0$, $D_z = 0$, but the extension is along the x-axis, say. Here we have $D_z = \varepsilon_{zz}E_z + 4\pi\gamma_{z,xx}\sigma_{xx} = 0$, $u_{xx} = \mu_{xxxx}\sigma_{xx} + \gamma_{z,xx}E_z$. Eliminating E_z, we obtain $1/E = \mu_{xxxx} - 4\pi\gamma_{z,xx}{}^2/\varepsilon_{zz}$.

PROBLEM 4. Obtain an equation for the velocity of sound in a piezoelectric medium.

SOLUTION. In this problem it is more convenient to use u_{ik} as the independent variables, instead of σ_{ik}. We write $\tilde{F}$ in the form

$$\tilde{F} = F_0 + \tfrac{1}{2}\lambda_{iklm}u_{ik}u_{lm} - \frac{1}{8\pi}\varepsilon_{ik}E_iE_k - \frac{1}{4\pi}E_iD_{0i} + \beta_{i,kl}E_iu_{kl},$$

whence

$$\sigma_{ik} = \partial\tilde{F}/\partial u_{ik} = \lambda_{iklm}u_{lm} + \beta_{l,ik}E_l.$$

The equations of motion from the theory of elasticity are

$$\rho\ddot{u}_i = \frac{\partial\sigma_{ik}}{\partial x_k} = \lambda_{iklm}\frac{\partial u_{lm}}{\partial x_k} + \beta_{l,ik}\frac{\partial E_l}{\partial x_k}, \tag{4}$$

where ρ is the density of the medium, and **u** is the displacement vector, related to u_{ik} by

$$u_{ik} = \frac{1}{2}\left(\frac{\partial u_i}{\partial x_k} + \frac{\partial u_k}{\partial x_i}\right).$$

The equation div **D** = 0 gives

$$\varepsilon_{ik}\frac{\partial E_k}{\partial x_i} - 4\pi\beta_{i,kl}\frac{\partial u_{kl}}{\partial x_i} = 0, \tag{5}$$

† It is not assumed to coincide with any particular crystallographic direction.

and the field can be expressed in terms of the field potential: $E_i = -\partial\phi/\partial x_i$, which takes into account the equation **curl E** = 0.

In a plane sound wave, **u** and ϕ are proportional to $\exp[i(\mathbf{k}\cdot\mathbf{r}-\omega t)]$, and we find from the above equations that

$$\rho\omega^2 u_i = \lambda_{iklm}k_k k_l u_m - \beta_{l,ik}k_k k_l \phi,$$
$$\varepsilon_{ik}k_i k_k \phi + 4\pi\beta_{i,kl}k_i k_k u_l = 0.$$

Eliminating ϕ, we can write the condition of compatibility of the resulting equations for u_i as

$$\det|\rho\omega^2\delta_{ik} - \lambda_{iklm}k_l k_m - 4\pi(\beta_{l,mi}k_l k_m)(\beta_{p,qk}k_p k_q)/\varepsilon_{rs}k_r k_s| = 0.$$

For any given direction of the wave vector **k**, this equation determines three phase velocities of sound ω/k, which are in general different. A characteristic property of a piezoelectric medium is the involved relation between the velocity and direction of the wave.

PROBLEM 5. A piezoelectric crystal of the class C_{6v} has a plane boundary (the xz-plane) which passes through the axis of symmetry (the z-axis). Find the speed of surface waves propagated at right angles to the symmetry axis (in the x-direction), in which there are oscillations of the displacement u_z and the electric field potential ϕ (J. L. Bleustein, 1968; Yu. V. Gulyaev, 1969).

SOLUTION. Under the conditions considered, two equations involving only u_z and ϕ separate out from (4) and (5); these quantities depend on the coordinates x and y, and on the time t, but not on z. The non-zero components of the stress tensor and the induction vector are

$$\sigma_{zx} = \beta E_x + 2\lambda u_{zx}, \qquad \sigma_{zy} = \beta E_y + 2\lambda u_{zy},$$
$$D_x = -8\pi\beta u_{zx} + \varepsilon E_x, \qquad D_y = -8\pi\beta u_{zy} + \varepsilon E_y,$$

with

$$u_{zx} = \tfrac{1}{2}\partial u_z/\partial x, \quad u_{zy} = \tfrac{1}{2}\partial u_z/\partial y,$$
$$E_x = -\partial\phi/\partial x, \quad E_y = -\partial\phi/\partial y,$$

and writing for brevity $\beta_{x,xz} = \beta_{y,yz} \equiv \beta$, $\lambda_{xzxz} = \lambda_{yzyz} \equiv \lambda$, $\varepsilon_{xx} = \varepsilon_{yy} \equiv \varepsilon$; the constant pyroelectric induction $D_z = D_0$ does not appear in the equations or the boundary conditions.

Equation (5) and the z-component of equation (4) give, in the region occupied by the piezoelectric medium (the half-space $y > 0$),

$$4\pi\beta\triangle u_z + \varepsilon\triangle\phi^{(i)} = 0, \qquad \rho\ddot{u}_z = -\beta\triangle\phi^{(i)} + \lambda\triangle u_z,$$

where $\triangle \equiv \partial^2/\partial x^2 + \partial^2/\partial y^2$; these may be rewritten as

$$\rho\ddot{u}_z = \bar{\lambda}\triangle u_z, \quad \triangle\psi = 0, \tag{6}$$

where

$$\bar{\lambda} = \lambda + 4\pi\beta^2/\varepsilon, \quad \psi = (4\pi\beta/\varepsilon)u_z + \phi^{(i)}.$$

In the vacuum (the half-space $y < 0$), the potential $\phi^{(e)}$ satisfies the equation

$$\triangle\phi^{(e)} = 0. \tag{7}$$

These equations are to be solved with the following boundary conditions: at the surface of the medium,

$$\phi^{(i)} = \phi^{(e)}, \quad \sigma_{zy} = 0, \quad D^{(i)}{}_y = -\partial\phi^{(e)}/\partial y \text{ for } y = 0, \tag{8}$$

and far from the surface

$$u_z \to 0 \text{ as } y \to \infty; \qquad \phi \to 0 \text{ as } y \to \pm\infty.$$

We seek the solution in the form

$$u_z = Ae^{-\kappa y}e^{i(kx-\omega t)}, \quad \psi = Be^{-ky}e^{i(kx-\omega t)}, \quad \phi^{(e)} = Ce^{ky}e^{i(kx-\omega t)},$$

with

$$\rho\omega^2 = \bar{\lambda}(k^2 - \kappa^2). \tag{9}$$

Equations (6) and (7) and the conditions at infinity are then satisfied, and the conditions (8) give three linear homogeneous equations for A, B and C; the condition for these to have a solution is

$$\kappa = 4\pi\beta^2 k/\bar{\lambda}\varepsilon(1+\varepsilon) \equiv \Lambda k.$$

Finally, substitution in (9) gives the phase velocity of the waves:

$$\omega/k = [(\bar{\lambda}/\rho)(1-\Lambda^2)]^{\frac{1}{2}}.$$

The surface propagation of these waves is restricted to piezoelectric media. As $\beta \to 0$, the penetration depth $1/\kappa \to \infty$, and a bulk wave is formed.

§18. Thermodynamic inequalities

According to the formulae of §10, the total free energy can be written as the integral

$$\mathscr{F} = \int F(T, \rho, \mathbf{D})\,\mathrm{d}V, \tag{18.1}$$

taken over all space. We shall suppose that the function $\mathbf{D}(\mathbf{r})$ which appears in the integrand satisfies only the equation

$$\operatorname{div} \mathbf{D} = 0 \tag{18.2}$$

inside a dielectric and the condition

$$\oint \mathbf{D} \cdot \mathrm{d}\mathbf{f} = 4\pi e \tag{18.3}$$

on the surface of a conductor which carries a given charge. These equations establish the relation between the field and its sources. Otherwise we regard the function $\mathbf{D}(\mathbf{r})$ as arbitrary, and in particular we do not require it to satisfy the second field equation $\mathbf{curl}\,\mathbf{E} = 0$ (where $\mathbf{E} = 4\pi \partial F/\partial \mathbf{D}$) or the boundary condition ϕ = constant on the surface of a conductor. We shall show that these equations can then be obtained from the condition that the integral (18.1) be a minimum with respect to changes in the function $\mathbf{D}(\mathbf{r})$ which satisfy equations (18.2) and (18.3). It should be emphasized that the possibility of this derivation is not *a priori* evident, since the field distributions which come into consideration in determining the minimum of the integral (18.1) do not necessarily correspond to physically possible states (because they do not satisfy all the field equations), whereas, in the thermodynamic condition that the free energy be a minimum, only the various physically possible states are considered.

The problem of finding the minimum of the integral (18.1) with the subsidiary conditions (18.2) and (18.3) is solved by Lagrange's method of multipliers. We multiply the variation of the condition (18.2) by some as yet undetermined function $-\phi/4\pi$ of the coordinates, and that of the condition (18.3) by some undetermined constant $\phi_0/4\pi$, and then equate to zero the sum of variations

$$\int \delta F\,\mathrm{d}V - \frac{1}{4\pi}\int \phi \operatorname{div} \delta\mathbf{D}\,\mathrm{d}V + \frac{\phi_0}{4\pi}\oint \delta\mathbf{D} \cdot \mathrm{d}\mathbf{f} = 0.$$

In the first term we write†

$$\delta F = (\partial F/\partial \mathbf{D})_{T,\rho} \cdot \delta\mathbf{D} = \mathbf{E} \cdot \delta\mathbf{D}/4\pi,$$

and the second can be integrated by parts: $\int \phi \operatorname{div} \delta\mathbf{D}\,\delta V = \oint \phi\delta\mathbf{D} \cdot \mathrm{d}\mathbf{f} - \int \delta\mathbf{D} \cdot \mathbf{grad}\,\phi\,\mathrm{d}V$. The result is

$$\int (\mathbf{E} + \mathbf{grad}\,\phi) \cdot \delta\mathbf{D}\,\mathrm{d}V + \oint (\phi_0 - \phi)\delta\mathbf{D} \cdot \mathrm{d}\mathbf{f} = 0.$$

† The free energy is the minimum for a given temperature. The variation is with respect to two independent quantities $\mathbf{D}$ and ρ. Here we are interested only in the result of varying with respect to $\mathbf{D}$. The variation of the integral (18.1) with respect to density (with the subsidiary condition of constant total mass of the body) gives one of the usual conditions of thermal equilibrium, namely the constancy of the chemical potential ζ.

Hence we conclude that, throughout the volume, we must have $\mathbf{E} = -\mathbf{grad}\,\phi$ (and so $\mathbf{curl\,E} = 0$), and on the surface of a conductor $\phi = \phi_0 =$ constant. These are the correct equations for the field, and the Lagrange multiplier ϕ is its potential.

Similarly it can be shown that the equations for the electric induction are obtained from the condition that the integral $\tilde{\mathscr{F}} = \int \tilde{F}(T, \rho, \mathbf{E})\,\mathrm{d}V$ be a maximum, in which the function $\mathbf{E}(\mathbf{r})$ is varied with the subsidiary conditions $\mathbf{E} = -\mathbf{grad}\,\phi$ and $\phi =$ constant on the surface of a conductor.† For

$$\delta\tilde{\mathscr{F}} = \int (\partial\tilde{F}/\partial\mathbf{E})\cdot\delta\mathbf{E}\,\mathrm{d}V = \int \mathbf{D}\cdot\mathbf{grad}\,\delta\phi\,\mathrm{d}V/4\pi$$

$$= \oint \delta\phi\mathbf{D}\cdot\mathrm{d}\mathbf{f}/4\pi - \int \delta\phi\,\mathrm{div}\,\mathbf{D}\,\mathrm{d}V/4\pi = 0.$$

The first integral is zero because $\delta\phi = 0$ on the surface, and from the second we find the required equation $\mathrm{div}\,\mathbf{D} = 0$, since $\delta\phi$ is arbitrary in the volume.

If the body is not in an external electric field (in particular, if there are no charged conductors), it may be possible to formulate the condition of themodynamic equilibrium as the condition that the total free energy (18.1) have an absolute (unconditional) minimum. This amounts to the condition that the free energy density F be a minimum as a function of the independent variable $\mathbf{D}$: $\partial F/\partial\mathbf{D} = \mathbf{E}/4\pi = 0$, i.e. the field must be zero in all space. If it is possible to find a distribution of the induction such that $\mathrm{div}\,\mathbf{D} = 0$, this state will correspond to thermodynamic equilibrium.‡

Equating to zero the first variation of the free energy, we find necessary but not sufficient conditions for this energy to be a minimum. The determination of the sufficient conditions requires a discussion of the second variation. These conditions take the form of certain inequalities (called *thermodynamic inequalities*) and are the conditions which ensure the stability of the state of the body (see *SP* 1, §21).

When $\mathbf{D} = \varepsilon\mathbf{E}$, the situation is much simplified, and the thermodynamic inequality of interest here (relating to the dielectric properties of the body) becomes evident. The total free energy is $\mathscr{F}_0 + \int (D^2/8\pi\varepsilon)\,\mathrm{d}V$. It is clear that this can have a minimum only if $\varepsilon > 0$, since otherwise the integral could be made to take any large negative value by making D^2 large enough. Thus in this case nothing new is learnt, since we know already that the permittivity must in fact be not only positive but greater than unity (see §14).

In the general case of an arbitrary relation between $\mathbf{D}$ and $\mathbf{E}$, however, it is necessary to consider the second variation of the integral (18.1), and to vary simultaneously both $\mathbf{D}$ and ρ (leaving only the temperature constant). In an isotropic body, $F(T, \rho, \mathbf{D})$ depends only on the magnitude of the vector $\mathbf{D}$, but its three components vary independently. We take the direction of the vector $\mathbf{D}$ before variation as the z-axis. Then the change in the magnitude of $\mathbf{D}$ is given in terms of the changes in its components, as far as the second-order terms, by $\delta D = \delta D_z + (\delta D_x)^2/2D + (\delta D_y)^2/2D$. The first and second variations of the integral (18.1)

† The point that the thermodynamic potential $\tilde{\mathscr{F}}$ has a maximum, not a minimum like $\mathscr{F}$, with respect to the variable $\mathbf{E}$ or $\mathbf{D}$ is a general one, and is accounted for as follows. Let the equilibrium value of a variable x, say $x = 0$, be determined by the condition of thermodynamic equilibrium. Then the free energy $\mathscr{F}$ has, for a given T and V, a minimum at $x = 0$. We thus have, at $x = 0$, $X \equiv (\partial\mathscr{F}/\partial x)_{V,T} = 0$, and near that point $X = \alpha x$, $\mathscr{F} = \mathscr{F}_0 + \frac{1}{2}\alpha x^2$ with $\alpha > 0$. With the thermodynamic potential $\tilde{\mathscr{F}} = \mathscr{F} - xX$, this gives $\tilde{\mathscr{F}} = \mathscr{F}_0 - \frac{1}{2}\alpha x^2 = \mathscr{F}_0 - X^2/2\alpha$, so that in equilibrium $\tilde{\mathscr{F}}$ has a maximum with respect to x or X. But both $\mathscr{F}$ and $\tilde{\mathscr{F}}$ have minima with respect to any other variables y that are independent of x.

‡ Here we are considering bodies in which $\mathbf{D}$ need not be zero even if $\mathbf{E} = 0$ (see §19). Otherwise we have simply the trivial result $\mathbf{E} = \mathbf{D} = 0$ in all space.

are both contained in the expression

$$\int \left\{ \frac{\partial F}{\partial D} \delta D + \frac{\partial F}{\partial \rho} \delta \rho + \frac{1}{2} \frac{\partial^2 F}{\partial D^2} (\delta D)^2 + \frac{\partial^2 F}{\partial D \partial \rho} \delta D \delta \rho + \frac{1}{2} \frac{\partial^2 F}{\partial \rho^2} (\delta \rho)^2 \right\} \mathrm{d} V.$$

Substituting δD and collecting the second-order terms, we find the second variation

$$\int \frac{1}{2D} \frac{\partial F}{\partial D} [(\delta D_x)^2 + (\delta D_y)^2] \mathrm{d} V + \int \left\{ \frac{1}{2} \frac{\partial^2 F}{\partial D^2} (\delta D_z)^2 + \frac{\partial^2 F}{\partial D \partial \rho} \delta D_z \delta \rho + \frac{1}{2} \frac{\partial^2 F}{\partial \rho^2} (\delta \rho)^2 \right\} \mathrm{d} V. \tag{18.4}$$

These two terms are independent. The first is positive if $(1/D)\partial F/\partial D > 0$. But $\partial F/\partial \mathbf{D} = \mathbf{E}/4\pi$, so that the derivative $\partial F/\partial D$ is positive or negative according as the vectors $\mathbf{D}$ and $\mathbf{E}$ are in the same or opposite directions. Thus these vectors must be in the same direction.

The conditions for the second term in (18.4) to be positive are

$$\partial^2 F/\partial \rho^2 > 0, \tag{18.5}$$

$$\frac{\partial^2 F}{\partial \rho^2} \frac{\partial^2 F}{\partial D^2} - \left(\frac{\partial^2 F}{\partial \rho \partial D} \right)^2 > 0. \tag{18.6}$$

Since $\partial F/\partial \rho = \zeta$, $\partial F/\partial D = E/4\pi$, the first of these gives

$$(\partial \zeta/\partial \rho)_{D,T} > 0, \tag{18.7}$$

and the second can be rewritten as a Jacobian:

$$\frac{\partial(\partial F/\partial D, \partial F/\partial \rho)}{\partial(D, \rho)} = \frac{1}{4\pi} \frac{\partial(E, \zeta)}{\partial(D, \rho)} > 0.$$

Changing from the variables D, ρ to D, ζ, we have

$$\frac{\partial(E, \zeta)}{\partial(D, \rho)} = \frac{\partial(E, \zeta)}{\partial(D, \zeta)} \frac{\partial(D, \zeta)}{\partial(D, \rho)} = \left(\frac{\partial E}{\partial D} \right)_\zeta \left(\frac{\partial \zeta}{\partial \rho} \right)_D > 0;$$

by (18.7), this gives

$$(\partial E/\partial D)_{\zeta,T} > 0. \tag{18.8}$$

Thus we have derived the required thermodynamic inequalities. In the absence of a field, the inequality (18.7) becomes the usual condition that the isothermal compressibility is positive: $(\partial P/\partial \rho)_T > 0$.† The inequality (18.8) gives $\varepsilon > 0$, since when $E \to 0$ the induction $D \to \varepsilon E$.

Of the two inequalities (18.5), (18.6) the latter is the stronger; it may be violated while the

† It should be recalled that, in the absence of a field, ζ is the thermodynamic potential of unit mass and, by the ordinary thermodynamic relations, its differential

$$\mathrm{d}\zeta = \mathrm{d}P/\rho - (S/\rho)\mathrm{d}T,$$

so that $(\partial \zeta/\partial \rho)_T = (1/\rho)(\partial P/\partial \rho)_T$. In the above derivation the second of the ordinary thermodynamic inequalities (that the specific heat is positive) is ignored.

first is not, whereas the reverse is impossible. The equation

$$\frac{\partial^2 F}{\partial \rho^2}\frac{\partial^2 F}{\partial D^2} - \left(\frac{\partial^2 F}{\partial \rho \partial D}\right)^2 = \frac{1}{4\pi}\frac{\partial(E,\zeta)}{\partial(D,\rho)} = 0$$

corresponds to what is called the *critical state* (see *SP* 1, §83). This condition is more conveniently written in a different form by multiplying it by the non-zero factor $\partial(D,\rho)/\partial(E,\rho)$:

$$\partial(E,\zeta)/\partial(E,\rho) = (\partial\zeta/\partial\rho)_{E,T} = 0. \tag{18.9}$$

The critical states occupy a curve in the *ET*-plane, which is a singularity of the thermodynamic functions of the body, just as the critical point is a singularity in the absence of the field.

§19. Ferroelectrics

The various crystalline modifications of a given substance may include some which are pyroelectric and some which are not. If the change from one to the other takes place by means of a second-order phase transition, then near the transition point the substance has a number of unusual properties which distinguish it from ordinary pyroelectrics; these are called *ferroelectric* substances.

In an ordinary pyroelectric crystal, a change in the direction of the spontaneous polarization involves a considerable reconstruction of the crystal lattice. Even if the final result of this reconstruction is energetically favourable, its realization may still be impossible because it would require the surmounting of very high energy barriers.

In a ferroelectric body, however, the situation is quite different because, near a second-order phase transition point, the arrangement of the atoms in the crystal lattice of the pyroelectric phase is only slightly different from the arrangement in the non-pyroelectric lattice (and so the spontaneous polarization also is small). For this reason the change in direction of the spontaneous polarization here requires only a slight reconstruction of the lattice and can occur quite easily.

The actual nature of the ferroelectric properties of a body depends on its crystal symmetry. The direction of the spontaneous polarization of the pyroelectric phase (which we shall call the *ferroelectric axis*) is determined by the structure of the non-pyroelectric phase beyond the transition point. In some cases it is uniquely determined, in the sense that the ferroelectric axis can lie in only one, crystallographically determinate, direction; the direction of the spontaneous polarization is then determined apart from sign, since in the non-pyroelectric phase the two opposite directions parallel to the ferroelectric axis must be equivalent (otherwise this form of the crystal would also be pyroelectric). In other cases, the symmetry of the non-pyroelectric phase may be such as to allow spontaneous polarization in any of several crystallographically equivalent directions.†

---

† An instance of the first type is sodium potassium tartrate, whose non-pyroelectric phase has orthorhombic symmetry. The ferroelectric axis appears in it in a completely definite crystallographic direction (one of the twofold axes), and the lattice becomes monoclinic.

An instance of the second type is barium titanate. Its non-pyroelectric modification has a cubic lattice, and any of the three cubic axes may become the ferroelectric axis. After the spontaneous polarization has appeared at the transition point, these three directions, of course, are no longer equivalent. The ferroelectric axis becomes the only fourfold axis, and the lattice becomes tetragonal.

The occurrence of polarization is always associated with a reduction in the symmetry of the crystal. We can therefore refer to pyroelectric and non-pyroelectric phases as the unsymmetrical and symmetrical phases respectively, using the terminology of *SP* 1, §142.

We shall show how the theory of ferroelectricity can be developed in terms of the general theory of Landau second-order phase transitions, as was first done by V. L. Ginzburg (1945).†

We take the dielectric polarization vector **P** of the substance as the order parameter, whose magnitude determines the difference between the unsymmetrical and symmetrical phase lattice structures. This means that **P** will be regarded as an independent thermodynamic variable whose actual value (as a function of the temperature, the field, etc.) is then determined from the condition of thermal equilibrium, namely that the thermodynamic potential be a minimum.

Let us consider first the case where the position of the ferroelectric axis, which we take as the z-axis, is uniquely determined. The dielectric properties of the crystal in the x and y directions then exhibit no anomalies, and to investigate the properties in the z-direction we need consider only those terms in the thermodynamic potential which contain P_z. Near the transition point, the order parameter P_z is small and the thermodynamic potential Φ can be expanded in powers of P_z. Since the two directions of the z-axis are equivalent, the expansion cannot depend on the sign of P_z, and therefore contains only even powers. As far as the fourth-order terms, we have

$$\Phi = \Phi_0 + AP_z^2 + BP_z^4. \tag{19.1}$$

In the symmetrical phase, $A > 0$, and $P_z = 0$ corresponds to a minimum of the thermodynamic potential. For spontaneous polarization to occur, A must be negative; it is therefore zero at the phase transition point. The Landau theory assumes that $A(T)$ can be expanded in integral powers of $T - T_c$, where T_c is the phase transition temperature; near this point, we write $A = a(T - T_c)$, a being a constant (independent of the temperature). We shall take the specific case where $a > 0$, so that the unsymmetrical phase corresponds to temperatures $T < T_c$. The condition for the state to be stable at the point $T = T_c$ itself is that the coefficient B be positive at that point and therefore throughout a neighbourhood of it. In what follows, B will denote $B(T_c)$.

If the electric field in the body is not zero, further terms appear in the thermodynamic potential. To find these, we start from the relation

$$4\pi \partial \tilde{\Phi} / \partial \mathbf{E} = -\mathbf{D} = -\mathbf{E} - 4\pi \mathbf{P}. \tag{19.2}$$

Integration with a fixed value of the independent variable **P**, using the fact that Φ and $\tilde{\Phi}$ are the same when $\mathbf{E} = 0$, gives

$$\tilde{\Phi}(\mathbf{P}, \mathbf{E}) = \Phi(\mathbf{P}, 0) - \mathbf{E} \cdot \mathbf{P} - \mathbf{E}^2/8\pi.$$

With the electric field in the z-direction, and $\Phi(\mathbf{P}, 0)$ from (19.1), we have

$$\tilde{\Phi} = \Phi_0 + a(T - T_c)P_z^2 + BP_z^4 - E_z P_z - E_z^2/8\pi. \tag{19.3}$$

---

† The Landau theory certainly becomes invalid in a neighbourhood of the transition point. The question of when this happens in ferroelectrics needs a specific analysis of experimental results, and lies outside the scope of the present book. Actually, many ferroelectric transitions are not second-order, but first-order ones close to being second-order. This seems to be due to the fluctuation effect mentioned in *SP* 1, end of §146.

The presence of the term $-E_zP_z$ has the result that in any field E_z, however weak, the order parameter P_z becomes different from zero at every temperature; the field polarizes the non-pyroelectric phase and thus reduces its symmetry. The qualitative difference between the two phases thereby disappears, and accordingly so does the discrete transition point; the transition is "smoothed out".†

The thermodynamic potential $\tilde{\Phi}$ in equilibrium must be a minimum, for any given field **E**. Differentiation of (19.3) at constant E_z gives

$$2P_z a(T-T_c)+4BP_z^3 = E_z. \tag{19.4}$$

This is the basic relation between the field and the polarization in a ferroelectric.‡

When $T > T_c$, in the non-pyroelectric phase, P_z vanishes with E_z. As E_z increases, the polarization at first increases linearly, $P_z = \kappa E_z$, with susceptibility

$$\kappa = 1/2a(T-T_c),\ T > T_c, \tag{19.5}$$

which increases without limit as $T \to T_c$. The induction $D_z = (1+4\pi\kappa)E_z$ also increases linearly with P_z. Near the transition point, κ is large, and we have to the same accuracy

$$\varepsilon \cong 4\pi\kappa = 2\pi/a(T-T_c). \tag{19.6}$$

In sufficiently strong fields, the polarization increases according to $P_z = (E_z/4B)^{1/3}$.

When $T < T_c$, in the pyroelectric phase, $P_z = 0$ cannot correspond to a stable state. For $E_z = 0$, we find from (19.4) the spontaneous polarization of this phase,

$$P_{z0} = \pm\sqrt{[a(T_c-T)/2B]}. \tag{19.7}$$

The dielectric susceptibility of the phase can be found as the derivative dP_z/dE_z as $E_z \to 0$. From (19.4),

$$[-2(T_c-T)a+12BP_z^2]dP_z/dE_z = 1, \tag{19.8}$$

and substitution of (19.7) gives

$$\kappa = [dP_z/dE_z]_{E_z=0} = 1/4a(T_c-T),\ T < T_c. \tag{19.9}$$

This is half the susceptibility of the non-pyroelectric phase for the same value of $|T_c-T|$. In sufficiently weak fields, the polarization is $P_z = P_{z0}+\kappa E_z$, the induction is $D_z = D_{z0} + \varepsilon E_z$, where $D_{z0} = 4P_{z0}$, and the permittivity is

$$\varepsilon \cong 4\pi\kappa = \pi/a(T_c-T). \tag{19.10}$$

Figure 14 (p. 80) shows the function $P_z(E_z)$ given by (19.4) for $T < T_c$. First of all, it should be noted that the dashed part cc' does not correspond to stable states: from (19.8) written in the form

$$(\mathrm{d}P_z/\mathrm{d}E_z)(\partial^2\tilde{\Phi}/\partial P_z^2)_{E_z} = 1$$

we see that $\mathrm{d}P_z/\mathrm{d}E_z < 0$ implies that $\partial^2\Phi/\partial P_z^2 < 0$, i.e., the thermodynamic potential $\tilde{\Phi}$ has a maximum, not a minimum. The ordinates of c and c' are given by the equation $\mathrm{d}E_z/\mathrm{d}P_z = 0$, and we conclude that the possible values of $|P_z|$ in the pyroelectric phase are

† Cf. *SP* 1, §144. The discussion below is largely a repetition of the one given there.

‡ Expressing **P(E)** by means of (19.4) and substituting in (19.3), we find the potential $\tilde{\Phi}(\mathbf{E})$ as a function of **E** only. From the condition $\partial\tilde{\Phi}(\mathbf{P}, \mathbf{E})/\partial\mathbf{P} = 0$, the equation $\mathbf{D} = -4\pi\partial\tilde{\Phi}/\partial\mathbf{E}$ is valid both for $\tilde{\Phi}(\mathbf{E})$ and for $\tilde{\Phi}(\mathbf{P}, \mathbf{E})$ (which is differentiated at constant **P**).

restricted by the condition

$$P_z^2 > (T_c - T)a/6B. \tag{19.11}$$

If we consider states of a ferroelectric with given values of E_z, there is still an ambiguity in the value of P_z, in the range of abscissae between c and c', and the question arises of the physical significance of the two values. We shall assume the ferroelectric to be a homogeneous flat slab, with the ferroelectric axis normal to it, lying between the plates of a capacitor, which are maintained at given potentials, i.e. which set up a given uniform field $E = E_z$.

For given potentials on the conductors, the condition of stability requires that the thermodynamic potential $\tilde{\Phi}$ be a minimum. In particular, for $\mathbf{E} = 0$ there are two states in which P_z has opposite signs (the points a and a' in Fig. 14) but $\tilde{\Phi}$ $(= \Phi)$ is the same. These two states, therefore, are equally stable, i.e. they are two phases which can coexist in contact.

Hence it is clear that the portions ac and $a'c'$ of the curve correspond to states which are metastable but not absolutely stable. It is easy to see directly that the values of $\tilde{\Phi}$ on ac and on a' c' are in fact greater than its values on $a'b'$ and ab for the same value of E_z. The ordinates of a and a' are given by formula (19.7). Thus the range of metastability is

$$(T_c - T)a/6B < P_z^2 < (T_c - T)a/2B. \tag{19.12}$$

The existence of these two phases with $\mathbf{E} = 0$ is very important, since it means that a ferroelectric body can be divided into a number of separate regions or *domains* in which the polarization is in opposite directions. On the surfaces separating these domains, the normal component of $\mathbf{D}$ and the tangential component of $\mathbf{E}$ must be continuous. The latter condition is satisfied identically, because $\mathbf{E} = 0$. From the former condition it follows that the domain boundaries must be parallel to the z-axis.

The actual shapes and sizes of the domains are determined by the condition that the total thermodynamic potential of the body should be a minimum.†

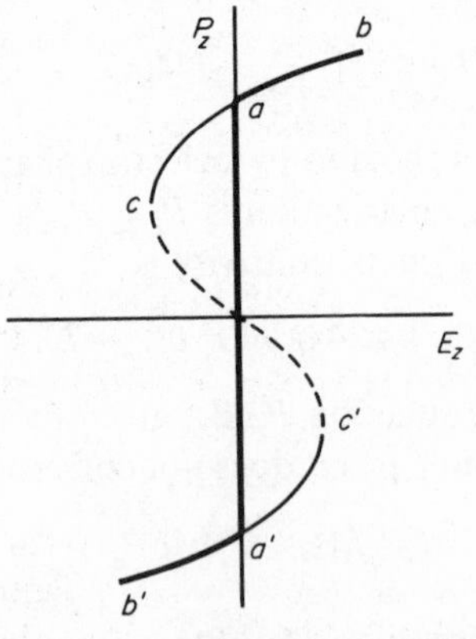

FIG. 14

† It should be emphasized that complete thermodynamic equilibrium is under consideration here. This can occur in ferroelectrics, but practically never does so in ordinary pyroelectrics because of the already mentioned difficulty of reorienting the polarization (and so of forming domains) in these materials. The shape and size of the domains will be discussed in §44 for the case of ferromagnets, which is in many respects analogous. We shall not pause to consider the specific features of the domain structure in ferroelectrics. They are due mainly to the rigid coupling of the direction of polarization to particular crystal axes, to the high dielectric susceptibility in comparison with the magnetic susceptibility of a ferromagnet, and to the greater role of striction effects.

If we are not interested in the details of the structure, and consider portions of the body which are large compared with the domains, we can use the polarization $\overline{\mathbf{P}}$ averaged over such portions. Its component $\overline{P}_z$ can evidently take values in the range between the ordinates of a and a' in Fig. 14, i.e.

$$-\sqrt{[(T_c-T)a/2B]} < \overline{P}_z < \sqrt{[(T_c-T)a/2B]}. \qquad (19.13)$$

In other words, if P_z in Fig. 14 is taken as the polarization averaged in this way, the vertical segment aa' corresponds to the region of domain structure, and the thick curve $baa'b'$ gives all stable states of the body.

Let us consider ferroelectrics which belong (in the non-pyroelectric phase) to the cubic system.† The cubic symmetry admits two independent fourth-order invariants formed from the components of the vector $\mathbf{P}$, which may be taken as $(P_x^2+P_y^2+P_z^2)^2$ and $P_x^2P_y^2+P_x^2P_z^2+P_y^2P_z^2$. Hence the expansion of the thermodynamic potential near the transition point (when $\mathbf{E}=0$) is of the form

$$\Phi = \Phi_0 + a(T-T_c)(P_x^2+P_y^2+P_z^2)+B(P_x^2+P_y^2+P_z^2)^2$$
$$+C(P_x^2P_y^2+P_x^2P_z^2+P_y^2P_z^2), \qquad (19.14)$$

where a, B, C are constants, and the x, y, z axes are along the three fourfold axes of symmetry.

The sum of the fourth-order terms in (19.14) must be essentially positive. Hence we must have

$$B>0, \qquad 3B+C>0. \qquad (19.15)$$

The spontaneous polarization of a ferroelectric when $\mathbf{E}=0$ is determined by the condition that Φ should be a minimum as a function of $\mathbf{P}$. In particular, since the second-order term and the first of the fourth-order terms are independent of the direction of $\mathbf{P}$, the direction of the spontaneous polarization is determined by the condition that the next term be a minimum for a given absolute value P. Two cases are possible. If $C>0$, the minimum of this term corresponds to $\mathbf{P}$ being along any one of the axes x, y, z, i.e. along any of the three edges of the cube. If, however, $C<0$, the minimum value occurs when $\mathbf{P}$ is along any one of the spatial diagonals of the cube, i.e. when $P_x^2 = P_y^2 = P_z^2 = \frac{1}{3}P^2$. In the former case the pyroelectric phase of the ferroelectric has tetragonal symmetry, and in the latter case it has rhombohedral symmetry.

Let us consider in more detail, for example, the first case ($C>0$), and take as the z-axis the direction of the spontaneous polarization below the transition point. The magnitude P_0 of this polarization is determined (when $\mathbf{E}=0$) by the minimum of the expression $-a(T_c-T)P^2+BP^4$, whence

$$P_0^2 = a(T_c-T)/2B. \qquad (19.16)$$

To find the polarization as a function of the field $\mathbf{E}$, we must add to (19.14) a term $-\mathbf{P}\cdot\mathbf{E}$ (thus changing to the potential $\tilde{\Phi}$) and equate to zero the derivative $\partial\tilde{\Phi}/\partial\mathbf{P}$.

When the field $\mathbf{E}$ is weak, P_x, P_y, and P_z-P_0 are small. Omitting from the equations the

† The crystal classes $\boldsymbol{T}_h$ and $\boldsymbol{O}_h$ are envisaged here. The cubic classes $\boldsymbol{T}$ and $\boldsymbol{T}_d$ allow also a third-order invariant $P_xP_yP_z$; under such conditions, the state with $\mathbf{P}=0$ certainly could not satisfy the stability condition (minimum Φ), and a second-order phase transition is therefore impossible. The symmetry of the class $\boldsymbol{O}$ (and that of $\boldsymbol{T}$) allows an invariant $\mathbf{P}\cdot\mathbf{curl}\,\mathbf{P}$ that is linear in the derivatives, and this gives rise to an incommensurate structure (cf. §52).

terms of the second and higher orders of smallness, and substituting $P_{0z} = P_0$ from (19.16), we find the longitudinal polarization

$$P_z - P_0 = E_z/4a(T_c - T) \tag{19.17}$$

and the transverse polarization

$$P_x = BE_x/aC(T_c - T) \tag{19.18}$$

(and similarly for P_y). Above the transition point, in the non-pyroelectric phase, the dielectric susceptibility of a cubic ferroelectric is the same in all directions:

$$\mathbf{P} = \mathbf{E}/2a(T - T_c). \tag{19.19}$$

Finally, let us briefly consider the elastic properties of ferroelectrics. According to its crystal class, the non-pyroelectric phase of a ferroelectric may or may not be piezoelectric.† Let us begin with the former case, and suppose that the symmetry admits a piezoelectric (linear) relation between the deformation and the polarization in the direction of the ferroelectric (z) axis. These include the classes $\boldsymbol{D}_2$, $\boldsymbol{D}_{2d}$ and $\boldsymbol{S}_4$; in each case the polarization P_z appears in the piezoelectric part of the thermodynamic potential through a term $-\gamma_{z,xy}P_z\sigma_{xy}$.

In the elastic energy of these crystals, the component σ_{xy} appears in a term $-\mu_{xyxy}\sigma_{xy}^2$. Thus the thermodynamic potential near the transition point is

$$\tilde{\Phi} = \Phi_0 + a(T - T_c)P_z^2 + BP_z^4 - \gamma P_z\sigma_{xy} - \mu\sigma_{xy}^2 - E_zP_z - E_z^2/8\pi, \tag{19.20}$$

where for brevity we have put $\gamma_{z,xy} = \gamma$, $\mu_{xyxy} = \mu$.‡ The terms involving the other components of $\mathbf{P}$ and σ_{ik} are of no interest, since they lead to no anomaly of the piezoelectric properties near the transition point.

Equating to zero the derivative $\partial\tilde{\Phi}/\partial P_z$ with E_z constant, we obtain

$$E_z = 2a(T - T_c)P_z + 4BP_z^3 - \gamma\sigma_{xy}. \tag{19.21}$$

The components of the strain tensor are found by differentiating the thermodynamic potential (19.20) with respect to the corresponding components σ_{ik} (see 17.4)):§

$$u_{xy} = \tfrac{1}{2}\gamma P_z + \mu\sigma_{xy}. \tag{19.22}$$

In the non-pyroelectric phase when $\mathbf{E}$ is small we can neglect the term in P_z^3 in (19.21):

$$E_z = 2a(T - T_c)P_z - \gamma\sigma_{xy}.$$

Substituting P_z in (19.22), we find

$$u_{xy} = \frac{\gamma}{4a(T - T_c)}E_z + \left[\mu + \frac{\gamma^2}{4a(T - T_c)}\right]\sigma_{xy}.$$

The coefficient of σ_{xy} in this formula represents the modulus of elasticity for deformations in which the field E_z is kept constant, while μ in formula (19.22) is the modulus for constant

† The non-pyroelectric phase of a ferroelectric is piezoelectric if it belongs to one of eight out of the ten classes listed at the end of §17: $\boldsymbol{D}_2$, $\boldsymbol{D}_4$, $\boldsymbol{D}_{2d}$, $\boldsymbol{S}_4$, $\boldsymbol{D}_3$, $\boldsymbol{D}_6$, $\boldsymbol{C}_{3h}$, $\boldsymbol{D}_{3h}$.

‡ Because the expansion is of a different type, the definitions of the tensors $\gamma_{i,kl}$ and μ_{iklm} are not the same as those denoted by the same letters in §17, but their symmetry properties are, of course, unchanged.

§ See the first footnote to §17, Problem 1, concerning differentiation with respect to the components of a tensor.

polarization P_z. Hence we can write

$$\mu^{(E)} = \mu^{(P)} + \gamma^2/4a(T-T_c), \tag{19.23}$$

where the superscripts indicate the nature of the deformation. We see that the two coefficients behave entirely differently: whereas $\mu^{(P)}$ is a finite constant, $\mu^{(E)}$ increases without limit as the transition point is approached.†

In the pyroelectric phase, formula (19.22) shows that the spontaneous polarization results in a certain deformation of the body. If there are no internal stresses and the field **E** is zero, the strain u_{xy} is proportional to P_{z0}, i.e. varies with temperature as $\sqrt{(T_c-T)}$.

If the symmetry (cubic, for example) of the non-pyroelectric phase of a ferroelectric does not admit a linear piezoelectric effect, then the first non-vanishing terms in an expansion of the thermodynamic potential in powers of σ_{ik} and **P** are quadratic in the components of **P**, i.e. they are of the form

$$-\gamma_{iklm}P_iP_k\sigma_{lm}, \tag{19.24}$$

where γ_{iklm} is a tensor of rank four, symmetrical with respect to the pairs of suffixes *i*, *k* and *l*, *m*. In such cases, the strain in the pyroelectric phase due to the spontaneous polarization is quadratic in $\mathbf{P}_0$, and accordingly varies with temperature as T_c-T.

Doubt might be cast on the legitimacy of using the expression (19.24) in the thermodynamic potential, on the grounds that, as stated in §17, this potential can be used only when quadratic effects are neglected. However, the ferroelectrics form an exception because, near the transition point, the field **E** is small compared with the polarization **P** or induction **D**, because of the unlimited increase in the dielectric susceptibility. The use of the thermodynamic potential involves the neglect of the quantities of the order of EDu_{ik} (or, what is the same thing, $ED\sigma_{ik}$), whereas the expression (19.24) is of the order of $D^2\sigma_{ik}$.

§20. Improper ferroelectrics

The theory of ferroelectrics given in §19 is based on identifying the polarization vector of the crystal with the order parameter which determines the change in the crystal symmetry in the phase transition. This is not always permissible, however. It may happen that the occurrence of spontaneous polarization does not in itself entirely determine the nature of the change in the crystal structure.

It is known (see *SP* 1, §145) that the order parameter in a second-order phase transition is a quantity, or a set of quantities, transformed by some irreducible (not the unit) representation of the symmetry group of the original ("symmetrical") phase. The transformation properties of the order parameter determine the nature of the change (the decrease) in the symmetry at the phase transition. The specific physical nature of the change is unimportant; the order parameter may be taken as any of various physical quantities, provided that they are related to one another by linear expressions and therefore have the same transformation properties.

The choice of the vector **P** as the order parameter is equivalent to assuming that this is transformed by the same representation as the components of a (polar) vector. If the phase transition occurs with no change in the unit cell of the lattice (or with only a strain), the irreducible representations concerned are those of the point symmetry groups (the crystal

† The modulus $\mu^{(D)} = \mu + \gamma^2/8\pi$, which determines the strain for a constant induction D_z, is also a constant.

classes). In the biaxial classes (§13), each component of a vector is transformed by one of the one-dimensional representations. The same applies to a component of a vector along the principal axis of symmetry (threefold, fourfold or sixfold) in uniaxial crystals. For all these representations, the order parameter can be the corresponding component of the vector **P**, and the theory based on the thermodynamic potential (19.1) is applicable to them. The components of **P** in a plane perpendicular to the principal axis of symmetry in a uniaxial crystal are transformed by a two-dimensional irreducible representation and can serve as the order parameter for that representation. Lastly, in crystals with cubic symmetry, all three components of the vector are transformed by a single three-dimensional representation. This case corresponds to the theory of ferroelectricity based on the thermodynamic potential (19.14).

There are also, however, ferroelectric transitions in which the order parameter is transformed by an irreducible representation of the symmetrical phase which does not correspond to the components of a vector. In such cases, the order parameter is not the polarization but a physically different quantity; the spontaneous polarization arises as a secondary effect (it is assumed, of course, that the symmetry of the unsymmetrical phase allows pyroelectricity). These substances are called *improper ferroelectrics*; they differ considerably from ordinary ferroelectrics as regards the nature of the dielectric anomalies.† Here belong all ferroelectric transitions in which the unit cell changes, i.e. in which the translational symmetry of the lattice changes (the corresponding irreducible representations are certainly not realized by vector quantities invariant under translations)‡, but they may also be transitions without change of translational symmetry (the order parameter is transformed by an irreducible representation of the point group, not corresponding to the components of a vector).

In an ordinary ferroelectric transition, when the change in symmetry is entirely determined by the polarization vector, the transition is to a higher sub-group (that allows pyroelectricity) of the space group of the original (non-pyroelectric) phase. In an improper ferroelectric transition, the pyroelectric phase belongs to a sub-group of lower symmetry.

The specific thermodynamic properties of improper ferroelectrics may vary considerably, like the transformational properties of quantities that are transformed by different irreducible representations of the space groups. Let us now consider (again in terms of the Landau theory of phase transitions) one formal example to illustrate some important fundamental points.

We shall take a transition (without change in the unit cell) from a non-pyroelectric crystal of the class $\boldsymbol{C}_{3h}$ to the class $\boldsymbol{C}_1$, which allows spontaneous polarization, the order parameter having two components η_1, η_2 and being transformed by the irreducible representation E_u of the group C_{3h}; the components P_x, P_y of the polarization vector in the plane perpendicular to the C_3 axis being transformed by the representation E_g.

The thermodynamic potential $\tilde{\Phi}$ near the transition point is to be expanded in powers of the order parameter η_1, η_2 and the polarization P_x, P_y. For ferroelectricity to occur, there must be mixed invariants formed from the same quantities, and linear in the vector **P**. There are two such invariants in this case: the real and imaginary parts of the product

† The possible existence of such ferroelectrics was noted by V. L. Indenbom (1960).

‡ All known improper ferroelectrics are in fact of this type.

$(\eta_1+i\eta_2)^2(P_x+iP_y)$. We thus obtain an expansion in the form

$$\tilde{\Phi} = \Phi_0 + a(T-T_c)\eta^2 + B\eta^4 + \kappa\mathbf{P}^2 + C_1\eta^2[P_x(\gamma_1{}^2-\gamma_2{}^2)-2P_y\gamma_1\gamma_2)]$$
$$+C_2\eta^2[P_y(\gamma_1{}^2-\gamma_2{}^2)+2P_x\gamma_1\gamma_2]-\mathbf{E}\cdot\mathbf{P}-\mathbf{E}^2/8\pi, \tag{20.1}$$

where $\eta^2 = \eta_1{}^2+\eta_2{}^2$, $\gamma_i = \eta_i/\eta$; the vectors **E** and **P** are in the xy-plane.

The order parameter and the polarization are determined by the condition that $\tilde{\Phi}$ be a minimum (for constant **E**). Here we shall give only some characteristic results that are evident without actually making the calculation. The order parameter in the unsymmetrical phase is found to be proportional to $(T_c-T)^{1/2}$, as for any second-order transition in the Landau theory. The polarization arises as an effect of the second order in η, and is therefore proportional to T_c-T. The dielectric susceptibility does not tend to infinity as $T \to T_c$ as in ordinary ferroelectrics, because it is not here determined by a coefficient of η^2 that tends to zero. It does, however, have a finite discontinuity at the transition point. This is because, in the symmetrical phase, the order parameter $\eta = 0$ and is not affected by the field **E**; in the unsymmetrical phase it does change, and this gives a further contribution to the susceptibility.

An improper ferroelectric transition is possible only when the order parameter has more than one component. With a one-component parameter η, the only possible mixed invariant linear in **P** is ηP_z, where P_z is one component of the vector **P** (since η^2 is an invariant for a one-dimensional representation). This would mean, however, that η and P_z had the same transformational properties, and therefore that P_z itself could be chosen as the order parameter.

CHAPTER III

STEADY CURRENT

§21. The current density and the conductivity

LET US now consider the steady motion of charges in conductors, i.e. steady electric currents. We shall denote by $\mathbf{j}$ the mean charge flux density or *electric current density*.† In a steady current, the spatial distribution of $\mathbf{j}$ is independent of time, and satisfies the equation

$$\operatorname{div} \mathbf{j} = 0, \tag{21.1}$$

which states that the mean total charge in any volume of the conductor remains constant.

The electric field in the conductor in which a steady current flows is constant, and therefore satisfies the equation

$$\mathbf{curl}\, \mathbf{E} = 0, \tag{21.2}$$

i.e. it is a potential field.

Equations (21.1) and (21.2) must be supplemented by an equation relating $\mathbf{j}$ and $\mathbf{E}$. This equation depends on the properties of the conductor, but in the great majority of cases it may be supposed linear (*Ohm's law*). If the conductor is homogeneous and isotropic, the linear relation is a simple proportionality:

$$\mathbf{j} = \sigma \mathbf{E}. \tag{21.3}$$

The coefficient σ depends on the nature and state of the conductor; it is called the *electrical conductivity*.

In a homogeneous conductor, σ = constant and, substituting (21.3) in (21.1), we have $\operatorname{div} \mathbf{E} = 0$. In this case the electric field potential satisfies Laplace's equation: $\triangle \phi = 0$.

At a boundary between two conducting media, the normal component of the current density must, of course, be continuous. Moreover, by the general condition that the tangential field component be continuous (which follows from $\mathbf{curl}\, \mathbf{E} = 0$; cf. (1.7) and (6.9)), the ratio $\mathbf{j}_t/\sigma$ must be continuous. Thus the boundary conditions on the current density are

$$j_{n1} = j_{n2}, \qquad \mathbf{j}_{t1}/\sigma_1 = \mathbf{j}_{t2}/\sigma_2, \tag{21.4}$$

or, as conditions on the field,

$$\sigma_1 E_{n1} = \sigma_2 E_{n2}, \qquad \mathbf{E}_{t1} = \mathbf{E}_{t2}. \tag{21.5}$$

† In this chapter we ignore the magnetic field due to the current, and therefore the reaction of that field on the current. If this effect is to be taken into account, the definition of the current density must be refined, which we do in §30.

At a boundary between a conductor and a non-conductor we have simply $j_n = 0$, or $E_n = 0$.†

An electric field in the presence of a current does mechanical work on the current-carrying particles moving in the conductor; the work done per unit time and volume is evidently equal to the scalar product $\mathbf{j}\cdot\mathbf{E}$. This work is dissipated into heat in the conductor. Thus the quantity of heat evolved per unit time and volume in a homogeneous conductor is

$$\mathbf{j}\cdot\mathbf{E} = \sigma E^2 = j^2/\sigma. \tag{21.6}$$

This is *Joule's law*.‡

The evolution of heat results in an increase in the entropy of the body. When an amount of heat $dQ = \mathbf{j}\cdot\mathbf{E}\,dV$ is evolved, the entropy of the volume element dV increases by dQ/T. The rate of change of the total entropy of the body is therefore

$$d\mathscr{S}/dt = \int (\mathbf{j}\cdot\mathbf{E}/T)\,dV. \tag{21.7}$$

Since the entropy must increase, this derivative must be positive. Putting $\mathbf{j} = \sigma\mathbf{E}$, we see that the conductivity σ must therefore be positive.

In an anisotropic body (a single crystal), the directions of the vectors $\mathbf{j}$ and $\mathbf{E}$ are in general different, and the linear relation between them is

$$j_i = \sigma_{ik} E_k, \tag{21.8}$$

where the quantities σ_{ik} form a tensor of rank two, the *conductivity tensor*, which is symmetrical (see below).

The following remark should be made here. The symmetry of the crystal would admit also an inhomogeneous term in the linear relation between $\mathbf{j}$ and $\mathbf{E}$, giving $j_i = \sigma_{ik}E_k + j_i^{(0)}$, with $\mathbf{j}^{(0)}$ a constant vector. The presence of this term would mean that the conductor was "pyroelectric", there being a non-zero field in it when $\mathbf{j} = 0$. In reality, however, this is impossible, because the entropy must increase: the term $\mathbf{j}^{(0)}\cdot\mathbf{E}$ in the integrand in (21.7) could take either sign, and so $d\mathscr{S}/dt$ could not be invariably positive.

Just as, for an isotropic medium, $d\mathscr{S}/dt > 0$ leads to $\sigma > 0$, so for an anisotropic medium this condition means that the principal values of the tensor σ_{ik} must be positive.

The dependence of the number of independent components of the tensor σ_{ik} on the symmetry of the crystal is the same as for any symmetrical tensor of rank two (see §13): for biaxial crystals, all three principal values are different, for uniaxial crystals two are equal, and for cubic crystals all three are equal, i.e. a cubic crystal behaves as an isotropic body as regards its conductivity.

The symmetry of the conductivity tensor

$$\sigma_{ik} = \sigma_{ki} \tag{21.9}$$

is a consequence of the *symmetry of the kinetic coefficients*. This general principle, due to L. Onsager, may be conveniently formulated, for use here and in §§26–28, as follows (see *SP* 1, §120).

† It should be noticed that the equations $\mathbf{curl}\,\mathbf{E} = 0$, $\operatorname{div}(\sigma\mathbf{E}) = 0$ and the boundary conditions (21.5) thereon are formally identical with the equations for the electrostatic field in a dielectric, the only difference being that ε is replaced by σ. This enables us to solve problems of the current distribution in an infinite conductor if the solutions of the corresponding electrostatic problems are known. When the conductor is bounded by a non-conductor this analogy does not serve, because in electrostatics there is no medium for which $\varepsilon = 0$.

‡ In Russian "Joule and Lenz's law".

Let $x_1, x_2, \ldots$ be some quantities which characterize the state of the body at every point. We define also the quantities

$$X_a = -\partial S/\partial x_a, \tag{21.10}$$

where S is the entropy of unit volume of the body, and the derivative is taken at constant energy of the volume. In a state close to equilibrium, the quantities x_a are close to their equilibrium values, and the X_a are small. Processes will occur in the body which tend to bring it into equilibrium. The rates of change of the quantities x_a at each point are usually functions only of the values of the x_a (or X_a) at that point. Expanding these functions in powers of X_a and taking only the linear terms, we have

$$\partial x_a/\partial t = -\sum_b \gamma_{ab} X_b. \tag{21.11}$$

Then we can assert that the coefficients γ_{ab} (the *kinetic coefficients*) are symmetrical with respect to the suffixes a and b:†

$$\gamma_{ab} = \gamma_{ba}. \tag{21.12}$$

In order to make practical use of this principle, it is necessary to choose the quantities x_a (or their derivatives $\dot{x}_a$) in some manner, and then to determine the X_a. This can usually be done very simply by means of the formula for the rate of change of the total entropy of the body

$$\frac{\mathrm{d}\mathscr{S}}{\mathrm{d}t} = -\int \sum_a X_a \frac{\partial x_a}{\partial t} \mathrm{d}V, \tag{21.13}$$

where the integration is extended over the whole volume of the body.

When a current flows in a conductor, $\mathrm{d}\mathscr{S}/\mathrm{d}t$ is given by (21.7). Comparing this with (21.13), we see that, if the components of the current density vector $\mathbf{j}$ are taken as the quantities $\dot{x}_a$, then the quantities X_a will be the components of the vector $-\mathbf{E}/T$. A comparison of formulae (21.8) and (21.11) shows that the kinetic coefficients in this case are the components of the conductivity tensor, multiplied by T. Thus the symmetry of this tensor follows immediately from the general relation (21.12).

PROBLEMS

PROBLEM 1. A system of electrodes maintained at constant potentials ϕ_a is immersed in a conducting medium. A current J_a flows from each electrode. Determine the total amount of Joule heat evolved in the medium per unit time.

SOLUTION. The required amount of heat Q is given by the integral

$$Q = \int \mathbf{j}\cdot\mathbf{E}\,\mathrm{d}V = -\int \mathbf{j}\cdot\mathbf{grad}\,\phi\,\mathrm{d}V = -\int \mathrm{div}\,(\phi\mathbf{j})\,\mathrm{d}V,$$

taken over the volume of the medium. We transform this into a surface integral, using the fact that $j_n = 0$ at the outer boundary of the medium, while on the surfaces of the electrodes $\phi = \text{constant} \equiv \phi_a$. The result is $Q = \Sigma\phi_a J_a$.

PROBLEM 2. Determine the potential distribution in a conducting sphere with a current J entering at a point O and leaving at the point O' diametrically opposite to O.

SOLUTION. Near O and O' (Fig. 15) the potential must be of the forms $\phi = J/2\pi\sigma R_1$ and $\phi = -J/2\pi\sigma R_2$ respectively, R_1 and R_2 being the distances from O and O'. These functions satisfy Laplace's equation, and the

† It is assumed that x_a and x_b behave in the same way under time reversal.

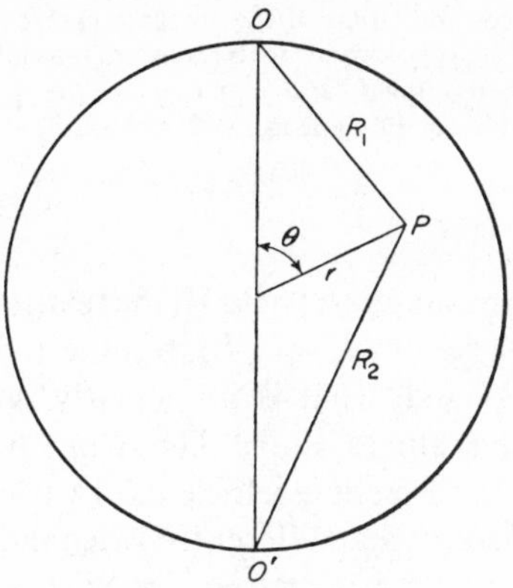

FIG. 15

integrals $-\sigma\int \mathbf{grad}\,\phi\cdot d\mathbf{f}$ over infinitesimal caps about O and O' are equal to $\pm J$. We seek the potential at an arbitrary point P in the sphere in the form

$$\phi = \frac{J}{2\pi\sigma}\left\{\frac{1}{R_1} - \frac{1}{R_2} + \psi\right\},$$

where ψ is a solution of Laplace's equation having no poles in or on the sphere. It is evident from symmetry that ψ, like ϕ, is a function of the spherical polar coordinates r and θ only.

On the surface of the sphere ($r = a$) we must have $\partial\phi/\partial r = 0$. Differentiating, we find the boundary condition on ψ:

$$\frac{\partial\psi}{\partial r} = \frac{1}{2a}\left(\frac{1}{R_1} - \frac{1}{R_2}\right) \text{ for } r = a.$$

If $f(r, \theta)$ is any solution of Laplace's equation, then the function

$$\int_0^r \frac{f(r,\theta)}{r}\,dr$$

is also a solution.† Comparing this with the above boundary condition, we see that the condition is met by the solution

$$\psi = \tfrac{1}{2}\int_0^r \left(\frac{1}{R_1} - \frac{1}{R_2}\right)\frac{dr}{r}.$$

Substituting $R_{1,2} = \sqrt{(a^2 + r^2 \mp 2ar\cos\theta)}$ and effecting the integration, we have finally

$$\phi = \frac{J}{2\pi\sigma}\left\{\frac{1}{R_1} - \frac{1}{R_2} + \frac{1}{2a}\left(\sinh^{-1}\frac{a + r\cos\theta}{r\sin\theta} - \sinh^{-1}\frac{a - r\cos\theta}{r\sin\theta}\right)\right\},$$

if $\phi = 0$ when $r = 0$.

PROBLEM 3. Show that the current distribution in a conductor is such that the energy dissipated is a minimum.

† This is easily seen either by direct calculation or from the fact that any solution $f(r, \theta)$ of Laplace's equation depending only on r and θ can be written $f = \Sigma c_n r^n P_n(\cos\theta)$, where the c_n are constants and the P_n are Legendre polynomials.

SOLUTION. The minimum concerned is that of the integral $\int \mathbf{j}\cdot\mathbf{E}\,dV = \int (j^2/\sigma)\,dV$, with the subsidiary condition $\operatorname{div}\mathbf{j} = 0$ (conservation of charge). Varying with respect to $\mathbf{j}$ the integral $\int [(j^2/\sigma) - 2\phi \operatorname{div}\mathbf{j}]\,dV$, where 2ϕ is an undetermined Lagrange multiplier, and equating the result to zero, we obtain the equation $\mathbf{j} = -\sigma\,\mathbf{grad}\,\phi$ or $\mathbf{curl}\,(\mathbf{j}/\sigma) = 0$, which is the same as (21.2) and (21.3).

§22. The Hall effect

If a conductor is in an external magnetic field **H**, the relation between the current density and the electric field is again given by $j_i = \sigma_{ik}E_k$, but the components of the conductivity tensor σ_{ik} are functions of **H** and, what is particularly important, they are no longer symmetrical with respect to the suffixes i and k. The symmetry of this tensor was proved in §21 from the symmetry of the kinetic coefficients. In a magnetic field, however, this principle must be formulated somewhat differently: when the suffixes are interchanged, the direction of the magnetic field must be reversed (see *SP* 1, §120). Hence we now have for the components $\sigma_{ik}(\mathbf{H})$ the relations

$$\sigma_{ik}(\mathbf{H}) = \sigma_{ki}(-\mathbf{H}). \tag{22.1}$$

The quantities $\sigma_{ik}(\mathbf{H})$ and $\sigma_{ki}(\mathbf{H})$ are not equal.

Like any tensor of rank two, σ_{ik} can be divided into symmetrical and antisymmetrical parts, which we denote by s_{ik} and a_{ik}:

$$\sigma_{ik} = s_{ik} + a_{ik}. \tag{22.2}$$

By definition

$$s_{ik}(\mathbf{H}) = s_{ki}(\mathbf{H}), \quad a_{ik}(\mathbf{H}) = -a_{ki}(\mathbf{H}), \tag{22.3}$$

and from (22.1) it follows that

$$\left.\begin{aligned} s_{ik}(\mathbf{H}) &= s_{ki}(-\mathbf{H}), = s_{ik}(-\mathbf{H}), \\ a_{ik}(\mathbf{H}) &= a_{ki}(-\mathbf{H}) = -a_{ik}(-\mathbf{H}). \end{aligned}\right\} \tag{22.4}$$

Thus the components of the tensor s_{ik} are even functions of the magnetic field, and those of a_{ik} are odd functions.

Any antisymmetrical tensor a_{ik} of rank two corresponds to some axial vector, whose components are

$$a_x = a_{yz}, \qquad a_y = -a_{xz}, \qquad a_z = a_{xy}. \tag{22.5}$$

In terms of this vector, the components of the product $a_{ik}E_k$ can be written as those of the vector product $\mathbf{E}\times\mathbf{a}$:

$$j_i = \sigma_{ik}E_k = s_{ik}E_k + (\mathbf{E}\times\mathbf{a})_i, \tag{22.6}$$

The Joule heat generated by the passage of the current is given by the product $\mathbf{j}\cdot\mathbf{E}$. Since the vectors $\mathbf{E}\times\mathbf{a}$ and $\mathbf{E}$ are perpendicular, their scalar product is zero identically, and so

$$\mathbf{j}\cdot\mathbf{E} = s_{ik}E_iE_k, \tag{22.7}$$

i.e. the Joule heat is determined (for a given field **E**) only by the symmetrical part of the conductivity tensor.

If the external magnetic field is sufficiently weak, the components of the conductivity tensor may be expanded in powers of that field. Since the function $\mathbf{a}(\mathbf{H})$ is odd, the expansion of this vector will involve only odd powers. The first terms are linear in the field, i.e. they are of the form

$$a_i = \alpha_{ik} H_k. \tag{22.8}$$

The vectors **a** and **H** are both axial, and the constants α_{ik} therefore form an ordinary (polar) tensor. The expansion of the even functions $s_{ik}(\mathbf{H})$ will involve only even powers. The first term is the conductivity $\sigma^{(0)}{}_{ik}$ in the absence of the field, and the next terms are quadratic in the field:

$$s_{ik} = \sigma^{(0)}{}_{ik} + \beta_{iklm} H_l H_m. \tag{22.9}$$

The tensor β_{iklm} is symmetrical with respect to i, k and l, m.

Thus the principal effect of the magnetic field is linear in the field and is given by the term $\mathbf{E}\times\mathbf{a}$; it is called the *Hall effect*. As we see, it gives rise to a current perpendicular to the electric field, whose magnitude is proportional to the magnetic field. It should be borne in mind, however, that, for an arbitrary anisotropic medium, the Hall current is not the only current perpendicular to **E**; the current $s_{ik}E_k$ also has a component in such a direction.

The Hall effect may be differently regarded if we use the inverse formulae which express **E** in terms of the current density: $E_i = \sigma^{-1}{}_{ik} j_k$. The inverse tensor $\sigma^{-1}{}_{ik}$, like σ_{ik} itself, can be resolved into a symmetrical part ρ_{ik} and an antisymmetrical part which may be represented by an axial vector **b**:

$$E_i = \rho_{ik} j_k + (\mathbf{j}\times\mathbf{b})_i. \tag{22.10}$$

The tensor ρ_{ik} and the vector **b** have the same properties as s_{ik} and **a**. In particular, in weak magnetic fields the vector **b** is linear in the field. In formula (22.10) the Hall effect is represented by the term $\mathbf{j}\times\mathbf{b}$, i.e. by an electric field perpendicular to the current and proportional to the magnetic field and to the current **j**.

The above relations are much simplified if the conductor is isotropic. The vectors **a** and **b** must then be parallel to the magnetic field, by symmetry. The only non-zero components of the tensor ρ_{ik} are $\rho_{xx} = \rho_{yy}$ and ρ_{zz}, the field being in the z-direction. Denoting these two quantities by $\rho_\perp$ and $\rho_\parallel$ and taking the current to lie in the xz-plane, we have

$$E_x = \rho_\perp j_x, \qquad E_y = -b j_x, \qquad E_x = \rho_\parallel j_z. \tag{22.11}$$

Hence we see that, in an isotropic conductor, the Hall field is the only electric field which is perpendicular to both the current and the magnetic field.

In weak magnetic fields, the vectors **b** and **H** are related (in an isotropic body) simply by

$$\mathbf{b} = -R\mathbf{H}. \tag{22.12}$$

The constant R (called the *Hall constant*) may be either positive or negative. The form of the terms quadratic in **H** in the relation between **E** and **j**, which enter through the tensor ρ_{ik}, is easily seen from the fact that the only vectors linear in **j** and quadratic in **H** which can be constructed from **j** and **H** are $(\mathbf{j}\cdot\mathbf{H})\mathbf{H}$ and $H^2\mathbf{j}$. Hence the general form of the relation between **E** and **j** in an isotropic body, as far as the terms quadratic in **H**, is

$$\mathbf{E} = \rho^{(0)}\mathbf{j} + R\mathbf{H}\times\mathbf{j} + \beta_1 H^2\mathbf{j} + \beta_2(\mathbf{j}\cdot\mathbf{H})\mathbf{H}. \tag{22.13}$$

PROBLEM

Express the components of the inverse tensor $\sigma^{-1}{}_{ik}$ in terms of those of s_{ik} and **a**.

SOLUTION. The calculations are most simply effected by taking a system of coordinates in which the axes are the principal axes of the tensor s_{ik}; the form of the results in an arbitrary coordinate system can easily be deduced from their form in this particular case. The determinant $|\sigma|$ is

$$|\sigma| = \begin{vmatrix} s_{xx} & a_z & -a_y \\ -a_z & s_{yy} & a_x \\ a_y & -a_x & s_{zz} \end{vmatrix}$$

$$= s_{xx}s_{yy}s_{zz} + s_{xx}a_x^2 + s_{yy}a_y^2 + s_{zz}a_z^2.$$

In the general case we evidently have

$$|\sigma| = |s| + s_{ik}a_i a_k.$$

From the minors of this determinant we find the components of the inverse tensor:

$$\sigma^{-1}{}_{xx} = \rho_{xx} = (s_{yy}s_{zz} + a_x^2)/|\sigma|,$$

$$\sigma^{-1}{}_{xy} = \rho_{xy} + b_z = (a_x a_y - a_z s_{zz})/|\sigma|, \ldots.$$

The general expressions which give these results for the particular system of coordinates chosen are

$$\rho_{ik} = \{s^{-1}{}_{ik}|s| + a_i a_k\}/|\sigma|, \qquad b_i = -s_{ik}a_k/|\sigma|.$$

This completes the solution.

§23. The contact potential

In order to remove a charged particle through the surface of a conductor, work must be done. The work required for a thermodynamically reversible removal of the particle is called the *work function*. This quantity is always positive; this follows immediately from the fact that a point charge is attracted to any neutral body, and therefore to any conductor (see §14). The work in question will be denoted by eW, where e is the charge on the particle; the sign of the *work potential* W thus defined is the same as that of the charge on the particle removed.

The work function depends both on the nature of the conductor (and its thermodynamic state, i.e. its temperature and density) and on that of the charged particle. For example, the work function for a given metal is different for the removal of a conduction electron and for the removal of an ion from the surface. It must also be emphasized that the work function is characteristic of the surface of the conductor. It therefore depends, for instance, on the treatment of the surface and the "contamination" of it. If the conductor is a single crystal, then the work function is different for different faces.

To ascertain the physical nature of the dependence of the work function on the properties of the surface, let us establish its relation to the electric structure of the surface layer. If $\rho(x)$ is the charge density *not* averaged over physically infinitesimal segments of the x-axis (perpendicular to the layer), we can write Poisson's equation in the layer as $\mathrm{d}^2\phi/\mathrm{d}x^2 = -4\pi\rho$. Let the conductor occupy the region $x < 0$. Then a first integration gives

$$\frac{\mathrm{d}\phi}{\mathrm{d}x} = -4\pi \int_{-\infty}^{x} \rho\,\mathrm{d}x,$$

and a second integration (by parts) gives

$$\phi - \phi(-\infty) = -4\pi x \int_{-\infty}^{x} \rho\,\mathrm{d}x + 4\pi \int_{-\infty}^{x} x\rho\,\mathrm{d}x.$$

For $x \to \infty$, the integral

$$\int_{-\infty}^{x} \rho\,\mathrm{d}x$$

tends very rapidly to zero (since the surface of an uncharged conductor is electrically neutral). Hence

$$\phi(+\infty)-\phi(-\infty)=4\pi\int_{-\infty}^{\infty} x\rho\,dx.$$

The integral on the right is the dipole moment of the charges near the surface of the body. These charges form a "double layer", in which charges of opposite sign are separated and the dipole moment is non-zero. The structure of the double layer, of course, depends on the properties of the surface (its crystallographic direction, contamination, etc.). The difference in the work potential for different surfaces of a given conductor is determined by the difference in the dipole moments.

If two different conductors are placed in contact, an exchange of charged particles may occur between them. Charges pass from the body with the smaller work function to that with the greater until a potential difference between them is set up which prevents further movement of charge. This is called a *contact potential.*

Fig. 16 shows a cross-section of two conductors in contact (a and b) near their surfaces AO and OB. Let the potentials of these surfaces be ϕ_a and ϕ_b respectively. Then the contact potential is $\phi_{ab}=\phi_b-\phi_a$. The quantitative relation between this potential and the work functions is given by the condition of thermodynamic equilibrium. Let us consider the work which must be done on a particle with charge e to remove it from the conductor a through the surface AO, transfer it to the surface OB, and finally carry it into the conductor b. In a state of thermodynamic equilibrium, this work must be zero.† The work done on the particle in the three stages mentioned is eW_a, $e(\phi_b-\phi_a)$, and $-eW_b$ respectively. Putting the sum of these equal to zero, we find the required relation:

$$\phi_{ab}=W_b-W_a. \tag{23.1}$$

Thus the contact potential of the neighbouring free surfaces of two conductors in contact is equal to the difference in their work functions.

The existence of the contact potential results in the appearance of an electric field in the space outside the conductors. It is easy to determine this field near the line of contact of the surfaces. In a small region near this line (the point O in Fig. 16), the surfaces may be

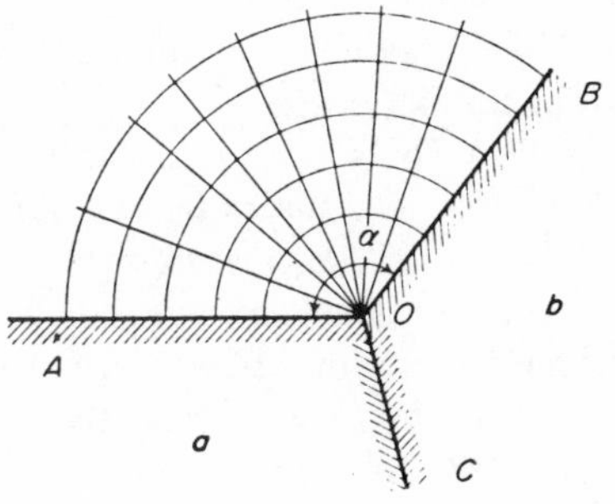

FIG. 16

† Of course, in reality a particle can pass from one conductor to another only through their surface of contact, and not through the space adjoining them, but the work done is independent of the path.

regarded as plane. The field potential outside the conductors satisfies the equation

$$\triangle\phi \equiv \frac{1}{r}\frac{\partial}{\partial r}\left(r\frac{\partial\phi}{\partial r}\right)+\frac{1}{r^2}\frac{\partial^2\phi}{\partial\theta^2}=0,$$

where r and θ are polar coordinates with origin at O; on AO and OB the potential takes given constant values. We are interested in the solution which contains the lowest power of r; this is the leading term in an expansion of the potential in powers of the small distance r. The solution concerned is $\phi = \text{constant} \times \theta$. Measuring the angle θ from AO and arbitrarily taking the potential on AO as zero, we have

$$\phi = \phi_{ab}\theta/\alpha, \tag{23.2}$$

where α is the angle AOB. Thus the equipotential lines in the plane of the diagram are straight lines diverging from O. The lines of force are arcs of circles centred at O. The field is

$$E = -\frac{1}{r}\frac{\partial\phi}{\partial\theta} = -\frac{\phi_{ab}}{\alpha}\frac{1}{r}; \tag{23.3}$$

it decreases inversely as the distance from O.

As has been said above, "contact" potentials also exist between the various faces of a single crystal of metal. Hence an electric field of the kind just described must exist near the edges of the crystal.†

If several metallic conductors (at equal temperatures) are connected together, the potential between the extreme conductors is, as we easily deduce from formula (23.1), simply the difference of their work potentials, as it is for two conductors in direct contact. In particular, if the metal at each end is the same, the contact potential between the ends is zero. This is evident, however, because if there were a potential difference between two like conductors, a current would flow when they were connected, in contradiction to the second law of thermodynamics.

§24. The galvanic cell

The statement at the end of §23 ceases to be valid if the circuit includes conductors in which the current is carried by different means (e.g. metals and solutions of electrolytes). Because the work function of a conductor is different for different charged particles (electrons and ions), the total contact potential in the circuit is not zero even when the conductors at each end are similar. This total potential difference is called the *electromotive force* or *e.m.f.* in the circuit; it is just the potential difference between the two like conductors before the circuit is closed. When the circuit is closed, a current flows in it; this is the basis of the operation of what are called *galvanic cells*. The energy which maintains the current in the circuit is supplied by chemical transformations occurring in the cell.

When we go completely round any closed circuit the field potential must, of course, return to its original value, i.e. the total change in the potential must be zero. Let us consider, for example, a contour on the surface of the conductors. When we pass from one conductor to another, the potential has a discontinuity ϕ_{ab}. The potential drop across any

† In reality, all such fields are usually compensated by the field of ions from the atmosphere which "adhere" to the surface of the crystal.

conductor is RJ, where J is the total current flowing through it and R is its resistance. Hence the total change in the potential round the circuit is $\Sigma\phi_{ab} - \Sigma JR$. Putting this equal to zero and using the facts that J is the same at every point in the circuit and $\Sigma\phi_{ab}$ is the electromotive force $\mathscr{E}$, we find

$$J\Sigma R = \mathscr{E}, \tag{24.1}$$

so that the current in a circuit containing a galvanic cell is equal to the e.m.f. divided by the total resistance of all the conductors in the circuit (including, of course, the internal resistance of the cell itself).

Although the e.m.f. of a galvanic cell can be expressed as a sum of contact potentials, it is very important to note that it is in reality a thermodynamic quantity, determined entirely by the states of the conductors and independent of the properties of the surfaces separating them. This is clear, because $\mathscr{E}$ is just the work per unit charge which must be done on a charged particle when it is carried reversibly along the closed circuit.

To illustrate this, let us consider a galvanic cell consisting of two electrodes of metals A and B immersed in solutions of electrolytes AX and BX, X^- being any anion. Let ζ_A and ζ_B be the chemical potentials of the metals A and B, and ζ_{AX} and ζ_{BX} those of the electrolytes in solution.† If an elementary charge e is carried along the closed circuit, an ion A^+ passes into solution from the electrode A and an ion B^+ passes out of solution to the electrode B, the change in the charges on the electrodes being compensated by the passage of an electron from A to B through the external circuit. The result is that the electrode A loses one neutral atom, the electrode B gains one, and in the electrolyte solution one molecule of BX is replaced by one of AX. Since the work done in a reversible process (at constant temperature and pressure) is equal to the change in the thermodynamic potential of the system, we have

$$e\mathscr{E}_{AB} = (\zeta_B - \zeta_{BX}) - (\zeta_A - \zeta_{AX}), \tag{24.2}$$

which expresses the e.m.f. of the cell in terms of the properties of the material of the electrodes and of the electrolyte solution.

From (24.2) we can also draw the following conclusion. If the solution contains three electrolytes AX, BX, CX and three metallic electrodes A, B, C, then the e.m.f.s between each pair of them are related by

$$\mathscr{E}_{AB} + \mathscr{E}_{BC} = \mathscr{E}_{AC}. \tag{24.3}$$

Using the general formulae of thermodynamics, we can relate the e.m.f. of a galvanic cell to the heat evolved when a current flows, which of course is actually an irreversible phenomenon. Let Q be the amount of heat generated (both in the cell itself and in the external circuit) when the unit charge passes along the circuit; Q is just the heat of the reaction which occurs in the cell when a current flows. By a well-known formula of thermodynamics (see *SP* 1, §91), it is related to the work $\mathscr{E}$ by

$$Q = -T^2 \frac{\partial}{\partial T}\left(\frac{\mathscr{E}}{T}\right). \tag{24.4}$$

The definition of the partial derivative with respect to temperature depends on the conditions under which the process occurs. For example, if the current flows at constant pressure (as usually happens), then the differentiation is effected at constant pressure.

---

† In this section we use the ordinary chemical potentials, i.e. those defined with respect to one particle.

§25. Electrocapillarity

The presence of charges on the boundary between two conducting media affects the surface tension there. This phenomenon is called *electrocapillarity*. In practice, the media concerned are both liquids; usually one is a liquid metal (mercury) and the other is a solution of an electrolyte.

Let ϕ_1, ϕ_2 be the potentials of the two conductors, and e_1, e_2 the charges at the surface of separation. These charges are equal in magnitude and opposite in sign, and thus form a double layer on the surface.

The differential of the potential $\tilde{\mathcal{F}}$ of a system of two conductors at given temperature and pressure is, taking into account the surface of separation,

$$d\tilde{\mathcal{F}} = \alpha dS - e_1 d\phi_1 - e_2 d\phi_2, \tag{25.1}$$

where the term αdS is the work done in a reversible change dS in the area S of the surface of separation; α is the surface-tension coefficient (see *SP* 1, §154).

The thermodynamic potential $\tilde{\mathcal{F}}$ in (25.1) may be replaced by its "surface part" $\tilde{\mathcal{F}}_s$, since the volume part is constant for given temperature and pressure, and is therefore of no interest here. Putting $e_1 = -e_2 \equiv e$ and the potential difference $\phi_1 - \phi_2 = \phi$, we can write (25.1) as

$$d\tilde{\mathcal{F}}_s = \alpha dS - e\, d\phi. \tag{25.2}$$

Hence

$$(\partial\tilde{\mathcal{F}}_s/\partial S)_\phi = \alpha, \tag{25.3}$$

α being expressed as a function of ϕ. Integrating, we find that $\tilde{\mathcal{F}}_s = \alpha S$. Substitution in (25.2) gives $d(\alpha S) = \alpha dS - e d\phi$, or $S d\alpha = -e d\phi$, whence

$$\sigma = -(\partial\alpha/\partial\phi)_{P,T}, \tag{25.4}$$

where $\sigma = e/S$ is the charge per unit area of the surface. The relation (25.4), derived by G. Lippmann and J. W. Gibbs, is the fundamental formula in the theory of electrocapillarity.

In a state of equilibrium, the thermodynamic potential $\tilde{\mathcal{F}}$ must be a minimum for given values of the electric potentials on the conductors. Regarding it as a function of the surface charges e, we can write the necessary conditions for a minimum as

$$\partial\tilde{\mathcal{F}}_s/\partial e = 0, \qquad \partial^2\tilde{\mathcal{F}}_s/\partial e^2 > 0, \tag{25.5}$$

where the derivatives are taken at constant area S. To calculate these, we express $\tilde{\mathcal{F}}_s$ in terms of the thermodynamic potential $\mathcal{F}_s = \mathcal{F}_s(e)$:

$$\tilde{\mathcal{F}}_s = \mathcal{F}_s(e) - e_1\phi_1 - e_2\phi_2 = \mathcal{F}_s(e) - e\phi. \tag{25.6}$$

The vanishing of the first derivative gives

$$\frac{\partial\tilde{\mathcal{F}}_s}{\partial e} = \frac{\partial\mathcal{F}_s}{\partial e} - \phi = 0,$$

and then the condition for the second derivative to be positive becomes

$$\frac{\partial^2\tilde{\mathcal{F}}_s}{\partial e^2} = \frac{\partial^2\mathcal{F}_s}{\partial e^2} = \frac{\partial\phi}{\partial e} = \frac{1}{S}\frac{\partial\phi}{\partial\sigma} > 0,$$

or

$$\partial\sigma/\partial\phi > 0. \tag{25.7}$$

This result was to be expected, since the double layer on the surface may be regarded as a capacitor with capacitance $\partial e/\partial \phi$.

Differentiating equation (25.4) with respect to ϕ and using (25.7), we find that

$$\partial^2 \alpha/\partial \phi^2 < 0. \tag{25.8}$$

This means that the point where $\partial \alpha/\partial \phi = -\sigma = 0$ is a maximum of α as a function of ϕ.

§26. Thermoelectric phenomena

The condition that there should be no current in a metal is that there is thermodynamic equilibrium with respect to the conduction electrons. This means not only that the temperature must be constant throughout the body, but also that the sum $e\phi + \zeta_0$ should be constant, where ζ_0 is the chemical potential of the conduction electrons in the metal (for $\phi = 0$).† If the metal is not homogeneous, ζ_0 is not constant throughout the body even if the temperature is constant. Hence the constancy of the electric potential ϕ in this case does not mean the absence of a current in the metal, although the field $\mathbf{E} = -\mathbf{grad}\,\phi$ is zero. This makes the ordinary definition of ϕ (as the average of the true potential) inconvenient, if we wish to take inhomogeneous conductors into consideration.

It is natural to redefine the potential as $\phi + \zeta_0/e$, and we shall write this henceforward as ϕ simply.‡ In a homogeneous metal, the change amounts to the adding of an unimportant constant to the potential. Accordingly, the "field" $\mathbf{E} = -\mathbf{grad}\,\phi$ (which we shall use henceforward) is the same as the true mean field only in a homogeneous metal, and in general the two differ by the gradient of some function of the state.§

With this definition, the current and field are both zero in a state of thermodynamic equilibrium with respect to the conduction electrons, and the relation between them is $\mathbf{j} = \sigma\mathbf{E}$ (or $j_i = \sigma_{ik} E_k$) even if the metal is not homogeneous.

Let us now consider a non-uniformly heated metal, which cannot be in thermodynamic equilibrium (with respect to the electrons). Then the field $\mathbf{E}$ is not zero even if the current is zero. In general, when both the current density $\mathbf{j}$ and the temperature gradient $\mathbf{grad}\,T$ are not zero, the relation between these quantities and the field can be written

$$\mathbf{E} = \mathbf{j}/\sigma + \alpha\,\mathbf{grad}\,T. \tag{26.1}$$

Here σ is the ordinary conductivity, and α is another quantity which is an electrical characteristic of the metal. Here we suppose for simplicity that the substance is isotropic (or of cubic symmetry), and therefore write the proportionality coefficients as scalars. The linear relation between $\mathbf{E}$ and $\mathbf{grad}\,T$ is, of course, merely the first term of an expansion, but it is sufficient in view of the smallness of the temperature gradients occurring in practice.

The same formula (26.1), in the form

$$\mathbf{j} = \sigma(\mathbf{E} - \alpha\,\mathbf{grad}\,T), \tag{26.2}$$

---

† See *SP* 1, §25. Here we take ζ to be the chemical potential defined in the usual manner, viz. per unit particle (electron).

‡ This definition can also be formulated as follows: the new $e\phi$ is the change in the free energy when one electron is isothermally brought into the metal. In other words, $\phi = \partial F/\partial \rho$, where F is the free energy of the metal and ρ the charge on the conduction electrons per unit volume.

§ It must be emphasized that $e\mathbf{E}$ is then not the force on the charge e. Consequently, this definition of $\mathbf{E}$, which is suitable in a phenomenological theory, may be inconvenient in the microscopic theory when calculating the kinetic coefficients (cf. *PK*, §44).

shows that a current can flow in a non-uniformly heated metal even if the field $\mathbf{E}$ is zero.

As well as the electric current density $\mathbf{j}$, we can consider the energy flux density $\mathbf{q}$. First of all, this quantity contains an amount $\phi\mathbf{j}$ resulting simply from the fact that each charged particle (electron) carries with it an energy $e\phi$. The difference $\mathbf{q}-\phi\mathbf{j}$, however, does not depend on the potential, and can be generally written as a linear function of the gradients $\mathbf{grad}\,\phi = -\mathbf{E}$ and $\mathbf{grad}\,T$, similarly to formula (26.2) for the current density. We shall for the present write this as

$$\mathbf{q}-\phi\mathbf{j} = \beta\mathbf{E}-\gamma\,\mathbf{grad}\,T.$$

The symmetry of the kinetic coefficients gives a relation between the coefficient β and the coefficient α in (26.2). To derive this, we calculate the rate of change of the total entropy of the conductor. The amount of heat evolved per unit time and volume is $-\operatorname{div}\mathbf{q}$. Hence we can put

$$\frac{\mathrm{d}\mathscr{S}}{\mathrm{d}t} = -\int\frac{\operatorname{div}\mathbf{q}}{T}\,\mathrm{d}V.$$

Using the equation $\operatorname{div}\mathbf{j} = 0$, we have

$$\frac{\operatorname{div}\mathbf{q}}{T} = \frac{1}{T}\{\operatorname{div}(\mathbf{q}-\phi\mathbf{j})+\operatorname{div}\phi\mathbf{j}\} = \frac{1}{T}\operatorname{div}(\mathbf{q}-\phi\mathbf{j})-\frac{\mathbf{E}\cdot\mathbf{j}}{T}.$$

The first term is integrated by parts, giving

$$\frac{\mathrm{d}\mathscr{S}}{\mathrm{d}t} = \int\frac{\mathbf{E}\cdot\mathbf{j}}{T}\,\mathrm{d}V-\int\frac{(\mathbf{q}-\phi\mathbf{j})\cdot\mathbf{grad}\,T}{T^2}\,\mathrm{d}V. \tag{26.3}$$

This formula shows that, if we take as the quantities $\partial x_a/\partial t$ (see §21) the components of the vectors $\mathbf{j}$ and $\mathbf{q}-\phi\mathbf{j}$, then the corresponding quantities X_a are the components of the vectors $-\mathbf{E}/T$ and $\mathbf{grad}\,T/T^2$. Accordingly in the relations

$$\mathbf{j} = \sigma T\frac{\mathbf{E}}{T}-\sigma\alpha T^2\frac{\mathbf{grad}\,T}{T^2},$$

$$\mathbf{q}-\phi\mathbf{j} = \beta T\frac{\mathbf{E}}{T}-\gamma T^2\frac{\mathbf{grad}\,T}{T^2},$$

the coefficients $\sigma\alpha T^2$ and βT must be equal. Thus $\beta = \sigma\alpha T$, so that $\mathbf{q}-\phi\mathbf{j} = \sigma\alpha T\mathbf{E} -\gamma\,\mathbf{grad}\,T$. Finally, expressing $\mathbf{E}$ in terms of $\mathbf{j}$ and $\mathbf{grad}\,T$ by (26.1), we have the result

$$\mathbf{q} = (\phi+\alpha T)\mathbf{j}-\kappa\,\mathbf{grad}\,T, \tag{26.4}$$

where $\kappa = \gamma - T\alpha^2\sigma$ is simply the ordinary thermal conductivity, which gives the heat flux in the absence of an electric current.

It should be pointed out that the condition that $\mathrm{d}\mathscr{S}/\mathrm{d}t$ should be positive places no new restriction on the thermoelectric coefficients. Substituting (26.1) and (26.4) in (26.3), we obtain

$$\frac{\mathrm{d}\mathscr{S}}{\mathrm{d}t} = \int\left(\frac{j^2}{\sigma T}+\frac{\kappa\,(\mathbf{grad}\,T)^2}{T^2}\right)\mathrm{d}V > 0, \tag{26.5}$$

whence we find only that the coefficients of thermal and electrical conductivity must be positive.

In the above formulae it was tacitly assumed that an inhomogeneity of pressure (or density) at constant temperature cannot cause a field (or current) to appear in the conductor, and consequently no term in $\mathbf{grad}\, p$ was included in (26.2) or (26.4). The existence of such terms would, in fact, contradict the law of the increase of entropy: the integrand in (26.5) would then contain terms in the products $\mathbf{j}\cdot\mathbf{grad}\, p$ and $\mathbf{grad}\, T\cdot\mathbf{grad}\, p$, which could be of either sign, and so the integral could not be necessarily positive.

The relations (26.1) and (26.4) indicate various thermoelectric effects. Let us consider the amount of heat $-\operatorname{div}\mathbf{q}$ evolved per unit time and volume in the conductor. Taking the divergence of (26.4), we have

$$\begin{aligned} Q &= -\operatorname{div}\mathbf{q} \\ &= \operatorname{div}(\kappa\,\mathbf{grad}\, T) + \mathbf{E}\cdot\mathbf{j} + \mathbf{j}\cdot\mathbf{grad}\,(\alpha T), \end{aligned}$$

or, substituting (26.1),

$$Q = \operatorname{div}(\kappa\,\mathbf{grad}\, T) + \frac{j^2}{\sigma} - T\mathbf{j}\cdot\mathbf{grad}\,\alpha. \tag{26.6}$$

The first term on the right pertains to ordinary thermal conduction, and the second term, proportional to the square of the current, is the Joule heat. The term of interest here is the third, which gives the thermoelectric effects.

Let us assume the conductor to be homogeneous. Then the change in α is due only to the temperature gradient, and $\mathbf{grad}\,\alpha = (\mathrm{d}\alpha/\mathrm{d}T)\,\mathbf{grad}\, T$; if, as usually happens, the pressure is constant through the body, $\mathrm{d}\alpha/\mathrm{d}T$ must be taken as $(\partial\alpha/\partial T)_p$. Thus the amount of heat evolved (called the *Thomson effect*) is

$$\rho\mathbf{j}\cdot\mathbf{grad}\, T, \quad \text{where } \rho = -T\,\mathrm{d}\alpha/\mathrm{d}T. \tag{26.7}$$

The coefficient ρ is called the *Thomson coefficient*. It should be noticed that this effect is proportional to the first power of the current, and not to the second power like the Joule heat. It therefore changes sign when the current is reversed. The coefficient ρ may be either positive or negative. If $\rho > 0$, the Thomson heat is positive (i.e. heat is emitted) when the current flows in the direction of increasing temperature, and heat is absorbed when it flows in the opposite direction; if $\rho < 0$ the reverse is true.

Another thermal effect, called the *Peltier effect*, occurs when a current passes through a junction of two different metals. At the surface of contact, the temperature, the potential and the normal components of the current density and energy flux density are all continuous. Denoting by the suffixes 1 and 2 the values of quantities for the two metals and equating the normal components of $\mathbf{q}$ (26.4) on the two sides, we have, since ϕ, T and j_x are continuous,

$$[-\kappa\partial T/\partial x]_1^2 = -j_x T(\alpha_2 - \alpha_1),$$

the x-axis being taken along the normal to the surface. If the positive direction of this axis is from metal 1 to metal 2, then the expression on the left-hand side of this equation is the amount of heat taken from the surface per unit time and area by thermal conduction. This heat loss is balanced by the evolution at the junction of an amount of heat given by the right-hand side of the equation. Thus the amount of heat generated per unit time and area is

$$j\Pi_{12}, \quad \text{where } \Pi_{12} = -T(\alpha_2 - \alpha_1). \tag{26.8}$$

The quantity Π_{12} is called the *Peltier coefficient*. Like the Thomson effect, the Peltier effect is proportional to the first power of the current, and changes sign when the direction of the current is reversed. The Peltier coefficient is additive: $\Pi_{13} = \Pi_{12} + \Pi_{23}$, where the suffixes 1, 2, 3 refer to three different metals.

A comparison of formulae (26.7) and (26.8) shows that the Thomson and Peltier coefficients are related by

$$\rho_2 - \rho_1 = T \frac{d}{dT}\left(\frac{\Pi_{12}}{T}\right). \tag{26.9}$$

Next, let us consider an open circuit containing two junctions, the two end conductors being of the same metal (1 in Fig. 17). We suppose that the junctions b and c are at different temperatures T_1 and T_2, while the temperature at each end (a and d) is the same. Then there is a potential difference called a *thermoelectromotive force*, which we denote by $\mathscr{E}_T$, between the ends.

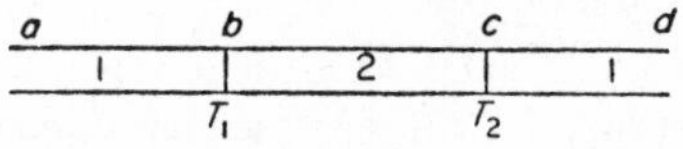

FIG. 17

To calculate this force, we put $\mathbf{j} = 0$ in (26.1) and integrate the field $\mathbf{E} = \alpha\, \mathbf{grad}\, T$ along the circuit (taken to be the x-axis):

$$\mathscr{E}_T = \int_a^d \alpha \frac{dT}{dx} dx = \int_a^d \alpha\, dT.$$

The integrations from a to b and from c to d are over temperatures from T_2 to T_1 in metal 1, and that from b to c is over temperatures from T_1 to T_2 in metal 2. Thus

$$\mathscr{E}_T = \int_{T_1}^{T_2} (\alpha_2 - \alpha_1)\, dT. \tag{26.10}$$

Comparing this with (26.8), we see that the thermo-e.m.f. is related to the Peltier coefficient by

$$\mathscr{E}_T = -\int_{T_1}^{T_2} \frac{\Pi_{12}}{T} dT \tag{26.11}$$

Formulae (26.9) and (26.11) are called *Thomson's relations* (W. Thomson, 1854).

To conclude this section, we shall give the formulae for the current and heat flux in an anisotropic conductor. These are derived from the symmetry of the kinetic coefficients in the same way as formulae (26.1) and (26.4), and the results are

$$\left.\begin{aligned} E_i &= \sigma^{-1}{}_{ik} j_k + \alpha_{ik} \partial T/\partial x_k, \\ q_i - \phi j_i &= T\alpha_{ki} j_k - \kappa_{ik} \partial T/\partial x_k. \end{aligned}\right\} \tag{26.12}$$

Here $\sigma^{-1}{}_{ik}$ is the tensor inverse to the conductivity tensor σ_{ik}, and the tensors σ_{ik} and κ_{ik} are symmetrical. The thermoelectric tensor α_{ik}, however, is in general not symmetrical.

§27. Thermogalvanomagnetic phenomena

There is a still greater variety of phenomena which occur when a current flows in the simultaneous presence of an electric field, a magnetic field, and a temperature gradient.

The discussion is entirely similar to that given in §26 for thermoelectric effects. It will be given here in tensor form, so as to be applicable to both isotropic and anisotropic conductors. We write the electric current density **j** and the heat flux **q** as

$$\left.\begin{aligned} j_i &= a_{ik}\frac{E_k}{T} + b_{ik}\frac{\partial}{\partial x_k}\left(\frac{1}{T}\right), \\ q_i - \phi j_i &= c_{ik}\frac{E_k}{T} + d_{ik}\frac{\partial}{\partial x_k}\left(\frac{1}{T}\right), \end{aligned}\right\} \tag{27.1}$$

where all the coefficients are functions of the magnetic field. The symmetry of the kinetic coefficients gives

$$\left.\begin{aligned} a_{ik}(\mathbf{H}) = a_{ki}(-\mathbf{H}), \qquad d_{ik}(\mathbf{H}) = d_{ki}(-\mathbf{H}), \\ b_{ik}(\mathbf{H}) = c_{ki}(-\mathbf{H}). \end{aligned}\right\} \tag{27.2}$$

Expressing **E** and $\mathbf{q}-\phi\mathbf{j}$ in terms of **j** and $\mathbf{grad}\,T$ from (27.1), we have

$$\left.\begin{aligned} E_i &= \sigma^{-1}{}_{ik} j_k + \alpha_{ik}\partial T/\partial x_k, \\ q_i - \phi j_i &= \beta_{ik} j_k - \kappa_{ik}\partial T/\partial x_k, \end{aligned}\right\} \tag{27.3}$$

where the tensors σ^{-1}, α, β, κ are certain functions of the tensors a, b, c, d, and have the following symmetry properties resulting from (27.2):

$$\left.\begin{aligned} \sigma^{-1}{}_{ik}(\mathbf{H}) = \sigma^{-1}{}_{ki}(-\mathbf{H}), \\ \kappa_{ik}(\mathbf{H}) = \kappa_{ki}(-\mathbf{H}), \qquad \beta_{ik}(\mathbf{H}) = T\alpha_{ki}(-\mathbf{H}). \end{aligned}\right\} \tag{27.4}$$

These are the required relations in their most general form. They generalize those found in §26 for the case where there is no magnetic field and in §22 for the case where there is no temperature gradient. It must be emphasized that in an anisotropic conductor the tensors α_{ik} and β_{ik} are in general not symmetrical even when there is no magnetic field.

The tensors σ^{-1}, κ, and $\beta + T\alpha$ can be resolved into symmetrical and antisymmetrical parts (cf. §22). In a weak magnetic field, the symmetrical parts may be regarded as constants independent of **H**, while the antisymmetric parts are linear in **H**. For an isotropic conductor we have, to this accuracy,

$$\mathbf{E} = \mathbf{j}/\sigma + \alpha\,\mathbf{grad}\,T + R\mathbf{H}\times\mathbf{j} + N\mathbf{H}\times\mathbf{grad}\,T, \tag{27.5}$$

$$\mathbf{q} - \phi\mathbf{j} = \alpha T\mathbf{j} - \kappa\,\mathbf{grad}\,T + NT\mathbf{H}\times\mathbf{j} + L\mathbf{H}\times\mathbf{grad}\,T. \tag{27.6}$$

Here σ and κ are the ordinary coefficients of electrical and thermal conductivity, α is the thermoelectric coefficient which appears in (26.1), R is the Hall coefficient, and N and L are new coefficients. The term $N\mathbf{H}\times\mathbf{grad}\,T$ may be regarded as representing the effect of the magnetic field on the thermo-e.m.f. (called the *Nernst effect*), and the term $L\mathbf{H}\times\mathbf{grad}\,T$ as representing the effect of this field on the thermal conduction (called the *Leduc–Righi effect*).

At a boundary between media, the normal components of the vectors **j** and **q** are continuous, and therefore so is that of the vector $-\kappa\,\mathbf{grad}\,T+\alpha T\mathbf{j}+NT\mathbf{H}\times\mathbf{j}+L\mathbf{H}\times\mathbf{grad}\,T$. The term $NT\mathbf{H}\times\mathbf{j}$ gives the influence of the magnetic field on the Peltier effect (called the *Ettingshausen effect*).

The amount of heat evolved in the conductor per unit time and volume is $Q = -\operatorname{div}\mathbf{q}$. Here we must substitute $\mathbf{q}$ from (27.6) and replace $-\mathbf{grad}\,\phi = \mathbf{E}$ in accordance with (27.5). If the conductor is homogeneous, then the quantities α, N, L, etc. are functions of temperature alone, and so their gradients are proportional to $\mathbf{grad}\,T$. In the calculation we neglect all quantities of the second order in $\mathbf{H}$, and to the same approximation we can take $\mathbf{curl}\,(\mathbf{j}/\sigma) \cong \mathbf{curl}\,\mathbf{E} = 0$. We also note that the external field $\mathbf{H}$ (arising from sources outside the conductor under consideration) is such that $\mathbf{curl}\,\mathbf{H} = 0$.† Finally, $\operatorname{div}\mathbf{j} = 0$, as for any steady current. The result is

$$Q = \frac{j^2}{\sigma} + \operatorname{div}(\kappa\,\mathbf{grad}\,T) - T\mathbf{j}\cdot\mathbf{grad}\,\alpha + \frac{1}{\sigma T}\frac{\mathrm{d}}{\mathrm{d}T}(\sigma N T^2)(\mathbf{j}\times\mathbf{H})\cdot\mathbf{grad}\,T.$$

The third term here gives the Thomson effect (26.7), and the last term gives the change in the Thomson effect resulting from the presence of the magnetic field.

§28. Diffusion phenomena

The presence of diffusion causes certain phenomena in electrolyte solutions which do not occur in solid conductors. We shall assume, for simplicity, that the temperature is the same everywhere in the solution, and so consider only pure diffusion phenomena, uncomplicated by thermoelectric effects.

Instead of the pressure P and the concentration c, it is more convenient to take as independent variables the pressure and the chemical potential ζ. We here define ζ as the derivative of the thermodynamic potential of unit mass of the solution with respect to its concentration c (at constant P and T); by the concentration we mean the ratio of the mass of electrolyte in a volume element to the total mass of fluid in the same volume.‡ It may be recalled that the constancy of the chemical potential is (like that of the pressure and the temperature) one of the conditions of thermodynamic equilibrium.

The definition of the electric field potential given in §26 has to be somewhat modified in this case, since the current is now carried by the ions of the dissolved electrolyte, and not by the conduction electrons. A suitable definition is (cf. the third footnote to §26) $\phi = (\partial\Phi/\partial\rho)_c$, where Φ is the thermodynamic potential and ρ the sum of the ion charges in unit volume of the solution (after differentiating we put $\rho = 0$, of course, because the

† This neglects the weak effect on the evolution of heat resulting from the magnetic fields of the currents themselves.

‡ The chemical potentials are usually defined as $\zeta_1 = \partial\Phi/\partial n_1$, $\zeta_2 = \partial\Phi/\partial n_2$, where Φ is the thermodynamic potential of any mass of the solution, and n_1, n_2 the numbers of particles of solute and solvent in that mass of solution. If Φ is the thermodynamic potential per unit mass, then the numbers n_1 and n_2 are related by $n_1 m_1 + n_2 m_2 = 1$ (where m_1, m_2 are the masses of the two kinds of particle), and the concentration $c = n_1 m_1$. Hence we have

$$\zeta = \frac{\partial\Phi}{\partial c} = \frac{\partial\Phi}{\partial n_1}\frac{\partial n_1}{\partial c} + \frac{\partial\Phi}{\partial n_2}\frac{\partial n_2}{\partial c} = \frac{\zeta_1}{m_1} - \frac{\zeta_2}{m_2},$$

where ζ is the chemical potential as here defined.

solution is electrically neutral). The derivative is taken at constant mass concentration, i.e. at a given sum of the masses of ions of both signs in unit volume.†

When a gradient of the chemical potential is present, a term proportional to it is added to the expression for the current density:

$$\mathbf{j} = \sigma(\mathbf{E} - \beta\,\mathbf{grad}\,\zeta), \tag{28.1}$$

in analogy with the added term in (26.2). We shall see below that, for a given gradient of the chemical potential (and of the temperature), $\mathbf{j}$ must be independent of the pressure gradient, and so no term in $\mathbf{grad}\,P$ appears in (28.1).‡

As well as the electric current, we have to consider the transport of the mass of the electrolyte which takes place at the same time. It must be borne in mind that the passage of a current through the solution may be accompanied by a macroscopic motion of the fluid. The mass flux density of the electrolyte resulting from this motion is $\rho c\mathbf{v}$, where $\mathbf{v}$ is the velocity and ρ the density of the solution. The electrolyte is also transported by molecular diffusion. We denote the diffusion flux density by $\mathbf{i}$, so that the total flux density is $\rho c\mathbf{v} + \mathbf{i}$. The irreversible processes of diffusion cause a further increase in entropy; the rate of change of the total entropy is§

$$\frac{d\mathscr{S}}{dt} = \int \frac{\mathbf{E}\cdot\mathbf{j}}{T}\,dV - \int \frac{\mathbf{i}\cdot\mathbf{grad}\,\zeta}{T}\,dV. \tag{28.2}$$

Like the electric current density, the diffusion flux may be written as a linear combination of $\mathbf{E}$ and $\mathbf{grad}\,\zeta$, or of $\mathbf{j}$ and $\mathbf{grad}\,\zeta$. Using the symmetry of the kinetic coefficients, we can relate one of the coefficients in this combination to the coefficient β in (28.1), in exactly the same way as we did for $\mathbf{j}$ and $\mathbf{q} - \phi\mathbf{j}$ in §26. The result is

$$\mathbf{i} = -\frac{\rho D}{(\partial\zeta/\partial c)_{P,T}}\,\mathbf{grad}\,\zeta + \beta\mathbf{j}. \tag{28.3}$$

The coefficient of $\mathbf{grad}\,\zeta$ is here expressed in terms of the ordinary diffusion coefficient, ρ being the density of matter. For $\mathbf{j} = 0$ and constant pressure and temperature we have $\mathbf{i} = -\rho D\,\mathbf{grad}\,c$.

The inadmissibility in (28.1) and (28.3) of terms proportional to the pressure gradient follows, as in §26, from the law of the increase of entropy: such terms would make the derivative of the total entropy (28.2) a quantity of variable sign.

Formulae (28.1) and (28.3) give all the diffusion phenomena in electrolytes, but we shall not pause here to examine them more closely.

PROBLEM

Two parallel plates of a metal A are immersed in a solution of an electrolyte AX. Find the current density as a function of the potential difference applied between the plates.

---

† In a strong electrolyte, the solute is completely dissociated, and so the mass concentration may be written as $c = m_+n_+ + m_-n_-$, where m_+ and m_- are the cation and anion masses, n_+ and n_- their number densities. With the above definition of the potential, $\phi = 0$ corresponds to $\zeta_+/m_+ = \zeta_-/m_-$ for the cation and anion chemical potentials, which are also related by $\zeta_+ + \zeta_- = \zeta_1$.

‡ It should be emphasized, however, that, for a given concentration gradient $\mathbf{j}$ does depend on the pressure gradient:

$$\mathbf{grad}\,\zeta = (\partial\zeta/\partial c)_{P,T}\,\mathbf{grad}\,c + (\partial\zeta/\partial P)_{c,T}\,\mathbf{grad}\,P.$$

§ The derivation of the second term is given in *FM*, §57.

SOLUTION. When the current passes, metal is dissolved from one plate and deposited on the other. The solvent (water) remains at rest, and a mass flux of metal of density† $\rho v = jm/e$ occurs in the solution, where j is the electric current density, and m and e are the mass and charge of an ion A^+. This flux is also given by $i + \rho vc$, where i is as shown in (28.3); assuming the pressure constant throughout the liquid,‡ we have

$$\rho D \frac{dc}{dx} = \left[\beta - \frac{m}{e}(1-c)\right] j, \tag{1}$$

where x is the coordinate in the direction of a line joining the electrodes. Since j = constant in the solution, this gives

$$jl = \int_{c_1}^{c_2} \frac{\rho D\,dc}{\beta - m(1-c)/e}, \tag{2}$$

where c_1, c_2 are the concentrations at the surfaces of the plates, and l is the distance between them.

The potential difference $\mathscr{E}$ between the plates is most simply found from the total amount of energy Q dissipated per unit time and unit area of the plates, which must equal $j\mathscr{E}$. By (28.1), (28.2) we have

$$Q = T\frac{d\mathscr{S}}{dt} = \int \left\{ \frac{j^2}{\sigma} + \rho D \frac{\partial \zeta}{\partial c}\left(\frac{dc}{dx}\right)^2 \right\} dx = j\mathscr{E},$$

and therefore, using (1),

$$\mathscr{E} = \int_{c_1}^{c_2} \frac{\rho D\,dc}{\sigma(\beta - m(1-c)/e)} + \int_{c_1}^{c_2} \frac{\partial \zeta}{\partial c}\left[\beta - \frac{m}{e}(1-c)\right] dc. \tag{3}$$

Formulae (2) and (3) implicitly solve the problem.

If the current j is small, the concentration difference $c_2 - c_1$ is also small. Replacing the integrals by $c_2 - c_1$ times the integrands, we find the effective specific resistance of the solution:

$$\frac{\mathscr{E}}{lj} = \frac{1}{\sigma} + \frac{1}{\rho D}\frac{\partial \zeta}{\partial c}\left[\beta - \frac{m}{e}(1-c)\right]^2.$$

The first term in (3) gives the potential drop ($\int (j/\sigma)\,dx$) due to the passage of the current. The second term is the e.m.f. due to the concentration gradient in the solution (in a certain sense analogous to the thermo-e.m.f.). This latter expression is independent of the conditions of the particular one-dimensional problem considered, and is the general expression for the e.m.f. of a "concentration cell".

† It may be recalled that the hydrodynamic velocity $\mathbf{v}$ in a solution is defined so that $\rho\mathbf{v}$ is the momentum of unit volume of the liquid; see *FM*, §57. Hence the fact that in this case only the dissolved metal is moving (relative to the electrodes) does not affect the calculation of $\rho\mathbf{v}$.

‡ The change in pressure due to the motion of the liquid gives only terms of a higher order of smallness.

CHAPTER IV

STATIC MAGNETIC FIELD

§29. Static magnetic field

A STATIC magnetic field in matter satisfies two of Maxwell's equations, obtained by averaging the microscopic equations

$$\operatorname{div} \mathbf{h} = 0, \qquad \mathbf{curl}\, \mathbf{h} = \frac{1}{c}\frac{\partial \mathbf{e}}{\partial t} + \frac{4\pi}{c}\rho \mathbf{v}. \tag{29.1}$$

The mean magnetic field is usually called the *magnetic induction* and denoted by **B**:

$$\overline{\mathbf{h}} = \mathbf{B}. \tag{29.2}$$

Hence the result of averaging the first equation (29.1) is

$$\operatorname{div} \mathbf{B} = 0. \tag{29.3}$$

In the second equation, the time derivative gives zero on averaging, since the mean field is supposed constant, and so we have

$$\mathbf{curl}\, \mathbf{B} = (4\pi/c)\overline{\rho \mathbf{v}}. \tag{29.4}$$

The mean value of the microscopic current density is in general not zero in either conductors or dielectrics. The only difference between these two classes is that in dielectrics we always have

$$\int \overline{\rho \mathbf{v}} \cdot d\mathbf{f} = 0, \tag{29.5}$$

where the integral is taken over the area of any cross-section of the body; in conductors, this integral need not be zero. Let us suppose to begin with that there is no net current in the body if it is a conductor, i.e. that (29.5) holds.

The vanishing of the integral in (29.5) for every cross-section of the body means that the vector $\overline{\rho \mathbf{v}}$ can be written as the curl of another vector, usually denoted by $c\mathbf{M}$:

$$\overline{\rho \mathbf{v}} = c\, \mathbf{curl}\, \mathbf{M}, \tag{29.6}$$

where **M** is zero outside the body; compare the similar discussion in §6. For, integrating over a surface bounded by a curve which encloses the body and nowhere enters it, we have $\int \overline{\rho \mathbf{v}} \cdot d\mathbf{f} = c \int \mathbf{curl}\, \mathbf{M} \cdot d\mathbf{f} = c \oint \mathbf{M} \cdot d\mathbf{l} = 0$. The vector **M** is called the *magnetization* of the body. Substituting it in (29.4), we find

$$\mathbf{curl}\, \mathbf{H} = 0, \tag{29.7}$$

where the vector **H** and the magnetic induction **B** are related by

$$\mathbf{B} = \mathbf{H} + 4\pi \mathbf{M}, \tag{29.8}$$

which is analogous to the relation between the electric field **E** and induction **D**. Although **H** is, by analogy with **E**, usually called the *magnetic field*, it must be remembered that the true mean field is really **B** and not **H**.

To see the physical significance of the quantity **M**, let us consider the total magnetic moment due to all the charged particles moving in the body. By the definition of the magnetic moment (see *Fields*, §44), this is†

$$\int \mathbf{r} \times \overline{\rho \mathbf{v}}\, dV/2c = \tfrac{1}{2}\int \mathbf{r} \times \mathbf{curl}\, \mathbf{M}\, dV.$$

Since $\rho\mathbf{v} \equiv 0$ outside the body, the integral can be taken over any volume which includes the body. We transform the integral as follows:

$$\int \mathbf{r} \times \mathbf{curl}\, \mathbf{M}\, dV = -\oint \mathbf{r} \times (\mathbf{M} \times d\mathbf{f}) - \int (\mathbf{M} \times \mathbf{grad}) \times \mathbf{r}\, dV.$$

The integral over the surface outside the body is zero. In the second term we have $(\mathbf{M} \times \mathbf{grad}) \times \mathbf{r} = -\mathbf{M}\, \mathrm{div}\, \mathbf{r} + \mathbf{M} = -2\mathbf{M}$. Thus we obtain

$$\frac{1}{2c}\int \mathbf{r} \times \overline{\rho \mathbf{v}}\, dV = \int \mathbf{M}\, dV. \tag{29.9}$$

We see that the magnetization vector is the magnetic moment per unit volume.‡

The equations (29.3) and (29.7) must be supplemented by a relation between **H** and **B** in order to complete the system of equations. For example, in non-ferromagnetic bodies in fairly weak magnetic fields, **B** and **H** are linearly related. In isotropic bodies, this linear relation becomes a simple proportionality:

$$\mathbf{B} = \mu \mathbf{H}. \tag{29.10}$$

The coefficient μ is called the *magnetic permeability*. We also have $\mathbf{M} = \chi \mathbf{H}$, where the coefficient

$$\chi = (\mu - 1)/4\pi \tag{29.11}$$

is called the *magnetic susceptibility*.

Unlike the permittivity ε, which always exceeds unity, the magnetic permeability may be either greater or less than unity. (It is, however, always positive, as we shall prove in §31. The reason for the differing behaviour of μ and ε is discussed in §32.) The magnetic susceptibility χ may correspondingly be either positive or negative.

Another, quantitative, difference is that the magnetic susceptibility of the great majority of bodies is very small in comparison with the dielectric susceptibility. This difference arises because the magnetization of a (non-ferromagnetic) body is a relativistic effect, of order v^2/c^2, where v is the velocity of the electrons in the atoms.§

In anisotropic bodies (crystals), the simple proportionality (29.10) is replaced by the linear relations

$$B_i = \mu_{ik} H_k. \tag{29.12}$$

---

† For clarity, it should be emphasized that **r** here is a variable coordinate of integration, not the position vector of a microscopic particle; it therefore does not come under the averaging sign.

‡ The quantity **M** is completely determined only when this relation is established. The relation (29.6) inside the body, and $\mathbf{M} = 0$ outside it, do not uniquely define **M**: the gradient of any scalar could be added to **M** inside the body without affecting (29.6) (cf. the similar remark in the first footnote to §6).

§ The ratio v/c appears with **H** in the Hamiltonian of the interaction of the body with the magnetic field, and again in the magnetic moments of the atoms or molecules.

The magnetic permeability tensor μ_{ik} is symmetrical; this follows from the thermodynamic relations to be derived in §31, in exactly the same way as for ε_{ik} (§13).

From the equations $\operatorname{div} \mathbf{B} = 0$, $\mathbf{curl}\,\mathbf{H} = 0$ it follows (cf. §6) that at a boundary between two different media we must have

$$B_{1n} = B_{2n}, \qquad \mathbf{H}_{1t} = \mathbf{H}_{2t}. \tag{29.13}$$

This system of equations and boundary conditions is formally identical with those for the electrostatic field in a dielectric in the absence of free charges, differing only in that **E** and **D** are replaced by **H** and **B** respectively. Since $\mathbf{curl}\,\mathbf{H} = 0$, we can put $\mathbf{H} = -\mathbf{grad}\,\psi$; the equations for the potential ψ are the same as those for the electrostatic potential. Thus the solutions of the various problems of electrostatics discussed in Chapter II can be immediately applied to problems with a static magnetic field. In particular, the formulae derived in §8 for a dielectric ellipsoid in a uniform electric field hold also, with appropriate substitutions, for a magnetic ellipsoid in a uniform magnetic field. For example, the magnetic field $\mathbf{H}^{(i)}$ and induction $\mathbf{B}^{(i)}$ inside the ellipsoid are related to the external field $\mathfrak{H}$ by

$$H_i^{(i)} + n_{ik}(B_k^{(i)} - H_k^{(i)}) = \mathfrak{H}_i, \tag{29.14}$$

where n_{ik} is the demagnetizing factor tensor. This relation is valid, whatever the relation between **B** and **H**.

The tangential component of the magnetic induction, unlike its normal component, is discontinuous at a surface separating two media. The magnitude of the discontinuity can be related to the current density on the surface. To do this, we integrate both sides of equation (29.4) over a small interval Δl crossing the surface along the normal. We then let Δl tend to zero; the integral $\int \overline{\rho \mathbf{v}}\, dl$ may tend to some finite limit. The quantity

$$\mathbf{g} = \int \overline{\rho \mathbf{v}}\, dl \tag{29.15}$$

may be called the *surface current density*; it gives the charge passing per unit time across unit length of a line in the surface. We take the direction of **g** at a given point on the surface as the y-axis, and the direction of the normal from medium 1 to medium 2 as the x-axis. Then the integration of equation (29.4) gives

$$\int \left(\frac{\partial B_x}{\partial z} - \frac{\partial B_z}{\partial x} \right) dx = \frac{4\pi}{c} g_y = \frac{4\pi}{c} g.$$

Since B_x is continuous, the derivative $\partial B_x/\partial z$ is finite, and so its integral tends to zero with Δl. The integral of $\partial B_z/\partial x$ gives the difference in the values of B_z on the two sides of the surface. Thus $B_{2z} - B_{1z} = -4\pi g/c$. This can be written in vector form:

$$4\pi \mathbf{g}/c = \mathbf{n} \times (\mathbf{B}_2 - \mathbf{B}_1) = 4\pi \mathbf{n} \times (\mathbf{M}_2 - \mathbf{M}_1), \tag{29.16}$$

where **n** is a unit vector along the normal into region 2; the last member of (29.16) is obtained by using the continuity of the tangential component of **H**.

§30. The magnetic field of a steady current

If a conductor carries a non-zero total current, the mean current density in it can be written as $\overline{\rho \mathbf{v}} = c\,\mathbf{curl}\,\mathbf{M} + \mathbf{j}$. The first term, resulting from the magnetization of the medium, makes no contribution to the total current, so that the net charge transfer through a cross-section of the body is given by the integral $\int \mathbf{j}\cdot d\mathbf{f}$ of the second term. The quantity **j** is called the *conduction current density*.† The statements made in §21 apply to this current; in particular, the energy dissipated per unit time and volume is $\mathbf{E}\cdot\mathbf{j}$.

The distribution of the current **j** over the volume of the conductor is given by the equations of §21, which do not involve the magnetic field due to **j** itself, if we neglect the effect of this field on the conductivity of the body. Hence the magnetic field of the currents must be determined for a given current distribution. The equations satisfied by this field differ from those in §29 by the presence of a term $4\pi\mathbf{j}/c$ on the right-hand side of (29.7):

$$\operatorname{div}\mathbf{B} = 0, \tag{30.1}$$

$$\mathbf{curl}\,\mathbf{H} = 4\pi\mathbf{j}/c. \tag{30.2}$$

The conduction current density **j**, which is proportional to the electric field, does not become infinite, and in particular is finite on a surface separating two media. Hence the term on the right of (30.2) does not affect the boundary condition that the tangential component of **H** be continuous.

To solve equations (30.1), (30.2), it is convenient to use the *vector potential* **A**, defined by

$$\mathbf{B} = \mathbf{curl}\,\mathbf{A}, \tag{30.3}$$

so that equation (30.1) is satisfied identically. Equation (30.3) does not uniquely define the vector potential, to which the gradient of any scalar may be added without affecting (30.3). For this reason we can impose on **A** a further condition, which we take to be

$$\operatorname{div}\mathbf{A} = 0. \tag{30.4}$$

The equation for **A** is obtained by substituting (30.3) in (30.2). If the linear relation $\mathbf{B} = \mu\mathbf{H}$ holds we have

$$\mathbf{curl}\left(\frac{1}{\mu}\,\mathbf{curl}\,\mathbf{A}\right) = 4\pi\mathbf{j}/c. \tag{30.5}$$

In this form the equation is valid for any medium, homogeneous or not.

In a homogeneous medium, μ = constant, and since

$$\mathbf{curl}\,\mathbf{curl}\,\mathbf{A} = \mathbf{grad}\operatorname{div}\mathbf{A} - \triangle\mathbf{A} = -\triangle\mathbf{A}$$

we find from (30.5)

$$\triangle\mathbf{A} = -4\pi\mu\mathbf{j}/c. \tag{30.6}$$

If we have two or more adjoining media with different magnetic permeability μ, the general equation (30.5) has the form (30.6) in each homogeneous medium, while at the interfaces the tangential component of the vector $(1/\mu)$ **curl A** must be continuous. Moreover, the tangential component of **A** itself must be continuous, since a discontinuity would mean that the induction **B** was infinite at the boundary.

The field equations are simpler in the two-dimensional problem of finding the magnetic field in a medium infinite and homogeneous in one direction (which we take as the z-direction), the currents which produce the field being everywhere in that direction, with the

† The quantity c **curl M** is sometimes called the *molecular current density*. This name, however, is not in complete accordance with the actual physical picture of motion of charges in a conductor. For example, in a metal the conduction electrons, as well as those moving in the atoms, contribute to the magnetization.

current density $j_z = j$ depending only on x and y. We make the plausible assumption (to be confirmed by the result) that the vector potential of such a field is also in the z-direction: $A_z = A(x, y)$. The condition (30.4) is then satisfied identically; the magnetic field is everywhere parallel to the xy-plane. We denote by $\mathbf{k}$ a unit vector in the z-direction; then

$$\mathbf{curl\,A} = \mathbf{curl}\,A\mathbf{k} = \mathbf{grad}\,A \times \mathbf{k},$$

$$\mathbf{curl}\left(\frac{1}{\mu}\,\mathbf{curl\,A}\right) = \mathbf{curl}\left(\frac{\mathbf{grad}\,A}{\mu}\times\mathbf{k}\right) = -\mathbf{k}\,\mathrm{div}\frac{\mathbf{grad}\,A}{\mu}.$$

Hence equation (30.5) becomes

$$\mathrm{div}\frac{\mathbf{grad}\,A}{\mu} = -\frac{4\pi}{c}j(x, y), \tag{30.7}$$

i.e. we in fact obtain one equation for the one scalar quantity $A(x, y)$. For a piecewise homogeneous medium, (30.7) becomes

$$\triangle A = -4\pi\mu j(x, y)/c, \tag{30.8}$$

with the boundary condition that A and $(1/\mu)\partial A/\partial n$ be continuous at an interface.†

The magnetic field is easily found if the current distribution is symmetrical about the z-axis: $j_z = j(r)$ (where r is the distance from that axis). In this case the lines of magnetic force are evidently the circles r = constant. The magnitude of the field is found at once from the formula

$$\oint \mathbf{H}\cdot d\mathbf{l} = \frac{4\pi}{c}\int \mathbf{j}\cdot d\mathbf{f}, \tag{30.9}$$

which is the integral form of (30.2). Thus

$$H(r) = 2J(r)/cr, \tag{30.10}$$

where $J(r)$ is the total current within the radius r.

The reduction of the vector equation (30.5) to a single scalar equation is possible also if the current distribution is axially symmetrical and has in cylindrical polar coordinates r, ϕ, z the form $j_r = j_z = 0$, $j_\phi = j(r, z)$. We seek the vector potential in the form $A_r = A_z = 0$, $A_\phi = A(r, z)$. The components of the magnetic induction $\mathbf{B} = \mathbf{curl\,A}$ are $B_r = -\partial A/\partial z$, $B_z = (1/r)\partial(rA)/\partial r$, $B_\phi = 0$, and the ϕ-component of equation (30.2) gives

$$\frac{\partial}{\partial z}\left(\frac{1}{\mu}\frac{\partial A}{\partial z}\right) + \frac{\partial}{\partial r}\left\{\frac{1}{\mu r}\frac{\partial}{\partial r}(rA)\right\} = -\frac{4\pi}{c}j(r, z). \tag{30.11}$$

The equations for the magnetic field of the currents can be solved in a general form in the important case where the magnetic properties of the medium may be neglected, i.e. where we can put $\mu = 1$. The vector potential then satisfies in all space the equation

† It should be noticed that the two-dimensional problem with a static magnetic field is equivalent to the two-dimensional electrostatic problem of determining the electric field due to extraneous charges with density $\rho_{ex}(x, y)$ in a dielectric medium. The equation to be solved in the latter problem is $\mathrm{div}\,(\varepsilon\,\mathbf{grad}\,\phi) = -4\pi\rho_{ex}$, where ϕ is the field potential; this differs from (30.7) only in that A, j/c and μ are replaced by ϕ, ρ_{ex} and $1/\varepsilon$ respectively. The boundary conditions on A and ϕ are the same. A difference occurs, however, on passing to $\mathbf{E}$ and $\mathbf{B}$ from ϕ and $\mathbf{A}$ respectively. The vectors $\mathbf{E} = -\mathbf{grad}\,\phi$ and $\mathbf{B} = \mathbf{curl\,A}$ are the same in magnitude but in perpendicular directions at any given point.

$\triangle \mathbf{A} = -4\pi \mathbf{j}/c$ with no conditions at the interfaces between different media (including the surface of the conductor on which the current flows). The solution of this equation which vanishes at infinity is

$$\mathbf{A} = \frac{1}{c}\int \frac{\mathbf{j}}{R}\,\mathrm{d}V, \tag{30.12}$$

where R is the distance from the volume element $\mathrm{d}V$ to the point at which $\mathbf{A}$ is to be calculated (see *Fields*, §43). In taking the curl of this equation, we must remember that the integrand $\mathbf{j}/R$ is to be differentiated with respect to the coordinates of this point, of which $\mathbf{j}$ is independent, so that

$$\mathbf{curl}\,(\mathbf{j}/R) = \mathbf{grad}\,(1/R)\times\mathbf{j} = -\mathbf{R}\times\mathbf{j}/R^3,$$

where the radius vector $\mathbf{R}$ is from $\mathrm{d}V$ to the point under consideration. Thus

$$\mathbf{B} = \mathbf{H} = \frac{1}{c}\int \frac{\mathbf{j}\times\mathbf{R}}{R^3}\,\mathrm{d}V. \tag{30.13}$$

If the conductor on which the current flows is sufficiently thin (a thin wire), and if we are interested only in the field in the surrounding space, the thickness of the wire may be neglected. In what follows we shall often discuss such *linear currents*. The integration over the volume of the conductor is then replaced by an integration along its length: the formulae for linear currents are obtained from those for volume currents by making the substitution $\mathbf{j}\,\mathrm{d}V \to J\,\mathrm{d}\mathbf{l}$, where J is the total current in the conductor. For example, from formulae (30.12) and (30.13) we have

$$\mathbf{A} = \frac{J}{c}\oint \frac{\mathrm{d}\mathbf{l}}{R}, \qquad \mathbf{H} = \frac{J}{c}\oint \frac{\mathrm{d}\mathbf{l}\times\mathbf{R}}{R^3}. \tag{30.14}$$

The latter formula is *Biot and Savart's law*.

This simple formula for the magnetic field of a linear current does not depend on the assumption that $\mu = 1$. Since we neglect the thickness of the conductor, no boundary conditions at its surface need be applied, and the magnetic properties of the conducting material are of no importance (it may even be ferromagnetic). The solution of equation (30.6) for the field in the medium surrounding the conductor is therefore

$$\mathbf{A} = \frac{\mu J}{c}\int \frac{\mathrm{d}\mathbf{l}}{R}, \quad \mathbf{B} = \frac{\mu J}{c}\int \frac{\mathrm{d}\mathbf{l}\times\mathbf{R}}{R^3}, \tag{30.15}$$

whatever the magnetic susceptibility of that medium. Thus the presence of the medium simply changes the magnetic induction by a factor μ. The field $\mathbf{H} = \mathbf{B}/\mu$ is unchanged.

The problem of determining the magnetic field of linear currents can also be solved as a problem of potential theory. Since we neglect the volume of conductors, we are in fact determining the field in a region containing no currents except along certain line singularities. In the absence of currents, a static magnetic field has a scalar potential, which in a homogeneous medium satisfies Laplace's equation. There is, however, an important difference between the magnetic field potential and the electrostatic potential: the latter is always a one-valued function, because $\mathbf{curl}\,\mathbf{E} = 0$ in all space (including charged regions) and so the change in the potential in going round any closed contour (i.e. the circulation of $\mathbf{E}$ round that contour) is zero. The circulation of the magnetic field round a contour enclosing a linear current is not zero, but $4\pi J/c$. Hence the potential changes by this amount on each passage round a contour enclosing a linear current, i.e. it is a many-valued function.

If the currents lie in a finite region of space (and $\mu = 1$ everywhere), the vector potential of the magnetic field at a great distance from the conductors is

$$\mathbf{A} = \mathscr{M}\times\mathbf{R}/R^3, \tag{30.16}$$

where

$$\mathscr{M} = \int \mathbf{r}\times\mathbf{j}\,\mathrm{d}V/2c \tag{30.17}$$

is the total magnetic moment of the system.†

For a linear current, this becomes

$$\mathscr{M} = J\oint \mathbf{r}\times\mathrm{d}\mathbf{l}/2c,$$

and can be transformed into an integral over a surface bounded by the line of the current. The product $\mathrm{d}\mathbf{f} = \frac{1}{2}\mathbf{r}\times\mathrm{d}\mathbf{l}$ is equal in magnitude to the area of the triangular surface element formed by the vectors $\mathbf{r}$ and $\mathrm{d}\mathbf{l}$. The vector $\int \mathrm{d}\mathbf{f}$ is independent of the particular surface (bounded by the current) over which it is taken. Thus the magnetic moment of a closed linear current is

$$\mathscr{M} = J\oint \mathrm{d}\mathbf{f}/c. \tag{30.18}$$

In particular, for a plane closed linear current the magnetic moment is simply JS/c, where S is the area of the plane enclosed by the current.

To conclude this section, we may briefly discuss the energy flux in a conductor. The energy dissipated as Joule heat in the conductor is derived from the energy of the electromagnetic field. In a steady state, the equation of continuity which expresses the law of conservation of energy is

$$-\operatorname{div}\mathbf{S} = \mathbf{j}\cdot\mathbf{E}, \tag{30.19}$$

where $\mathbf{S}$ is the energy flux density, given in a conductor by

$$\mathbf{S} = c\mathbf{E}\times\mathbf{H}/4\pi, \tag{30.20}$$

which is formally the same as the expression for the Poynting vector for the field in a vacuum. This is easily verified directly by calculating div $\mathbf{S}$ from the equations $\mathbf{curl}\,\mathbf{E} = 0$ and (30.2), when we obtain (30.19).

Formula (30.20) also follows independently from the obvious condition that the normal component of $\mathbf{S}$ must be continuous at the surface of a conductor, if we use the continuity of $\mathbf{E}_t$ and $\mathbf{H}_t$ and the validity of (30.20) in the vacuum outside the body.

PROBLEMS‡

PROBLEM 1. Determine the scalar potential of the magnetic field of a closed linear current.

SOLUTION. We transform the curvilinear integral into one over a surface bounded by the curve, obtaining

$$\mathbf{A} = \frac{J}{c}\oint\frac{\mathrm{d}\mathbf{l}}{R} = \frac{J}{c}\int \mathrm{d}\mathbf{f}\times\mathbf{grad}\frac{1}{R},$$

$$\mathbf{B} = \mathbf{curl}\,\mathbf{A} = -\frac{J}{c}\int(\mathrm{d}\mathbf{f}\cdot\mathbf{grad})\,\mathbf{grad}\frac{1}{R}$$

† See *Fields*, §44. In the derivation there given, we use explicitly the idea of a current as the result of the motion of individual charged particles. Such a derivation is, of course, quite general, but formula (30.16) can also be obtained by macroscopic arguments (see Problem 4).

‡ In Problems 1–4, $\mu = 1$.

(where we have used the fact that $\triangle(1/R) = 0$). Since $\mathbf{B} = -\mathbf{grad}\,\psi$, we have for the scalar potential

$$\psi = \frac{J}{c}\int \mathrm{d}\mathbf{f}\cdot\mathbf{grad}\frac{1}{R} = -\frac{J}{c}\int\frac{\mathrm{d}\mathbf{f}\cdot\mathbf{R}}{R^3}.$$

The integral is, geometrically, the solid angle Ω subtended by the closed contour at the point considered. The above-mentioned many-valuedness of the potential is seen from the fact that, as this point describes a closed path embracing the wire, the angle Ω changes suddenly from 2π to -2π.

PROBLEM 2. Find the magnetic field of a linear current flowing in a circle with radius a.

SOLUTION. We take the origin of cylindrical polar coordinates r, ϕ, z at the centre of the circle, with the angle ϕ measured from the plane which passes through the z-axis and the point at which the field is calculated. The vector potential has only one component, $A_\phi = A(r, z)$, and by formula (30.14) we have

$$A_\phi = \frac{J}{c}\oint\frac{\cos\phi\,\mathrm{d}l}{R}$$

$$= \frac{2J}{c}\int_0^\pi \frac{a\cos\phi\,\mathrm{d}\phi}{\sqrt{(a^2+r^2+z^2-2ar\cos\phi)}}.$$

Putting $\theta = \frac{1}{2}(\phi-\pi)$, we find

$$A_\phi = \frac{4J}{ck}\sqrt{\frac{a}{r}}\,[(1-\tfrac{1}{2}k^2)K - E],$$

where $k^2 = 4ar/[(a+r)^2+z^2]$, and K and E are complete elliptic integrals of the first and second kinds:

$$K = \int_0^{\frac{1}{2}\pi}\frac{\mathrm{d}\theta}{\sqrt{(1-k^2\sin^2\theta)}},\qquad E = \int_0^{\frac{1}{2}\pi}\sqrt{(1-k^2\sin^2\theta)}\,\mathrm{d}\theta.$$

The components of the induction are

$$B_\phi = 0,$$

$$B_r = -\frac{\partial A_\phi}{\partial z} = \frac{J}{c}\frac{2z}{r\sqrt{[(a+r)^2+z^2]}}\left[-K + \frac{a^2+r^2+z^2}{(a-r)^2+z^2}E\right],$$

$$B_z = \frac{1}{r}\frac{\partial}{\partial r}(rA_\phi) = \frac{J}{c}\frac{2}{\sqrt{[(a+r)^2+z^2]}}\left[K + \frac{a^2-r^2-z^2}{(a-r)^2+z^2}E\right].$$

Here we have used the easily verified formulae

$$\frac{\partial K}{\partial k} = \frac{E}{k(1-k^2)} - \frac{K}{k},\qquad \frac{\partial E}{\partial k} = \frac{E-K}{k}.$$

On the axis ($r = 0$) we have $B_r = 0$, $B_z = 2\pi a^2 J/c(a^2+z^2)^{3/2}$, as can also be found by a straightforward calculation.

PROBLEM 3. Determine the magnetic field in a cylindrical hole in a cylindrical conductor of infinite length carrying a current uniformly distributed over its cross-section (Fig. 18).

SOLUTION. If there were no hole, the field in the cylinder would be given by $H'_x = -2\pi jy/c$, $H'_y = 2\pi jx/c$. The dimensions and axes are as shown in Fig. 18. If a current with density $-j$ were to flow in the inner cylinder, it would produce a field $H''_x = 2\pi jy'/c$, $H''_y = -2\pi jx'/c$. The required field in the hole is obtained by superposing these two fields. Since $x - x' = OO' = h$, and $y = y'$, we have $H_x = 0$, $H_y = 2\pi jh/c = 2hJ/(b^2-a^2)c$, i.e. a uniform field in the y-direction.

PROBLEM 4. Derive from (30.12) the formula (30.16) for the vector potential of the field far from the currents.

SOLUTION. We write $\mathbf{R} = \mathbf{R}_0 - \mathbf{r}$, where $\mathbf{R}_0$ and $\mathbf{r}$ are the radius vectors from the origin (situated somewhere among the currents) to the point considered and to the volume element $\mathrm{d}V$ respectively. Expanding the integrand

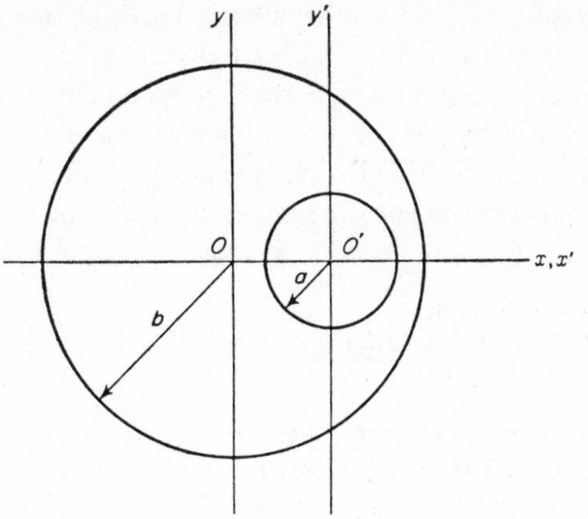

FIG. 18

in powers of **r** and using the fact that $\int \mathbf{j}\,dV \equiv 0$, we have $A_i \cong (R_k/cR^3)\int x_k j_i\,dV$. The suffix 0 to R is omitted. Integrating by parts the identity $\int x_i x_k \operatorname{div}\mathbf{j}\,dV = 0$ gives $\int (j_i x_k + j_k x_i)\,dV = 0$. Hence we can write

$$A_i = (R_k/2cR^3)\int (x_k j_i - x_i j_k)\,dV,$$

which agrees with (30.16).

PROBLEM 5. Determine the magnetic field produced by a linear current in a magnetically anisotropic medium (A. S. Viglin, 1954).

SOLUTION. In the anisotropic medium surrounding the conductor we have

$$\operatorname{div}\mathbf{B} = \mu_{ik}\partial H_k/\partial x_i = 0, \tag{1}$$

where μ_{ik} is the magnetic permeability tensor of the medium. Instead of introducing the vector potential by $\mathbf{B} = \mathbf{curl}\,\mathbf{A}$, we use another vector $\mathbf{C}$ defined by

$$H_i = e_{ikl}\mu_{km}\partial C_l/\partial x_m, \tag{2}$$

where e_{ikl} is the antisymmetrical unit tensor. Then equation (1) is again satisfied identically. We can also impose on the vector **C** thus defined the condition

$$\operatorname{div}\mathbf{C} \equiv \partial C_l/\partial x_l = 0. \tag{3}$$

Substituting (2) in $\mathbf{curl}\,\mathbf{H} = 4\pi\mathbf{j}/c$, we obtain $e_{ikl}\,\partial H_l/\partial x_k = -\mu_{kp}\partial^2 C_i/\partial x_k\partial x_p = 4\pi j_i/c$ (using the condition (3) and the fact that $e_{ikl}e_{lmn} = \delta_{im}\delta_{kn} - \delta_{in}\delta_{km}$). The equation thus obtained for **C** is the same in form as that for the electric field potential resulting from charges in an anisotropic medium (§13, Problem 2). The solution is

$$\mathbf{C} = \frac{1}{c}\int \frac{\mathbf{j}\,dV}{\sqrt{(|\mu|\mu^{-1}{}_{ik}R_i R_k)}},$$

where $|\mu|$ is the determinant of the tensor μ_{ik}, and **R** the radius vector from the point considered to dV. For a linear current we have

$$\mathbf{C} = \frac{J}{c\sqrt{|\mu|}}\oint \frac{d\mathbf{l}}{\sqrt{(\mu^{-1}{}_{ik}R_i R_k)}}.$$

§31. Thermodynamic relations in a magnetic field

The thermodynamic relations for a magnetic substance in a magnetic field are, as we shall see, very similar to the corresponding relations for a dielectric in an electric field. Their derivation, however, is quite different from that given in §10. This difference is ultimately due to the fact that a magnetic field, unlike an electric field, does no work on charges moving in it (since the force acting on a charge is perpendicular to its velocity). Hence, to calculate the change in the energy of the medium when a magnetic field is applied, we must examine the electric fields induced by the change in the magnetic field and determine the work done by these fields on the currents which produce the magnetic field.

Thus the equation which relates electric and variable magnetic fields must be used. This equation is

$$\mathbf{curl}\,\mathbf{E} = -\frac{1}{c}\frac{\partial \mathbf{B}}{\partial t}, \tag{31.1}$$

it follows immediately on averaging the microscopic equation (1.3).

During a time δt, the field $\mathbf{E}$ does work $\delta t \int \mathbf{j}\cdot\mathbf{E}\,\mathrm{d}V$ on the currents $\mathbf{j}$. This quantity with the opposite sign is the work δR "done on the field" by the external e.m.f. which maintains the currents. Substituting

$$\mathbf{j} = c\,\mathbf{curl}\,\mathbf{H}/4\pi,$$

we have

$$\begin{aligned}\delta R &= -\delta t \frac{c}{4\pi}\int \mathbf{E}\cdot\mathbf{curl}\,\mathbf{H}\,\mathrm{d}V \\ &= \delta t\frac{c}{4\pi}\int \operatorname{div}(\mathbf{E}\times\mathbf{H})\,\mathrm{d}V - \delta t\frac{c}{4\pi}\int \mathbf{H}\cdot\mathbf{curl}\,\mathbf{E}\,\mathrm{d}V.\end{aligned}$$

The first integral, on being transformed to an integral over an infinitely distant surface, is seen to be zero. In the second integral we substitute $\mathbf{curl}\,\mathbf{E}$ from (31.1) and put $\delta\mathbf{B} = \delta t\,\partial\mathbf{B}/\partial t$ for the change in the magnetic induction, obtaining finally

$$\delta R = \int \mathbf{H}\cdot\delta\mathbf{B}\,\mathrm{d}V/4\pi. \tag{31.2}$$

This formula appears entirely analogous to the expression (10.2) for the work done in an infinitesimal change in the electric field. It must be pointed out, however, that the physical analogy between the two formulae is actually not complete, since $\mathbf{H}$, unlike $\mathbf{E}$, is not the mean value of the microscopic field.

Having derived formula (31.2), we can write down all the thermodynamic relations for a magnetic substance in a magnetic field by analogy with those given in §10 for a dielectric in an electric field, simply replacing $\mathbf{E}$ and $\mathbf{D}$ by $\mathbf{H}$ and $\mathbf{B}$ respectively. We shall give some of these formulae here for purposes of reference. The differentials of the total free energy and the total internal energy are

$$\left.\begin{aligned}\delta\mathscr{F} &= -\mathscr{S}\delta T + \int \mathbf{H}\cdot\delta\mathbf{B}\,\mathrm{d}V/4\pi, \\ \delta\mathscr{U} &= T\delta\mathscr{S} + \int \mathbf{H}\cdot\delta\mathbf{B}\,\mathrm{d}V/4\pi,\end{aligned}\right\} \tag{31.3}$$

and those of the corresponding quantities per unit volume are

$$\left.\begin{aligned}\mathrm{d}F &= -S\,\mathrm{d}T + \zeta\,\mathrm{d}\rho + \mathbf{H}\cdot\mathrm{d}\mathbf{B}/4\pi, \\ \mathrm{d}U &= T\,\mathrm{d}S + \zeta\,\mathrm{d}\rho + \mathbf{H}\cdot\mathrm{d}\mathbf{B}/4\pi.\end{aligned}\right\} \tag{31.4}$$

We need also the thermodynamic potentials

$$\tilde{U} = U - \mathbf{H}\cdot\mathbf{B}/4\pi, \qquad \tilde{F} = F - \mathbf{H}\cdot\mathbf{B}/4\pi, \tag{31.5}$$

for which

$$\left.\begin{aligned} d\tilde{F} &= -S\,dT + \zeta\,d\rho - \mathbf{B}\cdot d\mathbf{H}/4\pi, \\ d\tilde{U} &= T\,dS + \zeta\,d\rho - \mathbf{B}\cdot d\mathbf{H}/4\pi. \end{aligned}\right\} \tag{31.6}$$

If the linear relation $\mathbf{B} = \mu\mathbf{H}$ holds, we can write the expressions for all these quantities in the form

$$\left.\begin{aligned} U &= U_0(S,\rho) + B^2/8\pi\mu, \quad & F &= F_0(T,\rho) + B^2/8\pi\mu, \\ \tilde{U} &= U_0(S,\rho) - \mu H^2/8\pi, \quad & \tilde{F} &= F_0(T,\rho) - \mu H^2/8\pi. \end{aligned}\right\} \tag{31.7}$$

The work δR (or, what is the same thing, the change $\delta\mathscr{F}$ at constant temperature) can be written in a different form, in terms of the current density and the vector potential of the magnetic field. For this purpose we put $\delta\mathbf{B} = \mathbf{curl}\,\delta\mathbf{A}$ and

$$\begin{aligned} (\delta\mathscr{F})_T &= \frac{1}{4\pi}\int \mathbf{H}\cdot\mathbf{curl}\,\delta\mathbf{A}\,dV \\ &= -\frac{1}{4\pi}\int \operatorname{div}(\mathbf{H}\times\delta\mathbf{A})\,dV + \frac{1}{4\pi}\int \delta\mathbf{A}\cdot\mathbf{curl}\,\mathbf{H}\,dV. \end{aligned}$$

The first integral is again zero, and the second gives

$$(\delta\mathscr{F})_T = \int \mathbf{j}\cdot\delta\mathbf{A}\,dV/c. \tag{31.8}$$

A similar transformation gives

$$(\delta\tilde{\mathscr{F}})_T = -\int \mathbf{A}\cdot\delta\mathbf{j}\,dV/c. \tag{31.9}$$

It is useful to note that in macroscopic electrodynamics the currents (sources of the magnetic field) are mathematically analogues of the potentials, not of the charges (the sources of the electric field). This is seen by comparing formulae (31.8) and (31.9) with the corresponding results for an electric field:

$$(\delta\mathscr{F})_T = \int \phi\,\delta\rho\,dV, \quad (\delta\tilde{\mathscr{F}})_T = -\int \rho\,\delta\phi\,dV \tag{31.10}$$

(see (10.13), (10.14)). We observe that the charges and potentials appear in these formulae in the opposite order to the currents and potentials in formulae (31.8), (31.9).†

On account of the complete formal correspondence between the thermodynamic relations (expressed in terms of field and induction) for electric and magnetic fields, the thermodynamic inequalities derived in §18 can also be applied to magnetic fields. In particular, we have seen that it follows from these inequalities that $\varepsilon > 0$. In the electric case this result was of no interest, because it was weaker than the inequality $\varepsilon > 1$ which follows on other grounds. In the magnetic case, however, the corresponding inequality $\mu > 0$ is very important, as it is the only restriction on the values which can be taken by the magnetic permeability.

† The significance of this difference is further discussed in the last footnote to §33.

§32. The total free energy of a magnetic substance

In §11 expressions have been derived for the total free energy $\mathscr{F}$ of a dielectric in an electric field. One of the thermodynamic properties of this quantity is that the change in it gives the work done by the electric field on the body when the charges producing the field remain constant. In a magnetic field a similar part is played by the free energy $\tilde{\mathscr{F}}$, since for given currents producing the field the change in $\tilde{\mathscr{F}}$ is the work done on the body.

The following derivation is entirely analogous to that given in §11. The "total" quantity $\tilde{\mathscr{F}}$ is defined as

$$\tilde{\mathscr{F}} = \int \left(\tilde{F} + \frac{\mathfrak{H}^2}{8\pi} \right) \mathrm{d}V, \tag{32.1}$$

where $\mathfrak{H}$ is the magnetic field which would be produced by the given currents in the absence of the magnetizable medium. The plus sign appears in the parenthesis (instead of the minus sign as in (11.1)) because the value of $\tilde{\mathscr{F}}$ for a magnetic field in a vacuum is $-\int(\mathfrak{H}^2/8\pi)\,\mathrm{d}V$ (see (31.7)). The integration in (32.1) is taken over all space, including the volume occupied by the conductors in which flow the currents producing the field.†

Let us calculate the change in $\tilde{\mathscr{F}}$ (for a given temperature and no departure from thermodynamic equilibrium in the medium) corresponding to an infinitesimal change in the field. Since $\delta\tilde{F} = -\mathbf{B}\cdot\delta\mathbf{H}/4\pi$, we have

$$\begin{aligned} \delta\tilde{\mathscr{F}} &= -\int(\mathbf{B}\cdot\delta\mathbf{H} - \mathfrak{H}\cdot\delta\mathfrak{H})\,\mathrm{d}V/4\pi \\ &= -\int(\mathbf{H}-\mathfrak{H})\cdot\delta\mathfrak{H}\,\mathrm{d}V/4\pi - \int\mathbf{B}\cdot(\delta\mathbf{H}-\delta\mathfrak{H})\,\mathrm{d}V/4\pi - \int(\mathbf{B}-\mathbf{H})\cdot\delta\mathfrak{H}\,\mathrm{d}V/4\pi. \end{aligned} \tag{32.2}$$

Introducing the vector potential $\mathfrak{A}$ of the field $\mathfrak{H}$, we can write in the first term

$$\begin{aligned} (\mathbf{H}-\mathfrak{H})\cdot\delta\mathfrak{H} &= (\mathbf{H}-\mathfrak{H})\cdot\mathbf{curl}\,\delta\mathfrak{A} \\ &= \mathrm{div}\,[\delta\mathfrak{A}\times(\mathbf{H}-\mathfrak{H})] + \delta\mathfrak{A}\cdot\mathbf{curl}\,(\mathbf{H}-\mathfrak{H}). \end{aligned}$$

By definition, the fields $\mathbf{H}$ and $\mathfrak{H}$ are produced by the same currents $\mathbf{j}$, the distribution of which over the volume of the conductors is (see §30) independent of the field which they produce, i.e. is independent of the presence or absence of magnetic substances in the surrounding medium. Hence $\mathbf{curl}\,\mathbf{H}$ and $\mathbf{curl}\,\mathfrak{H}$ are both equal to $4\pi\mathbf{j}/c$, and so $\mathbf{curl}(\mathbf{H}-\mathfrak{H}) = \mathbf{0}$. The integral of $\mathrm{div}\,[\delta\mathfrak{A}\times(\mathbf{H}-\mathfrak{H})]$ is transformed into an integral over an infinitely distant surface, and so vanishes.

Similarly, we see that the second term on the right of (32.2) is zero; thus

$$\begin{aligned} \partial\tilde{\mathscr{F}} &= -\int(\mathbf{B}-\mathbf{H})\cdot\delta\mathfrak{H}\,\mathrm{d}V/4\pi \\ &= -\int\mathbf{M}\cdot\delta\mathfrak{H}\,\mathrm{d}V. \end{aligned} \tag{32.3}$$

The expression which we have obtained for $\delta\tilde{\mathscr{F}}$ is exactly similar to (11.3) for the electrostatic problem. In particular, in a uniform magnetic field $\mathfrak{H}$ we have for $\mathrm{d}\tilde{\mathscr{F}}$ an

† In §11 we took the integration in (11.1) over all space except the volume occupied by the charged conductors producing the field. This was possible because there is no electric field in a conductor, charged or not. There is a magnetic field, however, inside the conductors which carry the currents, and it cannot be excluded in calculating the total free energy.

expression analogous to (11.5):

$$\mathrm{d}\tilde{\mathscr{F}} = -\mathscr{S}\,\mathrm{d}T - \mathscr{M}\cdot\mathrm{d}\mathfrak{H}, \tag{32.4}$$

where $\mathscr{M}$ is the total magnetic moment of the body.

Without repeating the subsequent calculations, we shall write down the following formulae by analogy with those in §11. If the linear relation $\mathbf{B} = \mu\mathbf{H}$ holds, we have

$$\tilde{\mathscr{F}} - \mathscr{F}_0(V,T) = -\int \tfrac{1}{2}\mathfrak{H}\cdot\mathbf{M}\,\mathrm{d}V. \tag{32.5}$$

In particular, if the external field is homogeneous, then

$$\tilde{\mathscr{F}} - \mathscr{F}_0(V,T) = -\tfrac{1}{2}\mathfrak{H}\cdot\mathscr{M}. \tag{32.6}$$

In the general case of an arbitrary relation between **B** and **H**, $\tilde{\mathscr{F}}$ can be calculated from the formula

$$\begin{aligned}\tilde{\mathscr{F}} &= \int\left(\tilde{F} + \frac{\mathbf{H}\cdot\mathbf{B}}{8\pi} - \tfrac{1}{2}\mathbf{M}\cdot\mathfrak{H}\right)\mathrm{d}V \\ &= \int\left(F - \frac{\mathbf{H}\cdot\mathbf{B}}{8\pi} - \tfrac{1}{2}\mathbf{M}\cdot\mathfrak{H}\right)\mathrm{d}V,\end{aligned} \tag{32.7}$$

which is analogous to (11.12) for dielectrics.

In §11 we gave also the simpler formulae obtained when the dielectric susceptibility is small. The analogous case for the magnetic problem is especially important because, as mentioned above, the magnetic susceptibility of the majority of bodies is indeed small. In this case

$$\tilde{\mathscr{F}} - \mathscr{F}_0 = -\tfrac{1}{2}\chi\int\mathfrak{H}^2\,\mathrm{d}V. \tag{32.8}$$

We can also derive results for the magnetic field analogous to those obtained in §14. These concern the change in the thermodynamic quantities resulting from an infinitesimal change in the magnetic permeability μ, the field sources being assumed unchanged. It is clear from the foregoing that we must consider the change in $\tilde{\mathscr{F}}$, and not that in $\mathscr{F}$ as in §14. We shall not repeat the derivation, which is similar to that of (14.1), but merely give the result:

$$\delta\tilde{\mathscr{F}} = -\int\delta\mu H^2\,\mathrm{d}V/8\pi. \tag{32.9}$$

In §14 we used the formula (11.7), an analogue of (32.5), to deduce that the dielectric susceptibility of any substance is positive. In the magnetic case we cannot draw this conclusion, and the magnetic susceptibility may be of either sign. The reason for this marked difference is that the Hamiltonian of a system of charges moving in a magnetic field contains not only terms linear in the field (as in the electric case) but also quadratic terms. Hence, in determining the change in the free energy of the body in the magnetic field by means of perturbation theory as in (14.2), we have a contribution in the first approximation as well as the second. In such a case no general conclusion can be drawn concerning the sign of the variation. It is positive for paramagnetic bodies and negative for diamagnetic ones.

In §14 we also drew conclusions concerning the direction of motion of bodies in an electric field. Similar conclusions follow from (32.9), but, since μ may be either greater or less than 1, there is no universal result. For example, in an almost uniform field paramagnetic bodies ($\mu > 1$) move in the direction of H increasing, and diamagnetic bodies ($\mu < 1$) in the opposite direction.

§33. The energy of a system of currents

Let us consider a system of conductors with currents flowing in them and assume that neither the conductors nor the medium surrounding them are ferromagnetic, so that $\mathbf{B} = \mu\mathbf{H}$ everywhere. According to §31, the total free energy of the system is given in terms of the magnetic field of the currents by

$$\mathscr{F} = \int \mathbf{H}\cdot\mathbf{B}\,\mathrm{d}V/8\pi. \tag{33.1}$$

Here we omit the quantity $\mathscr{F}_0$, which is a constant (at a given temperature) and is not related to the currents. The integration in (33.1) is taken over all space, both inside and outside the conductors.

The same energy can also be expressed in terms of the currents by means of the integral

$$\mathscr{F} = \int \mathbf{A}\cdot\mathbf{j}\,\mathrm{d}V/2c; \tag{33.2}$$

cf. the derivation of (31.8) from (31.2). Here the integration extends only over the conductors, because $\mathbf{j} = 0$ outside them.

Since the field equations are linear, the magnetic field can be written as the sum of the fields resulting from each current alone with no current in the other conductors: $\mathbf{H} = \Sigma\mathbf{H}_a$. Then the total free energy (33.1) is

$$\mathscr{F} = \sum_a \mathscr{F}_{aa} + \sum_{a>b} \mathscr{F}_{ab}, \tag{33.3}$$

where

$$\mathscr{F}_{aa} = \int \mathbf{H}_a\cdot\mathbf{B}_a\,\mathrm{d}V/8\pi, \quad \mathscr{F}_{ab} = \int \mathbf{H}_a\cdot\mathbf{B}_b\,\mathrm{d}V/4\pi. \tag{33.4}$$

We have put $\mathscr{F}_{ab} = \mathscr{F}_{ba}$, since $\mathbf{H}_a\cdot\mathbf{B}_b = \mu\mathbf{H}_a\cdot\mathbf{H}_b = \mathbf{H}_b\cdot\mathbf{B}_a$, where μ is the magnetic permeability at any point. The quantity $\mathscr{F}_{aa}$ may be called the *free self-energy* of the current in the ath conductor, and $\mathscr{F}_{ab}$ the *interaction energy* of the ath and bth conductors. It should be borne in mind, however, that these names are strictly correct only if the magnetic properties of both the conductors and the medium are neglected. Otherwise the field, and therefore the energy, of each current depend on the position and magnetic permeability of the other conductors.

The quantities (33.4) can also be expressed in terms of the currents $\mathbf{j}_a$ in each conductor, in accordance with formula (33.2):

$$\mathscr{F}_{aa} = \int \mathbf{j}_a\cdot\mathbf{A}_a\,\mathrm{d}V_a/2c, \quad \mathscr{F}_{ab} = \int \mathbf{j}_a\cdot\mathbf{A}_b\,\mathrm{d}V_a/c = \int \mathbf{j}_b\cdot\mathbf{A}_a\,\mathrm{d}V_b/c. \tag{33.5}$$

The integral in $\mathscr{F}_{aa}$ is here taken only over the volume of the ath conductor; $\mathscr{F}_{ab}$ can be written as either of the two expressions, in which the integration is over the volume of the ath and bth conductor respectively.

When the distribution of the current density over the volume of the conductor is given, $\mathscr{F}_{aa}$ depends only on the total current J_a passing through a cross-section. Both the current density $\mathbf{j}$ and the field which it produces will be proportional to J_a. Hence the integral $\mathscr{F}_{aa}$ is proportional to J_a^2, and we write it

$$\mathscr{F}_{aa} = L_{aa}J_a^2/2c^2, \tag{33.6}$$

where L_{aa} is called the *self-inductance* of the conductor. Similarly, the interaction energy of two currents is proportional to the product J_aJ_b:

$$\mathscr{F}_{ab} = L_{ab}J_aJ_b/c^2. \tag{33.7}$$

The quantity L_{ab} is called the *mutual inductance* of the conductors. Thus the total free energy of a system of currents is

$$\mathscr{F} = \frac{1}{2c^2}\sum_a L_{aa}J_a^2 + \frac{1}{c^2}\sum_{a>b} L_{ab}J_aJ_b = \frac{1}{2c^2}\sum_a\sum_b L_{ab}J_aJ_b. \tag{33.8}$$

The condition that this quadratic form should be positive definite places certain restrictions on the values of the coefficients. In particular $L_{aa} > 0$ for all a, and $L_{aa}L_{bb} > L_{ab}^2$.

The calculation of the energy of currents in the general case of arbitrary three-dimensional conductors requires a complete solution of the field equations, and is a difficult problem. It becomes simpler if the magnetic permeability of both the conductors and the surrounding medium can be taken as unity. It should be noted that the energy of the currents is then no longer dependent on the thermodynamic state (in particular, on the temperature) of the bodies, and hence the free energy in the above formulae may be referred to simply as the energy.

For $\mu = 1$ the vector field potential due to the currents $\mathbf{j}$ is given by formula (30.12). Hence the self-energy of the ath conductor is

$$\mathscr{F}_{aa} = \frac{1}{2c^2}\int\int \frac{\mathbf{j}\cdot\mathbf{j}'}{R}\,\mathrm{d}V\,\mathrm{d}V', \tag{33.9}$$

where both integrations are taken over the volume of the conductor considered, and R is the distance between $\mathrm{d}V$ and $\mathrm{d}V'$. Similarly, the mutual energy of two conductors is

$$\mathscr{F}_{ab} = \frac{1}{c^2}\int\int \frac{\mathbf{j}_a\cdot\mathbf{j}_b}{R}\,\mathrm{d}V_a\,\mathrm{d}V_b, \tag{33.10}$$

where $\mathrm{d}V_a$ and $\mathrm{d}V_b$ are volume elements in the two conductors.

The mutual energy of two linear currents is particularly easy to calculate. In formula (33.10) we change from volume currents to linear ones by replacing $\mathbf{j}_a\mathrm{d}V_a$ and $\mathbf{j}_b\mathrm{d}V_b$ by $J_a\mathrm{d}\mathbf{l}_a$ and $J_b\mathrm{d}\mathbf{l}_b$ respectively, and we find that the mutual inductance is $L_{ab} = \oint\oint \mathrm{d}\mathbf{l}_a\cdot\mathrm{d}\mathbf{l}_b/R$. In this approximation, therefore, L_{ab} depends only on the shape, size and relative position of the two currents, and not on the distribution of current over the cross-section of each wire. It must be emphasized that this simple formula can be obtained for linear currents without imposing the condition that $\mu = 1$. In the approximation where the thickness of the wires is neglected, their magnetic properties have no effect on the field which they produce, and therefore no effect on their mutual energy. If the magnetic permeability μ of the medium surrounding the wires is different from unity, the vector potential is, by (30.15), simply multiplied by μ, and therefore so is the magnetic induction. The mutual inductance is therefore multiplied by the same factor, so that

$$L_{ab} = \mu\oint\oint \mathrm{d}\mathbf{l}_a\cdot\mathrm{d}\mathbf{l}_b/R. \tag{33.11}$$

The self-inductance of linear conductors is much more difficult to calculate; we shall discuss it in §34.

The total energy of a system of linear currents can be written in still another form. To do this, we return to the integral (33.2), which for linear currents becomes

$$\mathscr{F} = \frac{1}{2c}\sum_a J_a \oint \mathbf{A}\cdot\mathrm{d}\mathbf{l}_a, \tag{33.12}$$

where $\mathbf{A}$ is the vector potential of the total field at the element $\mathrm{d}\mathbf{l}_a$ of the ath conductor. The main error in going from (33.2) to (33.12) arises from neglecting the change in the field (including the field of the current considered) over the cross-section of the wire. Each of the contour integrals in (33.12) can be transformed into a surface integral:

$$\oint \mathbf{A}\cdot\mathrm{d}\mathbf{l}_a = \int \mathbf{curl}\,\mathbf{A}\cdot\mathrm{d}\mathbf{f}_a = \int \mathbf{B}\cdot\mathrm{d}\mathbf{f}_a,$$

i.e. it is the flux of the magnetic induction or *magnetic flux* through the circuit of the ath current. We denote this flux by Φ_a. Then

$$\mathscr{F} = \frac{1}{2c}\sum_a J_a\Phi_a. \tag{33.13}$$

Similarly, the free energy $\mathscr{F}$ of a linear current J in an external magnetic field, i.e. the energy without the self-energy of the field sources, can be expressed in terms of the magnetic flux. Evidently

$$\mathscr{F} = J\Phi/c, \tag{33.14}$$

where Φ is the flux of the external field through the circuit of the current J. If the external field is uniform, and $\mu = 1$ in the external medium, then $\Phi = \mathfrak{H}\cdot\int\mathrm{d}\mathbf{f}$. Introducing the magnetic moment of the current in accordance with (30.18), we have $\mathscr{F} = \mathscr{M}\cdot\mathfrak{H}$.

Knowing the energy of a system of currents as a function of their shape, size and relative position, we can determine the forces on the conductors by simply differentiating with respect to the appropriate coordinates. Here, however, the question arises which characteristics of the currents should be kept constant in the differentiation. It is most convenient to differentiate at constant current. In this case the free energy is represented by $\tilde{\mathscr{F}}$, and so the generalized force F_q in the direction of a generalized coordinate q is $F_q = -(\partial\tilde{\mathscr{F}}/\partial q)_{J,T}$. The suffixes show that the differentiation is effected at constant current and constant temperature. Since we omit the term independent of the currents in the free energy, $\mathscr{F}$ and $\tilde{\mathscr{F}}$ differ only in sign, and so

$$F_q = -\left(\frac{\partial\tilde{\mathscr{F}}}{\partial q}\right)_J = \left(\frac{\partial\mathscr{F}}{\partial q}\right)_J = \frac{1}{2c^2}\sum_{a,b} J_aJ_b\frac{\partial L_{ab}}{\partial q}; \tag{33.15}$$

here and henceforward the suffix T to the derivatives is omitted, for brevity.

In particular, the forces exerted on a conductor by its own magnetic field are given by the formula

$$F_q = \frac{1}{2c^2}J^2\frac{\partial L}{\partial q}, \tag{33.16}$$

where L is the self-inductance of the conductor. The nature of these forces can be seen as follows. For given current (and temperature), $\tilde{\mathscr{F}}$ tends to a minimum. Since $\tilde{\mathscr{F}} = -LJ^2/2c^2$, this means that the forces on the conductor will tend to increase its self-inductance. The latter, having the dimensions of length, must be proportional to the dimension of the conductor. Thus the effect of the magnetic field is to increase the size of the conductor.

For a current in an external magnetic field we have†

$$\tilde{\mathscr{F}} = -\mathscr{F} = -\mathscr{M}\cdot\mathfrak{H}. \tag{33.17}$$

† The factor $\frac{1}{2}$ which appears in (32.6) is absent in (33.17) because the magnetic moment of the current in the latter equation is independent of the field, whereas the magnetic moment in (32.6) is itself due to the field.

In all the above formulae for the energy it is assumed that there is a linear relation between the magnetic field and induction. In the general case where this relation is arbitrary, analogous differential relations can be set up. The change in the free energy resulting from an infinitesimal change in the field (at constant temperature) is, by (31.8), $\delta\mathscr{F} = \int \mathbf{j}.\delta\mathbf{A}\, dV/c$ or, for a system of linear currents,

$$\delta\mathscr{F} = \frac{1}{c}\sum_a J_a \oint \delta\mathbf{A}\cdot d\mathbf{l}_a.$$

Proceeding as in the derivation of (33.13) from (33.12), we have†

$$\delta\mathscr{F} = \frac{1}{c}\sum_a J_a \delta\Phi_a. \tag{33.18}$$

Similarly, we find from (31.9)

$$\delta\tilde{\mathscr{F}} = -\frac{1}{c}\sum_a \Phi_a \delta J_a. \tag{33.19}$$

Thus we can say that, for a system of linear currents, $\mathscr{F}$ is the thermodynamic potential with respect to the magnetic fluxes, and $\tilde{\mathscr{F}}$ with respect to the currents, the two potentials being related by

$$\tilde{\mathscr{F}} = \mathscr{F} - \frac{1}{c}\sum_a J_a \Phi_a. \tag{33.20}$$

Whatever the magnetic properties of the substance, therefore, the thermodynamic relations

$$J_a/c = \partial\mathscr{F}/\partial\Phi_a, \quad \Phi_a/c = -\partial\tilde{\mathscr{F}}/\partial J_a \tag{33.21}$$

hold. If these formulae are applied to the case where the field and induction are linearly related, so that $\mathscr{F}$ is given by (33.8), we obtain

$$\Phi_a = \frac{1}{c}\sum_a L_{ab} J_b. \tag{33.22}$$

Thus the inductances are the coefficients of proportionality between the magnetic fluxes and the currents which produce the magnetic field. The product $L_{ab} J_b/c$ is the magnetic flux through the circuit of the current J_a due to the current $J_b\,(b \neq a)$, and $L_{aa} J_a/c$ is that due to the current J_a itself.

§34. The self-inductance of linear conductors

In calculating the self-inductance of a linear conductor its thickness cannot be entirely neglected as it was in calculating the mutual inductance of two conductors. If it were, we

† There is an obvious analogy between formulae (33.18) and (10.13) in the magnetic and electric cases respectively. In the magnetic case, the induction fluxes play the role of charges. The analogy has a clear physical interpretation. Just as the electric field can be maintained without any external energy supply, by charges on insulated conductors, the magnetic field can be maintained, without any external energy supply, by superconducting solenoids across which the magnetic fluxes remain constant. It is therefore not surprising that the change in the free energy $\mathscr{F}$ in the electric and magnetic cases is governed by the changes in the charges and in the induction fluxes respectively.

should obtain from (33.9) the self-inductance $L = \oint\oint \mathrm{d}\mathbf{l}\cdot\mathrm{d}\mathbf{l}'/R$, where both integrals are taken along the same circuit, and this integral is logarithmically divergent as $R \to 0$.

The exact value of the self-inductance of a conductor depends on the distribution of current in it, which may vary with the manner of excitation of the current, i.e. with the manner of application of the electromotive force. For a linear conductor, however, the self-inductance does not, to a fairly high accuracy, depend on the distribution of current over the cross-section.†

Let us write the self-inductance as $L = L_e + L_i$, where L_e and L_i result from the magnetic field energy outside and inside the conductor respectively. For a linear conductor, the "external" part L_e makes the main contribution to the self-inductance. This is because most of the magnetic energy of a closed linear circuit resides in the field at distances from the wire large compared with its thickness. For the energy per unit length of an infinite straight wire is

$$(\mu_e/8\pi)\int H^2 \cdot 2\pi r\,\mathrm{d}r = (\mu_e/8\pi)\int (2J/cr)^2 \cdot 2\pi r\,\mathrm{d}r = (\mu_e J^2/c^2)\int \mathrm{d}r/r,$$

where r is the distance from the axis of the wire and μ_e the magnetic permeability of the external medium. This integral diverges logarithmically for large r. For a closed linear circuit, of course, this divergence disappears, because the integral is "cut off" at distances of the order of the dimension of the circuit. We obtain an approximate value for the energy on multiplying this integral by the total length l of the wire, and taking l as the upper limit and the radius a of the wire as the lower limit. The result is $(\mu_e J^2 l/c^2)\log(l/a)$, and hence the self-inductance is

$$L = 2\mu_e l \log(l/a). \tag{34.1}$$

This expression is said to be of *logarithmic accuracy*: its relative error is of the order of $1/\log(l/a)$, and the ratio l/a is assumed to be so large that its logarithm is large.‡

A particular case of a linear conductor is a *solenoid*, which consists of a wire wound in a helix, with the turns very close together. Neglecting the thickness of the wire and the distance between the turns, we have simply a conducting cylindrical surface with a "surface" conduction current on it. The equation $\mathbf{curl}\,\mathbf{H} = 4\pi\mathbf{j}/c$ within the conductor is here replaced by the boundary condition.

$$\mathbf{n}\times(\mathbf{H}_2 - \mathbf{H}_1) = 4\pi\mathbf{g}/c, \tag{34.2}$$

where $\mathbf{g}$ is the surface current density, $\mathbf{H}_1$ and $\mathbf{H}_2$ the fields on each side of the surface, and $\mathbf{n}$ the unit normal vector into medium 2; cf. the derivation of (29.16).

If the solenoid is an infinitely long cylinder, the magnetic field which it produces can be found very simply. The surface currents flow in circles, and their density $g = nJ$, where J is the current in the wire and n the number of turns per unit length of the solenoid. The field outside the cylinder is zero; the field inside is uniform and along the axis of the cylinder, and

† More precisely, it is independent of the distribution of current provided that the current density varies appreciably only over distances comparable with the thickness a of the wire. If, however, the distribution is such that the current density varies appreciably over distances small compared with a (as happens, for particular reasons, in the *skin effect* and in superconductors), then the self-inductance does depend on the distribution.

‡ The assertion made above that the self-inductance is independent of the current distribution actually applies not only to the approximation (34.1) but also to the next approximation, in which terms not containing the large logarithm are included (or, what is the same thing, the coefficient of l/a is included in the argument of the logarithm); see the Problems at the end of this section.

is $H = 4\pi nJ/c$. For this field evidently satisfies the equations div $\mathbf{H} = 0$, **curl** $\mathbf{H} = 0$ in all space outside the conducting surface, and also the boundary condition (34.2) at that surface.

Accordingly, the field energy per unit length of the cylinder is

$$\mu_e H^2 \pi b^2/8\pi = 2\pi^2 n^2 b^2 \mu_e J^2/c^2,$$

where b is the radius of the cylinder and μ_e pertains to the material within the solenoid. Neglecting the end effects, we can apply this formula also to a solenoid whose length h is finite, but large compared with b. Then the self-inductance is

$$L = 4\pi^2 n^2 b^2 h\mu_e = 2\pi\mu_e nbl, \tag{34.3}$$

where $l = 2\pi bnh$ is the total length of the wire. The greater self-inductance of a solenoid as compared with that of a straight wire of equal length (cf. (34.1)) is, of course, due to the mutual induction between adjoining turns.

PROBLEMS†

PROBLEM 1. Determine the self-inductance of a closed circuit of thin wire of circular cross-section.

SOLUTION. The magnetic field in the wire can be taken to be the same as that inside an infinite straight cylinder: $H = 2Jr/ca^2$, where r is the distance from the axis of the wire and a its radius. Hence we find the internal part of the self-inductance:

$$L_i = \frac{2c^2}{J^2}\frac{\mu_i}{8\pi}\int H^2 \mathrm{d}V = \tfrac{1}{2} l\mu_i, \tag{1}$$

where l is the length of the wire.

To calculate L_e, we notice that the field outside a thin wire is independent of the distribution of current over its cross-section. In particular, the energy $\mathscr{F}_e$ of the external magnetic field is unchanged if we assume that the current flows only on the surface of the wire. The field inside the wire is then zero, and $\mathscr{F}_e$ may be calculated as the total energy from formula (33.2). On account of the assumed surface distribution of the current, the integral in this formula becomes a line integral along the axis of the wire, and so the external part of the self-inductance is

$$L_e = \frac{2c^2}{J^2}\frac{J}{2c}\oint [\mathbf{A}]_{r=a}\cdot \mathrm{d}\mathbf{l},$$

where the value of $\mathbf{A}$ in the integrand is taken at the surface of the wire. In obtaining this formula we also use the fact that, in the approximation used here, the field is constant along the perimeter of a cross-section.

Having reduced the problem to that of calculating $\mathbf{A}$ for $r = a$, we now make a different assumption concerning the current distribution, namely that the whole current J flows along the axis of the wire. The field on the surface of the wire is, in the approximation considered, unchanged by this assumption (nor would it be changed for a straight wire of circular cross-section). Then, by formula (30.14), we have

$$[\mathbf{A}]_{r=a} = \frac{J}{c}\left[\oint \frac{\mathrm{d}\mathbf{l}}{R}\right]_{r=a},$$

where R is the distance from the element $\mathrm{d}\mathbf{l}$ of the axis to a given point on the surface of the wire. We divide the integral into two parts, one for which $R > \Delta$ and the other for which $R < \Delta$, where Δ is a distance small compared with the dimension of the circuit but large compared with the radius a of the wire.‡ In the integral where $R > \Delta$, a may be neglected and R taken simply as the distance between two points on the circuit. The vector integral where $R < \Delta$ may be assumed to be along the tangent at the point considered. Denoting by $\mathbf{t}$ the unit vector in that direction, we have

$$\left[\int\limits_{R<\Delta} \frac{\mathrm{d}\mathbf{l}}{R}\right]_{r=a} \cong \mathbf{t}\int\limits_{-\Delta}^{\Delta} \frac{\mathrm{d}l}{\sqrt{(a^2+l^2)}} = 2\mathbf{t}\sinh^{-1}(\Delta/a)$$

$$\cong 2\mathbf{t}\log(2\Delta/a).$$

† In Problems 1–6 we assume $\mu_e = 1$.

‡ A similar procedure was used to calculate the capacitance of a thin ring in §2, Problem 4.

This expression can be written as the integral

$$\int_{\Delta > R > \frac{1}{2}a} d\mathbf{l}/R,$$

where R is again the distance between points on the circuit. Adding the two integrals for $R > \Delta$ and $R < \Delta$, we obtain

$$[\mathbf{A}]_{r=a} = \frac{J}{c} \int_{R > \frac{1}{2}a} \frac{d\mathbf{l}}{R},$$

from which the arbitrary parameter Δ has disappeared, as it should.

The final result is therefore

$$L_e = \iint_{R > \frac{1}{2}a} \frac{d\mathbf{l} \cdot d\mathbf{l}'}{R}. \tag{2}$$

The integration here extends over all pairs of points on the circuit whose distance apart exceeds $\frac{1}{2}a$.

PROBLEM 2. Determine the self-inductance of a thin wire ring (with radius b) of circular cross-section (with radius a).

SOLUTION. The integrand in (2), Problem 1, depends only on the central angle ϕ subtended by the chord R, and $R = 2b \sin \frac{1}{2}\phi$, while $d\mathbf{l} \cdot d\mathbf{l}' = dl\,dl' \cos \phi$. Hence

$$L_e = 2 \int_{\phi_0}^{\pi} \frac{\cos \phi \cdot 2\pi b \cdot b\,d\phi}{2b \sin \frac{1}{2}\phi} = 4\pi b\,[-\log \tan \tfrac{1}{4}\phi_0 - 2 \cos \tfrac{1}{2}\phi_0].$$

The lower limit of integration is determined from $2b \sin \frac{1}{2}\phi_0 = \frac{1}{2}a$, whence $\phi_0 \cong a/2b$. Substituting this value and adding $L_i = \pi b \mu_i$, we have to the required accuracy

$$L = 4\pi b\,[\log(8b/a) - 2 + \tfrac{1}{4}\mu_i].$$

In particular, for $\mu_i = 1$ we obtain

$$L = 4\pi b\,[\log(8b/a) - (7/4)].$$

PROBLEM 3. Determine the extension of a ring of wire (with $\mu_i = 1$) under the action of the magnetic field of a current flowing in it.

SOLUTION. The internal stresses parallel and perpendicular to the axis are, by (33.16), given by

$$\pi a^2 \sigma_\parallel = \frac{J^2}{2c^2} \frac{\partial L}{\partial(2\pi b)}, \qquad 2\pi a b \sigma_\perp = \frac{J^2}{2c^2} \frac{\partial L}{\partial a}.$$

Substituting L from Problem 2, we have

$$\sigma_\parallel = \frac{J^2}{\pi a^2 c^2} \left[\log \frac{8b}{a} - \frac{3}{4} \right], \qquad \sigma_\perp = -\frac{J^2}{a^2 c^2}.$$

Hence the required relative extension of the ring is

$$\frac{\Delta b}{b} = \frac{1}{E}(\sigma_\parallel - 2\sigma\sigma_\perp) = \frac{J^2}{\pi a^2 c^2 E} \left(\log \frac{8b}{a} - \frac{3}{4} + 2\pi\sigma \right).$$

where E is Young's modulus and σ Poisson's ratio for the wire; see *TE*, §5.

PROBLEM 4. Determine the self-inductance per unit length of a system of two parallel straight wires (with $\mu_i = 1$) having circular cross-sections of radii a and b, with their axes a distance h apart, and carrying equal currents J in opposite directions (Fig. 19).

SOLUTION. The vector potential of the magnetic field of each current is parallel to the axes of the wires, and so the two vector potentials can be added algebraically. For the magnetic field of wire 1, with a uniformly distributed

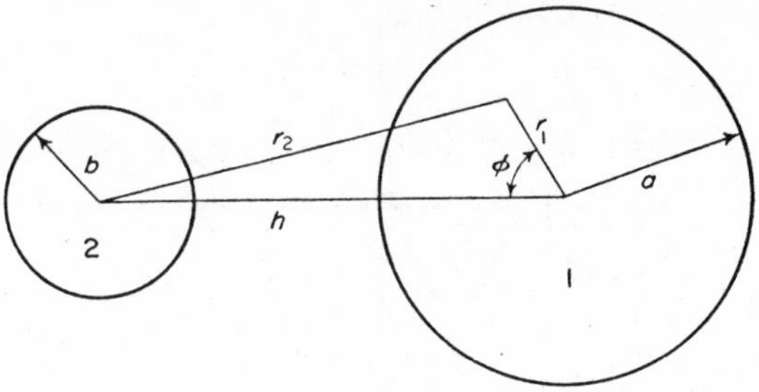

FIG. 19

current $+J$, we have in cylindrical polar coordinates

$$A_z = \frac{J}{c}\left(C - \frac{r^2}{a^2}\right) \text{ for } r < a,$$

$$A_z = \frac{J}{c}\left(C - 1 - 2\log\frac{r}{a}\right) \text{ for } r > a,$$

where C is an arbitrary constant; A_z is continuous at the surface of the wire. The formulae for wire 2 are obtained by substituting b for a and changing the sign of J. Integration over the cross-section of wire 1 in formula (33.2) gives

$$\frac{J^2}{2c^2\pi a^2}\int\left\{\left(C - \frac{r_1{}^2}{a^2}\right) - \left(C - 1 - 2\log\frac{r_2}{b}\right)\right\}\mathrm{d}f_1$$

$$= \frac{J^2}{2c^2\pi a^2}\int_0^a\int_0^{2\pi}\left\{1 - \frac{r_1{}^2}{a^2} + \log\frac{h^2 + r_1{}^2 - 2hr_1\cos\phi}{b^2}\right\}r_1\,\mathrm{d}\phi\,\mathrm{d}r_1 = \frac{J^2}{2c^2}\left(\frac{1}{2} + 2\log\frac{h}{b}\right).$$

The integration over the cross-section of wire 2 gives the same thing with a in place of b. The required self-inductance per unit length of the double wire is therefore

$$L = 1 + 2\log(h^2/ab).$$

PROBLEM 5. Determine the self-inductance of a toroidal solenoid.

SOLUTION. We regard the solenoid as a toroidal conducting surface carrying surface currents with density $g = NJ/2\pi r$, where N is the total number of turns and J the current; the coordinates and dimensions are as shown in Fig. 20 (p. 126). The magnetic field outside the solenoid is zero, and inside the solenoid $H_{ir} = H_{iz} = 0$, $H_{i\phi} = 2NJ/cr$, where r, z, ϕ are cylindrical polar coordinates; for this solution satisfies the equations $\operatorname{div}\mathbf{H} = 0$, $\mathbf{curl}\,\mathbf{H} = 0$ and the boundary condition (34.2).† The energy of the magnetic field in the solenoid is

$$\int(H_i{}^2/8\pi)\mathrm{d}V = (N^2J^2/c^2)\oint z\,\mathrm{d}r/r,$$

where the integration is taken along the perimeter of the cross-section, and is easily effected by putting $z = a\sin\theta$, $r = b + a\cos\theta$. The self-inductance is found to be

$$L = 4\pi N^2[b - \sqrt{(b^2 - a^2)}].$$

PROBLEM 6. Determine the end-effect correction of order l/h to the expression (34.3) (with $\mu_e = 1$) for the self-inductance of a cylindrical solenoid.

SOLUTION. The self-inductance is calculated as a double integral over the surface of the solenoid:

$$L = \frac{1}{J^2}\iint\frac{\mathbf{g}_1\cdot\mathbf{g}_2}{R}\,\mathrm{d}f_1\,\mathrm{d}f_2,$$

† It is valid also for an annular solenoid with any cross-section.

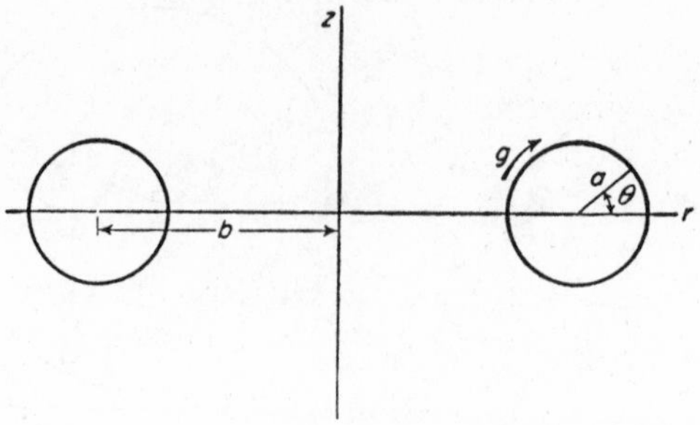

FIG. 20

where **g** is the surface current density ($g = nJ$). In cylindrical polar coordinates

$$L = 2\pi b^2 n^2 \int_0^h \int_0^h \int_0^{2\pi} \frac{\cos\phi \, d\phi \, dz_1 \, dz_2}{\sqrt{[(z_2 - z_1)^2 + 4b^2 \sin^2 \frac{1}{2}\phi]}}$$

$$= 8\pi b^2 n^2 \int_0^h \int_0^\pi \frac{(h-\zeta)\cos\phi \, d\phi \, d\zeta}{\sqrt{(\zeta^2 + 4b^2 \sin^2 \frac{1}{2}\phi)}},$$

where ϕ is the angle between the diametral planes through df_1 and df_2, and $\zeta = z_2 - z_1$.
Effecting the integration with respect to ζ, we have for $h \gg b$

$$L \cong 8\pi b^2 n^2 \int_0^\pi [h \log \frac{h}{b \sin \frac{1}{2}\phi} - h + 2b \sin \tfrac{1}{2}\phi] \cos\phi \, d\phi,$$

and finally

$$L = 4\pi^2 b^2 n^2 [h - 8b/3\pi].$$

PROBLEM 7. Determine the factor by which the self-inductance of a plane circuit changes when it is placed on the surface of a half-space of magnetic permeability μ_e. The internal part of the self-inductance is neglected.

SOLUTION. It is evident from symmetry that, in the absence of the half-space, the magnetic field of the current is symmetrical about the plane of the circuit, and the lines of magnetic force cross that plane normally. Let this field be $\mathbf{H}_0$. We can satisfy the field equations and the boundary conditions on the surface of the half-space by putting $\mathbf{H} = 2\mu_e \mathbf{H}_0/(\mu_e + 1)$ in the vacuum and $\mathbf{B} = \mu_e \mathbf{H} = 2\mu_e \mathbf{H}_0/(\mu_e + 1)$ in the medium: B_n and $\mathbf{H}_t$ are then continuous at the boundary, and the circulation of **H** along any line of force is equal to that of $\mathbf{H}_0$. Hence we easily see that, when the medium is present, the total energy of the field, and therefore the self-inductance of the circuit, are multiplied by $2\mu_e/(\mu_e + 1)$.

§35. Forces in a magnetic field

To determine the forces on matter in a magnetic field hardly any further calculations are necessary, on account of the complete analogy with electrostatics. The analogy is due mainly to the fact that the expressions for the thermodynamic quantities in a magnetic field differ from those for an electric field only in that **E** and **D** are replaced by **H** and **B** respectively. In calculating the stress tensor in §15 we used the fact that the electric field satisfies the equation **curl E** = 0, and is therefore a potential field. The magnetic field satisfies the equation

$$\mathbf{curl\,H} = 4\pi \mathbf{j}/c, \tag{35.1}$$

which reduces to **curl H** = 0 only in the absence of conduction currents. In calculating the stress tensor, however, we must always put **j** = 0. Since **j** involves the derivatives of the magnetic field, an allowance for the currents in calculating the stresses would amount to adding to the stress tensor σ_{ik} the very small corrections due to the non-uniformity of the field; cf. the second footnote to §15.

Thus all the formulae obtained in §§15 and 16 for the stress tensor can be applied immediately for a magnetic field. For example, in a fluid medium with $\mathbf{B} = \mu\mathbf{H}$ we have

$$\sigma_{ik} = -P_0(\rho, T)\delta_{ik} - \frac{H^2}{8\pi}\left[\mu - \rho\left(\frac{\partial\mu}{\partial\rho}\right)_T\right]\delta_{ik} + \frac{\mu H_i H_k}{4\pi}. \tag{35.2}$$

From this the body forces are calculated by the formula $f_i = \delta\sigma_{ik}/\partial x_k$. If the medium is a conductor carrying a current, the calculation differs from that in §15 in that the equation **curl H** = 0 is replaced by (35.1).

Differentiating (35.2) and using also the equation div **B** = div (μ**H**) = 0, we find

$$\mathbf{f} = -\mathbf{grad}\,P_0 + \frac{1}{8\pi}\mathbf{grad}\left[H^2\rho\left(\frac{\partial\mu}{\partial\rho}\right)_T\right] - \frac{H^2}{8\pi}\mathbf{grad}\,\mu - \frac{\mu}{8\pi}\mathbf{grad}\,H^2 + \frac{\mu}{4\pi}(\mathbf{H}\cdot\mathbf{grad})\mathbf{H}.$$

By a well-known formula of vector analysis,

$$\begin{aligned}(\mathbf{H}\cdot\mathbf{grad})\mathbf{H} &= \tfrac{1}{2}\mathbf{grad}\,H^2 - \mathbf{H}\times\mathbf{curl}\,\mathbf{H}\\ &= \tfrac{1}{2}\mathbf{grad}\,H^2 + 4\pi\mathbf{j}\times\mathbf{H}/c.\end{aligned}$$

Thus

$$\mathbf{f} = -\mathbf{grad}\,P_0 + \frac{1}{8\pi}\mathbf{grad}\left[H^2\rho\left(\frac{\partial\mu}{\partial\rho}\right)_T\right] - \frac{H^2}{8\pi}\mathbf{grad}\,\mu + \frac{\mu}{c}\mathbf{J}\times\mathbf{H}. \tag{35.3}$$

The last term does not appear in the corresponding formula (15.12). It would, however, be incorrect to suppose that the presence of this term means that a force can be isolated in **f** which is due to the conduction current. The reason is that, by (35.1), the current **j** is inseparable from non-uniformity of the field, and another term in (35.3) also involves the space derivatives of the field. When the magnetic permeability of the medium is appreciably different from unity, all the terms in (35.3) are in general of the same order of magnitude.

If, however, as usually happens, μ is close to 1, the last term in (35.3) gives the main contribution to the force when a conduction current is present, and the remaining terms form only a small correction. In calculating the forces we can then put $\mu = 1$, obtaining simply

$$\mathbf{f} = \mathbf{j}\times\mathbf{H}/c. \tag{35.4}$$

The term $-\mathbf{grad}\,P_0$ is of no interest henceforward, and we omit it. For $\mu = 1$ the properties of the substance have no effect on the magnetic phenomena, and the expression (35.4) for the force is equally valid for fluid and for solid conductors. The total force exerted by a magnetic field on a conductor carrying a current is given by the integral

$$\mathbf{F} = \int \mathbf{j}\times\mathbf{H}\,dV/c. \tag{35.5}$$

Formula (35.4) can, of course, be very easily obtained from the familiar expression for the Lorentz force. The macroscopic force on a body at rest in a magnetic field is just the averaged Lorentz force exerted on the charged particles in the body by the microscopic field **h**: $\mathbf{f} = \rho\mathbf{v}\times\mathbf{h}/c$. For $\mu = 1$ the field **h** is equal to the mean field **H**, and the mean value of $\rho\mathbf{v}$ is the conduction current density.

When a conductor moves, the forces (35.4) do mechanical work on it. At first sight it might appear that this contradicts the result that the Lorentz forces do no work on moving charges. In reality, of course, there is no contradiction, since the work done by the Lorentz forces in a moving conductor includes not only the mechanical work but also the work done by the electromotive forces induced in the conductor during its motion. These two quantities of work are equal and opposite; see the second footnote to §63.

In the expression (35.4) **H** is the true value of the magnetic field due both to external sources and to the currents themselves on which the force (35.4) acts. In calculating the total force from (35.5), however, we can take **H** to be simply the external field $\mathfrak{H}$ in which the conductor carrying a current is placed. The field of the conductor itself cannot, by the law of conservation of momentum, contribute to the total force acting on the conductor.

The calculation of the forces is particularly simple for a linear conductor. Its magnetic properties are of no significance, and, if $\mu = 1$ in the surrounding medium, the total force on the conductor is given by the line integral

$$\mathbf{F} = J \oint \mathbf{dl} \times \mathfrak{H}/c. \tag{35.6}$$

This expression can be written as an integral over a surface bounded by the current circuit. Using Stokes' theorem, we replace $\mathbf{dl}$ by the operator $\mathbf{df} \times \mathbf{grad}$, obtaining $\oint \mathbf{dl} \times \mathfrak{H} = \int (\mathbf{df} \times \mathbf{grad}) \times \mathfrak{H}$. Now

$$\begin{aligned}(\mathbf{df} \times \mathbf{grad}) \times \mathfrak{H} &= -\mathbf{df}\,\mathrm{div}\,\mathfrak{H} + \mathbf{grad}\,(\mathbf{df} \cdot \mathfrak{H}) \\ &= -\mathbf{df}\,\mathrm{div}\,\mathfrak{H} + \mathbf{df} \times \mathbf{curl}\,\mathfrak{H} + (\mathbf{df} \cdot \mathbf{grad})\,\mathfrak{H}.\end{aligned}$$

But $\mathrm{div}\,\mathfrak{H} = 0$, and in the space outside the currents $\mathbf{curl}\,\mathfrak{H} = 0$ also. Thus

$$\mathbf{F} = J \int (\mathbf{df} \cdot \mathbf{grad})\,\mathfrak{H}/c. \tag{35.7}$$

In particular, in an almost uniform external field $\mathfrak{H}$ can be taken outside the integral, together with the operator **grad**. With the magnetic moment of the current given by (30.18), we then have the obvious result

$$\mathbf{F} = (\mathscr{M} \cdot \mathbf{grad})\,\mathfrak{H}. \tag{35.8}$$

Since $\mathscr{M}$ in this formula is constant, we can also write

$$\mathbf{F} = \mathbf{grad}\,(\mathscr{M} \cdot \mathfrak{H}), \tag{35.9}$$

in agreement with the expression (33.17) for the energy of the current. The torque acting on a current in an almost uniform field is easily seen to be given by the usual expression

$$\mathbf{K} = \mathscr{M} \times \mathfrak{H}. \tag{35.10}$$

PROBLEM

Determine the force on a straight wire carrying a current J and parallel to an infinite circular cylinder with magnetic permeability μ, radius a and axis at a distance b from the wire.

SOLUTION. On account of the relation, mentioned in the second footnote to §30, between two-dimensional problems of electrostatics and magnetostatics, the field of the current is obtained from the result in §7, Problem 3, by changing the notation. The field in the space round the cylinder is the same as that produced in a vacuum by the current J and currents $+J'$ and $-J'$ through A and O' (Fig. 12, §7) respectively, where $J' = (\mu - 1)J/(\mu + 1)$. The field within the cylinder is the same as that due to a current $J'' = 2J/(\mu + 1)$ through O. The force per unit length of the conductor is

$$F = JB/c = \frac{2JJ'}{c^2}\left(\frac{1}{OA} - \frac{1}{OO'}\right)$$
$$= \frac{2J^2a^2(\mu-1)}{b(b^2-a^2)(\mu+1)c^2}.$$

Similarly we find (see §7, Problem 4) that a linear conductor passing through a cylindrical hole in a magnetic medium is attracted to the nearest surface of the hole by a force

$$F = 2J^2 b(\mu-1)/(a^2-b^2)(\mu+1)c^2.$$

§36. Gyromagnetic phenomena

Uniform rotation of a body (having no magnetic structure) causes a magnetization which is linearly dependent on the angular velocity $\mathbf{\Omega}$ (the *Barnett effect*). Phenomenologically, a linear relation between the vectors $\mathscr{M}$ and $\mathbf{\Omega}$ is possible because both change sign under time reversal. Since both are axial vectors, the relation can hold even in an isotropic body, where it reduces to a simple proportionality between $\mathscr{M}$ and $\mathbf{\Omega}$.

There must also be an inverse effect: a freely suspended body, on being magnetized, begins to rotate (the *Einstein–de Haas effect*). There is a simple thermodynamic relation between the two effects; it can be derived as follows.

As we know (see *SP* 1, §26), the thermodynamic potential with respect to the angular velocity (for given temperature and volume of the body) is the free energy $\mathscr{F}'$ of the body in a system of coordinates rotating with it. The angular momentum $\mathbf{L}$ of the body is

$$\mathbf{L} = -\partial\mathscr{F}'/\partial\mathbf{\Omega}. \tag{36.1}$$

The gyromagnetic phenomena are described by adding to the free energy a further term, which is the first term containing both $\mathbf{\Omega}$ and $\mathbf{M}$ in an expansion in powers of $\mathbf{\Omega}$ and of the magnetization $\mathbf{M}$ at each point in the body. This term is linear in both, i.e. it is

$$\mathscr{F}'_{\text{gyro}} = -\int \lambda_{ik}\Omega_i M_k \, dV = -\lambda_{ik}\Omega_i\mathscr{M}_k, \tag{36.2}$$

where λ_{ik} is a constant tensor, in general unsymmetrical.

According to (36.1) and (36.2) the angular momentum acquired by the body as a result of magnetization is related to its total magnetic moment by $L_{\text{gyro},i} = \lambda_{ik}\mathscr{M}_k$. It is usual to replace λ_{ik} by the inverse tensor, defined as $g_{ik} = (2mc/e)\lambda^{-1}{}_{ik}$, where e and m are the electron charge and mass. The dimensionless quantities g_{ik} are called *gyromagnetic coefficients*. Then

$$\mathscr{M}_i = (e/2mc)g_{ik}L_{\text{gyro},k}. \tag{36.3}$$

The expression (36.2) also shows that, as regards its magnetic effect, the rotation of the body is equivalent to an external field $\mathfrak{H}_i = \lambda_{ki}\Omega_k$ or

$$\mathfrak{H}_i = (2mc/e)g^{-1}{}_{ki}\Omega_k. \tag{36.4}$$

We thus have the possibility, in principle, of calculating the magnetization caused by the rotation. For example, if the magnetic susceptibility χ_{ik} of the body is small, the magnetic moment which it acquires is independent of its shape and is

$$\mathscr{M}_i = \chi_{ik}\mathfrak{H}_k = (2mc/e)\chi_{ik}g^{-1}{}_{lk}\Omega_l.$$

Formulae (36.3) and (36.4) represent respectively the Einstein–de Haas and Barnett effects. We see that both effects are determined by the same tensor g_{ik}.

CHAPTER V

FERROMAGNETISM AND ANTIFERROMAGNETISM

§37. Magnetic symmetry of crystals

THERE is a profound difference between the electric properties of crystals and their magnetic properties, which results from a difference in the behaviour of charges and currents with respect to time reversal.

The invariance of the equations of motion with respect to this change means that the formal substitution $t \to -t$, on being applied to any state of thermodynamic equilibrium of a body, must give some possible equilibrium state. There are then two possibilities: either the state obtained by changing the sign of t is the same as the original state, or it is not.

In this section we denote by $\rho(x, y, z)$ and $\mathbf{j}(x, y, z)$ the true (microscopic) charge and current densities at any given point in the crystal, averaged only over time, and not over "physically infinitesimal" volumes as in the macroscopic theory. These are the functions which determine the electric and magnetic structure of the crystal respectively.

When t is replaced by $-t$, $\mathbf{j}$ changes sign. If the state of the body remains unchanged, it follows that $\mathbf{j} = -\mathbf{j}$, i.e. $\mathbf{j} = 0$. Thus there is a reason why bodies can exist in which the function $\mathbf{j}(x, y, z)$ is identically zero. In such bodies, not only the current density but also the (time) average magnetic field and magnetic moment vanish at every point (we are speaking, of course, of states in the absence of an external magnetic field). Such bodies may be said to have no "magnetic structure", and indeed the great majority of bodies fall into this category.

The charge density ρ, on the other hand, is unchanged when $t \to -t$. There is therefore no reason why this function should be identically zero. In other words, there are no crystals without "electric structure", and herein lies the essential difference, mentioned at the beginning of this section, between the electric and the magnetic properties of crystals.

Let us now consider crystals for which the change from t to $-t$ results in a change of state, so that $\mathbf{j} \neq 0$. We shall say that such bodies have a magnetic structure. First of all, we note that, although $\mathbf{j}$ is not zero, there can be no total current in an equilibrium state of the body, i.e. the integral $\int \mathbf{j}\, dV$ taken over a unit cell must always be zero.† Otherwise the current would produce a macroscopic magnetic field, and the crystal would have a magnetic energy per unit volume increasing rapidly with its dimensions. Since such a state is energetically unfavourable, it could not correspond to thermodynamic equilibrium.

The currents $\mathbf{j}$ may, however, produce a non-zero macroscopic magnetic moment, i.e.

† It should be emphasized that the cell spoken of here is the true unit cell, whose definition involves the magnetic structure of the crystal, and this "magnetic cell" may be different from the purely crystallographic cell, which relates only to the symmetry of the charge distribution in the lattice (cf. §38 below).

the integral $\int \mathbf{r} \times \mathbf{j}\, dV$, again taken over a unit cell, need not be zero. Accordingly, the bodies for which $\mathbf{j} \neq 0$ may be divided into two types: those in which the macroscopic magnetic moment is not zero, called *ferromagnets*, and those in which it is zero, called *antiferromagnets*.

What are the possible symmetry groups of the current distribution $\mathbf{j}(x, y, z)$? This symmetry contains, first of all, the usual rotations, reflections and translations, and so the possible symmetry groups of $\mathbf{j}$ always include the usual 230 crystallographic space symmetry groups. These, however, are by no means all. As has already been mentioned, the substitution $t \to -t$ changes the sign of the vector $\mathbf{j}$. For this reason a new symmetry element comes in, namely that resulting from the reversal of all currents; we shall denote this transformation by R. If the current distribution itself has the symmetry R, it follows that $\mathbf{j} = -\mathbf{j}$, i.e. $\mathbf{j} = 0$, and the body has no magnetic structure. A non-zero function $\mathbf{j}(x, y, z)$ may, however, be symmetrical with respect to various combinations of R with the other symmetry elements (rotations, reflections and translations). Thus the problem of determining the possible types of symmetry of the current distribution (the *magnetic space groups*) amounts to the enumeration of all possible groups containing both the transformations of the ordinary space groups and the combinations of these with R.

If the symmetry of the current distribution is given, the crystallographic symmetry of the particle distribution, which is also the symmetry of the function $\rho(x, y, z)$, is determined. It is the symmetry of the space group which is obtained from the symmetry group of $\mathbf{j}$ by formally regarding the transformation R as the identity (as it is with respect to the function ρ).

If only the macroscopic properties of the body are of interest, however, it is not necessary to know the complete symmetry group of the function $\mathbf{j}(x, y, z)$. These properties depend only on the direction in the crystal, and the translational symmetry of the lattice does not affect them. As regards crystallographic structure, the "symmetry of directions" is specified by the 32 crystal classes. These are the symmetry groups consisting of rotations and reflections only, and are obtained from the space groups by regarding every translation as the identity, and the screw axes and glide planes as simple axes and planes of symmetry. As regards the magnetic properties, the macroscopic symmetry can be classified by groups (consisting of rotations, reflections and combinations of these with R) which may be called the *magnetic crystal classes*. They are related to the magnetic space groups in the same way as are the ordinary crystal classes to the ordinary space groups. They include, firstly, the usual 32 classes, and those classes augmented by the element R. These augmented classes are, in particular, the macroscopic symmetry groups for all bodies having no magnetic structure, but they occur also in bodies with magnetic structure. This happens if the magnetic space symmetry group of such bodies includes R only in combination with translations, and not alone.

There are also 58 classes in which R enters only in combination with rotations or reflections. Each of these becomes one of the ordinary crystal classes if R is replaced by the identity.

It should be noted that the occurrence of magnetic structure (ferromagnetic or antiferromagnetic) always involves comparatively weak interactions.† Hence the crystal

† The exchange interaction between the magnetic moments of atoms usually results in the saturation of the valency bonds and the formation of non-magnetic structures. A magnetic structure results only from the relatively weak exchange interactions between deep-lying d and f electrons of atoms of elements in the intermediate groups of the periodic system.

structure of a magnetic body is only a slight modification of that in the non-magnetic phase, which usually changes into the magnetic phase when the temperature is reduced. In this respect ferromagnets, in particular, differ from ordinary pyroelectrics, but are analogous to ferroelectrics.

If the magnetic crystal class of a body is specified, its macroscopic magnetic properties are qualitatively determined. The most important of these is the presence or absence of a macroscopic magnetic moment, i.e. of spontaneous magnetization in the absence of an external field. The magnetic moment **M** is a vector, behaving as an axial vector (the vector product of two polar vectors) under rotation and reflection, and changing sign under the operation R. The crystal will possess spontaneous magnetization if it has one or more directions such that a vector **M** in that direction and having the above-mentioned properties is invariant under all transformations belonging to the magnetic crystal class concerned.

We must again emphasize the difference between these (macroscopic) properties and the corresponding ones in electrostatics. The latter are qualitatively determined by the ordinary crystal class. In particular, a body is pyroelectric if its crystal class admits the existence of a polar vector **P** (the polarization). It would, however, be entirely wrong to base conclusions about the existence or otherwise of a macroscopic magnetic moment on the behaviour of the axial vector **M** with respect to the transformations of the purely structural crystal class of the body concerned, which corresponds to the symmetry of the function $\rho(x, y, z)$; we shall return to this problem in §38, after actually constructing the magnetic classes.

Instead of the symmetry of the function $\mathbf{j}(x, y, z)$, we can use that of the microscopic magnetic moment density distribution $\mathbf{M}(x, y, z) = \mathbf{r} \times \mathbf{j}(x, y, z)$. This in turn may be regarded as the symmetry of the configuration and orientation of the (time) averaged values of the magnetic moments $\boldsymbol{\mu}$ of the atoms or ions in the lattice. In a body without magnetic symmetry these averaged values are zero. In a ferromagnet the sum of the atomic moments in each unit cell is non-zero, but in an antiferromagnet it is zero.

A set of atoms in the lattice having equal values of $\boldsymbol{\mu}$ is called a *magnetic sub-lattice*. An antiferromagnet will evidently contain at least two sub-lattices with equal and opposite values of $\boldsymbol{\mu}$. If the directions of the moments $\boldsymbol{\mu}$ are parallel or antiparallel for all sub-lattices, such a body is said to be *collinear*; in the contrary case, it is a *non-collinear* antiferromagnet.

A ferromagnet also may contain more than one sub-lattice. In the narrow sense, the term "ferromagnetic" applies to bodies in which all the mean atomic magnetic moments are parallel. A crystal is said to be *ferrimagnetic* if it contains two or more sub-lattices with atomic moments that differ in direction or magnitude. In contrast to the antiferromagnetic case, in these substances the vector sum **M** of the magnetic moments of the sub-lattices is not zero. A ferromagnet may be collinear, or not, according as the magnetic moments of all its sub-lattices are parallel or antiparallel, or not.

§38. Magnetic classes and space groups

We shall now show how the magnetic symmetry groups are constructed, beginning with the magnetic classes.

As already mentioned in §37, the magnetic classes fall into three types. Type I includes the 32 ordinary crystal classes, which do not involve the element R. Type II includes the same 32 classes augmented by R. Each such class contains all the elements of the ordinary class (of the point group $\boldsymbol{G}$) and also all these elements multiplied by R; if the magnetic class is denoted by $\boldsymbol{M}$, then we can write

$$\boldsymbol{M} = \boldsymbol{G} + R\boldsymbol{G}. \tag{38.1}$$

The transformation R commutes, of course, with all spatial rotations and reflections; hence $RG = GR$, where G is any element of $\boldsymbol{G}$.

These two types of class are, in a sense, trivial. The non-trivial type III includes the 58 magnetic classes, in which R enters only in combination with rotations or reflections. Each of these becomes one of the ordinary crystal classes of $\boldsymbol{G}$ if R is replaced by the identity. The construction of all magnetic classes of this type is carried out as follows.

Let $\boldsymbol{H}$ denote the set of elements of the group $\boldsymbol{G}$ which are not multiplied by R when the magnetic class $\boldsymbol{M}$ is formed. Such a set includes, by definition, the unit element E, since otherwise $\boldsymbol{M}$ would contain the element R itself, i.e., would belong to type II. The products of any two of the elements in this set are also members of the set. Thus $\boldsymbol{H}$ is a sub-group of $\boldsymbol{G}$. All other elements of $\boldsymbol{G}$ appear in $\boldsymbol{M}$ multiplied by R; since $R^2 = E$, all products of pairs of these elements belong to $\boldsymbol{H}$. It follows that $\boldsymbol{H}$ is a sub-group of $\boldsymbol{G}$ (and therefore of $\boldsymbol{M}$) with index 2.† The structure of a type III magnetic class $\boldsymbol{M}$ may therefore be written as

$$\boldsymbol{M} = \boldsymbol{H} + RG_1\boldsymbol{H}, \tag{38.2}$$

where G_1 is any element of $\boldsymbol{G}$ that is not an element of $\boldsymbol{H}$. It is evident that the groups $\boldsymbol{M}$ and $\boldsymbol{G} = \boldsymbol{H} + G_1\boldsymbol{H}$ are isomorphous.

The problem of constructing all the magnetic classes thus amounts to finding the sub-groups with index 2 in all the crystal classes. This is easily done by means of the character tables for the irreducible representations of the point groups. Every one-dimensional (other than the unit) representation of the group contains an equal number of characters $+1$ and -1; the elements with characters $+1$ form a sub-group with index 2. On changing to the magnetic class, these elements remain the same, while all others are multiplied by R.

The procedure may be illustrated with reference to the point group C_{4v}. The character table for its irreducible representations is (see *QM*, §95)

	E	C_2	$2C_4$	$2\sigma_v$	$2\sigma'_v$
A_1	1	1	1	1	1
A_2	1	1	1	-1	-1
B_1	1	1	-1	1	-1
B_2	1	1	-1	-1	1
E	2	-2	0	0	0

The one-dimensional (other than the unit) representations are A_2, B_1, B_2. In the representation A_2, the elements with characters $+1$ form the sub-group C_4. The corresponding magnetic class, denoted by $C_{4v}(C_4)$, consists of the elements E, C_2, $2C_4$, $2R\sigma_v$, $2R\sigma'_v$. In the representations B_1 and B_2, the elements with characters $+1$ form the sub-groups C_{2v}, which differ only in the position of the planes σ_v relative to fixed coordinates. These sub-groups are crystallographically indistinguishable, and correspond to the same magnetic class $C_{4v}(C_{2v})$, whose elements are E, C_2, $2RC_4$, $2\sigma_v$, $2R\sigma'_v$.

† This means that $\boldsymbol{H}$ contains half as many elements as $\boldsymbol{G}$. The statement is a consequence of a general theorem which is fairly obvious: a sub-group $\boldsymbol{H}$ of $\boldsymbol{G}$ has index 2 if and only if the product of any two elements of $\boldsymbol{G}$ that do not belong to $\boldsymbol{H}$ is an element of $\boldsymbol{H}$.

TABLE 1 *Magnetic classes*

$C_i(C_1)$	$C_{3v}(C_3)$
$C_s(C_1)$	$D_3(C_3)$
$C_2(C_1)$	$D_{3d}(D_3, S_6, C_{3v})$
$C_{2h}(C_i, C_2, C_s)$	$C_{3h}(C_3)$
$C_{2v}(C_s, C_2)$	$C_6(C_3)$
$D_2(C_2)$	$D_{3h}(C_{3h}, C_{3v}, D_3)$
$D_{2h}(D_2, C_{2h}, C_{2v})$	$C_{6h}(C_6, S_6, C_{3h})$
$C_4(C_2)$	$C_{6v}(C_6, C_{3v})$
$S_4(C_2)$	$D_6(C_6, D_3)$
$D_{2d}(S_4, D_2, C_{2v})$	$D_{6h}(D_6, C_{6h}, C_{6v}, D_{3d}, D_{3h})$
$D_4(C_4, D_2)$	$T_h(T)$
$C_{4v}(C_4, C_{2v})$	$O(T)$
$C_{4h}(C_4, C_{2h}, S_4)$	$T_d(T)$
$D_{4h}(D_4, C_{4h}, D_{2h}, C_{4v}, D_{2d})$	$O_h(O, T_h, T_d)$
$S_6(C_3)$	

On going through all the 32 crystal classes in this way, we obtain the 58 magnetic classes of type III, as listed in Table 1. Each class $\boldsymbol{G}(\boldsymbol{H})$ is defined by an original point group $\boldsymbol{G}$ and a sub-group $\boldsymbol{H}$ thereof which is one of those listed in parentheses after the symbol for the particular group $\boldsymbol{G}$. The crystal classes $\boldsymbol{C}_1$, $\boldsymbol{C}_3$ and $\boldsymbol{T}$ have no sub-groups with index 2, and so they give rise to no magnetic classes. The rotation C_3, moreover, never appears multiplied by R in a (non-trivial) magnetic class: the rotation C_3R thrice repeated would give R, which is not a member of such classes.†

It has been mentioned in §37 that the crystal class concerned does not determine whether or not ferromagnetism can exist. To illustrate this, let us consider a tetragonal lattice of identical atoms, with magnetic moments parallel to the tetragonal axis.‡ Its magnetic crystal class is $\boldsymbol{D}_{4h}(\boldsymbol{D}_4)$, containing the transformations§ $E, C_2, 2C_4, 2U_2R, 2U'_2R, I, \sigma_h, 2S_4, 2\sigma_vR, 2\sigma'_vR$. All these transformations leave invariant the axial vector $\mathbf{M}$, which is parallel to the fourfold axis. The crystal class $\boldsymbol{D}_{4h}$ itself, however, would not allow the existence of an axial vector: all the components M_x, M_y, M_z would change sign in a rotation about some twofold axis, for example.

Let us now turn to the magnetic space groups. These are in the same relation to the ordinary crystal space groups as the magnetic classes are to the crystal classes, reducing to these if R is replaced by the identity. The total number of magnetic space groups is 1651; like the magnetic classes, they fall into three types.

Type I contains 230 groups which coincide with the crystal groups and do not involve R; type II has the same 230 groups augmented by R.

The non-trivial type III contains 1191 groups in which R enters only in combination with rotations, reflections or translations. It has the structure (38.2), where $\boldsymbol{H}$ is any sub-group

† In abstract symmetry theory, magnetic symmetry is called antisymmetry. This concept was independently proposed by H. Heesch (1929) and A. V. Shubnikov (1945). The antisymmetry classes were found by Shubnikov (1951) as the symmetry groups of geometric figures (polyhedra) with faces of two colours; the element R then corresponds to changing the colour of each face. These classes were derived as magnetic symmetry groups by B. A. Tavger and V. M. Zaïtsev (1956). The method of derivation given here is due to V. L. Indenbom (1959).

‡ Such, for instance, is the lattice of iron in its ferromagnetic phase. Crystallographically, it is a cubic lattice slightly distorted along one of the fourfold symmetry axes. The distortion is the result of magnetostriction arising simply from the presence of the magnetic structure.

§ The notation for the symmetry elements is everywhere as in *QM*, §§93 and 94. In particular, U_2 and U'_2 are rotations through 180° about horizontal axes perpendicular to the fourfold axis, σ_h is a reflection in the horizontal plane, σ_v and σ'_v are reflections in vertical planes passing through the axes C_4 and U_2, U'_2. The symbols U_2R, σR, etc. denote symmetry planes and axes which appear in this class in combination with R.

with index 2 of the crystal space group G, and G_1 is an element of G that is not an element of H. The sub-group H must evidently coincide with the original space group G as regards either translational symmetry or class, having respectively either half as many "rotational" elements (rotations and reflections) as G, or half as many translations. Accordingly, type III may be subdivided into two sub-types.

Sub-type IIIa contains the magnetic space groups for which G_1 in (38.2) is a "rotational" transformation of the crystal group $G = H + G_1 H$ which does not belong to H. The translational symmetry (Bravais lattice) of a space group M of this type is the same as that of G; that is, the unit cell of the magnetic structure is the same as the purely crystallographic one. These magnetic space groups, 674 in number, belong to magnetic classes of type III.

Sub-type IIIb contains the magnetic space groups for which G_1 in (38.2) may be taken to be a pure translation through one of the basic periods of G. The unit cell of the magnetic structure has a volume twice that of the crystallographic unit cell. The set of pure translations and translations multiplied by R forms the *magnetic Bravais lattice*; there are 22 different lattices of this kind. The magnetic space groups of sub-type IIIb, 517 in number, belong to magnetic classes of type II.†

PROBLEM

List the magnetic classes that allow ferromagnetism.

SOLUTION. The classes of type II do not allow ferromagnetism. It must be emphasized that the same is therefore true of every space group of type II or IIIb that contains translations multiplied by R; these transformations certainly change the sign of $\mathbf{M}$. In other words, for ferromagnetism to occur, it is always necessary that the magnetic unit cell should coincide with the crystallographic one.‡

Of the classes of type I (which are the same as the ordinary crystal classes), the following allow the existence of an axial vector $\mathbf{M}$: C_1 and C_i with $\mathbf{M}$ in any direction; C_s with $\mathbf{M}$ in the plane of symmetry; C_2, C_{2h}, C_4, S_4, C_{4h}, C_3, C_6, S_6, C_{3h}, C_{6h} with $\mathbf{M}$ parallel to the axis of symmetry.

For ferromagnetism to exist in classes of type III, they must never contain the element IR, which changes the sign of the vector $\mathbf{M}$ whatever the direction of the latter. Of such classes, the following allow ferromagnetism: $C_2(C_1)$, $C_{2h}(C_i)$ with $\mathbf{M}$ perpendicular to the C_2R axis; $C_s(C_1)$ with $\mathbf{M}$ in the σR plane; $C_{2v}(C_s)$ with $\mathbf{M}$ in the $\sigma_v R$ plane and perpendicular to the C_2R axis; $D_2(C_2)$, $C_{2v}(C_2)$, $D_{2h}(C_{2h})$ with $\mathbf{M}$ parallel to the C_2 axis; $D_4(C_4)$, $C_{4v}(C_4)$, $D_{2d}(S_4)$, $D_{4h}(C_{4h})$, $D_3(C_3)$, $C_{3v}(C_3)$, $D_{3d}(S_6)$, $D_{3h}(C_{3h})$, $D_6(C_6)$, $C_{6v}(C_6)$, $D_{6h}(C_{6h})$ with $\mathbf{M}$ parallel to the C_4, C_3 or C_6 axis.

§39. Ferromagnets near the Curie point

There is a close analogy between the magnetic properties of ferromagnets and the electric properties of ferroelectrics. Both exhibit spontaneous polarization, magnetic or electric, in macroscopic volumes. In each case, this polarization vanishes at a temperature corresponding to a second-order phase transition (the transition point between the ferromagnetic and paramagnetic phases is called the *Curie point*).

There are also, however, important differences between ferromagnetic and ferroelectric phenomena, arising from the difference in the microscopic interaction forces which bring

† The magnetic space groups were constructed (as antisymmetry groups) by A. M. Zamorzaev (1953) and by N. V. Belov, N. N. Neronova, and T. S. Smirnova (1955); the latter authors' tables are given by A. V. Shubnikov and N. V. Belov, *Colored Symmetry*, Pergamon Press, Oxford, 1964. The most complete tables of the magnetic space groups and their properties are those given by V. A. Koptsik, *Shubnikov Groups* (*Shubnikovskie gruppy*), Moscow State University Press, 1966.

‡ This refers (see the last footnote but two) to the symmetry of a lattice already distorted by the very existence of the magnetic structure.

about the spontaneous polarization. In ferroelectrics, the interaction between the molecules in the crystal lattice is essentially anisotropic, and consequently the spontaneous polarization vector is fairly closely related to certain directions in the crystal. The formation of a magnetic (including ferromagnetic) structure, on the other hand, is due mainly to the exchange interaction of the atoms, which is quite independent of the direction of the total magnetic moment relative to the lattice.† It is true that, together with the exchange interaction, there is also a direct magnetic interaction between the magnetic moments of the atoms. This interaction, however, is an effect of order v^2/c^2 (v being the atomic velocities), since the magnetic moments themselves contain a factor v/c. The effects of this order include also the interaction of the magnetic moments of the atoms with the electric field of the crystal lattice. All these interactions, which may be called *relativistic* by virtue of the factor $1/c^2$ in them, are weak in comparison with the exchange interaction, so that they can result only in a comparatively slight dependence of the energy of the crystal on the direction of magnetization. This relationship between the exchange and relativistic interactions will be assumed throughout the rest of the chapter.‡

Consequently, the magnetization of a ferromagnet is a quantity which, in the first approximation (i.e. on the basis of the exchange interaction), is conserved. This fact endows with greater physical significance the thermodynamic theory, in which the magnetization **M** is regarded as an independent variable, the actual value of which (as a function of temperature, field, etc.) is afterwards determined by the appropriate conditions of thermal equilibrium.§

We denote by $\Phi(\mathbf{M}, \mathbf{H})$ the thermodynamic potential per unit volume of the substance, regarded as a function of the independent variable **M** (and of the other thermodynamic variables). We shall, for the present, neglect the relativistic interactions, i.e. we shall take into account only the exchange interaction. Then $\Phi(\mathbf{M}, 0)$ may be a function of the magnitude of **M**, but not of its direction.

In order to find the thermodynamic quantities when the field **H** is not zero, we proceed exactly as in the derivation of (19.3), starting from the relation $\partial\tilde{\Phi}/\partial\mathbf{H} = -\mathbf{B}/4\pi$. This gives

$$\tilde{\Phi}(\mathbf{M}, \mathbf{H}) = \Phi(M, 0) - \mathbf{M}\cdot\mathbf{H} - H^2/8\pi. \tag{39.1}$$

Hence the potential Φ is

$$\begin{aligned}\Phi(\mathbf{M}, \mathbf{B}) &= \tilde{\Phi} + \mathbf{H}\cdot\mathbf{B}/4\pi \\ &= \Phi(M, 0) + H^2/8\pi \\ &= \Phi(M, 0) + (\mathbf{B} - 4\pi\mathbf{M})^2/8\pi.\end{aligned} \tag{39.2}$$

---

† The exchange interaction is a quantum effect resulting from the symmetry of the wave functions of the system with respect to interchanges of the particles. The interchange symmetry of the wave functions, and therefore the exchange interaction, depend only on the total spin of the system, and not on the direction of the spin; see *QM*, §60. The importance of the exchange interaction in ferromagnets was first pointed out by Ya. I. Frenkel', Ya. G. Dorfman and W. Heisenberg (1928).

‡ The order of magnitude of the ratio of the relativistic and exchange interactions is given by the ratio U_{aniso}/NT_c, where U_{aniso} is the *magnetic anisotropy energy* (see §40), N the number of atoms per unit volume, and T_c the temperature of the Curie point. For ferromagnets this ratio is usually between 10^{-2} and 10^{-5}. In some ferromagnets (rare-earth metals and their compounds), however, it may be much greater and even reach values of the order of unity, both because of the "anomalously" large anisotropy energy and because of the relative weakness of the exchange interactions. There are, of course, only limited possibilities of applying the macroscopic theory to such substances. A detailed discussion of the microscopic mechanisms of the various interactions in specific magnetic materials is outside the scope of this book.

§ The collinear ferrimagnets are indistinguishable from ferromagnets in the narrow sense (see the end of §37) as regards macroscopic magnetic symmetry and as regards their behaviour in fairly weak magnetic fields. The theory given below relates to both these types of material.

When the magnetic anisotropy of the ferromagnet is neglected, the directions of the vectors **M** and **H** are, of course, the same, and so the vectors in formulae (39.1) and (39.2) are replaced by their magnitudes.

Near the Curie point, the magnetization **M** is small. We shall consider the properties of a ferromagnet in this range in the general Landau theory of second-order phase transitions.† In accordance with this theory, we expand $\Phi(M, 0)$ as a series in powers of the vector **M**, which acts as an order parameter. The expansion of an isotropic function in powers of a vector quantity can contain only even powers:

$$\tilde{\Phi} = \Phi_0 + AM^2 + BM^4 - MH - H^2/8\pi, \tag{39.3}$$

where Φ_0, A and B are functions only of temperature and pressure.‡

The Curie point $T = T_c(P)$ is given by the vanishing of the coefficient A: $A > 0$ for $T > T_c$ and $A < 0$ for $T < T_c$. (This relation of the phases occurs in all known ferromagnets, although it is not thermodynamically necessary.) Near the Curie point,

$$A = a(T - T_c) \tag{39.4}$$

where a is a positive quantity independent of temperature. The expressions (39.3) and (39.4) differ from (19.3) only in that M and H replace P_z and E_z. The conclusions which follow from these expressions will therefore be stated without repeating the arguments given in §19.

The spontaneous magnetization in the ferromagnetic phase varies with temperature according to

$$M = \sqrt{[a(T_c - T)/2B]}. \tag{39.5}$$

Above the Curie point there is no spontaneous magnetization, and the magnetic susceptibility is

$$\chi = 1/2a(T - T_c), \tag{39.6}$$

i.e., there is paramagnetism with susceptibility inversely proportional to $T - T_c$ (the *Curie–Weiss law*). Below the Curie point,

$$\chi = (\partial M/\partial H)_{H \to 0} = 1/4a(T_c - T). \tag{39.7}$$

It should be pointed out, however, that this quantity is here not the susceptibility in the ordinary sense of the word (i.e. the coefficient of proportionality between M and H), since $M \neq 0$ even when $H = 0$.

The susceptibility (39.7) can actually attain values of the order of unity only in the immediate neighbourhood of the Curie point. Except in this region, we may suppose that the magnetization M changes only very slightly with the magnetic field and may be regarded as a constant for any given temperature. In the following sections we shall assume this to be true.

This constitutes a further difference between ferromagnets and ferroelectrics: for the latter, $\partial P/\partial E$ is in general not small even far from the Curie point. The reason again lies in the smallness of the magnetic moments of the atoms in comparison with the electric dipole moments of the molecules.

It has been mentioned in §19 that the application of an electric field blurs the discrete

---

† The properties of a ferromagnet in the fluctuation range near the Curie point will be discussed in §47.

‡ The fact that the expansion contains only even powers of the components of **M** has also another more profound origin which does not depend on the exchange approximation: **M** is an odd function under time reversal, whereas the thermodynamic potential must of course be invariant under time reversal.

second-order phase transition point in ferroelectrics. The same is true, of course, for a ferromagnet in a magnetic field. Since **M** and **H** have the same direction in the exchange approximation, the blurring of the transition occurs in that approximation whatever the crystallographic direction of **H**.

§40. The magnetic anisotropy energy

As already mentioned, the anisotropy of the magnetic properties of ferromagnets is due to the relativistic interactions between their atoms, and these interactions are comparatively weak. In the macroscopic theory, the anisotropy is described by the addition to the thermodynamic potential of the *magnetic anisotropy energy*, which depends on the direction of magnetization.

The calculation of the anisotropy energy from the microscopic theory would require the use of quantum perturbation theory, the energy of the perturbation being represented by the terms in the Hamiltonian of the crystal which pertain to the relativistic interactions. The general form of the desired expressions, however, can be deduced without such calculations, from simple arguments concerning symmetry.

The Hamiltonian of the relativistic interactions contains terms of the first and second powers in the electron spin vector operators; these are respectively the *spin–orbit* and *spin–spin* interactions. Both kinds are small in proportion to v^2/c^2, where v is of the order of magnitude of the velocities of atomic electrons, and c is the velocity of light. The perturbation-theory series for the anisotropy energy is an expansion in powers of this small quantity, but, because of the above-mentioned dependence of the perturbation operator on the spin operators, the anisotropy energy is automatically obtained as a series of powers of the direction cosines of the magnetization vector, i.e. the components of a unit vector **m** in the direction of **M**. The anisotropy energy U_{aniso}, like the potential Φ itself (cf. the last footnote), is invariant under time reversal, whereas the magnetization **M** changes sign. Hence it follows that the anisotropy energy must be an even function of the components of **m**.

For uniaxial and biaxial crystals, the expansion of the anisotropy energy begins with squares of these components, and may be written

$$U_{\text{aniso}} = K_{ik} m_i m_k, \tag{40.1}$$

where K_{ik} is a symmetrical tensor of rank two, whose components, like U_{aniso} itself, have the dimensions of energy density. In uniaxial and biaxial crystals, such a tensor has respectively two and three independent components. In the present case, however, it must also be borne in mind that one quadratic combination, namely $m_x{}^2 + m_y{}^2 + m_z{}^2 = 1$, is independent of the direction of **m** and so cannot appear in the anisotropy energy. Hence the expression (40.1) for uniaxial and biaxial crystals contains only one and two independent coefficients respectively.

For example, in uniaxial crystals the anisotropy energy can be written

$$U_{\text{aniso}} = K(m_x{}^2 + m_y{}^2) = K \sin^2\theta \tag{40.2}$$

or, equivalently,†

$$U_{\text{aniso}} = -Km_z{}^2 = -K\cos^2\theta,$$

† These two expressions differ by the quantity K, which is independent of direction. The change from one to the other implies that this quantity is included in the isotropic part of Φ; in particular, it alters the coefficient A in the expansion (39.3).

where θ is the angle between $\mathbf{m}$ and the z-axis, taken to be along the principal axis of symmetry of the crystal. If K (a function of temperature) is positive, the anisotropy energy is least for magnetization in the z-direction; this will then be the *direction of easy magnetization*. Such a ferromagnet is said to be of the "easy-axis" type. If K is negative, the direction of easy magnetization lies in the xy-plane (the basal plane of the crystal); such a ferromagnet is said to be of the "easy-plane" type.† The expression (40.2) is isotropic in the xy-plane. This isotropy, however, does not occur in the higher-order terms in the expansion of U_{aniso}, and these determine (when $K < 0$) the direction of easy magnetization in the xy-plane. The form of such terms depends on the particular crystal system to which the crystal belongs.

For tetragonal crystals, the fourth-order terms include two independent invariants, $(m_x^2+m_y^2)^2$ and $m_x^2 m_y^2$ (the x and y axes being taken along the two twofold axes in the basal plane). The second of these leads to anisotropy in the basal plane.

In a hexagonal crystal, the anisotropy energy contains only one fourth-order term, proportional to $(m_x^2+m_y^2)^2$; in this approximation,

$$U_{\text{aniso}} = K_1 \sin^2\theta + K_2 \sin^4\theta, \tag{40.3}$$

the coefficient K in (40.2) being here denoted by K_1. The anisotropy in the basal plane, however, occurs only in the sixth-order terms; the anisotropic invariant of this order is

$$\tfrac{1}{2}[(m_x+im_y)^6+(m_x-im_y)^6] = \sin^4\theta \cos 6\phi, \tag{40.4}$$

the x-axis being taken along one of the twofold axes in the basal plane, from which the azimuthal angle ϕ is measured.

Lastly, rhombohedral symmetry allows two fourth-order terms, with the invariants

$$(m_x^2+m_y^2)^2 = \sin^4\theta, \tag{40.5}$$

$$\tfrac{1}{2}m_z[(m_x+im_y)^3+(m_x-im_y)^3] = \cos\theta \sin^3\theta \cos 3\phi, \tag{40.6}$$

the y-axis being along one of the twofold axes, and ϕ being measured from the x-axis. The presence of the factor m_z in the second invariant means that the direction of easy magnetization is at a small angle to the basal plane; since m_z is small, the determination of this angle would require the simultaneous inclusion of both fourth-order and sixth-order terms, one of the latter being the term with the invariant (40.4).

Let us now consider ferromagnetic crystals of the cubic system. Their properties differ considerably from those of uniaxial (and biaxial) crystals. This is because the only quadratic combination which is invariant under the cubic symmetry transformations and which can be formed from the components of the vector $\mathbf{m}$ is $\mathbf{m}^2 = 1$. The first non-vanishing term in the expansion of the anisotropy energy for a cubic crystal is therefore of the fourth, not the second, order. For this reason, the magnetic anisotropy effects in cubic crystals are in general weaker than in uniaxial and biaxial crystals.

Cubic symmetry allows only one independent fourth-order invariant which depends on the direction of $\mathbf{m}$. The anisotropy energy of a cubic ferromagnet can therefore be expressed as

$$U_{\text{aniso}} = K(m_x^2 m_y^2 + m_x^2 m_z^2 + m_y^2 m_z^2) \tag{40.7}$$

---

† An example of a uniaxial ferromagnet is hexagonal cobalt, in which K changes sign, being positive below $\sim 530\,°\text{K}$ and negative above this temperature. The value extrapolated to $0\,°\text{K}$ is $\approx 0.8\times10^7$ erg/cm^3. The value of NT_c is $\sim 2\times10^{10}$ erg/cm^3.

or

$$U_{\text{aniso}} = -\tfrac{1}{2}K(m_x^4 + m_y^4 + m_z^4);$$

these two expressions are evidently equivalent, since their difference is $\frac{1}{2}K$, which is independent of direction.

If $K > 0$, as for example in iron, the anisotropy energy has equal minimum values for three positions of the vector **m**, namely parallel to the edges of the cube (the x, y, z axes, with crystallographic directions [100], [010], [001]). In this case, therefore, the crystal has three equivalent axes of easy magnetization.

If, on the other hand, $K < 0$, as for example in nickel, then the anisotropy energy is least when $m_x^2 = m_y^2 = m_z^2 = \frac{1}{3}$, i.e. when **m** is parallel to any of the four spatial diagonals of the cube, with crystallographic directions [111], $[\bar{1}11]$, etc. These are then the directions of easy magnetization.†

The next approximation after (40.7) for the anisotropy energy of a cubic crystal corresponds to the sixth-order terms. Omitting from these the direction-independent invariant $(\mathbf{m}^2)^3$ and an expression which differs from (40.7) only by the factor $\mathbf{m}^2$, we are left with just one invariant, which may be taken as $m_x^2 m_y^2 m_z^2$. Then

$$U_{\text{aniso}} = K_1(m_x^2 m_y^2 + m_x^2 m_z^2 + m_y^2 m_z^2) + K_2 m_x^2 m_y^2 m_z^2. \tag{40.8}$$

It should be noted that, strictly speaking, a ferromagnetic cubic crystal, when spontaneously magnetized along one of the directions of easy magnetization, ceases to possess cubic symmetry, and so there is a displacement of the atoms, i.e. a distortion of the crystal lattice. Such a crystal, when magnetized parallel to an edge of the cube, becomes slightly tetragonal, while one magnetized parallel to a spatial diagonal becomes rhombohedral. In this respect cubic crystals differ from uniaxial crystals with the direction of easy magnetization along the principal axis of symmetry, where a magnetization in this direction evidently does not change the symmetry of the crystal.

We must again emphasize that the above expansion of the anisotropy energy of a ferromagnet in powers of the components of the unit vector **m** is not one in powers of the magnetization **M** itself (which is not small far from the Curie point); the series converges only because the relativistic interactions are weak. Near the Curie point, however, where **M** is small, it does become an expansion in powers of **M**. In the Landau theory, this would imply that the ratio K/M^2 in a uniaxial crystal, with K from (40.2), must tend to a non-zero limit as $T \to T_c$. To illustrate this statement, let us consider, for example, the transition from a paramagnetic phase to a ferromagnetic phase of the easy-axis type. According to (39.3) and (40.2), the quadratic terms in the expansion of the thermodynamic potential in powers of the components of **M** are $AM_z^2 + (A + K/M^2)(M_x^2 + M_y^2)$. The transition point is given by the vanishing of the coefficient of M_z^2; the coefficient of $M_x^2 + M_y^2$ does not become zero.

Similarly, in a cubic ferromagnet the ratio K/M^4, with K from (40.7), must tend to a non-zero limit.

In the fluctuation range, however, this behaviour of the anisotropy coefficients does not in general occur.

† As an example, it may be noted that in iron and nickel the value of $\frac{1}{3}K$ (the difference in U_{aniso} for the directions of easiest and most difficult magnetization), extrapolated to 0 °K, is about 2×10^5 erg/cm^3.

§41. The magnetization curve of ferromagnets

Let us examine the relation between the magnetization of a uniaxial ferromagnet and the magnetic field in it, taking the particular case of an easy-axis material. The anisotropy energy may here be conveniently written in the form

$$U_{\text{aniso}} = \tfrac{1}{2}\beta M^2 \sin^2\theta, \tag{41.1}$$

with the dimensionless coefficient $\beta > 0$ (and $K = \frac{1}{2}\beta M^2$).†

It should be recalled that we suppose the magnitude of **M** to be independent of **H**, so that only rotations of **M** are considered.‡ It is evident from symmetry that the vector **M** lies in a plane through the z-axis and the direction of **H** (if terms of higher order, anisotropic in the xy-plane, are neglected in the anisotropy energy). We take this as the xz-plane. The thermodynamic potential, including the anisotropy energy, is§

$$\begin{aligned}\tilde{\Phi} &= \Phi_0(M) + \tfrac{1}{2}\beta M_x{}^2 - \mathbf{M}\cdot\mathbf{H} - H^2/8\pi \\ &= \Phi_0(M) + \tfrac{1}{2}\beta M^2 \sin^2\theta - M(H_x \sin\theta + H_z\cos\theta) - H^2/8\pi.\end{aligned} \tag{41.2}$$

The dependence of **M** on **H** is given by the equilibrium condition $\partial\tilde{\Phi}/\partial\theta = 0$, whence

$$\beta M \sin\theta\cos\theta = H_x\cos\theta - H_z\sin\theta. \tag{41.3}$$

This is an algebraic equation of the fourth degree for $\xi = \sin\theta$:

$$(\beta M\xi - H_x)^2(1-\xi^2) = H_z{}^2\xi^2,$$

in which the coefficients of odd powers of ξ are not zero. This equation has either two or four real roots (all less than unity). Since all such roots correspond to extrema of $\tilde{\Phi}(\theta)$, it is clear that, in the first case, this function has one minimum and one maximum and, in the second case, two minima and two maxima. In other words, the number of possible directions of the magnetization **M** for a given field **H** is one and two respectively. In the second case, one direction (corresponding to the lower minimum of $\tilde{\Phi}$) is thermodynamically completely stable, while the other (corresponding to the higher minimum) is thermodynamically metastable.

Either of the two cases can occur, depending on the values of H_x and H_z. When these two parameters vary continuously, one case passes into the other at the point where one maximum and one minimum coalesce. The curve of $\tilde{\Phi}(\theta)$ then has a point of inflection instead of an extremum, i.e. both $\partial\tilde{\Phi}/\partial\theta$ and $\partial^2\tilde{\Phi}/\partial\theta^2$ are zero. Writing equation (41.3) in the form

$$\frac{H_x}{\sin\theta} - \frac{H_z}{\cos\theta} = \beta M$$

† The analysis given below is based on the expression (41.1) for the anisotropy energy. It should be mentioned, however, that the expansion of which (41.1) is the first term is usually not rapidly convergent in practice. For a satisfactory quantitative description of the phenomena, it is therefore necessary to include also the term of the next (fourth) order.

‡ In addition to the rotation of **M** considered here, another process can occur in ferrites in very strong fields, where the antiparallel magnetic moments rotate towards each other and become parallel. This is found, however, only in "exchange" fields $H \sim T_c/\mu$; for instance, in the ferrite $FeO.Fe_2O_3$ ($T_c \sim 580\,°K$, $\mu \sim \mu_B$), these fields are $H \sim 10^7$ Oe.

§ By including U_{aniso} in the thermodynamic potential $\tilde{\Phi}$, we imply that the anisotropy constants are defined for given elastic stresses.

and differentiating with respect to θ, we have $H_x/\sin^3\theta = -H_z/\cos^3\theta$. Eliminating θ from these two equations gives

$$H_x^{2/3} + H_z^{2/3} = (\beta M)^{2/3}. \tag{41.4}$$

In the H_xH_z-plane equation (41.4) represents a closed astroid curve of the kind shown in Fig. 21. It divides the plane into two parts, in one of which metastable states can exist, while in the other they cannot. It is evident without further investigation that the region where metastable states do not exist is that outside the curve, because for $H \to \infty$ only one direction of **M** can be stable, namely that of the field **H**.

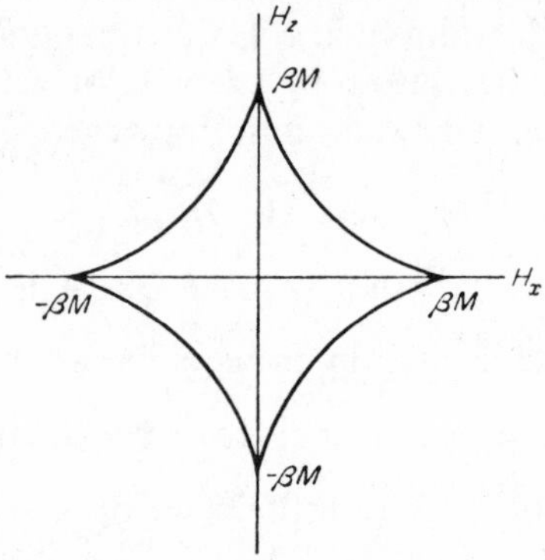

FIG. 21

The existence of metastable states means that what is called *hysteresis* can occur; this is an irreversible change of magnetization on passing through these states when the external magnetic field is varied. The curve shown in Fig. 21 is therefore the absolute limit of hysteresis; this phenomenon cannot occur for fields outside the curve.†

States in which the field **H** is perpendicular to the direction of easy magnetization ($H_x = H$, $H_z = 0$) require special consideration. The thermodynamic potential is

$$\tilde{\Phi} = \Phi_0 - \frac{H^2}{8\pi} + \tfrac{1}{2}\beta M^2 \sin^2\theta - HM\sin\theta. \tag{41.5}$$

If $H > \beta M$, $\tilde{\Phi}$ has only one minimum, at $\theta = \frac{1}{2}\pi$, i.e. the magnetization is parallel to the field. If, however, $H < \beta M$, then $\tilde{\Phi}$ has a minimum when

$$M_x = M\sin\theta = H/\beta, \tag{41.6}$$

to which there correspond two possible positions of the vector **M** (at angles θ and $\pi - \theta$), symmetrical about the x-axis. Thus in this case there are two equilibrium states, which have the same value of $\tilde{\Phi}$ and are therefore equally stable.

This result is very important, since it means that two phases can exist in contact in which the field **H** is the same but the magnetization **M** (and therefore the induction **B**) is different. Thus a new possibility appears for reducing the total thermodynamic potential of the body: its volume may be divided into separate regions, in each of which the magnetization has one of its two possible directions. These regions are called *regions of spontaneous*

† The whole of the discussion in this chapter concerns only thermodynamic equilibrium states in ferromagnets and therefore reversible processes in them. In particular, we entirely ignore the mechanism of hysteresis phenomena; these may arise from defects in the crystal, internal stresses, a polycrystalline state, and so on.

magnetization or *domains*. The actual determination of the thermodynamic equilibrium structure of a ferromagnet requires a consideration of the shape and size of the body as a whole. We shall return to this problem in §44.

Let us consider a portion of the body which is small compared with the total volume but large compared with the domains. The field H_x can be regarded as constant in this portion; we denote by $\overline{\mathbf{M}}$ and $\overline{\mathbf{B}}$ the values of **M** and **B** averaged over its volume. As well as H_x, the transverse component $M_x = H_x/\beta$ of the magnetization is constant. The longitudinal component M_z, however, has opposite signs in different domains, so that its mean value certainly cannot exceed $|M_z|$. Since $H_z = 0$ everywhere, the mean induction is therefore

$$\overline{B}_x = H_x\left(1+\frac{4\pi}{\beta}\right), \quad \overline{B}_z < 4\pi\sqrt{\left(M^2-\frac{H_x{}^2}{\beta^2}\right)}. \tag{41.7}$$

These formulae give the range of values of the mean induction corresponding to the domain structure of a uniaxial ferromagnet.

The relation between **M** and **H** for a cubic crystal can in principle be investigated in the same way as was done above for a uniaxial crystal. However, we shall not pause to discuss this, because the equations are more complex, and explicit analytical formulae cannot be obtained.

PROBLEMS

PROBLEM 1. A uniaxial ferromagnetic crystal is in the shape of a spheroid, the axis of easy magnetization being the axis of revolution, and is placed in an external magnetic field $\mathfrak{H}$. Determine the range of values of $\mathfrak{H}$ for which the body has a domain structure.

SOLUTION. According to the general properties of an ellipsoid in a uniform external field (§8), the induction $\overline{\mathbf{B}}$ and field $\overline{\mathbf{H}}$ (= **H**) averaged over the domain structure are related to $\mathfrak{H}$ by

$$n\overline{B}_z+(1-n)H_z = \mathfrak{H}_z, \qquad \tfrac{1}{2}(1-n)\overline{B}_x+\tfrac{1}{2}(1+n)H_x = \mathfrak{H}_x,$$

where n is the demagnetizing factor in the direction of the axis of revolution (taken as the z-axis). Putting $H_z = 0$ and using formulae (41.7), we obtain

$$H_x = \frac{\mathfrak{H}_x}{1+2\pi(1-n)/\beta}, \qquad \overline{B}_z = \frac{\mathfrak{H}_z}{n} < 4\pi\sqrt{\left(M^2-\frac{H_x{}^2}{\beta^2}\right)}.$$

Elimination of H_x gives the required inequality

$$\frac{\mathfrak{H}_z{}^2}{(4\pi n)^2}+\frac{\mathfrak{H}_x{}^2}{[\beta+2\pi(1-n)]^2} < M^2$$

for the range in which there is a domain structure.

PROBLEM 2. Determine the magnetization averaged over the crystallites (which have uniaxial symmetry) for a polycrystalline body in a strong magnetic field ($H \gg 4\pi M$).

SOLUTION. In a particular crystallite, let θ and ψ be the angles between the direction of easy magnetization and, respectively, **M** and **H**. It is evident that, in a strong field, **M** and **H** will be in almost the same direction, i.e. the angle $\vartheta \equiv \theta-\psi$ is small. Putting in (41.2) $\mathbf{M}\cdot\mathbf{H} = MH\cos(\theta-\psi)$ and equating to zero the derivative $\partial\tilde{\Phi}/\partial\theta$, we have $\vartheta \cong \sin\vartheta = -(\beta M/H)\sin\theta\cos\theta$. The average magnetization is clearly parallel to **H**, and is

$$\overline{M} = M\,\overline{\cos\vartheta} = M(1-\tfrac{1}{2}\overline{\vartheta^2}) = M\left(1-\frac{\beta^2M^2}{2H^2}\overline{\sin^2\theta\cos^2\theta}\right).$$

The bar denotes averaging over the crystallites. Assuming that all directions of the axis of easy magnetization of the crystallites are equally probable, we have

$$\overline{M} = M\left(1-\frac{\beta^2M^2}{15H^2}\right).$$

Thus the mean magnetization approaches saturation in the manner $\overline{M}-M \propto 1/H^2$.

PROBLEM 3. The same as Problem 2, but for the case where the crystallites have cubic symmetry.

SOLUTION. The conditions for a minimum of the expression

$$-\tfrac{1}{4}\beta(M_x{}^4+M_y{}^4+M_z{}^4)-(H_xM_x+H_yM_y+H_zM_z),$$

with $K=\frac{1}{2}\beta M^2$ in (40.7) and the subsidiary condition $M_x{}^2+M_y{}^2+M_z{}^2=$ constant, are

$$\beta M_x{}^3+H_x=\lambda M_x,\ \beta M_y{}^3+H_y=\lambda M_y,\ \beta M_z{}^3+H_z=\lambda M_z,$$

where λ is an undetermined Lagrange multiplier. For large H, we therefore have

$$M_x\cong\frac{1}{\lambda}H_x+\frac{1}{\lambda^4}\beta H_x{}^3+\ldots,\ \text{etc.};$$

adding the squares of the three equations, we obtain $M^2\cong H^2/\lambda^2$, i.e. $\lambda\cong H/M$. The angle ϑ between **M** and **H** is found from

$$\vartheta^2\cong\sin^2\vartheta=\frac{(\mathbf{M}\times\mathbf{H})^2}{M^2H^2}$$

$$=\frac{\beta^2M^6}{H^{10}}\sum H_x{}^2H_y{}^2(H_x{}^2-H_y{}^2)^2,$$

where the summation is over cyclic permutations of the suffixes x, y, z. Averaging this expression over the orientations of the crystallites is equivalent to averaging over directions of the vector **H**. The latter averaging is effected by integrating over the angles which specify the direction of **H**, and the result is

$$\bar{M}=M(1-\tfrac{1}{2}\overline{\vartheta^2})=M\left(1-\frac{2\beta^2M^6}{105H^2}\right).$$

§42. Magnetostriction of ferromagnets

A change in the magnetization of a ferromagnet in a magnetic field causes a deformation in it; this phenomenon is called *magnetostriction*, and may be due to either exchange interactions or relativistic interactions in the body. Since the exchange energy depends only on the magnitude M of the magnetization, its value can change only when M changes in the magnetic field. Although the latter change is, in general, very small, the exchange energy is large compared with the anisotropy energy. Hence the magnetostriction effects from each type of interaction may be of comparable magnitude.

This happens, for instance, in uniaxial crystals. Marked deformations resulting from a change in the direction of **M** occur in fields $H\sim\beta M$; the change in the magnitude M is considerable when $H\sim4\pi M$. If these two values of H are almost the same, it is in general necessary to take account of both effects in discussing the magnetostriction of uniaxial ferromagnets. We shall not pause here to derive the formulae, which are fairly complex.

In cubic crystals the situation is different, because the anisotropy energy is of the fourth order and therefore relatively small. A considerable magnetostriction, due to the change in the direction of **M**, occurs even in comparatively weak fields, where the change in the magnitude M may be entirely neglected. Let us consider these effects.

The change in the relativistic interaction energy in the deformed body is described by the inclusion in the thermodynamic potential $\tilde{\Phi}$ of *magnetoelastic terms* depending on the components of the elastic stress tensor σ_{ik} and the direction of the vector **M** (N. S. Akulov, 1928). The first such terms which do not vanish are linear in σ_{ik}, and quadratic in the direction cosines of **M** because of the symmetry with respect to time reversal. In general, therefore, the magnetoelastic energy is given by an expression of the form

$$U_{\text{m-el}}=-a_{iklm}\sigma_{ik}m_lm_m, \qquad (42.1)$$

where a_{iklm} is a dimensionless tensor of rank four, symmetrical with respect to the pairs of suffixes i, k and l, m (but not with respect to interchange of the two pairs). Near the Curie point, where the expansion in powers of the direction cosines of the vector **M** is equivalent to one in powers of its components, the quantities a_{iklm}/M^2 tend to constants.

In calculating the number of independent components of the tensor a_{iklm} it must be borne in mind that the terms in (42.1) which involve the components of **m** in the form $m_x^2+m_y^2+m_z^2$ are independent of the direction of **m**, and so may be omitted from the magnetoelastic energy.† Thus we find that, in a cubic crystal, the magnetoelastic energy contains two independent coefficients; we shall write it as

$$U_{\text{m-el}} = -a_1(\sigma_{xx}m_x^2+\sigma_{yy}m_y^2+\sigma_{zz}m_z^2) - 2a_2(\sigma_{xy}m_xm_y+\sigma_{xz}m_xm_z+\sigma_{yz}m_ym_z). \tag{42.2}$$

The strain tensor is obtained by differentiating $\tilde{\Phi}$ with respect to the various components σ_{ik}: $u_{ik} = -\partial\tilde{\Phi}/\partial\sigma_{ik}$, where $\tilde{\Phi}$ includes also the ordinary elastic energy (with reversed sign; see the first footnote to §17). For a cubic crystal, the latter energy involves three independent elastic coefficients, and can be put in the form

$$U_{\text{el}} = \tfrac{1}{2}\mu_1(\sigma_{xx}^2+\sigma_{yy}^2+\sigma_{zz}^2)+\tfrac{1}{2}\mu_2(\sigma_{xx}+\sigma_{yy}+\sigma_{zz})^2 + \mu_3(\sigma_{xy}^2+\sigma_{xz}^2+\sigma_{yz}^2), \tag{42.3}$$

where μ_1, μ_2 and μ_3 are positive. The strain tensor is‡

$$\left.\begin{aligned} u_{xx} &= (\mu_1+\mu_2)\sigma_{xx}+\mu_2(\sigma_{yy}+\sigma_{zz})+a_1m_x^2, \\ u_{xy} &= \mu_3\sigma_{xy}+a_2m_xm_y, \end{aligned}\right\} \tag{42.4}$$

and similarly for the other components.

These formulae give all the magnetostriction effects in the range of fields considered. In particular, if there are no internal stresses the change in the deformation resulting from a change in the direction of magnetization is given by

$$u_{xx} = a_1m_x^2, \qquad u_{xy} = a_2m_xm_y, \text{ etc.} \tag{42.5}$$

It should be recalled that the magnitude of the deformation itself is to some extent arbitrary, because the direction of **m** for which the deformation is supposed zero is arbitrarily chosen.

The stress tensor found by solving a specific problem (e.g. for a clamped crystal) is in order of magnitude $\sigma \sim a/\mu$, where a and μ are the orders of magnitude of the coefficients a_{iklm} and the elastic coefficients respectively. In this sense, the magnetoelastic energy (per unit volume, as usual) is of the order of a^2/μ. The coefficients a are of the first order in the relativistic spin–spin interaction, and the magnetoelastic energy is therefore of the second order. In a uniaxial crystal, the anisotropy energy is of the first order in the relativistic interaction, and is therefore usually much larger than the magnetoelastic energy. In cubic crystals, the anisotropy energy is of the second order in this interaction, and is therefore in

† There is consequently some arbitrariness in the choice of the a_{iklm}, which simply reflects the arbitrariness involved in choosing the direction of **m** for which (applied mechanical forces being absent) we regard the crystal as undeformed.

‡ In differentiating $\tilde{\Phi}$ the third footnote to §17 should be recalled.

general comparable with the magnetoelastic energy.† It may therefore be necessary to take into account simultaneously both forms of energy (for instance, in considering the magnetization curve), which greatly complicates the problem.

Let us now consider magnetostriction in fields so strong ($H \gg 4\pi M$) that the anisotropy energy is unimportant and there is no domain structure, so that the directions of **M** and **H** may be assumed to coincide.

Since the anisotropy energy is neglected, the particular symmetry of the crystal is of no importance, and the formulae given below are valid for any ferromagnet.

Let the body be placed in a uniform external magnetic field $\mathfrak{H}$. Its total thermodynamic potential $\tilde{\Phi}$ is‡

$$\tilde{\Phi} = -\mathscr{M}\cdot\mathfrak{H} = -MV\mathfrak{H}, \tag{42.6}$$

where $\mathscr{M} = MV$ is the total magnetic moment of a body uniformly magnetized in the direction of the field; we omit the term $\tilde{\Phi}_0$ which is unrelated to the magnetic field. The strain tensor averaged over the volume of the body is $\bar{u}_{ik} = -(1/V)\partial\tilde{\Phi}/\partial\sigma_{ik}$, whence

$$\bar{u}_{ik} = \frac{\mathfrak{H}}{V}\frac{\partial(MV)}{\partial\sigma_{ik}}. \tag{42.7}$$

Thus the deformation is determined by the dependence of the magnetization on the internal stresses.

For cubic symmetry, any symmetrical tensor of rank two characterizing the properties of the crystal reduces to a scalar multiple of δ_{ik}. This is true, in particular, of the tensor $\partial(MV)/\partial\sigma_{ik}$, so that the magnetostriction deformation amounts in this case to a uniform compression or extension.

If we are interested only in the change δV in the total volume of the body, we can obtain it by simply differentiating $\tilde{\Phi}$ with respect to the pressure:

$$\delta V = \partial\tilde{\Phi}/\partial P = -\mathfrak{H}\partial(MV)/\partial P, \tag{42.8}$$

where P is to be regarded as a uniform pressure applied to the suface of the body.

PROBLEMS

PROBLEM 1. Find the relative extension of a ferromagnetic cubic crystal as a function of the direction of the magnetization **m** and the direction of measurement **n**.

SOLUTION. The relative extension in the direction of the unit vector **n** is given in terms of the strain tensor by $\delta l/l = u_{ik}n_i n_k$. Substituting u_{ik} (in the absence of internal stresses) from (42.5), we have

$$\delta l/l = a_1(m_x^2 n_x^2 + m_y^2 n_y^2 + m_z^2 n_z^2) + a_2(m_x m_y n_x n_y + m_x m_z n_x n_z + m_y m_z n_y n_z).$$

Only the difference in the values of this quantity for different directions **m** and **n** has absolute significance. For instance, if **m** is along the x-axis, the difference in the values of $\delta l/l$ along the x and y axes is a_1. If **m** is along one of the spatial diagonals, the difference in the values of $\delta l/l$ in that direction and along the other three spatial diagonals is $4a_2/9$.

PROBLEM 2. Determine the change in volume in magnetostriction of a ferromagnetic ellipsoid in an external field $\mathfrak{H} \sim 4\pi M$ parallel to one of its axes. The ferromagnet is assumed to be a cubic crystal.§

† In cubic crystals too, however, the magnetoelastic energy may be much smaller than the anisotropy energy. In iron at room temperature, for example, the ratio is $\sim 10^{-2}$.

‡ Here the definition of $\tilde{\Phi}$ is that given in §12, which is applicable except when the deformation of the body is appreciably inhomogeneous.

§ In a uniaxial ferromagnet with $\mathfrak{H} \sim 4\pi M$ the anisotropy energy would have to be taken into account, but this is not necessary in a cubic crystal.

SOLUTION. When the anisotropy energy is neglected, the range for which the domain structure exists is given by $\bar{B} < 4\pi M$ when $H = 0$. The bar denotes averaging over the volume of the body; cf. §41. In an ellipsoid $n\bar{B} + (1-n)\bar{H} = \mathfrak{H}$; putting $H = 0$, we find that the domain structure exists when $\mathfrak{H} < 4\pi nM$. Since $n\bar{B} = 4\pi nM = \mathfrak{H}$, the mean magnetization is $\bar{M} = \mathfrak{H}/4\pi n$. Hence the thermodynamic potential is

$$\tilde{\Phi} = -V\int_0^{\mathfrak{H}} \bar{M}\,d\mathfrak{H} = -\mathfrak{H}^2 V/8\pi n. \qquad (1)$$

If $\mathfrak{H} > 4\pi nM$, the ellipsoid is magnetized entirely in the direction of the field, and $\bar{M} = M$. Then

$$\tilde{\Phi} = -M\mathfrak{H}V + 2\pi M^2 Vn. \qquad (2)$$

The expressions (1) and (2) are the same for $\mathfrak{H} = 4\pi Mn$.

The required change in volume is obtained by differentiating $\tilde{\Phi}$ with respect to pressure:

$$\delta V = -\frac{\mathfrak{H}^2}{8\pi n}\frac{\partial V}{\partial P} \quad \text{for } \mathfrak{H} < 4\pi nM,$$

$$\delta V = -\mathfrak{H}\frac{\partial(MV)}{\partial P} + 2\pi n\frac{\partial(M^2 V)}{\partial P} \quad \text{for } \mathfrak{H} > 4\pi nM.$$

For $\mathfrak{H} \gg 4\pi nM$ we obtain (42.8).

§43. Surface tension of a domain wall

As already mentioned in §41, there exists a wide range of states in which a ferromagnet must have a *domain structure*, i.e. must consist of regions with different directions of magnetization. This is true, in particular, of a ferromagnet which is not in an external magnetic field.

Thermodynamically, adjacent domains are different phases of a ferromagnet, with different directions of their spontaneous magnetization. Let us consider, first of all, the properties of the phase boundaries (*domain walls*) as such, and calculate their surface tension (L. D. Landau and E. M. Lifshitz, 1935).

The phase boundaries are in reality fairly narrow transition layers in which the direction of the magnetization varies continuously between its directions in the two adjoining domains. The "width" of such a layer and the manner in which **M** varies within it are given by the conditions of thermodynamic equilibrium. The additional energy due to the non-uniformity of the magnetization must be taken into account. The largest contribution to this *non-uniformity energy* is given by the exchange interaction. Macroscopically, this energy can be expressed in terms of the derivatives of **M** with respect to the coordinates. This can be done in a general form if the gradient of the direction of **M** is supposed relatively small, i.e. if the change in the direction of the magnetic moments occurs over distances large compared with the interatomic distances. In the present case, this condition is evidently fulfilled, because a considerable difference in the directions of the magnetic moments of adjoining atoms would lead to a very large increase in the exchange energy, and is therefore thermodynamically unfavourable.

We denote the non-uniformity energy density by $U_{\text{non-u}}$. There can be no terms of the form $a_{ik}(M)\partial M_i/\partial x_k$ linear in the first derivatives, because of the requirement of symmetry under time reversal, which changes the sign of **M** but leaves the energy unchanged. The crystal symmetry might allow the existence of terms containing products of the derivatives $\partial M_i/\partial x_k$ with the components of **M** itself. We are interested here in the non-uniformity exchange energy; the corresponding terms in $U_{\text{non-u}}$ must be invariant when the vectors **M** are simultaneously rotated by the same amount throughout all space (with no change in the coordinates used).† The terms involving such products can then only have the form

$$a_i(M)M_l\partial M_l/\partial x_i = \tfrac{1}{2}\mathbf{a}(M)\cdot\mathbf{grad}\,M^2.$$

Such terms cannot, however, actually represent the non-uniformity energy, even if the magnitude as well as the direction of **M** is assumed to change within the crystal. The reason is that not $U_{\text{non-u}}$ itself but only its integral over the volume of the body has a real thermodynamic significance, and such an integration would reduce terms of the above form to expressions depending only on the values of the magnetization on the surface of the body, not on its variation within the body.‡

For the same reason, we need not include in $U_{\text{non-u}}$, among the terms of the next order of smallness, those linear in the second derivatives of **M** with respect to the coordinates, since on integration over the volume of the body they become expressions quadratic in the first derivatives. The quadratic expressions are the principal non-vanishing terms in the expansion of the non-uniformity exchange energy.

The most general form of these terms is

$$U_{\text{non-u}} = \tfrac{1}{2}\alpha_{ik}(\partial M_l/\partial x_i)(\partial M_l/\partial x_k), \tag{43.1}$$

where α_{ik} is a symmetrical tensor. For the ferromagnetic order to be stable, this expression must be positive definite, i.e. the principal values of the tensor α_{ik} must be positive. In a cubic crystal, this tensor reduces to a scalar ($\alpha_{ik} = \alpha\delta_{ik}$, $\alpha > 0$), so that §

$$U_{\text{non-u}} = \tfrac{1}{2}\alpha(\partial\mathbf{M}/\partial x_i)\cdot(\partial\mathbf{M}/\partial x_i). \tag{43.2}$$

In a uniaxial crystal, α_{ik} has two independent components, and the non-uniformity energy is of the form

$$U_{\text{non-u}} = \tfrac{1}{2}\alpha_1[(\partial\mathbf{M}/\partial x)^2 + (\partial\mathbf{M}/\partial y)^2] + \tfrac{1}{2}\alpha_2(\partial\mathbf{M}/\partial z)^2. \tag{43.3}$$

It must be emphasized that, far from the Curie point, the expressions (43.1)–(43.3) are to be regarded as the first terms in an expansion in powers of the derivatives of the unit vector $\mathbf{m} = \mathbf{M}/M$, not of the magnetization **M** itself. Only near the Curie point do they become an expansion in powers of the derivatives of **M**. Accordingly, in the Landau theory the coefficients α_{ik} in these expressions should tend, as $T \to T_c$, to non-zero limits. (See §47 regarding the role of fluctuations here.)

As an example, let us consider the boundary between phases in a uniaxial crystal of the easy-axis type, assuming that the vector **M** is parallel (or antiparallel) to the direction of easy magnetization (the *z*-axis).

The structure of the transition layer is determined by the condition that the total free energy of the layer be a minimum.‖ Here the exchange energy tends to increase the width of

† This means that the scalar expression $U_{\text{non-u}}$ must be so constructed that the magnetic and coordinate vector suffixes are each interchanged among themselves but not with each other.

‡ The symmetry of the crystal may, however, allow the existence of terms having the form $a_{ikl}M_i\partial M_k/\partial x_l$ and not of an exchange nature. This would alter the nature of the ferromagnetic order in the crystal; see §52.

§ In order of magnitude (in iron, for example), $\alpha \sim 10^{-12}$ cm.

‖ When the magnetostriction is neglected, there is no need to distinguish the free energy $\mathscr{F}$ from the thermodynamic potential $\tilde{\Phi}$. If, however, we intend to take account of the elastic and magnetoelastic energies of the non-uniform deformation (as may sometimes be necessary; see Problem 2), the total free energy must be used. Here it should be remembered that the equilibrium equations of the medium are found by varying its total free energy with respect to the components of the displacement vector **u**; cf. the derivation for a fluid at the end of §15, the equilibrium equation $\mathbf{f} = 0$ being obtained by equating to zero the variation $\delta\mathscr{F}$.

the layer (i.e. to make the direction of **M** vary less rapidly). The anisotropy energy has the opposite effect, because any deviation of **M** from the direction of easy magnetization increases this energy.

We take the x-axis perpendicular to the plane of the layer; the direction of **M** depends only on x. The rotation of the vector **M** across the layer must take place in the yz-plane, i.e. $M_x = 0$ everywhere. This is seen as follows. The non-uniformity and anisotropy energies in a uniaxial crystal are independent of the plane in which the rotation of the magnetization takes place. The presence of a non-zero component M_x would necessarily result in a magnetic field which was thermodynamically unfavourable, because of the additional magnetic energy. For it follows from the equation $\operatorname{div} \mathbf{B} = dB_x/dx = 0$ that $B_x =$ constant throughout the transition layer, and, since $M_x = 0$ and $H_x = 0$ within the domains, we find that $B_x = 0$ everywhere. Hence, together with the component $M_x \neq 0$, a field $H_x = -4\pi M_x$ must occur. Correspondingly, the free energy $\tilde{F}$ contains a term† $-M_x H_x - H_x^2/8\pi = H_x^2/8\pi > 0$.

Let θ be the angle between **M** and the z-axis. Then the components of **M** are $M_x = 0$, $M_y = M \sin\theta$, $M_z = M \cos\theta$. The sum of the non-uniformity energy (43.3) and the anisotropy energy (41.1) is the integral

$$\int_{-\infty}^{\infty} [\tfrac{1}{2}\alpha_1 (M_y'^2 + M_z'^2) + \tfrac{1}{2}\beta M_y^2]\,dx = \tfrac{1}{2}M^2 \int_{-\infty}^{\infty} (\alpha_1 \theta'^2 + \beta \sin^2\theta)\,dx, \tag{43.4}$$

where the prime denotes differentiation with respect to x. The remaining terms in the free energy are independent of the structure of the layer, and so can be omitted here.

To determine the function $\theta(x)$ which makes this integral a minimum, we write down the corresponding Euler's equation $\alpha_1 \theta'' - \beta \sin\theta \cos\theta = 0$. Assuming the width of the transition layer small compared with that of the domains themselves, we can write the boundary conditions on this equation as

$$\theta(+\infty) = 0, \quad \theta(-\infty) = \pi, \quad \theta'(\pm\infty) = 0. \tag{43.5}$$

These state that adjoining domains are magnetized in opposite directions. The first integral of Euler's equation which satisfies these conditions is‡

$$\theta'^2 - (\beta/\alpha_1)\sin^2\theta = 0. \tag{43.6}$$

A second integration gives

$$\cos\theta = \tanh [x\sqrt{(\beta/\alpha_1)}], \tag{43.7}$$

which describes the manner of variation of the magnetization in the transition layer. The width of this layer is $\delta \sim \sqrt{(\alpha_1/\beta)}$.

---

† More precisely, we should use here the thermodynamic potential $\mathscr{F}'$; cf. the end of §44. For a given value of $B_n = B_x$, this potential must have a minimum with respect to M_x or H_x. When $B_x = 0$, however, $\tilde{F}$ and F' are the same.

‡ This expression can, of course, be written down directly (without using Euler's equation) if we notice that the integral in (43.4) has the form of the action integral for a one-dimensional motion of a particle in a field whose potential energy is $-\beta \sin^2\theta$, with θ acting as the coordinate and x as the time. Then (43.6) expresses the conservation of energy for the particle.

With (43.6), the integral (43.4) becomes

$$M^2\alpha_1 \int_{-\infty}^{\infty} \theta'^2\,dx = M^2\sqrt{(\alpha_1\beta)} \int_0^{\pi} \sin\theta\,d\theta$$
$$= 2M^2\sqrt{(\alpha_1\beta)}.$$

If the boundary between domains is regarded as a geometrical surface, this quantity is the surface tension which must be ascribed to the boundary in order to take account of the energy needed to create it. The surface tension of a domain wall may be denoted by $M^2\Delta$, where Δ has the dimensions of length. Then

$$\Delta = 2\sqrt{(\alpha_1\beta)}. \tag{43.8}$$

PROBLEMS

PROBLEM 1. Determine the surface tension of a domain wall in a cubic ferromagnet with the directions of easy magnetization parallel to the cube edges (the x, y, z axes). The domains are magnetized parallel and antiparallel to the z-axis, and the domain wall is (a) parallel to the (100) plane, (b) parallel to the (110) plane (E. M. Lifshitz, 1944; L. Néel, 1944).

SOLUTION. (a) The domain wall is parallel to the yz-plane, all quantities in it depend only on the coordinate x, and the rotation of the vector **M** takes place in the yz-plane (as well as the argument in the text of §43 for uniaxial crystals, note that a departure of **M** from the yz-plane in this case would increase the anisotropy energy). Neglecting the magnetostriction energy, and using formula (43.2) for the non-uniformity energy and (40.7) for the anisotropy energy (with $K = \frac{1}{2}\beta M^2$), we find as the free energy of the wall

$$\tfrac{1}{2}M^2 \int_{-\infty}^{\infty} (\alpha\theta'^2 + \beta\sin^2\theta\cos^2\theta)dx,$$

where θ is the angle between **M** and the z-axis. The first integral of Euler's equation for the problem of minimizing this functional, which satisfies the boundary conditions (43.5), is $\alpha\theta'^2 - \beta\sin^2\theta\cos^2\theta = 0$, or $\theta' = \sqrt{(\beta/a)}\sin\theta\,|\cos\theta|$; by writing $|\cos\theta|$, we ensure that the angle θ varies monotonically in the transition layer. This equation has no solution that could describe the structure of a domain wall of finite width (which would demand an allowance for the magnetostriction energy; see Problem 2), but it is adequate for calculating the surface tension, which is not zero even with the above approximations:

$$\Delta_{(100)} = \sqrt{(\alpha\beta)}.\,2\int_0^{\frac{1}{2}\pi} \sin\theta\cos\theta\,d\theta = \sqrt{(\alpha\beta)}.$$

(b) The domain wall passes through the z-axis at 45° to the x and y axes. The need to avoid the presence of a considerable magnetic field again tends to keep the vector **M** in the wall plane. The magnetic anisotropy in this case, however, moves **M** slightly out of that plane. Nevertheless, since the anisotropy energy in a cubic crystal is assumed small, this deviation will be small and may with sufficient accuracy be neglected. Then $M_x = M_y = (M/\sqrt{2})\sin\theta$, $M_z = M\cos\theta$, where θ is again the angle between **M** and the z-axis, and the anisotropy energy is

$$U_{\text{aniso}} = \frac{1}{8}\beta M^2\sin^2\theta\,(3\cos^2\theta + 1).$$

We can write down at once the first integral of Euler's equation for the variational problem:

$$\theta'^2 = A\sin^2\theta\,(\cos^2\theta + B), \tag{1}$$

where $A = 3\beta/4\alpha$, $B = \frac{1}{3}$, and the prime denotes differentiation with respect to the coordinate (ξ, say) normal to the wall plane. Hence, again using the conditions (43.5), we find the equation for the wall structure:

$$\sinh\xi\sqrt{[A(1+B)]} = \sqrt{\frac{1+B}{B}}\cot\theta, \tag{2}$$

i.e.

$$\sinh\sqrt{(\beta/\alpha)}\,\xi = 2\cot\theta.$$

The surface tension is

$$\Delta = \alpha\sqrt{A}\{\sqrt{(1+B)}+B\sinh^{-1}B^{-\frac{1}{2}}\}, \tag{3}$$

or, with the above values of A and B,

$$\Delta_{(110)} = 1.38\sqrt{(\alpha\beta)}.$$

PROBLEM 2. Find the structure of the domain wall in the (100) plane whose surface tension was calculated in Problem 1(a) (E. M. Lifshitz, 1944).

SOLUTION. As already mentioned, a non-zero width of this wall occurs only when the magnetostriction energy is taken into account. The structure of the wall is determined by the condition for the free energy $\mathscr{F}$ to be a minimum; the free energy density $\tilde{F}$ is to be expressed in terms of u_{ik} (cf. the fourth footnote to §43). The corresponding expressions for the magnetoelastic and elastic energies are analogous to (42.2) and (42.3) with different coefficients:

$$U_{\text{m-el}} = b_1(u_{yy}m_y^2+u_{zz}m_z^2)+2b_2u_{yz}m_ym_z,$$
$$U_{\text{el}} = \tfrac{1}{2}\lambda_1(u_{xx}^2+u_{yy}^2+u_{zz}^2)+\tfrac{1}{2}\lambda_2(u_{xx}+u_{yy}+u_{zz})^2+\lambda_3(u_{xy}^2+u_{xz}^2+u_{yz}^2);$$

here we have put $m_x = 0$.

Not only the magnetization distribution but also the strain in the transition layer must be a function of x only. It follows that the y and z components of the displacement vector $\mathbf{u}$ must have the form $u_y = \text{constant}\times y$, $u_z = \text{constant}\times z$; if the constants here were replaced by functions of x, then u_{xy} and u_{xz} would depend on y or z. Thus u_{yy}, u_{zz}, and u_{yz} are constants. From the general equations of elastic equilibrium, $\partial\sigma_{ik}/\partial x_k = 0$, it follows that $\sigma_{ix}' = 0$; at $x = \pm\infty$, where the strain is zero, we must have $\sigma_{ik} = 0$, and therefore $\sigma_{xx} = \sigma_{xy} = \sigma_{xz} = 0$ everywhere. Calculating these components of the stress tensor as the derivatives $\sigma_{ik} = \partial\tilde{F}/\partial u_{ik}$, we find that $u_{xy} = u_{xz} = 0$, $u_{xx} = \text{constant}$. Thus all the u_{ik} are in fact constant. It is therefore sufficient to calculate their values at infinity, where all the $\sigma_{ik} = 0$, $m_y = 0$, $m_z = \pm 1$. The equations $\sigma_{yz} = 0$, $\sigma_{zz}-\sigma_{yy} = 0$ give

$$u_{yz} = 0,\quad u_{yy}-u_{zz} = b_1/\lambda_1.$$

Omitting the constant terms from $U_{\text{m-el}}$ and U_{el}, we find that the further term $U_{\text{m-el}} = (b_1^2/\lambda_1)\sin^2\theta$ is to be added to $U_{\text{non-u}}+U_{\text{aniso}}$. The determination of $\theta(x)$ thus amounts to solving equation (1) with $A = \beta/\alpha$, $B = 2b_1^2/\lambda_1\beta M^2$. The constant B, which represents the ratio of the magnetostriction and anisotropy energies, is small.† Putting $B = 0$ in (3), we obtain the value of $\Delta_{(100)}$ already found in Problem 1(a). From (2) we find as the distribution of magnetization in the wall

$$\sinh\sqrt{(\beta/\alpha)}x = \sqrt{(\lambda_1\beta M^2/2b_1^2)}\cot\theta.$$

The width of this distribution,

$$\delta \sim \sqrt{(\alpha/\beta)}\log(\lambda_1\beta M^2/b_1^2),$$

depends considerably on the magnetostriction constant.‡

PROBLEM 3. In a crystal of the same kind as in Problem 2, find the surface tension of a domain wall between domains magnetized in the [001] and [010] (z and y) directions, in two cases: (a) the wall is parallel to the (100) plane, (b) the wall is parallel to the (011) plane (S. V. Vonsovskiĭ, 1944; L. Néel, 1944).§

SOLUTION. In both cases, the magnetoelastic energy may be neglected.

(a) Here the vector $\mathbf{M}$ remains in the wall (yz) plane as it rotates. The only difference from Problem 1(a) lies in the boundary conditions:

$$\theta(-\infty) = 0,\quad \theta(+\infty) = \tfrac{1}{2}\pi,\quad \theta'(\pm\infty) = 0.$$

The wall structure is given by the solution

$$\tan\theta = \exp\sqrt{(\beta/\alpha)}x,$$

---

† For example, in iron at room temperature $B \sim 2\times 10^{-3}$.

‡ As $b_1 \to 0$, this 180-*degree wall* (as regards the angle of rotation of the vector $\mathbf{M}$ in it) decays, as it were, into two 90-*degree walls*, separated by a region with $\theta = \frac{1}{2}\pi$ which tends to infinity.

§ In a cubic crystal (unlike a uniaxial one; see the second footnote to §44), a 90-degree wall is a true phase boundary, since both domains are stable phases and each is magnetized in one of the easy directions.

and the surface tension is

$$\Delta^{90^\circ}_{(100)} = \tfrac{1}{2}\sqrt{(\alpha\beta)},$$

half the value for a 180-degree wall.

(b) Together with the crystallographic x, y, z axes, we use axes x, η, ζ as shown in Fig. 22; the x-axis is perpendicular to the plane of the diagram, and the arrows show the directions of $\mathbf{M}$ in domains separated by the plane $\eta = 0$. In the transition layer, $\mathbf{M}$ rotates over half a circular cone with its axis along the η-axis; M_η = constant $= 1/\sqrt{2}$, so that div $\mathbf{M} = M_\eta' = 0$, as it should be (the prime denotes differentiation with respect to η). Let ϕ denote the angle between the projection of $\mathbf{M}$ on the $x\zeta$-plane and the ζ-axis; ϕ varies from 0 to π. Then $m_\eta = 1/\sqrt{2}$, $m_\zeta = (1/\sqrt{2})\cos\phi$, $m_x = (1/\sqrt{2})\sin\phi$, $m_y = \tfrac{1}{2}(1-\cos\phi)$, $m_z = \tfrac{1}{2}(1+\cos\phi)$. The non-uniformity and anisotropy energies are

$$U_{\text{non-u}} = \tfrac{1}{4}\alpha M^2 \phi'^2, \quad U_{\text{aniso}} = \tfrac{1}{4}\beta M^2 (\sin^2\phi - \tfrac{3}{8}\sin^4\phi).$$

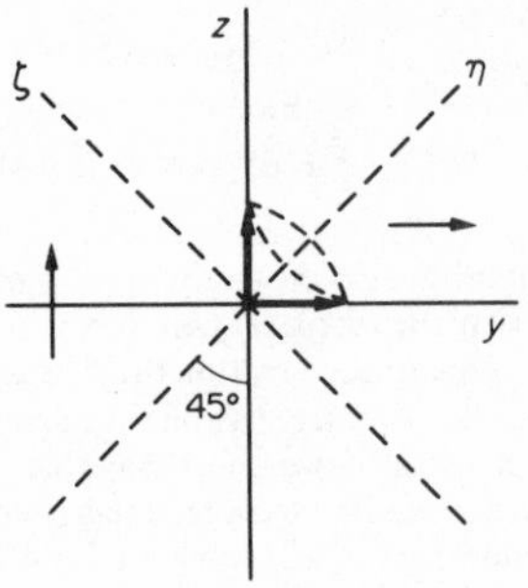

FIG. 22

The surface tension is

$$\Delta^{90^\circ}_{(110)} = \tfrac{1}{4}\sqrt{(\alpha\beta)}.\, 2\int_0^\pi \sin\theta\,[1-\tfrac{3}{8}\sin^2\theta]^{\frac{1}{2}}\,d\theta$$

$$= \tfrac{1}{2}\sqrt{(\alpha\beta)}\left[1+\frac{5}{2\sqrt{6}}\sinh^{-1}\sqrt{\tfrac{3}{5}}\right]$$

$$= 1.73 \times \tfrac{1}{2}\sqrt{(\alpha\beta)}.$$

PROBLEM 4. Find the surface tension of a domain wall in a uniaxial crystal when the transition between domains takes place by a change in the magnitude of $\mathbf{M}$ without rotation, its direction being reversed as $\mathbf{M}$ passes through zero. The dependence of the free energy on M (when $\mathbf{H} = 0$) is taken in the form of the expansion (39.3), appropriate near the Curie point (V. A. Zhirnov, 1958).

SOLUTION. Throughout the transition layer, M_z is equal to M, and varies in the x-direction, which is perpendicular to the wall plane. The free energy density is, taking into account the non-uniformity energy,

$$F = F_0 - |A|M^2 + BM^4 + \tfrac{1}{2}\alpha_1 M'^2. \tag{1}$$

The equilibrium value of the magnetization within the domains is here denoted by M_0: $M_0{}^2 = |A|/2B$; see (39.5). With $\mathbf{m} = \mathbf{M}/M_0$ ($m \neq 1$), we can write the free energy of the wall as

$$\tfrac{1}{2}|A|M_0{}^2 \int_{-\infty}^{\infty} [(1-m^2)^2 + (\alpha_1/|A|)m'^2]\,dx;$$

the additive constant in F is chosen so that F is zero within the domains. This integral is to be minimized with the boundary conditions $m(+\infty) = 1$, $m(-\infty) = -1$, $m'(\pm\infty) = 0$. The first integral of Euler's equation for this variational problem is $(\alpha_1/|A|)m'^2 = (1-m^2)^2$. Hence

$$m(x) = \tanh \sqrt{(|A|/\alpha_1)}\,x,$$

and the calculation of the integral gives for the surface tension $M_0{}^2\Delta$ the value

$$\Delta = \frac{4}{3}\sqrt{(\alpha_1|A|)}. \tag{2}$$

This wall structure can in principle occur sufficiently close to the Curie point (if $\beta/|A|$ tends to infinity as $T \to T_c$), where a change in the magnitude of **M** becomes energetically more favourable than a deviation of **M** from the direction of easy magnetization.

§44. The domain structure of ferromagnets

Let us now ascertain the actual shape and size of the domains.†

Some conclusions concerning the shape of the surfaces separating the domains may be obtained directly from the boundary conditions on the magnetic field. Since the field **H** is the same in adjoining domains, the condition of continuity of the normal induction B_n reduces to the continuity of M_n. In uniaxial crystals, the sign of M_z is different in different domains, but M_x and M_y are the same. Under these conditions the continuity of M_n means that the surface of separation must be parallel to the z-axis, i.e. to the direction of easy magnetization.

The shape and size of the domains in thermodynamic equilibrium are given by the condition that the total free energy should be a minimum. They depend considerably on the actual shape and size of the body. In the simplest case, that of a ferromagnet in the form of a flat plate, the domains may in principle form parallel layers across the body from one surface to the other. In what follows we shall take this case.

The formation of an entire new boundary between domains results in an increase in the total surface tension energy. This energy consequently tends to reduce the number of domains, i.e. to increase their thickness. The excess energy near the outer surface of the body, where the domains emerge, has the opposite effect. In the body the magnetic field $\mathbf{H} = 0$, and the anisotropy energy is also zero, because the vector **M** is in a direction of easy magnetization. Near the surface, however, this is not so.

The way in which the domains emerge at the surface of the body is different in the limiting cases of large and small magnetic anisotropy. A natural measure of this quantity in the present case is not the coefficient β itself in (41.1) but $\beta/4\pi$. This is seen from the expression $\mu_{xx} = 1 + 4\pi/\beta$ (41.7) for the transverse magnetic permeability of a uniaxial ferromagnet.

When the magnetic anisotropy is large, the layers must emerge at the surface with no change in the direction of **M** (Fig. 23a, p. 154, where for simplicity we suppose that the surface is perpendicular to the direction of easy magnetization). Near the surface there is a magnetic field which penetrates into the surrounding space, and into the body, to distances of the order of the layer thickness a.

In the opposite case of weak anisotropy, a more favourable disposition is that where there is no magnetic field, and **M** deviates from the direction of easy magnetization. For $\mathbf{H} = 0$ we must have everywhere $\operatorname{div} \mathbf{B} = 4\pi \operatorname{div} \mathbf{M} = 0$, and M_n must be continuous at all

† The concept of domains was first put forward by P. Weiss (1907). The thermodynamic theory of domains was given by L. D. Landau and E. M. Lifshitz (1935).

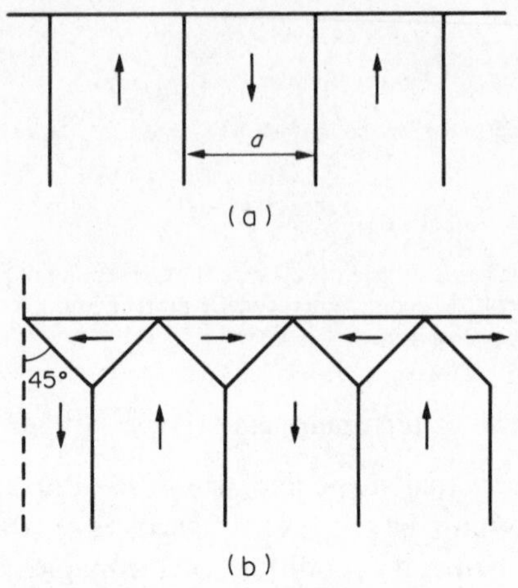

FIG. 23

domain boundaries and at the free surface. This is achieved by the setting up of *domains of closure* with triangular cross-section (Fig. 23b), in which the magnetization is parallel to the surface of the body. The total volume of these regions, and therefore the anisotropy energy in them, are proportional to the layer thickness a.†

Thus in all cases the emergence of the domains at the surface of the body results in an excess energy which increases with the thickness of the domains. This effect therefore tends to reduce the domain thickness.

The actual thickness of the domains is determined by the equilibrium of the two oppositely acting effects, namely the surface energy of the domain walls and the *energy of emergence* of the domains at the surface of the body. The number of plane-layer domains in a plate is proportional to $1/a$, while the surface-tension energy at the surfaces separating them is proportional to their total area, i.e. to l/a, where l is the thickness of the plate. The energy of emergence is proportional to a. The sum of these two energies has, as a function of a, a minimum when a has a value proportional to $\sqrt{l}$.

For example, in the case of weak anisotropy (Fig. 23a), the energy of emergence per unit area of the two sides of the plate is $\sim a\beta M^2$; the surface energy is $M^2\Delta l/a$ (the plate thickness l is, of course, assumed much greater than the domain thickness). Hence

$$a \sim \sqrt{(l\Delta/\beta)} \sim \sqrt{(l\delta)}. \tag{44.1}$$

Thus the thickness of the domains increases with the dimension of the body, but the quantitative law $a \propto \sqrt{l}$ of this increase is based on the assumption that the domains are of

† It should be emphasized that the boundary between the principal domains and the domains of closure, which separates regions where the directions of $\mathbf{M}$ differ by 90°, is not in this case (i.e. in uniaxial crystals) a phase boundary in the literal sense, as is seen from the fact that a state with magnetization perpendicular to the easy axis (when $\mathbf{H} = 0$) is in itself unstable and not a possible phase of matter. Strictly speaking, the distribution of magnetization shown in Fig. 23b relates only to the limit $\beta \to 0$. Even in the first approximation with respect to the small quantity $\beta/4\pi$, deviations from this pattern and transitions between principal domains and domains of closure occur at distances less than the domain thickness a by this small factor, and fields $\sim \beta M$ occur (although this does not affect the estimate of the energy of domain emergence at the surface of the body).

constant thickness, and clearly cannot be valid for all values of l. The reason is that the thickness of the domains at the surface of the body cannot exceed some limiting value a_k which depends on the properties of the ferromagnet but not on the shape and size of the body. The value of a_k is determined by the point at which, as a increases, the splitting of the domain near the surface to a depth $\sim a$ becomes thermodynamically favourable. Such a point must necessarily be reached, since the energy of the emergence of one domain increases as a^2, whereas the excess surface-tension energy resulting from the splitting of the domain increases only as a.

Thus we conclude that, as the size of the body, and therefore the domain thickness, increase, a progressive branching of the domains occurs as they approach the surface of the body (E. M. Lifshitz, 1944).† In principle, when l becomes sufficiently great, the branching continues until the thickness of the branches at the surface itself becomes comparable with the domain wall width δ.

The function $a(l)$ can be found for this limiting case. In making estimates, we shall take the particular case of weak anisotropy.

Figure 24 shows the domain branching pattern. The energy of emergence of the domains is here made up of the additional surface energy of wedge-shaped domains and the anisotropy energy due to the deviation of **M** from the z-axis as a result of branching; here there are no triangular domains of closure. Assuming rapid convergence of the sum over successive branchings, it is sufficient to consider the first wedge, whose length is denoted by h. It is evident from energy favourability arguments that h is much greater than the thickness of the wedge (or, equivalently as regards order of magnitude, than a), i.e. the angle ϑ between the wedge boundary and the z-axis is $\vartheta \sim a/h \ll 1$. The surface energy of the wedge is $\sim hM^2\Delta$, and the anisotropy energy due to the wedge is $\sim ha\beta M^2 \vartheta^2 \sim a^3\beta M^2/h$; the energy of emergence, the sum of these two, is least when $h^2 \sim a^3\beta/\Delta$, i.e. when

$$a/h \sim \sqrt{(\Delta/a\beta)} \sim \sqrt{(\delta/a)} \ll 1 \tag{44.2}$$

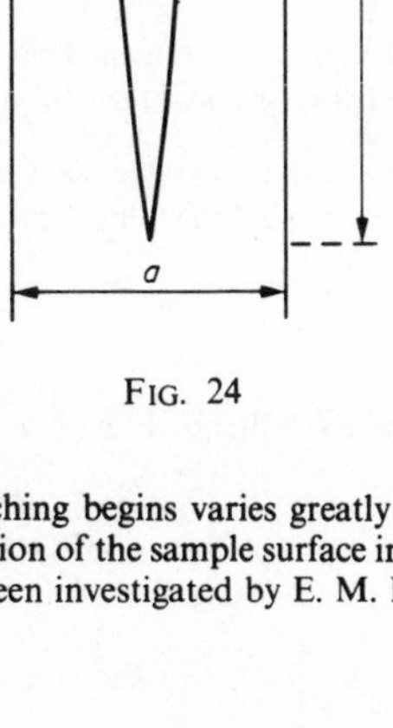

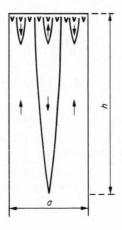

FIG. 24

† The critical value of l at which branching begins varies greatly according to the nature of the magnetic anisotropy and according to the configuration of the sample surface in relation to the crystallographic axes. The initial stages of domain branching have been investigated by E. M. Lifshitz (*Journal of Physics* **8**, 337, 1944).

(with the assumption, of course, that $a \gg \delta$). The energy of emergence per unit area of the plate surface is then $\sim a^{3/2}M^2\sqrt{(\beta\Delta)/a}$. The surface energy of the principal domain boundaries is $\sim M^2\Delta l/a$. Minimizing the sum of these two expressions gives

$$a \sim l^{2/3}(\Delta/\beta)^{1/3} \sim (l^2\delta)^{1/3} \qquad (44.3)$$

(I. A. Privorotskiĭ, 1970).†

To conclude this section, let us consider the phase equilibrium conditions in a ferromagnet from a somewhat more general standpoint than was needed for the above discussion, by no longer assuming that all the components of **H** are equal in the two phases; this formulation may be appropriate, for instance, in the treatment of curved domain boundaries.

First of all, the general magnetostatic conditions (29.13) must be satisfied:

$$B_{1n} = B_{2n}, \qquad \mathbf{H}_{1t} = \mathbf{H}_{2t}; \qquad (44.4)$$

these follow from Maxwell's equations div $\mathbf{B} = 0$, **curl** $\mathbf{H} = 0$. In addition, the thermodynamic condition must be satisfied which expresses the equilibrium with regard to a movement of the interface (in the direction normal to it), i.e. with regard to a transfer of matter from one phase to the other. This condition is represented by the equality in the two phases of the thermodynamic potentials relative to B_n and $\mathbf{H}_t$.‡ To find these, it is sufficient to rewrite the product $-\mathbf{B}\cdot\mathrm{d}\mathbf{H}$ in the expression for the differential $\mathrm{d}\tilde{F}$ (31.6) (again neglecting magnetostriction) as

$$\begin{aligned} -\mathbf{B}\cdot\mathrm{d}\mathbf{H} &= -\mathbf{B}_t\cdot\mathrm{d}\mathbf{H}_t - B_n\mathrm{d}H_n \\ &= -\mathrm{d}(B_nH_n) - \mathbf{B}_t\cdot\mathrm{d}\mathbf{H}_t + H_n\mathrm{d}B_n. \end{aligned}$$

From this it is clear that the required thermodynamic potential F' is

$$F' = \tilde{F} + B_nH_n/4\pi, \qquad (44.5)$$

and the boundary condition sought is

$$F'_1 = F'_2 \qquad (44.6)$$

(I. A. Privorotskiĭ and M.Ya. Azbel', 1969).

PROBLEM

Determine the energy of the magnetic field near the surface of a ferromagnet at which plane-parallel domains perpendicular to the surface emerge without change in the direction of magnetization (Fig. 23a).

SOLUTION. The problem of determining the magnetic field near such a surface is equivalent to the electrostatic problem of the field due to a plane divided into strips charged alternately positively and negatively with surface charge density $\sigma = \pm M$.

Let the surface of the body be the plane $z = 0$, and let the x-axis be perpendicular to the plane of the domains. The "surface charge density" $\sigma(x)$ is a periodic function with period $2a$ (a being the width of the domains), and its

† The idea of domain branching and the result that in the limit $a \propto l^{2/3}$ were first put forward by L. D. Landau for the intermediate state of superconductors (§57).

‡ This statement is justified in a similar way to the derivation of the ordinary condition of phase equilibrium, the equality of the chemical potentials of the two phases, in which the independent variables are the temperature and the pressure, these being equal in the two phases; see *SP* 1, §81.

value in a typical period is $\sigma = -M$ for $-a < x < 0$, $\sigma = +M$ for $0 < x < a$. Its expansion in Fourier series is

$$\sigma(x) = \sum_{n=0}^{\infty} c_n \sin\frac{(2n+1)\pi x}{a}, \qquad c_n = 4M/(2n+1)\pi.$$

The field potential satisfies Laplace's equation

$$\frac{\partial^2\phi}{\partial x^2} + \frac{\partial^2\phi}{\partial z^2} = 0;$$

we seek ϕ as a series

$$\phi(x, z) = \sum_{n=0}^{\infty} b_n \sin\frac{(2n+1)\pi x}{a} e^{\mp(2n+1)\pi z/a},$$

where the two signs in the exponent relate to the half-spaces $z > 0$ and $z < 0$. The coefficients b_n are given by the boundary condition

$$-[\partial\phi/\partial z]_{z=0+} + [\partial\phi/\partial z]_{z=0-} = 4\pi\sigma,$$

whence $b_n = 2ac_n/(2n+1)$.

The required field energy can be calculated as the integral $\frac{1}{2}\int \sigma\phi\, \mathrm{d}f$ over the "charged surface". The energy per unit area is

$$\frac{1}{2}\frac{1}{2a}\int_{-a}^{a} [\sigma\phi]_{z=0}\, \mathrm{d}x = \frac{1}{4}\sum_{n=0}^{\infty} c_n b_n$$

$$= \frac{8aM^2}{\pi^2}\sum_{n=0}^{\infty}\frac{1}{(2n+1)^3} = \frac{7aM^2}{\pi^2}\zeta(3).$$

With the zeta function $\zeta(3) = 1.202$, the result is equal to $0.852\, aM^2$. The width of the domains in the plate is found by minimizing the sum $1.7\, aM^2 + M^2\Delta l/a$, the first term being the energy of emergence of domains on both sides of the plate, and the second term the surface energy. Hence $a = 0.8\sqrt{(\Delta l)}$ (C. Kittel, 1946).

§45. Single-domain particles

As the dimensions of the body decrease, the formation of any domains at all ultimately becomes thermodynamically unfavourable, so that sufficiently small ferromagnetic particles are uniformly magnetized single domains. The criterion giving their dimension l is obtained by comparing the magnetic energy of a uniformly magnetized particle with the non-uniformity energy which would result if there were considerable non-uniformity in the distribution of the magnetization over its volume. The former energy is of the order of $M^2 V$, and the latter of the order of $\alpha M^2 V/l^2$. The condition for a single domain to be formed is therefore†

$$l \lesssim \sqrt{\alpha}. \tag{45.1}$$

In order to ascertain the behaviour of a uniformly magnetized particle in an external magnetic field, we must consider its total free energy, substituting in (32.7) for $\tilde{F}$ the sum of (39.1) and the anisotropy energy:‡

$$\mathscr{F} = VU_{\text{aniso}} - \tfrac{1}{2}\mathbf{M}\cdot\int(\mathbf{H} + \mathfrak{H})\,\mathrm{d}V, \tag{45.2}$$

the integration being taken only over the volume of the body, and the unimportant

† The term *micromagnetism* is sometimes used to refer to the properties of assemblies of such particles.

‡ In neglecting the magnetostriction, we make no distinction between the thermodynamic potential and the free energy, considering the latter for a given volume V of the body.

constant VF_0 being omitted. Let the particle be ellipsoidal. Then the field **H** within it is given by (29.14), or

$$H_i = \mathfrak{H}_i - 4\pi n_{ik} M_k; \tag{45.3}$$

the second term here is the "demagnetizing field" due to the body. We thus find

$$\mathcal{F} = 2\pi n_{ik} M_i M_k V - V\mathbf{M} \cdot \mathfrak{H} + VU_{\text{aniso}}. \tag{45.4}$$

The first term is called the intrinsic *magnetostatic energy* of the magnetized particle; the second term is the energy of the particle in the external field.

The direction of magnetization of the particle in the external field $\mathfrak{H}$ is determined by the condition that $\mathcal{F}$ be a minimum as a function of the direction of **M**. For a cubic crystal, the anisotropy energy may be neglected in (45.4). For a uniaxial crystal, writing the anisotropy energy in the form $\frac{1}{2}\beta_{ik} M_i M_k$, we have

$$\mathcal{F} = \tfrac{1}{2} V (4\pi n_{ik} + \beta_{ik}) M_i M_k - V \mathfrak{H} \cdot \mathbf{M}. \tag{45.5}$$

The problem thus formulated is mathematically identical with the one in §41 concerning the dependence of the local magnetization **M** on the local field **H**, the only difference being that **H** is replaced by $\mathfrak{H}$, and β_{ik} by $4\pi n_{ik}$ or by $4\pi n_{ik} + \beta_{ik}$.

Lastly, let us derive the equation which must be satisfied by the distribution of the magnetization in a single-domain sample, under conditions such that this distribution cannot be supposed uniform. To do so, we must require that the total free energy of the body be a minimum; we write it as the integral

$$\begin{aligned}\mathcal{F} &= \int \tilde{F}\, \mathrm{d}V \\ &= \int \{F_0(M) + U_{\text{non-u}} + U_{\text{aniso}} - \mathbf{M} \cdot \mathbf{H} - H^2/8\pi\}\, \mathrm{d}V,\end{aligned} \tag{45.6}$$

taken over all space. The variation is with respect to **M**, which is here a function of the coordinates, the value of **H** being specified at every point; the magnitude of **M** is fixed, and only its direction varies. Omitting from the integrand terms which depend only on M or on **H**, we vary the integral

$$\int \left\{ \frac{1}{2} \alpha_{ik} M^2 \frac{\partial \mathbf{m}}{\partial x_i} \cdot \frac{\partial \mathbf{m}}{\partial x_k} + U_{\text{aniso}} - M\, \mathbf{m} \cdot \mathbf{H} \right\} \mathrm{d}V,$$

which is now taken only over the volume of the body (where $\mathbf{M} \neq 0$). Integrating the first term by parts after the variation, we find

$$\begin{aligned}\delta \mathcal{F} = &- \int \left\{ \alpha_{ik} M^2 \frac{\partial^2 \mathbf{m}}{\partial x_i\, \partial x_k} - \frac{\partial U_{\text{aniso}}}{\partial \mathbf{m}} + M\mathbf{H} \right\} \cdot \delta \mathbf{m}\, \mathrm{d}V \\ &+ \oint \alpha_{ik} M^2 \frac{\partial \mathbf{m}}{\partial x_k} \cdot \delta \mathbf{m}\, \mathrm{d}f_i;\end{aligned} \tag{45.7}$$

the second integral is taken over the surface of the body. Since $\mathbf{m}^2 = 1$, we have $\mathbf{m} \cdot \delta\mathbf{m} = 0$, i.e. the variation is of the form $\delta\mathbf{m} = \delta\boldsymbol{\omega} \times \mathbf{m}$, where $\delta\boldsymbol{\omega}$ is an arbitrary (small) function of the coordinates. The condition $\delta\mathcal{F} = 0$ gives the required equation† when the

† It is, of course, the same as the equation of motion (precession) of the magnetic moment in a ferromagnet, if the rate of change $\partial \mathbf{M}/\partial t$ is put equal to zero (see *SP* 2, §69).

coefficient of $\delta\omega$ in the volume integrand is put equal to zero:

$$\mathbf{m} \times \left(\alpha_{ik} M^2 \frac{\partial^2 \mathbf{m}}{\partial x_i \partial x_k} - \frac{\partial U_{\text{aniso}}}{\partial \mathbf{m}} + M \mathbf{H} \right) = 0. \tag{45.8}$$

Equating the surface integral to zero, we obtain the boundary condition on this equation; for example, when $\alpha_{ik} = \alpha \delta_{ik}$ the condition is

$$\mathbf{m} \times \partial \mathbf{m}/\partial n = 0, \tag{45.9}$$

where **n** is the direction of the normal to the surface of the body.

Together with (45.8), Maxwell's equations

$$\text{div}\,(\mathbf{H} + 4\pi M \mathbf{m}) = 0, \quad \mathbf{curl}\,\mathbf{H} = 0 \tag{45.10}$$

must of course be satisfied in all space, with the usual boundary conditions on the surface of the body and the condition $\mathbf{H} \to \mathfrak{H}$ at infinity.†

For a uniformly magnetized (ellipsoidal) body the first term in the parentheses in (45.8) is zero. The remaining equation, with **H** from (45.3), is the same as the condition for a minimum of the free energy (45.5).

§46. Orientational transitions

The anisotropy constant of a ferromagnet is a function of temperature, and so can change sign at any point. The direction of the spontaneous magnetization then changes also, and therefore so does the symmetry of the magnetic structure. The resulting transitions between different phases of a magnetic substance are called *orientational transitions.* Let us see how such transitions occur in a uniaxial hexagonal ferromagnetic crystal (H. Horner and C. M. Varma, 1968).

Since we have in mind to consider the neighbourhood of a point where the anisotropy constant K_1 becomes zero, it is necessary to take into account also the next term in the expansion of the anisotropy energy; for a hexagonal ferromagnet, such an expression for U_{aniso} is given by (40.3).

Let us first suppose that $K_2 > 0$. The minimum of U_{aniso} then corresponds to the following phases, depending on the values of K_1 and K_2:

$$\left.\begin{aligned} &\text{(I)}\ \theta = 0, \pi \text{ for } K_1 > 0, \\ &\text{(II)}\ \sin\theta = \pm\sqrt{(-K_1/2K_2)} \text{ for } -2K_2 < K_1 < 0, \\ &\text{(III)}\ \theta = \tfrac{1}{2}\pi \text{ for } K_1 < -2K_2. \end{aligned}\right\} \tag{46.1}$$

Phases I and III are of the easy axis and easy plane type respectively. In phase II, the magnetization vector does not have a fixed orientation as in phases I and III; as the temperature varies, its direction changes continuously between $\theta = 0$ (or π) and $\theta = \frac{1}{2}\pi$. The symmetry of this phase (sometimes called the *angular phase*) is lower than that of both

† The question may arise whether it is correct to vary the integral (45.6) with respect to **m** for constant **H**, even though these are related by the first equation (45.10). In fact, however, if we put (from the second equation) $\mathbf{H} = -\mathbf{grad}\,\phi$ and calculate the variation of the integral with respect to ϕ, the result is zero by virtue of the first equation. Thus the variation with respect to **H** gives no contribution to $\delta\mathcal{F}$.

phase I and phase III. The transitions between phases I and II, and II and III, take place as second-order phase transitions at temperatures T_1 and T_2 given by the conditions

$$K_1(T_1) = 0, \quad K_1(T_2) + 2K_2(T_2) = 0. \tag{46.2}$$

We shall take the particular case where $K_1 > 0$ for $T > T_1$. Then $T_2 < T_1$ if $K_2 > 0$.

In the presence of a magnetic field, the thermodynamic potential is

$$\tilde{\Phi} = K_1 \sin^2\theta + K_2 \sin^4\theta - \mathbf{H}\cdot\mathbf{M}; \tag{46.3}$$

Only the terms which depend on the direction of **M** are shown here. Near the transition point between phases I and II, the order parameter is the small quantity $\sin\theta \cong \theta \equiv \eta$. In this range, the thermodynamic potential in a weak transverse magnetic field $H = H_x$ is

$$\tilde{\Phi} = K_1\eta^2 + K_2\eta^4 - MH_x\eta,$$

with $K_1 = \text{constant} \times (T - T_1)$. In the usual way (cf. §39), we deduce that at the transition point the transverse susceptibility

$$\chi_\perp = (\partial M_x/\partial H_x)_{H_x\to 0} \cong M(\partial\theta/\partial H_x)_{H_x\to 0}$$

tends to infinity in a similar manner to that given by (39.6) and (39.7). Analogously, near the transition point between phases II and III, the order parameter is the small angle $\eta = \frac{1}{2}\pi - \theta$. In a longitudinal field $H = H_z$, the thermodynamic potential contains a term $-MH_z\eta$, and at the transition point the longitudinal susceptibility $\chi_\parallel$ becomes infinite.

The above argument is based on the Landau theory of phase transitions. For orientational transitions, this theory is valid almost without restrictions. The closeness to the transition point that is allowable in the Landau theory is determined by a condition (see (47.1) below) having in the denominator the cube of the coefficient α in the term (43.1) of the thermodynamic potential, which arises from the non-uniform distribution of the order parameter. In the present case, this term is due to the exchange interactions in the ferromagnet, whereas the expansion terms relative to the parameter η itself are due to the relativistic interactions. This is the cause of the extreme narrowness of the temperature range near the transition point within which the Landau theory is not valid.

Now let $K_2 < 0$. Then phase II is unstable (U_{aniso} has a maximum and not a minimum), so that the easy-axis phase I must pass directly into the easy-plane phase III. It is clear from the start that this cannot be a second-order transition, since neither of the symmetry groups of phases I and III is a sub-group of the other. The transition takes place as a first-order phase transition at the point $T = T_0$ which is given by the equality of the thermodynamic potentials in the two phases (i.e., of the values of U_{aniso}):

$$K_1(T_0) + K_2(T_0) = 0. \tag{46.4}$$

The point T_0 lies between T_1 and T_2 as given by equations (46.2) (where now $T_2 > T_1$). In this case, the temperatures T_1 and T_2 determine the limits of metastability of phases I and II respectively; beyond these limits, U_{aniso} has a maximum and not a minimum at $\theta = 0$ or $\frac{1}{2}\pi$.

Orientational transitions can occur not only when the temperature varies (spontaneous transitions) but also when the magnetic field applied to the body changes (field-induced transitions). The transition points occupy curves in the (H, T) phase diagrams (for a given crystallographic orientation of **H**). As an example, let us consider the phase diagrams of the same uniaxial ferromagnet in a field H_z parallel to the hexagonal axis.

A longitudinal field does not affect the symmetry of the easy-axis phase (*EAP* in Fig. 25).

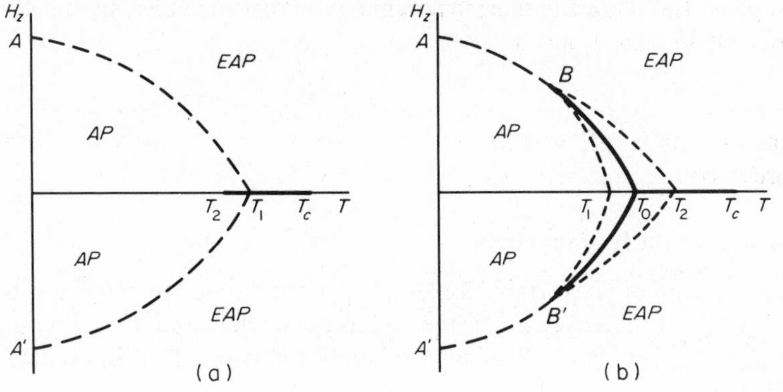

FIG. 25

The easy-plane phase, however, becomes an angular phase (*AP*), since the field moves the magnetization **M** out of the basal plane.

Let us first consider the case where $K_2 > 0$. The regions of the two phases are separated by second-order phase transition curves beginning from T_1 on the abscissa axis (dashed curves in Fig. 25a). The upper and lower parts of the diagram correspond to two opposite directions of the longitudinal field, and accordingly to opposite signs of the longitudinal component M_z. Near the curve AT_1, the angle θ is small (near $A'T_1$, $\pi - \theta$ is small). As far as terms of the fourth order in θ, we have from (46.3)

$$\tilde{\Phi} = (K_1 + \tfrac{1}{2} MH_z)\,\theta^2 + (K_2 - \tfrac{1}{3} K_1 - \tfrac{1}{24} MH_z)\,\theta^4. \tag{46.5}$$

The equation of the curve AT_1 is found by equating to zero the coefficient of θ^2:

$$K_1\,(T) + \tfrac{1}{2} MH_z = 0 \tag{46.6}$$

($K_1 < 0$ when $T < T_1$); the curve $A'T_1$ is obviously given by the same equation with the sign of the second term reversed, and is symmetrical with respect to the curve AT_1.

The segment T_2T_c of the abscissa axis is a first-order phase transition line, on which two phases with opposite signs of M_z are in equilibrium. The second-order phase transition which occurs at T_2 in the absence of the field disappears when the field is present; T_2 is a critical point in the (H,T) phase diagram, at which a first-order transition line terminates. Another termination point of this line is the Curie point T_c (at which the magnetization vanishes when $H_z = 0$).

When $K_2 < 0$ (Fig. 25b), the initial part of the boundary ABT_0 between the two phases in the (H,T) diagram (and similarly $A'B'T_0$) is a first-order phase transition curve (the continuous curve BT_0), on either side of which are regions of metastability of the two phases bounded by the dashed curves BT_1 and BT_2.† At B (a tricritical point) the first-order transition curve becomes a second-order one (the dashed curve BA, whose equation is (46.6)). The coordinates of this point are given by the simultaneous vanishing of the

† It is assumed that the interface between two coexistent phases is parallel to the magnetic field, so that H_z is continuous at this interface.

coefficients of θ^2 and θ^4 in the thermodynamic potential (46.5), i.e. by the equations†

$$MH_z = 2\,|K_1(T)|, \quad K_1(T) = 4K_2(T). \tag{46.7}$$

Lastly, the segment $T_0 T_c$ is a line of first-order transitions between phases with oppositely directed magnetizations $M_z = \pm M$.

§47. Fluctuations in ferromagnets

The Landau theory of phase transitions, on which the discussion in §39 was based, does not take account of fluctuations of the order parameter, and thus becomes invalid sufficiently near the Curie point. The range of applicability of this theory is determined by the condition

$$|t|/T_c \gg T_c B^2/a\alpha^3, \tag{47.1}$$

where $t = T - T_c$, a and B are the coefficients in the expansion (39.3), (39.4), and α is of the order of the components of the tensor α_{ik} in (43.1). We must also, of course, have $|t| \ll T_c$ as the condition of closeness to the Curie point.‡

When the inequality sign in (47.1) is reversed, i.e. in the *fluctuation region*, the order parameter fluctuations are of decisive importance. The rigorous statistical theory for the Curie point of a ferromagnet in this region would have to be based on the effective Hamiltonian.

$$\mathscr{H}_{\text{eff}} = \int \left\{ at\mathbf{M}^2 + B(\mathbf{M}^2)^2 + \tfrac{1}{2}\bar{\alpha}\frac{\partial \mathbf{M}}{\partial x_i}\cdot\frac{\partial \mathbf{M}}{\partial x_i} - \mathbf{H}\cdot\mathbf{M} \right\} dV. \tag{47.2}$$

The function $\mathbf{M}(\mathbf{r})$ is here assumed to vary only slowly, in the sense that its Fourier expansion contains only wave numbers much less than the characteristic reciprocal atomic distance. The coefficients a, B, and $\bar{\alpha}$ are the same as the expansion coefficients in the Landau theory, i.e. are independent of temperature; for simplicity, the derivative term is written on the assumption that the crystal has cubic symmetry, the coefficient being denoted by $\bar{\alpha}$ to distinguish it from the true coefficient α (see below). The effective Hamiltonian (47.2) corresponds to the exchange approximation, neglecting the anisotropy energy; in that approximation, we should also neglect the fluctuations of the magnetic field that are due to those of the magnetization (i.e. the magnetostatic energy of the fluctuations; cf. *SP* 2, §70). The exchange approximation is characterized by a "degeneracy": the order parameter $\mathbf{M}$ has three components, but the effective Hamiltonian is invariant when this vector is rotated through the same angle in all space.

To describe the behaviour of the thermodynamic quantities near a second-order phase transition, use is made of *critical indices*, defined in a way which will be repeated here as it applies to the Curie point in a ferromagnet. The index α describes the temperature dependence of the specific heat C_p on either side of the Curie point in the absence of a

† At the tricritical point in the (H,T) phase diagram, we can apply the results derived in *SP* 1, §150 (within the Landau theory) for the tricritical point in the *PT*-plane.

‡ In this section, we use the results given in *SP* 1, §§146–149. The reader is assumed to be familiar with these, and some of them are mentioned afresh only to make the discussion more coherent. The appropriate references will not be given on every occasion.

magnetic field:†

$$C_p \propto |t|^{-\alpha} \text{ for } \mathbf{H} = 0. \tag{47.3}$$

The index β describes the temperature dependence of the spontaneous magnetization below the Curie point:

$$M \propto (-t)^{\beta} \text{ for } t < 0, \mathbf{H} = 0. \tag{47.4}$$

The index γ describes the dependence on t of the magnetic susceptibility in the paramagnetic phase:

$$\chi \propto t^{-\gamma} \text{ for } t > 0, \mathbf{H} = 0; \tag{47.5}$$

see below concerning the behaviour of the susceptibility when $t < 0$. The field dependence of the magnetization at the Curie point itself is written

$$M \propto H^{1/\delta} \text{ for } t = 0. \tag{47.6}$$

The temperature dependence of the correlation radius of the magnetization fluctuations is described by the index ν:

$$r_c \propto |t|^{-\nu} \text{ for } \mathbf{H} = 0. \tag{47.7}$$

At the Curie point itself, the correlation function is a decreasing power function of the distance:‡

$$G_{ik}(r) = \langle \delta M_i(0)\, \delta M_k(\mathbf{r}) \rangle \propto r^{-(1+\zeta)} \text{ for } t = 0, \mathbf{H} = 0. \tag{47.8}$$

The critical indices are connected by certain relations, some of which follow from the hypothesis of scale invariance. These relations are universal ones, and in particular do not depend on the number of components of the order parameter; they enable all the above-mentioned indices to be expressed in terms of any two of them.

Table 2 gives the values of the critical indices for a ferromagnetic transition. For comparison, we give also the values for other degrees of degeneracy n (where n is the number of components of the order parameter $\mathbf{M}$ in the effective Hamiltonian in the form (47.2); $n = 3$ for a three-dimensional ferromagnet in the exchange approximation).§

In the paramagnetic phase, the correlation function decreases exponentially at distances $r \gg r_c$. In the ferromagnetic phase, we have to distinguish fluctuations with and without a change in the magnitude of $\mathbf{M}$. The degeneracy of the Curie point problem in the exchange approximation makes its appearance here.

The difference does not arise when the change in $\mathbf{M}$, including the change in its direction, takes places over short distances $r \ll r_c$; the correlation function then has the same behaviour (47.8) for all fluctuations. At distances $r \gg r_c$ (wave numbers such that $kr_c \ll 1$),

TABLE 2

n	α	β	γ	ν	δ	ζ
1	0.110	0.325	1.240	0.630	0.48	0.031
2	−0.007	0.346	1.315	0.669	0.48	0.033
3	−0.115	0.364	1.387	0.705	0.48	0.033

† This refers to the "singular" part of the specific heat: when $\alpha < 0$, the specific heat $C_p = C_{p0} + \text{constant} \times |t|^{|\alpha|}$. The index α is not to be confused with the coefficient in (47.1), (47.2) and below.

‡ We are everywhere considering only ordinary three-dimensional bodies.

§ The values of the small indices α and ζ are taken from numerical calculations (J. C. Le Guillou and J. Zinn-Justin, *Physical Review* B **21**, 3976, 1980). The others have been calculated from these two.

the fluctuations of the direction of **M** increase "anomalously" because of the decreased amount of energy needed for this departure from equilibrium; for a rotation of the magnetization uniform throughout the crystal, $k \to 0$, and no energy at all is needed. The correlation function of the fluctuations of the direction of **M** at such distances can be derived thermodynamically.

Since **M** varies slowly, we can separate in the thermodynamic potential the terms related to this variation:

$$\Phi = \Phi_0 (T, M) + \tfrac{1}{2} \alpha \int \frac{\partial \mathbf{M}}{\partial x_i} \cdot \frac{\partial \mathbf{M}}{\partial x_i} \, \mathrm{d}V, \tag{47.9}$$

where Φ_0 applies to a uniformly magnetized body.† It must be emphasized that we are here discussing the true thermodynamic potential, and the coefficient α, a function of temperature, is not the same as $\bar{\alpha}$ in the effective Hamiltonian.

In a small rotation of **M** (without change in magnitude), the term $\Phi_0(T, M)$ is unaltered. If some particular direction of **M** is chosen as the z-axis, a small deviation from this direction can be described by a small two-dimensional vector $\delta \mathbf{M}_\perp \equiv \mathbf{M}_\perp$ in the xy-plane. The corresponding change in the thermodynamic potential is‡

$$\delta \Phi = \tfrac{1}{2} \alpha \int \frac{\partial \delta \mathbf{M}_\perp}{\partial x_i} \cdot \frac{\partial \delta \mathbf{M}_\perp}{\partial x_i} \, \mathrm{d}V. \tag{47.10}$$

Substituting $\delta \mathbf{M}_\perp$ as a Fourier series

$$\delta \mathbf{M}_\perp = \sum_{\mathbf{k}} \delta \mathbf{M}_{\perp \mathbf{k}} e^{i\mathbf{k}\cdot\mathbf{r}}, \quad \delta \mathbf{M}_{\perp \mathbf{k}} = \delta \mathbf{M}^*_{\perp, -k},$$

we find

$$\delta \Phi = \tfrac{1}{2} V \alpha \sum_{\mathbf{k}} k^2 |\delta \mathbf{M}_{\perp \mathbf{k}}|^2,$$

and the mean square fluctuation

$$\langle \delta M_{\alpha \mathbf{k}} \, \delta M_{\beta \mathbf{k}} \rangle = (T/V\alpha k^2) \, \delta_{\alpha\beta}, \tag{47.11}$$

where α and β are vector suffixes in the xy-plane. The corresponding coordinate correlation function is §

$$G_{\alpha\beta}(r) = (T/4\pi\alpha r) \, \delta_{\alpha\beta}. \tag{47.12}$$

Thus the correlation function of the fluctuations of the direction of **M** in the ferromagnetic phase decreases according to a low power law ($\sim 1/r$) at distances $r \gg r_c$, whereas the correlation function of the fluctuations of the magnitude M decreases more rapidly.

† Strictly speaking, we should use here the thermodynamic potential Ω, not Φ; cf. *SP* 1, §146. However, according to the theorem of small increments, the derivative term concerned here has the same form for either potential, and the potential will therefore be written as Φ.

‡ The calculations below are exactly analogous to those in *SP* 1, §146.

§ Formula (47.12) can also be derived in microscopic spin-wave theory. It is valid independently of the closeness to the Curie point; far from the latter, the only requirement is that r be much greater than atomic dimensions (see *SP* 2, §71, Problem 4). Neglecting the magnetic anisotropy energy, however, places an upper limit on the range of validity of (47.12). For example, in a uniaxial crystal with the anisotropy energy (41.1), we must have $r \ll \sqrt{(\beta/\alpha)}$.

Using the hypothesis of scale invariance, we can now determine the temperature dependence of the thermodynamic quantity α, expressing it in terms of the critical indices defined above. As already mentioned, when $r \ll r_c$ the correlation functions of all fluctuations of the magnetization, including those of its direction, have the same behaviour (47.8). One feature of scale invariance is precisely that the characteristic distance at which one limiting form changes to the other is r_c. That is, when $r \sim r_c$ the two forms should give the same order of magnitude of G. When $r \sim r_c$, the temperature dependence of G from (47.7) and (47.8) is

$$G \propto r_c^{-(1+\zeta)} \propto (-t)^{\nu(1+\zeta)}.$$

According to (47.12),

$$G \propto (\alpha r_c)^{-1} \sim \alpha^{-1}(-t)^{\nu}.$$

Comparison of these two expressions gives

$$\alpha \sim (-t)^{-\nu\zeta} \tag{47.13}$$

(P. C. Hohenberg and P. C. Martin, 1965). Thus, as $T \to T_c$, α tends to infinity, though only slowly (since ζ is small). In the Landau theory, it tends to a finite non-zero limit.†

When $\mathbf{H} \neq 0$, the thermodynamic potential (47.9) must contain a further term $-\int HM_z \, dV$ (when a field is present, the mean direction of $\mathbf{M}$, i.e. the z-axis, is of course the same as the direction of $\mathbf{H}$). In a fluctuation of the direction of $\mathbf{M}$ with $\mathbf{M}^2$ = constant, we have

$$2M\delta M_z + \delta \mathbf{M}_\perp{}^2 = 0.$$

The fluctuation therefore causes the expression (47.10) for the change in $\mathcal{G}$ to have an additional term

$$-\int H\,\delta M_z \, dV = -(H/2M)\int (\delta\mathbf{M}_\perp)^2 \, dV.$$

It is clear that this causes αk^2 in (47.11) to be replaced by $\alpha k^2 + H/M$, so that

$$\langle \delta M_\alpha \delta M_\beta \rangle = \frac{T}{V(\alpha k^2 + H/M)}\delta_{\alpha\beta}. \tag{47.14}$$

These formulae yield the susceptibility of the ferromagnetic phase, i.e. the derivative

$$\chi = \frac{\partial}{\partial H}\langle \delta M_z \rangle = -\frac{1}{2M}\frac{\partial}{\partial H}\langle (\delta\mathbf{M}_\perp)^2 \rangle. \tag{47.15}$$

Substituting $\langle (\delta\mathbf{M}_\perp)^2 \rangle$ from (47.14) and changing from summation over $\mathbf{k}$ to integration, we find

$$\chi = \frac{T}{VM^2}\int \frac{1}{(\alpha k^2 + H/M)^2}\frac{V\,d^3k}{(2\pi)^3},$$

and finally, after evaluating the integral,‡

$$\chi = T/8\pi\,(\alpha M)^{3/2} H^{1/2}. \tag{47.16}$$

† It must be emphasized that the possibility of determining the function $\alpha(t)$ depends on the degeneracy of the problem. For the same reason it is possible to determine the temperature dependence of the superfluid density ρ_s of liquid helium near the λ-point; in that case, the degree of degeneracy $n = 2$ (see *SP* 2, §28). The quantities αM^2 and ρ_s play an analogous role in the two cases.

‡ The integral is governed by the range $k^2 \sim H/M\alpha$; for sufficiently small H, this is compatible with the condition $kr_c \ll 1$ for (47.14) to be valid.

We see that, in the exchange approximation, the fluctuations of the direction of $\mathbf{M}$ deprive the susceptibility χ in the ferromagnetic phase of its literal significance, as it increases without limit as $H \to 0$. Formula (47.16) is applicable not only near the Curie point.†

By M in the denominator of (47.16) is to be understood its value for $H = 0$. In the fluctuation range (near the Curie point), taking the temperature dependences of M and α from (47.4) and (47.13), we find that in the ferromagnetic phase

$$\chi \propto (-t)^{-3(\beta - \nu\zeta)/2} H^{-1/2}, \qquad t < 0. \tag{47.17}$$

In the (temperature) range where the Landau theory is valid, there is a range of values of H for which the ordinary term (39.7) predominates in the susceptibility. For, substituting into (47.16) $M \sim (-at/B)^{1/2}$, we obtain as the condition for this predominance

$$\sqrt{H} \gg T_c B^{3/4} (a|t|)^{1/4} \alpha^{-3/2}.$$

In the Landau theory, H may be considered weak if $MH \ll AM^2$, i.e. $H \ll (a|t|)^{3/2} B^{-1/2}$. The condition for these two inequalities to be compatible is the same as the condition (47.1) for the Landau theory to be valid.

The exchange approximation becomes invalid in the immediate neighbourhood of the Curie point, where the anisotropy energy becomes considerable on account of the decrease in the exchange energy. The number of order parameter components then varies; for instance, it decreases to one in a uniaxial ferromagnet of the easy-axis type (M_z in place of $\mathbf{M}$; cf. the end of §40). In the Landau theory, this circumstance does not affect the form of the temperature dependence of the spontaneous magnetization and the magnetic susceptibility. In the fluctuation range, however, it is important: the fluctuations of M_z increase without limit, but those of M_x and M_y remain finite. This alters the values of the critical indices and causes temperature dependence of the anisotropy coefficients. The problem is further complicated by the fact that it may be necessary to take into account also the magnetostatic energy of the magnetic field fluctuations; we shall not pause here to discuss this complex situation.

§48. Antiferromagnets near the Curie point

Like ferromagnetism, an antiferromagnetic structure is established basically by the isotropic exchange interaction of electrons; the weaker relativistic interactions determine the crystallographic orientation of the magnetizations of the sub-lattices.‡ The antiferromagnets now known are extremely varied in structure and therefore in their particular magnetic properties.

As an illustration, we shall consider only a simple but typical and important case, that of a uniaxial antiferromagnet with two antiparallel magnetic sub-lattices. The atoms in these sub-lattices occupy equivalent lattice sites (that is, the symmetry elements of the magnetic space group include rotations or reflections which interchange the atoms of the different

† It can also be derived in spin-wave theory (see *SP* 2, §71, Problem 2). Neglecting the magnetic anisotropy and the magnetostatic energy of the fluctuations, however, restricts its validity in accordance with the condition (for uniaxial crystals) $H \gg \beta M$ or $H \gg 4\pi M$.

‡ The idea that the exchange interaction may produce a state in which the sub-lattices have antiparallel magnetic moments was first put forward by L. Neel (1932). Independently, a similar idea was suggested by L. D. Landau (1933), who formulated the concept of the antiferromagnetic state as a thermodynamic phase different from the paramagnetic one, and of the necessary existence of a point of transition between them. This will be called the *antiferromagnetic Curie point*.

sub-lattices); otherwise, symmetry would not require strict equality of the magnitudes of the sub-lattice magnetic moments, and the crystal would be ferromagnetic.

Let $\mathbf{M}_1$ and $\mathbf{M}_2$ be the magnetic moments of the two sub-lattices per unit volume. We take as the order parameter, equal to zero in the paramagnetic phase but not in the antiferromagnetic phase, the difference

$$\mathbf{L} = \mathbf{M}_1 - \mathbf{M}_2, \qquad (48.1)$$

called the *antiferromagnetic vector*. The magnetization, equal to zero in the absence of a magnetic field, is the sum $\mathbf{M} = \mathbf{M}_1 + \mathbf{M}_2$.

In relation to $\mathbf{L}$, the symmetry transformations fall into two classes: those which interchange atoms only within the same sub-lattice, and those which interchange atoms in different sub-lattices. Then $\mathbf{L}$ is transformed simply as an axial vector for the first class, but in addition changes sign for the second class.

Near the Curie point, $\mathbf{L}$ is small. In the Landau theory, the thermodynamic potential $\tilde{\Phi}$ in this range is expanded in powers of $\mathbf{L}$ and $\mathbf{M}$, as first discussed by Landau in 1933. However, since the magnetization occurs only when the field $\mathbf{H}$ is present, it would be more correct to expand in terms of $\mathbf{L}$ and $\mathbf{H}$ immediately. For a uniaxial crystal, such an expansion is

$$\begin{aligned}\tilde{\Phi} &= \Phi_0 + AL^2 + BL^4 + D(\mathbf{H}\cdot\mathbf{L})^2 + D'H^2L^2 \\ &\quad -\tfrac{1}{2}\chi_p H^2 + \tfrac{1}{2}\beta(L_x^2 + L_y^2) - \tfrac{1}{2}\gamma(H_x^2 + H_y^2) - H^2/8\pi. \end{aligned} \qquad (48.2)$$

The first five terms (after Φ_0) are independent of the crystallographic orientation of the vectors $\mathbf{H}$ and $\mathbf{L}$; they are of exchange origin. The next two terms are due to the relativistic interactions; the z-axis is, as usual, taken to be along the principal axis of symmetry of the crystal. A term $\mathbf{H}\cdot\mathbf{L}$ linear in $\mathbf{H}$ is forbidden by the requirement of invariance under transformations that change the sign of $\mathbf{L}$.

If $\beta > 0$, the antiferromagnetic vector $\mathbf{L}$ is along the z-axis (an antiferromagnet of the easy-axis type). If $\beta < 0$, $\mathbf{L}$ lies in the basal plane (an antiferromagnet of the easy-plane type). In the former case, the antiferromagnetism corresponds to the vanishing of the function $A(T)$ (the coefficient of L_z^2); in the latter case, to the vanishing of $A + \frac{1}{2}\beta$ (the coefficient of $L_x^2 + L_y^2$).

We shall consider an easy-axis antiferromagnet ($\beta > 0$). Near the Curie point, we put as usual $A = a(T - T_c)$, and take B to have its value for $T = T_c$; then $B > 0$ is the condition for the state with $L = 0$ to be stable at $T = T_c$. In the paramagnetic phase, $A > 0$ and $L = 0$. In the antiferromagnetic phase, $A < 0$, and minimizing the potential $\tilde{\Phi}$ for $H = 0$ gives the usual Landau-theory temperature dependence of L:

$$L = \sqrt{[a(T_c - T)/2B]}. \qquad (48.3)$$

Differentiating the potential (48.2), using the formula

$$\partial\tilde{\Phi}/\partial\mathbf{H} = -\mathbf{B}/4\pi = -\mathbf{H}/4\pi - \mathbf{M},$$

we find for $L = 0$, i.e. in the paramagnetic phase,†

$$M_x = (\chi_p + \gamma)H_x, \quad M_y = (\chi_p + \gamma)H_y, \quad M_z = \chi_p H_z. \qquad (48.4)$$

† The term $-H^2/8\pi$ is separated in (48.2) so that the differentiation of the remaining terms with respect to $\mathbf{H}$ gives just $\mathbf{M}$.

The constant γ makes the magnetic susceptibility anisotropic in this phase. Because of its relativistic origin, we have $|\gamma| \ll \chi_p$. This constant will be neglected in what follows, so that χ_p will be the isotropic susceptibility for $T > T_c$.†

In the antiferromagnetic phase, when $H \to 0$ (i.e. when the field dependence of the equilibrium value of **L** is neglected). we have

$$\mathbf{M} = \chi_p \mathbf{H} - 2D\,\mathbf{L}(\mathbf{L}\cdot\mathbf{H}) - 2D' L^2 \mathbf{H}. \tag{48.5}$$

If the field is perpendicular to **L**, then

$$\begin{aligned} \mathbf{M} = \chi_\perp \mathbf{H}, \quad \chi_\perp &= \chi_p - 2D' L^2 \\ &= \chi_p - (D'a/B)(T_c - T). \end{aligned} \tag{48.6}$$

In a longitudinal field,

$$\begin{aligned} \mathbf{M} = \chi_\parallel \mathbf{H}, \quad \chi_\parallel &= \chi_\perp - 2DL^2 \\ &= \chi_p - (D+D')a\,(T_c - T)/B. \end{aligned} \tag{48.7}$$

The susceptibility is anisotropic and remains so even when the relativistic interactions are neglected, i.e. the anisotropy is of exchange origin.

It must also be emphasized that at the Curie point itself the susceptibility remains finite and continuous, but its first derivatives are discontinuous. There is an important difference here from a ferromagnet, for which the susceptibility becomes infinite at the transition point. This difference between the ferromagnetic and antiferromagnetic Curie points is closely related to the difference in the way they are affected by the presence of a magnetic field. In a ferromagnet, any field, however weak, blurs the transition, since by magnetizing the paramagnetic phase it removes the difference in symmetry of the two phases. Antiferromagnetic order, however, cannot be created by a magnetic field; the difference in symmetry between the two phases is maintained even in the presence of a field, and the transition remains sharp.

With the accuracy allowed by the expansion (48.2), the coefficients D and D' in (48.6) and (48.7) may be taken as having their values at $T = T_c$. The discontinuities in the derivatives of the susceptibility are therefore‡

$$\frac{\partial \chi_p}{\partial T} - \frac{\partial \chi_\perp}{\partial T} = -\frac{aD'}{B}, \quad \frac{\partial \chi_p}{\partial T} - \frac{\partial \chi_\parallel}{\partial T} = -\frac{a(D+D')}{B}. \tag{48.8}$$

Let us now return to the expression (48.2) for $\tilde{\Phi}$. The terms with the coefficients D and β depend on the direction of **L**. We shall assume that $D > 0$ (i.e. $\chi_\perp > \chi_\parallel$), and as before that $\beta > 0$ (an easy-axis antiferromagnet). If the magnetic field is perpendicular to the z-axis, the form of these terms shows that the minimum of $\tilde{\Phi}$ corresponds to $L_x = L_y = 0$, i.e. **L** is always in the z-direction. If, however, the field is also along the z-axis, we see that, when the magnetic energy in the field (the first of the terms mentioned) becomes comparable in magnitude with the anisotropy energy, there should be a change in the direction of **L**,

† In all known cases, the substance becomes paramagnetic at the antiferromagnetic Curie point, i.e. $\chi_p > 0$. The sign of χ_p cannot, however, be deduced from thermodynamic arguments alone.

‡ No general conclusion as to the sign of D and D' can be drawn from purely thermodynamic arguments. In practice, the discontinuity of $\partial\chi_\parallel/\partial T$ is always negative; this means that $D + D' > 0$. That of $\partial\chi_\perp/\partial T$ is usually negative also, and near the Curie point $\chi_\perp > \chi_\parallel$; this means that $D > 0$ and $D' > 0$.

which rotates into the basal plane. This *spin flop* of the sub-lattices occurs abruptly at a certain field value $H = H_f$ (L. Néel, 1936), a result proved as follows. The terms in question in (48.2) may be written in the form

$$H_z^2 DL^2 + L^2(-DH_z^2 + \tfrac{1}{2}\beta)\sin^2\theta,$$

where θ is the angle between **L** and the z-axis. It is evident that the minimum of $\tilde{\Phi}$ occurs for $\theta = 0$ if $H_z < H_f$, where

$$H_f^2 = \beta/2D = \beta L^2/(\chi_\perp - \chi_\parallel). \tag{48.9}$$

If $H_z > H_f$, however, the value corresponding to equilibrium is $\theta = \frac{1}{2}\pi$, and **L** is perpendicular to the z-axis. A similar situation continues to exist far from the Curie point. The flop field H_f depends on the temperature, and the above discussion shows that the curve $H = H_f(T)$ in the TH-plane is a curve of first-order phase transitions.†

In sufficiently strong magnetic fields, the antiferromagnetic structure cannot be thermodynamically stable, since the parallel orientation along the field becomes energetically favourable for the magnetic moments of the two sub-lattices. The disappearance of the antiferromagnetic structure involves a change in symmetry and takes place by a second-order phase transition. In the TH-plane, therefore, the region in which the antiferromagnetic phase exists is bounded by a curve $H = H_c(T)$. The loss of antiferromagnetism must occur when the magnetic energy in the field becomes comparable with the exchange energy. Far from the Curie point, the critical field can be estimated in order of magnitude as $\mu H_c \sim T_c$, where μ is the atomic magnetic moment. As the Curie point is approached, H_c decreases, and becomes zero at that point. The function $H_c(T)$ in this range is easily determined by means of the same expression (48.2) for the thermodynamic potential.

It has been shown above that, if the field **H** is perpendicular to the z-axis, **L** is always parallel to that axis. The L-dependent terms in the thermodynamic potential are

$$AL^2 + BL^4 + D'H^2L^2. \tag{48.10}$$

From this we see that the presence of the magnetic field replaces the coefficient A by $A + D'H^2$, and the latter is zero at the new transition point. The critical field is therefore given by

$$H_c^2 = -A/D' = a(T_c - T)/D'. \tag{48.11}$$

If **H** is parallel to the z-axis, for $H < H_f$ the vector **L** is again along that axis, but the L-dependent terms in (48.2) differ from (48.10) in that D' is replaced by $D + D'$. In this case, therefore,

$$H_c^2 = \frac{a}{D + D'}(T_c - T). \tag{48.12}$$

Lastly, if $H > H_f$ the vector **L** is perpendicular to the z-axis; in this case, we similarly find for the critical field‡

$$H_c^2 = \frac{a}{D'}(T_c - T - \beta/2a). \tag{48.13}$$

† As has already been stressed more than once, the anisotropic terms in expressions such as (48.2) remain significant far from the transition point, whether or not L is small, since they are part of the expansion of the relativistic interactions in powers of the components of the unit vector $\mathbf{l} = \mathbf{L}/L$. Formula (48.9) also is therefore meaningful far from the Curie point, provided that we regard βL^2 as the coefficient in the $\frac{1}{2}(l_x^2 + l_y^2)$ term in the thermodynamic potential.

‡ The derivation of formulae (48.11)–(48.13) in the Landau theory is due to A. S. Borovik-Romanov (1959).

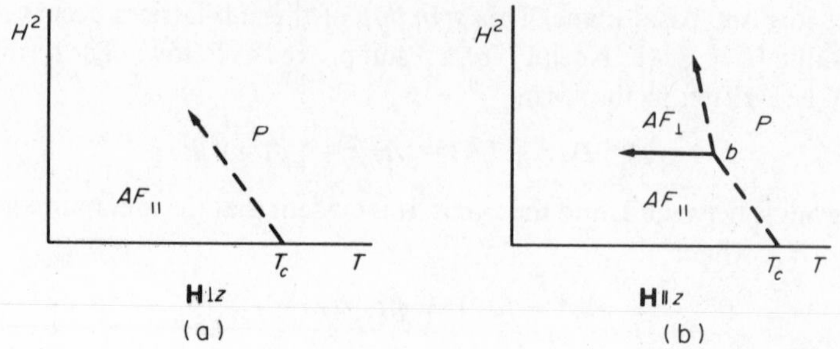

FIG. 26

Figure 26 shows the (T, H^2) phase diagram for an antiferromagnet near the Curie point for both the field directions considered. The dashed lines are second-order transitions, and the continuous line a first-order one; P denotes the paramagnetic phase, $AF_{\parallel}$ and $AF_{\perp}$ the antiferromagnetic phases with $\mathbf{L}$ respectively parallel and perpendicular to the z-axis. In the Landau theory (which is used for the whole treatment here) the diagrams consist of straight lines. For a longitudinal field there is a bicritical point (b in Fig. 26b) where the first-order phase transition line ends on a second-order one. This point is analogous to the bicritical point in the PT-plane (*SP* 1, §150, Fig. 67). Its coordinates are

$$T_b = T_c - \frac{\beta}{2a}\frac{D+D'}{D}, \quad H_b{}^2 = \beta/2D \tag{48.14}$$

where β, D, D' are the values of the coefficients for $T = T_b$.

§49. The bicritical point for an antiferromagnet

The Landau theory used in §48 becomes inapplicable, as usual, in the immediate neighbourhood of second-order transition curves, i.e. in the fluctuation range. Let us consider this range near a bicritical point in the phase diagram of a uniaxial antiferromagnet (of the easy-axis type) in a longitudinal magnetic field.

The effective Hamiltonian for this problem is

$$\mathscr{H}_{\text{eff}} = \int \Big\{ aL^2 \Big[T - T_b - \frac{D+D'}{a}(H_f{}^2 - H_z{}^2) \Big] + BL^4 + DL^2(H_f{}^2 - H_z{}^2)\sin^2\theta + \frac{1}{2}\bar{\alpha}\frac{\partial \mathbf{L}}{\partial x_i}\cdot\frac{\partial \mathbf{L}}{\partial x_i} \Big\} \, dV, \tag{49.1}$$

The integrand consists of the $\mathbf{L}$-dependent terms in the expansion (48.2), with $A = (T-T_c)a$, the notation T_b and H_f from (48.9) and (48.14), and the addition of a gradient term. This Hamiltonian has the same form as for a uniaxial ferromagnet (not in a magnetic field), differing only in the notation: $\mathbf{L}$ replaces $\mathbf{M}$, the expression in the square brackets replaces $T - T_c$, and

$$u = 2D(H_f{}^2 - H_z{}^2) \tag{49.2}$$

replaces the anisotropy constant β of the ferromagnet. When $H_z = H_f$, this constant is zero, and (49.1) reduces to

$$\mathscr{H}_{\text{eff}} = \int \left\{ aL^2(T-T_b) + BL^4 + \frac{1}{2}\bar{\alpha}\frac{\partial \mathbf{L}}{\partial x_i}\cdot\frac{\partial \mathbf{L}}{\partial x_i} \right\} dV, \tag{49.3}$$

which is formally the same as the effective Hamiltonian of a ferromagnet in the exchange approximation (47.2). These analogies enable us to elucidate a number of properties of an antiferromagnet near a bicritical point, by using the known results for a ferromagnet (M. E. Fisher and D. R. Nelson, 1974).

The equation $H_z = H_f$ corresponds to a first-order phase transition curve (spin flop). The analogy leads to the conclusion that the thermodynamic properties of the antiferromagnet on this curve near the bicritical point are the same (with appropriate reinterpretation of the quantities) as those of a purely exchange ferromagnet near its Curie point. In particular, as the bicritical point is approached along this curve, the antiferromagnetic vector tends to zero in accordance with

$$L \propto (T_b - T)^{\beta}, \tag{49.4}$$

where β is the same as in (47.4).†

Near the bicritical point but off the first-order transition curve, the parameter u is small but not zero. However small the value of u, its importance increases as the second-order transition curve is approached (see the comment at the end of §47): the number of components of the order parameter decreases to one (L_z in the $AF_{\parallel}$ phase) or two (L_x and L_y in the $AF_{\perp}$ phase). As u becomes smaller, so does the region in which it is important. The change from $n = 3$ to $n = 1$ or 2 takes place via an intermediate range. It is plausible to assume scale invariance in that range: only the scale of measurement of u changes as the bicritical point is approached. Accordingly, there is a new *cross-over index* ϕ: as the scale of $t = T - T_b$ changes, that of u varies as $|t|^{\phi}$. The index ϕ must be positive, since any value of u, however small, affects the nature of the transition.

In using this hypothesis, we must assume that the phase transition curves near the bicritical point are given by constant values of $x = u/|t|^{\phi}$. The first-order transition curves correspond to $x = 0$, the second-order ones between P and $AF_{\parallel}$ to some $x_1 > 0$, and those between P and $AF_{\perp}$ to some $x_2 < 0$. By H_f in the definition (49.2) we must now understand, of course, the actual function $H_f(T)$ that would give the exact solution of the statistical problem with the effective Hamiltonian (49.1). Expanding it in powers of t and omitting from u the constant factor $2H_b$, we find the variable u near the bicritical point as

$$u = H_b - H_z + c(T - T_b), \tag{49.5}$$

where c is a constant. The equation of the first-order transition curve is

$$H_z - H_b = c(T - T_b),$$

and those of the two second-order transition curves are

$$H_z - H_b = c(T - T_b) \pm c_{1,2}(T - T_b)^{\phi}, \tag{49.6}$$

where c_1 and c_2 are positive constants. When $\phi > 1$ (numerical estimates give $\phi \cong 1.25$),

† The index γ in this case has no direct physical significance, since it is related to the field $\mathbf{h}$, which would appear as a term $-\mathbf{h}\cdot\mathbf{L}$ in the effective Hamiltonian, but has no real existence in an antiferromagnet.

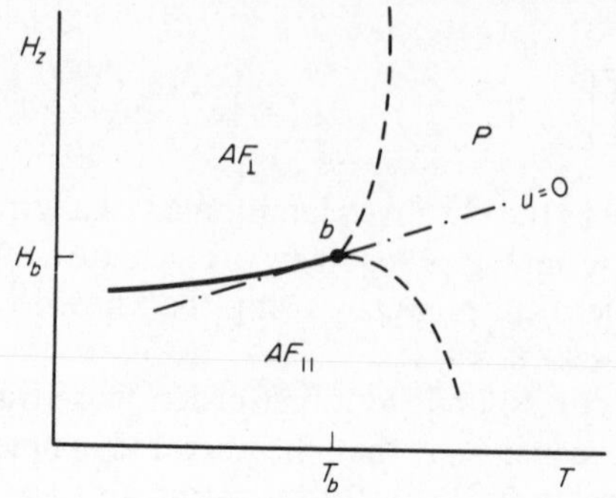

FIG. 27

these curves have the form shown in Fig. 27 in the fluctuation range; all the curves have a common tangent at b.

The hypothesis of scale invariance near the bicritical point also leads to a number of conclusions as to the manner of variation of the magnitude of the antiferromagnetic vector **L** as that point is approached from various directions in the TH_z-plane, but we shall not pause to discuss these.†

§50. Weak ferromagnetism

There are crystals in which the exchange interaction establishes an antiferromagnetic structure but the comparatively weak relativistic interactions cause a slight distortion of this structure, as a result of which a magnetization **M** appears which is "anomalously" small in proportion to the ratio of relativistic to exchange interactions. This is called *weak ferromagnetism.*‡

The exchange interaction itself allows any orientation of the antiferromagnetic vector **L** in the crystal. A particular crystallographic orientation of **L** is established only by the relativistic interactions described by terms anisotropic with respect to **L** in the expansion of the thermodynamic potential. It may happen that the symmetry of the resulting structure would in itself allow the existence of a ferromagnetic moment **M** also.§ It is in such cases that weak ferromagnetism occurs: among the relativistic terms in the expansion of the thermodynamic potential, there are some which give rise to the required distortion of the antiferromagnetic structure.‖ This will be shown by means of a characteristic example.

Let us consider rhombohedral crystals belonging to the space group $\boldsymbol{D}_{3d}^6$. As we know (see *QM*, §93), the crystal class $\boldsymbol{D}_{3d}$ contains the following symmetry elements: a threefold axis of symmetry C_3 (trigonal axis), three twofold axes perpendicular to it (denoted by U_2), and a centre of inversion I; in consequence, there are three planes of symmetry σ_d, each passing through the C_3 axis and perpendicular to one of the U_2 axes (and thus bisecting the

---

† See M. E. Fisher and D. R. Nelson, *Physical Review Letters* **32**, 1350, 1974.

‡ With the terminology described at the end of §37, it would be more correctly named "weak ferrimagnetism".

§ For this, it is always necessary that the magnetic unit cell should be the same as the crystallographic one; see the end of §38.

‖ The theory of weak ferromagnetism is due to I. E. Dzyaloshinskiĭ (1957).

angle between the other two). In the space group D_{3d}^6, the σ_d planes become glide planes with a translation of half a period along the trigonal axis. This gives an arrangement of the axes and centre of inversion in each unit cell as shown in Fig. 28. The vertical segment is one period along the trigonal axis (a spatial diagonal of the rhombohedral cell), whose length is arbitrarily taken as unity. The twofold axes pass through the points $\frac{1}{4}$ and $\frac{3}{4}$. The centre of inversion occurs at 0 and $\frac{1}{2}$ (marked by crosses in Fig. 28). The vertical planes σ_d are not shown.

In the antiferromagnets $FeCO_3$ and $MnCO_3$, each unit cell contains two magnetic ions (Fe^{++} or Mn^{++}), which occupy positions at the equivalent points 0 and $\frac{1}{2}$ on the trigonal axis. The exchange interaction establishes a magnetic structure in which the moments of these two ions are antiparallel. In $FeCO_3$ the moments of the Fe^{++} ions are along the trigonal axis (Fig. 29). It is easy to see that such a structure is invariant under all transformtions of the class $\boldsymbol{D}_{3d}$, and therefore does not allow ferromagnetism: the existence of a vector $\mathbf{M}$ along the trigonal axis is excluded by the presence of the U_2 axes, and of one in the basal plane by that of the C_3 axis.

In the antiferromagnet $MnCO_3$, the magnetic moments of the ions lie in the basal (xy) plane, which is perpendicular to the trigonal (z) axis (Fig. 30). If the moments lie in one of the σ_d planes, which we then take as the xz-plane, the magnetic structure has the symmetry elements (in addition to the unit element) $U_2^{(y)}, I, \sigma_d^{(xz)}$, i.e. belongs to the magnetic class which coincides with the ordinary class $\boldsymbol{C}_{2h}$; it allows the existence of a vector $\mathbf{M}$ in the y-direction. If, on the other hand, the moments lie along one of the U_2 axes, which we then take as the x-axis, the magnetic structure has the symmetry elements $U_2^{(x)}R, I, \sigma_d^{(xz)}R$, i.e. belongs to the magnetic class $\boldsymbol{C}_{2h}(\boldsymbol{C}_i)$; it too allows the existence of a vector $\mathbf{M}$ in the y-direction. In either case, $\mathbf{M}$ arises from the turning of the moments of the two ions in each unit cell to lie oppositely in the xy-plane (Fig. 31).

Going on to a quantitative theory, we again define vectors $\mathbf{M} = \mathbf{M}_1 + \mathbf{M}_2$ and $\mathbf{L} = \mathbf{M}_1 - \mathbf{M}_2$, where the suffixes 1 and 2 refer to the two magnetic sub-lattices. The unit vector in the direction of $\mathbf{L}$ is denoted by $\mathbf{l}$.

Let us consider the expansion of the thermodynamic potential Φ (for $\mathbf{H} = 0$) in powers of $\mathbf{M}$ and $\mathbf{l}$. The expansion in powers of $\mathbf{M}$ is permissible because this quantity is small in a

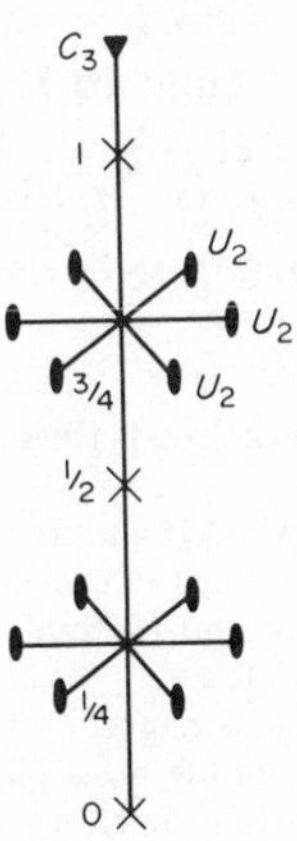

FIG. 28

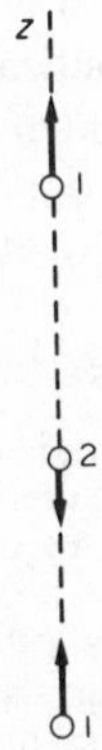

FIG. 29

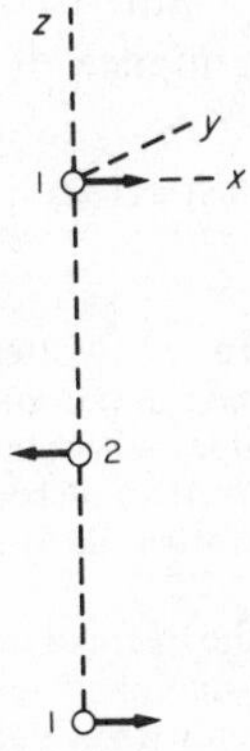

FIG. 30

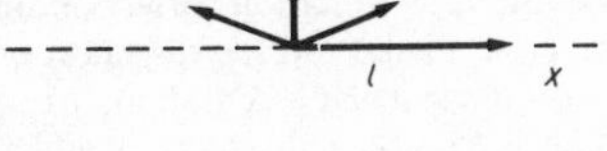

FIG. 31

weak ferromagnet. The expansion of the anisotropy energy in powers of $\mathbf{l}$ is based, as usual, on the relative smallness of the relativistic interactions. There is thus no assumption of nearness to a second-order phase transition point (L small); the theory given here is therefore not subject to the limitations inherent in the Landau theory.

The expansion terms must be invariant under all transformations in the group $\boldsymbol{D}_{3d}^{6}$. The first terms in such an expansion are

$$\Phi = \Phi_0(L) + BM^2 + D(\mathbf{l}\cdot\mathbf{M})^2 - \tfrac{1}{2}\beta M_z{}^2 - \tfrac{1}{2}\gamma L^2 l_z{}^2 + \zeta L(M_x l_y - M_y l_x), \qquad (50.1)$$

where $\Phi_0(L)$ is a function isotropic with respect to $\mathbf{L}$. The first two terms (after Φ_0) are of exchange origin; here $B > 0$ (otherwise, there would exist a spontaneous magnetization independent of the antiferromagnetism, i.e. the body would be an ordinary exchange ferromagnet). The next three terms are of the first order ($\sim v^2/c^2$) in the relativistic interactions.† The last of these may be put in the form $L\boldsymbol{\zeta}\cdot\mathbf{M}\times\mathbf{l}$, where ζ is a vector parallel to the z-axis.‡

It is obvious that all the terms in (50.1) except the last are invariant. To test the invariance of the last term, it is sufficient to do so with respect to the C_3 axis, one of the U_2 axes, and the inversion I. The invariance under rotations about the trigonal (z) axis is evident from the form which shows the z-component of $\mathbf{M}\times\mathbf{l}$; here it is important that the rotations do not interchange atoms between different sub-lattices, so that $\mathbf{M}$ and $\mathbf{l}$ are transformed in the same manner. The invariance under inversion follows from that of $\mathbf{M}$ and $\mathbf{l}$ separately; for $\mathbf{M}$ this follows from its being an axial vector, for $\mathbf{l}$ we need also to note that in the structure considered the inversion interchanges atoms only within each sub-lattice. The transformation $U_2{}^{(x)}$ interchanges atoms with oppositely directed moments; under such a rotation, therefore, $M_x, M_y \to M_x, -M_y; l_x, l_y \to -l_x, l_y$; and the invariance of $M_x l_y - M_y l_x$ is obvious.

We shall suppose that $\gamma < 0$. Then the vector $\mathbf{l}$ is in the basal plane ($l_z = 0$). Taking as the xz-plane the plane containing $\mathbf{l}$, and minimizing Φ with respect to $\mathbf{M}$ for a given $\mathbf{L}$, we find for the ferromagnetic moment

$$M_x = 0,\; M_y = (\zeta/2B)L_x,\; M_z = 0. \qquad (50.2)$$

Since $|\zeta| \ll B$, M is in fact small. We see that the occurrence of weak ferromagnetism is due to the last term in (50.1), which is bilinear in $\mathbf{M}$ and $\mathbf{l}$. The close relation between the direction of $\mathbf{M}$ and the antiferromagnetic structure is characteristic of weak ferromagnetism; in the present case, $\mathbf{M}$ is in the same basal plane and is perpendicular to $\mathbf{L}$.§

When a field is present, the dependence of the magnetization on $\mathbf{H}$ is given by the conditions for a minimum of the thermodynamic potential $\tilde{\Phi} = \Phi - \mathbf{M}\cdot\mathbf{H} - H^2/8\pi$. The minimization is to be effected with respect to the orientation of the structure in the basal

† The factors L^2 and L in the last two terms are written only in order to make all the coefficients in (50.1) dimensionless, and do not signify an expansion in powers of L.

‡ The microscopic origin of such a term is due to the spin-antisymmetric interaction which occurs as a second-order effect in perturbation theory, the perturbation being the mixed terms bilinear in the exchange and relativistic spin–orbit interactions. See T. Moriya, in *Magnetism* (ed. by G. T. Rado and H. Suhl), Vol. 1, Academic Press, New York, 1963, p. 85.

§ In the approximation corresponding to the thermodynamic potential (50.1), there is no magnetic anisotropy in the basal plane, and the direction of $\mathbf{L}$ in that plane remains arbitrary. Anisotropy in the basal plane, and therefore a particular orientation of $\mathbf{L}$, occur only when higher-order terms (up to the sixth order) are taken into account. Terms then also come in which are mixed as regards the z and the x or y components of the vectors, so that the magnetic moments deviate from the basal plane by a small angle.

plane and with respect to the components of **M**. It is evident that, when the magnetic anisotropy in the basal plane is neglected, the magnetization rotates to a position such that its component $\mathbf{M}_\perp$ in the basal plane is parallel to the field $\mathbf{H}_\perp$, and **L** is accordingly perpendicular to $\mathbf{H}_\perp$.† The minimization of $\tilde{\Phi}$ with respect to $\mathbf{M}_\perp$ and M_z then gives

$$M_z = \chi_\parallel H_z, \quad M_\perp = \chi_\perp(H_D + H_\perp), \tag{50.3}$$

where the susceptibilities are

$$\chi_\parallel = 1/(2B-\beta), \chi_\perp = 1/2B \tag{50.4}$$

and

$$H_D = |\zeta| L \tag{50.5}$$

is the *Dzyaloshinskiĭ field*, i.e. the "effective field" which determines the spontaneous magnetization of a weak ferromagnet. Since $|\beta| \ll B$, we have $\chi_\parallel \cong \chi_\perp$.

Another property of these substances is one which appears when a field **H** is applied, and is found near the point of transition to the paramagnetic phase. In the Landau theory, we expand the function $\Phi_0(L)$ in this region in powers of L:

$$\Phi_0(L) = \Phi_0(0) + AL^2 + CL^4.$$

Let the vector **L** be in the positive x-direction. We take the particular case where $\zeta > 0$‡; then **M** is in the positive y-direction. The field **H** is assumed to be in that direction also. The thermodynamic potential is

$$\tilde{\Phi} = \Phi_0(0) + AL^2 + CL^4 + BM^2 - \zeta LM - HM - H^2/8\pi; \tag{50.6}$$

near the Curie point, the expansion of the anisotropy energy in powers of the unit vector **l** becomes one in powers of **L** itself. In the absence of a field, $M = \zeta L/2B$, and the expansion (50.6) becomes

$$\Phi = \Phi_0 + (A - \zeta^2/4B)L^2 + CL^4.$$

The Curie point is given by the vanishing of the coefficient of L^2; near this point, therefore,

$$A - \zeta^2/4B = a(T - T_c)$$

($a > 0$). The remaining coefficients are taken to have their values at $T = T_c$ (with $C > 0$).

When the field is present, elimination of M from the equations $\partial\tilde{\Phi}/\partial L = 0$, $\partial\tilde{\Phi}/\partial M = 0$ gives the following equation for L:

$$2CL^3 + a(T-T_c)L - \zeta H/4B = 0. \tag{50.7}$$

This shows that the magnetic field blurs the phase transition in a weak ferromagnet, as in an ordinary one.§ The antiferromagnetic vector also is non-zero on both sides of T_c; the

† It is assumed here that $D > 0$; when $D < 0$, the presence of the term $-|D|(\mathbf{l}\cdot\mathbf{M})^2$ with a large coefficient $|D|$ may have the result that **l** and $\mathbf{H}_\perp$ are no longer at right angles.

The neglect of the anisotropy in the basal plane imposes certain lower limits on the fields concerned.

‡ The sign of ζ is arbitrary in the sense that it depends on the definition of $\mathbf{L} = \mathbf{M}_1 - \mathbf{M}_2$ according to which magnetic atoms in the crystal are taken as the sub-lattices 1 and 2. But when the choice has been made in a given crystal the sign of ζ is definite, and determines the direction of **M** with respect to that of **L** and the crystallographic axes.

§ Equation (50.7) has the same form as (19.4) for a ferroelectric in an electric field and the corresponding equation for an ordinary ferromagnet in a magnetic field.

magnetic field creates antiferromagnetic order in the paramagnetic phase, thus removing the difference between the symmetry properties of the two phases (A. S. Borovik-Romanov and V. I. Ozhogin, 1960). When $T > T_c$, L decreases beyond a certain distance from T_c according to

$$L = \zeta H/4aB(T - T_c).$$

§51. Piezomagnetism and the magnetoelectric effect

The phenomena of piezomagnetism and the magnetoelectric effect in antiferromagnets are closely related to the magnetic symmetry.

Piezomagnetism is the occurrence of magnetization when elastic stresses are applied to the crystal; it is analogous to piezoelectricity, and is represented by the presence in the thermodynamic potential of the crystal of a term linear both in the field and in the elastic stress tensor:

$$\tilde{\Phi}_{pm} = -\lambda_{i,\,kl} H_i \sigma_{kl}, \tag{51.1}$$

where $\lambda_{i,\,kl}$ is a tensor symmetrical in the suffixes k and l; cf. (17.7). This term causes the induction $B_i = -4\pi\partial\tilde{\Phi}/\partial H_i$ to contain a further term $4\pi\lambda_{i,\,kl}\sigma_{kl}$. Thus, when $\mathbf{H} = 0$, there is a magnetization linear in the stress:

$$M_i = \lambda_{i,\,kl} \sigma_{kl}. \tag{51.2}$$

Another manifestation of the same property is *linear magnetostriction*, i.e. the occurrence of a strain linear in the field applied to the crystal:

$$u_{kl} = -\partial\tilde{\Phi}_{pm}/\partial\sigma_{kl} = \lambda_{i,\,kl} H_i. \tag{51.3}$$

Time reversal changes the sign of the field $\mathbf{H}$ (and of the magnetization $\mathbf{M}$) while leaving the tensor σ_{kl} unchanged; the thermodynamic potential must also remain unchanged, of course. The piezomagnetic tensor $\lambda_{i,\,kl}$ therefore changes sign under time reversal. In turn, it follows that piezomagnetism is possible only in bodies having a magnetic structure; without this, their properties are invariant under the transformation R, and therefore $\lambda_{i,kl} = -\lambda_{i,kl} = 0$. Piezomagnetism is possible in antiferromagnets belonging to certain magnetic symmetry classes which contain R only in combination with rotations or reflections, or else do not contain R at all (B. A. Tavger and V. M. Zaĭtsev, 1956).

The *magnetoelectric effect* consists in a linear relation between the magnetic and electric fields in matter; it causes, for example, a magnetization proportional to the electric field present (L. D. Landau and E. M. Lifshitz, 1956). This effect is described by a term in the thermodynamic potential that is linear in both the magnetic field and the electric field:

$$\tilde{\Phi}_{m-e} = -\alpha_{ik} E_i H_k, \tag{51.4}$$

where α_{ik} is an unsymmetrical tensor. When $\mathbf{H} = 0$, the electric field generates in the substance a magnetization

$$M_k = \alpha_{ik} E_i, \tag{51.5}$$

and when $\mathbf{E} = 0$ the magnetic field generates an electric polarization

$$P_i = \alpha_{ik} H_k. \tag{51.6}$$

Like piezomagnetism, the magnetoelectric effect is possible only for certain magnetic symmetry classes; the magnetoelectric tensor α_{ik} is odd under time reversal and zero in bodies without a magnetic structure. Since **E** is a polar vector and **H** an axial one, α_{ik} is always zero if the crystal symmetry contains the inversion I; for the magnetoelectric effect to exist, the inversion can occur only in the combination IR.

PROBLEMS†

PROBLEM 1. Find the non-zero components of the piezomagnetic tensor in the antiferromagnet $FeCO_3$ (with the structure shown in Fig. 29).

SOLUTION. As already mentioned in §50, the magnetic class of this structure is the same as the crystal class $\boldsymbol{D}_{3d}$ and does not contain the element R at all. The transformations of this class leave invariant the expression

$$\tilde{\Phi}_{pm} = -\lambda_1 [(\sigma_{xx}-\sigma_{yy}) H_x - 2\sigma_{xy} H_y] - \lambda_2 (\sigma_{xz} H_y - \sigma_{yz} H_x),$$

the z-axis being along the threefold symmetry axis, and the x-axis along one of the horizontal twofold axes. (This expression for the class $\boldsymbol{D}_3$ in the piezoelectric case has been found in §17, Problem 1; it remains valid in the piezomagnetic case for the class $\boldsymbol{D}_{3d} = \boldsymbol{D}_3 \times \boldsymbol{C}_i$ also, since the spatial inversion I leaves unchanged both the tensor σ_{ik} of rank two and the axial vector **H**.) The magnetization is therefore

$$M_x = \lambda_1 (\sigma_{xx}-\sigma_{yy}) - \lambda_2 \sigma_{yz}, \quad M_y = -2\lambda_1 \sigma_{xy} + \lambda_2 \sigma_{xz}.$$

PROBLEM 2. The same as Problem 1, but for the magnetic class $\boldsymbol{D}_{4h}(\boldsymbol{D}_{2h})$.‡

SOLUTION. This class has the elements (besides the unit element) C_2, $2C_4R$, $2U_2$, $2U'_2R$, I, σ_h, $2\sigma_v$, $2\sigma'_vR$; the notation is everywhere as in *QM*, §95. These transformations leave invariant the expression

$$\tilde{\Phi}_{pm} = -\lambda_1 (\sigma_{xz} H_y + \sigma_{yz} H_x) - \lambda_2 \sigma_{xy} H_z,$$

the z-axis being along the fourfold symmetry axis, the x and y axes along two horizontal twofold axes. The magnetization is therefore

$$M_x = \lambda_1 \sigma_{yz}, \quad M_y = \lambda_1 \sigma_{xz}, \quad M_z = \lambda_2 \sigma_{xy}.$$

PROBLEM 3. Find the non-zero components of the magnetoelectric tensor for the antiferromagnet chromium sesquioxide (Cr_2O_3). This crystal belongs to the space group $\boldsymbol{D}_{3d}^6$ (see §50) and contains in each unit cell four chromium atoms at the equivalent points $u, \frac{1}{2}-u, \frac{1}{2}+u, 1-u$ ($u < \frac{1}{4}$) on the trigonal axis; their magnetic moments are along this axis (Fig. 32).

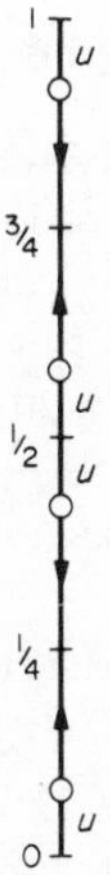

FIG. 32

† The examples given below are due to I. E. Dzyaloshinskiĭ (1957, 1959).

‡ The antiferromagnets manganese difluoride (MnF_2) and cobalt difluoride (CoF_2) are of this type.

SOLUTION. The antiferromagnet in question belongs to the magnetic class $\boldsymbol{D}_{3d}\,(\boldsymbol{D}_3)$, which contains the elements $2C_3$, $3U_2$, IR, $2S_6R$, $3\sigma_dR$. These transformations leave invariant the expression

$$\Phi_{m-e} = -\alpha_{\parallel} E_z H_z - \alpha_{\perp}(E_x H_x + E_y H_y),$$

the z-axis being along the trigonal axis; the non-zero components are therefore $\alpha_{xx} = \alpha_{yy} = \alpha_{\perp}$, $\alpha_{zz} = \alpha_{\parallel}$.

§52. Helicoidal magnetic structures

A special category of magnetic structures consists of those in which the periods of the "magnetic lattice" are incommensurate with the periods of the original crystal lattice. Various mechanisms are possible for the formation of such structures; here we shall consider one mechanism noted by I. E. Dzyaloshinskiĭ (1964), which allows a simple formulation in macroscopic terms. This will be done for a specific example, that of a crystal of the cubic class $\boldsymbol{T}$, in which the exchange interaction alone would give a purely ferromagnetic order of the magnetic moments (V. G. Bar'yakhtar and E. P. Stefanovskiĭ, 1969).†

For the structure in fact to exist, it must be stable under small perturbations which remove the macroscopic spatial uniformity of the crystal. This condition has been tacitly assumed to be satisfied throughout the discussion up to now. The additional non-uniformity energy resulting from the perturbation was given by (43.1); since this is positive definite, the necessary stability does exist.

The expression (43.2) was derived in §43 as the first non-vanishing term in the expansion of the non-uniformity energy in a cubic crystal in powers of the derivatives of the magnetization **M**. There, however, we were concerned only with the energy of exchange origin. It was also noted that the crystal symmetry may allow the existence of non-exchange (relativistic) terms containing products of the derivatives $\partial M_i/\partial x_k$ and the components M_i; for this to occur, the symmetry must not include a centre of inversion. The crystal class $\boldsymbol{T}$ is such a class; it allows the existence of a term of the form $\mathbf{M}\cdot\mathbf{curl}\,\mathbf{M}$ that is invariant under the symmetry operations in it. The non-uniformity energy thus has the form

$$U_{\text{non-u}} = \gamma\,\mathbf{M}\cdot\mathbf{curl}\,\mathbf{M} + \tfrac{1}{2}\alpha\,[(\mathbf{grad}\,M_x)^2 + (\mathbf{grad}\,M_y)^2 + (\mathbf{grad}\,M_z)^2]. \qquad (52.1)$$

The condition $\alpha > 0$ does not here guarantee the stability of the uniform state.

We must take into account, however, the fact that the first-order term in the derivatives in (52.1) contains an extra small factor ($\sim v^2/c^2$) in comparison with the second-order term, because of its relativistic origin‡; this means that $\gamma \ll \alpha/a$, where a is the lattice constant. It is thus possible to find a new stable state while remaining within the range of validity of the expansion (52.1): the simple ferromagnetic structure is distorted in a non-uniform manner, but only over distances much greater than a, so that the derivatives remain small.

The value of M is determined by the basic exchange interactions (which are not related to the non-uniformity). The "large-scale" structure is determined by the minimization of

† Manganese silicide (MnSi) and iron germanide (FeGe) crystals are of this kind, with magnetic ions Mn and Fe which belong to the space group $\boldsymbol{T}^4$, with a simple cubic Bravais lattice.

‡ The first term in (52.1) is not invariant under a simultaneous equal rotation of the vectors **M** at every point in space; it does not satisfy the condition stated in the second footnote to §43.

The relativistic terms may also occur as quadratic ones; for example, cubic symmetry allows a term of the form

$$\tfrac{1}{2}\alpha'\,[(\partial M_x/\partial_x)^2 + (\partial M_y/\partial y)^2 + (\partial M_z/\partial z)^2].$$

the energy (52.1). For a given M, this structure consists in a slow change in the direction of **M**.†

We seek $\mathbf{M}(\mathbf{r})$ as a periodic function

$$\mathbf{M}(\mathbf{r}) = \frac{M}{\sqrt{2}}(\mathbf{m}e^{i\mathbf{k}\cdot\mathbf{r}} + \mathbf{m}^* e^{-i\mathbf{k}\cdot\mathbf{r}}); \tag{52.3}$$

if $\mathbf{M}^2 = M^2$ is constant, the square of the complex unit vector $\mathbf{m}$ ($\mathbf{m}\cdot\mathbf{m}^* = 1$) must be zero: $\mathbf{m}^2 = 0$. Such a vector may be expressed as $\mathbf{m} = (\mathbf{m}_1 + i\mathbf{m}_2)/\sqrt{2}$, where $\mathbf{m}_1$ and $\mathbf{m}_2$ are two mutually perpendicular real unit vectors. Then

$$\mathbf{M}(\mathbf{r}) = \mathbf{M}\,(\mathbf{m}_1 \cos \mathbf{k}\cdot\mathbf{r} - \mathbf{m}_2 \sin \mathbf{k}\cdot\mathbf{r}). \tag{52.3}$$

Substitution of (52.2) in (52.1) gives

$$U_{\text{non-u}} = iM^2 \gamma\, \mathbf{k}\cdot\mathbf{m}\times\mathbf{m}^* + \tfrac{1}{2}\alpha M^2 k^2$$
$$= M^2 \gamma\, \mathbf{k}\cdot\mathbf{m}_1\times\mathbf{m}_2 + \tfrac{1}{2}\alpha M^2 k^2.$$

This expression has a minimum as a function of $\mathbf{k}$ if the vectors $\mathbf{k}$ and $\mathbf{m}_1\times\mathbf{m}_2$ are collinear (parallel if $\gamma < 0$ and antiparallel if $\gamma > 0$) and

$$k = \gamma/\alpha \ll 1/a. \tag{52.4}$$

Thus the presence of a small additional term in $U_{\text{non-u}}$, linear in the derivatives, causes the occurrence of a *helicoidal* magnetic structure superimposed on the basic ferromagnetic structure: the magnetic moments of the atoms lie in planes perpendicular to the direction of **k**, and the directions of the moments rotate slowly in successive layers of atoms. The ends of the atomic moment vectors on a straight line parallel to **k** form a helix. The pitch $2\pi/k$ of the helix is the period of the superlattice; it is large in comparison with the crystallographic periods, and is in general incommensurate with them.‡ Phases having structures of this type are said to be *incommensurate*.

In the approximation considered here, the direction of **k** relative to the crystallographic axes remains indeterminate. It is determined by the minimization of the sum of the anisotropy energy (40.7) and the relativistic part of the non-uniformity energy. We shall not pause to discuss this topic.

† To avoid complications having no fundamental significance, we assume that $U_{\text{non-u}}$ is much less than U_{exch} (so that M is determined only by the latter), but is also much greater than the relativistic anisotropy energy U_{aniso} (so that the latter may be neglected in determining the change in the direction of **M**). These conditions may be satisfied in the neighbourhood of the Curie point, neither too close to it to satisfy the first inequality nor too far from it to satisfy the second. We shall not pause to set out the explicit limits of this range.

‡ The situation discussed here is exactly analogous to that in cholesteric liquid crystals (see *SP* 1, § 140). In these, the presence of a term $\mathbf{n}\cdot\mathbf{curl}\,\mathbf{n}$ in the energy (where $\mathbf{n}$ is the "director" unit vector of the crystal) again leads to the occurrence of a helicoidal structure.

CHAPTER VI

SUPERCONDUCTIVITY

§53. The magnetic properties of superconductors

AT TEMPERATURES close to absolute zero many metals enter a peculiar state whose most striking property, discovered by Kamerlingh Onnes in 1911, is what is called *superconductivity*, i.e. the complete absence of electric resistance to a steady current. Superconductivity first occurs at a definite temperature for each metal, called the *superconductivity transition point*, which is a second-order phase transition point.

From the standpoint of phenomenological theory, however, the change in the magnetic, not the electrical, properties is more fundamental as regards the transition to the superconducting state; as we shall see, the change in its electric properties is an inevitable consequence of its magnetic properties.

The magnetic properties of a superconducting metal can be described as follows. The magnetic field does not penetrate into the superconductor; since the mean magnetic field in the medium is, by definition, the magnetic induction **B**, we can say that throughout a superconductor

$$\mathbf{B} = 0 \tag{53.1}$$

(W. Meissner and R. Ochsenfeld, 1933). This property holds whatever the conditions under which the transition to the superconducting state occurs. For example, if the metal is cooled in a magnetic field, then at the transition point the lines of magnetic force cease to enter the body.

However, it should be mentioned that the equation $\mathbf{B} = 0$ is not valid in a thin surface layer. The magnetic field penetrates into a superconductor to a depth large compared with the interatomic distances and usually of the order of 10^{-5} cm, but depending on the metal concerned and on the temperature. For the same reason, the equation $\mathbf{B} = 0$ does not hold at all in thin films of metal or small particles whose thickness or dimension is of the order of the penetration depth.

In what follows, we shall consider only thick superconductors, and neglect the penetration of the magnetic field into a thin surface layer.†

As we know, the normal component of the induction must be continuous at any boundary between two media; this condition follows from the equation $\operatorname{div} \mathbf{B} = 0$, which is

† We shall not give here the theory of phenomena related to the depth of penetration of the magnetic field into the superconductor (the Londons' theory and the Ginzburg–Landau theory). Although these are macroscopic theories, the quantities which appear in them have a clear significance only in relation to the microscopic theory. These theories are dealt with in *SP* 2.

It must also be emphasized that the present chapter is concerned with "type I" superconductors, which include the metallic elements and stoichiometric compounds. In a "type II" superconductor (including the superconducting alloys) the Meissner effect is fully seen only in fairly weak fields. A sufficiently strong field penetrates into a type II superconductor without completely destroying its superconducting properties (see *SP* 2, Chapter 5).

universally valid. Since $\mathbf{B} = 0$ in a superconductor, the normal component of the external field must be zero on the surface, i.e. the field outside a superconductor must be everywhere tangential to its surface, the lines of magnetic force having the surface as their envelope.

Using this result, we can easily find the forces acting on a superconductor in a magnetic field. As in §5 for an ordinary conductor in an electric field, we calculate the force per unit surface area as $\sigma_{ik}n_k$, where $\sigma_{ik} = (H_i H_k - \frac{1}{2}H^2\delta_{ik})/4\pi$ is the Maxwell stress tensor for a magnetic field in a vacuum. Since in the present case $\mathbf{n}\cdot\mathbf{H}_e = 0$, where $\mathbf{H}_e$ is the field just outside the body, we find

$$\mathbf{F}_s = -H_e^{\,2}\mathbf{n}/8\pi, \tag{53.2}$$

i.e. the surface is subject to a compression, of magnitude equal to the field energy density.

According to equation (29.4)

$$\mathbf{curl}\,\mathbf{B} = 4\pi\overline{\rho\mathbf{v}}/c, \tag{53.3}$$

and from the equation $\mathbf{B} = 0$ it follows that the mean current density is also zero everywhere inside the superconductor. That is, no macroscopic volume currents can flow in a superconductor. It should be emphasized that in a superconductor the conduction current cannot meaningfully be isolated from $\overline{\rho\mathbf{v}}$ as it can in an ordinary conductor. For the same reason the magnetization $\mathbf{M}$, and therefore $\mathbf{H}$, have no physical significance here.

Thus any electric current which flows in a superconductor must be a surface current. The surface current density $\mathbf{g}$ is given, according to (29.16), by the discontinuity in the tangential component of the induction at the boundary of the body. Since $\mathbf{B} = 0$ inside the superconductor, and $\mathbf{B} = \mathbf{H}$ outside it, we have

$$\mathbf{g} = c\mathbf{n}\times\mathbf{H}_e/4\pi. \tag{53.4}$$

The presence of surface currents is not peculiar to superconductors. Similar currents can occur in any magnetized body, and their density is $\mathbf{g} = c\mathbf{n}\times(\mathbf{H}_e - \mathbf{B})/4\pi$. Since the tangential component of $\mathbf{H} = \mathbf{B}/\mu$ is continuous on the surface of a normal (not superconducting) body, we have $\mathbf{n}\times\mathbf{H}_e = \mathbf{n}\times\mathbf{B}/\mu$, and so the expression for $\mathbf{g}$ can be written

$$\mathbf{g} = \mathbf{n}\times\mathbf{B}\frac{c}{4\pi}\frac{1-\mu}{\mu}. \tag{53.5}$$

A fundamental difference between superconductors and other bodies, however, appears when we consider the total current through a cross-section of the body. In a non-superconductor the surface currents always balance, and the total current is zero. This is seen from the condition (53.5) which relates the current density $\mathbf{g}$ to the magnetic induction inside the body, and so to the current $\mathbf{g}$ at every point on the surface. In superconductors, however, the condition (53.5) has no meaning. For the transition from the ordinary state (with magnetic permeability μ) to the superconducting state corresponds formally to the limit $B \to 0$ and $\mu \to 0$ simultaneously. The right-hand side of (53.5) then becomes indeterminate, and there is no condition which restricts the possible values of the current.

Thus we have the result that the currents flowing on the surface of a superconductor may amount to a non-zero total current. Of course, this can occur only in a multiply-connected body (a ring, for example), or in a simply-connected superconductor forming part of a closed circuit which includes also a source of the electromotive force needed to maintain the currents in the parts of the circuit which are not superconducting.

It is very important to note that a steady flow of current on a superconductor is possible even if no electric field is present. This means that no dissipation of energy occurs, whose replacement would involve work being done by an external field. This property of a superconductor may also be described by saying that it has no electric resistance, a result which is thus a necessary consequence of its magnetic properties.

§54. The superconductivity current

Let us consider in more detail some properties of superconductors which depend on their shape.

If a superconductor is a simply-connected body, then no steady distribution of surface currents on it can exist in the absence of an external magnetic field. This can be seen as follows. The surface currents would produce in the surrounding space a static magnetic field vanishing at infinity. Like any static magnetic field in a vacuum, this field would have a potential ϕ, and by the boundary conditions on the superconductor the normal derivative $\partial\phi/\partial n$ would vanish at the surface. We know from potential theory, however, that if $\partial\phi/\partial n = 0$ on the surface of a simply-connected body and at infinity, then ϕ is a constant in all space outside the body. Thus a magnetic field of this kind cannot exist, and therefore neither can the assumed surface currents.

An external magnetic field, on the other hand, causes currents to flow on the surface of a simply-connected superconductor, and these currents can be observed through the appearance of a magnetic moment of the whole body. This "magnetization" is easily calculated for an ellipsoidal superconductor.†

Let $\mathfrak{H}$ be the external field, parallel to one of the principal axes of the ellipsoid. The relation $(1-n)\mathbf{H}+n\mathbf{B} = \mathfrak{H}$ holds for the magnetic field $\mathbf{H}$ inside a non-superconducting ellipsoid, n being the demagnetizing factor for the axis in question (see (29.14)). In a superconductor there is no physical "field" $\mathbf{H}$, as we have already shown, and so the magnetization $\mathbf{M} = (\mathbf{B}-\mathbf{H})/4\pi$ also lacks its usual significance. Nevertheless, it is here convenient to introduce $\mathbf{H}$ and $\mathbf{M}$ as formal auxiliary quantities in the calculation of the total magnetic moment $\mathscr{M} = \mathbf{M}V$ (V being the volume of the ellipsoid), which retains its usual physical meaning. Putting $\mathbf{B} = 0$ in the superconducting ellipsoid, we find

$$\mathbf{H} = \mathfrak{H}/(1-n), \tag{54.1}$$

and

$$\mathscr{M} = -V\mathbf{H}/4\pi = -V\mathfrak{H}/4\pi(1-n). \tag{54.2}$$

In particular, for a long cylinder in a longitudinal field $n = 0$, so that $\mathbf{H} = \mathfrak{H}$ and $\mathscr{M} = -V\mathfrak{H}/4\pi$.‡ These values of $\mathscr{M}$ are the same as would be found if the body had a diamagnetic volume susceptibility of $-1/4\pi$.

The magnetic field $\mathbf{H}_e$ just outside the ellipsoid is everywhere tangential to it, and so its magnitude can be determined at once from the condition that the tangential component of $\mathbf{H}$ be continuous. Within the ellipsoid $\mathbf{H} = \mathfrak{H}/(1-n)$; taking the tangential component, we

† In the present section we always assume that the magnetic field does not exceed the value at which superconductivity ceases (see §55).

‡ These relations for a cylinder follow immediately from the continuity condition on $\mathbf{H}$, and are therefore valid for a cylinder of any cross-section, circular or otherwise.

have

$$(1-n)H_e = \mathfrak{H} \sin\theta, \tag{54.3}$$

where θ is the angle between the direction of the external field $\mathfrak{H}$ and the normal to the surface at the point considered. The greatest value of H_e occurs on the equator of the ellipsoid, and is $\mathfrak{H}/(1-n)$.

It may be pointed out once more that there is no fundamental difference between the currents which cause the "magnetization" of a superconductor and those which produce the total current in it: their physical nature is the same. This important fact makes possible, in particular, an immediate determination of the gyromagnetic coefficients for any superconductor. The momentum density of the electrons which form the "magnetizing" currents differs from the current density only by a factor m/e, e and m being the charge and mass of the electron. From the definition of the gyromagnetic coefficients (see (36.3)) it follows at once that for a superconductor $g_{ik} = \delta_{ik}$.

Let us now consider multiply-connected superconductors. Their properties are very different from those of simply-connected ones, mainly because it is no longer true that a steady distribution of surface currents is impossible in the absence of an external magnetic field. Moreover, the surface currents need not balance out, and may result in a steady total superconductivity current on the body, even if no external e.m.f. is applied.

Let us consider a doubly-connected body (i.e. a ring), with no external magnetic field. We shall show that the state of such a body is entirely determined if the total current J on it is given. The problem of determining the field of the ring can again be solved as a problem of potential theory, but the potential ϕ is now a many-valued function, which changes by $4\pi J/c$ when we go round any closed path interlinked with the ring (cf. §30). In order to state the problem in mathematically precise terms, we must draw some open surface which spans the ring. Then the problem is to solve Laplace's equation with the boundary conditions $\partial\phi/\partial n = 0$ on the surface of the ring, $\phi = 0$ at infinity, and $\phi_2 - \phi_1 = 4\pi J/c$ on the chosen surface, where ϕ_1 and ϕ_2 are the values of the potential on the two sides of that surface. Such a problem is known from potential theory to have a unique solution, which does not depend on the form of the chosen surface. From the field near the surface of the ring, we can in turn uniquely determine the surface current distribution.

The self-inductance of a superconducting ring is entirely determinate together with the current distribution. Here there is a marked difference from ordinary conductors, where the current distribution, and therefore the precise value of the self-inductance, depend on the manner of excitation of the current (§34).†

In §33 we introduced the concept of the magnetic flux Φ through a linear conductor circuit, and showed that $\Phi = LJ/c$, where L is the self-inductance of the conducting circuit. For a superconducting ring, the magnetic flux is meaningful for any thickness, not necessarily small, of the ring. For, since the magnetic field is tangential, the magnetic flux through any part of the surface of the ring is zero; the magnetic flux through every surface

† The self-inductance of a superconducting ring with radius b, made of wire whose cross-section is a circle with small radius a, is the same as the external part of the inductance of a non-superconducting ring, namely $L = 4\pi b\,[\log(8b/a) - 2]$; see §34, Problem 2. The exact solution of the problem of a current in a superconducting circular ring was first given by V. A. Fok, *Physikalische Zeitschrift der Sowjetunion* **1**, 215, 1932.

spanning the ring is therefore the same. Moreover, the formula

$$\Phi = LJ/c \tag{54.4}$$

remains valid, the self-inductance L being again defined in terms of the total energy of the magnetic field of the current. The total energy of the magnetic field of the superconductor is given by the integral $\int (H^2/8\pi)\,\mathrm{d}V$, taken over all space outside the body. Again spanning the ring by a surface C, we can use the field potential and write

$$\begin{aligned}\int H^2\,\mathrm{d}V/8\pi &= -\int \mathbf{H}\cdot\mathbf{grad}\,\phi\,\mathrm{d}V/8\pi \\ &= \int \phi\,\mathrm{div}\,\mathbf{H}\,\mathrm{d}V/8\pi - \oint H_n\phi\,\mathrm{d}f/8\pi.\end{aligned}$$

The first term is zero, because $\mathrm{div}\,\mathbf{H} = 0$. The surface integral is taken over an infinitely remote surface, the surface of the ring, and the two sides of the surface C. The first two of these give zero, so that

$$\begin{aligned}\int H^2\,\mathrm{d}V/8\pi &= \oint_C H_n(\phi_2-\phi_1)\,\mathrm{d}f/8\pi \\ &= (J/2c)\oint_C H_n\,\mathrm{d}f = J\Phi/2c,\end{aligned}$$

where Φ is the magnetic flux through the surface C. Comparing this with the definition of the self-inductance, we have $J\Phi/2c = LJ^2/2c^2$, which gives (54.4).

If the ring is in an external magnetic field, the total magnetic flux Φ is composed of the flux LJ/c and the flux Φ_e of the external field. A very important property of a superconducting ring is that, even if the external field and the current vary, the magnetic flux through the ring remains constant:

$$LJ/c + \Phi_e = \Phi_0, \text{ a constant.} \tag{54.5}$$

This follows immediately from the integral form of Maxwell's equation in the space outside the body:

$$\frac{1}{c}\frac{\partial}{\partial t}\oint_C \mathbf{H}\cdot\mathbf{df} = -\oint \mathbf{E}\cdot\mathbf{dl}.$$

If the integration on the left-hand side is taken over a surface C which spans the ring, the contour of integration on the right-hand side is a line on the surface of the ring. On the surface of a superconductor, the tangential component of $\mathbf{E}$ is zero (since $\mathbf{E} = 0$ inside a superconductor and $\mathbf{E}_t$ is continuous on the surface). Hence the right-hand side is zero, and therefore $\mathrm{d}\Phi/\mathrm{d}t = 0$.

The relation (54.5) gives the variation of the current in the ring when the external field changes. For example, if the ring is made superconducting in an external field with flux Φ_0, which is then removed, a steady current $J = c\Phi_0/L$ flows in the ring.

The constancy of the magnetic flux through a superconducting ring holds not only when the external field changes but also when the shape of the ring or its position in space is altered.† An intuitive statement of this result is that the lines of force can never intersect the surface of the superconductor, and so cannot escape from the aperture of the ring.

The above results can be immediately generalized to the case of multiply-connected superconducting bodies, including sets of rings. The state of an n-ply connected system in the absence of an external field is completely determined by the $n-1$ total currents J_a. The relation (54.5) becomes the system of equations

$$\sum_b L_{ab} J_b + \Phi_a^{(e)} = \Phi_{a0}. \tag{54.6}$$

These equations hold, not only for any external field, but also for any change in shape or relative position.

PROBLEM

Determine the magnetic moment of a superconducting disc in an external magnetic field perpendicular to its plane.‡

SOLUTION. The problem of a superconductor in a static magnetic field is identical with the electrostatic problem of a dielectric with permittivity $\varepsilon = 0$. Regarding the disc as the limit of a spheroid as $c \to 0$ (cf. §4, Problem 4), and using (8.10), we find with appropriate change of notation (the field $\mathfrak{H}$ being along the z-axis) $\mathscr{M} = -2a^3 \mathfrak{H}/3\pi$.

§55. The critical field

A cylindrical superconductor in a longitudinal magnetic field has an additional magnetic energy $-\frac{1}{2}\mathfrak{H}\cdot\mathscr{M} = \mathfrak{H}^2 V/8\pi$. For a non-superconducting cylinder, on the other hand, the total energy would be almost unchanged when the external field was applied (we shall neglect the slight diamagnetism or paramagnetism of a non-superconducting metal, i.e. take $\mu = 1$). Thus it is clear that, in sufficiently strong magnetic fields, the superconducting state must be thermodynamically less favourable than the normal state, and so the superconductivity must be destroyed.

The value of the longitudinal magnetic field at which the superconductivity of a cylindrical body is destroyed depends on the metal concerned and on the temperature (and pressure). This value is called the *critical field* H_{cr}, and is one of the most important characteristics of a superconductor.§

When the critical field is reached, the superconductivity is destroyed throughout the cylinder, because of the uniformity of the field over the surface. In bodies of other shapes, however, the destruction of superconductivity is a more complex process, in which the volume occupied by matter in the normal state gradually extends as $\mathfrak{H}$ increases over some range (§56).

---

† This statement follows at once from the relation between the induced e.m.f. and the change in the magnetic flux through the circuit when it is moved (§63).

‡ We consider this problem principally with a view to using the result elsewhere (see §95, Problem 2). For a superconducting disc the magnetic fields must in reality be very weak, since its superconductivity is very easily destroyed (see §55).

§ There is a sharp transition between the superconducting and normal states only in type I superconductors (see the footnote to §53), which are the only ones considered here. In type II superconductors, the destruction of superconductivity and the penetration of the magnetic field occur gradually over a fairly wide range of fields, so that there is no critical field in the sense here defined.

Thus, at any temperature below the transition point, the metal can exist in either the superconducting or the normal state, denoted by the suffixes s and n respectively. We denote by $\mathscr{F}_{s0}(V,T)$ and $\mathscr{F}_n(V,T)$ the total free energies of the superconducting and normal body in the absence of an external magnetic field; these quantities depend on the substance concerned and on the volume, but not on the shape, of the body. The free energy in the normal state does not change when the external field is applied, and so we omit the suffix 0 in $\mathscr{F}_{n0}$.† In the superconducting state, however, the magnetic field considerably affects the free energy.

For a superconducting cylinder, with given V and T, the free energy in a longitudinal external field $\mathfrak{H}$ is

$$\mathscr{F}_s = \mathscr{F}_{s0}(V,T) + \mathfrak{H}^2 V/8\pi. \tag{55.1}$$

From this we can derive all the other thermodynamic quantities. Differentiating (55.1) with respect to the volume, we find the pressure on the body:

$$P = P_0(V,T) - \mathfrak{H}^2/8\pi, \tag{55.2}$$

where $P_0(V,T)$ is the pressure (for given V and T) in the absence of the field. The equation (55.2) gives the relation between P, V and T, i.e. it is the equation of state for a superconducting cylinder in an external magnetic field. We see that the volume $V(P,T)$ in the presence of the magnetic field is the same as the volume with no magnetic field but a pressure $P + \mathfrak{H}^2/8\pi$. This result accords, of course, with formula (53.2) for the force on the surface of a superconductor in a magnetic field.

The thermodynamic potential‡ of the superconducting cylinder is $\Phi_s = \mathscr{F}_s + PV = \mathscr{F}_{s0}(V,T) + P_0 V$, the volume V being expressed in terms of P and T by (55.2). Hence we can write

$$\Phi_s(P,T) = \Phi_{s0}\left(P + \frac{\mathfrak{H}^2}{8\pi}, T\right), \tag{55.3}$$

where $\Phi_{s0}(P,T)$ is the thermodynamic potential in the absence of the field. Differentiating this equation with respect to T and to P, we obtain analogous relations for the entropy and the volume:

$$\mathscr{S}_s(P,T) = \mathscr{S}_{s0}\left(P + \frac{\mathfrak{H}^2}{8\pi}, T\right), \tag{55.4}$$

$$V_s(P,T) = V_{s0}\left(P + \frac{\mathfrak{H}^2}{8\pi}, T\right). \tag{55.5}$$

We can now write down the condition which determines the critical field. The transition of the cylinder from the superconducting to the normal state occurs when Φ_n becomes less than Φ_s (for given P and T). At the transition point we have $\Phi_s = \Phi_n$, i.e.

$$\Phi_{s0}\left(P + \frac{H_{cr}^2}{8\pi}, T\right) = \Phi_n(P,T). \tag{55.6}$$

This is an exact thermodynamic relation.§ The change in the thermodynamic potential in

---

† By definition, the "total" quantities $\mathscr{F}$ and Φ do not include the energy of the magnetic field that would exist in the absence of the body.

‡ Here defined as in §12.

§ We give here calculations more accurate than is usually necessary, so as to exhibit more clearly the interrelation between the various thermodynamic quantities.

the magnetic field is usually a small correction to $\mathscr{G}_{s0}(P,T)$. We can then expand the left-hand side of equation (55.6) in series, taking the first two terms:

$$\mathscr{G}_{s0}(P,T)+\frac{H_{cr}^2}{8\pi}V_{s0}(P,T)=\mathscr{G}_n(P,T), \tag{55.7}$$

where $V_{s0}(P,T)=\partial\mathscr{G}_{s0}(P,T)/\partial P$ is the volume of the superconducting cylinder in the absence of the field. Thus, in this approximation, we can say that the thermodynamic potential per unit volume is greater by $H_{cr}^2/8\pi$ in the normal state than its value in the superconducting state.

We denote by $T_{cr}=T_{cr}(P)$ the transition temperature in the absence of the magnetic field. The transition concerned is a second-order phase transition. Hence, in particular, $H_{cr}(T)$ must tend continuously to zero at $T=T_{cr}$. We know from the general theory of second-order phase transitions† that the change in the thermodynamic potential near the transition point is proportional to the square of $T-T_{cr}$. We can therefore deduce from (55.7) that the critical field in this temperature range varies as the temperature difference $T-T_{cr}$:

$$H_{cr}=\text{constant}\times(T_{cr}-T). \tag{55.8}$$

Differentiating both sides of equation (55.6) with respect to temperature (for given pressure), remembering that H_{cr} is a function of T, and using (55.4), (55.5), we have

$$\mathscr{S}_n-\mathscr{S}_s=-V_s\frac{\partial}{\partial T}\left(\frac{H_{cr}^2}{8\pi}\right), \tag{55.9}$$

where all the quantities $\mathscr{S}_n$, $\mathscr{S}_s$, V_s are for the point of transition between the two states of the body (i.e. for $H=H_{cr}$). Multiplying by T, we obtain the heat of the transition:

$$Q=T(\mathscr{S}_n-\mathscr{S}_s)=-\frac{V_sH_{cr}T}{4\pi}\left(\frac{\partial H_{cr}}{\partial T}\right)_P \tag{55.10}$$

(W. H. Keesom, 1924). When the transition occurs at $T=T_{cr}$ (i.e. in the absence of the magnetic field), the quantity Q vanishes with H_{cr}, in accordance with the fact that we have a second-order phase transition. A transition at $T<T_{cr}$ (in a magnetic field) involves absorption or evolution of heat, i.e. it is a first-order phase transition. In practice, H_{cr} increases monotonically with decreasing temperature throughout the range from T_{cr} to zero. Hence the derivative $\partial H_{cr}/\partial T$ is always negative, and we see from (55.10) that $Q>0$, i.e. heat is absorbed in the (isothermal) transition from the superconducting to the normal state.

As $T\to 0$, the entropy of the whole body must vanish, by Nernst's theorem. Hence it follows from (55.9) that $\partial H_{cr}/\partial T=0$ for $T=0$, i.e. the curve of $H_{cr}(T)$ intersects the H_{cr}-axis at right angles.

We may differentiate the difference $\mathscr{S}_n-\mathscr{S}_s$ (55.9) again with respect to temperature, and again use equations (55.4), (55.5). Since also $(\partial S/\partial P)_T=-(\partial V/\partial T)_P$, the result is

$$\frac{\partial\mathscr{S}_n}{\partial T}-\frac{\partial\mathscr{S}_s}{\partial T}=-V_s\frac{\partial^2}{\partial T^2}\left(\frac{H_{cr}^2}{8\pi}\right)-2\frac{\partial V_s}{\partial T}\frac{\partial}{\partial T}\left(\frac{H_{cr}^2}{8\pi}\right)-\frac{\partial V_s}{\partial P}\left[\frac{\partial}{\partial T}\left(\frac{H_{cr}^2}{8\pi}\right)\right]^2. \tag{55.11}$$

† For the superconductivity transition, the Landau theory may be regarded as valid with no practical restrictions, up to the transition point itself; see *SP* 2, §45.

Multiplying both sides of this equation by T, we obtain the difference of the specific heats (at constant pressure) of the two phases. The terms involving the thermal-expansion coefficient and the compressibility are usually very small in comparison with the remaining terms; neglecting them, we have

$$\mathscr{C}_s - \mathscr{C}_n = \frac{V_s T}{4\pi} H_{cr} \frac{\partial^2 H_{cr}}{\partial T^2} + \frac{V_s T}{4\pi}\left(\frac{\partial H_{cr}}{\partial T}\right)^2. \tag{55.12}$$

This formula could also be obtained by direct differentiation of the approximate relation (55.7). In this approximation the difference between V_s and V_{s0}, and between $\mathscr{C}_s$ and $\mathscr{C}_{s0}$, may be neglected.

For $T = T_{cr}$, the first term in (55.12) is zero, and we obtain the following formula, which relates the change in specific heat in the second-order phase transition (in the absence of an external magnetic field) to the temperature dependence of H_{cr}:

$$\mathscr{C}_s - \mathscr{C}_n = \frac{V_s T_{cr}}{4\pi}\left(\frac{\partial H_{cr}}{\partial T}\right)^2 \tag{55.13}$$

(A. J. Rutgers, 1933). Hence we see that in this case $\mathscr{C}_s > \mathscr{C}_n$. As the temperature falls, i.e. when the superconductivity is destroyed by the magnetic field, the difference $\mathscr{C}_s - \mathscr{C}_n$ changes sign, because the difference $\mathscr{S}_n - \mathscr{S}_s$ is zero for $T = 0$ and for $T = T_{cr}$, and must have a maximum in between.

We can similarly discuss effects related to the change in volume in the transition. To do so, we differentiate equation (55.6) with respect to pressure (for given temperature), H_{cr} being a function of P. This gives

$$V_n = V_s \frac{\partial}{\partial P}\left(P + \frac{H_{cr}^2}{8\pi}\right)$$

or

$$V_n - V_s = \frac{V_s H_{cr}}{4\pi}\frac{\partial H_{cr}}{\partial P}. \tag{55.14}$$

which determines the change in volume at the transition point.† For $T = T_{cr}$ this difference is zero, like the entropy difference. The transition at temperatures $T < T_{cr}$, however, is accompanied by a change in volume, which may be of either sign, depending on the sign of the derivative $(\partial H_{cr}/\partial P)_T$. For $T = T_{cr}$ there is no change in volume, but the compressibility is discontinuous; the discontinuity is easily found by differentiating equation (55.14). It may be noted that, if we substitute in (55.14)

$$\left(\frac{\partial H_{cr}}{\partial P}\right)_T = -\left(\frac{\partial H_{cr}}{\partial T}\right)_P \left(\frac{\partial T}{\partial P}\right)_{H_{cr}}$$

(obtained by differentiating the equation $H_{cr}(P, T) = \text{constant}$), we obtain the Clapeyron–Clausius equation

$$\left(\frac{\partial P}{\partial T}\right)_{H_{cr}} = \frac{Q}{T(V_n - V_s)}, \tag{55.15}$$

† This difference must of course be distinguished from the change in volume (magnetostriction) of the superconductor when the field changes from zero to H_{cr}. This can be found from (55.5): $V_s(P,T) - V_{s0}(P,T) \cong (H_{cr}^2/8\pi)(\partial V_s/\partial P)_T$.

where the derivative $(\partial P/\partial T)_{H_{cr}}$ defines the change in pressure needed to keep the applied external field critical when the temperature changes.

The physical significance of the critical field H_{cr} is much wider than would appear from its definition in terms of the behaviour of a superconducting cylinder. The equation $H = H_{cr}$ is a condition of equilibrium which must be fulfilled at every point of a surface separating normal and superconducting phases in the same body. This is evident from the following simple arguments. If a cylinder is in a longitudinal magnetic field H_{cr}, then both the boundary conditions on the magnetic field and the conditions of thermodynamic stability are satisfied for all states in which an interior cylindrical part is in the superconducting state and the rest of the body is in the normal state, and the field at the boundary between these parts is H_{cr}. Thus the surface of separation, on which $H = H_{cr}$, is in neutral equilibrium with respect to its location. This is a characteristic property of phase equilibrium.

In a variable magnetic field, the boundary between the superconducting and normal phases changes its position. The kinetics of this process is very complex, and its discussion requires a simultaneous solution of the equations of electrodynamics and of thermal conduction, taking into account the heat evolved in the phase transition. We shall not pause to carry out this investigation,† but merely give the boundary condition which must be satisfied at the moving boundary between the normal and superconducting phases.

To derive this condition we take a coordinate system K' moving with the velocity $\mathbf{v}$ of the boundary between the phases. By the formulae for transformation of fields, the electric field $\mathbf{E}'$ in the system K' is related to the fields $\mathbf{E}$ and $\mathbf{B}$ in a fixed system K by $\mathbf{E}' = \mathbf{E} + \mathbf{v} \times \mathbf{B}/c$; see (63.1). Since the boundary is at rest in the system K', the usual condition of continuity of the tangential component of $\mathbf{E}'$ holds, i.e. $\mathbf{n} \times \mathbf{E}' = \mathbf{n} \times \mathbf{E} - v\mathbf{B}/c$ must be continuous, where $\mathbf{n}$ is a unit vector normal to the surface, in the direction of the velocity $\mathbf{v}$. In the superconducting phase $\mathbf{E} = \mathbf{B} = 0$, and in the normal phase $B = H_{cr}$ at the boundary. We therefore find that a tangential electric field appears on the moving boundary, its direction being perpendicular to that of the magnetic field and its magnitude being

$$E = vH_{cr}/c. \tag{55.16}$$

§56. The intermediate state

If a superconducting body of any shape is in an external magnetic field $\mathfrak{H}$ which is gradually increased, a stage is finally reached where the field at some point on the surface of the body becomes equal to the critical field H_{cr}, but $\mathfrak{H}$ itself is still less than H_{cr}. For example, on the surface of an ellipsoid placed in a field $\mathfrak{H}$ parallel to one of its axes, the greatest value of the field occurs on the equator (see (54.3)), and is equal to H_{cr} when $\mathfrak{H} = H_{cr}(1-n)$.

When $\mathfrak{H}$ increases further, the body cannot remain entirely in the superconducting state. Nor can it pass entirely into the normal state, because then the field would become $\mathfrak{H}$ everywhere. Hence the superconductivity must be lost only in part.

At first sight one might imagine that this process occurs as follows. As $\mathfrak{H}$ increases, the

† See I. M. Lifshitz, *Zhurnal éksperimental'noĭ i teoreticheskoĭ fiziki* **20**, 834, 1950; *Doklady Akademii Nauk SSSR* **90**, 363, 1953.

superconductivity is lost in a gradually increasing part of the body, while a gradually decreasing part remains superconducting, and the whole body becomes normal when $\mathfrak{H} = H_{cr}$. It is a easy to see, however, that such states of the body are thermodynamically unstable. On the surface separating the superconducting and normal phases the magnetic field is, as we know, tangential to the surface, and its magnitude is H_{cr}. That is, the lines of force are on the surface. If the boundary is convex to the normal phase, the equipotential surfaces of the field, being at right angles to the lines of force, will diverge into the normal region, as shown by dashed lines in Fig. 33a. The field decreases, however, in the direction in which the equipotential surfaces diverge, so that we should have $H < H_{cr}$ in the shaded region, contrary to the supposition that this region is in the normal state. If, on the other hand, the boundary of the superconducting phase is concave, then the lines of force on that boundary must have a bend on the free surface of the superconducting region, to which the field is tangential (at the point O, in Fig. 33b). At a bend in a line of force, however, the field becomes infinite, which again contradicts the boundary conditions at the surface of the superconductor.

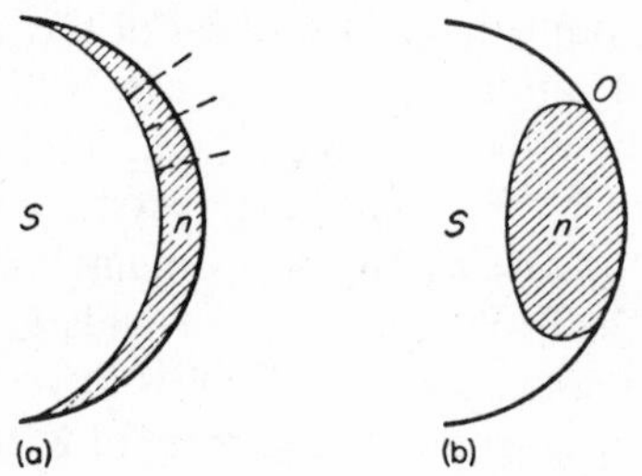

FIG. 33

The above arguments represent essentially another form of the situation which leads to the domain structure of ferroelectrics and ferromagnets. Here also the conditions of thermodynamic stability have the result that, if the magnetic field reaches the value H_{cr} at even one point on the surface, the body is divided into numerous parallel alternating thin layers of normal and superconducting matter (L. D. Landau, 1937). This state of the superconductor is called the *intermediate* state. As $\mathfrak{H}$ increases, the total volume of the normal layers increases, and when $\mathfrak{H} = H_{cr}$ the body becomes entirely normal.

In general, a body of arbitrary shape need not be entirely in the intermediate state. There may also be regions in the purely superconducting and purely normal states; these must be separated by the region which is in the intermediate state. A simpler case is the ellipsoid already considered. In a field parallel to the axis the intermediate state occurs in the range

$$H_{cr}(1-n) < \mathfrak{H} < H_{cr}, \tag{56.1}$$

and the whole volume of the ellipsoid is in this state. For a sphere (for example), $n = \frac{1}{3}$, and the intermediate state exists in the range $\frac{2}{3}H_{cr} < \mathfrak{H} < H_{cr}$. For a cylinder in a transverse field, $n = \frac{1}{2}$, and the corresponding range is $\frac{1}{2}H_{cr} < \mathfrak{H} < H_{cr}$. For a cylinder in a longitudinal field, $n = 0$; there is no intermediate state, and the superconductivity is totally destroyed at $\mathfrak{H} = H_{cr}$. Finally, for a flat plate in a transverse field $n = 1$, and it is in the intermediate state for any field $\mathfrak{H} < H_{cr}$.

The intermediate state can also be described in an averaged manner if the thickness of the regions under consideration is large compared with the layer thickness (R. E. Peierls, and F. London, 1936). In this description it is assumed that there is inside the body a magnetic induction $\overline{\mathbf{B}}$ which varies from zero in the purely superconducting state to H_{cr} in the purely normal state. If we ascribe a non-zero induction to the matter in the intermediate state, we must also ascribe to it a definite magnetic "field" $\overline{\mathbf{H}}$. To determine the relation between these quantities, we must consider the true structure of the intermediate state.

The magnetic field in a normal layer at its boundary with a superconducting layer is H_{cr}, and by virtue of the assumed smallness of the layer thickness we can suppose that the field has this value everywhere in the normal layers. In the superconducting layers $\mathbf{B} = 0$. Hence, averaging the magnetic field over a volume large compared with the layer thickness, we find that the mean induction $\overline{B} = x_n H_{cr}$, where x_n is the fraction of the volume that is in the normal state. Next, we determine the thermodynamic potential per unit volume of the body, taking as zero the value for the purely superconducting state. In the absence of a magnetic field, unit volume of the normal phase has an excess thermodynamic potential $H_{cr}{}^2/8\pi$.† When a magnetic field is present, a further $H_{cr}{}^2/8\pi$ is added as magnetic energy, giving altogether $H_{cr}{}^2/4\pi$. The mean thermodynamic potential per unit volume in the intermediate state is therefore

$$\Phi = x_n H_{cr}{}^2/4\pi = H_{cr}\overline{B}/4\pi. \tag{56.2}$$

The relation between $\overline{\mathbf{B}}$ and $\overline{\mathbf{H}}$ is obtained from the general thermodynamic relation $\mathbf{H} = 4\pi\partial\Phi/\partial\mathbf{B}$. In the present case we find that $\overline{\mathbf{H}}$ is parallel to $\overline{\mathbf{B}}$ and its magnitude is

$$\overline{H} = H_{cr}, \tag{56.3}$$

i.e. it is independent of the induction.

If the relation between $\overline{B}$ and $\overline{H}$ is shown graphically (Fig. 34), then the segment OA of the axis of abscissae corresponds to the superconducting state, and the line BC ($\overline{B} = \overline{H}$) to the normal state. The vertical line AB ($\overline{H} = H_{cr}$) corresponds to the intermediate state.

Let $\mathbf{n}$ be a unit vector in the direction of the lines of force of the averaged magnetic field. Putting $\overline{\mathbf{H}} = H_{cr}\mathbf{n}$ and substituting in the equation $\mathbf{curl}\,\overline{\mathbf{H}} = 0$ (which holds in the absence of a volume current), we find that $\mathbf{curl}\,\mathbf{n} = 0$. Since $\mathbf{n}^2 = 1$, we have

$$\mathbf{grad}\,\mathbf{n}^2 = 2(\mathbf{n}\cdot\mathbf{grad})\mathbf{n} + 2\mathbf{n}\times\mathbf{curl}\,\mathbf{n} = 0,$$

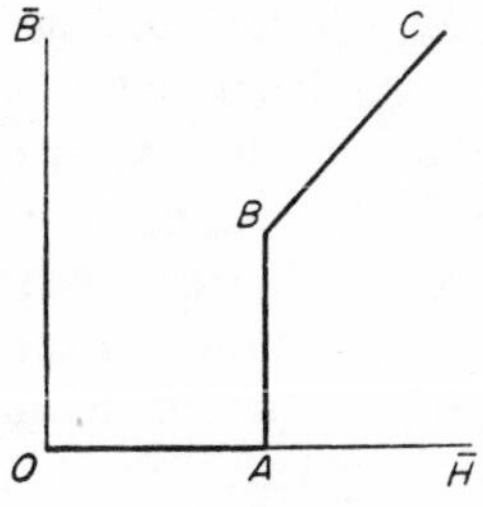

FIG. 34

† Here we neglect all magnetostriction effects. Instead of the change in the thermodynamic potential we could therefore speak of the (equal) change in the free energy.

and therefore $(\mathbf{n}\cdot\mathbf{grad})\mathbf{n} = 0$. This means that the direction of the vector $\mathbf{n}$ is constant. Thus the lines of force of the mean field are straight lines.

Let us apply these results to an ellipsoid in the intermediate state. For a uniform field inside the ellipsoid, the relation $(1-n)\bar{H}+n\bar{B} = \mathfrak{H}$ holds, whatever the relation between $\mathbf{B}$ and $\mathbf{H}$. Putting $\bar{H} = H_{cr}$, we have

$$\bar{B} = \frac{\mathfrak{H}}{n} - \frac{1-n}{n} H_{cr}. \tag{56.4}$$

Thus the mean induction in the ellipsoid varies linearly with the external field, from zero when $\mathfrak{H} = (1-n)H_{cr}$ to H_{cr} when $\mathfrak{H} = H_{cr}$.

We may also write down an expression for the total thermodynamic potential $\tilde{\mathscr{F}}$ of an ellipsoid in the intermediate state. To do so, we start from the general formula

$$\tilde{\mathscr{F}} = \int \left[\Phi - \frac{\mathbf{H}\cdot\mathbf{B}}{8\pi} - \frac{(\mathbf{B}-\mathbf{H})\cdot\mathfrak{H}}{8\pi} \right] dV$$

(cf. (32.7)), which is also valid whatever the relation between $\mathbf{B}$ and $\mathbf{H}$. Substituting Φ, H and B from (56.2)–(56.4), we obtain

$$\tilde{\mathscr{F}}_t = \frac{V}{8\pi}\left[H_{cr}^{\,2} - \frac{1}{n}(H_{cr} - \mathfrak{H})^2 \right], \tag{56.5}$$

V being the volume of the ellipsoid; $\tilde{\mathscr{F}}_t$ is taken to be zero in the purely superconducting state of the ellipsoid, in the absence of a magnetic field. For a superconducting ellipsoid in an external field $\mathfrak{H}$ we have

$$\tilde{\mathscr{F}}_s = -\tfrac{1}{2}\mathscr{M}\cdot\mathfrak{H} = V\,\mathfrak{H}^2/8\pi(1-n), \tag{56.6}$$

in accordance with (32.6) and (54.2). For $\mathfrak{H} = H_{cr}(1-n)$, the thermodynamic potential and its first derivative with respect to the temperature are continuous; in this sense, the transition from the superconducting to the intermediate state is analogous to a second-order phase transition.†

It should be emphasized that the averaged description of the intermediate state given here is in reality not very accurate, because of the comparatively large thickness of the layers. For the same reason, this description fails to reproduce certain phenomena related to the properties of the layer structure. These include the fact that the transition from the superconducting to the intermediate state actually occurs only when $\mathfrak{H}$ slightly exceeds $(1-n)H_{cr}$. The reason for this "delay" is as follows. The passage into the intermediate state occurs when that state becomes thermodynamically stable, i.e. when $\tilde{\mathscr{F}}_t = \tilde{\mathscr{F}}_s$. The layer structure, however, has not only the "volume" energy (56.5) allotted to it in the averaged description but also additional energy resulting from the existence of the boundaries between the layers and their change in shape near the surface of the body. This results in some displacement of the transition point towards stronger fields.

As mentioned in the footnote to §53, this chapter deals with type I superconductors. Nevertheless, a minor comment may be made here concerning the thermodynamics of the

† It is therefore not surprising that $\tilde{\mathscr{F}}_t(\mathfrak{H}) < \tilde{\mathscr{F}}_s(\mathfrak{H})$ on both sides of the point $\mathfrak{H} = H_{cr}(1-n)$. In a second-order phase transition, neither phase exists beyond the transition point, and hence it is meaningless to compare the thermodynamic potentials of the two phases.

"magnetization curve" for a cylindrical type II superconductor. Such bodies are characterized by a gradual penetration of the magnetic field into them. For example, with a long cylindrical superconductor in a longitudinal magnetic field $\mathfrak{H}$, the penetration begins when the field reaches some value $H_{cr,1}(T)$, and the superconductor changes continuously into the normal state only at a field $H_{cr,2} > H_{cr,1}$.†

We start from the relation $\partial\tilde{\mathscr{G}}/\partial H = -\mathscr{M}$; cf. (32.4). Integrating both sides over H from 0 to $H_{cr,2}$, we find

$$-\int_0^{H_{cr,2}} \mathscr{M}\,d\mathfrak{H} = \mathscr{G}_n - \mathscr{G}_{s0},$$

where $\mathscr{G}_{s0}$ relates to the superconductor in the absence of the field, and $\mathscr{G}_n$ is independent of the external field (so that the tilde can be omitted from both). In the field range $0 \leqslant H \leqslant H_{cr,1}$, the field does not penetrate into the cylinder, whose magnetic moment is therefore $\mathscr{M} = -V\mathfrak{H}/4\pi$. Separating this part of the integral gives

$$\int_{H_{cr,1}}^{H_{cr,2}} \mathscr{M}\,d\mathfrak{H} = -(\mathscr{G}_n - \mathscr{G}_{s0}) + \frac{VH_{cr,1}^2}{8\pi}.$$

If H_{cr} is defined as before, but purely formally, for a type II superconductor:‡

$$\mathscr{G}_n - \mathscr{G}_{s0} = VH_{cr}^2/8\pi,$$

the relation obtained may be written in the final form

$$\int_{H_{cr,1}}^{H_{cr,2}} \mathscr{M}\,d\mathfrak{H} = -\frac{V}{8\pi}(H_{cr}^2 - H_{cr,1}^2). \tag{56.7}$$

PROBLEM

Determine the heat capacity of an ellipsoid in the intermediate state.

SOLUTION. The entropy, and thence the heat capacity, are found by differentiating the thermodynamic potential (56.5) with respect to temperature. Neglecting the terms containing the thermal-expansion coefficient, we obtain

$$\mathscr{C}_t - \mathscr{C}_s = \frac{VT}{4\pi n}[(1-n)(H_{cr}'^2 + H_{cr}H_{cr}'') - \mathfrak{H}H_{cr}''],$$

the prime denoting differentiation with respect to T; $\mathscr{C}_s$ is the heat capacity of the body in the superconducting state, whose slight dependence on $\mathfrak{H}$ we here neglect. Hence it follows that, as $\mathfrak{H}$ varies (at constant temperature), the heat capacity changes discontinuously at the point $\mathfrak{H} = (1-n)H_{cr}$ from $\mathscr{C}_s$ to

$$\mathscr{C}_s + \frac{VT(1-n)}{4\pi n}H_{cr}'^2,$$

and thereafter varies linearly with $\mathfrak{H}$, reaching the value

$$\mathscr{C}_s - \frac{VT}{4\pi}(H_{cr}'^2 + H_{cr}H_{cr}'') + \frac{VT}{4\pi n}H_{cr}'^2 = \mathscr{C}_n + \frac{VT}{4\pi n}H_{cr}'^2$$

for $\mathfrak{H} = H_{cr}$, whence it falls discontinuously to $\mathscr{C}_n$.

† See *SP* 2, §§47, 48. The state of the superconductor in the field range between $H_{cr,1}$ and $H_{cr,2}$ is called the *mixed state*. It must be emphasized that this is not the same as the intermediate state of a type I superconductor. In the mixed state, the magnetic field penetrates into the sample in the form of what are called vortex filaments.

‡ The value of H_{cr} lies between those of $H_{cr,1}$ and $H_{cr,2}$. It has, however, no distinctive property in a type II superconductor.

§57. Structure of the intermediate state

The shape and size of the normal and superconducting layers in the intermediate state are governed by the conditions of thermodynamic equilibrium for the body as a whole, in the same manner as was the shape of the domains in a ferromagnet (§44). As there, the resulting thickness of the layers is determined by two oppositely acting factors. The surface tension at the normal and superconducting phase boundaries tends to reduce the number of layers, i.e. to increase their thickness. The energy of emergence of the layers at the free surface of the body has the opposite tendency. The layer thickness increases with the size of the body, and consequently (for the same reasons as in the case of ferromagnetic domains) the layers must ultimately undergo branching near the surface.†

The problem of determining the shape and size of unbranched layers in the intermediate state in a flat plate can be solved exactly. We shall do this on the assumption that the external field $\mathfrak{H}$ is perpendicular to the plate (L. D. Landau, 1937).

The layers are parallel to the field, except near the surface of the plate. The lines of magnetic force (shown dashed in Fig. 35, p. 196) pass only through the normal layers, and the boundaries of the superconducting layers are also lines of force, since $B_n = 0$ there. Since also $H = H_{\rm cr}$ on the boundary between the normal and superconducting phases, the conditions at the boundaries of a superconducting layer are

$$\left.\begin{array}{ll}\text{on } BC & H_x = 0,\\ \text{on } BA \text{ and } CD & H_x^2 + H_y^2 = H_{\rm cr}^2,\end{array}\right\} \tag{57.1}$$

the coordinate axes being taken as shown in Fig. 35. Far from the plate, the field **H** must be the same as the external field $\mathfrak{H}$, i.e.

$$\text{for } x \to -\infty \qquad H_x = \mathfrak{H}, \qquad H_y = 0. \tag{57.2}$$

We use the scalar and vector potentials: $H_x = -\partial\phi/\partial x = \partial A/\partial y$, $H_y = -\partial\phi/\partial y = -\partial A/\partial x$, and the complex potential $w = \phi - iA$ (cf. §3).

On a line of force A = constant. We put $A = 0$ on the line of force which reaches O and then branches into OCD and OBA, forming the boundary of one superconducting layer. The difference between the values of A at the boundaries of two successive superconducting layers is equal to the magnetic flux across the segment $a = a_s + a_n$, namely $\mathfrak{H}a$. Hence the value of A at the boundary of any superconducting layer is an integral multiple of $\mathfrak{H}a$. Using also the "complex field" $\eta = H_x - iH_y = -dw/dz$, $z = x + iy$, we can write the conditions (57.1) as

$$\left.\begin{array}{ll}\text{on } BC & \mathrm{re}\,\eta = 0,\\ \text{on } BA \text{ and } CD & |\eta| = H_{\rm cr}.\end{array}\right\} \tag{57.3}$$

We introduce a new variable

$$\zeta = \exp(-2\pi w/\mathfrak{H}a) - 1 \tag{57.4}$$

and regard η as a function of ζ. ζ is real on all boundary lines of force and on their continuations beyond the plate: $\zeta = \exp(-2\pi\phi/\mathfrak{H}a) - 1$.

Since ϕ is determined apart from a constant, its value at any one point can be chosen

† In certain conditions, when the external field is near zero or $H_{\rm cr}$, a filamentary structure may be thermodynamically more favourable than the layer structure; see E. R. Andrew, *Proceedings of the Royal Society* A **194**, 98, 1948.

arbitrarily. Let $\phi = 0$ at O. Then $\zeta = 0$ there also. On the limiting line of force considered, far from the plate, $\zeta = -1$ (since for $x \to -\infty$ we have $\phi \to -\mathfrak{H}x \to +\infty$). The value of ζ at B or C, where the line of force enters the plate, is ζ_0, say. On CD and BA, ζ varies from ζ_0 to ∞. Then the conditions (57.1) and (57.3) can be written

$$\text{for } \zeta = -1 \qquad \eta = \mathfrak{H}, \tag{57.5}$$

$$\left.\begin{aligned} &\text{for } 0 < \zeta < \zeta_0 \qquad \operatorname{re}\eta = 0,\\ &\text{for } \zeta_0 < \zeta \qquad |\eta| = H_{\text{cr}}. \end{aligned}\right\} \tag{57.6}$$

The function $\eta(\zeta)$ must, furthermore, be everywhere finite.

The conditions (57.6) are satisfied by the function

$$\eta = H_{\text{cr}}\left[\sqrt{\left(1-\frac{\zeta_0}{\zeta}\right)} - \sqrt{\left(-\frac{\zeta_0}{\zeta}\right)}\right]. \tag{57.7}$$

For real negative ζ the two roots are real, and are taken with the signs shown. For $0 < \zeta < \zeta_0$ both are imaginary, and we take

$$\eta = \mp iH_{\text{cr}}\left[\sqrt{\frac{\zeta_0}{\zeta}} - \sqrt{\left(\frac{\zeta_0}{\zeta}-1\right)}\right],$$

with the minus and plus signs on OC and OB respectively. For $\zeta > \zeta_0$

$$\eta = H_{\text{cr}}\left[\sqrt{\left(1-\frac{\zeta_0}{\zeta}\right)} \mp i\sqrt{\frac{\zeta_0}{\zeta}}\right],$$

with the minus and plus signs on CD and BA respectively. The value of ζ_0 is found from the condition (57.5), and is

$$\zeta_0 = \frac{1}{4}\left(\frac{1}{h} - h\right)^2, \tag{57.8}$$

where $h = \mathfrak{H}/H_{\text{cr}}$.

The shape of the layer, i.e. the equation of the limiting line of force, is obtained by integrating the relation $dz = -dw/\eta$ over real ζ:

$$z = -\int \frac{dw}{\eta} = \frac{ah}{2\pi}\int \frac{d\zeta}{\eta(\zeta+1)}.$$

Substituting $\eta(\zeta)$, taking real and imaginary parts, and choosing appropriately the constants of integration, we obtain the following parametric equations of the line CD:

$$\begin{aligned} x &= \frac{ah}{2\pi}\int_{\zeta_0}^{\zeta} \sqrt{\left(1-\frac{\zeta_0}{\zeta}\right)}\frac{d\zeta}{\zeta+1}\\ &= \frac{ah}{\pi}\left[\cosh^{-1}\sqrt{\frac{\zeta}{\zeta_0}} - \sqrt{(\zeta_0+1)}\cosh^{-1}\sqrt{\frac{\zeta(\zeta_0+1)}{\zeta_0(\zeta+1)}}\right], \end{aligned} \tag{57.9}$$

$$y = Y - \frac{ah}{2\pi}\int_{\zeta}^{\infty}\sqrt{\frac{\zeta_0}{\zeta}}\frac{d\zeta}{\zeta+1} = Y - \frac{ah}{\pi}\sqrt{\zeta_0}\left(\tfrac{1}{2}\pi - \tan^{-1}\sqrt{\zeta}\right),$$

where $Y = \frac{1}{2}a_s$ is the value of y for $x \to \infty$; see Fig. 35.

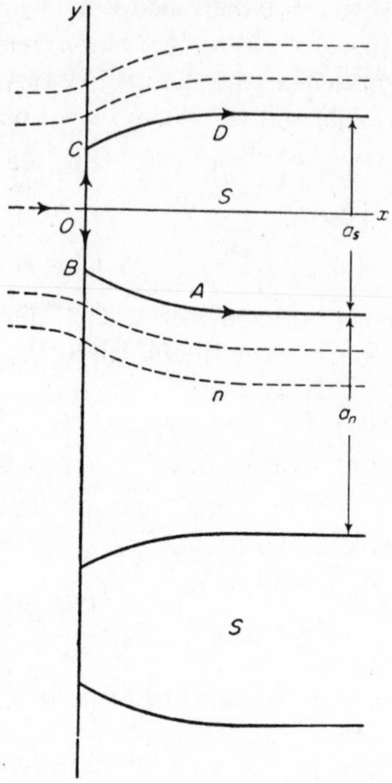

FIG. 35

The period a of the layer structure is related to the thicknesses a_s and a_n of the superconducting and normal layers by $a = a_s + a_n$, $a\mathfrak{H} = a_n H_{cr}$. The latter equation follows from the continuity of the magnetic flux, which passes entirely through normal layers. Hence $a_s = a(1-h)$, $a_n = ah$.

The period a is determined by the condition that the total thermodynamic potential of the plate be a minimum. The existence of surface tension at the boundaries between the normal and superconducting phases gives a term

$$\wp_1 = 2lH_{cr}^2\Delta/8\pi a \tag{57.10}$$

in the thermodynamic potential per unit area of the surface of the plate. Here l is the thickness of the plate, and the surface-tension coefficient is written as $H_{cr}^2\Delta/8\pi$, where Δ has the dimensions of length. In calculating this part of the energy we can, of course, neglect the curvature of the layers near the surface of the plate.

The energy of emergence of the layers at the surface of the plate can be written as the sum of two parts. First, the increase in the volume of the normal layers as compared with the volume they would occupy if they were everywhere plane-parallel gives an additional energy

$$\wp_2 = \frac{4}{a}\int_0^\infty (Y-y)\,dx\frac{H_{cr}^2}{8\pi}, \tag{57.11}$$

where the factor 4 takes into account the presence of four angles (such as B and C in Fig. 35) on the two sides of each of the $1/a$ superconducting layers.

Second, the emergence of the layers at the surface of the plate changes the energy of the system in the external field, i.e. the energy $-\frac{1}{2}\mathscr{M}\cdot\mathfrak{H}$. The magnetic moment of the plate is due to currents on the surfaces of the superconducting layers. When the tangential component of the induction changes discontinuously from H to zero, the surface current density is $g = \pm cH/4\pi$. Hence the magnetic moment per unit length in the z-direction and per boundary surface of the superconducting layer is

$$-\int_{OCD} \frac{H}{4\pi} y \, ds, \qquad ds = \sqrt{(dx^2 + dy^2)}.$$

If the layer did not emerge at the surface, there would be no segment OC, and on CD we should have $y = Y$. Hence the excess magnetic moment for each of the four angles is

$$-\int_{OCD} \frac{H}{4\pi} y \, ds + \int_0^\infty \frac{H_{cr}}{4\pi} Y dx.$$

Accordingly, the excess energy is

$$\begin{aligned}\mathscr{P}_3 &= -\tfrac{1}{2}\mathfrak{H}\cdot\frac{4}{a}\left[\int_0^\infty \frac{H_{cr}}{4\pi} Y dx - \int_{OCD} \frac{H}{4\pi} y \, ds\right] \\ &= -\frac{\mathfrak{H}}{2\pi a}\left[H_{cr}\int_{CD}(-Y dx + y\, ds) + \int_{OC} H_y dy\right].\end{aligned} \tag{57.12}$$

The coordinates x and y, expressed in terms of ζ, are proportional to a. Hence all the integrals in $\mathscr{P}_1 + \mathscr{P}_2$ are proportional to a^2, and this part of the thermodynamic potential is therefore proportional to a. The sum $\mathscr{P}_1 + \mathscr{P}_2 + \mathscr{P}_3$ is therefore of the form

$$\mathscr{P} = \frac{{H_{cr}}^2}{4\pi}\left[\frac{l\Delta}{a} + af(h)\right]. \tag{57.13}$$

The condition that this be a minimum gives

$$a = \sqrt{[l\Delta/f(h)]}. \tag{57.14}$$

The integrals in (57.11) and (57.12) can be calculated†, and the result obtained for $f(h)$ is

$$\begin{aligned}f(h) = \frac{1}{4\pi}\{&(1+h)^4 \log(1+h) + (1-h)^4 \log(1-h) \\ &-(1+h^2)^2 \log(1+h^2) - 4h^2 \log 8h\}.\end{aligned} \tag{57.15}$$

† See A. Fortini and E. Paumier, *Physical Review* B **5**, 1850, 1972.

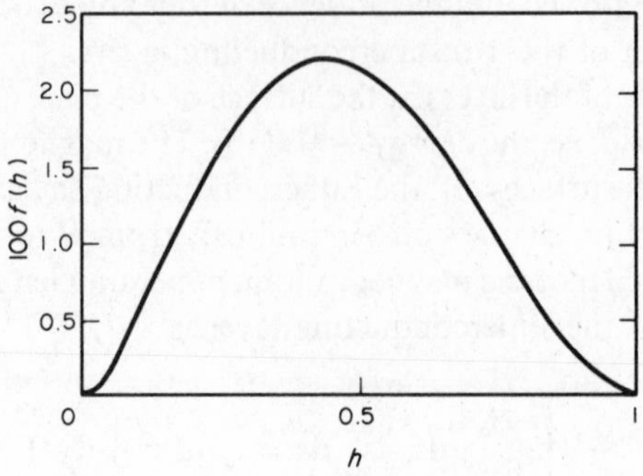

FIG. 36

The limiting forms of this are

$$\left.\begin{aligned} f(h) &= \frac{h^2}{\pi}\log\frac{0.56}{h^2} \quad \text{for } h \ll 1; \\ f(h) &= \frac{\log 2}{\pi}(1-h)^2 \quad \text{for} \quad 1-h \ll 1. \end{aligned}\right\} \tag{57.16}$$

Figure 36 shows a graph of $f(h)$.

It should be noted that, in normal layers near the surface of the plate, the magnetic field may be considerably less than H_{cr}, i.e. there is a situation corresponding to that shown in Fig. 33a.† In this case the unfavourable thermodynamic state is made possible by the surface-tension energy, which prevents further reduction in the layer thickness.

As already mentioned, as the plate thickness increases, branching of the layers must occur. This in turn affects the dependence of the period a of the structure on l; in the limit of multiple branching, $a \propto l^{2/3}$. Numerical calculations show, however, that branching should begin only at a comparatively late stage‡.

† For instance, when $h = \frac{1}{2}$ the field on the surface in the centre of the normal layer is only $0.73 H_{cr}$, and as $h \to 0$ it tends to $0.65 H_{cr}$.

‡ The calculation for a model with multiply branched layers is given by L. D. Landau, *Journal of Physics* 7, 99, 1943 (*Collected Papers*, Pergamon Press, Oxford, 1965, p. 365).

CHAPTER VII

QUASI-STATIC ELECTROMAGNETIC FIELD

§58. Equations of the quasi-static field

SO FAR we have discussed only static electric and magnetic fields, and have used Maxwell's equation

$$\mathbf{curl}\,\mathbf{E} = -\frac{1}{c}\frac{\partial \mathbf{B}}{\partial t} \tag{58.1}$$

only as a step in deriving the expression for the energy of a magnetic field (§31).

The nature of the variable electromagnetic fields in matter depends greatly on the kind of matter concerned and on the order of magnitude of the frequency of the field. In the present section we shall consider the phenomena which occur in extended conductors placed in a variable external magnetic field. We shall assume that the rate of change of the field is not too large, and therefore satisfies various conditions which will be derived below. Electromagnetic fields and currents which satisfy these conditions are said to be *quasi-static*.

We shall first of all suppose that the wavelength $\lambda \sim c/\omega$ which corresponds (in the vacuum or dielectric surrounding the conductor) to the field frequency ω is large compared with the dimension l of the body: $\omega \ll c/l$. Then the magnetic field distribution outside the conductor at any instant can be described by the equations of a static field:

$$\mathrm{div}\,\mathbf{B} = 0, \qquad \mathbf{curl}\,\mathbf{H} = 0, \tag{58.2}$$

all effects due to the finite velocity of propagation of electromagnetic disturbances being neglected. Of course, this neglect is permissible only at distances from the body which are small compared with λ; these are the only distances which need be considered in determining the field inside the body.

The complete system of field equations inside the conductor consists of (58.1) together with

$$\mathrm{div}\,\mathbf{B} = 0, \tag{58.3}$$

$$\mathbf{curl}\,\mathbf{H} = 4\pi\mathbf{j}/c, \quad \mathbf{j} = \sigma\mathbf{E}; \tag{58.4}$$

in an electrically anisotropic (non-cubic) crystal, we must write $j_i = \sigma_{ik}E_k$. The second of these equations has been derived, strictly speaking, only for steady currents and static magnetic fields. It is therefore necessary to specify conditions under which this equation can reliably be used for variable fields. In equation (58.4) the current has been written in terms of the electric field in accordance with the relation $\mathbf{j} = \sigma\mathbf{E}$ with constant σ, which holds for a steady state. This relation remains valid if the period of the field is large compared with the characteristic times of microscopic conduction. That is, the field frequency must be small compared with the reciprocal mean free time of the electrons in the conductor. For typical metals at room temperature, the limiting frequencies given by this condition lie in the infra-red region of the spectrum.†

There is another condition which restricts the applicability of the equations in this case. Equation (58.4) assumes that the relation between the current and the field is a *local* one, i.e. that the current density at a point in the conductor depends on the field at that point only. This in turn presupposes that the electron mean free path is small compared with the distances over which the field changes appreciably. We shall return to this condition in §59.

In equations (58.1) and (58.4), **E** is the induced electric field resulting from the variation of the magnetic field. When **H** is known, the field **E** can be immediately determined by equation (58.4). The equation for **H** is obtained by eliminating **E** from (58.1) and (58.4):

$$\frac{4\pi}{c^2}\frac{\partial \mathbf{B}}{\partial t} = -\mathbf{curl}\frac{\mathbf{curl\, H}}{\sigma}. \qquad (58.5)$$

In a homogeneous medium with constant conductivity σ and constant magnetic permeability μ, the factor $1/\sigma$ can be taken in front of the curl operator, and by (58.3) we have div $\mathbf{B} = \mu$ div $\mathbf{H} = 0$. Hence **curl curl H** $= -\triangle\mathbf{H}$, and we obtain the equation

$$\triangle\mathbf{H} = \frac{4\pi\mu\sigma}{c^2}\frac{\partial \mathbf{H}}{\partial t}. \qquad (58.6)$$

With the equation div $\mathbf{H} = 0$ this suffices to determine the magnetic field. It may be noted that equation (58.6) is a heat-conduction equation, the thermometric conductivity χ being represented by $c^2/4\pi\mu\sigma$.

The boundary conditions on the magnetic field at the surface of a conductor are evident from the form of the equations, and are as before

$$B_{n1} = B_{n2}, \qquad \mathbf{H}_{t1} = \mathbf{H}_{t2}. \qquad (58.7)$$

The expression on the right-hand side of equation (58.4), being bounded, does not affect the second of (58.7). For $\mu = 1$ we can put simply‡

$$\mathbf{H}_1 = \mathbf{H}_2. \qquad (58.8)$$

From (58.4) we have div $\mathbf{j} = 0$; the boundary condition on this equation is that $j_n = 0$ at the surface of the conductor. In an electrically isotropic conductor, it follows (since $\mathbf{j} = \sigma\mathbf{E}$) that $E_n^{(i)} = 0$ at the boundary, the index (i) denoting the field inside the conductor; in the general case of an anisotropic conductor, the normal component of the field in the conductor at its surface is in general not zero.

The boundary condition (58.8) is insufficient for a complete formulation of the problem if the conductor is composite and its parts have different conductivities. At the interfaces

† For poor conductors (e.g. semiconductors), equation (58.4) is valid only if a further condition, which may be more stringent, is satisfied. For such bodies it may be possible to define both a conductivity and a permittivity. Then a term $-(\varepsilon/c)\,\partial\mathbf{E}/\partial t$ is added to the right-hand side of (58.4), and the condition for this term to be small in comparison with $4\pi\sigma\mathbf{E}/c$ is $\sigma/\omega \gg \varepsilon$. In good conductors (e.g. metals), on the other hand, $\sigma/\omega \gg 1$ throughout the frequency range in which the conductivity can be regarded as constant (see also the second footnote to §59).

‡ For ordinary diamagnetic and paramagnetic bodies, μ is very nearly 1, and the inclusion of μ in the following formulae would be a pointless refinement. Values of μ differing considerably from 1 occur in ferromagnetic metals, whose magnetic properties (in sufficiently weak fields) can be described in terms of a large constant permeability. For quite moderate frequencies, however, such substances exhibit a dispersion of μ (i.e. a dependence of μ on the frequency ω), together with a decrease of μ almost to 1. We shall therefore put $\mu = 1$ in the present chapter.

between the parts we must use both the continuity of **H** and that of $\mathbf{E}_t$; the latter implies the condition

$$(\mathbf{curl\ H})_{t1}/\sigma_1 = (\mathbf{curl\ H})_{t2}/\sigma_2 \tag{58.9}$$

on the magnetic field.

Suppose that a conductor is placed in an external magnetic field which is suddenly removed. The field in and around the conductor does not vanish immediately; the manner of its decay with time is given by equation (58.6). To solve a problem of this kind, we use the following procedure, a general technique in mathematical physics. We seek solutions of equation (58.6) which have the form $\mathbf{H} = \mathbf{H}_m(x, y, z)e^{-\gamma_m t}$, where γ_m is a constant. The equation for the function $\mathbf{H}_m(x, y, z)$ is then

$$(c^2/4\pi\sigma)\triangle \mathbf{H}_m = -\gamma_m \mathbf{H}_m. \tag{58.10}$$

For a conductor of given shape, this equation has non-zero solutions (satisfying the necessary boundary conditions) only for certain γ_m, the *eigenvalues* of (58.10), all of which are real and positive.† The corresponding functions $\mathbf{H}_m(x,y,z)$ form a complete set of orthogonal vector functions. Let the field distribution at the initial instant be $\mathbf{H}_0(x,y,z)$. On expanding this in terms of the functions $\mathbf{H}_m$:

$$\mathbf{H}_0(x,y,z) = \sum_m c_m \mathbf{H}_m(x,y,z),$$

we obtain the solution of the problem:

$$\mathbf{H}(x,y,z,t) = \sum_m c_m e^{-\gamma_m t}\mathbf{H}_m(x,y,z) \tag{58.11}$$

gives the manner of decay of the field with time.

The rate of decay is determined principally by the term in the sum for which γ_m is least; let this be γ_1. The decay time of the field may be defined as $\tau = 1/\gamma_1$. The order of magnitude of τ is evident from equation (58.10). Since $\triangle\mathbf{H} \sim \mathbf{H}/l^2$, where l is the dimension of the conductor, we have

$$\tau \sim 4\pi\sigma l^2/c^2. \tag{58.12}$$

§59. Depth of penetration of a magnetic field into a conductor

Let us consider a conductor in an external magnetic field which varies with frequency ω. The magnetic field penetrates into the conductor and induces in it a variable electric field, which in turn causes currents to appear; these are called *eddy currents*.‡ A general idea of

---

† This is easily seen as follows. So as to avoid having to take account of the boundary conditions at the surface of the body, we start from equation (58.5) and suppose σ to vanish continuously outside the body. Multiplying both sides of the equation

$$-4\pi\gamma_m\mathbf{H}_m/c^2 = -\mathbf{curl}\,[(1/\sigma)\,\mathbf{curl}\,\mathbf{H}_m]$$

by $\mathbf{H}_m{}^*$ and integrating over all space, we have

$$\frac{4\pi}{c^2}\gamma_m\int|\mathbf{H}_m|^2 dV = \int \mathbf{H}_m{}^*\cdot\mathbf{curl}\frac{\mathbf{curl}\,\mathbf{H}_m}{\sigma}dV = \int\frac{1}{\sigma}|\,\mathbf{curl}\ \mathbf{H}_m|^2 dV,$$

whence it is evident that the γ_m are real and positive.

‡ In Russian "Foucault currents".

the way in which the field penetrates into the conductor can be obtained from the analogy already mentioned between equation (58.6) and the equation of thermal conduction. It is known from the theory of thermal conduction that a quantity which satisfies such an equation is propagated through a distance $\sim \sqrt{(\chi t)}$ in time t. We can therefore immediately conclude that the magnetic field penetrates into the conductor to a distance δ, given in order of magnitude by $\delta \sim \sqrt{(c^2/\sigma\omega)}$. The same is true, of course, of the induced electric field and currents.

In a variable field with frequency ω, all quantities depend on the time through a factor $e^{-i\omega t}$. Equation (58.6) then becomes

$$\triangle \mathbf{H} = -4\pi i\sigma\omega\mathbf{H}/c^2. \tag{59.1}$$

Let us consider two limiting cases. If the *penetration depth* δ is large compared with the dimension of the body (low frequencies), we can put the right-hand side of (59.1) equal to zero as a first approximation. Then the magnetic field distribution at any instant will be the same as it would be in a steady state with the same external field far from the body. Let this solution be $\mathbf{H}_{\text{st}}$; it is independent of the frequency (or rather involves the frequency only in the time factor $e^{-i\omega t}$). The induced electric field appears only in the next approximation, being absent in the steady state. This corresponds to the fact that $\mathbf{curl}\, \mathbf{H}_{\text{st}} = 0$, and so the value of $\mathbf{E}$ obtained from (58.4) is zero. To calculate $\mathbf{E}$, therefore, we must use equation (58.1), according to which

$$\mathbf{curl}\, \mathbf{E} = i\omega\mathbf{H}_{\text{st}}/c. \tag{59.2}$$

This equation, together with div $\mathbf{E} = 0$ (which follows from (58.4) when σ is constant in the body), entirely determines the electric field distribution. It is seen to be proportional to the frequency ω.

The opposite limiting case is that where $\delta \ll l$ (high frequencies). The condition for the field equations to be local (§58) requires that δ should still be large compared with the mean free path of the conduction electrons.†

When $\delta \ll l$ the magnetic field penetrates only into a thin surface layer of the conductor. In calculating the field outside the conductor we can neglect the thickness of this layer, i.e. assume that the magnetic field does not penetrate into the conductor at all. In this sense a conductor in a high-frequency magnetic field behaves like a superconductor in a constant field, and the field outside it must be calculated by solving the corresponding steady-state problem for a superconductor of the same shape.

The true field distribution in the surface layer of the conductor can be investigated in a general manner by regarding small regions of the surface as plane. It is necessary to solve equation (59.1) for a conducting medium bounded by a plane surface, outside which the field has a given value $\mathbf{H}_0 e^{-i\omega t}$, say. This vector is obtained as shown above, by solving the problem for a semi-infinite medium, and is parallel to the surface of the conductor. The boundary condition (58.8) shows that the magnetic field in the conductor is also $\mathbf{H}_0 e^{-i\omega t}$ at the surface.

We take the surface of the conductor as the xy-plane, the conducting medium being in $z > 0$. Since the conditions of the problem are independent of x and y, the required field $\mathbf{H}$ depends only on the z coordinate (and on the time). We therefore have div $\mathbf{H} = \partial H_z/\partial z$

† This condition is, in fact, the first to be violated in metals as the frequency increases. The condition $\omega \ll 1/\tau$, where τ is the mean free time, may, however, be the more stringent for semiconductors of low conductivity.

$= 0$, and since $H_z = 0$ at the boundary it must be zero everywhere. By (59.1), the equation for **H** is $\partial^2\mathbf{H}/\partial z^2 + k^2\mathbf{H} = 0$, where $k = \sqrt{(4\pi i\sigma\omega/c^2)} = (1+i)\sqrt{(2\pi\sigma\omega)}/c$. The solution of this equation which vanishes far from the surface ($z \to \infty$) is proportional to e^{ikz}. Using the boundary condition at $z = 0$, we obtain

$$\mathbf{H} = \mathbf{H}_0 e^{-z/\delta} e^{iz/\delta} e^{-i\omega t} \tag{59.3}$$

where the penetration depth δ is

$$\delta = c/\sqrt{(2\pi\sigma\omega)} \text{ and } k = (1+i)/\delta. \tag{59.4}$$

The electric field is now determined by means of equation (58.4). If **n** is a unit vector in the z-direction, we have

$$\mathbf{E} = \sqrt{(\omega/8\pi\sigma)}(1-i)\mathbf{H}\times\mathbf{n}. \tag{59.5}$$

Thus $E \sim H\delta/\lambda$.

If the field $\mathbf{H}_0 e^{-i\omega t}$ is linearly polarized, then $\mathbf{H}_0$ can be made real by a suitable choice of the origin of time. We then take the direction of $\mathbf{H}_0$ as the y-axis. Taking the real part in (59.4) and (59.5), we have

$$\left.\begin{aligned} H = H_y &= H_0 e^{-z/\delta}\cos\left(\frac{z}{\delta} - \omega t\right), \\ E = E_x &= H_0\sqrt{(\omega/4\pi\sigma)}e^{-z/\delta}\cos\left(\frac{z}{\delta} - \omega t - \frac{1}{4}\pi\right). \end{aligned}\right\} \tag{59.6}$$

The eddy current density $\mathbf{j} = \sigma\mathbf{E}$ has the same distribution as **E**.

The relation (59.5) is valid in the present case for the field throughout the half-space $z > 0$. In more general cases, a relation

$$\mathbf{E}_t = \zeta\mathbf{H}_t\times\mathbf{n} \tag{59.7}$$

is in general valid only at the surface of the conductor for the field components tangential to the surface; since these components are continuous, (59.7) applies to the field on either side of the surface. The coefficient ζ is called the *surface impedance* of the conductor; we shall return in §87 to more general aspects of this concept.† In the present case,

$$\zeta = \sqrt{(\omega/8\pi\sigma)}(1-i). \tag{59.8}$$

The presence of eddy currents implies a dissipation of the field energy, which appears as Joule heat. The time average energy Q dissipated in the conductor per unit time is $Q = \int \overline{\mathbf{j}\cdot\mathbf{E}}\,dV = \int \sigma\overline{\mathbf{E}^2}\,dV$. It can also be calculated as the mean field energy entering the conductor per unit time:

$$Q = \oint \overline{\mathbf{S}}\cdot d\mathbf{f} = (c/4\pi)\oint \overline{\mathbf{E}\times\mathbf{H}}\cdot d\mathbf{f}, \tag{59.9}$$

† In electrically anisotropic media, the surface impedance is a two-dimensional tensor:

$$\mathbf{E}_{t\alpha} = \zeta_{\alpha\beta}(\mathbf{H}_t\times\mathbf{n})_\beta, \tag{59.7a}$$

where α and β are tensor suffixes in the plane perpendicular to **n**. This tensor may be affected by the heat fluxes that result from the thermoelectric effect (see Problem 4).

the integral being taken over the surface of the conductor.†

We have already seen that, in the limiting case $\delta \gg l$, the amplitude of the magnetic field inside the conductor is independent of the frequency, while that of the electric field is proportional to ω. The energy dissipation Q at low frequencies is therefore proportional to ω^2. When $\delta \ll l$, on the other hand, the magnetic and electric fields on the surface of the conductor are given by formulae (59.3) and (59.5) with $z = 0$. The Poynting vector is normal to the surface, and its mean value is $\overline{S} = (c/16\pi)\sqrt{(\omega/2\pi\sigma)}|\mathbf{H}_0|^2$, the variation of $\mathbf{H}_0$ over the surface being given by the solution of the problem of the field outside a superconductor of the same shape (cf. above). The energy dissipation is

$$Q = \frac{c}{16\pi}\sqrt{\left(\frac{\omega}{2\pi\sigma}\right)}\oint|\mathbf{H}_0|^2\,\mathrm{d}f. \tag{59.10}$$

Thus at high frequencies it is proportional to $\sqrt{\omega}$.

The energy dissipation can also be expressed in terms of the total magnetic moment $\mathscr{M}$ acquired by the conductor in the magnetic field. In a periodic field, the magnetic moment is likewise a periodic function of time, with the same frequency. According to formula (32.4), the rate of variation of the free energy is given by $-\mathscr{M}\cdot\mathrm{d}\mathfrak{H}/\mathrm{d}t$, where $\mathfrak{H}$ is a uniform external field in which the conductor is placed. This expression does not immediately give the required energy dissipation, because the energy of the body changes not only on account of dissipation but also by the periodic movement of energy between the body and the surrounding field. If we average over time, however, the latter contribution vanishes, and the mean dissipation of energy per unit time is

$$Q = -\overline{\mathscr{M}\cdot\mathrm{d}\mathfrak{H}/\mathrm{d}t}. \tag{59.11}$$

If $\mathscr{M}$ and $\mathfrak{H}$ are written in complex form, then $\mathrm{d}\mathfrak{H}/\mathrm{d}t = -i\omega\mathfrak{H}$, and Q can be calculated as

$$Q = -\tfrac{1}{2}\mathrm{re}(i\omega\mathscr{M}\cdot\mathfrak{H}^*) = \tfrac{1}{2}\omega\,\mathrm{im}(\mathscr{M}\cdot\mathfrak{H}^*). \tag{59.12}$$

The components of the magnetic moment $\mathscr{M}$ are linear functions of the external field:

$$\mathscr{M}_i = V\alpha_{ik}\mathfrak{H}_k, \tag{59.13}$$

where the dimensionless coefficients $\alpha_{ik}(\omega)$ depend on the shape of the body and on its orientation in the external field, but not on its volume V. In this formula we assume that $\mathscr{M}$ and $\mathfrak{H}$ are written in complex form, so that the α_{ik} are also in general complex. The tensor $V\alpha_{ik}$ may be called the *magnetic polarizability tensor* for the body as a whole. This tensor is a generalized susceptibility, and has the properties common to all such quantities. In particular, it is symmetrical (see *SP* 1, §125):

$$\alpha_{ik} = \alpha_{ki}. \tag{59.14}$$

We can therefore write

$$\begin{aligned}\mathscr{M}\cdot\mathfrak{H}^* = V\alpha_{ik}\mathfrak{H}_i^*\mathfrak{H}_k &= \tfrac{1}{2}V\alpha_{ik}(\mathfrak{H}_i^*\mathfrak{H}_k + \mathfrak{H}_i\mathfrak{H}_k^*)\\ &= V\alpha_{ik}\,\mathrm{re}(\mathfrak{H}_i\mathfrak{H}_k^*).\end{aligned}$$

† If any two quantities $a(t)$ and $b(t)$ are written in complex form (proportional to $e^{-i\omega t}$), the real parts must of course be taken before calculating their product. If, however, we are interested only in the time average value of the product, it may be calculated as $\frac{1}{2}$ re ab^*. The terms containing $e^{\pm 2i\omega t}$ give zero on averaging, and so $\frac{1}{4}(a+a^*)(b+b^*) = \frac{1}{4}(ab^*+a^*b)$. In particular, $\overline{\mathbf{S}}$ can be calculated as the real part of the "complex Poynting vector":

$$\overline{\mathbf{S}} = \mathrm{re}\left[\frac{c}{4\pi}\cdot\tfrac{1}{2}\mathbf{E}\times\mathbf{H}^*\right]. \tag{59.9a}$$

If we also write the complex quantities α_{ik} as $\alpha_{ik}' + i\alpha_{ik}''$, the energy dissipation (59.12) becomes

$$Q = \tfrac{1}{2} V\omega\alpha_{ik}'' \operatorname{re}(\mathfrak{H}_i \mathfrak{H}_k^*). \qquad (59.15)$$

Thus the energy dissipation is determined by the imaginary part of the magnetic polarizability. We have already seen that Q is proportional to ω^2 for low frequencies, and to $\sqrt{\omega}$ for high frequencies. We can therefore conclude that the quantities α_{ik}'' in these two limiting cases are proportional to ω and to $1/\sqrt{\omega}$ respectively. Since they decrease both as $\omega \to 0$ and $\omega \to \infty$, they must have a maximum in between.

The magnetic moment of a conductor in a variable magnetic field is due mainly to the conduction currents set up in the body; it is not zero even if $\mu = 1$, when the moment in a static field vanishes. The latter can be obtained from $\mathscr{M}(\omega)$ by taking the limit as $\omega \to 0$. Hence it follows that the real part α_{ik}' of the polarizability tends to a constant limit as $\omega \to 0$ (the limit being zero for $\mu = 1$), corresponding to magnetization in a static field. In the limit $\omega \to \infty$, when the magnetic field does not penetrate into the body, α_{ik}' tends to a different constant limit, corresponding to the static magnetization of a superconductor of the same shape.

PROBLEMS

PROBLEM 1. Determine the magnetic polarizability of an isotropic conducting sphere with radius a in a uniform periodic external field.

SOLUTION. The field $\mathbf{H}^{(i)}$ inside the sphere satisfies the equations $\triangle \mathbf{H}^{(i)} + k^2 \mathbf{H}^{(i)} = 0$, $\operatorname{div} \mathbf{H}^{(i)} = 0$, where $k = (1+i)/\delta$. We write this field in the form $\mathbf{H}^{(i)} = \mathbf{curl}\, \mathbf{A}$, where $\mathbf{A}$ satisfies the equation $\triangle \mathbf{A} + k^2 \mathbf{A} = 0$; since $\mathbf{H}$ is an axial vector, $\mathbf{A}$ is a polar vector. By symmetry, the only constant vector on which the required solution can depend is the external field $\mathfrak{H}$. We denote by f the spherically symmetrical solution, finite for $r = 0$, of the scalar equation $\triangle f + k^2 f = 0$, namely $f = (1/r) \sin kr$. Then the polar vector $\mathbf{A}$, which satisfies the vector equation $\triangle \mathbf{A} + k^2 \mathbf{A} = 0$ and depends linearly on the constant axial vector $\mathfrak{H}$, can be written as $\mathbf{A} = \beta\, \mathbf{curl}(f\mathfrak{H})$, where β is a constant. Thus we have

$$\mathbf{H}^{(i)} = \beta\, \mathbf{curl}\, \mathbf{curl}(f\mathfrak{H})$$
$$= \beta\left(\frac{f'}{r} + k^2 f\right)\mathfrak{H} - \beta\left(\frac{3f'}{r} + k^2 f\right)(\mathbf{n}\cdot\mathfrak{H})\mathbf{n},$$

where $\mathbf{n}$ is a unit vector in the direction of $\mathbf{r}$; the second derivative f'' has been eliminated by means of the equation $\triangle f + k^2 f = 0$.

The field $\mathbf{H}^{(e)}$ outside the sphere satisfies the equations $\mathbf{curl}\, \mathbf{H}^{(e)} = 0$, $\operatorname{div} \mathbf{H}^{(e)} = 0$. We put $\mathbf{H}^{(e)} = -\mathbf{grad}\, \phi + \mathfrak{H}$; ϕ satisfies the equation $\triangle \phi = 0$ and vanishes at infinity. Since ϕ depends linearly on the constant vector $\mathfrak{H}$, we have $\phi = -V\alpha\, \mathfrak{H}\cdot\mathbf{grad}(1/r)$, where $V = 4\pi a^3/3$. Thus

$$\mathbf{H}^{(e)} = V\alpha\, \mathbf{grad}[(\mathfrak{H}\cdot\mathbf{grad})(1/r)] + \mathfrak{H}$$
$$= \frac{V\alpha}{r^3}[3(\mathbf{n}\cdot\mathfrak{H})\mathbf{n} - \mathfrak{H}] + \mathfrak{H}.$$

It is evident that $V\alpha\mathfrak{H}$ is the magnetic moment of the sphere, so that $V\alpha$ is its magnetic polarizability (by symmetry, the tensor α_{ik} reduces to a scalar $\alpha\delta_{ik}$).

On the surface of the sphere ($r = a$), all the components of $\mathbf{H}$ must be continuous. Equating separately the components parallel and perpendicular to $\mathbf{n}$, we obtain two equations to determine α and β. The polarizability per unit volume is found to be

$$\alpha = \alpha' + i\alpha'' = -\frac{3}{8\pi}\left[1 - \frac{3}{a^2 k^2} + \frac{3}{ak}\cot ak\right],$$

$$\alpha' = -\frac{3}{8\pi}\left[1 - \frac{3}{2}\frac{\delta}{a}\frac{\sinh(2a/\delta) - \sin(2a/\delta)}{\cosh(2a/\delta) - \cos(2a/\delta)}\right],$$

$$\alpha'' = -\frac{9\delta^2}{16\pi a^2}\left[1 - \frac{a}{\delta}\frac{\sinh(2a/\delta) + \sin(2a/\delta)}{\cosh(2a/\delta) - \cos(2a/\delta)}\right].$$

In the limit of low frequencies ($\delta \gg a$),

$$\alpha' = -\frac{1}{105\pi}\left(\frac{a}{\delta}\right)^4 = -\frac{4\pi}{105}\frac{a^4\sigma^2\omega^2}{c^4},$$

$$\alpha'' = \frac{1}{20\pi}\left(\frac{a}{\delta}\right)^2 = \frac{a^2\sigma\omega}{10c^2}.$$

For high frequencies ($\delta \ll a$),

$$\alpha' = -\frac{3}{8\pi}\left[1-\frac{3\delta}{2a}\right] = -\frac{3}{8\pi}\left[1-\frac{3c}{2a\sqrt{(2\pi\sigma\omega)}}\right],$$

$$\alpha'' = \frac{9}{16\pi}\frac{\delta}{a} = \frac{9c}{16\pi a\sqrt{(2\pi\sigma\omega)}}.$$

The limiting value $V\alpha' = -\frac{1}{2}a^3$ corresponds to the magnetic moment of a superconducting sphere; the corresponding value of α'' could be found from formula (59.10), using the expression (54.3) for the field at the surface of a superconducting sphere.

The external field is assumed to be written in the complex form $\mathfrak{H} = \mathfrak{H}_0 e^{-i\omega t}$ with an arbitrary constant complex vector $\mathfrak{H}_0$. The analysis therefore includes both the "linearly polarized" variable field whose direction is constant, and the elliptically or circularly polarized fields which rotate in some plane.

PROBLEM 2. The same as Problem 1, but for a conducting cylinder (with radius a) in a uniform periodic mangnetic field perpendicular to its axis.

SOLUTION. This problem is the "two-dimensional analogue" of Problem 1. In what follows all vector operations are two-dimensional operations in a plane perpendicular to the axis of the cylinder, and $\mathbf{r}$ is the position vector in that plane. The field inside the cylinder is of the form

$$\begin{aligned}\mathbf{H}^{(i)} &= \beta\,\mathbf{curl}\,\mathbf{curl}(f\,\mathfrak{H})\\ &= \beta\left(\frac{f'}{r}+k^2f\right)\mathfrak{H} - \beta\left(\frac{2f'}{r}+k^2f\right)(\mathbf{n}\cdot\mathfrak{H})\mathbf{n},\end{aligned}$$

where $f = J_0(kr)$ is the symmetrical solution of the two-dimensional equation $\triangle f + k^2 f = 0$ which is finite for $r = 0$. The field outside the cylinder is

$$\begin{aligned}\mathbf{H}^{(e)} &= -2V\alpha\,\mathbf{grad}\,[(\mathfrak{H}\cdot\mathbf{grad})\log r] + \mathfrak{H}\\ &= \frac{2V\alpha}{r^2}\,[2(\mathbf{n}\cdot\mathfrak{H})\mathbf{n} - \mathfrak{H}] + \mathfrak{H},\end{aligned}$$

where $V = \pi a^2$. The magnetic moment per unit length of the cylinder is $V\alpha\mathfrak{H}$ (see §3, Problem 2). From the condition $\mathbf{H}^{(i)} = \mathbf{H}^{(e)}$ for $r = a$, as in Problem 1, we obtain

$$\alpha = -\frac{1}{2\pi}\left[1-\frac{2}{ka}\frac{J_1(ka)}{J_0(ka)}\right],$$

using the relation $J_0'(kr) = -kJ_1(kr)$.

For $\delta \gg a$, expanding the Bessel functions in powers of ka, we have

$$\alpha' = -\frac{1}{24\pi}\left(\frac{a}{\delta}\right)^4 = -\frac{\pi a^4\sigma^2\omega^2}{6c^4},$$

$$\alpha'' = \frac{1}{8\pi}\left(\frac{a}{\delta}\right)^2 = \frac{a^2\sigma\omega}{4c^2}.$$

For $\delta \ll a$, we use the asymptotic expressions for the Bessel functions, obtaining

$$\alpha' = -\frac{1}{2\pi}\left(1-\frac{\delta}{a}\right) = -\frac{1}{2\pi}\left(1-\frac{c}{a\sqrt{(2\pi\sigma\omega)}}\right),$$

$$\alpha'' = \frac{1}{2\pi}\frac{\delta}{a} = \frac{c}{2\pi a\sqrt{(2\pi\sigma\omega)}}.$$

PROBLEM 3. The same as Problem 2, but for a magnetic field parallel to the axis of the cylinder.

SOLUTION. The magnetic field is everywhere parallel to the axis of the cylinder. Outside the cylinder we have $\mathbf{H}^{(e)} = \mathfrak{H}$, and inside it $\mathbf{H}^{(i)} = f\,\mathfrak{H}$, where f is the symmetrical solution of the two-dimensional equation $\triangle f + k^2 f = 0$ which is 1 for $r = a$ and finite for $r = 0$: $\mathbf{H}^{(i)} = \mathfrak{H}\, J_0(kr)/J_0(ka)$. The eddy currents in the cylinder are azimuthal (i.e. the only non-zero component is j_ϕ), and are given in terms of the field $H_z = H$ by $4\pi j/c = -\partial H/\partial r$. The magnetic moment generated per unit length of the cylinder by the conduction currents is $\mathscr{M} = \pi a^2 \alpha\, \mathfrak{H}$ $= (1/2c)\int jr\, dV = -\frac{1}{4}\int (\partial H/\partial r) r^2\, dr$; it is parallel to the axis. Evaluating the integral, we have

$$\alpha = -\frac{1}{4\pi}\left[1 - \frac{2}{ka}\frac{J_1(ka)}{J_0(ka)}\right].$$

Thus the longitudinal polarizability of the cylinder is half the transverse polarizability derived in Problem 2.

PROBLEM 4. Determine the least decay coefficient for the magnetic field in a conducting sphere.

SOLUTION. The solutions of equations (58.10) for a sphere include functions with various symmetries. The most symmetrical solution is that which is defined by an arbitrary constant scalar. This solution is inapplicable, however, for the following reason: it would be spherically symmetrical ($H = H_r(r)$) and would have to be $H = \text{constant}/r$ in order to satisfy the equation $\operatorname{div} \mathbf{H} = (1/r)\partial(rH)/\partial r = 0$, which is valid both outside and inside the sphere; but this function is not finite at the centre of the sphere.

The least value of γ corresponds to one of the solutions defined by an arbitrary constant vector. The form of these solutions is evidently the same as has been found in Problem 1, the only difference being that the constant term in the field $\mathbf{H}^{(e)}$ must be omitted so as to have $\mathbf{H} = 0$ at infinity. The quantity k is now real ($= \sqrt{(4\pi\sigma\gamma/c^2)}$), and the vector $\mathbf{H}$ is an arbitrary constant vector. From the boundary condition $\mathbf{H}^{(i)} = \mathbf{H}^{(e)}$ at $r = a$ we obtain two equations, and on eliminating α and β we find $\sin ka = 0$. The smallest non-zero root of this equation is $ka = \pi$, and so the smallest value of γ is $\gamma_1 = \pi c^2/4\sigma a^2$.

PROBLEM 5. The plane surface of a uniaxial metal crystal is cut so that the normal is at an angle θ to the principal axis of symmetry of the crystal. Determine the surface impedance, taking account of the thermoelectric effect (M. I. Kaganov and V. M. Tsukernik, 1958).

SOLUTION. We take the crystal surface as the xy-plane, and the z-axis along the inward normal to the surface; let the principal axis of symmetry of the crystal be in the xz-plane at an angle θ to the z-axis. Let the magnetic field at the surface be in the y-direction: $H_y = H_0 e^{-i\omega t}$. Then it is in this direction everywhere within the metal also. Since all quantities depend only on z (and on the time as $e^{-i\omega t}$), we find that Maxwell's equations (58.1), (58.4) become

$$-H_y' = 4\pi j_x/c, \quad j_y = j_z = 0, \quad E_x' = i\omega H_y/c, \tag{1}$$

and $E_y' = 0$, whence $E_y = 0$; the prime denotes differentiation with respect to z. To take account of the thermoelectric effect, we must include also the equation of thermal conduction, $C\partial T/\partial t + \operatorname{div} \mathbf{q} = 0$, or

$$-i\omega C\tau + q_z' = 0, \tag{2}$$

where τ is a variable increment of the mean temperature: $T = \overline{T} + \tau$, C the specific heat of the metal per unit volume, and $\mathbf{q}$ the heat flux density; $\mathbf{j}$ and $\mathbf{q}$ are related to the field $\mathbf{E}$ and the temperature gradient by (26.12).

The tensors $\rho_{ik} = \sigma^{-1}{}_{ik}$ and κ_{ik} are symmetrical. We shall suppose the symmetry of the crystal to be such that α_{ik} is symmetrical also. With the above choice of the x, y and z axes, we have

$$\rho_{xx} = \rho_\parallel \sin^2\theta + \rho_\perp \cos^2\theta, \quad \rho_{yy} = \rho_\perp,$$
$$\rho_{zz} = \rho_\parallel \cos^2\theta + \rho_\perp \sin^2\theta, \quad \rho_{xy} = \rho_{yz} = 0,$$
$$\rho_{xz} = (\rho_\parallel - \rho_\perp)\sin\theta\cos\theta,$$

where $\rho_\parallel$ and $\rho_\perp$ are the principal values of the tensor ρ_{ik} along the crystal axis and in the plane perpendicular to it; similar expressions are valid for κ_{ik} and α_{ik}. Then, from (26.12),

$$E_x = \rho_{xx} j_x + \alpha_{xz}\tau', \; q_z = T\alpha_{zx} j_x - \kappa_{zz}\tau', \tag{3}$$
$$E_z = \rho_{zx} j_x + \alpha_{zz}\tau'. \tag{4}$$

Elimination of H_y from equations (1)–(3) gives

$$E_x'' + k^2(E_x + E_T) = 0,$$

$$(1+a)E_T'' + k^2[(b-a)E_T - aE_x] = 0,$$

with $E_T = -\alpha_{xz}\tau'$ and the parameters

$$k^2 = 4\pi i\omega/c^2\rho_{xx}, \; a = T\alpha_{xz}{}^2/\rho_{xx}\kappa_{zz}, \; b = c^2 C\rho_{xx}/4\pi\kappa_{zz}.$$

The solution of these two equations for the half-space $z > 0$ occupied by the metal is

$$E_x = Ae^{ik_1 z} + Be^{ik_2 z},$$

$$E_T = -(1-k_1^2/k^2)Ae^{ik_1 z} - (1-k_2^2/k^2)Be^{ik_2 z},$$

where

$$k_{1,2} = k\left[\frac{1+b \pm \sqrt{\{(1-b)^2-4ab\}}}{2(1+a)}\right]^{\frac{1}{2}};$$

the imaginary parts of k_1 and k_2 must be positive.

The relation between the coefficients A and B is found from the boundary conditions for the temperature; formula (4) gives the field E_z in the metal; the normal component of **E** is not required to be continuous at the surface of the conductor. According to the definition (59.7a), the surface impedance is

$$\zeta_{xx} = [E_x/H_y]_{z=0}$$
$$= \frac{\omega}{c}\frac{A+B}{k_1 A + k_2 B},$$
$$\zeta_{yx} = 0.$$

The final expressions are as follows for the impedance when $a \ll 1$ (as is in fact true for ordinary metals) in two cases. With an isothermal boundary ($\tau = 0$),

$$\zeta_{xx}^{(is)} = \zeta_0[1 + a/2(1+\sqrt{b})^2];$$

with an adiabatic boundary ($q_z = 0$),

$$\zeta_{xx}^{(ad)} = \zeta_0[1 + a(1+2\sqrt{b})/2(1+\sqrt{b})^2],$$

where $\zeta_0 = (\omega\rho_{yy}/8\pi)^{\frac{1}{2}}(1-i)$ is the impedance when the thermoelectric effect is neglected. The parameter a, and therefore the correction to the impedance, are zero when $\theta = \frac{1}{2}\pi$ or $\theta = 0$, i.e. when the principal axis is either in the plane of the surface or perpendicular to it.

If the field **H** is in the x-direction, then $E_x = 0$, $j_x = j_y = 0$, and there is no temperature gradient, so that $\zeta_{yy} = \zeta_0$.

§60. The skin effect

Let us consider the distribution of current density over the cross-section of a conductor in which a non-zero and variable total current is flowing. From the results of §59 we should expect that, as the frequency increases, the current will tend to be concentrated near the surface of the conductor. This phenomenon is called the *skin effect*.†

The exact solution of the problem of the skin effect depends, in general, not only on the shape of the conductor but also on the manner of excitation of the current in it, i.e. the nature of the variable external magnetic field which induces the current. An important particular case, however, is that where the current flows in a wire of thickness small compared with its length; here the current distribution is independent of the manner of excitation.

In calculating the current distribution over the cross-section of a thin wire, the latter may be regarded as straight. The electric field is parallel to the axis of the wire, and the magnetic field vector **H** is in a plane perpendicular to the axis.

Let us consider a wire of circular cross-section. This is a particularly simple case, because the form of the field outside the wire is immediately obvious. By symmetry, **E** = constant over the surface of the wire (though the value of the constant varies with time). With this

† The term *skin effect* is used, more generally, in all cases where a variable electromagnetic field (and therefore the currents due to it) penetrates only to a relatively small depth in a conductor.

boundary condition, the only solution of the equations div $\mathbf{E} = 0$, **curl** $\mathbf{E} = 0$ outside the wire is $\mathbf{E} =$ constant. Similarly, the magnetic field outside the wire must be the same as it would be outside a wire carrying a constant current equal to the instantaneous value of the variable current.

Inside the wire, the electric field satisfies the equation $\triangle \mathbf{E} = (4\pi\sigma/c^2)\partial\mathbf{E}/\partial t$, which is the same as equation (58.6) for $\mathbf{H}$; it is obtained by eliminating $\mathbf{H}$ from (58.1) and (58.4), just as (58.6) was obtained by eliminating $\mathbf{E}$. In cylindrical polar coordinates, with the z-axis along the axis of the wire, the only non-zero component of $\mathbf{E}$ is E_z, which depends only on r. For a periodic field with frequency ω we have

$$\frac{1}{r}\frac{\partial}{\partial r}\left(r\frac{\partial E}{\partial r}\right)+k^2E=0, \quad k=\frac{\sqrt{(2i)}}{\delta}=\frac{1+i}{\delta}, \tag{60.1}$$

where δ is the penetration depth (59.4). The solution of this equation which remains finite at $r = 0$ is

$$E = E_z = \text{constant} \times J_0(kr)e^{-i\omega t}, \tag{60.2}$$

where J_0 is the Bessel function. The current density $j = \sigma E$ is similarly distributed.

The magnetic field $H_\phi = H$ is found from the electric field by equation (58.1):

$$i\omega H_\phi/c = (\mathbf{curl}\ \mathbf{E})_\phi = -\partial E_z/\partial r. \tag{60.3}$$

Since $J_0'(u) = -J_1(u)$, we obtain

$$H = H_\phi = -\text{constant} \times i\sqrt{(4\pi\sigma i/\omega)}\, J_1(kr)e^{-i\omega t}, \tag{60.4}$$

the constant being the same as in (60.2); it is easily determined from the condition that $H = 2I/ca$ on the surface of the wire, a being the radius of the wire and I the total current in it.

In the limiting case of low frequencies ($a/\delta \ll 1$) we can take just the first few terms of the expansions of the Bessel functions at every point in the cross-section:

$$\left.\begin{aligned} E_z &= \text{constant} \times \left[1-\frac{1}{2}i(r/\delta)^2-\frac{1}{16}(r/\delta)^4\right]e^{-i\omega t}, \\ H_\phi &= \text{constant} \times \frac{2\pi\sigma}{c} r\left[1-\frac{1}{4}i(r/\delta)^2-\frac{1}{48}(r/\delta)^4\right]e^{-i\omega t}. \end{aligned}\right\} \tag{60.5}$$

The amplitude of E, and therefore that of the current density, increase as $1+(r/2\delta)^4$ with increasing distance r from the axis.

In the opposite limiting case of high frequencies ($a/\delta \gg 1$) we can use the asymptotic formula

$$J_0[u\sqrt{(2i)}] \sim u^{-\frac{1}{2}}e^{(1-i)u}, \tag{60.6}$$

which is valid for large values of the argument of the Bessel function, over most of the cross-section. Retaining only the rapidly varying exponential factor, we have

$$\left.\begin{aligned} E_z &= \text{constant} \times e^{-(a-r)/\delta}e^{i(a-r)/\delta}e^{-i\omega t}, \\ H_\phi &= \text{constant} \times (1+i)\sqrt{\frac{2\pi\sigma}{\omega}}\, e^{-(a-r)/\delta}e^{i(a-r)/\delta}e^{-i\omega t}. \end{aligned}\right\} \tag{60.7}$$

These formulae are, of course, the same as (59.3)–(59.5), which are valid near the surface of a conductor of any shape when the skin effect is strong.

In the general case of a wire whose cross-section is not circular, the exact calculation of the skin effect is considerably more involved, since the fields inside and outside the wire must be determined simultaneously. Only in the limiting case of strong skin effect is the problem again simplified, because the field outside the wire may then be determined as the static field outside a superconductor of the same shape.

§61. The complex resistance

If the frequency of the variable current is low, the instantaneous current $J(t)$ in a linear circuit is determined by the instantaneous e.m.f. $\mathscr{E}$:

$$\mathscr{E}(t) = RJ(t), \tag{61.1}$$

where R is the resistance of the wire to a constant current.

There is no reason, however, to expect a direct relation between the values of $\mathscr{E}$ and J at the same instant for all frequencies. We can say only that the value of $J(t)$ must be a linear function of the values of $\mathscr{E}(t)$ at all previous instants. This relation may be symbolically written as $J = \hat{Z}^{-1}\mathscr{E}$ or, conversely.

$$\mathscr{E} = \hat{Z}J, \tag{61.2}$$

where $\hat{Z}$ is some linear operator.† If the functions $\mathscr{E}(t)$ and $J(t)$ are expanded as Fourier integrals, then for each monochromatic component (depending on time through a factor $e^{-i\omega t}$), the effect of the linear operator $\hat{Z}$ is simply multiplication by a quantity Z which depends on the frequency:

$$\mathscr{E} = Z(\omega)\,J. \tag{61.3}$$

The function $Z(\omega)$ is in general complex. It is called the *complex resistance* or *impedance* of the conductor.

It is evident from a comparison of (61.3) and (61.1) that the ordinary resistance R is the zero-order term in an expansion of the function $Z(\omega)$ in powers of ω. To find the next term, we must take account both of R and of the self-inductance L of the conductor.‡

Let us consider a linear circuit containing a variable e.m.f. $\mathscr{E}(t)$. By the definition of $\mathscr{E}$, the work done per unit time by the electric field on the charges moving in the wire is $\mathscr{E}J$. This work goes partly into Joule heat and partly to change the energy of the magnetic field of the current. By the definition of R and L, the Joule heat evolved in the wire per unit time is RJ^2, and the magnetic energy of the current is $LJ^2/2c^2$. The law of conservation of energy therefore gives the equation

$$\mathscr{E}J = RJ^2 + \frac{d}{dt}\frac{LJ^2}{2c^2} = RJ^2 + \frac{1}{c^2}LJ\frac{dJ}{dt},$$

or

$$\mathscr{E} = RJ + \frac{1}{c^2}L\frac{dJ}{dt}. \tag{61.4}$$

† We shall not pause here to discuss the general properties of this operator, since they are entirely analogous to those of the operator $\hat{\varepsilon}$, which will be examined in detail in §§77 and 82.

‡ Here, and in what follows, R and L denote the values for a steady current.

In order to use the quadratic expressions $\mathscr{E} J$ and J^2 we must write $\mathscr{E}$ and J as real functions. Having derived the linear equation (61.4), however, we can take complex monochromatic forms: $\mathscr{E} = \mathscr{E}_0 e^{-i\omega t}$, $J = J_0 e^{-i\omega t}$. Then equation (61.4) gives the algebraic relation

$$\mathscr{E} = ZJ, \qquad Z = R - \frac{i}{c^2}\omega L. \tag{61.5}$$

Taking the real part in $J = \mathscr{E}/Z$, we have

$$J(t) = \frac{\mathscr{E}_0}{\sqrt{(R^2 + \omega^2 L^2/c^4)}} \cos(\omega t - \phi), \qquad \tan\phi = \omega L/c^2 R, \tag{61.6}$$

which determines the amplitude of the current and the phase difference between the current and the e.m.f.

The real part of the expression (61.5) is the resistance R, which determines the energy dissipation in the circuit. It is easy to see that, whatever the function $Z(\omega)$, a similar relation holds between re Z and the energy dissipation for a given current. On averaging with respect to time the power $\mathscr{E} J$ required to maintain the periodic current in the circuit, we obtain the part of this power which continually makes good the dissipative losses. The energy dissipation in the circuit per unit time is therefore $Q = \frac{1}{2}\mathrm{re}(\mathscr{E} J^*)$, where $\mathscr{E}$ and J are expressed in complex form; see the last footnote to §59. Substituting $\mathscr{E} = ZJ$ and denoting the real and imaginary parts of Z by Z' and Z'' respectively:†

$$Z = Z' + iZ'', \tag{61.7}$$

we obtain $Q = \frac{1}{2}Z'|J|^2$ or, in terms of the real function $J(t)$,

$$Q = Z'(\omega)\overline{J^2}, \tag{61.8}$$

which gives the required relation.

It may be noted that, since Q is necessarily positive, Z' is also positive:

$$Z' > 0. \tag{61.9}$$

We may calculate $Z(\omega)$ for a wire of circular cross-section for any frequency which satisfies the quasi-steady condition, i.e. without neglecting the skin effect. To do so, we again use the law of conservation of energy, but in a different form. We divide the power $\mathscr{E} J$ (where $\mathscr{E}$ and J are real) into two parts, one being the change in the magnetic field energy outside the wire, and the other the total energy consumed inside the wire (both in changing the field and in evolution of heat). The second part can be calculated as the total energy flux entering the conductor through its surface per unit time. Thus we have

$$\mathscr{E} J = \frac{\mathrm{d}}{\mathrm{d}t}\left(\frac{L_e J^2}{2c^2}\right) + \frac{cEH}{4\pi}\cdot 2\pi a l = \frac{L_e}{c^2} J \frac{\mathrm{d}J}{\mathrm{d}t} + \frac{1}{2} cEHal,$$

where L_e is the external part of the self-inductance of the wire, E and H the electric and magnetic fields at its surface, a its radius, and l its length. The field H is related to the current J by $H = 2J/ca$. Hence, dividing the above equation by J, we have

$$\mathscr{E} = \frac{1}{c^2} L_e \frac{\mathrm{d}J}{\mathrm{d}t} + El.$$

† Sometimes called the *resistance* and *reactance* (in Russian: *active* and *reactive* resistances).

This is a linear equation, and hence we can use complex quantities. Then

$$\mathscr{E} = ZJ = -\frac{i\omega L_e}{c^2} J + El,$$

whence

$$Z = -\frac{i\omega}{c^2} L_e + \frac{El}{J} = -\frac{i\omega}{c^2} L_e + \frac{2El}{caH}. \tag{61.10}$$

For general frequencies, E and H are given by (60.2) and (60.4), and we have

$$Z = -\frac{i\omega}{c^2} L_e + \frac{1}{2} Rka \frac{J_0(ka)}{J_1(ka)}, \tag{61.11}$$

where $R = l/\pi a^2 \sigma$. When the skin effect is weak, we use the expansions (60.5); taking terms as far as $(a/\delta)^4$ and separating the real part, we find

$$Z' = R\left[1 + \frac{1}{48}\left(\frac{a}{\delta}\right)^4\right] = R\left[1 + \frac{1}{12}\left(\frac{\pi\sigma\omega a^2}{c^2}\right)^2\right]. \tag{61.11a}$$

In the opposite case of a strong skin effect we use the expressions (60.7), obtaining

$$\begin{aligned} Z' &= Ra/2\delta = (l/ca)\sqrt{(\omega/2\pi\sigma)}, \\ Z'' &= -\frac{\omega}{c^2}\left[L_e + \frac{2\delta}{a} L_i\right] = -\frac{\omega}{c^2}\left[L_e + \frac{lc}{a\sqrt{(2\pi\sigma\omega)}}\right]. \end{aligned} \tag{61.12}$$

It is seen from (61.11a) that we can put $Z' = R$ if $(\pi\sigma\omega a^2/c^2)^2 \ll 12$. We also have $Z''|Z' = \omega L/c^2 R = (\pi\sigma\omega a^2/c^2)/2\log(l/a)$, where L is given by (34.1). Comparing this with the inequality just given, we see that the range of frequencies in which the expression (61.5) can be used to take the self-inductance into account depends on the ratio l/a and is fairly narrow.

In practice, however, the most important case is that in which the self-inductance of the circuit is due mainly to coils in it, whose self-inductance is large compared with that of an uncoiled wire (see §34). In such circuits formula (61.5) (i.e. equation (61.4) with constant R and L) can be used over a fairly wide range of frequencies.

Let us consider a circuit in a variable external magnetic field $\mathbf{H}_e$, which may be generated in any manner. We denote by $\mathbf{E}_e$ the electric field which would be induced by the variable field $\mathbf{H}_e$ in the absence of conductors. Both $\mathbf{H}_e$ and $\mathbf{E}_e$ vary only very slightly over the thickness of a thin wire (unlike the field of the currents in the wire). We can therefore discuss the circulation of $\mathbf{E}_e$ round the current circuit without specifying the exact position of the contour of integration in the wire. This circulation is just the e.m.f. $\mathscr{E}$ induced in the circuit by the variable external magnetic field. By the integral form of Maxwell's equation we have

$$\mathscr{E} = \oint \mathbf{E}_e \cdot d\mathbf{l} = -\frac{1}{c}\frac{\partial}{\partial t}\int \mathbf{H}_e \cdot d\mathbf{f} = -\frac{1}{c}\frac{d\Phi_e}{dt}, \tag{61.13}$$

where Φ_e is the flux of the external field through the circuit. Substituting this expression in equation (61.4), we obtain

$$RJ + \frac{1}{c^2} L \frac{dJ}{dt} = -\frac{1}{c}\frac{d\Phi_e}{dt}.$$

Taking the self-inductance term to the right-hand side, we have

$$RJ = -\frac{1}{c}\frac{d\Phi_e}{dt} - \frac{L}{c^2}\frac{dJ}{dt} = -\frac{1}{c}\frac{d\Phi}{dt}, \tag{61.14}$$

where $\Phi = \Phi_e + LJ/c$ is the total magnetic flux from the external magnetic field and the field of the current. In this form the equation gives Ohm's law for the whole circuit, i.e. the equality of RJ to the total e.m.f. in the circuit.

The formulation of equation (61.14) as expressing Ohm's law makes possible a generalization of it to the case where the shape of the circuit also varies with time. The self-inductance L is then a function of time, and (61.14) becomes

$$RJ = -\frac{1}{c^2}\frac{d}{dt}(LJ) - \frac{1}{c}\frac{d\Phi_e}{dt}. \tag{61.15}$$

In deriving this from the law of conservation of energy we should have to take into account also the work done in deforming the conductor.

If there are several circuits in proximity, carrying currents J_a, then for each of them Φ_e in equation (61.14) is the sum of the magnetic fluxes due to all the other circuits (and to the external field, if any). The magnetic flux through the ath circuit due to the current J_b is $L_{ab} J_b/c$, where L_{ab} is the mutual inductance of the two circuits. We therefore have the following set of equations for the variable currents in the circuits:

$$R_a J_a + \frac{1}{c^2}\sum_b L_{ab}\frac{dJ_b}{dt} = \mathscr{E}_a. \tag{61.16}$$

The sum over b includes the self-inductance term ($b = a$), and $\mathscr{E}_a$ is the e.m.f. produced in the ath circuit by sources external to the system of currents considered.

For monochromatic periodic currents, the system of differential equations (61.16) becomes a set of algebraic equations:

$$\sum_b Z_{ab} J_b = \mathscr{E}_a, \tag{61.17}$$

where the quantities

$$Z_{ab} = \delta_{ab} R_a - \frac{i\omega}{c^2} L_{ab} \tag{61.18}$$

form the *impedance matrix*. Like (61.5), the expressions (61.18) represent the first terms in an expansion of the functions $Z_{ab}(\omega)$ in powers of the frequency.

It should be noted that, in this approximation, the circuits have no mutual effect on the real parts of their impedances. Such an effect arises because the magnetic field of the variable current in one conductor generates eddy currents, and therefore an additional dissipation of energy, in another conductor. For linear conductors this effect is negligible, but it may become important if extended conductors are located near them.

Finally, let us consider how the equations of variable currents in linear circuits obtained in this section are related to the general equations of a variable magnetic field in arbitrary conductors. We shall take the simple example of the current set up in a circuit when a constant e.m.f. $\mathscr{E}_0$ is removed at time $t = 0$. From equation (61.4) we have†

$$\left.\begin{aligned} J &= \mathscr{E}_0/R \text{ for } t < 0, \\ J &= (\mathscr{E}_0/R)e^{-c^2Rt/L} \text{ for } t > 0. \end{aligned}\right\} \tag{61.19}$$

We see that, after the removal of the e.m.f., the current decays exponentially with time, the decrement being

$$\gamma = c^2R/L. \tag{61.20}$$

If the problem is exactly formulated, this γ is the smallest of the γ_m obtained by solving the exact equation (58.10) for the conductor in question. Among the γ_m for a linear conductor there is one, the smallest, which is less than the others by a factor of the order of $\log(l/a)$, and this is (61.20).

§62. Capacitance in a quasi-steady current circuit

A variable current, unlike a steady one, can flow in an open circuit as well as in a closed one. Let us consider a linear circuit whose ends are connected to the plates of a capacitor, which are at a small distance apart. When a variable current flows in the circuit, the capacitor plates will be periodically charged and discharged, thereby acting as sources and sinks of current in the open circuit.

Since the distance between the capacitor plates is small, the magnetic energy of the current can again be taken as $LJ^2/2c^2$, where L is the self-inductance of the closed circuit which would be obtained by joining the capacitor plates by a short piece of wire.‡ The only change in equation (61.4) is then to add to the voltage drop RJ across the resistance the potential difference e/C between the capacitor plates, where C is the capacitance and $\pm e(t)$ the charges on the plates. We obtain

$$\mathscr{E} = RJ + \frac{1}{c^2}L\frac{\mathrm{d}J}{\mathrm{d}t} + \frac{e}{C}.$$

The current J is equal to the rate of decrease and increase of the charges on the two plates: $J = \mathrm{d}e/\mathrm{d}t$. Expressing J in terms of e, we have

$$\frac{1}{c^2}L\frac{\mathrm{d}^2e}{\mathrm{d}t^2} + R\frac{\mathrm{d}e}{\mathrm{d}t} + \frac{e}{C} = \mathscr{E}. \tag{62.1}$$

This is the required equation for a variable current in a circuit with a capacitance.

If $\mathscr{E}$ is a periodic function of time having frequency ω, then equation (62.1) reduces to an algebraic relation between $\mathscr{E}$ and the charge e, or between $\mathscr{E}$ and the current $J = -i\omega e$. We have, in fact, $JZ = \mathscr{E}$, where the impedance Z is defined by

$$Z = R - i\left(\frac{\omega L}{c^2} - \frac{1}{\omega C}\right). \tag{62.2}$$

† Strictly speaking, these formulae are invalid for very small t, when the high-frequency terms in the Fourier expansions of the functions are important and so equation (61.4) cannot be used. During this short interval of time, however, the current J cannot change significantly, and so formula (61.19) gives the current at subsequent times with sufficient accuracy.

‡ In the present section we neglect the skin effect.

Taking real parts in the relation $J = \mathscr{E}/Z$, we obtain

$$\left.\begin{aligned} J(t) &= \frac{\mathscr{E}_0 \cos(\omega t - \phi)}{\sqrt{\left[R^2 + \left(\dfrac{\omega L}{c^2} - \dfrac{1}{\omega C}\right)^2\right]}}, \\ \tan\phi &= \left(\frac{\omega L}{c^2} - \frac{1}{\omega C}\right)\frac{1}{R}, \end{aligned}\right\} \tag{62.3}$$

which give the current in a circuit to which an external e.m.f. $\mathscr{E} = \mathscr{E}_0 \cos\omega t$ is applied.

If $\mathscr{E} = 0$, the current in the circuit consists of free electric oscillations. The (complex) frequency of these oscillations is given by $Z = 0$, whence

$$\omega = -i\frac{Rc^2}{2L} \pm \sqrt{\left[\frac{c^2}{LC} - \left(\frac{Rc^2}{2L}\right)^2\right]}. \tag{62.4}$$

We may have either periodic oscillations damped with decrement $Rc^2/2L$ or an aperiodically damped discharge, depending on the sign of the radicand. In the limit as $R \to 0$ we have undamped oscillations whose frequency is given by *Thomson's formula*:

$$\omega = c/\sqrt{(LC)} \tag{62.5}$$

(W. Thomson, 1853).

Equation (62.1) can be immediately generalized to a system of several inductively coupled circuits containing capacitors. The current J_a in the ath circuit is related to the charges $\pm e_a$ on the corresponding capacitor by $J_a = \mathrm{d}e_a/\mathrm{d}t$, and equation (62.1) is replaced by the set of equations

$$\sum_b \frac{1}{c^2} L_{ab} \frac{\mathrm{d}^2 e_b}{\mathrm{d}t^2} + R_a \frac{\mathrm{d}e_a}{\mathrm{d}t} + \frac{e_a}{C_a} = \mathscr{E}_a. \tag{62.6}$$

For periodic (monochromatic) currents, these give the algebraic equations

$$\sum_b Z_{ab} J_b = \mathscr{E}_a, \tag{62.7}$$

the matrix elements Z_{ab} being given by the formulae

$$Z_{ab} = \delta_{ab}\left(R_a + \frac{i}{\omega C_a}\right) - \frac{i\omega}{c^2} L_{ab}. \tag{62.8}$$

The eigenfrequencies of the current system are given by the condition of compatibility of equations (62.7) when $\mathscr{E}_a = 0$, i.e. by the condition

$$\det |Z_{ab}| = 0. \tag{62.9}$$

If the resistances R are not zero, all the "frequencies" have a non-zero imaginary part, and the electric oscillations are therefore damped.

It should be noticed that equations (62.6) are formally identical with the mechanical equations of motion of a system with several degrees of freedom which executes small damped oscillations. The generalized coordinates are represented by the charges e_a, and the generalized velocities by the currents $J_a = de_a/dt$. The Lagrangian of the system is

$$\mathscr{L} = \sum_{a,b} \frac{1}{2c^2} L_{ab} \dot{e}_a \dot{e}_b - \sum_a \frac{e_a^{\,2}}{2C_a} + \sum_a e_a \mathscr{E}_a. \tag{62.10}$$

The kinetic and potential energies of the mechanical system are represented by the magnetic and electric energies of the current system, and the quantities $\mathscr{E}_a$ correspond to the externally applied forces which cause the forced oscillations of the system. The quantities R_a appear in the *dissipative function*

$$\mathscr{R} = \sum_a \tfrac{1}{2} R_a \dot{e}_a^{\,2}. \tag{62.11}$$

Equations (62.6) are the analogues of Lagrange's equations

$$\frac{d}{dt}\frac{\partial \mathscr{L}}{\partial \dot{e}_a} - \frac{\partial \mathscr{L}}{\partial e_a} = -\frac{\partial \mathscr{R}}{\partial \dot{e}_a}. \tag{62.12}$$

PROBLEMS

PROBLEM 1. Determine the eigenfrequencies of electric oscillations in two inductively coupled circuits containing self-inductances L_1 and L_2 and capacitances C_1 and C_2, neglecting the resistances R_1 and R_2.

SOLUTION. The required frequencies are determined from the condition

$$\det|Z_{ab}| = Z_{11}Z_{22} - Z_{12}^{\,2} = 0,$$

where

$$Z_{11} = -i\left(\frac{\omega}{c^2}L_1 - \frac{1}{\omega C_1}\right), \quad Z_{22} = -i\left(\frac{\omega}{c^2}L_2 - \frac{1}{\omega C_2}\right), \quad Z_{12} = -\frac{i\omega}{c^2}L_{12}.$$

Calculation gives

$$\omega_{1,2}^{\,2} = c^2 \frac{L_1C_1 + L_2C_2 \mp \sqrt{[(L_1C_1 - L_2C_2)^2 + 4C_1C_2L_{12}^{\,2}]}}{2C_1C_2(L_1L_2 - L_{12}^{\,2})}.$$

Both frequencies are purely real, owing to the fact that R_1 and R_2 have been neglected. As $L_{12} \to 0$, ω_1 and ω_2 tend to $c/\sqrt{(L_1C_1)}$ and $c/\sqrt{(L_2C_2)}$. These are the frequencies for the two circuits separately.

PROBLEM 2. The same as Problem 1, but for a circuit consisting of a resistance R, a capacitance C and an inductance L connected in parallel.

SOLUTION. The impedances of the three branches are $Z_1 = R$, $Z_2 = i/\omega C$, $Z_3 = -i\omega L/c^2$, and the currents in them are such that $J_1 + J_2 + J_3 = 0$, $Z_1J_1 = Z_2J_2 = Z_3J_3$. Hence we have $1/Z_1 + 1/Z_2 + 1/Z_3 = 0$, whence

$$\omega = -\frac{i}{2RC} \pm \sqrt{\left[\frac{c^2}{LC} - \frac{1}{4R^2C^2}\right]}.$$

PROBLEM 3. Discuss the propagation of electric oscillations in a circuit consisting of an infinite succession of identical meshes containing impedances

$$Z_1 = -i\left(\frac{\omega}{c^2}L_1 - \frac{1}{\omega C_1}\right), \qquad Z_2 = -i\left(\frac{\omega}{c^2}L_2 - \frac{1}{\omega C_2}\right),$$

as shown in Fig. 37. Find the range of frequencies which can be propagated in the circuit without damping.†

† The condition for the quasi-steady theory to be applicable to such a periodic circuit is that the dimension of one mesh should be small compared with the "wavelength" c/ω.

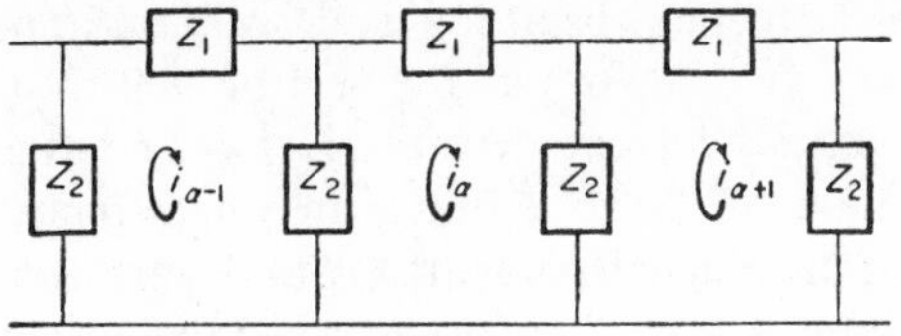

FIG. 37

SOLUTION. The current in mesh α is denoted by i_α, as shown in Fig. 37. Kirchhoff's second law gives for this mesh $Z_1 i_\alpha + Z_2(2i_\alpha - i_{\alpha-1} - i_{\alpha+1}) = 0$. This is a linear difference equation in the integral variable α, with constant coefficients. We seek the solution in the form $i_\alpha = \text{constant} \times q^\alpha$, obtaining for the parameter q the equation

$$q^2 - \left(2 + \frac{Z_1}{Z_2}\right) q + 1 = 0. \tag{1}$$

Let $-4 \leqslant Z_1/Z_2 \leqslant 0$, corresponding to values of ω^2 lying between c^2/L_1C_1 and $c^2(4/C_2 + 1/C_1)/(4L_2 + L_1)$. Then equation (1) has two complex conjugate roots with moduli $|q| = 1$. This means that the current does not decrease from one mesh to the next, i.e. the electric oscillations are propagated in the circuit without damping. Putting $q = e^{ikl}$, where l is the length of one mesh and k is the "wave number" of the oscillations propagated in the circuit, we can calculate the velocity of propagation u from the general result $u = d\omega/dk$.

If, however, ω is outside the range mentioned, equation (1) has two real roots q_1 and q_2, say; since $q_1 q_2 = 1$, one root (q_1, say) is less than 1 in absolute magnitude, while q_2 is greater. It is easy to see that the propagation of undamped oscillations in the circuit is then impossible. To elucidate the reason for this, let us consider a circuit of large but finite length. An initial oscillatory impulse is given to one end of the circuit, the other end being closed in some manner. This closure corresponds mathematically to a certain boundary condition, by means of which we can determine the ratio of the coefficients c_1 and c_2 in the general solution $i_\alpha = c_1 q_1^{-(A-\alpha)} + c_2 q_2^{-(A-\alpha)}$, where A is the "coordinate" of the end of the circuit. This ratio is of the order of unity. As $A - \alpha$ increases, the second term in the solution rapidly becomes very small compared with the first term, because $|q_2| > 1$. Thus the solution is $i_\alpha = c_1 q_1^{-(A-\alpha)}$ everywhere except for a small part near the end of the circuit, and $|i_\alpha|$ decreases towards the end of the circuit.

It should be emphasized that this damping does not involve dissipative absorption, because there is no resistance in the circuit; it can be imagined as being the result of reflection of the oscillatory impulse from each successive mesh of the circuit.

§63. Motion of a conductor in a magnetic field

Hitherto we have tacitly assumed that a conductor in an electromagnetic field is at rest in the frame of reference K in which $\mathbf{E}$, $\mathbf{H}$, etc. are defined. In particular, the relation $\mathbf{j} = \sigma\mathbf{E}$ between the current and the field is generally valid only for conductors at rest.

To determine the corresponding relation in a moving conductor, we change from the frame K to another frame K' in which the conductor, or some part of it, is at rest at the instant considered. In this frame we have $\mathbf{j} = \sigma\mathbf{E}'$, where $\mathbf{E}'$ is the electric field in K'. The well-known formula for the transformation of fields† gives $\mathbf{E}'$ in terms of the field in K:

$$\mathbf{E}' = \mathbf{E} + \mathbf{v} \times \mathbf{B}/c, \tag{63.1}$$

where $\mathbf{v}$ is the velocity of K' relative to K, i.e. in this case the velocity of the conductor, which we of course suppose small compared with the velocity of light. Thus we find

$$\mathbf{j} = \sigma(\mathbf{E} + \mathbf{v} \times \mathbf{B}/c). \tag{63.2}$$

This gives the relation between the current and the field in moving conductors. The

† See *Fields*, §24. The microscopic values of the electric and magnetic fields are replaced by their averaged values $\bar{\mathbf{e}} = \mathbf{E}$, $\bar{\mathbf{h}} = \mathbf{B}$.

following remark should be made concerning its derivation. In going from one frame of reference to the other we have transformed the field but left the current **j** unaltered. The correct transformation of the current density gives only terms of a higher order of smallness if $v \ll c$. In formula (63.2) the second term, which appears as a result of the field transformation, is in general not small compared with the first term, despite the factor v/c. For example, if the electric field is due to electromagnetic induction from a variable magnetic field, its order of magnitude contains a factor $1/c$ as compared with the magnetic field.

The energy dissipation in a conductor when a given current flows in it cannot, of course, depend on the motion of the conductor. The rate of evolution of Joule heat per unit volume in a moving conductor is therefore given in terms of the current density by the same expression j^2/σ as for a conductor at rest. The expression $\mathbf{j}\cdot\mathbf{E}$, however, is replaced by† $j^2/\sigma = \mathbf{j}\cdot(\mathbf{E}+\mathbf{v}\times\mathbf{B}/c)$.

Thus, in a moving conductor, the sum $\mathbf{E}+\mathbf{v}\times\mathbf{B}/c$ acts as an "effective" electric field producing the conduction current. Hence the e.m.f. acting in a closed linear circuit C is given by the integral

$$\mathscr{E} = \oint_C (\mathbf{E}+\mathbf{v}\times\mathbf{B}/c)\cdot d\mathbf{l}. \tag{63.3}$$

This expression can be transformed as follows. According to Maxwell's equation, $\mathbf{curl}\,\mathbf{E} = -(1/c)\partial\mathbf{B}/\partial t$, and so

$$\oint_C \mathbf{E}\cdot d\mathbf{l} = \int_S \mathbf{curl}\,\mathbf{E}\cdot d\mathbf{f} = -\frac{1}{c}\frac{\partial}{\partial t}\int_S \mathbf{B}\cdot d\mathbf{f}$$

or, denoting by Φ the magnetic flux through the surface S, which spans the circuit C,

$$\oint_C \mathbf{E}\cdot d\mathbf{l} = -\frac{1}{c}\left(\frac{\partial\Phi}{\partial t}\right)_{\mathbf{v}=0}.$$

The time derivative with the suffix $\mathbf{v} = 0$ denotes the rate of change of the magnetic flux due to the time variation of the magnetic field, the position of the contour C remaining unchanged.

In the second term in (63.3), we put $\mathbf{v} = d\mathbf{u}/dt$, where $d\mathbf{u}$ is an infinitesimal displacement of the circuit element dl. Then

$$\oint_C \mathbf{v}\times\mathbf{B}\cdot d\mathbf{l} = \oint_C d\mathbf{u}\times\mathbf{B}\cdot d\mathbf{l}/dt = -\oint \mathbf{B}\cdot d\mathbf{f}/dt,$$

where $d\mathbf{f} = d\mathbf{u}\times d\mathbf{l}$ is an element of area on the "side" surface between two infinitely close positions C and C' of the current circuit, which it occupies at times t and $t+dt$ (Fig. 38). Since the total magnetic flux through any closed surface is zero, the flux through the side

† It is seen from this formula that the additional heat evolved in time δt in a conductor moving in a magnetic field is

$$\delta t\int \mathbf{j}\cdot\mathbf{v}\times\mathbf{B}\,dV/c = -\int \delta\mathbf{u}\cdot\mathbf{j}\times\mathbf{B}\,dV/c,$$

where $\delta\mathbf{u} = \mathbf{v}\delta t$ is the displacement in time δt. This expression is equal and opposite to the work done on the conductor in time δt by the volume forces $\mathbf{f} = \mathbf{j}\times\mathbf{B}/c$. This explains the apparent contradiction mentioned in §35.

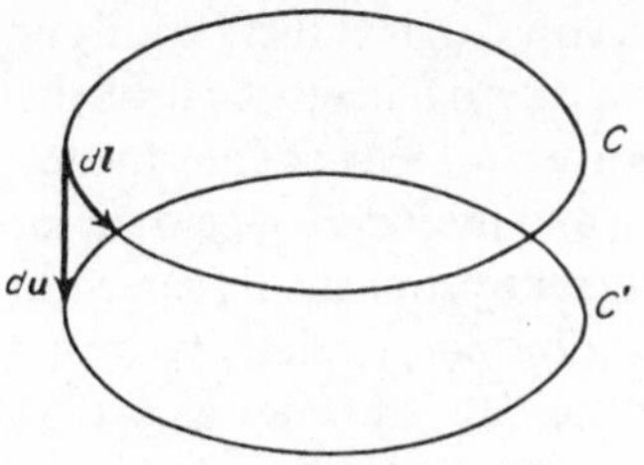

FIG. 38

surface must evidently equal the difference of the fluxes through surfaces spanning C and C'. Thus

$$\oint_C \mathbf{v}\times\mathbf{B}\cdot d\mathbf{l} = -(\partial\Phi/\partial t)_{\mathbf{B}=\text{constant}},$$

where the time derivative denotes the rate of change of the magnetic flux due to the motion of the conductor in a constant field.

Adding the two terms, we have finally

$$\mathscr{E} = -(1/c)\,d\Phi/dt, \tag{63.4}$$

where the time derivative now denotes the total rate of change of the magnetic flux through the moving circuit. Thus the expression (63.4), which is *Faraday's law*, is valid whatever the reason for the change in the magnetic flux, whether variation of the field itself (already discussed in §61, formula (61.13)) or motion of the conductor.

In a static magnetic field, the change in the flux may be due entirely to the motion of the circuit. If the circuit moves in such a way that every point of it moves along a line of force, then the flux through the circuit does not vary. This is an obvious result of the fact that the magnetic flux through any closed surface is zero, and the flux through the side surface described by the moving circuit is in this case identically zero (since $B_n = 0$ on this surface). Thus we can say that, to induce an e.m.f., the conductor must certainly move so as to cross lines of magnetic force.

The electromagnetic field in a moving conductor is given by the equations

$$\left.\begin{aligned}\mathbf{curl}\,\mathbf{E} &= -(1/c)\partial\mathbf{B}/\partial t,\\ \mathbf{curl}\,\mathbf{H} = 4\pi\mathbf{j}/c &= (4\pi\sigma/c)(\mathbf{E}+\mathbf{v}\times\mathbf{B}/c),\\ \operatorname{div}\mathbf{B} &= 0.\end{aligned}\right\} \tag{63.5}$$

Expressing $\mathbf{E}$ in terms of $\mathbf{H}$ by means of the second equation and substituting in the first, we obtain

$$\frac{\partial\mathbf{B}}{\partial t} - \mathbf{curl}\,(\mathbf{v}\times\mathbf{B}) = -\frac{c^2}{4\pi}\mathbf{curl}\left(\frac{\mathbf{curl}\,\mathbf{H}}{\sigma}\right). \tag{63.6}$$

In a homogeneous conductor with constant conductivity σ and constant magnetic permeability μ, we have

$$\frac{\partial\mathbf{H}}{\partial t} - \mathbf{curl}\,(\mathbf{v}\times\mathbf{H}) = \frac{c^2}{4\pi\sigma\mu}\triangle\mathbf{H}, \qquad \operatorname{div}\mathbf{H} = 0. \tag{63.7}$$

These equations generalize those obtained in §58.

It should be pointed out, however, that, if there is only one conductor moving as a whole (without change of shape) in an external magnetic field, then the solution of the problem is considerably simplified if we use a system of coordinates fixed in the conductor. In this system the conductor is at rest, and the external field varies with time in a given manner, so that we return to the eddy-current problems discussed in §59. This possibility does not depend on Galileo's (or on Einstein's) relativity principle, since the new system of coordinates is in general not inertial. The equivalence of the problems results from the above-mentioned fact that the electromagnetic induction is independent of the cause of the change in the magnetic flux. This equivalence can also be demonstrated mathematically. To do so, we expand the expression **curl** $(\mathbf{v}\times\mathbf{B})$, using the facts that div $\mathbf{B} = 0$ and (for motion of the body as a whole) div $\mathbf{v} = 0$ (i.e. the body is incompressible). Then the left-hand side of equation (63.6) becomes

$$\partial\mathbf{B}/\partial t + (\mathbf{v}\cdot\mathbf{grad})\mathbf{B} - (\mathbf{B}\cdot\mathbf{grad})\mathbf{v}. \tag{63.8}$$

This sum is just the time derivative of **B** with respect to axes fixed in a rotating body. For the sum of the first two terms is the "substantial" time derivative (derivative following the motion) $d\mathbf{B}/dt$, which gives the rate of change of **B** at a point moving with velocity **v**. The third term takes into account the change in the direction of **B** relative to the body; it is zero for pure translation ($\mathbf{v}$ = constant) and equals $-\boldsymbol{\Omega}\times\mathbf{B}$ for rotation ($\mathbf{v} = \boldsymbol{\Omega}\times\mathbf{r}$, where $\boldsymbol{\Omega}$ is the angular velocity).

To conclude this section, let us consider the phenomenon of *unipolar induction*, which occurs when a magnetized conductor rotates. If a stationary wire is connected to the rotating magnet by means of two sliding contacts *A* and *B* (Fig. 39) then a current flows in the wire. It is not difficult to calculate the e.m.f. which produces the current; the simplest procedure is to use a system of coordinates rotating with the magnet. If $\boldsymbol{\Omega}$ is the angular velocity of rotation of the magnet, then in the new system the wire rotates with angular velocity $-\boldsymbol{\Omega}$, while the magnet is at rest. Thus we have a conductor moving in a given static magnetic field **B** due to a fixed magnet. We neglect the distortion of the field by the wire itself. According to formula (63.3), the e.m.f. between the ends of the wire is

$$\mathscr{E} = \frac{1}{c}\int\limits_{ACB} \mathbf{v}\times\mathbf{B}\cdot d\mathbf{l} = \frac{1}{c}\int\limits_{ACB} \mathbf{B}\times(\mathbf{r}\times\boldsymbol{\Omega})\cdot d\mathbf{l}, \tag{63.9}$$

taken along the wire. This is the required solution.

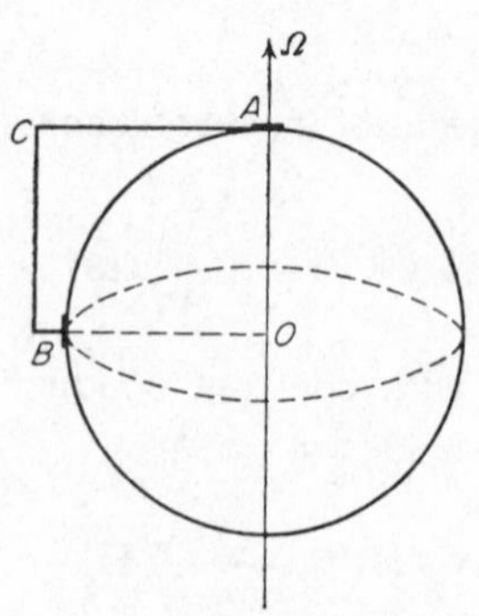

FIG. 39

PROBLEMS

PROBLEM 1. Determine the magnetic moment of a conducting sphere (with $\mu = 1$) rotating uniformly in a uniform static magnetic field, and the torque on the sphere.

SOLUTION. Let the external field have components $\mathfrak{H}_x$, 0, $\mathfrak{H}_z$ in a fixed system of coordinates with the z-axis in the direction of the angular velocity vector $\mathbf{\Omega}$. In a coordinate system ξ, η, z which rotates with the sphere, the field components are $\mathfrak{H}_\xi = \mathfrak{H}_x \cos \Omega t$, $\mathfrak{H}_\eta = -\mathfrak{H}_x \sin \Omega t$, $\mathfrak{H}_z$, or, in complex form, $\mathfrak{H}_\xi = \mathfrak{H}_x e^{-i\Omega t}$, $\mathfrak{H}_\eta = -i\mathfrak{H}_x e^{-i\Omega t}$, $\mathfrak{H}_z$.

Thus variable fields with frequency Ω act along the ξ and η axes, and the magnetic moment which they induce is

$$\mathscr{M}_\xi = V \operatorname{re}(\alpha \mathfrak{H}_\xi) = V\mathfrak{H}_x(\alpha' \cos \Omega t + \alpha'' \sin \Omega t),$$
$$\mathscr{M}_\eta = V \operatorname{re}(\alpha \mathfrak{H}_\eta) = V\mathfrak{H}_x(-\alpha' \sin \Omega t + \alpha'' \cos \Omega t),$$

where $V\alpha$ is the complex magnetic polarizability of the sphere, which has been determined in §59, Problem 1. Along the z-axis, on the other hand, the magnetic field is constant, and therefore causes no magnetic moment (if $\mu = 1$). The components of the magnetic moment in the fixed system of coordinates are $\mathscr{M}_x = V\alpha' \mathfrak{H}_x$, $\mathscr{M}_y = V\alpha'' \mathfrak{H}_x$, $\mathscr{M}_z = 0$. Thus in this problem α' and α'' give the components of the magnetic moment of the sphere respectively parallel and perpendicular to the plane of the vectors $\mathbf{\Omega}$ and $\mathfrak{H}$.

The torque on the sphere is $\mathbf{K} = \mathscr{M} \times \mathfrak{H}$. Its components relative to the fixed axes are

$$K_x = V\alpha'' \mathfrak{H}_x \mathfrak{H}_z, \qquad K_y = -V\alpha' \mathfrak{H}_x \mathfrak{H}_z, \qquad K_z = -V\alpha'' \mathfrak{H}_x^2.$$

The above reduction of the problem of a sphere rotating in a field to that of a sphere at rest in a variable field is perfectly natural, in view of the comment at the end of the solution in §59, Problem 1. An interesting feature of the analogy is that, as the frequency ω of the variable magnetic field increases, the field is squeezed out of the sphere: in the limit $\omega \to \infty$, all the lines of force pass round the sphere, not through it. In a similar way, a magnetic field perpendicular to the axis of rotation is squeezed out of a rapidly rotating sphere.

PROBLEM 2. Determine the e.m.f. due to unipolar induction between the pole and the equator (Fig. 39) of a uniformly magnetized sphere rotating uniformly about the direction of magnetization.

SOLUTION. When the sphere rotates about its direction of magnetization, it generates a static field, and, since no currents flow within the sphere, we find from (63.6) that $\mathbf{curl}\,(\mathbf{v} \times \mathbf{B}) = 0$. Hence the integral of $\mathbf{v} \times \mathbf{B}$ along the closed contour $OACBO$ (Fig. 39) is zero, and so the integration along ACB in formula (63.9) may be replaced by one along the path AOB, which lies inside the sphere. The integral along the segment AO of the axis of rotation is zero, since $\mathbf{\Omega}$ and $\mathbf{r}$ are parallel; the integral along the radius OB gives, since $\mathbf{B}$ and $\mathbf{\Omega}$ are parallel within the sphere,

$$\mathscr{E} = \frac{1}{c}\int_0^a B_0 \Omega r dr = B_0 \Omega a^2/2c,$$

where a is the radius of the sphere and B_0 the magnetic induction in it. In a uniformly magnetized sphere (in the absence of an external field) the induction is related to the magnetization by $B_0 + 2H = 0$ (cf. (8.1)) and $B_0 - H = 4\pi M$, whence $B_0 = 8\pi M/3$. In terms of the total magnetic moment $\mathscr{M}$ of the sphere we have finally $\mathscr{E} = \Omega\mathscr{M}/ca$.

PROBLEM 3. Determine the total charge which flows along a closed linear circuit when the magnetic flux through the circuit changes for any reason from one constant value (Φ_1) to another (Φ_2).

SOLUTION. The required total charge is the integral

$$\int_{-\infty}^{\infty} J\,\mathrm{d}t,$$

where $J(t)$ is the induction current in the circuit. Mathematically, this integral is the Fourier component of the function $J(t)$ that has the frequency $\omega = 0$. It is therefore related to the corresponding component of the e.m.f. by

$$\int_{-\infty}^{\infty} \mathscr{E}\,\mathrm{d}t = Z(0)\int_{-\infty}^{\infty} J\,\mathrm{d}t;$$

see (61.3). Putting $Z(0) = R$, where R is the resistance of the circuit to a steady current and $\mathscr{E} = -(1/c)\mathrm{d}\Phi/\mathrm{d}t$, we have

$$\int_{-\infty}^{\infty} J\,\mathrm{d}t = \frac{1}{cR}(\Phi_1 - \Phi_2).$$

§64. Excitation of currents by acceleration

In discussing the motion of a conductor in §63 we have neglected possible effects of the acceleration, if any. The accelerated motion of a metal, however, is equivalent to the action of additional inertia forces on the conduction electrons. If $\dot{\mathbf{v}}$ is the acceleration of the conductor and m the mass of the electron, then the force on an electron is $-m\dot{\mathbf{v}}$. It affects the electron in the same way as an electric field $m\dot{\mathbf{v}}/e$, where $-e$ is the charge on the electron. Thus the effective electric field on the conduction electrons in an accelerated metal is

$$\mathbf{E}' = \mathbf{E} + m\dot{\mathbf{v}}/e. \tag{64.1}$$

The current density is accordingly

$$\mathbf{j} = \sigma\mathbf{E}' = \sigma(\mathbf{E} + m\dot{\mathbf{v}}/e). \tag{64.2}$$

Expressing $\mathbf{E}$ in terms of $\mathbf{E}'$ from (64.1), we substitute in the equation

$$\mathbf{curl\,E} = -(1/c)\partial\mathbf{H}/\partial t$$

(as usual, we put $\mu = 1$). Then

$$\mathbf{curl\,E}' = -\frac{1}{c}\frac{\partial\mathbf{H}}{\partial t} + \frac{m}{e}\,\mathbf{curl}\,\dot{\mathbf{v}}. \tag{64.3}$$

We write $\mathbf{v}$ as a sum $\mathbf{v} = \mathbf{u} + \boldsymbol{\Omega}\times\mathbf{r}$, where $\mathbf{u}$ is the translational velocity and $\boldsymbol{\Omega}$ the angular velocity of rotation of the body. Differentiating with respect to time, we find the acceleration to be $\dot{\mathbf{v}} = \dot{\mathbf{u}} + \boldsymbol{\Omega}\times\mathbf{v} + \dot{\boldsymbol{\Omega}}\times\mathbf{r} = \dot{\mathbf{u}} + \boldsymbol{\Omega}\times\mathbf{u} + \boldsymbol{\Omega}\times(\boldsymbol{\Omega}\times\mathbf{r}) + \dot{\boldsymbol{\Omega}}\times\mathbf{r}$. The first two terms are independent of $\mathbf{r}$, and therefore give zero on differentiation with respect to the coordinates. The third term can be written as $\boldsymbol{\Omega}\times(\boldsymbol{\Omega}\times\mathbf{r}) = -\frac{1}{2}\,\mathbf{grad}\,(\boldsymbol{\Omega}\times\mathbf{r})^2$, and its curl is therefore zero. Finally, $\mathbf{curl}\,(\dot{\boldsymbol{\Omega}}\times\mathbf{r}) = 2\dot{\boldsymbol{\Omega}}$. Thus, substituting for $\dot{\mathbf{v}}$ in equation (64.3), we have $\mathbf{curl\,E}' = -(1/c)\partial\mathbf{H}/\partial t + 2m\dot{\boldsymbol{\Omega}}/e$ or

$$\mathbf{curl\,E}' = -\frac{1}{c}\frac{\partial\mathbf{H}'}{\partial t}, \tag{64.4}$$

where

$$\mathbf{H}' = \mathbf{H} - 2mc\boldsymbol{\Omega}/e. \tag{64.5}$$

Since $\boldsymbol{\Omega}$ is independent of the coordinates, the equation $\mathbf{curl\,H} = 4\pi\mathbf{j}/c$ is still valid if $\mathbf{H}$ is replaced by $\mathbf{H}'$:

$$\mathbf{curl\,H}' = 4\pi\sigma\mathbf{E}'/c. \tag{64.6}$$

Eliminating $\mathbf{E}'$ from equations (64.4) and (64.6), we obtain for $\mathbf{H}'$ the equation

$$\triangle\mathbf{H}' = (4\pi\sigma/c^2)\partial\mathbf{H}'/\partial t, \tag{64.7}$$

which is the same as the equation for $\mathbf{H}$ in a conductor at rest.

Outside the body, the field satisfies the equation $\triangle \mathbf{H} = 0$ (the wavelength being supposed large compared with the dimension of the body), and $\mathbf{H}'$ satisfies the same equation.

Finally, on the surface of the conductor $\mathbf{H}'$, like $\mathbf{H}$, is continuous. The only difference is in the condition at infinity, where $\mathbf{H}$ tends to zero but $\mathbf{H}'$ tends to the limit $-2mc\mathbf{\Omega}/e$.

Thus the problem of determining the variable magnetic field $\mathbf{H}$ near a non-uniformly rotating body is equivalent to that of determining the field $\mathbf{H}'$ near a body at rest in a uniform external magnetic field

$$\mathfrak{H} = -2mc\mathbf{\Omega}/e. \tag{64.8}$$

The required field $\mathbf{H}^{(e)}$ outside the conductor is obtained by subtracting $\mathfrak{H}$ from the solution $\mathbf{H}'$ of this latter problem.

The magnetic field thus produced, like any variable field, induces electric currents in the conductor itself. In a simply-connected body, these currents appear in the form of a magnetic moment. In a non-uniformly rotating ring, the effect appears as an e.m.f.—the *Stewart–Tolman effect*.

Misunderstanding may arise from the appearance of the angular velocity $\mathbf{\Omega}$ itself, and not its time derivative, in formula (64.8). We may therefore emphasize that the above discussion, and therefore the significance here attached to the quantity (64.8), pertain only to non-uniform rotation. When $\mathbf{\Omega}$ is constant, equation (64.7) with the appropriate boundary condition at infinity is identically satisfied by $\mathbf{H}' = \mathfrak{H}$, and the definition (64.5) then gives $\mathbf{H} = 0$. The magnetic field which arises from the gyromagnetic effect (§36) in uniform rotation is a small quantity which is here neglected.

The derivation has also ignored the deformation of the body which results from non-uniform rotation. It can be seen that including this deformation would not affect the result if the characteristic time of variation of the angular velocity is (as we assume) much longer than the relaxation time of the conduction electrons in the deformation: the electric current in the conductor is due to the gradient of $\phi + \zeta_0/e$, where ϕ is the field potential and ζ_0 the chemical potential of the conduction electrons (see §26). A non-uniform deformation produces a gradient of ζ_0, but this is compensated by the electric field which results from the thermodynamic equilibrium condition $e\phi + \zeta_0 = \text{constant}$.

PROBLEMS

PROBLEM 1. Determine the magnetic moment of a non-uniformly rotating sphere with radius a. The rate of rotation is assumed so small that the penetration depth $\delta \gg a$.

SOLUTION. The magnetic moment of the sphere in the field $\mathfrak{H}(t)$ (64.8) is $\mathscr{M} = V\hat{\alpha}\mathfrak{H}$, where $\hat{\alpha}$ is an operator whose action on the Fourier components of the function $\mathfrak{H}(t)$ is given by the formulae of §59, Problem 1. For the components with frequencies ω such that $\delta \gg a$ we have $\mathscr{M} = V\alpha(\omega)\mathfrak{H} \cong -4\pi m a^5 \sigma i\omega\mathbf{\Omega}/15ce$. This formula, when written $\mathscr{M} = (4\pi m a^5\sigma/15ce)\,d\mathbf{\Omega}/dt$, does not contain ω explicitly, and is therefore valid also for the functions $\mathscr{M}(t)$ and $\mathbf{\Omega}(t)$ themselves, as well as their individual Fourier components (on the assumption that the Fourier expansion contains chiefly terms whose frequencies satisfy the above condition).

PROBLEM 2. Determine the total charge which flows along a thin circular ring when it ceases a uniform rotation about an axis perpendicular to its plane.

SOLUTION. In the formula obtained in §63, Problem 3, $\mathbf{\Phi}$ must be taken as the flux of the field $\mathfrak{H}$ (64.8). The total charge transferred when the angular velocity changes from $\mathbf{\Omega}$ to zero is

$$\int_{-\infty}^{\infty} J\,\mathrm{d}t = \frac{2mc}{eRc}\Omega\pi b^2 = \frac{m\sigma V}{2\pi e}\Omega,$$

where b is the radius of the ring and V its volume.

PROBLEM 3. Determine the current in a superconducting circular ring which ceases to rotate uniformly.

SOLUTION. From the condition that the total magnetic flux through the ring be constant (see (54.5)), we have

$$J = \frac{2mc^2}{eL}\Omega\pi b^2 = \frac{mc^2 b\Omega}{2e[\log(8b/a)-2]}.$$

See the third footnote to §54 concerning the value of L.

CHAPTER VIII

MAGNETOHYDRODYNAMICS

§65. The equations of motion for a fluid in a magnetic field

If a conducting fluid moves in a magnetic field, electric fields are induced in it and electric currents flow. The magnetic field exerts forces on these currents which may considerably modify the flow. Conversely, the currents themselves modify the magnetic field. Thus we have a complex interaction between the magnetic and the fluid-dynamic phenomena, and the flow must be examined by combining the field equations with those of fluid dynamics.

The applications of magnetohydrodynamics cover a very wide range of physical objects, from liquid metals to cosmic plasmas. We shall not discuss here the specific conditions that exist in various particular objects, but simply mention that for magnetohydrodynamics to be strictly applicable it is of course necessary that the characteristic distances and time intervals for the motion in question should be much greater than the mean free path and mean free time of the current-carriers (electrons or ions). In some cases, however, equations that are formally identical with those of magnetohydrodynamics in an ideal fluid may describe also the motion of a medium with a long mean free path. Such a situation occurs, for example, in a non-equilibrium plasma in which the electron temperature considerably exceeds the ion temperature (cf. *PK*, §38).

The magnetic premeability of the media considered in magnetohydrodynamics differs only slightly from unity, and the difference is unimportant as regards the phenomena under discussion. We shall therefore take $\mu = 1$ throughout the present chapter.†

Let us first set up the equations of magnetohydrodynamics for conditions such that all dissipative processes may be neglected, i.e. for an ideal fluid. This means that no account is taken of viscosity and thermal conduction, or of the finite electrical conductivity σ of the medium, which is regarded as being indefinitely great.

Putting $\sigma \to \infty$ in equations (63.7), we write

$$\operatorname{div} \mathbf{H} = 0, \tag{65.1}$$

$$\partial \mathbf{H}/\partial t = \mathbf{curl}\,(\mathbf{v} \times \mathbf{H}). \tag{65.2}$$

The equations of fluid dynamics are the equation of continuity

$$\partial \rho/\partial t + \operatorname{div}(\rho \mathbf{v}) = 0, \tag{65.3}$$

where ρ is the fluid density, and the Navier–Stokes (Euler) equation

$$\frac{\partial \mathbf{v}}{\partial t} + (\mathbf{v}\cdot\mathbf{grad})\mathbf{v} = -\frac{1}{\rho}\,\mathbf{grad}\,P + \frac{\mathbf{f}}{\rho},$$

† In the literature on magnetohydrodynamics, the magnetic field under these conditions is often denoted by **B** to emphasize that it is the averaged microscopic field, $\overline{\mathbf{h}} = \mathbf{B}$. We shall here use the notation **H**, however, for the sake of uniformity with the other chapters, in which non-magnetic media are discussed.

where **f** is the volume density of the external (in this case, electromagnetic) forces. By formula (35.4) we have $\mathbf{f} = \mathbf{j} \times \mathbf{H}/c = (\mathbf{curl}\,\mathbf{H}) \times \mathbf{H}/4\pi$. Thus the equation of motion of the fluid is

$$\frac{\partial \mathbf{v}}{\partial t} + (\mathbf{v} \cdot \mathbf{grad})\mathbf{v} = -\frac{1}{\rho}\mathbf{grad}\, P - \frac{1}{4\pi\rho}\mathbf{H} \times \mathbf{curl}\,\mathbf{H}. \tag{65.4}$$

To these equations we must add the equation of state

$$P = P(\rho, T), \tag{65.5}$$

which relates the pressure, density and temperature of the fluid, and the equation of conservation of entropy, which expresses the fact that the motion is adiabatic in the absence of dissipation:

$$\mathrm{d}s/\mathrm{d}t \equiv \partial s/\partial t + \mathbf{v} \cdot \mathbf{grad}\, s = 0, \tag{65.6}$$

where s is the entropy per unit mass of the fluid and $\mathrm{d}/\mathrm{d}t \equiv \partial/\partial t + \mathbf{v} \cdot \mathbf{grad}$ denotes the "substantial" derivative giving the rate of change of a quantity at a point moving with the fluid particle. Equations (65.1)–(65.6) form a complete system of equations of magnetohydrodynamics for an ideal fluid.

The Navier–Stokes equation can be written (using the equation of continuity) in a form expressing the law of conservation of momentum:

$$\partial(\rho v_i)/\partial t = -\partial \Pi_{ik}/\partial x_k, \tag{65.7}$$

where Π_{ik} is the momentum flux density tensor (see *FM*, §7). In the absence of external forces, $\Pi_{ik} = \rho v_i v_k + P\delta_{ik}$. Transforming the last term in (65.4) by means of $\mathbf{H} \times \mathbf{curl}\,\mathbf{H} = \frac{1}{2}\mathbf{grad}\,\mathbf{H}^2 - (\mathbf{H} \cdot \mathbf{grad})\mathbf{H}$, and $\operatorname{div} \mathbf{H} \equiv \partial H_k/\partial x_k = 0$, we have in magnetohydrodynamics

$$\Pi_{ik} = \rho v_i v_k + P\delta_{ik} - (H_i H_k - \tfrac{1}{2}H^2\delta_{ik})/4\pi. \tag{65.8}$$

The added term is the Maxwell stress tensor, as it should be.

The equation of conservation of energy in ordinary fluid dynamics is

$$\partial(\tfrac{1}{2}\rho v^2 + \rho\varepsilon)/\partial t = -\operatorname{div}\mathbf{q},$$

where

$$\mathbf{q} = \rho\mathbf{v}(\tfrac{1}{2}v^2 + w);$$

ε and $w = \varepsilon + P/\rho$ are respectively the internal energy and heat function per unit mass of fluid; this necessarily follows from the equation of motion (see *FM*, §6). When a magnetic field is present in the conducting medium, the energy density includes also the magnetic energy $H^2/8\pi$, and the energy flux density includes also the Poynting vector $\mathbf{S} = c\mathbf{E} \times \mathbf{H}/4\pi$. Expressing **E** in the latter in terms of **H** by

$$\mathbf{E} = -\mathbf{v} \times \mathbf{H}/c, \tag{65.9}$$

which follows from (63.2) when $\sigma \to \infty$ and **j** is finite†, we find that the equation of conservation of energy in magnetohydrodynamics is

$$\frac{\partial}{\partial t}\left(\tfrac{1}{2}\rho v^2 + \rho\varepsilon + \frac{H^2}{8\pi}\right) = -\operatorname{div}\mathbf{q}, \tag{65.10}$$

† This expression corresponds to a zero field **E**′ (63.1) in a frame of reference moving with the fluid volume element concerned: in a perfectly conducting medium, the electric field is completely screened.

where the energy flux density is

$$\mathbf{q} = \rho\mathbf{v}(\tfrac{1}{2}v^2 + w) + \mathbf{H}\times(\mathbf{v}\times\mathbf{H})/4\pi. \tag{65.11}$$

It is not difficult to verify this by direct calculation.

The above equations of magnetohydrodynamics are based on neglecting the displacement current in Maxwell's equations. We thus assume that

$$(1/c)\,|\partial\mathbf{E}/\partial t| \ll |\mathbf{curl\,H}|. \tag{65.12}$$

Expressing $\mathbf{E}$ in terms of $\mathbf{H}$ by (65.9) then gives

$$vl/c^2\tau \ll 1, \tag{65.13}$$

where l and τ are length and time parameters characteristic of the motion concerned. From (65.2), $l/\tau \sim v$, and we then find from (65.13) the condition $v \ll c$: the motion must be non-relativistic (as we have assumed from the start). Equation (65.4) gives $\rho v/\tau \sim H^2/l$; in combination with (65.13), this yields a condition on the magnetic field:

$$H^2 \ll \rho c^2. \tag{65.14}$$

It should be noted that the left-hand side of (65.10) does not include the electrical energy $E^2/8\pi$, and that of (65.7) does not include the electromagnetic field momentum $\mathbf{S}/c^2$. This is a necessary consequence of neglecting the displacement current. The smallness of the electrical energy in comparison with the magnetic energy corresponds to the inequality $E \sim vH/c \ll H$, and that of $S/c^2 \sim EH/c \sim vH^2/c^2$ in comparison with ρv corresponds to the inequality (65.14).

Let us return to the equation (65.2), which has an important physical interpretation (H. Alfvén, 1942). We expand the right-hand side, using the fact that div $\mathbf{H} = 0$:

$$\partial\mathbf{H}/\partial t = (\mathbf{H}\cdot\mathbf{grad})\,\mathbf{v} - (\mathbf{v}\cdot\mathbf{grad})\mathbf{H} - \mathbf{H}\,\mathrm{div}\,\mathbf{v}.$$

Substituting from the equation of continuity (65.3)

$$\mathrm{div}\,\mathbf{v} = -\frac{1}{\rho}\frac{\partial\rho}{\partial t} - \frac{\mathbf{v}\cdot\mathbf{grad}\,\rho}{\rho},$$

we obtain after a simple rearrangement of terms

$$\left(\frac{\partial}{\partial t} + \mathbf{v}\cdot\mathbf{grad}\right)\frac{\mathbf{H}}{\rho} \equiv \frac{\mathrm{d}}{\mathrm{d}t}\frac{\mathbf{H}}{\rho} = \left(\frac{\mathbf{H}}{\rho}\cdot\mathbf{grad}\right)\mathbf{v}. \tag{65.15}$$

Let us now consider some "fluid line", i.e. a line which moves with the fluid particles composing it. Let $\delta\mathbf{l}$ be an element of length of this line; we shall determine how $\delta\mathbf{l}$ varies with time. If $\mathbf{v}$ is the fluid velocity at one end of the element $\delta\mathbf{l}$, then the fluid velocity at the other end is $\mathbf{v} + (\delta\mathbf{l}\cdot\mathbf{grad})\mathbf{v}$. During a time interval $\mathrm{d}t$, the length of $\delta\mathbf{l}$ therefore changes by $\mathrm{d}t(\delta\mathbf{l}\cdot\mathbf{grad})\mathbf{v}$, i.e. $\mathrm{d}(\delta\mathbf{l})/\mathrm{d}t = (\delta\mathbf{l}\cdot\mathbf{grad})\mathbf{v}$. We see that the rates of change of the vectors $\delta\mathbf{l}$ and $\mathbf{H}/\rho$ are given by identical formulae. Hence it follows that, if these vectors are initially in the same direction, they will remain parallel, and their lengths will remain in the same ratio. In other words, if two infinitely close fluid particles are on the same line of force at any time, then they will always be on the same line of force, and the value of H/ρ will be proportional to the distance between the particles.

Passing now from particles at an infinitesimal distance apart to those at any distance apart, we conclude that every line of force moves with the fluid particles which lie on it. We can picture this by saying that (in the limit $\sigma \to \infty$) the lines of magnetic force are "frozen" in the fluid and move with it. The quantity H/ρ varies at every point proportionally to the extension of the corresponding "fluid line". If the fluid may be supposed incompressible ρ = constant, and the field H varies as the extension of the lines of force.

These results can be viewed in another way: as any closed fluid contour moves about in the course of time, it cuts no line of force. This means (cf. §63) that the flux of the magnetic field through any surface spanning the fluid contour does not vary with time.

§66. Dissipative processes in magnetohydrodynamics

In ordinary fluid dynamics the dissipative processes are governed by three quantities, namely two coefficients of viscosity and the thermal conductivity. In magnetohydrodynamics the number is considerably greater, both because new electrical quantities occur, and because there is at each point a distinctive direction, that of **H**, which makes the fluid no longer isotropic. We shall, however, take only the simple case where all the kinetic coefficients may be regarded as constant throughout the medium, and in particular as being independent of the magnitude and direction of the magnetic field. There is then only one quantity, the electrical conductivity σ, to be added to the usual viscosity coefficients η and ζ and the thermal conductivity κ.†

The assumption that the kinetic coefficients are independent of the magnetic field implies that certain conditions are satisfied which considerably restrict the range of validity of the equations in comparison with that of the equations of magnetohydrodynamics for an ideal fluid. The mean free path of the current-carriers must be much less than the radius of curvature of their orbits in the magnetic field; that is, the collision frequency must be large compared with the Larmor frequency of the carriers. This condition is not satisfied in a highly rarefied medium or in a strong magnetic field.‡

When viscosity and electrical conduction are taken into account, equation (65.2) is replaced by the complete equation (63.7):

$$\partial \mathbf{H}/\partial t = \mathbf{curl}\,(\mathbf{v}\times\mathbf{H}) + (c^2/4\pi\sigma)\triangle\mathbf{H}, \tag{66.1}$$

and equation (65.4) by the (complete) Navier–Stokes equation

$$\frac{\partial \mathbf{v}}{\partial t} + (\mathbf{v}\cdot\mathbf{grad})\mathbf{v} = -\frac{1}{\rho}\,\mathbf{grad}\,P + \frac{\eta}{\rho}\triangle\mathbf{v} + \frac{1}{\rho}(\zeta + \tfrac{1}{3}\eta)\,\mathbf{grad}\,\mathrm{div}\,\mathbf{v} - \frac{1}{4\pi\rho}\mathbf{H}\times\mathbf{curl}\,\mathbf{H}. \tag{66.2}$$

Equation (66.1) does not involve the viscosity, and so the "freezing" of the lines of force as $\sigma \to \infty$ continues to occur even in a viscous perfectly conducting fluid.

The adiabatic equation (65.6) is replaced by the equation of heat transfer. In ordinary fluid dynamics the latter is

$$\rho T\left(\frac{\partial s}{\partial t} + \mathbf{v}\cdot\mathbf{grad}\,s\right) = \sigma'_{ik}\frac{\partial v_i}{\partial x_k} + \mathrm{div}(\kappa\,\mathbf{grad}\,T);$$

† The relation between the current and the electric field in a thermodynamically non-uniform medium isotropic at every point involves also the thermoelectric coefficient α (§26). However, if this coefficient is constant, it does not appear in the equations of motion.

‡ The problem of the equations of magnetohydrodynamics in a plasma where these conditions are not satisfied is dealt with elsewhere; see *PK*, §§58, 59.

see *FM*, §49. The left-hand side of the equation is the quantity of heat generated per unit time and volume in a moving fluid particle. The right-hand side is the energy dissipated per unit time and volume. The first term is due to viscosity; σ'_{ik} is the viscous stress tensor:

$$\sigma'_{ik} = \eta\left(\frac{\partial v_i}{\partial x_k} + \frac{\partial v_k}{\partial x_i} - \tfrac{2}{3}\delta_{ik}\operatorname{div}\mathbf{v}\right) + \zeta\delta_{ik}\operatorname{div}\mathbf{v}.$$

The second term gives the dissipation due to thermal conduction. In a conducting fluid, a term giving the Joule heat must be added. The rate of evolution of this heat per unit volume is $\mathbf{j}^2/\sigma = (c^2/16\pi^2\sigma)(\mathbf{curl}\,\mathbf{H})^2$. The equation of heat transfer in magnetohydrodynamics is therefore

$$\rho T\left(\frac{\partial s}{\partial t} + \mathbf{v}\cdot\mathbf{grad}\,s\right) = \sigma'_{ik}\frac{\partial v_i}{\partial x_k} + \kappa\triangle T + \frac{c^2}{16\pi^2\sigma}(\mathbf{curl}\,\mathbf{H})^2. \tag{66.3}$$

The momentum flux density tensor contains in addition the viscous stress tensor:

$$\Pi_{ik} = \rho v_i v_k + P\delta_{ik} - \sigma'_{ik} - (H_i H_k - \tfrac{1}{2}H^2\delta_{ik})/4\pi. \tag{66.4}$$

The heat flux density is now

$$\mathbf{q} = \rho\mathbf{v}(\tfrac{1}{2}v^2 + w) - (\mathbf{v}\sigma') - \kappa\,\mathbf{grad}\,T + \frac{1}{4\pi}\mathbf{H}\times(\mathbf{v}\times\mathbf{H}) - \frac{c^2}{16\pi^2\sigma}\mathbf{H}\times\mathbf{curl}\,\mathbf{H}, \tag{66.5}$$

where $\mathbf{v}\sigma'$ is a vector whose components are $\sigma'_{ik}v_k$. This includes extra terms due to viscosity, thermal conduction and electrical conduction, the last of which is obtained by substituting in the Poynting vector the field $\mathbf{E}$ from (63.2):

$$\begin{aligned}\mathbf{E} &= \mathbf{j}/\sigma - \mathbf{v}\times\mathbf{H}/c\\ &= (c/4\pi\sigma)\,\mathbf{curl}\,\mathbf{H} - \mathbf{v}\times\mathbf{H}/c.\end{aligned} \tag{66.6}$$

The equations are somewhat simplified if the moving fluid can be supposed incompressible. The equation of continuity then reduces to $\operatorname{div}\mathbf{v} = 0$, and in equation (66.2) the penultimate term is zero. We shall write out here the appropriate system of equations (in equations (66.1) and (66.2) it is convenient to transform the terms $\mathbf{curl}(\mathbf{v}\times\mathbf{H})$ and $\mathbf{H}\times\mathbf{curl}\,\mathbf{H}$ by means of the appropriate formulae of vector analysis):

$$\operatorname{div}\mathbf{H} = 0, \qquad \operatorname{div}\mathbf{v} = 0, \tag{66.7}$$

$$\partial\mathbf{H}/\partial t + (\mathbf{v}\cdot\mathbf{grad})\mathbf{H} = (\mathbf{H}\cdot\mathbf{grad})\mathbf{v} + (c^2/4\pi\sigma)\triangle\mathbf{H}, \tag{66.8}$$

$$\frac{\partial\mathbf{v}}{\partial t} + (\mathbf{v}\cdot\mathbf{grad})\mathbf{v} = -\frac{1}{\rho}\mathbf{grad}\left(P + \frac{H^2}{8\pi}\right) + \frac{1}{4\pi\rho}(\mathbf{H}\cdot\mathbf{grad})\mathbf{H} + \nu\triangle\mathbf{v}, \tag{66.9}$$

where $\nu = \eta/\rho$ is the kinematic viscosity. Equation (66.3) is not needed in solving the problem of incompressible flow unless we are interested in the temperature distribution.

In ordinary fluid dynamics, the Reynolds number describes the role of viscosity terms in the equations of motion, in relation to the convection terms: $\mathrm{R} = ul/\nu$, where l and $u \sim l/\tau$ are the characteristic parameters of length and velocity for a particular motion of the fluid. In magnetohydrodynamics, we can supplement this by the *magnetic Reynolds number*

$$\mathrm{R}_m = ul/\nu_m, \qquad \nu_m = c^2/4\pi\sigma, \tag{66.10}$$

which describes the role of the conduction term in equation (66.1). This term is analogous to the term $\nu \triangle \mathbf{v}$ in the Navier–Stokes equation, and ν_m acts as a "diffusion coefficient" of the magnetic field. When $\mathrm{R}_m \gg 1$, this term may prove to be negligible. However, there is no general answer to the question of when in fact dissipative processes in the fluid may be neglected, since the relevant conditions depend greatly on the nature of the motion and are, for instance, completely different for steady and non-steady flows.

In the opposite limiting case of a poorly conducting fluid, $\mathrm{R}_m \ll 1$, the equations of magnetohydrodynamics can be considerably simplified (S. I. Braginskiĭ, 1959). The reason is that in this case the magnetic field is only slightly perturbed by the motion of the fluid. If the unperturbed field $\mathfrak{H}$ is independent of time (as we shall assume), the change $\mathbf{H}'$ in this field in the moving fluid may be estimated by comparing the two terms on the right of (66.1): $\mathbf{curl}\,(\mathbf{v}\times\mathfrak{H}) \sim \nu_m \triangle \mathbf{H}'$, whence $H' \sim \mathrm{R}_m \mathfrak{H}$, and when $\mathrm{R}_m \ll 1$ we in fact have $H' \ll \mathfrak{H}$. Neglecting this change, we can suppose that the magnetic field $\mathbf{H}$ is the same as the field $\mathfrak{H}$ that would result from the external sources in vacuum. Since $\mathfrak{H}$ is constant, $\mathbf{curl}\,\mathbf{E} = -(1/c)\partial\mathfrak{H}/\partial t = 0$, i.e. the electric field has a potential: $\mathbf{E} = -\mathbf{grad}\,\phi$. An equation for the potential ϕ can be derived from $\operatorname{div}\mathbf{j} = 0$, which is satisfied identically when the displacement current is neglected (i.e. because of the relation $\mathbf{curl}\,\mathbf{H} = 4\pi\mathbf{j}/c$). Substituting the current density $\mathbf{j} = \sigma(-\mathbf{grad}\,\phi + \mathbf{v}\times\mathfrak{H}/c)$ and noting that $\mathbf{curl}\,\mathfrak{H} = 0$ for the unperturbed field, we find (if σ = constant)

$$\triangle\phi = \mathfrak{H}\cdot\mathbf{curl}\,\mathbf{v}/c. \tag{66.11}$$

The second equation is the Navier–Stokes equation

$$\frac{\partial\mathbf{v}}{\partial t} + (\mathbf{v}\cdot\mathbf{grad})\mathbf{v} = -\frac{1}{\rho}\,\mathbf{grad}\,P + \nu\triangle\mathbf{v} + \mathbf{f} \tag{66.12}$$

(for an incompressible fluid), in which the volume density of external forces is

$$\begin{aligned}\mathbf{f} &= \mathbf{j}\times\mathfrak{H}/c \\ &= (\sigma/c)\,[\,\mathfrak{H}\times\mathbf{grad}\,\phi + \mathfrak{H}\times(\mathfrak{H}\times\mathbf{v})/c\,].\end{aligned} \tag{66.13}$$

Equations (66.11)–(66.13) are the appropriate ones for this case.

§67. Magnetohydrodynamic flow between parallel planes

An instructive example of the magnetohydrodynamic flow of a viscous conducting fluid occurs in steady flow in the space between two parallel solid planes when a uniform magnetic field $\mathfrak{H}$ is applied perpendicular to the planes (J. Hartmann, 1937). This is the simplest analogue of Poiseuille flow in ordinary fluid dynamics.

It is natural to assume that the velocity of the fluid is everywhere in the same direction (which we take as the x-direction), and depends only on the coordinate z (whose direction is perpendicular to the planes). The same is true of the longitudinal field H_x which results from the movement of the fluid. The pressure P, however, depends on x also, because there must be a pressure gradient in the direction of the motion in order to maintain a steady flow. The equation $\operatorname{div}\mathbf{v} = 0$ is satisfied identically, and from $\operatorname{div}\mathbf{H} = 0$ it follows that $H_z = \text{constant} = \mathfrak{H}$. The z-component of (66.9) shows that $P + H_x^2/8\pi$ is a function of x only. Since H_x is independent of x, the pressure gradient $\mathrm{d}P/\mathrm{d}x$ might be a function of x alone, but actually (because of the uniformity in the x-direction) it is a constant $-\Delta P/l$ (where ΔP is the pressure drop over a distance l).

The x-components of (66.8) and (66.9) give

$$\mathfrak{H}\frac{\mathrm{d}v}{\mathrm{d}z}+\frac{c^2}{4\pi\sigma}\frac{\mathrm{d}^2H_x}{\mathrm{d}z^2}=0, \tag{67.1}$$

$$\eta\frac{\mathrm{d}^2v}{\mathrm{d}z^2}+\frac{\mathfrak{H}}{4\pi}\frac{\mathrm{d}H_x}{\mathrm{d}z}=\text{constant}\equiv -\Delta P/l. \tag{67.2}$$

The boundary conditions on the solid surfaces are that the velocity of the viscous fluid be zero and that the tangential component of the magnetic field be continuous:

$$v=0,\quad H_x=0 \quad \text{for} \quad z=\pm a,$$

where $2a$ is the distance between the solid planes and $z=0$ lies half-way between them. The solution of (67.1) and (67.2) which satisfies these conditions is

$$\left.\begin{aligned} v &= v_0\frac{\cosh(a/\delta)-\cosh(z/\delta)}{\cosh(a/\delta)-1}, \qquad \delta=\frac{c}{\mathfrak{H}}\sqrt{\frac{\eta}{\sigma}},\\ H_x &= -v_0\frac{4\pi}{c}\sqrt{(\sigma\eta)}\frac{(z/a)\sinh(a/\delta)-\sinh(z/\delta)}{\cosh(a/\delta)-1}. \end{aligned}\right\} \tag{67.3}$$

The constant v_0 is the velocity of the fluid in the medial plane $z=0$. Its relation to the pressure gradient may be found by substituting (67.3) in (67.2). The velocity averaged over the cross-section is

$$\bar{v}=\frac{1}{2a}\int_{-a}^{a}v\,\mathrm{d}z=\frac{\Delta P}{l}\frac{a\delta}{\eta}\left(\coth\frac{a}{\delta}-\frac{\delta}{a}\right). \tag{67.4}$$

The effect of the magnetic field on the flow, in comparison with that of the viscosity, is described by the quantity

$$\mathrm{G}=a/\delta=(a\mathfrak{H}/c)\sqrt{(\sigma/\eta)}, \tag{67.5}$$

called the *Hartmann number*. When $\mathrm{G}\ll 1$ we have

$$v=v_0(1-z^2/a^2),\quad \bar{v}=(\Delta P/l)a^2/3\eta, \tag{67.6}$$

that is, ordinary Poiseuille flow. When $\mathrm{G}\gg 1$,

$$v=v_0\left[1-\exp\left(-\frac{a-|z|}{\delta}\right)\right],\quad \bar{v}=\frac{\Delta P}{l}\frac{ac}{\mathfrak{H}\sqrt{(\sigma\eta)}}. \tag{67.7}$$

When the magnetic field increases, the velocity profile is flattened over the greater part of the cross-section, and the mean velocity is reduced (for a given pressure gradient); the decrease in the velocity occurs mainly in layers of thickness $\sim\delta$ near the planes.

The motion of the fluid gives rise to an electric field in the y-direction. Since the motion is steady, **curl E** $=0$, and hence $E_y=$ constant. The current density in the fluid is $j_y=\sigma(E_y-v\mathfrak{H}/c)$. The total current through a cross-section of the fluid is zero: since also

$j_y = (c/4\pi)\,(\mathbf{curl\,H})_y$, we have

$$\int_{-a}^{a} j_y\,\mathrm{d}z = \frac{c}{4\pi}\int_{-a}^{a}\frac{\partial H_x}{\partial z}\mathrm{d}z$$

$$= \frac{c}{4\pi}\left[H_x(a) - H_x(-a)\right] = 0.$$

Hence

$$\int_{-a}^{a} j_y\,\mathrm{d}z = 2a\sigma E_y - \frac{\sigma}{c}\mathfrak{H}\int_{-a}^{a} v\,\mathrm{d}z = 0,$$

and so

$$E_y = \bar{v}\,\mathfrak{H}/c. \tag{67.8}$$

§68. Equilibrium configurations

The equilibrium of a perfectly conducting fluid (referred to here for clarity as a plasma) at rest in a constant magnetic field is described by the equations

$$\mathbf{grad}\,P = \mathbf{j}\times\mathbf{H}/c, \tag{68.1}$$

$$\mathbf{j} = c\,\mathbf{curl\,H}/4\pi, \tag{68.2}$$

$$\mathrm{div}\,\mathbf{H} = 0. \tag{68.3}$$

The first of these is equation (65.4) with $\mathbf{v} = 0$ and written in terms of the electric current density, which is related to the magnetic field by Maxwell's equation (68.2). In the present section, we shall consider some general properties of equilibrium configurations derivable from these equations, but without entering into the complex and manifold problems of the stability of such configurations.†

Scalar multiplication of (68.1) by $\mathbf{H}$ and by $\mathbf{j}$ gives

$$(\mathbf{H}\cdot\mathbf{grad})P = 0, \qquad (\mathbf{j}\cdot\mathbf{grad})P = 0; \tag{68.4}$$

that is, the pressure gradient is zero along the lines of magnetic force and along the current lines. Thus both sets of lines lie on surfaces

$$P(x, y, z) = \text{constant}, \tag{68.5}$$

called *magnetic surfaces*. In principle, every magnetic surface could be the boundary of an equilibrium configuration.‡

The equilibrium equations (68.1), (68.2) can also be put in the form

$$\partial\Pi_{ik}/\partial x_k = 0, \quad \Pi_{ik} = P\delta_{ik} - (H_iH_k - \tfrac{1}{2}H^2\delta_{ik})/4\pi, \tag{68.6}$$

if we start from the equation of motion in the momentum conservation form (65.7), (65.8). We multiply this equation by x_k, and integrate over some volume bounded by a closed

† The main results on this subject (within the scope of magnetohydrodynamics) are given by B. B. Kadomtsev, *Reviews of Plasma Physics* **2**, 153, 1966.

‡ The pressure P is determined by equations (68.1) and (68.2) only to within an arbitrary additive constant. Hence any of the magnetic surfaces can be $P = 0$.

surface. Integration by parts gives†, since $\partial x_k/\partial x_i = \delta_{ik}$,

$$\int \Pi_{ii}\,\mathrm{d}V = \oint \Pi_{ik} x_k\,\mathrm{d}f_i. \tag{68.7}$$

On substitution of Π_{ik} from (68.6), this equation becomes

$$\oint (3P + H^2/8\pi)\,\mathrm{d}V = \oint \{(P + H^2/8\pi)\mathbf{r} - (\mathbf{H}\cdot\mathbf{r})\mathbf{H}/4\pi\}\cdot \mathrm{d}\mathbf{f} \tag{68.8}$$

(S. Chandrasekhar and E. Fermi, 1953).

Let the plasma occupy some finite volume outside which $P = 0$, and let there be no fixed field sources (rigid current-carrying conductors) outside it. Then the field far from the plasma decreases as $1/r^3$, and, if the integration is taken over all space, the surface integral is zero. The integral of the positive quantity $3P + H^2/8\pi$ cannot, however, be zero. It is therefore not possible for an equilibrium configuration bounded in space to exist without being maintained by a magnetic field from external sources; when such sources *are* present, the right-hand side of (68.8) becomes an integral over their surfaces, and the condition can in principle be satisfied (V. D. Shafranov, 1957).

Let us consider the simplest unbounded configuration, a cylindrical plasma *pinch* of unlimited length and uniform along its length. In cylindrical polar coordinates r, ϕ, z, with the z-axis along the pinch, all quantities depend on the radial coordinate r only. The radial component H_r must be zero, since otherwise it would become infinite as $r \to 0$, in accordance with the equations

$$\operatorname{div}\mathbf{H} = \frac{1}{r}\frac{\mathrm{d}}{\mathrm{d}r}(rH_r) = 0, \quad H_r = \text{constant}/r.$$

The same is true of j_r because of the equation $\operatorname{div}\mathbf{j} = 0$ which necessarily follows from (68.2).

The components of equation (68.2) give

$$j_\phi = -\frac{c}{4\pi}\frac{\mathrm{d}H_z}{\mathrm{d}r}, \quad j_z = \frac{c}{4\pi r}\frac{\mathrm{d}}{\mathrm{d}r}(rH_\phi).$$

From the second of these,

$$H_\phi = 2J(r)/cr, \quad J(r) = \int_0^r j_z \cdot 2\pi r\,\mathrm{d}r. \tag{68.9}$$

Then equation (68.1) becomes

$$-\frac{\mathrm{d}P}{\mathrm{d}r} = \frac{1}{2\pi c^2 r^2}\frac{\mathrm{d}J^2(r)}{\mathrm{d}r} + \frac{1}{8\pi}\frac{\mathrm{d}H_z^2}{\mathrm{d}r}. \tag{68.10}$$

Two particular cases can occur here which are significantly different. In one, the *z pinch*, $H_z = 0$ and $j_\phi = 0$. Multiplying (68.10) by r^2 and integrating with respect to r from zero to the pinch radius a, with the boundary condition $P(a) = 0$, we find as the equilibrium condition

$$\int_0^a P(r)\cdot 2\pi r\,\mathrm{d}r = J^2(a)/2c^2, \tag{68.11}$$

† This derivation is similar to the familiar derivation of the virial theorem (*Fields*, §34).

where $J(a)$ is the total current through the pinch (W. Bennett, 1934). The equilibrium configuration is here contained by the field of the longitudinal current.

In the other case, the *theta pinch*†, $H_\phi = 0$ and $j_z = 0$. In this case, we have from (68.10)

$$P + H_z^2/8\pi = \mathfrak{H}^2/8\pi, \tag{68.12}$$

where $\mathfrak{H}$ is the longitudinal magnetic field outside the pinch. Here the plasma is contained by the external longitudinal field.

In an arbitrary space-bounded axially symmetric configuration, the radial components H_r and j_r may be non-zero (a toroidal configuration). Moreover, all quantities may then depend on z as well as on r.

Equations (68.1)–(68.3), written in components, are

$$j_\phi H_z - j_z H_\phi = c\partial P/\partial r, \quad j_r H_\phi - j_\phi H_r = c\partial P/\partial z, \quad j_z H_r = j_r H_z, \tag{68.13}$$

$$j_r = -\frac{c}{4\pi}\frac{\partial H_\phi}{\partial z}, \quad j_z = \frac{c}{4\pi r}\frac{\partial}{\partial r}(rH_\phi), \quad j_\phi = \frac{c}{4\pi}\left(\frac{\partial H_r}{\partial z} - \frac{\partial H_z}{\partial r}\right), \tag{68.14}$$

$$\frac{1}{r}\frac{\partial}{\partial r}(rH_r) + \frac{\partial H_z}{\partial z} = 0. \tag{68.15}$$

A consequence of these equations, which is already evident from the vector form (68.1), is that if the current density distribution is *azimuthal* ($j_r = j_z = 0, j_\phi \neq 0$) the magnetic field is *meridional* ($H_\phi = 0$). If the magnetic field is azimuthal, a stronger conclusion is possible: not only is the current density meridional, but the whole equilibrium configuration must be a z pinch ($j_r = 0$, H_ϕ and j_z independent of z), as is easily seen by eliminating P from the first two equations (68.13) and then using the other equations.

Equations (68.13)–(68.15) can be reduced to a single equation in the following way (V. D. Shafranov, 1957; H. Grad, 1958). We define the quantities

$$\psi(r, z) = \int_0^r H_z \cdot 2\pi r \, dr, \quad J(r, z) = \int_0^r j_z \cdot 2\pi r \, dr, \tag{68.16}$$

the magnetic flux and total current through a circle of radius r perpendicular to the z-axis. From these definitions and the equations div $\mathbf{H} = 0$, div $\mathbf{j} = 0$, we find the meridional components of the field and the current density:

$$\left.\begin{aligned} H_r &= -\frac{1}{2\pi r}\frac{\partial\psi}{\partial z}, \quad H_z = \frac{1}{2\pi r}\frac{\partial\psi}{\partial r}, \\ j_r &= -\frac{1}{2\pi r}\frac{\partial J}{\partial z}, \quad j_z = \frac{1}{2\pi r}\frac{\partial J}{\partial r}. \end{aligned}\right\} \tag{68.17}$$

These expressions show that the gradients of ψ and J are orthogonal to the line of magnetic force and the current line respectively. In view of the discussion of the surfaces (68.5) at the beginning of this section, we can conclude that ψ and J are constant on the magnetic surfaces, and therefore any two of ψ, J and P can be expressed as functions of the third

† The name comes from the angle in cylindrical polar coordinates, which is often denoted by θ.

alone. In particular,

$$P = P(\psi), \quad J = J(\psi). \tag{68.18}$$

The azimuthal components of the field and the current are expressed in terms of ψ and J by means of equations (68.14):

$$H_\phi = \frac{2J}{cr}, \quad j_\phi = -\frac{c}{8\pi^2 r}\left(\frac{\partial^2\psi}{\partial r^2} - \frac{1}{r}\frac{\partial\psi}{\partial r} + \frac{\partial^2\psi}{\partial z^2}\right). \tag{68.19}$$

Finally, substituting these results in the first equation (68.13), we find

$$\frac{\partial^2\psi}{\partial r^2} - \frac{1}{r}\frac{\partial\psi}{\partial r} + \frac{\partial^2\psi}{\partial z^2} = -16\pi^3 r^2\frac{\mathrm{d}P}{\mathrm{d}\psi} - \frac{8\pi^2}{c^2}\frac{\mathrm{d}J^2}{\mathrm{d}\psi}. \tag{68.20}$$

On specifying any particular functions $P(\psi)$ and $J(\psi)$ and solving this equation, we obtain some equilibrium configuration that is in principle possible; the field and current distributions in this configuration are given by (68.17) and (68.19), and the magnetic surfaces are $\psi(r, z) = \text{constant}$.

As an illustration, the expression

$$\psi/\psi_0 = \tfrac{1}{2}(bR^2 + r^2)z^2 + \tfrac{1}{8}(a-1)(r^2 - R^2)^2 \tag{68.21}$$

is the solution of (68.20) with $\mathrm{d}P/\mathrm{d}\psi = \text{constant}$ and $\mathrm{d}J^2/\mathrm{d}\psi = \text{constant}$; ψ_0, a, b and R are constants, with

$$16\pi^3\mathrm{d}P/\mathrm{d}\psi = -a\psi_0, \quad (8\pi^2/c^2)\mathrm{d}J^2/\mathrm{d}\psi = -bR^2\psi_0.$$

This solution describes a toroidal configuration consisting of nested toroidal magnetic surfaces $\psi = \text{constant}$, any of which may be taken as the plasma boundary $P = 0$. The innermost surface degenerates to a curve, the circle $r = R$, $z = 0$, called the *magnetic axis*. Near the magnetic axis,

$$\psi/\psi_0 \cong \tfrac{1}{2}R^2(b+1)z^2 + \tfrac{1}{2}R^2(a-1)(r-R)^2.$$

Thus, if $b+1 > 0$ and $a > 1$, the cross-sections of the magnetic surfaces near the axis are ellipses. With increasing distance from the axis, ψ increases and the pressure falls. Outside the surface $P = 0$ (the plasma boundary), the magnetic field needed to maintain equilibrium is given by equation (68.20) with zero on the right-hand side and the boundary conditions that ψ and its normal derivative be continuous.

§69. Hydromagnetic waves

Let us consider the propagation of small disturbances in a homogeneous conducting medium in a uniform constant magnetic field $\mathbf{H}_0$. We shall assume that the fluid is ideal, i.e. neglect all dissipation processes in it.†

We start from the equations of magnetohydrodynamics, (65.1)–(65.4). The adiabaticity equation (65.6) signifies only that, if the unperturbed medium is homogeneous, $s = \text{constant}$ in the perturbed medium also, i.e. the flow is isentropic.

We write

$$\mathbf{H} = \mathbf{H}_0 + \mathbf{h}, \quad \rho = \rho_0 + \rho', \quad P = P_0 + P',$$

† The condition for this approximation to be valid is that the wave damping coefficient (calculated in the Problem at the end of this section) should be small.

where the suffix 0 denotes the constant equilibrium value, and $\mathbf{h}$, ρ' and P' are the small variations in the wave. The velocity $\mathbf{v}$, which is zero in equilibrium, is a small quantity of the same order. Since the flow is isentropic, the changes in pressure and density are related by $P' = u_0^2\rho'$, where $u_0^2 = (\partial P/\partial \rho)_s$ is the square of the ordinary velocity of sound in the medium concerned. Neglecting terms of higher order than the first in equations (65.1)–(65.4), we obtain the linear equations

$$\left.\begin{aligned} \operatorname{div}\mathbf{h} = 0, \qquad \partial\mathbf{h}/\partial t &= \mathbf{curl}(\mathbf{v}\times\mathbf{h}),\\ \partial\rho'/\partial t + \rho \operatorname{div}\mathbf{v} &= 0,\\ \partial\mathbf{v}/\partial t = -(u_0^2/\rho)\,\mathbf{grad}\,\rho' &- (\mathbf{H}\times\mathbf{curl}\,\mathbf{h})/4\pi\rho. \end{aligned}\right\} \tag{69.1}$$

Here and in what follows we omit, for brevity, the suffix zero to the equilibrium values.

We shall seek solutions of these equations as plane waves proportional to $\exp[i(\mathbf{k}\cdot\mathbf{r}-\omega t)]$. The system of equations (69.1) then gives the algebraic equations

$$\left.\begin{aligned} -\omega\mathbf{h} = \mathbf{k}\times(\mathbf{v}\times\mathbf{H}), \qquad \omega\rho' = \rho\mathbf{k}\cdot\mathbf{v},\\ -\omega\mathbf{v} + (u_0^2/\rho)\rho'\mathbf{k} = -\mathbf{H}\times(\mathbf{k}\times\mathbf{h})/4\pi\rho; \end{aligned}\right\} \tag{69.2}$$

the equation $\mathbf{k}\cdot\mathbf{h} = 0$, which follows from $\operatorname{div}\mathbf{h} = 0$, is automatically satisfied and need not be considered separately.

The first of these equations shows that the vector $\mathbf{h}$ is perpendicular to the wave vector $\mathbf{k}$, which we shall take to be along the x-axis, with the plane of $\mathbf{k}$ and $\mathbf{H}$ as the xy-plane. We also introduce the *phase velocity* of the wave, $u = \omega/k$. Eliminating ρ' from the third equation by means of the second equation, and rewriting the result in components, we have

$$uh_z = -v_zH_x, \qquad uv_z = -H_xh_z/4\pi\rho, \tag{69.3}$$

$$\left.\begin{aligned} uh_y = v_xH_y - v_yH_x, \qquad uv_y &= -H_xh_y/4\pi\rho,\\ v_x\left(u - \frac{u_0^2}{u}\right) &= H_yh_y/4\pi\rho. \end{aligned}\right\} \tag{69.4}$$

We have here separated the equations into two groups, the first involving only h_z and v_z and the second only h_y, v_x and v_y. It therefore follows that perturbations of the two groups of variables are propagated independently. The density, and therefore the pressure, are propagated with the perturbations h_y, v_x and v_y, being related to v_x by

$$\rho' = \rho v_x/u. \tag{69.5}$$

The compatibility condition for the two equations (69.3) is

$$u = u_A \equiv |H_x|/\sqrt{(4\pi\rho)}; \tag{69.6}$$

we shall assume that $H_x > 0$, and omit the modulus sign. In these waves the component h_z of the magnetic field which is perpendicular to the directions of propagation and of the constant field $\mathbf{H}$ oscillates, and with it the velocity v_z, which is related to h_z by

$$v_z = -h_z/\sqrt{(4\pi\rho)}. \tag{69.7}$$

The relation between ω and $\mathbf{k}$ (the *dispersion relation*, as it is called) given by (69.6) involves the direction of the wave vector:

$$\omega = \mathbf{H}\cdot\mathbf{k}/\sqrt{(4\pi\rho)}. \tag{69.8}$$

The physical velocity of propagation of the waves is called the *group velocity* and is given by the derivative $\partial\omega/\partial\mathbf{k}$. In the present case we have

$$\partial\omega/\partial\mathbf{k} = \mathbf{H}/\sqrt{(4\pi\rho)}, \tag{69.9}$$

which does not involve the direction of **k**. The direction of propagation of the wave, in the sense of the direction of its group velocity, is the direction of **H**. These are called *Alfvén waves* (H. Alfvén, 1942), and (69.6) is the *Alfvén velocity*.

Let us now consider waves described by the equations (69.4), called *magnetosonic waves*. Equating to zero the determinant of these equations, we find an equation quadratic in u^2, whose roots are

$$u_{f,s}{}^2 = \tfrac{1}{2}\left\{\frac{H^2}{4\pi\rho} + u_0{}^2 \pm \left[\left(\frac{H^2}{4\pi\rho} + u_0{}^2\right)^2 - \frac{H_x{}^2}{\pi\rho}u_0{}^2\right]^{1/2}\right\}. \tag{69.10}$$

We thus obtain two further types of wave. The waves corresponding to the plus and minus signs in (69.10) are called respectively *fast* and *slow* magnetosonic waves.

In the limiting case where $H^2 \ll 4\pi\rho u_0{}^2$ we have $u_f \cong u_0$, and it follows from equations (69.4) that $v_y \ll v_x$. In other words, in the limit the fast magnetosonic waves become ordinary sound waves propagated with velocity u_0. The weak transverse field in the wave is related to v_x by $h_y \cong v_x H_y/u_0$. In the same limiting case, the velocity of the slow magnetosonic wave becomes the Alfvén velocity u_A, with $v_x \cong 0$, $v_y \cong -h_y/\sqrt{(4\pi\rho)}$, as in a wave of the first type, but with a different polarization: the vectors **v** and **h** are in a plane through **k** and **H**, not perpendicular to it.

In an incompressible fluid (corresponding formally to the limit $u_0 \to \infty$) only one type of wave remains, namely Alfvén waves with two independent directions of polarization. The dispersion relation for these waves is given by (69.8); the vectors **v** and **h** are perpendicular to the wave vector and are related by

$$\mathbf{v} = -\mathbf{h}/\sqrt{(4\pi\rho)}. \tag{69.11}$$

There is a simple interpretation of the fact that, in a longitudinal magnetic field, transverse displacements of the fluid are propagated in the form of waves. Because the lines of force are "frozen in", the transverse displacement of the particles results in a curvature of these lines, and therefore in their stretching and, at some points, in their compression. The forces in a magnetic field (expressed by the Maxwell stress tensor) are such as would occur if the lines of magnetic force tended to contract and at the same time to repel one another.† Hence a curvature of the lines results in quasi-elastic forces which tend to straighten them, leading to further oscillations.

Let us now return to formulae (69.4) and (69.10), and consider the opposite limiting case, where $H^2 \gg 4\pi\rho u_0{}^2$. We then have, in the first approximation, $u_f = H/\sqrt{(4\pi\rho)}$. Since this expression is independent of **k**, the group velocity is of magnitude u_f and its direction is that of **k**. In this wave the vector **v** is perpendicular to **H** (Fig. 40, p. 238), and its magnitude is given in terms of $h = |h_y|$ by $v = h/\sqrt{(4\pi\rho)}$. For u_s we have in this limiting case $u_s = u_0 H_x/H$. The group velocity is $\partial\omega/\partial\mathbf{k} = u_0\mathbf{H}/H$. The vector **v** in this case is anti-parallel to **H**, and its magnitude is given by $v = hH^2/4\pi\rho u_0 H_y$.

When the relation between H^2 and $\rho u_0{}^2$ is arbitrary, both u_f and u_s depend on the

---

† For, let a line of force be along the z-axis. Then the longitudinal stress Π_{zz} (65.8) contains a negative term $-H^2/8\pi$, and the transverse stresses Π_{xx} and Π_{yy} contain a positive term $H^2/8\pi$.

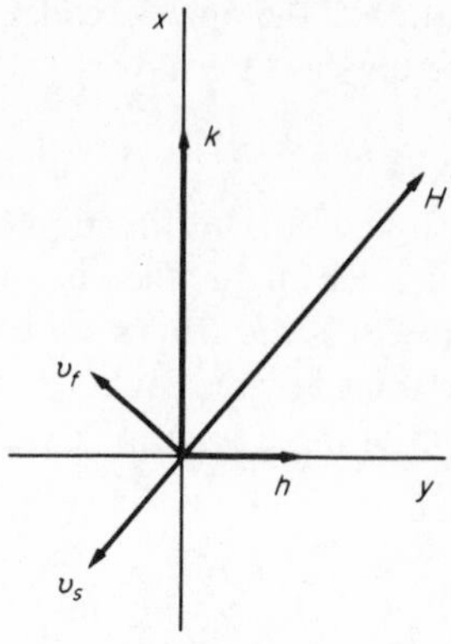

FIG. 40

direction of the wave vector. When the angle between **k** and **H** increases, u_f increases monotonically and u_s decreases monotonically. It is easy to see that the inequalities

$$u_s \leqslant u_A \leqslant u_f, \qquad u_f \geqslant u_0, \qquad u_s \leqslant u_0 \tag{69.12}$$

always hold. If **k** is parallel to **H**, u_f and u_s are respectively equal to the greater and the smaller of u_0 and $u_A = H/\sqrt{(4\pi\rho)}$. If **k** is perpendicular to **H**, then

$$u_f = \sqrt{\left(u_0^2 + \frac{H^2}{4\pi\rho}\right)}, \tag{69.13}$$

while u_A and u_s are zero, i.e. only the fast magnetosonic waves exist.

Lastly, let us consider two exact solutions of the equations of magnetohydrodynamics in the form of a plane wave with any amplitude (not necessarily small).

One of these is a plane Alfvén wave in an incompressible fluid, propagated with velocity u_A; that is, involving x and t only in the combination $x - u_A t$. To show this, we go back to the exact equations (65.1)–(65.4). The equation of continuity (65.3) in an incompressible fluid becomes div **v** = 0, whence v_x = constant; without loss of generality, we can put $v_x = 0$, by an appropriate choice of the frame of reference. The equation div **H** = 0 gives H_x = constant. Denoting the transverse components of **H** by **h**, we find from (65.2) and (65.4)

$$\frac{\partial \mathbf{h}}{\partial t} = H_x \frac{\partial \mathbf{v}}{\partial x}, \qquad \frac{\partial \mathbf{v}}{\partial t} = \frac{H_x}{\sqrt{(4\pi\rho)}} \frac{\partial \mathbf{h}}{\partial x},$$

i.e. the exact equations necessarily become linear equations describing a plane wave with the phase velocity (69.6) and **v** and **h** related by (69.11). The wave profile, i.e. the function $\mathbf{h}(x - u_A t)$, is arbitrary. The x-component of (65.4) gives

$$\frac{1}{\rho}\frac{\partial P}{\partial x} + \frac{1}{4\pi\rho}\mathbf{h}\cdot\frac{\partial \mathbf{h}}{\partial x} = 0,$$

whence

$$P + h^2/8\pi = \text{constant}, \tag{69.14}$$

which determines the variation of the pressure in the wave.

The other case is a simple wave propagated at right angles to the magnetic field (S. A. Kaplan and K. P. Stanyukovich, 1954). Let the field be in the y-direction, and the wave be again propagated in the x-direction. Then $H_x = 0$, $H_y = H$, and the equation $\operatorname{div} \mathbf{H} = 0$ is satisfied identically. Equations (65.2)–(65.4) give

$$\partial H/\partial t + \partial(v_x H)/\partial x = 0, \tag{69.15}$$

$$\partial \rho/\partial t + \partial(v_x \rho)/\partial x = 0, \tag{69.16}$$

$$\frac{\partial v_x}{\partial t} + v_x \frac{\partial v_x}{\partial x} + \frac{1}{8\pi\rho}\frac{\partial H^2}{\partial x} = -\frac{1}{\rho}\frac{\partial P}{\partial x}. \tag{69.17}$$

From the first two of these equations it is easily seen that the ratio $H/\rho = b$ satisfies the equation $\partial b/\partial t + v_x \partial b/\partial x = 0$ or $\mathrm{d}b/\mathrm{d}t = 0$, where the total derivative signifies the rate of change in a given fluid particle as it moves about. Hence, if the fluid is homogeneous at some initial instant, so that b is constant, then at all subsequent instants we have $b = \text{constant}$. Substituting $H = \rho b$ into the third equation, we obtain

$$\frac{\partial v_x}{\partial t} + v_x \frac{\partial v_x}{\partial x} = -\frac{1}{\rho}\frac{\partial}{\partial x}\left(P + \frac{b^2}{8\pi}\rho^2\right). \tag{69.18}$$

Thus the magnetic field has been eliminated from the equations, and the problem reduces to the solution of equations (69.16) and (69.18). These equations differ from those for one-dimensional motion in ordinary fluid dynamics only by a change in the equation of state of the gas: the true pressure $P = P(\rho)$ (for given entropy s) must be replaced by $P^*(\rho) = P(\rho) + b^2\rho^2/8\pi$. This fact enables us to apply the results of ordinary fluid dynamics to this case of magnetohydrodynamic flow. In particular, the formulae giving the exact solution for one-dimensional travelling waves (Riemann's solution; simple waves, see *FM*, §94) can be applied, the role of velocity of sound being played by

$$u^* = \sqrt{\left(\frac{\partial P^*}{\partial \rho}\right)_s} = \sqrt{\left(u_0{}^2 + \frac{b^2}{4\pi}\rho\right)}$$
$$= \sqrt{\left(u_0{}^2 + \frac{H^2}{4\pi\rho}\right)},$$

in accordance with formula (69.13).

PROBLEM

Determine the absorption coefficient (assumed small) for an Alfvén wave in an incompressible fluid.

SOLUTION. The absorption coefficient for a wave is defined as $\gamma = \overline{Q}/2\bar{q}$, where $\overline{Q}$ is the (time) average energy dissipated per unit time and volume, and $\bar{q}$ is the mean energy flux density in the wave. The amplitude of the wave decreases as $e^{-\gamma x}$ during its propagation. Q is given by the right-hand side of equation (66.3); in an incompressible fluid we have for a wave propagated in the x-direction (so that $v_x = 0$)

$$Q = \eta(\partial \mathbf{v}/\partial x)^2 + (c^2/16\pi^2\sigma)(\partial \mathbf{h}/\partial x)^2.$$

In the energy flux density (66.5), we can omit the small dissipative terms, leaving $q_x = -H_x \mathbf{h}\cdot\mathbf{v}/4\pi$. Using formulae (69.6) and (69.11), we have the result

$$\gamma = \frac{\omega^2}{2u_A{}^3}\left(\frac{\eta}{\rho} + \frac{c^2}{4\pi\sigma}\right).$$

§70. Conditions at discontinuities

The equations of motion for an ideal magnetohydrodynamic medium admit discontinuous flows as in ordinary fluid dynamics. To elucidate the conditions which must be satisfied on a surface of discontinuity, let us consider an element of the surface and use a system of coordinates in which it is at rest.†

First of all, the mass flux must be continuous at a surface of discontinuity: the mass of gas entering from one side must be equal to the mass leaving on the other side. Thus $\rho_1 v_{1n} = \rho_2 v_{2n}$, where the suffixes 1 and 2 refer to the two sides of the discontinuity, and the suffix n denotes the component of a vector normal to the surface. In what follows we shall denote the difference between the values of any quantity on the two sides of the surface of discontinuity by enclosing it in square brackets. Thus $[\rho v_n] = 0$.

Next, the energy flux must be continuous. Using the expression (65.11) and omitting the dissipative terms, we obtain

$$[q_n] = [\rho v_n(\tfrac{1}{2}v^2 + w) + v_n H^2/4\pi - H_n \mathbf{v}\cdot\mathbf{H}/4\pi] = 0.$$

The momentum flux must also be continuous. This condition means that $[\Pi_{ik} n_k] = 0$, where Π_{ik} is the momentum flux density tensor, and $\mathbf{n}$ is a unit vector normal to the surface. Using (65.8), we therefore have

$$[P + \rho v_n^2 + (\mathbf{H}_t^2 - H_n^2)/8\pi] = 0,$$

$$[\rho v_n \mathbf{v}_t - H_n \mathbf{H}_t/4\pi] = 0,$$

where the suffix t denotes the component tangential to the surface.

Finally, the normal component of the magnetic field and the tangential component of the electric field must be continuous. If the conductivity of the medium is infinite, the induced electric field is given by $\mathbf{E} = -\mathbf{v}\times\mathbf{H}/c$, and the condition $[\mathbf{E}_t] = 0$ leads to $[H_n \mathbf{v}_t - \mathbf{H}_t v_n] = 0$.

In what follows it is more convenient to use the specific volume of the gas ($V = 1/\rho$) in place of its density. The mass flux density through the discontinuity is denoted by

$$j = \rho v_n = v_n/V. \tag{70.1}$$

Since j and H_n are continuous, we can write the remaining boundary conditions in the following form:

$$j[w + \tfrac{1}{2}j^2V^2 + \tfrac{1}{2}\mathbf{v}_t^2 + V\mathbf{H}_t^2/4\pi] = H_n[\mathbf{H}_t\cdot\mathbf{v}_t]/4\pi, \tag{70.2}$$

$$[P] + j^2[V] + [\mathbf{H}_t^2]/8\pi = 0, \tag{70.3}$$

$$j[\mathbf{v}_t] = H_n[\mathbf{H}_t]/4\pi, \tag{70.4}$$

$$H_n[\mathbf{v}_t] = j[V\mathbf{H}_t]. \tag{70.5}$$

This is the fundamental system of equations of discontinuities in magnetohydrodynamics.

§71. Tangential and rotational discontinuities

In ordinary fluid dynamics, discontinuities of two different kinds are possible: shock waves and tangential discontinuities. Mathematically, the two types occur because some

† This condition fixes only the velocity of the coordinate system in the direction normal to the surface. Any constant vector may be added to its tangential velocity.

of the boundary conditions can be written as the vanishing of a product of two factors, and the two different solutions are obtained by equating the factors to zero in turn. This feature is not present in the equations (70.2)–(70.5) of magnetohydrodynamics, and it might therefore be supposed that only one type of discontinuity occurs. In reality, however, it is found that essentially different types of discontinuity again occur (F. de Hoffmann and E. Teller, 1950).

Let us consider, first of all, discontinuities for which $j = 0$. This means that $v_{1n} = v_{2n} = 0$, i.e. the fluid moves parallel to the surface of discontinuity. If $H_n \neq 0$, we see from equations (70.2)–(70.5) that the velocity, pressure and magnetic field must be continuous. The density (and therefore the entropy, temperature, etc.) may have any discontinuity. Such a surface may be called a *contact discontinuity*, and is simply the boundary between two media at rest which have different densities and temperatures.

If both j and H_n are zero, then three of the four equations (70.2)–(70.5) are satisfied identically, and therefore this is clearly a special case. We thus find a type of discontinuity which may be called a *tangential discontinuity*, as in ordinary fluid dynamics. At such a discontinuity the velocity and the magnetic field are tangential and can have any discontinuity in both magnitude and direction:

$$j = 0, \qquad H_n = 0, \qquad [\mathbf{v}_t] \neq 0, \qquad [\mathbf{H}_t] \neq 0. \tag{71.1}$$

The density discontinuity also can take any value, but the pressure discontinuity is related to that of $\mathbf{H}_t$ by equation (70.3):

$$[V] \neq 0, \qquad \left[P + \frac{\mathbf{H}_t^2}{8\pi}\right] = 0. \tag{71.2}$$

The discontinuities of the other thermodynamic quantities (entropy, temperature, etc.) are related to those of V and P by the equation of state.

Another type of discontinuity is one in which the gas density is continuous. Since the flux $j = v_n/V$ is continuous, the normal velocity component is therefore continuous also:

$$j \neq 0, \qquad [V] = 0, \qquad [v_n] = 0. \tag{71.3}$$

On the right-hand side of equation (70.5) we can take V outside the brackets and divide this equation by equation (70.4), obtaining

$$j = H_n/\sqrt{(4\pi V)} \tag{71.4}$$

and

$$[\mathbf{v}_t] = \sqrt{(V/4\pi)}\,[\mathbf{H}_t], \tag{71.5}$$

In equation (70.2) we put $w = \varepsilon + PV$; since V is continuous, this equation can be rewritten as

$$j[\varepsilon] + jV\left[P + \frac{\mathbf{H}_t^2}{8\pi}\right] + \tfrac{1}{2}j\left[\left(\mathbf{v}_t - \sqrt{\frac{V}{4\pi}}\mathbf{H}_t\right)^2\right] = 0,$$

H_n being replaced in accordance with (71.4). The second term is zero by (70.3) and the third term is zero by (71.5), so that we find $[\varepsilon] = 0$, i.e. the internal energy also is continuous. Every other thermodynamic quantity is determined if ε and V are given. Hence all the other thermodynamic quantities, including the pressure, are continuous. It then follows from (70.3) that $\mathbf{H}_t^2$ is continuous, i.e.

$$[P] = 0, \qquad [H_t] = 0. \tag{71.6}$$

The fact that H_t and H_n are both continuous means that the magnitude of **H** itself and its angle to the surface are likewise continuous.

Formulae (71.3)–(71.6) give all the properties of the discontinuities under consideration. The thermodynamic quantities are continuous, but the magnetic field is turned through an angle about the normal, its magnitude remaining unchanged. The vector $\mathbf{H}_t$, and therefore (by (71.5)) the tangential velocity component, are discontinuous, but the normal velocity component $v_n = jV$ is continuous, and its value is

$$v_n = H_n\sqrt{(V/4\pi)} = H_n/\sqrt{(4\pi\rho)}. \tag{71.7}$$

These are called *rotational* or *Alfvén discontinuities*.

It is useful to note that, by a suitable choice of the coordinate system, we can always ensure that the gas velocity is parallel to the field on each side of a rotational discontinuity. To achieve this, we use a coordinate system moving with velocity $\mathbf{v}_{1t} - \mathbf{H}_{1t}\sqrt{(V/4\pi)} = \mathbf{v}_{2t} - \mathbf{H}_{2t}\sqrt{(V/4\pi)}$. (Compare the footnote to §70.) In the new system the ratio of each component of **v** to the corresponding component of **H** on either side of the discontinuity is $\sqrt{(V/4\pi)}$, i.e.

$$\mathbf{v}_1 = \mathbf{H}_1\sqrt{(V/4\pi)}, \qquad \mathbf{v}_2 = \mathbf{H}_2\sqrt{(V/4\pi)}. \tag{71.8}$$

Thus in this system of coordinates the velocity is rotated with the magnetic field, its magnitude and angle to the normal remaining unchanged.

The velocity v_n is also minus the velocity of propagation of the discontinuity relative to the fluid. This is equal to the phase velocity u_A of the Alfvén wave. The occurrence of this equality for all rotational discontinuities is to some extent accidental, but when the discontinuities of the various quantities are small the equality must hold. For such a discontinuity is a weak perturbation, in which the velocity **v** and the magnetic field **H** receive small increments perpendicular to the plane through **H** and the normal **n**. This perturbation is of the type whose phase velocity is u_A. The physical velocity of propagation of the front of a small perturbation is the normal component of the group velocity, i.e. its component in the direction of the wave vector **k**. Since the relation between ω and **k** is linear, we have $\mathbf{k}\cdot\partial\omega/\partial\mathbf{k} = \omega$, and so this component is the same as the phase velocity $\omega/k = u_A$.

Although tangential and rotational discontinuities form two different types, there are also discontinuities having the properties of both. These discontinuities are such that **v** and **H** are tangential in direction and continuous in magnitude.

In ordinary fluid dynamics, tangential discontinuities are always unstable with respect to infinitesimal perturbations, and so are rapidly broadened into turbulent regions. A magnetic field, however, has a stabilizing effect on the motion of a conducting fluid, and in this case tangential discontinuities may be stable. This result is a natural consequence of the fact that a perturbation involving fluid displacements transverse to the field leads to a stretching of the lines of magnetic force "frozen" in it, and therefore to the appearance of forces which tend to restore the unperturbed flow. Let us ascertain the stability conditions for a tangential discontinuity in an incompressible fluid (S. I. Syrovatskiĭ, 1953). We write $\mathbf{v} = \mathbf{v}_{1,2} + \mathbf{v}'$, $P = P_{1,2} + P'$, $\mathbf{H} = \mathbf{H}_{1,2} + \mathbf{H}'$, where $\mathbf{v}_{1,2}$, $P_{1,2}$ and $\mathbf{H}_{1,2}$ are the unperturbed values constant on either side of the discontinuity; $\mathbf{v}'$, P' and $\mathbf{H}'$ are the small perturbations. Substituting in equations (66.7)–(66.9), we have for an ideal fluid

$$\operatorname{div}\mathbf{u}' = 0, \qquad \operatorname{div}\mathbf{v}' = 0, \tag{71.9}$$

$$\partial\mathbf{u}'/\partial t = (\mathbf{u}\cdot\mathbf{grad})\mathbf{v}' - (\mathbf{v}\cdot\mathbf{grad})\mathbf{u}', \tag{71.10}$$

$$\frac{\partial \mathbf{v}'}{\partial t}+(\mathbf{v}\cdot\mathbf{grad})\,\mathbf{v}' = -\frac{1}{\rho}\,\mathbf{grad}\;P'-\mathbf{u}\times\mathbf{curl}\;\mathbf{u}'$$

$$= -\frac{1}{\rho}\,\mathbf{grad}\;(P'+\rho\mathbf{u}\cdot\mathbf{u}')+(\mathbf{u}\cdot\mathbf{grad})\,\mathbf{u}'. \qquad (71.11)$$

For brevity, we omit here and henceforward the suffixes 1, 2, and write $\mathbf{u} \equiv \mathbf{H}/\sqrt{(4\pi\rho)}$. Taking the divergence of (71.11) and using (71.9), we find

$$\triangle\,(P'+\rho\mathbf{u}\cdot\mathbf{u}') = 0. \qquad (71.12)$$

Let the plane of the discontinuity be $x = 0$; the vectors $\mathbf{v}$ and $\mathbf{u}$ are parallel to it. In each of the half-spaces $x > 0$ and $x < 0$, we seek all quantities $\mathbf{v}'$, $\mathbf{u}'$, ρ in a form proportional to $\exp\{i(\mathbf{k}\cdot\mathbf{r}-\omega t)+\kappa x\}$, where $\mathbf{k}$ is a two-dimensional vector in the yz-plane. From equation (71.12), we find that $k^2 = \kappa^2$, so that we must take $\kappa = k$ for $x < 0$ and $\kappa = -k$ for $x > 0$. Next, we eliminate v'_x from the x-components of (71.10) and (71.11), obtaining

$$P'+\rho\mathbf{u}\cdot\mathbf{u}' = -u'_x\,\frac{i\rho}{\kappa\mathbf{k}\cdot\mathbf{u}}\,\{(\omega-\mathbf{k}\cdot\mathbf{v})^2-(\mathbf{k}\cdot\mathbf{u})^2\}. \qquad (71.13)$$

The case where the expression in the braces is zero is of no interest, since ω is then real, whereas instability can arise only from complex values of ω.

Let $\zeta\,(x, y, z)$ be the displacement, along the x-axis, of the discontinuity surface, as a result of the perturbation. On the displaced surface the conditions (71.1) and (71.2) must be satisfied:

$$[P+P'+\rho(\mathbf{u}+\mathbf{u}')^2] \cong [P'+\rho\mathbf{u}\cdot\mathbf{u}'] = 0,$$

$$u_{1n}+u'_{1n} \cong u'_{1x}-(\mathbf{u}_1\cdot\mathbf{grad})\zeta = 0,$$

$$u_{2n}+u'_{2n} \cong u'_{2x}-(\mathbf{u}_2\cdot\mathbf{grad})\zeta = 0;$$

the condition that there be no flux of the fluid through the discontinuity surface is automatically satisfied. Putting $\zeta = \text{constant}\times e^{i(\mathbf{k}\cdot\mathbf{r}-\omega t)}$ and eliminating ζ, u_{1x} and u_{2x} from these three equations, we get an equation determining the possible values of ω:

$$(\omega-\mathbf{k}\cdot\mathbf{v}_1)^2+(\omega-\mathbf{k}\cdot\mathbf{v}_2)^2 = (\mathbf{k}\cdot\mathbf{u}_1)^2+(\mathbf{k}\cdot\mathbf{u}_2)^2.$$

This quadratic equation has no complex root if

$$2(\mathbf{k}\cdot\mathbf{u}_1)^2+2(\mathbf{k}\cdot\mathbf{u}_2)^2-(\mathbf{k}\cdot\mathbf{v})^2 > 0$$

or

$$(2u_{1i}u_{1k}+2u_{2i}u_{2k}-v_iv_k)k_ik_k > 0,$$

where $\mathbf{v} = \mathbf{v}_2-\mathbf{v}_1$ is the discontinuity of the velocity.

This quadratic form is positive-definite if the trace and determinant of the rank-two tensor in the brackets are both positive. Hence we have the required stability conditions:†

$$\left.\begin{aligned} &H_1{}^2+H_2{}^2 > 2\pi\rho v^2,\\ &(\mathbf{H}_1\times\mathbf{H}_2)^2 \geqslant 2\pi\rho\{(\mathbf{H}_1\times\mathbf{v})^2+(\mathbf{H}_2\times\mathbf{v})^2\}.\end{aligned}\right\} \qquad (71.14)$$

† If the incompressible fluids on either side of the discontinuity have different densities, then ρ in these inequalities is replaced by $2\rho_1\rho_2/(\rho_1+\rho_2)$.

In reality, however, the existence of a small but finite viscosity and electrical resistance in the fluid means that such tangential discontinuities cannot exist indefinitely, even if the conditions (71.14) are fulfilled. Although no turbulence occurs, the sharp discontinuity is replaced by a gradually widening transitional region, in which the velocity and the magnetic field change smoothly from one value to another. This is easily seen from the equations of motion (66.8) and (66.9) if the dissipative terms are retained. We take the x-axis in the direction of the normal to the discontinuity. Assuming all quantities to depend only on x (and possibly on the time), we can write the transverse components of these equations as

$$\left.\begin{aligned}\partial \mathbf{H}_t/\partial t &= (c^2/4\pi\sigma)\partial^2 \mathbf{H}_t/\partial x^2,\\ \partial \mathbf{v}_t/\partial t &= \nu\partial^2 \mathbf{v}_t/\partial x^2,\end{aligned}\right\} \tag{71.15}$$

the fluid being supposed incompressible. If we assume steady flow, the left-hand sides of equations (71.15) are zero, and the only solution which remains finite as $x \to \pm\infty$ is $\mathbf{H}_t$ = constant, $\mathbf{v}_t$ = constant, which contradicts the assumption that these quantities undergo a change at the discontinuity. Thus a tangential discontinuity cannot have a constant width such as is found for (e.g.) a weak shock wave. Equations (71.15) have the form of heat-conduction equations. As we know from the theory of thermal conduction, a discontinuity in a quantity satisfying such an equation is gradually smoothed out into a transitional region, whose width increases as the square root of the time. Since the coefficients in the two equations (71.15) are different, the widths δ_v and δ_H of the transitional regions for the velocity and the magnetic field are also different:

$$\delta_v \sim \sqrt{(\nu t)}, \qquad \delta_H = \sqrt{(c^2 t/\sigma)}. \tag{71.16}$$

Rotational discontinuities in an incompressible fluid are stable with respect to infinitesimal perturbations, whatever the strength of the magnetic field (S. I. Syrovatskiĭ, 1953). Like tangential discontinuities, however, they cannot have constant widths, but are gradually smoothed out by the viscosity and electrical resistance of the fluid (see Problem).

PROBLEM

Find the manner of widening of a rotational discontinuity with time.

SOLUTION. Assuming all quantities to depend only on the coordinate x (and on the time), we find from the equations div $\mathbf{v} = 0$ and div $\mathbf{H} = 0$ that v_x = constant and H_x = constant. Let the coordinate system be such that the values of $\mathbf{v}$ and $\mathbf{H}$ on each side of the discontinuity (outside the transitional layer) are related by (71.8). Then $v_x = u_x$, where $\mathbf{u}$ has the same meaning as in (71.9)–(71.11). For the transverse components $\mathbf{u}_t$ and $\mathbf{v}_t$ we have from equations (66.8) and (66.9)

$$\left.\begin{aligned}\frac{\partial \mathbf{u}_t}{\partial t} + u_x\frac{\partial \mathbf{u}_t}{\partial x} &= u_x\frac{\partial \mathbf{v}_t}{\partial x} + \frac{c^2}{4\pi\sigma}\frac{\partial^2 \mathbf{u}_t}{\partial x^2},\\ \frac{\partial \mathbf{v}_t}{\partial t} + u_x\frac{\partial \mathbf{v}_t}{\partial x} &= u_x\frac{\partial \mathbf{u}_t}{\partial x} + \nu\frac{\partial^2 \mathbf{v}_t}{\partial x^2}.\end{aligned}\right\} \tag{1}$$

Since the difference $\mathbf{v}_t - \mathbf{u}_t$ tends to zero for $x \to \pm\infty$, because of the relations (71.8), this difference must be small in the transitional layer in comparison with the sum $\mathbf{v}_t + \mathbf{u}_t$. Adding the equations (1), we can therefore neglect a term in $\mathbf{v}_t - \mathbf{u}_t$, obtaining

$$\frac{\partial}{\partial t}(\mathbf{v}_t + \mathbf{u}_t) = \frac{1}{2}\left(\frac{c^2}{4\pi\sigma} + \nu\right)\frac{\partial^2}{\partial x^2}(\mathbf{v}_t + \mathbf{u}_t).$$

From this we see that the width of the discontinuity varies in a manner given by

$$\delta \sim \sqrt{\left\{\left(\frac{c^2}{4\pi\sigma}+\nu\right)t\right\}}.$$

§72. Shock waves

Let us now consider the type of discontinuity in which

$$j \neq 0, \qquad [V] \neq 0. \tag{72.1}$$

Such discontinuities are called *shock waves*, as in ordinary fluid dynamics. They are characterized by a discontinuity of density and by the fact that the gas moves through them (v_{n1} and v_{n2} being non-zero). The normal component of the magnetic field may or may not be zero.

On comparing equations (70.4) and (70.5) we see that, when $H_n \neq 0$, the vectors $\mathbf{H}_{t2}-\mathbf{H}_{t1}$ and $V_2\mathbf{H}_{t2}-V_1\mathbf{H}_{t1}$ are parallel to the same vector $\mathbf{v}_{t2}-\mathbf{v}_{t1}$, and therefore to each other. Hence it follows that $\mathbf{H}_{t1}$ and $\mathbf{H}_{t2}$ are collinear, i.e. the vectors $\mathbf{H}_1$, $\mathbf{H}_2$ and the normal to the surface are coplanar, unlike what happens (in general) in tangential and Alfvén discontinuities, in which the $\mathbf{H}_1\,\mathbf{n}$ and $\mathbf{H}_2\,\mathbf{n}$ planes in general do not coincide. This result holds also when $H_n = 0$; in this case, which we shall discuss later, it follows from (70.5) that $V_1\mathbf{H}_{t1} = V_2\mathbf{H}_{t2}$.

The velocity discontinuity $\mathbf{v}_{t1}-\mathbf{v}_{t2}$ lies in the same plane as $\mathbf{H}_1$ and $\mathbf{H}_2$. We can, without loss of generality, assume that the vectors $\mathbf{v}_1$ and $\mathbf{v}_2$ themselves lie in this plane, so that the motion in the shock wave is two-dimensional. Furthermore, it is easy to see that, if $H_n \neq 0$, a suitable transformation of the coordinates will always ensure that the vectors $\mathbf{v}$ and $\mathbf{H}$ are collinear on each side of the discontinuity. To achieve this, we use a coordinate system which moves with velocity $\mathbf{v}_t-(v_n/H_n)\mathbf{H}_t = \mathbf{v}_t-(jV/H_n)\mathbf{H}_t$ (the value of this expression is the same on each side of the discontinuity, by (70.5)). In the following formulae, however, the choice of this particular coordinate system is not implied.

Let us derive the relation for shock waves in magnetohydrodynamics which corresponds to the shock adiabatic (Hugoniot adiabatic) in ordinary fluid dynamics. Eliminating $[\mathbf{v}_t]$ from (70.4) and (70.5) we have

$$j^2[VH_t] = H_n^2[H_t]/4\pi; \tag{72.2}$$

here we have replaced $\mathbf{H}_t$ by H_t, since $\mathbf{H}_{t1}$ and $\mathbf{H}_{t2}$ are collinear.† In order to eliminate $[\mathbf{v}_t]$ from equation (70.2), we rewrite that equation as

$$[w]+\tfrac{1}{2}j^2[V^2]+\tfrac{1}{2}\left[\left(\mathbf{v}_t-\frac{H_n}{4\pi j}\mathbf{H}_t\right)^2\right]+[VH_t^2]/4\pi-H_n^2[H_t^2]/32\pi^2j^2=0.$$

The third term is zero by equation (70.4) and so $\mathbf{v}_t$ does not appear. In the last term we substitute j^2 from (72.2) and in the second term from (70.3), i.e.

$$j^2 = \{P_2-P_1+(H_{t2}^2-H_{t1}^2)/8\pi\}/(V_1-V_2). \tag{72.3}$$

† The vectors $\mathbf{H}_{t1}$ and $\mathbf{H}_{t2}$ here might, however, be either in the same or in opposite directions; in this sense, H_{t1} and H_{t2} might have either the same or different signs. It will not appear till later (§73) that for other reasons they must in fact have the same sign.

A simple calculation then gives

$$\varepsilon_2 - \varepsilon_1 + \tfrac{1}{2}(P_2 + P_1)(V_2 - V_1) + (V_2 - V_1)(H_{t2} - H_{t1})^2/16\pi = 0. \tag{72.4}$$

This is the equation of the shock adiabatic in magnetohydrodynamics. It differs from the ordinary equation by the presence of the third term.

We may also write out again equation (70.4), which gives the discontinuity of $\mathbf{v}_t$ in terms of that of $\mathbf{H}_t$:

$$\mathbf{v}_{t2} - \mathbf{v}_{t1} = H_n(\mathbf{H}_{t2} - \mathbf{H}_{t1})/4\pi j. \tag{72.5}$$

Equations (72.2)–(72.5) form a complete system of equations of shock waves. We shall conventionally assign the suffix 1 to the medium towards which the wave is propagated; thus the gas passes from side 1 in front of the shock wave to side 2 behind it. As already mentioned, the coordinates used are such that a particular surface element of the discontinuity is at rest and the gas passes through it.

In ordinary fluid dynamics, Zemplén's theorem (*FM*, §84) states that the pressure and density increase in the shock wave:

$$P_2 > P_1, \qquad \rho_2 > \rho_1; \tag{72.6}$$

the shock wave is therefore a compression wave. Here it is assumed that

$$(\partial^2 V/\partial P^2)_s > 0; \tag{72.7}$$

although this is not a thermodynamic inequality, it is satisfied in almost all cases. Zemplén's theorem follows from the law of increase of entropy.

It is easy to see that Zemplén's theorem remains valid in magnetohydrodynamics for weak shock waves, provided that the condition (72.7) is satisfied. In a weak shock wave, the discontinuities of all quantities are small. Expanding equation (72.4) in powers of the pressure and entropy discontinuities, we have

$$T(s_2 - s_1) = \tfrac{1}{12}(\partial^2 V/\partial P^2)_s(P_2 - P_1)^3 - \frac{1}{16\pi}\left(\frac{\partial V}{\partial P}\right)_s(P_2 - P_1)(H_{t2} - H_{t1})^2; \tag{72.8}$$

the first term is that found in ordinary fluid dynamics (see *FM*, §83). Since $-(\partial V/\partial P)_s > 0$ according to one of the thermodynamic inequalities, the condition $s_2 - s_1 > 0$ with (72.8) gives $P_2 > P_1$, and accordingly $V_2 < V_1$.

If, in addition to the inequality (72.7), the thermal expansion coefficient is positive, $(\partial V/\partial T)_P > 0$, Zemplén's theorem in magnetohydrodynamics can be proved as in ordinary fluid dynamics, without assuming that the discontinuities of all quantities are small (R. V. Polovin and G.Ya. Lyubarskiĭ, 1958; S. V. Iordanskiĭ, 1958).

Let P_1, V_1 be the specified initial state of the gas, and $s_2^{(0)}$ the entropy in the final state for a given value of V_2 in the absence of a magnetic field. The entropy in the final state of the gas in the presence of a magnetic field with the same values of P_1, V_1 and V_2 is denoted by $s_2^{(H)}$. In ordinary fluid dynamics, $V_2 > V_1$ implies that $s_2^{(0)} < s_1$, so that a rarefaction wave is impossible. We shall show that (under the conditions stated above) $s_2^{(H)} < s_2^{(0)}$, so that certainly $s_2^{(H)} < s_1$; this proves that rarefaction waves are impossible in magnetohydrodynamics also.

Differentiating equation (72.4) with respect to P_2 for a constant V_2, and using $(\partial\varepsilon/\partial P)_V = T(\partial s/\partial P)_V$, we find

$$T_2(\partial s_2^{(H)}/\partial P_2)_{V_2} + \tfrac{1}{2}(V_2 - V_1) + (\partial Q/\partial P_2)_{V_2} = 0, \tag{72.9}$$

where

$$Q = (V_2 - V_1)(H_{t2} - H_{t1})^2/16\pi.$$

Because of the thermodynamic relations

$$\left(\frac{\partial P}{\partial s}\right)_V = \frac{T}{c_v}\left(\frac{\partial P}{\partial T}\right)_V, \quad \left(\frac{\partial P}{\partial T}\right)_V = -\left(\frac{\partial P}{\partial V}\right)_T\left(\frac{\partial V}{\partial T}\right)_P,$$

the first term in (72.9) has the same sign as $(\partial V/\partial T)_P$, and is therefore, by hypothesis, positive. Hence, if $V_2 > V_1$, then $(\partial Q/\partial P_2)_{V_2} < 0$. The presence of a magnetic field increases Q ($Q = 0$ without a field, $Q > 0$ with one), and so decreases P_2 for a given V_2. Since, by hypothesis, $(\partial s/\partial P)_V > 0$, it follows in turn that $s_2^{(H)} < s_2^{(0)}$. This completes the proof.

Lastly, let us consider the case mentioned at the beginning of this section, when the magnetic field on each side of the discontinuity surface is in the tangential plane, $H_n = 0$ (a *perpendicular shock wave*). From (72.5) we then have $\mathbf{v}_{t2} = \mathbf{v}_{t1}$, i.e. the tangential velocity component is continuous. By a suitable choice of coordinates, therefore, we can always ensure that $\mathbf{v}_t = 0$ on either side of the discontinuity, i.e. the gas moves perpendicularly to the discontinuity, and we shall henceforth assume this. From equation (72.2) we have $V_2 H_2 = V_1 H_1$. This relation shows that equations (72.3) and (72.4) can be written

$$j^2 = (P_2{}^* - P_1{}^*)/(V_1 - V_2), \quad \varepsilon_2{}^* - \varepsilon_1{}^* + \tfrac{1}{2}(P_2{}^* + P_1{}^*)(V_2 - V_1) = 0,$$

which differ from the ordinary equations for shock waves in the absence of a magnetic field only by a change in the equation of state: the true equation of state $P = P(V, s)$ must be replaced by $P^* = P^*(V, s)$, where $P^* = P + b^2/8\pi V^2$, and b denotes the constant product HV. Accordingly ε^* must be defined so as to satisfy the thermodynamic relation $(\partial \varepsilon^*/\partial V)_s = -P^*$, whence $\varepsilon^* = \varepsilon + b^2/8\pi V$.

§73. Evolutionary shock waves

For an actual hydrodynamic discontinuity to exist, it must be stable with regard to disintegration into two or more others. This condition can also be formulated as stating that any infinitesimal perturbation of the initial state must lead only to infinitesimal changes in the discontinuity. A discontinuity that meets this requirement is said to be *evolutionary*. It must be emphasized that the property in question is not the opposite of instability in the ordinary sense. The latter signifies that an initial small perturbation will gradually increase, which leads ultimately to the breakdown of the flow concerned; but the perturbation remains small over a sufficiently short time ($t \lesssim 1/\gamma$) even when there is an exponential increase as $e^{\gamma t}$ with $\gamma > 0$. In a non-evolutionary discontinuity, the perturbation becomes large immediately, although when t is small it occupies only a small region in space. The situation is illustrated by Fig. 41 (p. 248), which shows the splitting of a discontinuity of the density $\rho(x)$ into two; the perturbation $\delta\rho$ is not small, although when t is small, and the two discontinuities have not moved apart appreciably, it occupies only a small range δx.

The condition for an evolutionary discontinuity can be derived by counting the number of independent parameters which define an arbitrary initial ($t = 0$) small perturbation of the discontinuity, and the number of equations (linearized boundary conditions at the discontinuity) which these parameters must satisfy. The discontinuity is evolutionary if the

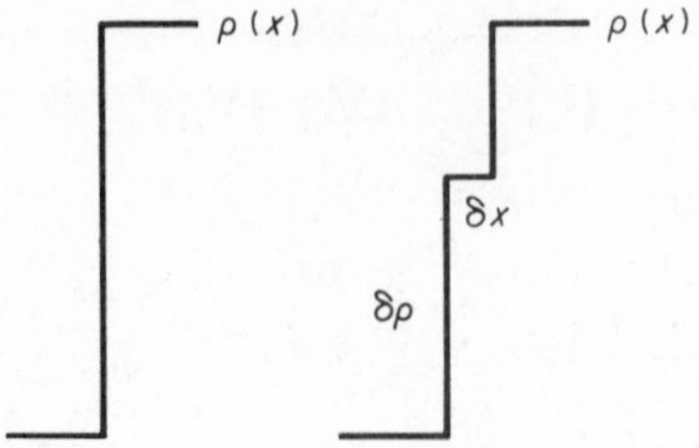

FIG. 41

two numbers are the same. The boundary conditions then uniquely determine the subsequent development of the perturbation, which remains small for small $t > 0$.† If the number of equations is greater or less than the number of unknown parameters, the problem of a small perturbation of the discontinuity has either no solution or an infinity of solutions. Either case may occur, and such a situation indicates that the discontinuity is not evolutionary, the initial assumption that the perturbation is small for small t being incorrect.

In ordinary fluid dynamics, the requirement that a shock wave be evolutionary does not impose any further restriction in comparison with the condition of increasing entropy: shock waves allowed by Zemplén's theorem are necessarily evolutionary (see *FM*, §84). In magnetohydrodynamics, this is not the case, and the requirement that a shock wave be evolutionary imposes significant new limitations on the way in which quantities vary in the shock wave (A. I. Akhiezer, G. Ya. Lyubarskiĭ and R. V. Polovin, 1958).

In order to ascertain the actual condition for shock waves to be evolutionary in magnetohydrodynamics, let us first count the number of equations to be satisfied by any small perturbation at the discontinuity surface.

The shock wave will be regarded as lying in the yz-plane. The positive x-direction is taken to be that of the flow across the discontinuity surface. The unperturbed fields H_1, H_2 and velocities v_1, v_2 on either side of the discontinuity are assumed to be in the xy-plane.

On each side of the discontinuity surface, seven quantities undergo perturbation: three fluid velocity components v_x, v_y, v_z; two magnetic field components H_y, H_z; the density $\rho = 1/V$, and the entropy s. The perturbations of the other thermodynamic quantities (P and w) are determined by those of ρ and s. The equation $\operatorname{div} \mathbf{H} = \partial H_x/\partial x = 0$ shows that the longitudinal field component H_x is constant along the x-axis and does not undergo any perturbation. In addition, the speed of propagation of the shock wave itself is perturbed, i.e. it acquires a small velocity δU relative to the chosen system of coordinates (in which the unperturbed discontinuity is at rest). This velocity, however, can be expressed immediately in terms of the perturbations of ρ and v_x, by means of the condition for the mass flux density j to be continuous across the discontinuity: the gas velocity relative to the discontinuity is $v_{x0} + \delta v_x - \delta U$, where v_{x0} is the unperturbed velocity and δv_x its perturbation; writing also $\rho = \rho_0 + \delta\rho$, linearizing the boundary condition $[j] = 0$, and

† Here the wave may be either unstable (if the eigenfrequencies of the equations include some which are complex with a positive imaginary part) or stable (if not).

then omitting the suffix zero to the unperturbed quantities, we find

$$[\delta j] = [\rho\delta v_x] + [v_x\delta\rho] - \delta U[\rho] = 0,$$

which determines δU.

Linearization of the boundary conditions of continuity of the momentum flux component Π_{zx} and the electric field component E_y (i.e. the z-components of equations (70.4) and (70.5)) gives the two equations

$$[\rho v_x \delta v_z - H_x \delta H_z/4\pi] = 0, \quad [H_x \delta v_z - v_x \delta H_z] = 0$$

(the unperturbed values are $v_z = 0$, $H_z = 0$). These equations involve the perturbations of only two quantities, namely

$$\delta v_z, \delta H_z. \tag{73.1}$$

The boundary conditions of continuity of the energy flux q_x, the momentum flux components Π_{xx} and Π_{yx}, and the electric field component E_z (i.e. equations (70.2) and (70.3) and the y-components of (70.4) and (70.5)) give four linear equations involving the perturbations

$$\delta v_x, \delta v_y, \delta H_y, \delta\rho, \delta s, \tag{73.2}$$

which will not be written out here.

Let us now count the number of parameters which define the perturbation of the shock wave. Perturbations varying with time as $e^{-i\omega t}$ are propagated in both directions from the discontinuity as hydromagnetic waves of three kinds (Alfvén, and fast and slow magnetosonic) and as an entropy wave; the latter is a small perturbation of the entropy transported by the adiabatic gas flow and moving with the gas. All these must of course be outgoing waves propagated to the left or right of the discontinuity. In each wave, the changes in all quantities are related in a specific way, as shown in §69, and therefore each wave is determined by just one parameter, the amplitude of some single quantity.

Magnetosonic and entropy waves transport the perturbations (73.2); Alfvén waves, the perturbations (73.1). Since the equations separate for these two groups of perturbations, the condition for an evolutionary discontinuity has to be satisfied for each separately (S. I. Syrovatskiĭ, 1958), and this makes the restrictions even stronger.

Let us first consider the conditions as regards Alfvén perturbations. This requires that the number of outgoing waves be equal to two, the number of equations. The Alfvén wave phase velocities relative to the discontinuity surface can have the values $v_{x1} \pm u_{A1}$, $v_{x2} \pm u_{A2}$, where u_A is the phase velocity (69.6) of the wave relative to the gas. Because of the direction chosen for the x-axis, $v_{x1}, v_{x2} > 0$. In region 1, in front of the discontinuity, the wave is outgoing if its phase velocity (relative to the discontinuity) is negative; in region 2, behind the discontinuity, a positive phase velocity is needed. The wave with velocity $v_{x1} + u_{A1}$ never satisfies this condition, and is always ingoing; that with velocity $v_{x1} - u_{A1}$ is outgoing if $v_{x1} < u_{A1}$. Similarly, the wave with velocity $v_{x2} + u_{A2}$ is always outgoing; that with $v_{x2} - u_{A2}$ is outgoing if $v_{x2} > u_{A2}$. There are therefore two ranges in which the condition for an evolutionary discontinuity is satisfied for Alfvén waves:

$$(1) \quad v_{x1} > u_{A1}, \qquad v_{x2} > u_{A2},$$

$$(2) \quad v_{x1} < u_{A1}, \qquad v_{x2} < u_{A2}.$$

These are shown by the vertical lines in Fig. 42 (p. 250), which is drawn for the case in which

$$u_s < u_A < u_f. \tag{73.3}$$

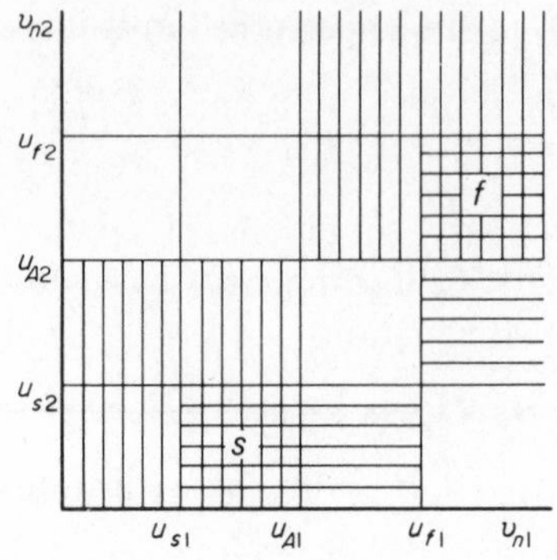

FIG. 42

The condition as regards magnetosonic and entropy perturbations requires that there be four outgoing waves. There is always an outgoing entropy wave moving with the gas, but only from side 2. The number of outgoing magnetosonic waves must therefore be three. Arguments similar to those given above for Alfvén waves indicate two ranges in which the condition for an evolutionary discontinuity is satisfied for this group of perturbations; they are shown by the horizontal lines in Fig. 42.†

The intersection of the vertically and horizontally lined areas defines two regions where the discontinuity is evolutionary with regard to all perturbations: those of (1) *fast shock waves*, with

$$v_{n1} > u_{f1}, \quad u_{f2} > v_{n2} > u_{A2}, \tag{73.4}$$

and (2) *slow shock waves*, with

$$u_{A1} > v_{n1} > u_{s1}, \quad u_{s2} > v_{n2}; \tag{73.5}$$

here the normal component of the gas velocity is again denoted by v_n instead of v_x. In the limit of a low-intensity wave (discontinuities of all quantities weak), the fast and slow shock waves are propagated with the respective velocities $u_{f2} \cong u_{f1}$ and $u_{s2} \cong u_{s1}$.

Let us apply these conditions to find how the magnetic field varies in a shock wave. We start from equation (72.2):

$$H_n^2 [\mathbf{H}_t]/4\pi j^2 = [\mathbf{H}_t/\rho] = [v_n \mathbf{H}_t]/j$$

or

$$\left(\frac{H_n^2}{4\pi j} - v_{n1}\right)\mathbf{H}_{t1} = \left(\frac{H_n^2}{4\pi j} - v_{n2}\right)\mathbf{H}_{t2}. \tag{73.6}$$

Since $H_n^2/4\pi\rho = u_A^2$, this can also be written as

$$\frac{u_{A1}^2 - v_{n1}^2}{v_{n1}}\mathbf{H}_{t1} = \frac{u_{A2}^2 - v_{n2}^2}{v_{n2}}\mathbf{H}_{t2}. \tag{73.7}$$

† The ranges where the condition is not satisfied include cases where the number of parameters is either greater or less than the number of equations, for each of the two groups of perturbations.

From the inequalities (73.4) and (73.5), with (73.7), we see that the tangential fields on the two sides of the shock wave are not only collinear but parallel.

In slow shock waves we have, on both sides of the discontinuity,

$$v_n < H_n^2/4\pi j = v_A^2/v_n.$$

Noting also that by the continuity of the mass flux $\rho_1 v_{n1} = \rho_2 v_{n2}$, we deduce from $\rho_1 < \rho_2$ that

$$v_{n1} > v_{n2}, \tag{73.8}$$

and hence, from (73.6), $H_{t2} < H_{t1}$: in a slow shock wave, the tangential magnetic field becomes weaker. In a fast wave, however, $v_n > H_n^2/4\pi j$, and from (73.6) it follows that $H_{t2} > H_{t1}$, the tangential magnetic field thus becoming stronger.

A particular case of shock waves is that where the magnetic field is parallel to the normal on both sides of the discontinuity surface. As mentioned at the beginning of §72, we can always choose the coordinates so as to make **v** and **H** parallel on both sides. In the case considered, we then have $\mathbf{H}_{t1} = \mathbf{H}_{t2} = 0$, $\mathbf{v}_{t1} = \mathbf{v}_{t2} = 0$ (a *parallel shock wave*). For such a shock wave, the boundary conditions do not involve the magnetic field, i.e. they are the same as in ordinary fluid dynamics. The presence of the magnetic field has the result, however, that over a certain range of shock parameter values the conditions for an evolutionary discontinuity are no longer satisfied, and such shock waves become impossible (see Problem 1).

All the perpendicular shock waves described at the end of §72 are evolutionary fast compression shock waves. They are seen to be fast from the fact that when $H_n = 0$ we have $u_A = u_s = 0$.

Having discussed the various types of discontinuity in magnetohydrodynamics, let us now consider the possible existence of forms transitional between these types, i.e. discontinuities having the properties of two types at the same time. Such possibilities are severely limited by conditions which result from the requirement of the evolutionary property.

An Alfvén discontinuity, first of all, cannot pass continuously into a shock wave, since for the latter the magnetic fields on either side are coplanar with the normal to the discontinuity surface. Such a shock wave can coincide with an Alfvén discontinuity only if the vector **H** turns through 180° in it. But the tangential field component would then change sign, whereas in an evolutionary shock wave it does not change sign.

A continuous transition between fast and slow shock waves would be possible only with $\mathbf{H}_{t1} = \mathbf{H}_{t2} = 0$, since in a fast shock the field $\mathbf{H}_t$ (if not zero) is strengthened, but in a slow one it is weakened. In other words, a continuous transition would be possible only between fast and slow parallel shocks. But the ranges where these are evolutionary meet only at $u_{A1} = u_{01}$, when the slow shock disappears (see Problem 1). Thus there cannot be a continuous transition between fast and slow shocks.

A fast shock cannot pass continuously into a tangential discontinuity, because of the inequalities (73.4).

There can therefore be continuous transitions only between a tangential discontinuity and either a contact discontinuity, an Alfvén discontinuity, or a slow shock wave.

PROBLEMS

PROBLEM 1. Find the range of values of v_1 in which a parallel shock wave is no longer evolutionary in a (thermodynamically) ideal monatomic gas with ratio of specific heats $c_p/c_v = 5/3$.

SOLUTION. For such a gas, the heat function $w = 5P/2\rho$, and the boundary conditions (70.1)–(70.3) become

$$\rho_1 v_1 = \rho_2 v_2, \quad P_1 + \rho_1 v_1^2 = P_2 + \rho_2 v_2^2, \quad v_1^2 + 5P_1/\rho_1 = v_2^2 + 5P_2/\rho_2.$$

Hence $v_2 = (v_1^2 + 3u_{01}^2)/4v_1$, where $u_0 = \sqrt{(5P/3\rho)}$ is the ordinary velocity of sound, and

$$u_{A2} = \sqrt{(\rho_1/\rho_2)}u_{A1} = \sqrt{(v_1^2 + 3u_{01}^2)}u_{A1}/2v_1,$$
$$u_{02} = \sqrt{\{(5v_1^2 - u_{01}^2)(v_1^2 + 3u_{01}^2)\}}/4v_1,$$
$$u_{f1} = \max(u_{01}, u_{A1}), \quad u_{s1} = \min(u_{01}, u_{A1}),$$
$$u_{f2} = \max(u_{02}, u_{A2}), \quad u_{s2} = \min(u_{02}, u_{A2}).$$

When $u_{A1} < u_{01}$, the shock wave is fast and always evolutionary. Figure 43 shows the approximate form of the functions $v_2(v_1)$ (thick curve) and $u_{A2}(v_1)$, $u_{f2}(v_1) = u_{02}(v_1)$ in this case; the slanting dot-and-dash line bisects the right angle.

Figure 44 shows a corresponding diagram for the case $u_{A1} > u_{01}$. The thin curves are again the functions $u_{A2}(v_1)$ and $u_{02}(v_1)$; the various parts of these curves are marked to show whether u_{f2} or u_{s2} coincides with them. The thick continuous curve is the function $v_2(v_1)$ for the evolutionary ranges; the left-hand and right-hand parts correspond to slow and fast shock waves respectively. The thick dashed curve is the non-evolutionary range $u_{A1} < v_1 < \sqrt{(4u_{A1}^2 - 3u_{01}^2)}$, ending on the right at the point $v_2 = u_{A2}$.†

PROBLEM 2. In front of a shock wave the tangential magnetic field $\mathbf{H}_{t1} = 0$, and behind it $\mathbf{H}_{t2} \neq 0$; this is called a *switch-on shock*. Find the range of values of v_{n1} which such a shock wave can have, in a gas with the same thermodynamic properties as in Problem 1.

SOLUTION. It follows from (72.2) that when $\mathbf{H}_{t1} = 0$ the speed of the shock wave relative to the gas behind it is $v_{n2} = H_n/\sqrt{(4\pi\rho_2)} = u_{A2}$, and relative to the gas in front $v_{n1} = \rho_2 v_{n2}/\rho_1 = u_{A1}\sqrt{(\rho_2/\rho_1)} > u_{A1}$. This is a fast shock; v_{n1} and v_{n2} are related by $v_{n1}v_{n2} = u_{A1}^2$.‡

For a switch-on shock in a gas with the specified thermodynamic properties, the boundary conditions give

$$H_{t2}^2/8\pi = (\rho_1/3u_{A1}^2)(v_{n1}^2 - u_{A1}^2)(4u_{A1}^2 - 3u_{01}^2 - v_{n1}^2).$$

Since the right-hand side must be positive and $v_{n1} > u_{A1}$, we see that the possible values of v_{n1} in a switch-on shock lie in the range

$$u_{A1} \leqslant v_{n1} \leqslant \sqrt{(4u_{A1}^2 - 3u_{01}^2)}.$$

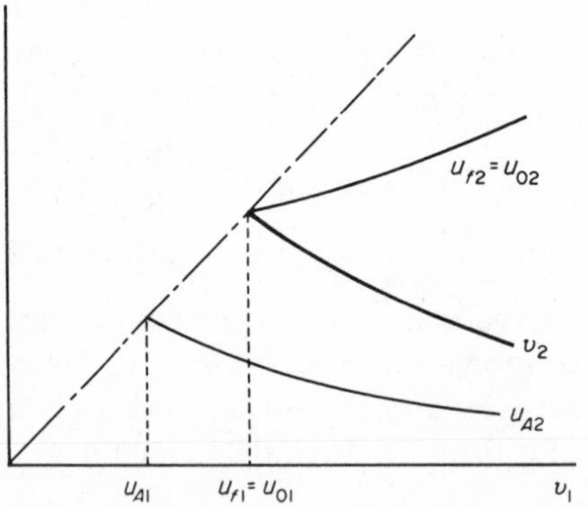

FIG. 43

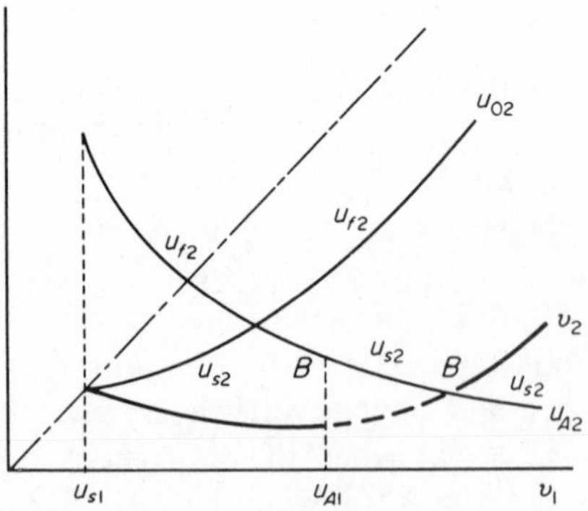

FIG. 44

† See Problem 2 as regards the segment BB.

‡ A *switch-off shock* is one in which $\mathbf{H}_{t1} \neq 0$ and $\mathbf{H}_{t2} = 0$. This is a slow shock, with $v_{n1} = u_{A1}$, $v_{n2} = \sqrt{(\rho_1/\rho_2)}u_{A2} < u_{A2}$, $v_{n1}v_{n2} = u_{A2}^2$.

There may be doubt as to whether the switch-on and switch-off shocks are evolutionary; for example, in a switch-off shock $v_{n1} > u_{A1}$, and it might seem that only one Alfvén perturbation wave can leave it, travelling backwards with $v_{n2} + u_{A2} = 2u_{A2}$. It must not be forgotten, however, that when a perturbation is imposed the tangential magnetic field becomes non-zero.

These inequalities show that switch-on shocks can occur only if $u_{A1} > u_{01}$. In Fig. 44 they correspond to the segment BB of the thin continuous curve.

§74. The turbulent dynamo

Turbulent motion of a conducting fluid has the remarkable property that it may lead to spontaneous magnetic fields which are quite strong. This is called the *turbulent dynamo* effect. There are always small perturbations in a conducting fluid, resulting from causes extraneous to the fluid motion itself (for example, the magnetomechanical effect in rotating parts of a fluid, or even thermal fluctuations), and accompanied by very weak electric and magnetic fields. The question is whether these perturbations are, on the average, amplified or damped by the turbulent motion in the course of time.

The manner of variation with time of magnetic field perturbations, once they have arisen, is determined by various physical agencies. The magnetic field tends to increase, by the purely hydromagnetic effect of the stretching of the lines of force. We have shown in §65 that, when a fluid with sufficiently high conductivity is in motion, the lines of magnetic force move as fluid lines "frozen" in, and the magnetic field varies proportionally to the stretching at each point on each line of force. In turbulent motion any two neighbouring particles move apart, on the average, in the course of time. As a result, the lines of force are stretched and the magnetic field is strengthened.

The dissipation of magnetic energy, which is converted into the Joule heat of the induced currents, tends to diminish the field. Since the energy dissipation is proportional to $(\mathbf{curl\,H})^2$, i.e. is quadratic in the spatial derivatives of the field, it is clear that the dissipation will be small in flow where the spatial scale of variation of the field is sufficiently great. This does not mean that the field will be strengthened over such distances. The reason is that the lines of force are not only stretched but also "tangled", which reduces the spatial scale. A situation may therefore occur where the strengthening of a field with a given scale is replaced merely by an energy flux from large-scale turbulent eddies to smaller-scale ones; when the scale becomes sufficiently small, the energy is dissipated.

Such a situation would occur in "two-dimensional" turbulence, where the fluid velocity $\mathbf{v}$ is always parallel to a fixed xy-plane (Ya. B. Zel'dovich, 1956). It must be stressed that the resulting field $\mathbf{H}$ is not assumed to be two-dimensional. The proof is as follows.

Let us first consider the time variation of the field component H_z perpendicular to the flow. With the equations $\operatorname{div}\mathbf{H} = 0$ and $\operatorname{div}\mathbf{v} = 0$ (the fluid being assumed incompressible in the present section), the z-component of equation (66.1) is

$$\partial H_z/\partial t = -(\mathbf{v}\cdot\mathbf{grad})H_z + \nu_m \triangle H_z, \tag{74.1}$$

which involves only H_z. The first term describes just the transport of a given value of H_z with the fluid particle to which it belongs. The second term describes the "diffusional" equalization of H_z values at various points in the fluid. It is evident that neither can cause H_z to increase. If the initial perturbation H_z occupies a finite region of space, it will disappear in the course of time because of the "diffusion".

To prove that the field components H_x and H_y decay, we can put $H_z = 0$, since what we have to exclude is the possibility that these components remain when H_z has decayed.† We shall do this under the further restriction that all quantities ($\mathbf{v}$ and $\mathbf{H}$) are independent of

† The arguments which follow do not exclude an increase of the field in the initial stage of the process.

the coordinate z.† Then the vector **curl H** is in the z-direction, and the same is true of $\mathbf{v} \times \mathbf{H}$, so that the electric field **E** is likewise in the z-direction; see (66.6). In such a case the electromagnetic field can be described by means of a vector potential **A** which is in the z-direction and independent of the coordinate z: $H_x = \partial A_z/\partial y$, $H_y = -\partial A_z/\partial x$, $E_z = -(1/c)\partial A_z/\partial t$, $\operatorname{div} \mathbf{A} = \partial A_z/\partial z = 0$. Substituting these expressions in (66.6), we find after a simple rearrangement the equation for A_z:

$$\partial A_z/\partial t = -(\mathbf{v}\cdot\mathbf{grad})A_z + \nu_m \triangle A_z, \tag{74.2}$$

which agrees exactly in form with (74.1). Hence it again follows that in the course of time the perturbations A_z decay, and therefore so do H_x and H_y.

The turbulent dynamo is thus an essentially three-dimensional effect. As an illustration, the following is an example of a flow which amplifies the field without altering its spatial scale (S. I. Vaĭnshteĭn and Ya. B. Zel'dovich, 1972). Let us consider a set of closed lines of magnetic force frozen inside a torus of fluid (Fig. 45a). Let the flow stretch the torus by a lengthwise factor of two, say (Fig. 45b); the cross-sectional area decreases by the same factor, and the magnetic field increases. Next, suppose that the torus is twisted by the flow (Fig. 45c), and that the loops are then superimposed (Fig. 45d). This gives a configuration of about the same size as the original one, but with twice the field in the torus. Repetitions of this cycle give an unlimited exponential increase of the field. Such a flow is evidently fundamentally three-dimensional.

This illustration is, of course, not a proof that the turbulent dynamo actually exists; there are also flows which reduce the scale of the field. To settle this point, we need to make a direct investigation of the stability of turbulent motion of a conducting fluid with respect to small initial perturbations of the magnetic field. This approach has made it seem very likely that, when the magnetic Reynolds number is sufficiently large, a magnetic field is in fact generated.‡ These fairly complicated analyses will not be given here; we shall discuss only the general form of the turbulence established by the process, on the assumption that the turbulent dynamo exists.

Turbulent flow may be regarded as an assembly of turbulent eddies of various sizes, from the basic "external" scale l to the smallest "internal" scale λ_0. The former is the same as the characteristic dimensions of the region in which the turbulent flow takes place; the latter gives the order of magnitude of the distances over which energy dissipation becomes important (see *FM*, §§31, 32). By "steady turbulence" we mean that the mean characteristics (averaged over times that are of the order of the periods of the corresponding fluctuations but of course are short compared with the total observation

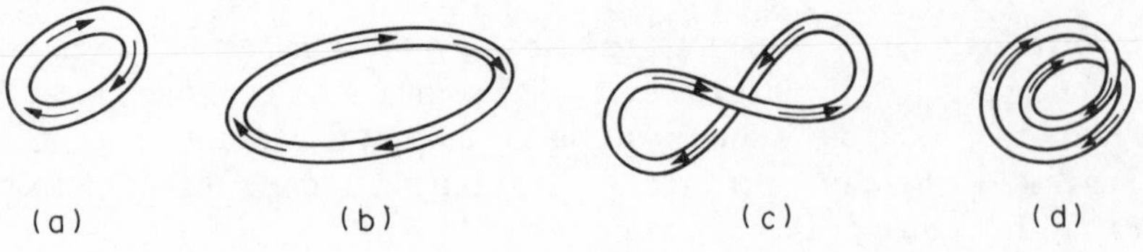

FIG. 45

† This assumption is made here only for purposes of simplification and is not fundamental. The same result can be reached without this restriction by somewhat more complex arguments.

‡ See S. I. Vaĭnshteĭn, *Zhurnal éksperimental'noĭ i teoreticheskoĭ fiziki* **79**, 2175, 1980; **83**, 161, 1982 (*Soviet Physics JETP* **52**, 1099, 1981; **56**, 86, 1983).

time) are constant. The suffix λ will denote mean values for eddies with scale λ; for example, v_λ and H_λ are the mean variations of the velocity and the field over distances $\sim \lambda$.

The statement that a turbulent dynamo exists means that on the basic scale l there are magnetic fields whose energy density $H_l^2/8\pi$ is comparable with the kinetic energy density $\frac{1}{2}\rho v_l^2$ of the fluid. That is, the Alfvén velocity in the field H_l,

$$u_A \sim H_l/\sqrt{(4\pi\rho)}, \tag{74.3}$$

is comparable with the basic scale of the fluid velocity $v_l \equiv u$ (the change in the mean velocity over distances $\sim l$). For scales $\lambda \ll l$, the fields $H_\lambda \ll H_l$.†

We must immediately emphasize the chief difference between magnetic and ordinary turbulence. In the latter, motion on the basic scale does not significantly affect the properties of the small-scale eddies, but simply transports them convectively. In the magnetic case, however, the basic-scale field H_l affects the motion on all smaller scales.

Since for scales $\lambda \ll l$ the field H_l may be regarded as locally uniform, and $H_\lambda \ll H_l$, the small-scale flow in this case is just an assembly of small-amplitude hydromagnetic waves with wave numbers $k \sim 1/\lambda$ and velocities $\sim u_A$. According to (69.11), in these waves the kinetic energy of the fluid is the same as the magnetic energy. That is, in the small-scale eddies there is almost exact equipartition between the magnetic and kinetic energies:

$$\rho v_\lambda^2 \sim H_\lambda^2/4\pi. \tag{74.4}$$

This relation, as an order-of-magnitude one, can be extended to the basic scale, for which it gives $u \sim u_A$, in agreement with the assumption made.

Let us consider the range of scales λ

$$l \gg \lambda \gg \lambda_0. \tag{74.5}$$

It must be remembered that the viscous and Joule dissipations will in general become important for different values of λ, and in this sense magnetic turbulence may have two internal scales, λ_0 in (74.5) being taken as the larger of the two, so that in the *inertial range* (74.5) there is no dissipation.

As in ordinary turbulence theory, we introduce the mean amount of energy ε dissipated per unit time and unit mass of fluid. This energy is derived from the large-scale flow, whence it is gradually transferred to smaller and smaller scales until it is dissipated in eddies of size $\lambda \lesssim \lambda_0$. It is evident that in the inertial range, where there is no dissipation, ε is also the constant (independent of λ) flux of energy towards smaller scales. In ordinary non-magnetic turbulence it was possible to assert that the local properties (i.e. over distances $\lambda \ll l$) of the turbulence must be determined only by ρ, ε and of course the distances λ themselves, but not by the scales l and u of the sizes and velocities as a whole; this was sufficient to give the dependence of v_λ on λ from dimensional arguments. In magnetic turbulence, the local properties may depend also on the field H_l (or, equivalently, on the velocity u_A). Dimensional arguments then do not suffice to determine v_λ, and it is necessary to bring in the actual mechanism by which the energy flux is established.

† Here we are considering the generation of a turbulent magnetic field whose spatial variation has characteristic dimensions $\lesssim l$. The generation of a "large-scale" field with dimensions $\gg l$ is found to be possible only if the averaged properties of the turbulent flow are not invariant under spatial inversion, a case not discussed here. The theory is given in the books by S. I. Vaĭnshteĭn, Ya. B. Zel'dovich and A. A. Ruzmaĭkin, *Turbulentnoe dinamo v astrofizike* (*The Turbulent Dynamo in Astrophysics*), Nauka, Moscow, 1980; H. K. Moffatt, *Magnetic Field Generation in Electrically Conducting Fluids*, Cambridge University Press, Cambridge, 1978.

This mechanism is the mutual interaction of small-amplitude hydromagnetic waves, which is described by non-linear terms in the equations of motion. The energy flux ε must therefore be expanded in powers of the small amplitudes v_λ, and this expansion must begin with terms above the second order (the quadratic terms would correspond to ordinary dissipation, which is absent here). The third-order terms would depend on the random phases of the interacting waves, and would vanish on averaging over these phases. Hence $\varepsilon \propto v_\lambda^4$. The proportionality factor can be found from dimensional arguments (the dimensions of ε are erg/g·s = cm^2/s^3):

$$\varepsilon \sim v_\lambda^4/u_A\lambda, \tag{74.6}$$

or

$$v_\lambda \sim (u_A\varepsilon\lambda)^{1/4} \tag{74.7}$$

(R. H. Kraichnan, 1965). This replaces Kolmogorov and Obukhov's law $v_\lambda \sim (\varepsilon\lambda)^{1/3}$ in ordinary turbulence. Extrapolation of (74.6) to the basic scale gives for ε the estimate $\varepsilon \sim u^4/u_A l \sim u^3/l$, as in non-magnetic fluid dynamics.

The internal scale of turbulence λ_0 can be estimated by regarding the small eddies as hydromagnetic waves against the background of the large-scale field H_l. The viscosity and conductivity of the medium cause absorption of these waves; the corresponding absorption coefficient γ has been found in §69, Problem. The dissipation becomes important when the absorption length $1/\gamma$ is comparable with the wavelength, i.e. with the scale λ. Since the hydromagnetic waves are propagated with velocity u_A (independent of their wavelength), the frequency $\omega \sim u_A/\lambda$. The absorption coefficient can be estimated as $\gamma \sim (\nu+\nu_m)/\lambda^2 u_A$. From the condition that $\gamma \sim 1/\lambda$ when $\lambda \sim \lambda_0$, we find the internal scale

$$\lambda_0 \sim (\nu+\nu_m)/u_A \sim (\nu+\nu_m)/u. \tag{74.8}$$

If $\nu \gg \nu_m$, then $\lambda_0 \sim \nu/u \sim l/\mathrm{R}$, where $\mathrm{R} \sim ul/\nu$ is the Reynolds number for the basic motion. Similarly, if $\nu_m \gg \nu$, then $\lambda_0 \sim l/\mathrm{R}_m$.

There is one further interesting property of the turbulent flow of a highly conducting medium: the magnetic field is displaced from the turbulent region. Let us consider a finite region occupied by the turbulence, with a magnetic field outside it. The lines of force of this field, on entering the turbulent region, become tangled, since they are "frozen"; the magnetic field becomes random in direction. This means that the time-average field **H** becomes almost zero, and the higher the conductivity, the more nearly exact is the vanishing of the field; finite conductivity gives rise to a "slippage" of the lines of force, so that the field does not become completely random. In other words, when a fairly weak magnetic field is applied to a fluid in turbulent motion (within a limited region), the fluid behaves diamagnetically with a small permeability $\mu \ll 1$ which decreases with increasing magnetic Reynolds number R_m.

If the magnetic field is sufficiently strong, however, it is unable to penetrate into the fluid. This does not mean that a strong field will suppress turbulence completely. Two-dimensional turbulence, with the fluid velocity everywhere parallel to the xy-plane and independent of the coordinate z, can occur in any uniform external magnetic field (in the z-direction), no matter how strong. For, in this case, $\mathbf{curl}\,(\mathbf{v}\times\mathbf{H}) = (\mathbf{H}\cdot\mathbf{grad})\mathbf{v} = 0$, and it follows from (65.2) that the flow does not perturb the external field, which remains uniform. No currents arise, therefore, and the Lorentz force is zero. We may say that the two-dimensional flow "does not see" a uniform field. In a strong external field, the turbulence degenerates into just this two-dimensional form.

CHAPTER IX

THE ELECTROMAGNETIC WAVE EQUATIONS

§75. The field equations in a dielectric in the absence of dispersion

In §58 we gave the equations for a variable electromagnetic field in a metal:

$$\mathbf{curl}\,\mathbf{H} = 4\pi\sigma\mathbf{E}/c, \qquad \mathbf{curl}\,\mathbf{E} = -(1/c)\,\partial\mathbf{B}/\partial t, \tag{75.1}$$

which hold when the field changes sufficiently slowly: the frequencies of the field must be such that the dependence of **j** on **E** (and of **B** on **H**, if needed) is that corresponding to the static case.†

We shall now examine the corresponding problem for a variable electromagnetic field in a dielectric, and shall formulate equations valid for frequencies such that the relations between **D** and **E**, and **B** and **H**, are the same as when the fields are constant. If, as usually happens, these relations are simple proportionalities, this means that we can put

$$\mathbf{D} = \varepsilon\mathbf{E}, \qquad \mathbf{B} = \mu\mathbf{H}, \tag{75.2}$$

with the static values of ε and μ.

These relations are not valid (or, as we say, ε and μ exhibit *dispersion*) at frequencies comparable with the eigenfrequencies of the molecular or electronic vibrations which lead to the electric or magnetic polarization of the matter. The order of magnitude of such frequencies depends on the substance concerned, and varies widely. It may also be entirely different for electric and for magnetic phenomena.‡

The equations

$$\operatorname{div}\mathbf{B} = 0, \tag{75.3}$$

$$\mathbf{curl}\,\mathbf{E} = -(1/c)\,\partial\mathbf{B}/\partial t \tag{75.4}$$

are obtained immediately by replacing **e** and **h** in the exact microscopic Maxwell's equations by their averaged values **E** and **B**, and are therefore always valid. The equation

$$\operatorname{div}\mathbf{D} = 0 \tag{75.5}$$

is obtained (§6) by averaging the exact microscopic equation $\operatorname{div}\mathbf{e} = 4\pi\rho$, using only the fact that the total charge on the body is zero. This result is evidently independent of the assumption made in §6 that the field is static, and equation (75.5) is therefore valid in variable fields also.

† The condition $l \ll \lambda$ does not relate to the validity of equations (75.1) as they stand. In the problems discussed in Chapter VII this condition was necessary in order to justify the neglect of retardation effects in the field outside the conductor.

‡ In diamond, for example, the electric polarization is due to the electrons, and the dispersion of ε begins only in the ultra-violet. In a polar liquid such as water, the polarization is due to the orientation of molecules with permanent dipole moments, and the dispersion of ε appears at frequencies $\omega \sim 10^{11}\,\text{sec}^{-1}$, i.e. in the centimetre wavelength range. The dispersion of μ in ferromagnets may begin at even lower frequencies.

A further equation is to be obtained by averaging the exact equation

$$\mathbf{curl\,h} = \frac{1}{c}\frac{\partial \mathbf{e}}{\partial t} + \frac{4\pi}{c}\rho\mathbf{v}. \tag{75.6}$$

A direct averaging gives

$$\mathbf{curl\,B} = \frac{1}{c}\frac{\partial \mathbf{E}}{\partial t} + \frac{4\pi}{c}\overline{\rho\mathbf{v}}. \tag{75.7}$$

When the macroscopic field depends on time, the establishment of the relation between the mean value $\overline{\rho\mathbf{v}}$ and the other quantities is fairly difficult. It is simpler to effect the averaging in the following more formal way.

Let us assume for the moment that extraneous charges of volume density ρ_{ex} are placed in the dielectric. The motion of these charges causes an "extraneous current" $\mathbf{j}_{ex}$, and the conservation of charge is expressed by an equation of continuity:

$$\partial \rho_{ex}/\partial t + \operatorname{div}\mathbf{j}_{ex} = 0.$$

Instead of equation (75.5) we have $\operatorname{div}\mathbf{D} = 4\pi\rho_{ex}$; see (6.8). Differentiating this equation with respect to time and using the equation of continuity, we obtain $\partial(\operatorname{div}\mathbf{D})/\partial t = 4\pi\partial\rho_{ex}/\partial t = -4\pi \operatorname{div}\mathbf{j}_{ex}$, or

$$\operatorname{div}\left(\frac{\partial \mathbf{D}}{\partial t} + 4\pi\mathbf{j}_{ex}\right) = 0.$$

Hence it follows that the vector in parentheses can be written as the curl of another vector, which we denote by $c\mathbf{H}$. Thus

$$\mathbf{curl\,H} = \frac{4\pi}{c}\mathbf{j}_{ex} + \frac{1}{c}\frac{\partial \mathbf{D}}{\partial t}. \tag{75.8}$$

Outside the body this must be the same as the exact Maxwell's equation for the field in a vacuum, and therefore $\mathbf{H}$ is the magnetic field. Inside the body, in the static case, the current $\mathbf{j}_{ex}$ is related to the magnetic field by the equation $\mathbf{curl\,H} = 4\pi\mathbf{j}_{ex}/c$, where $\mathbf{H}$ is the quantity introduced in §29 and related in a definite manner to the mean field $\mathbf{B}$. Hence it follows that, in the limit of zero frequency, the vector $\mathbf{H}$ in equation (75.8) is the static quantity $\mathbf{H}(\mathbf{B})$, and our present assumption that the field varies "slowly" means that the same relation $\mathbf{H}(\mathbf{B})$ holds between these variable fields. Thus $\mathbf{H}$ is a definite quantity, so that we can drop the auxiliary quantity $\mathbf{j}_{ex}$ and obtain the final equation

$$\mathbf{curl\,H} = (1/c)\,\partial\mathbf{D}/\partial t. \tag{75.9}$$

The quantity $\dot{\mathbf{D}}/4\pi$ is called the *displacement current*.

This equation replaces in dielectrics the first equation (75.1) for the field in metals. It might be supposed that the term in $\partial\mathbf{E}/\partial t$ ought to be included when this equation is used for variable fields in metals also, giving

$$\mathbf{curl\,H} = \frac{4\pi}{c}\sigma\mathbf{E} + \frac{\varepsilon}{c}\frac{\partial \mathbf{E}}{\partial t}, \tag{75.10}$$

with ε constant. In good conductors such as the true metals, however, the introduction of this term is pointless. The two terms on the right-hand side of (75.10) are essentially the first two terms in an expansion in powers of the field frequency. Since this frequency is assumed small, the second term must represent at most a small correction. In actual fact, in metals the corrections for the effect of the spatial non-uniformity of the field become important sooner than the frequency correction (see the first footnote to §59).

There are, however, substances, namely poor conductors, for which equation (75.10) may be meaningful. For such reasons as the small number of conduction electrons in semiconductors, or the small mobility of the ions in electrolyte solutions, these substances exhibit anomalously low conductivity, and hence the second term on the right of equation (75.10) may be comparable with, or even exceed, the first term at frequencies for which σ and ε may still be regarded as constants. In a field of a single frequency ω, the ratio of the second term to the first is $\varepsilon\omega/4\pi\sigma$. If this ratio is small, the body behaves as an ordinary conductor with conductivity σ. At frequencies $\omega \gg 4\pi\sigma/\varepsilon$, it behaves as a dielectric with permittivity ε.

In a homogeneous medium with constant ε and μ, equations (75.3)–(75.5) and (75.9) become

$$\operatorname{div} \mathbf{E} = 0, \qquad \operatorname{div} \mathbf{H} = 0, \tag{75.11}$$

$$\mathbf{curl}\, \mathbf{E} = -\frac{\mu}{c}\frac{\partial \mathbf{H}}{\partial t}, \qquad \mathbf{curl}\, \mathbf{H} = \frac{\varepsilon}{c}\frac{\partial \mathbf{E}}{\partial t}. \tag{75.12}$$

Eliminating $\mathbf{E}$ in the usual manner, we obtain

$$\mathbf{curl}\,\mathbf{curl}\, \mathbf{H} = \frac{\varepsilon}{c}\frac{\partial}{\partial t}\mathbf{curl}\, \mathbf{E} = -\frac{\varepsilon\mu}{c^2}\frac{\partial^2 \mathbf{H}}{\partial t^2},$$

and, since $\mathbf{curl}\,\mathbf{curl}\,\mathbf{H} = \mathbf{grad}\operatorname{div}\mathbf{H} - \triangle\mathbf{H} = -\triangle\mathbf{H}$, we reach the wave equation

$$\triangle \mathbf{H} - \frac{\varepsilon\mu}{c^2}\frac{\partial^2 \mathbf{H}}{\partial t^2} = 0.$$

A similar equation for $\mathbf{E}$ can be obtained by eliminating $\mathbf{H}$. We see that the velocity of propagation of electromagnetic waves in a homogeneous dielectric is

$$c/\sqrt{(\varepsilon\mu)}. \tag{75.13}$$

The energy flux density is made up of the electromagnetic field energy flux and the flux of energy carried directly by the movement of matter. In a medium at rest (as is considered here) the latter is zero, and the energy flux density in a dielectric is given by the same formula (30.20) as in a metal:

$$\mathbf{S} = c\mathbf{E} \times \mathbf{H}/4\pi. \tag{75.14}$$

This is easily seen by calculating $\operatorname{div}\mathbf{S}$. Using equations (75.4) and (75.9), we obtain

$$\begin{aligned}\operatorname{div}\mathbf{S} &= \frac{c}{4\pi}(\mathbf{H}\cdot\mathbf{curl}\,\mathbf{E} - \mathbf{E}\cdot\mathbf{curl}\,\mathbf{H}) \\ &= -\frac{1}{4\pi}\left(\mathbf{E}\cdot\frac{\partial \mathbf{D}}{\partial t} + \mathbf{H}\cdot\frac{\partial \mathbf{B}}{\partial t}\right) = -\frac{\partial U}{\partial t},\end{aligned} \tag{75.15}$$

in accordance with the expression $dU = (\mathbf{E}\cdot d\mathbf{D} + \mathbf{H}\cdot d\mathbf{B})/4\pi$ for the differential of the internal energy of a dielectric at given density and entropy.

The requirement of symmetry of the four-dimensional energy-momentum tensor for any closed system (in this case, a dielectric in an electromagnetic field) implies that the energy flux density must be the same, apart from a factor c^2, as the spatial density of momentum (see *Fields*, §§32, 94), which is therefore

$$\mathbf{E}\times\mathbf{H}/4\pi c. \tag{75.16}$$

This expression must, in particular, be used in determining the forces on a dielectric in a variable electromagnetic field. The force **f** per unit volume may be calculated from the stress tensor σ_{ik}: $f_i = \partial\sigma_{ik}/\partial x_k$. Here, however, it must be remembered that σ_{ik} is the momentum flux density, which includes the momentum of both the matter and the electromagnetic field. If **f** is taken as the force on the medium, the rate of change of the field momentum per unit volume must be subtracted:

$$f_i = \frac{\partial\sigma_{ik}}{\partial x_i} - \frac{\partial}{\partial t}\frac{(\mathbf{E}\times\mathbf{H})_i}{4\pi c}. \tag{75.17}$$

In a constant field the last term is zero, and so this question did not arise previously.

Since the field varies slowly, the stress tensor may be taken to have the same value as in a static field. For instance, in a fluid dielectric, σ_{ik} is given by the sum of the electric part (15.9) and the magnetic part (35.2). In differentiating these expressions with respect to the coordinates we must use the fact that the equations **curl E** = 0, **curl H** = 0 for a static field (in the absence of currents) are replaced by equations (75.12). This introduces new terms $-\varepsilon\mathbf{E}\times\mathbf{curl}\,\mathbf{E}/4\pi - \mu\mathbf{H}\times\mathbf{curl}\,\mathbf{H}/4\pi$, which are now not zero but equal to

$$\frac{\varepsilon\mu}{4\pi c}\mathbf{E}\times\dot{\mathbf{H}} - \frac{\varepsilon\mu}{4\pi c}\mathbf{H}\times\dot{\mathbf{E}} = \frac{\varepsilon\mu}{4\pi c}\frac{\partial}{\partial t}(\mathbf{E}\times\mathbf{H}).$$

The required force is therefore

$$\mathbf{f} = -\mathbf{grad}\,P_0(\rho, T) - \frac{E^2}{8\pi}\mathbf{grad}\,\varepsilon - \frac{H^2}{8\pi}\mathbf{grad}\,\mu$$

$$+\mathbf{grad}\left[\rho\left(\frac{\partial\varepsilon}{\partial\rho}\right)_T\frac{E^2}{8\pi} + \rho\left(\frac{\partial\mu}{\partial\rho}\right)_T\frac{H^2}{8\pi}\right] + \frac{\varepsilon\mu-1}{4\pi c}\frac{\partial}{\partial t}(\mathbf{E}\times\mathbf{H}). \tag{75.18}$$

The last term in this expression is called the *Abraham force* (M. Abraham, 1909).

§76. The electrodynamics of moving dielectrics

The motion of a medium results in an interaction between the electric and magnetic fields. Such phenomena for conductors have been discussed in §63; we shall now discuss them for dielectrics. Here we are in practice concerned with the phenomena occurring in moving media when external electric or magnetic fields are present. It should be emphasized that they are in no way related to the appearance of fields as a result of the motion itself (§§36, 64).

Our starting point in §63 was the formulae giving the transformation of the field when the frame of reference is changed. There it was sufficient to know the general formulae for the transformation of electric and magnetic fields in a vacuum, the averaging of which gives immediately the formulae for the transformation of **E** and **B**. In dielectrics the problem is considerably more complex, because the electromagnetic field is described by a greater number of quantities.

In the motion of macroscopic bodies, the velocities involved are usually small compared with the velocity of light. To obtain the necessary approximate transformation formulae, however, it is simplest to use the exact relativistic formulae which hold for all velocities.

In the electrodynamics of the field in a vacuum, the components of the electric and magnetic field vectors **e** and **h** are actually components of an antisymmetrical four-dimensional tensor (or four-tensor) of rank two (see *Fields*, §23). The same is true of **E** and **B**, which are the mean values of **e** and **h**. Thus there is a four-tensor $F_{\mu\nu}$ whose components are given by†

$$F_{\mu\nu} = \begin{pmatrix} 0 & E_x & E_y & E_z \\ -E_x & 0 & -B_z & B_y \\ -E_y & B_z & 0 & -B_x \\ -E_z & -B_y & B_x & 0 \end{pmatrix}, \quad F^{\mu\nu} = \begin{pmatrix} 0 & -E_x & -E_y & -E_z \\ E_x & 0 & -B_z & B_y \\ E_y & B_z & 0 & -B_x \\ E_z & -B_y & B_x & 0 \end{pmatrix}. \tag{76.1}$$

Using this tensor, the first two Maxwell's equations,

$$\operatorname{div} \mathbf{B} = 0, \qquad \mathbf{curl}\, \mathbf{E} = -(1/c)\,\partial \mathbf{B}/\partial t, \tag{76.2}$$

can be written in the four-dimensional form

$$\frac{\partial F_{\lambda\mu}}{\partial x^{\nu}} + \frac{\partial F_{\mu\nu}}{\partial x^{\lambda}} + \frac{\partial F_{\nu\lambda}}{\partial x^{\mu}} = 0. \tag{76.3}$$

This shows the relativistic invariance of the equations. The applicability of equations (76.2) to moving bodies is evident, since they are obtained directly from the exact microscopic Maxwell's equations by replacing **e** and **h** by their averaged values **E** and **B**.

The second pair of Maxwell's equations

$$\operatorname{div} \mathbf{D} = 0, \qquad \mathbf{curl}\, \mathbf{H} = (1/c)\,\partial \mathbf{D}/\partial t \tag{76.4}$$

also retain their form in moving media. This is seen from the arguments given in §75, in which we used only general properties of bodies (e.g. that the total charge is zero), equally valid for moving bodies and bodies at rest. However, the relations between **D** and **E**, and **B** and **H**, need not be the same as in bodies at rest.

Since they are valid for bodies both at rest and in motion, equations (76.4) must be unaltered by the Lorentz transformation. For a field in a vacuum, the vectors **D** and **H** are the same as **E** and **B**, and the relativistic invariance of the second pair of Maxwell's equations appears in the fact that they also can be written in four-dimensional form, using the same tensor $F_{\lambda\mu}$: $\partial F^{\lambda\mu}/\partial x^{\mu} = 0$ (see *Fields*, §30). Hence it is clear that, to ensure the relativistic invariance of equations (76.4), it is necessary that the components of the vectors **D** and **H** should be transformed as the components of a four-tensor exactly similar to $F_{\mu\nu}$, which we denote by $H_{\mu\nu}$:

$$H_{\mu\nu} = \begin{pmatrix} 0 & D_x & D_y & D_z \\ -D_x & 0 & -H_z & H_y \\ -D_y & H_z & 0 & -H_x \\ -D_z & -H_y & H_x & 0 \end{pmatrix}, \quad H^{\mu\nu} = \begin{pmatrix} 0 & -D_x & -D_y & -D_z \\ D_x & 0 & -H_z & H_y \\ D_y & H_z & 0 & -H_x \\ D_z & -H_y & H_x & 0 \end{pmatrix}. \tag{76.5}$$

† In the present section the four-dimensional tensor suffixes, which take the values 0, 1, 2, 3, are denoted by Greek letters λ, μ, ν.

Using this tensor, we can write equations (76.4) in the form

$$\partial H^{\lambda\mu}/\partial x^{\mu} = 0. \tag{76.6}$$

Having elucidated that the quantities **E**, **D**, **H**, **B** form four-dimensional tensors, we have also ascertained the law of their transformation from one frame of reference to another. However, we are interested rather in the relations between the quantities in a moving medium, which generalize the relations $\mathbf{D} = \varepsilon\mathbf{E}$ and $\mathbf{B} = \mu\mathbf{H}$ valid in a medium at rest.

We denote by u^{μ} the velocity four-vector of the medium; its components are related to the three-dimensional velocity **v** by

$$u^{\mu} = \left(\frac{1}{\sqrt{(1-v^2/c^2)}}, \frac{\mathbf{v}}{c\sqrt{(1-v^2/c^2)}}\right).$$

From this four-vector and the four-tensors $F^{\mu\nu}$ and $H^{\mu\nu}$ we form combinations which become **E** and **D** in a medium at rest. These combinations are the four-vectors $F^{\lambda\mu}u_{\mu}$ and $H^{\lambda\mu}u_{\mu}$; for $\mathbf{v} = 0$ their time components are zero and their space components are **E** and **D** respectively. The four-dimensional generalization of the equation $\mathbf{D} = \varepsilon\mathbf{E}$ is therefore evidently†

$$H^{\lambda\mu}u_{\mu} = \varepsilon F^{\lambda\mu}u_{\mu}. \tag{76.7}$$

Similarly, we see that the generalization of $\mathbf{B} = \mu\mathbf{H}$ is the four-dimensional equation

$$F_{\lambda\mu}u_{\nu} + F_{\mu\nu}u_{\lambda} + F_{\nu\lambda}u_{\mu} = \mu(H_{\lambda\mu}u_{\nu} + H_{\mu\nu}u_{\lambda} + H_{\nu\lambda}u_{\mu}). \tag{76.8}$$

Returning from the four-dimensional to the three-dimensional notation, we derive from these two equations the vector relations‡

$$\left.\begin{aligned} \mathbf{D} + \mathbf{v}\times\mathbf{H}/c &= \varepsilon(\mathbf{E} + \mathbf{v}\times\mathbf{B}/c), \\ \mathbf{B} + \mathbf{E}\times\mathbf{v}/c &= \mu(\mathbf{H} + \mathbf{D}\times\mathbf{v}/c). \end{aligned}\right\} \tag{76.9}$$

These formulae, first derived by H. Minkowski (1908), are exact in the sense that no assumption has yet been made concerning the magnitude of the velocity. If the ratio v/c is assumed small the equations can be solved for **D** and **B** as far as terms of the first order to give

$$\mathbf{D} = \varepsilon\mathbf{E} + (\varepsilon\mu - 1)\mathbf{v}\times\mathbf{H}/c, \tag{76.10}$$

$$\mathbf{B} = \mu\mathbf{H} + (\varepsilon\mu - 1)\mathbf{E}\times\mathbf{v}/c. \tag{76.11}$$

These formulae, together with Maxwell's equations (76.2) and (76.4), form the basis for the electrodynamics of dielectrics in motion.

The boundary conditions on Maxwell's equations are also somewhat modified. From the equations $\operatorname{div}\mathbf{D} = 0$, $\operatorname{div}\mathbf{B} = 0$ the continuity of the normal components of the inductions follows as before:

$$D_{n1} = D_{n2}, \qquad B_{n1} = B_{n2}. \tag{76.12}$$

The conditions on the tangential components of the fields are most simply obtained by

† It should be noted that, by writing down relations involving only the local value of the velocity, we neglect slight effects due to the possibility of a velocity gradient, such as gyromagnetic effects (§36).

‡ If either of the relations $\mathbf{D} = \varepsilon\mathbf{E}$ and $\mathbf{B} = \mu\mathbf{H}$ does not hold in the medium at rest, the corresponding relation (76.9) is replaced by a different functional relation between the vector sums on the two sides of the equation.

changing from the fixed frame of reference K to another, K', which moves with the surface element considered, whose velocity along the normal $\mathbf{n}$ we denote by v_n. The usual conditions, namely that $\mathbf{E}'_t$ and $\mathbf{H}'_t$ are continuous, hold in the frame K'. By the relativistic transformation formulae (see *Fields*, §24), these are equivalent to the continuity of the tangential components of the vectors $\mathbf{E}+\mathbf{v}\times\mathbf{B}/c$ and $\mathbf{H}-\mathbf{v}\times\mathbf{D}/c$. Taking the components perpendicular to $\mathbf{n}$ and using equations (76.12), we obtain the required boundary conditions:

$$\left.\begin{aligned}\mathbf{n}\times(\mathbf{E}_2-\mathbf{E}_1) &= v_n(\mathbf{B}_2-\mathbf{B}_1)/c,\\ \mathbf{n}\times(\mathbf{H}_2-\mathbf{H}_1) &= -v_n(\mathbf{D}_2-\mathbf{D}_1)/c.\end{aligned}\right\} \tag{76.13}$$

If we substitute here the expressions (76.10) and (76.11), and neglect terms of higher order in v/c, we obtain

$$\left.\begin{aligned}\mathbf{n}\times(\mathbf{E}_2-\mathbf{E}_1) &= v_n(\mu_2-\mu_1)\mathbf{H}_t/c,\\ \mathbf{n}\times(\mathbf{H}_2-\mathbf{H}_1) &= -v_n(\varepsilon_2-\varepsilon_1)\mathbf{E}_t/c.\end{aligned}\right\} \tag{76.14}$$

In this approximation the values of $\mathbf{H}$ and $\mathbf{E}$ on the two sides of the surface need not be distinguished on the right-hand sides of equations (76.14).

If the body moves so that its surface moves tangentially to itself (e.g. a solid of revolution rotating about its axis), then $v_n = 0$. Only in this case do the boundary conditions (76.13) or (76.14) reduce to the usual conditions that $\mathbf{E}_t$ and $\mathbf{H}_t$ be continuous.

PROBLEMS

PROBLEM 1. A dielectric sphere rotates uniformly in a vacuum in a uniform static magnetic field $\mathfrak{H}$. Determine the resulting electric field near the sphere.

SOLUTION. In calculating the resulting electric field, the magnetic field may be taken to be the same as for a sphere at rest, since an allowance for the reciprocal effect of the magnetic field variation would give corrections of a higher order of smallness. Within the sphere, the magnetic field has the uniform value $\mathbf{H}^{(i)} = 3\mathfrak{H}/(2+\mu)$; cf. (8.2).

Since the rotation is steady, the resulting electric field is constant and, like any static electric field, has a potential: $\mathbf{E} = -\,\mathbf{grad}\,\phi$. Outside the sphere, the potential satisfies the equation $\triangle\phi^{(e)} = 0$; inside the sphere, it satisfies

$$\triangle\phi^{(i)} = 2(\varepsilon\mu-1)\mathbf{\Omega}\cdot\mathbf{H}^{(i)}/c\varepsilon, \tag{1}$$

where $\mathbf{\Omega}$ is the angular velocity. The latter equation is obtained from div $\mathbf{D}=0$ by substituting for $\mathbf{D}$ the expression (76.10) with $\mathbf{v} = \mathbf{\Omega}\times\mathbf{r}$. The condition that the normal component of $\mathbf{D}$ be continuous at the surface of the sphere gives

$$-\varepsilon\left[\frac{\partial\phi^{(i)}}{\partial r}\right]_{r=a} + \frac{\varepsilon\mu-1}{c}a[\mathbf{\Omega}\cdot\mathbf{H}^{(i)} - (\mathbf{\Omega}\cdot\mathbf{n})(\mathbf{H}^{(i)}\cdot\mathbf{n})] = -\left[\frac{\partial\phi^{(e)}}{\partial r}\right]_{r=a}. \tag{2}$$

Here a is the radius of the sphere and $\mathbf{n}$ a unit radial vector.

From the symmetry of the sphere, the required electric field is determined by only two constant vectors, $\mathbf{\Omega}$ and $\mathfrak{H}$. From the components of these vectors we can form a bilinear scalar $\mathfrak{H}\cdot\mathbf{\Omega}$ and a bilinear tensor $\mathfrak{H}_i\Omega_k + \mathfrak{H}_k\Omega_i - \frac{2}{3}\delta_{ik}\mathfrak{H}\cdot\mathbf{\Omega}$, whose trace is zero. Accordingly, we seek the field potential outside the sphere in the form

$$\phi^{(e)} = \frac{1}{6}D_{ik}\frac{\partial^2}{\partial x_i\partial x_k}\left(\frac{1}{r}\right) = \frac{1}{2}D_{ik}\frac{n_i n_k}{r^3}, \tag{3}$$

where D_{ik} is a constant tensor (with $D_{ii}=0$), the electric quadrupole moment tensor of the sphere (see *Fields*, §41). No term of the form constant/r can appear in $\phi^{(e)}$, since such a term would give a non-zero total electric flux through a surface surrounding the sphere, whereas the sphere is uncharged. The field potential inside the sphere is sought in the form

$$\phi^{(i)} = \frac{r^2}{2a^5}D_{ik}n_i n_k + \frac{\varepsilon\mu-1}{3c\varepsilon}\mathbf{\Omega}\cdot\mathbf{H}^{(i)}(r^2-a^2). \tag{4}$$

The first term is the solution of the homogeneous equation $\triangle\phi = 0$, and the coefficient is chosen so as to give continuity of the potential, and therefore of $\mathbf{E}_t$, at the surface of the sphere. Substituting (3) and (4) in (2), we obtain

$$D_{ik} = -\frac{a^5}{c}\frac{3(\varepsilon\mu-1)}{(3+2\varepsilon)(2+\mu)}[\mathfrak{H}_i\Omega_k + \mathfrak{H}_k\Omega_i - \tfrac{2}{3}\delta_{ik}\mathfrak{H}\cdot\mathbf{\Omega}]. \tag{5}$$

Thus a quadrupole electric field is formed near the rotating sphere, and the quadrupole moment of the sphere is given by formula (5). In particular, if the axis of rotation (the z-axis) is parallel to the external field, D_{ik} has only the diagonal components

$$D_{zz} = -\frac{a^5}{c}\frac{4(\varepsilon\mu-1)}{(3+2\varepsilon)(2+\mu)}\mathfrak{H}\cdot\mathbf{\Omega}, \qquad D_{xx} = D_{yy} = -\tfrac{1}{2}D_{zz}.$$

Similarly, a quadrupole magnetic field occurs near a sphere rotating in a uniform electric field. The magnetic quadrupole moment is given by (5) if the sign is changed and ε, μ, $\mathfrak{H}$ are replaced by μ, ε, $\mathfrak{E}$ respectively.

PROBLEM 2. A magnetized sphere rotates uniformly in a vacuum about its axis, which is parallel to the direction of magnetization. Determine the resulting electric field near the sphere.†

SOLUTION. The magnetic field inside the sphere is uniform, and is expressed in terms of the constant magnetization $\mathbf{M}$ by the equations $\mathbf{B}^{(i)} + 2\mathbf{H}^{(i)} = 0$ (cf. (8.1)) and $\mathbf{B}^{(i)} - \mathbf{H}^{(i)} = 4\pi\mathbf{M}$, whence $\mathbf{B}^{(i)} = 8\pi\mathbf{M}/3$, $\mathbf{H}^{(i)} = -4\pi\mathbf{M}/3$. The second of formulae (76.9) does not hold in this case, because the formula $\mathbf{B} = \mu\mathbf{H}$ is not valid for a ferromagnet at rest; from the first of (76.9) we have, inside the sphere,

$$\begin{aligned}\mathbf{D} &= \varepsilon\mathbf{E} + \varepsilon\mathbf{v}\times\mathbf{B}/c - \mathbf{v}\times\mathbf{H}/c \\ &= \varepsilon\mathbf{E} + 4\pi(2\varepsilon+1)\,\mathbf{v}\times\mathbf{M}/3c.\end{aligned}$$

The potential of the resulting electric field outside the sphere satisfies the equation $\triangle\phi^{(e)} = 0$, and that inside the sphere satisfies $\triangle\phi^{(i)} = 8\pi(2\varepsilon+1)M\Omega/3c\varepsilon$.

The boundary condition that D_n be continuous at the surface of the sphere gives

$$-\varepsilon\left[\frac{\partial\phi^{(i)}}{\partial r}\right]_{r=a} + \frac{4\pi(2\varepsilon+1)}{3c}a\Omega M\sin^2\theta = -\left[\frac{\partial\phi^{(e)}}{\partial r}\right]_{r=a},$$

where θ is the angle between the normal $\mathbf{n}$ and the direction of $\mathbf{\Omega}$ and $\mathbf{M}$ (the z-axis). We seek $\phi^{(e)}$ and $\phi^{(i)}$ in the forms

$$\phi^{(e)} = \frac{D_{ik}n_in_k}{2r^3} = \frac{D_{zz}}{4r^3}(3\cos^2\theta - 1),$$

$$\phi^{(i)} = \frac{r^2}{4a^5}D_{zz}(3\cos^2\theta - 1) + \frac{4\pi(2\varepsilon+1)}{9c\varepsilon}M\Omega(r^2 - a^2).$$

From the boundary condition we obtain the following expressions for the electric quadrupole moment of the rotating sphere:

$$D_{zz} = -\frac{4(2\varepsilon+1)}{3c(2\varepsilon+3)}a^2\Omega\mathscr{M}, \qquad D_{xx} = D_{yy} = -\tfrac{1}{2}D_{zz},$$

where $\mathscr{M}$ is the total magnetic moment of the sphere. For a metal sphere we must take $\varepsilon \to \infty$, giving

$$D_{zz} = -4\Omega\mathscr{M}a^2/3c.$$

§77. The dispersion of the permittivity

Let us now go on to study the important subject of rapidly varying electromagnetic fields, whose frequencies are not restricted to be small in comparison with the frequencies

---

† If the direction of magnetization is not the same as that of the axis of rotation, the problem is considerably changed, since the sphere then emits electromagnetic waves.

which characterize the establishment of the electric and magnetic polarization of the substances concerned.

An electromagnetic field variable in time must necessarily be variable in space also. For a frequency ω, the spatial periodicity is characterized by a wavelength $\lambda \sim c/\omega$. As the frequency increases, λ eventually becomes comparable with the atomic dimensions a. The macroscopic description of the field is thereafter invalid.

The question may arise whether there is any frequency range in which, on the one hand, dispersion phenomena are important but, on the other hand, the macroscopic formulation still holds good. It is easy to see that such a range must exist. The most rapid manner of establishment of the electric or magnetic polarization in matter is the electronic mechanism. Its relaxation time is of the order of the atomic time a/v, where v is the velocity of the electrons in the atom. Since $v \ll c$, even the wavelength $\lambda \sim ac/v$ corresponding to these times is large compared with a.

In what follows we shall assume the condition $\lambda \gg a$ to hold.† It must be borne in mind, however, that this condition may not be sufficient: for metals at low temperatures there is a range of frequencies in which the macroscopic theory is inapplicable, although the inequality $c/\omega \gg a$ is satisfied (see §87).

The formal theory given below is equally applicable to metals and to dielectrics. At frequencies corresponding to the motion of the electrons within the atoms (*optical frequencies*) and at higher frequencies, there is, indeed, not even a quantitative difference in the properties of metals and dielectrics.

It is clear from the discussion in §75 that Maxwell's equations

$$\operatorname{div} \mathbf{D} = 0, \qquad \operatorname{div} \mathbf{B} = 0, \tag{77.1}$$

$$\mathbf{curl}\, \mathbf{E} = -(1/c)\, \partial \mathbf{B}/\partial t, \qquad \mathbf{curl}\, \mathbf{H} = (1/c)\, \partial \mathbf{D}/\partial t \tag{77.2}$$

remain formally the same in arbitrary variable electromagnetic fields. These equations are, however, largely useless until the relations between the quantities **D**, **B**, **E** and **H** which appear in them have been established. At the high frequencies at present under consideration, these relations bear no resemblance to those which are valid in the static case and which we have used for variable fields in the absence of dispersion.

First of all, the principal property of these relations, namely the dependence of **D** and **B** only on the values of **E** and **H** at the instant considered, no longer holds good. In the general case of an arbitrary variable field, the values of **D** and **B** at a given instant are not determined only by the values of **E** and **H** at that instant. On the contrary, they depend in general on the values of $\mathbf{E}(t)$ and $\mathbf{H}(t)$ at every previous instant. This expresses the fact that the establishment of the electric or magnetic polarization of the matter cannot keep up with the change in the electromagnetic field. The frequencies at which dispersion phenomena first appear may be completely different for the electric and the magnetic properties of the substance.

In the present section we shall refer to the dependence of **D** on **E**; the specific features of the dispersion of magnetic properties will be discussed in §79.

The polarization vector **P** has been introduced in §6 by means of the definition $\bar{\rho} = -\operatorname{div} \mathbf{P}$, ρ being the true (microscopic) charge density. This equation expresses the

† The effects (called the *natural optical activity*) resulting from terms of the next order in the small ratio a/λ will be considered in §§104–106.

electric neutrality of the body as a whole, and together with the condition $\mathbf{P} = 0$ outside the body it shows that the total electric moment of the body is $\int \mathbf{P}\, dV$. This derivation is evidently valid for variable as well as for static fields. Thus in any variable field, even if dispersion is present, the vector $\mathbf{P} = (\mathbf{D} - \mathbf{E})/4\pi$ retains its physical significance: it is the electric moment per unit volume.

In rapidly varying fields, the field strengths involved are usually fairly small. Hence the relation between **D** and **E** can always be taken to be linear.† The most general linear relation between $\mathbf{D}(t)$ and the values of the function $\mathbf{E}(t)$ at all previous instants can be written in the integral form

$$\mathbf{D}(t) = \mathbf{E}(t) + \int_0^\infty f(\tau)\, \mathbf{E}(t-\tau)\, d\tau. \tag{77.3}$$

It is convenient to separate the term $\mathbf{E}(t)$, for reasons which will become evident later. In equation (77.3) $f(\tau)$ is a function of time and of the properties of the medium. By analogy with the electrostatic formula $\mathbf{D} = \varepsilon \mathbf{E}$, we write the relation (77.3) in the symbolic form $\mathbf{D} = \hat{\varepsilon} \mathbf{E}$, where $\hat{\varepsilon}$ is a linear integral operator whose effect is shown by (77.3).

Any variable field can be resolved by a Fourier expansion into a series of monochromatic components, in which all quantities depend on time through the factor $e^{-i\omega t}$. For such fields the relation (77.3) between **D** and **E** becomes

$$\mathbf{D} = \varepsilon(\omega)\, \mathbf{E}, \tag{77.4}$$

where the function $\varepsilon(\omega)$ is defined as

$$\varepsilon(\omega) = 1 + \int_0^\infty f(\tau) e^{i\omega\tau}\, d\tau. \tag{77.5}$$

Thus, for periodic fields, we can regard the permittivity (the coefficient of proportionality between **D** and **E**) as a function of the field frequency as well as of the properties of the medium. The dependence of ε on the frequency is called its *dispersion relation.*

The function $\varepsilon(\omega)$ is in general complex. We denote its real and imaginary parts by ε' and ε'':

$$\varepsilon(\omega) = \varepsilon'(\omega) + i\varepsilon''(\omega). \tag{77.6}$$

From the definition (77.5) we see at once that

$$\varepsilon(-\omega) = \varepsilon^*(\omega). \tag{77.7}$$

Separating the real and imaginary parts, we have

$$\varepsilon'(-\omega) = \varepsilon'(\omega), \qquad \varepsilon''(-\omega) = -\varepsilon''(\omega). \tag{77.8}$$

Thus $\varepsilon'(\omega)$ is an even function of the frequency, and $\varepsilon''(\omega)$ is an odd function.

For frequencies which are small compared with those at which the dispersion begins, we can expand $\varepsilon(\omega)$ as a power series in ω. The expansion of the even function $\varepsilon'(\omega)$ contains only even powers, and that of the odd function $\varepsilon''(\omega)$ contains only odd powers. In the limit

† Here we assume that **D** depends linearly on **E** alone, and not on **H**. In a constant field, a linear dependence of **D** on **H** is excluded by the requirement of invariance with respect to time reversal. In a variable field, this condition no longer applies, and a linear relation between **D** and **H** is possible if the substance possesses symmetry of various kinds. It is, however, a small effect of the order of a/λ, of the kind mentioned in the last footnote.

as $\omega \to 0$, the function $\varepsilon(\omega)$ in dielectrics tends, of course, to the electrostatic permittivity, which we here denote by ε_0. In dielectrics, therefore, the expansion of $\varepsilon'(\omega)$ begins with the constant term ε_0, while that of $\varepsilon''(\omega)$ begins, in general, with a term in ω.

The function $\varepsilon(\omega)$ at low frequencies can also be discussed for metals, if it is defined in such a way that, in the limit $\omega \to 0$, the question

$$\mathbf{curl\, H} = (1/c)\,\partial\mathbf{D}/\partial t$$

becomes the equation

$$\mathbf{curl\, H} = 4\pi\sigma\,\mathbf{E}/c$$

for a static field in a conductor. Comparing the two equations, we see that for $\omega \to 0$ we must have $\partial\mathbf{D}/\partial t \to 4\pi\sigma\,\mathbf{E}$. But, in a periodic field, $\partial\mathbf{D}/\partial t = -i\omega\varepsilon\,\mathbf{E}$, and we thus obtain the following expression for $\varepsilon(\omega)$ in the limit of low frequencies:

$$\varepsilon(\omega) = 4\pi i\sigma/\omega. \qquad (77.9)$$

Thus the expansion of the function $\varepsilon(\omega)$ in conductors begins with an imaginary term in $1/\omega$, which is expressed in terms of the ordinary conductivity σ for steady currents.† The next term in the expansion of $\varepsilon(\omega)$ is a real constant, although for metals this constant does not have the same electrostatic significance as it does for dielectrics.‡

Moreover, this term of the expansion may again be devoid of significance if the effects of the spatial non-uniformity of the field of the electromagnetic wave appear before those of its periodicity in time.

§78. The permittivity at very high frequencies

In the limit as $\omega \to \infty$, the function $\varepsilon(\omega)$ tends to unity. This is evident from simple physical considerations: when the field changes sufficiently rapidly, the polarization processes responsible for the difference between the field **E** and the induction **D** cannot occur at all.

It is possible to establish the limiting form of the function $\varepsilon(\omega)$ at high frequencies, which is valid for all bodies, whether metals or dielectrics. The field frequency is assumed large compared with the frequencies of the motion of all, or at least the majority, of the electrons in the atoms forming the body. When this condition holds, we can calculate the polarization of the substance by regarding the electrons as free and neglecting their interaction with one another and with the nuclei of the atoms.

The velocities v of the motion of the electrons in the atoms are small compared with the velocity of light. Hence the distances v/ω which they traverse during one period of the electromagnetic wave are small compared with the wavelength c/ω. For this reason we can assume the wave field uniform in determining the velocity acquired by an electron in that field.

The equation of motion is $m\,d\mathbf{v}'/dt = e\mathbf{E} = e\mathbf{E}_0 e^{-i\omega t}$, where e and m are the electron charge and mass, and $\mathbf{v}'$ is the additional velocity acquired by the electron in the wave field.

† The imaginary part of the function $\varepsilon(\omega)$ is sometimes represented in the form (77.9) for all frequencies; this amounts to introducing a new function $\sigma(\omega)$, which has no physical significance apart from its relationship to $\varepsilon''(\omega)$.

‡ To avoid misunderstanding, we should point out a slight change in notation in comparison with §75. In equation (75.10) for poor conductors, $\varepsilon(\omega)$ is $(4\pi i\sigma/\omega)+\varepsilon$.

Hence $\mathbf{v}' = ie\mathbf{E}/m\omega$. The displacement $\mathbf{r}$ of the electron due to the field is given by $\dot{\mathbf{r}} = \mathbf{v}'$, and therefore $\mathbf{r} = -e\mathbf{E}/m\omega^2$. The polarization $\mathbf{P}$ of the body is the dipole moment per unit volume. Summing over all electrons, we find $\mathbf{P} = \Sigma e\mathbf{r} = -e^2 N\mathbf{E}/m\omega^2$, where N is the number of electrons in all the atoms in unit volume of the substance. By the definition of the electric induction, we have $\mathbf{D} = \varepsilon\mathbf{E} = \mathbf{E} + 4\pi\mathbf{P}$. We thus have the formula

$$\varepsilon(\omega) = 1 - 4\pi Ne^2/m\omega^2. \tag{78.1}$$

The range of frequencies over which this formula is applicable begins, in practice, at the far ultra-violet for light elements and at the X-ray region for heavier elements.

If $\varepsilon(\omega)$ is to retain the significance which it has in Maxwell's equations, the frequency must also satisfy the condition $\omega \ll c/a$. We shall see later (§124), however, that the expression (78.1) can be allotted a certain physical significance even at higher frequencies.

§79. The dispersion of the magnetic permeability

Unlike $\varepsilon(\omega)$, the magnetic permeability $\mu(\omega)$ ceases to have any physical meaning at relatively low frequencies. To take account of the deviation of $\mu(\omega)$ from unity would then be an unwarrantable refinement. To show this, let us investigate to what extent the physical meaning of the quantity $\mathbf{M} = (\mathbf{B} - \mathbf{H})/4\pi$, as being the magnetic moment per unit volume, is maintained in a variable field. The magnetic moment of a body is, by definition, the integral

$$\frac{1}{2c}\int \mathbf{r}\times\overline{\rho\mathbf{v}}\,\mathrm{d}V. \tag{79.1}$$

The mean value of the microscopic current density is related to the mean field by equation (75.7):

$$\mathbf{curl}\,\mathbf{B} = \frac{4\pi}{c}\overline{\rho\mathbf{v}} + \frac{1}{c}\frac{\partial\mathbf{E}}{\partial t}. \tag{79.2}$$

Subtracting the equation $\mathbf{curl}\,\mathbf{H} = (1/c)\,\partial\mathbf{D}/\partial t$, we obtain

$$\overline{\rho\mathbf{v}} = c\,\mathbf{curl}\,\mathbf{M} + \partial\mathbf{P}/\partial t. \tag{79.3}$$

The integral (79.1) can, as shown in §29, be put in the form $\int\mathbf{M}\,\mathrm{d}V$ only if $\overline{\rho\mathbf{v}} = c\,\mathbf{curl}\,\mathbf{M}$ and $\mathbf{M} = 0$ outside the body.

Thus the physical meaning of $\mathbf{M}$, and therefore of the magnetic susceptibility, depends on the possibility of neglecting the term $\partial\mathbf{P}/\partial t$ in (79.3). Let us see to what extent the conditions can be fulfilled which make this neglect permissible.

For a given frequency, the most favourable conditions for measuring the susceptibility are those where the body is as small as possible (to increase the space derivatives in **curl M**) and the electric field is as weak as possible (to reduce **P**). The field of an electromagnetic wave does not satisfy the latter condition, because $E \sim H$. Let us therefore consider a variable magnetic field, say in a solenoid, with the body under investigation placed on the axis. The electric field is due only to induction by the variable magnetic field, and the order of magnitude of E inside the body can be obtained by estimating the terms in the equation $\mathbf{curl}\,\mathbf{E} = -(1/c)\partial\mathbf{B}/\partial t$, whence $E/l \sim \omega H/c$ or $E \sim (\omega l/c)H$, where l is the dimension of the body. Putting $\varepsilon - 1 \sim 1$, we have $\partial P/\partial t \sim \omega E \sim \omega^2 lH/c$. For the space derivatives of the magnetic moment $\mathbf{M} = \chi \mathbf{H}$ we have $c\,\mathbf{curl}\,\mathbf{M} \sim c\chi \mathbf{H}/l$. If $|\partial \mathbf{P}/\partial t|$ is small compared with $|c\,\mathbf{curl}\,\mathbf{M}|$, we must have

$$l^2 \ll \chi c^2/\omega^2. \tag{79.4}$$

It is evident that the concept of magnetic susceptibility can be meaningful only if this inequality allows dimensions of the body which are (at least) just macroscopic, i.e. if it is compatible with the inequality $l \gg a$, where a is the atomic dimension. This condition is certainly not fulfilled for the optical frequency range; for such frequencies, the magnetic susceptibility is always $\sim v^2/c^2$, where v is the electron velocity in the atom;† but the optical frequencies themselves are $\sim v/a$, and therefore the right-hand side of the inequality (79.4) is $\sim a^2$.

Thus there is no meaning in using the magnetic susceptibility from optical frequencies onward, and in discussing such phenomena we must put $\mu = 1$. To distinguish between **B** and **H** in this frequency range would be an over-refinement. Actually, the same is true for many phenomena even at frequencies well below the optical range.‡

The presence of a considerable dispersion of the permeability makes possible the existence of quasi-steady oscillations of the magnetization in ferromagnetic bodies. In order to exclude the possible influence of the conductivity, we shall consider ferrites, which are non-metallic ferromagnets.

The term "quasi-steady" means, as usual (§58), that the frequency is assumed to satisfy the condition $\omega \ll c/l$, where l is the characteristic dimension of the body (or the "wavelength" of the oscillation). In addition, we shall neglect the exchange energy related to the inhomogeneity of the magnetization resulting from the oscillations; that is, the spatial dispersion (§103) of the permeability is assumed to be unimportant. For this, the dimensions l must be much greater than the characteristic length for the inhomogeneity energy: $l \gg \sqrt{\alpha}$, where α is of the order of the coefficients in (43.1).

We can put **H** and **B** in the forms $\mathbf{H} = \mathbf{H}_0 + \mathbf{H}'$, $\mathbf{B} = \mathbf{B}_0 + \mathbf{B}'$, where $\mathbf{H}_0$ and $\mathbf{B}_0$ are the field and induction in the statically magnetized body, $\mathbf{H}'$ and $\mathbf{B}'$ the variable parts in the oscillations. When the displacement current is neglected, these variable parts satisfy the equations

$$\mathbf{curl}\,\mathbf{H}' = 0, \quad \operatorname{div} \mathbf{B}' = 0, \tag{79.5}$$

which differ from the magnetostatic equations only in that the permeability is now (for a monochromatic field $\propto e^{-i\omega t}$) a function of the frequency, not a constant.§ A ferromagnetic medium is magnetically anisotropic, and its permeability is therefore a tensor $\mu_{ik}(\omega)$, which determines the linear relation between the variable parts of the induction and the field.

---

† This estimate relates to the diamagnetic susceptibility; the relaxation times of any paramagnetic or ferromagnetic processes are certainly long compared with the optical periods. It must be emphasized, however, that the estimates are made for an isotropic body, and must be used with caution when applied to ferromagnets. In particular, the gyrotropic terms in the tensor μ_{ik} which decrease only slowly (as $1/\omega$) with increasing frequency (see Problem 1) may be important even at fairly high frequencies.

‡ This is discussed from a somewhat different standpoint in §103 below; see the second footnote to that section.

§ These oscillations are therefore called *magnetostatic oscillations*. The theory has been given by C. Kittel (1947) for homogeneous (see below) magnetostatic oscillations and by L. R. Walker (1957) for inhomogeneous ones.

The first equation (79.5) shows that the magnetic field has a potential: $\mathbf{H}' = -\,\mathbf{grad}\,\psi$. Substituting in the second equation $B'_i = \mu_{ik}H'_k = -\mu_{ik}\,\partial\psi/\partial x_k$, we then obtain an equation for the potential within the body:

$$\mu_{ik}(\omega)\partial^2\psi/\partial x_i\,\partial x_k = 0. \tag{79.6}$$

Outside the body, the potential satisfies Laplace's equation $\triangle\psi = 0$; on the surface, $\mathbf{H}'_t$ and B'_n must as usual be continuous. The first condition is equivalent to the continuity of the potential ψ itself; the second implies the continuity of $\mu_{ik}n_i\partial\psi/\partial x_k$, where $\mathbf{n}$ is a unit vector along the normal to the surface. Far from the body, we must have $\psi \to 0$.

The problem thus formulated has non-trivial solutions only for certain values of the μ_{ik}, regarded as parameters. Equating the functions $\mu_{ik}(\omega)$ to these, we find the natural oscillation frequencies of the magnetization of the body, called the *inhomogeneous ferromagnetic resonance* frequencies.

The simplest type of magnetostatic oscillation of a uniformly magnetized ellipsoid consists in oscillations which maintain the uniformity, the magnetization oscillating as a whole. To find their frequencies, it is not necessary to obtain a new solution of the field equations; they can be derived directly from the relations (29.14):

$$H_i + n_{ik}(B_k - H_k) = \mathfrak{H}_i, \tag{79.7}$$

where n_{ik} is the demagnetizing factor tensor of the ellipsoid; $\mathbf{H}$ and $\mathbf{B}$ relate to the field within it, and $\mathfrak{H}$ is the external magnetic field. The latter is assumed to be uniform; in $\mathbf{H}$ and $\mathbf{B}$, we again separate the oscillatory parts $\mathbf{H}'$ and $\mathbf{B}'$, which are now uniform throughout the body. For these we have

$$H'_i + n_{ik}(B'_k - H'_k) = 0$$

or

$$(\delta_{ik} + 4\pi n_{il}\chi_{lk})H'_k = 0,$$

with the magnetic susceptibility tensor $\chi_{ik}(\omega)$ defined by $\mu_{ik} = \delta_{ik} + 4\pi\chi_{ik}$. Equating to zero the determinant of this system of linear homogeneous equations, we find

$$\det|\delta_{ik} + 4\pi n_{il}\chi_{lk}(\omega)| = 0, \tag{79.8}$$

the roots of which give the natural oscillation frequencies. These are called the *homogeneous ferromagnetic resonance* frequencies.

PROBLEMS

PROBLEM 1. Using the macroscopic equation of motion of the magnetic moment (the Landau–Lifshitz equation; see *SP* 2, (69.9)), derive the magnetic permeability tensor for a uniformly magnetized uniaxial ferromagnet of the easy-axis type, in the absence of dissipation (L. D. Landau and E. M. Lifshitz, 1935).

SOLUTION. The equation of motion of the magnetization in a ferromagnet is

$$\dot{\mathbf{M}} = \gamma(\mathbf{H} + \beta M_z \boldsymbol{\nu}) \times \mathbf{M},$$

where $\gamma = g|e|/2mc$ (g being the gyromagnetic ratio), $\beta > 0$ the anisotropy coefficient, and $\boldsymbol{\nu}$ a unit vector along the axis of easy magnetization (the z-axis). We write $\mathbf{H} = \mathbf{H}_0 + \mathbf{H}'$, where $\mathbf{H}'$ is a small variable field in any direction, and $\mathbf{H}_0$ a constant field which we take to be along the z-axis.† The transverse magnetization M_x, M_y due

† This field is used here with a view to applying the results in the subsequent Problems.

to $\mathbf{H}'$ is also small; $M_z \cong M = \text{constant}$. Neglecting second-order small quantities, we find the equations

$$-i\omega M_x = -\gamma(H_0+\beta M)M_y+\gamma MH'_y,$$
$$-i\omega M_y = \gamma(H_0+\beta M)M_x-\gamma MH'_x.$$

On determining M_x and M_y from these equations, we find the susceptibility (as the coefficients in the relations $M'_i = \chi_{ik}H'_k$), and hence the permeability

$$\left.\begin{aligned}\mu_{xx} = \mu_{yy} &= 1-\frac{4\pi}{\beta}\frac{\omega_M(\omega_M+\omega_H)}{\omega^2-(\omega_M+\omega_H)^2} \equiv \mu, \quad \mu_{zz}=1,\\ \mu_{xy} = -\mu_{yx} &= i\frac{4\pi}{\beta}\frac{\omega\omega_M}{\omega^2-(\omega_M+\omega_H)^2}, \quad \mu_{xz}=\mu_{yz}=0,\end{aligned}\right\} \tag{1}$$

where $\omega_M = \gamma\beta M$, $\omega_H = \gamma H_0$. A ferromagnetic medium is thus gyrotropic, in the sense defined in §101.

PROBLEM 2. Find the homogeneous ferromagnetic resonance frequencies for an ellipsoid with one principal axis along the axis of easy magnetization. An external field is applied in that direction (C. Kittel, 1947).†

SOLUTION. Within the ellipsoid there is a field $H_0 = \mathfrak{H} - 4\pi n^{(z)}M$ along the z-axis (the axis of easy magnetization); $n^{(x)}$, $n^{(y)}$, $n^{(z)}$ are the demagnetizing factors along the principal axes of the ellipsoid. A simple calculation of the determinant (79.8) gives

$$\frac{\omega^{(x)}\omega^{(y)}-\omega^2}{(\omega_M+\omega_H)^2-\omega^2} = 0,$$

where

$$\omega^{(x)} = \gamma[M\beta+\mathfrak{H}+4\pi M\,(n^{(x)}-n^{(z)})],$$
$$\omega^{(y)} = \gamma[M\beta+\mathfrak{H}+4\pi M\,(n^{(y)}-n^{(z)})].$$

The homogeneous resonance frequency is therefore

$$\omega = \sqrt{(\omega^{(x)}\omega^{(y)})}.$$

For example, in the case of a sphere $n^{(x)} = n^{(y)} = n^{(z)} = \frac{1}{3}$, and the resonance frequency is $\omega = \gamma(M\beta+\mathfrak{H})$. For a plane-parallel slab whose surface is perpendicular to the easy-magnetization axis we have $n^{(x)} = n^{(y)} = 0$, $n^{(z)} = 1$, and the resonance frequency is $\omega = \gamma(M\beta+\mathfrak{H}-4\pi M)$; the slab is magnetized if $M\beta+\mathfrak{H} > 4\pi M$.

PROBLEM 3. Find the dispersion relation for magnetostatic oscillations in an infinite medium.

SOLUTION. With the tensor μ_{ik} from (1), equation (79.6) becomes

$$\mu\left(\frac{\partial^2\psi}{\partial x^2}+\frac{\partial^2\psi}{\partial y^2}\right)+\frac{\partial^2\psi}{\partial z^2} = 0. \tag{2}$$

Putting $\psi \propto e^{i\mathbf{k}\cdot\mathbf{r}}$, we find $\mu(\omega) = -\cot^2\theta$, where θ is the angle between $\mathbf{k}$ and the easy-magnetization axis (the z-axis). With $\mu(\omega)$ from (1) ($\mathfrak{H} = 0$), the oscillation frequency is

$$\omega = \gamma M\,(\beta+4\pi\,\sin^2\theta)^{\frac{1}{2}}.$$

This depends only on the direction of the wave vector, not on its magnitude. The result agrees, as it should, with the limit (as $k \to 0$) of the dispersion relation for spin waves in a ferromagnet; see *SP* 2, §70.

PROBLEM 4. Find the inhomogeneous resonance frequencies in an infinite plane-parallel slab whose surface is perpendicular to the easy-magnetization axis. An external field $\mathfrak{H}$ is applied in that direction.

SOLUTION. It is necessary to solve equation (2) for the potential $\psi^{(i)}$ within the slab, and the equation $\triangle\psi^{(e)} = 0$ for the potential outside the slab, with the boundary conditions

$$\psi^{(i)} = \psi^{(e)}, \quad \partial\psi^{(i)}/\partial z = \partial\psi^{(e)}/\partial z \text{ for } z = \pm L,$$
$$\psi^{(e)} \to 0 \text{ as } |z| \to \infty;$$

the z-axis is perpendicular to the slab surface, the plane $z = 0$ bisects the slab, and $2L$ is the slab thickness. Such a

† In Problems 2–4 the permeability of the substance is assumed to be given by the equations (1).

solution may be an even or odd function of z. In the first case,

$$\psi^{(i)} = A \cos k_z z \, e^{ik_x x}, \quad \psi^{(e)} = Be^{-k_x|z|} \, e^{ik_x x},$$

with $\mu k_x^2 = -k_z^2$ (the wave vector is in the xz-plane); the boundary conditions give the relation.

$$\tan k_z L = k_x/k_z. \tag{3}$$

In the second case,

$$\psi^{(i)} = A \sin k_z z \, e^{ik_x x}, \quad \psi^{(e)} = \pm Be^{-k_x|z|} \, e^{ik_x x},$$

and the boundary conditions give

$$\tan k_z L = -k_z/k_x. \tag{4}$$

The demagnetizing factor of the slab is $n^{(z)} = 1$, and the demagnetizing field is therefore $-4\pi M$. With $\mu(\omega)$ from (1), we find the oscillation frequency

$$\omega^2 = \gamma^2 (M\beta + \mathfrak{H} - 4\pi M)(M\beta + \mathfrak{H} - 4\pi M \cos^2 \theta), \tag{5}$$

where θ is the angle between $\mathbf{k}$ and the z-axis. For any value of k_x there is a discrete infinity of k_z values given by the conditions (3) and (4). The corresponding frequencies are given by (5), and depend only on the ratio k_x/k_z. All possible frequency values lie in the range

$$\gamma(M\beta + \mathfrak{H} - 4\pi M) \leqslant \omega \leqslant \gamma[(M\beta + \mathfrak{H} - 4\pi M)(M\beta + \mathfrak{H})]^{\frac{1}{2}}.$$

As $k_z \to 0$, only symmetrical oscillations are possible, and from (3) $k_x L \sim (k_z L)^2$, i.e. is a second-order small quantity. Accordingly, we put $\theta = 0$ in (5) and obtain a frequency equal to the homogeneous resonance frequency, as it should be.

§80. The field energy in dispersive media

The formula

$$\mathbf{S} = c\mathbf{E} \times \mathbf{H}/4\pi \tag{80.1}$$

for the energy flux density remains valid in variable electromagnetic fields, even if dispersion is present. This is evident from the arguments given at the end of §30: on account of the continuity of the tangential components of **E** and **H**, formula (80.1) follows from the condition that the normal component of **S** be continuous at the boundary of the body and the validity of the same formula in the vacuum outside the body.

The rate of change of the energy in unit volume of the body is div **S**. Using Maxwell's equations, we can write this expression as

$$-\operatorname{div} \mathbf{S} = \frac{1}{4\pi}\left(\mathbf{E} \cdot \frac{\partial \mathbf{D}}{\partial t} + \mathbf{H} \cdot \frac{\partial \mathbf{B}}{\partial t}\right); \tag{80.2}$$

see (75.15). In a dielectric medium without dispersion, when ε and μ are real constants, this quantity can be regarded as the rate of change of the electromagnetic energy

$$U = (\varepsilon \mathbf{E}^2 + \mu \mathbf{H}^2)/8\pi, \tag{80.3}$$

which has an exact thermodynamic significance: it is the difference between the internal energy per unit volume with and without the field, the density and entropy remaining unchanged.

In the presence of dispersion, no such simple interpretation is possible. Moreover, in the general case of arbitrary dispersion, the electromagnetic energy cannot be rationally defined as a thermodynamic quantity. This is because the presence of dispersion in general signifies a dissipation of energy, i.e. a dispersive medium is also an absorbing medium.

To determine this dissipation, let us consider a monochromatic electromagnetic field. By averaging with respect to time the expression (80.2), we find the steady inflow of energy per unit time and volume from the external sources which maintain the field. Since the amplitude of the field is assumed constant, all of this energy goes to compensate the dissipation. Under the conditions stated, therefore, the time average of (80.2) is the mean quantity Q of heat evolved per unit time and volume.

Since the expression (80.2) is quadratic in the fields, all quantities must be written in real form. If, as is convenient for a monochromatic field, we take **E** and **H** to be complex, then in (80.2) we must substitute for **E** and $\partial\mathbf{D}/\partial t$ respectively $\frac{1}{2}(\mathbf{E}+\mathbf{E}^*)$ and $\frac{1}{2}(-i\omega\varepsilon\mathbf{E}+i\omega\varepsilon^*\mathbf{E}^*)$, and similarly for **H** and $\partial\mathbf{B}/\partial t$. On averaging with respect to time, the products $\mathbf{E}\cdot\mathbf{E}$ and $\mathbf{E}^*\cdot\mathbf{E}^*$, which contain factors $e^{\mp 2i\omega t}$, give zero, leaving

$$Q = \frac{i\omega}{16\pi}[(\varepsilon^*-\varepsilon)\mathbf{E}\cdot\mathbf{E}^*+(\mu^*-\mu)\mathbf{H}\cdot\mathbf{H}^*] = \frac{\omega}{8\pi}(\varepsilon''|\mathbf{E}|^2+\mu''|\mathbf{H}|^2). \tag{80.4}$$

This expression can also be written

$$Q = \omega(\varepsilon''\overline{\mathbf{E}^2}+\mu''\overline{\mathbf{H}^2})/4\pi, \tag{80.5}$$

where **E** and **H** are the real fields, and the bar denotes an average with respect to time; cf. the last footnote to §59.

It is easy to derive also a formula for the energy dissipation in a non-monochromatic field which tends sufficiently rapidly to zero as $t \to \pm\infty$. In this case it is appropriate to consider the dissipation not per unit time but over the whole duration of the existence of the field.

Expanding the field $\mathbf{E}(t)$ as a Fourier integral, we write

$$\mathbf{E}(t) = \int_{-\infty}^{\infty} \mathbf{E}_\omega e^{-i\omega t}\,\mathrm{d}\omega/2\pi,$$

$$\frac{\partial\mathbf{D}(t)}{\partial t} = -i\int_{-\infty}^{\infty} \omega\,\varepsilon(\omega)\mathbf{E}_\omega e^{-i\omega t}\,\mathrm{d}\omega/2\pi,$$

with $\mathbf{E}_{-\omega} = \mathbf{E}^*_\omega$. Writing the products of these quantities as a double integral and then integrating with respect to time, we have

$$\frac{1}{4\pi}\int_{-\infty}^{\infty}\mathbf{E}\cdot\frac{\partial\mathbf{D}}{\partial t} = -\frac{i}{4\pi}\int_{-\infty}^{\infty}\omega\varepsilon(\omega)\mathbf{E}_\omega\cdot\mathbf{E}_{\omega'}\,e^{-i(\omega+\omega')t}\frac{\mathrm{d}\omega\,\mathrm{d}\omega'}{(2\pi)^2}\,\mathrm{d}t.$$

The integration with respect to t is effected by means of the formula

$$\int_{-\infty}^{\infty} e^{-i(\omega+\omega')t}\,\mathrm{d}t = 2\pi\delta(\omega+\omega'),$$

and the delta function is then eliminated by the integration over ω'. The result is

$$-\frac{i}{4\pi}\int_{-\infty}^{\infty}\omega\varepsilon(\omega)|\mathbf{E}_\omega|^2\frac{\mathrm{d}\omega}{2\pi}.$$

On substituting $\varepsilon = \varepsilon' + i\varepsilon''$, the term in $\varepsilon'(\omega)$ vanishes in the integration, since the integrand is an odd function of ω. With a similar expression for the magnetic field, we have finally

$$\int_{-\infty}^{\infty} Q\,dt = \frac{1}{4\pi}\int_{-\infty}^{\infty} \omega\,[\varepsilon''(\omega)|\mathbf{E}_\omega|^2 + \mu''(\omega)|\mathbf{H}_\omega|^2]\frac{d\omega}{2\pi}; \tag{80.6}$$

the integral from $-\infty$ to ∞ can be replaced by twice the integral from 0 to ∞.

These formulae show that the absorption (dissipation) of energy is determined by the imaginary parts of ε and μ. The two terms in (80.5) are called the *electric* and *magnetic losses* respectively. On account of the law of increase of entropy, the sign of these losses is determinate: the dissipation of energy is accompanied by the evolution of heat, i.e. $Q > 0$. It therefore follows that the imaginary parts of ε and μ are always positive:

$$\varepsilon'' > 0, \qquad \mu'' > 0 \tag{80.7}$$

for all substances and at all (positive) frequencies.† The signs of the real parts of ε and μ for $\omega \neq 0$ are subject to no physical restriction.

Any non-steady process in an actual body is to some extent thermodynamically irreversible. The electric and magnetic losses in a variable electromagnetic field therefore always occur to some extent, however slight. That is, the functions $\varepsilon''(\omega)$ and $\mu''(\omega)$ are not exactly zero for any frequency other than zero. We shall see in § 81 that this statement is of fundamental importance, although it does not exclude the possibility of only very small losses in certain frequency ranges. Such ranges, in which ε'' and μ'' are very small in comparison with ε' and μ', are called *transparency ranges*. It is possible to neglect the absorption in these ranges and to introduce the concept of the internal energy of the body in the electromagnetic field, in the same sense as in a static field. To determine this quantity, it is not sufficient to consider a purely monochromatic field, since the strict periodicity results in no steady accumulation of electromagnetic energy. Let us therefore consider a field whose components have frequencies in a narrow range about some mean value ω_0. The field strengths can be written

$$\mathbf{E} = \mathbf{E}_0(t)e^{-i\omega_0 t}, \qquad \mathbf{H} = \mathbf{H}_0(t)e^{-i\omega_0 t}, \tag{80.8}$$

where $\mathbf{E}_0(t)$ and $\mathbf{H}_0(t)$ are functions of time which vary only slowly in comparison with the factor $e^{-i\omega_0 t}$. The real parts of these expressions are to be substituted on the right-hand side of (80.2), and we then average with respect to time over the period $2\pi/\omega_0$, which is small compared with the time of variation of the factors $\mathbf{E}_0$ and $\mathbf{H}_0$.

The first term in (80.2), with $\mathbf{E}$ written in complex form, is

$$\tfrac{1}{2}(\mathbf{E}+\mathbf{E}^*)\cdot\tfrac{1}{2}(\dot{\mathbf{D}}+\dot{\mathbf{D}}^*)/4\pi,$$

and similarly for the second term. The products $\mathbf{E}\cdot\dot{\mathbf{D}}$ and $\mathbf{E}^*\cdot\dot{\mathbf{D}}^*$ vanish when averaged

† This statement applies to bodies which, in the absence of the variable field, are in thermodynamic equilibrium; we assume this condition to hold. If the body is not in thermal equilibrium, then Q may in principle be negative. The second law of thermodynamics requires only a net increase in entropy as a result of the effects of the variable electromagnetic field and of the absence of thermodynamic equilibrium, the latter effect being independent of the presence of the field. An example of such a body is one in which all the atoms have been excited artificially (i.e. not by spontaneous thermal excitation, but by an external "pumping" field).

over time, and can therefore be ignored, leaving

$$\frac{1}{16\pi}\left(\mathbf{E}\cdot\frac{\partial \mathbf{D}^*}{\partial t}+\mathbf{E}^*\cdot\frac{\partial \mathbf{D}}{\partial t}\right). \tag{80.9}$$

We write the derivative $\partial\mathbf{D}/\partial t$ as $\hat{f}\mathbf{E}$, where $\hat{f}$ is the operator $\partial\hat{\varepsilon}/\partial t$, and ascertain the effect of this operator on a function of the form (80.8). If $\mathbf{E}_0$ were a constant, we should have simply $\hat{f}\mathbf{E}=f(\omega)\mathbf{E}$, where $f(\omega)=-i\omega\varepsilon(\omega)$. We expand the function $\mathbf{E}_0(t)$ as a series of Fourier components $\mathbf{E}_{0\alpha}e^{-i\alpha t}$, with constant $\mathbf{E}_{0\alpha}$. Since $\mathbf{E}_0(t)$ varies only slowly, this series will include only components with $\alpha \ll \omega_0$. We can therefore put

$$\begin{aligned}\hat{f}\mathbf{E}_{0\alpha}e^{-i(\omega_0+\alpha)t} &= f(\alpha+\omega_0)\mathbf{E}_{0\alpha}e^{-i(\omega_0+\alpha)t}\\ &\cong f(\omega_0)\mathbf{E}_{0\alpha}e^{-i(\omega_0+\alpha)t}+\alpha\frac{\mathrm{d}f(\omega_0)}{\mathrm{d}\omega_0}\mathbf{E}_{0\alpha}e^{-i(\omega_0+\alpha)t}.\end{aligned}$$

Summing the Fourier components, we have

$$\hat{f}\mathbf{E}_0(t)e^{-i\omega_0 t}=f(\omega_0)\mathbf{E}_0e^{-i\omega_0 t}+i\frac{\mathrm{d}f(\omega_0)}{\mathrm{d}\omega_0}\frac{\partial\mathbf{E}_0}{\partial t}e^{-i\omega_0 t}.$$

Omitting henceforward the suffix 0 to ω, we thus obtain

$$\frac{\partial\mathbf{D}}{\partial t}=-i\omega\varepsilon(\omega)\mathbf{E}+\frac{\mathrm{d}(\omega\varepsilon)}{\mathrm{d}\omega}\frac{\partial\mathbf{E}_0}{\partial t}e^{-i\omega t}. \tag{80.10}$$

Substituting this expression in (80.9) and neglecting the imaginary part of $\varepsilon(\omega)$ gives

$$\frac{1}{16\pi}\frac{\mathrm{d}(\omega\varepsilon)}{\mathrm{d}\omega}\left(\mathbf{E}_0{}^*\cdot\frac{\partial\mathbf{E}_0}{\partial t}+\mathbf{E}_0\cdot\frac{\partial\mathbf{E}_0{}^*}{\partial t}\right)=\frac{1}{16\pi}\frac{\mathrm{d}(\omega\varepsilon)}{\mathrm{d}\omega}\frac{\partial}{\partial t}(\mathbf{E}\cdot\mathbf{E}^*),$$

since $\mathbf{E}\cdot\mathbf{E}^*=\mathbf{E}_0\cdot\mathbf{E}_0{}^*$. Adding a similar expression involving the magnetic field, we conclude that the steady rate of change of the energy in unit volume is given by $\mathrm{d}\overline{U}/\mathrm{d}t$, where

$$\overline{U}=\frac{1}{16\pi}\left[\frac{\mathrm{d}(\omega\varepsilon)}{\mathrm{d}\omega}\mathbf{E}\cdot\mathbf{E}^*+\frac{\mathrm{d}(\omega\mu)}{\mathrm{d}\omega}\mathbf{H}\cdot\mathbf{H}^*\right]. \tag{80.11}$$

In terms of the real fields **E** and **H** this expression can be written

$$\overline{U}=\frac{1}{8\pi}\left[\frac{\mathrm{d}(\omega\varepsilon)}{\mathrm{d}\omega}\overline{E^2}+\frac{\mathrm{d}(\omega\mu)}{\mathrm{d}\omega}\overline{H^2}\right] \tag{80.12}$$

(L. Brillouin, 1921).

This is the required result: $\overline{U}$ is the mean value of the electromagnetic part of the internal energy per unit volume of a transparent medium. If there is no dispersion, ε and μ are constants, and (80.12) becomes the mean value of (80.3), as it should.

If the external supply of electromagnetic energy to the body is cut off, the absorption which is always present (even though very small) ultimately converts the energy $\overline{U}$ entirely into heat. Since, by the law of increase of entropy, there must be evolution and not absorption of heat, we must have $\overline{U}>0$. It therefore follows, by (80.11), that the inequalities

$$\mathrm{d}(\omega\varepsilon)/\mathrm{d}\omega>0, \quad \mathrm{d}(\omega\mu)/\mathrm{d}\omega>0 \tag{80.13}$$

must hold. In reality, these conditions are necessarily fulfilled, by virtue of more stringent inequalities always satisfied by the functions $\varepsilon(\omega)$ and $\mu(\omega)$ in transparency ranges (see the footnote to §84).

It should be emphasized once more that the expression (80.12) has been derived in the first approximation with respect to the frequencies α with which the amplitude $\mathbf{E}_0(t)$ varies. It is therefore valid only for fields whose amplitude varies sufficiently slowly with time. This comment applies also to the calculation of the stress tensor in §81.

§81. The stress tensor in dispersive media

Another topic of considerable interest is the determination of the (time) averaged stress tensor giving the forces on matter in a variable electric field. We shall show that, even when dispersion is present (but absorption is again absent), the expression for this tensor contains no derivatives with respect to the frequency, unlike the energy equation (80.12). In particular, for a transparent isotropic dispersive fluid in a monochromatic electric field, the mean value $\bar{\sigma}_{ik}$ is derived from (15.9) by simply replacing ε by $\varepsilon(\omega)$ and the products $E_i E_k$ and $\mathbf{E}^2$ by their mean values $\overline{E_i E_k}$ and $\overline{\mathbf{E}^2}$ (L. P. Pitaevskiĭ, 1960).

To prove this, we return to the proof given in §15 and reformulate it to some extent. We there considered a dielectric-filled plane capacitor and determined the stress tensor from the equality between the work done by the ponderomotive forces in a movement of the plates and the change in the appropriate thermodynamic potential. We now write this condition for total quantities (not those per unit area), in the form

$$A\sigma_{ik}\xi_i n_k = (\delta\mathscr{U})_{\mathscr{S},e}, \tag{81.1}$$

where A is the area of the capacitor plates. The potential $\tilde{\mathscr{F}}$ is here replaced by the ordinary energy $\mathscr{U}$, the change in which is taken for fixed values of the entropy $\mathscr{S}$ of the dielectric and of the total charges $\pm e$ on the capacitor plates (instead of a fixed potential ϕ), and the result $(\delta\tilde{\mathscr{F}})_{T,\phi} = (\delta\mathscr{U})_{\mathscr{S},e}$ given by the theorem of small increments is used.

In the form (81.1), this condition has a particularly simple significance: a thermally insulated capacitor with fixed charges on the plates is an electrically closed system. If mechanical work is done on it (to move the plates) by an external source, all the work goes to increase the energy of the capacitor. The latter is

$$\mathscr{U} = \mathscr{U}_0 + e^2/2C, \tag{81.2}$$

where $\mathscr{U}_0$ is the energy of the dielectric in the absence of the field (for the same value of the entropy $\mathscr{S}$) and C is the capacitance. For a plane capacitor, $C = \varepsilon A/4\pi h$, where h is the distance between the plates. Hence

$$(\delta\mathscr{U})_{\mathscr{S},e} = (\delta\mathscr{U}_0)_{\mathscr{S}} - \frac{e^2}{2C^2}(\delta C)_{\mathscr{S}}. \tag{81.3}$$

On expressing δC in terms of the displacement ξ of the plates (taking into account the dependence of ε on the density of the dielectric, which is changed by the displacement), we immediately find (15.9)†; since the result is obvious, it will not be given here.

† It is expressed in terms of different variables, with the adiabatic instead of the isothermal derivative $\partial\varepsilon/\partial\rho$ and function P_0. The two forms are, of course, equivalent.

When dispersion is present, the expression for $\mathscr{U}$ is different. We shall show that the relation (81.3), however, remains valid for the time-averaged variation $\overline{\delta \mathscr{U}}$, and this will complete the proof of the above statement concerning the averaged stress tensor.

Let the charge on the capacitor plates vary monochromatically with frequency ω. Then the capacitor itself is not an electrically closed system, since charge has to be supplied and withdrawn. The oscillatory circuit with natural frequency ω consisting of the capacitor and an appropriately chosen self-inductance *is* an electrically closed system, however†; the relation (81.1) is therefore valid for its energy.

In the absence of resistance, the potential difference ϕ between the capacitor plates is equal to the sum of the external e.m.f. and the self-inductance e.m.f.:

$$\phi = \mathscr{E} - \frac{1}{c^2} L \frac{\mathrm{d}J}{\mathrm{d}t}, \tag{81.4}$$

and the current J is related to the capacitor plate charge e by $J = \mathrm{d}e/\mathrm{d}t$. For quantities varying monochromatically, the capacitance $C(\omega)$ is by definition such that $\phi = e/C(\omega)$. Putting $\mathscr{E} = 0$, $J = -i\omega e$, in (81.4) we find first of all that even when there is dispersion of the capacitance the circuit frequency again satisfies Thomson's relation (62.5):

$$\omega = c/\sqrt{[LC(\omega)]}. \tag{81.5}$$

Next, multiplying (81.4) by $J = \mathrm{d}e/\mathrm{d}t$ and considering almost monochromatic quantities as in the derivation of (80.12), we easily find

$$\overline{\mathscr{E} J} = \frac{\mathrm{d}}{\mathrm{d}t}\left\{\frac{1}{2} L \overline{J^2} + \frac{\mathrm{d}(\omega C)}{\mathrm{d}\omega} \cdot \frac{1}{2}\overline{\phi^2}\right\}.$$

The form of this equation shows that the expression in the braces is the energy $\mathscr{U}$ of the oscillatory circuit. The first term in this expression can be transformed by substituting $J = -i\omega e$ and using (81.5):

$$\frac{1}{2c^2} L \overline{J^2} = \frac{1}{2c^2} L\omega^2 \overline{e^2} = \frac{1}{2c^2} LC^2 \omega^2 \overline{\phi^2} = \tfrac{1}{2} C \overline{\phi^2}.$$

The final expression for the circuit energy is‡

$$\overline{\mathscr{U}} = \frac{1}{\omega} \frac{\mathrm{d}(\omega^2 C)}{\mathrm{d}\omega} \cdot \tfrac{1}{2}\overline{\phi^2}. \tag{81.6}$$

We have to calculate the variation of this energy for a small displacement of the capacitor plates, i.e., for a small change in the capacitance. In a variable field, this displacement is to be regarded as taking place with infinite slowness. In such a change, the adiabatic invariant given (as in any linear oscillatory system) by the ratio of the oscillation energy and the frequency remains constant.§ We thus have $\delta(\overline{\mathscr{U}}/\omega) = 0$, i.e.

$$\delta \overline{\mathscr{U}} = \overline{\mathscr{U}} \delta\omega/\omega. \tag{81.7}$$

---

† To meet the conditions for a quasi-steady state, the circuit must be small in comparison with the wavelength c/ω. This limitation is not a fundamental one, and does not impair the generality of the proof given.

‡ From now on, to simplify the formulae, we omit the "non-electromagnetic" part $\mathscr{U}_0$ of the energy.

§ Cf. *Mechanics*, §49. The invariance of this quantity is particularly clear from the quantum theory: the ratio $\overline{\mathscr{U}}/\hbar\omega$ is the state quantum number, which remains unchanged when the conditions vary adiabatically.

From (81.5), for a small change in the capacitance,

$$\delta\omega/\omega = -\tfrac{1}{2}\delta C/C. \tag{81.8}$$

This change consists of two parts:

$$\delta C = (\delta C)_{st} + (\mathrm{d}C/\mathrm{d}\omega)\delta\omega. \tag{81.9}$$

The first term is the static part, related to the strain as in the static case; here it is important that, in the presence of dispersion, the capacitance $C(\omega)$ is expressed in terms of $\varepsilon(\omega)$ in the same way as in the static case. The second term depends only on the frequency change. From (81.8) and (81.9) we find as the static part

$$(\delta C)_{st} = -\frac{1}{\omega^2}\frac{\mathrm{d}(\omega^2 C)}{\mathrm{d}\omega}\delta\omega. \tag{81.10}$$

When (81.6) is substituted in (81.7), and (81.10) is used, $\mathrm{d}C/\mathrm{d}\omega$ disappears, and the energy variation becomes

$$\delta\overline{\mathscr{U}} = -\tfrac{1}{2}\overline{\phi^2}(\delta C)_{st} = -\tfrac{1}{2}(\overline{e^2}/C^2)(\delta C)_{st}, \tag{81.11}$$

which is in fact the same as the averaged second term in (81.3).

The disappearance from $\delta\overline{\mathscr{U}}$ of the terms involving the derivative with respect to ω is quite general and does not depend on a specific manner of change in the state of the body (in this case, the capacitor). In particular, for a dispersive medium formula (14.1) (with $\mathbf{E}^2$ replaced by $\overline{\mathbf{E}^2}$) remains valid for the change in the free energy due to a small change in ε:

$$\delta\mathscr{F} = -\int \delta\varepsilon(\omega)\overline{\mathbf{E}^2}\,\mathrm{d}V/8\pi. \tag{81.12}$$

Knowing the stress tensor, we can use (75.17) to find the force on unit volume of the dielectric. The terms containing spatial derivatives coincide with the corresponding terms in the time-averaged expression (75.18), in which we must put $\mu = 1$. The term in the time derivative (the Abraham force) is not the same. This term arises as the difference

$$\frac{1}{4\pi c}\left\{\frac{\partial}{\partial t}(\mathbf{D}\times\mathbf{H}) - \frac{\partial}{\partial t}(\mathbf{E}\times\mathbf{H})\right\},$$

which is now to be averaged over time. To do so, we express **D**, **E** and **H** in complex form (i.e. replace them by $\tfrac{1}{2}(\mathbf{D}+\mathbf{D}^*)$ and so on), and then use (80.10) for $\partial\mathbf{D}/\partial t$. This gives the Abraham force in the form

$$\frac{1}{8\pi c}(\varepsilon - 1)\,\mathrm{re}\frac{\partial}{\partial t}(\mathbf{E}\times\mathbf{H}^*) + \frac{1}{8\pi c}\omega\frac{\mathrm{d}\varepsilon}{\mathrm{d}\omega}\,\mathrm{re}\left(\frac{\partial\mathbf{E}}{\partial t}\times\mathbf{H}^*\right) \tag{81.13}$$

(H. Vashina and V. I. Karpman, 1976).

The stress tensor in a variable field is significant for absorbing as well as transparent media, unlike the internal energy, for which the problem can be formulated only by neglecting the dissipation. There is, however, reason to suppose that in an absorbing medium the stress tensor cannot be expressed in terms of the permittivity alone, and therefore cannot be derived in a general form by macroscopic methods.

§82. The analytical properties of $\varepsilon(\omega)$

The function $f(\tau)$ in (77.3) is finite for all values of τ, including zero.† For dielectrics it tends to zero as $\tau \to \infty$. This simply expresses the fact that the value of $\mathbf{D}(t)$ at any instant cannot be appreciably affected by the values of $\mathbf{E}(t)$ at remote instants. The physical agency underlying the integral relation (77.3) consists in the processes of the establishment of the electric polarization. Hence the range of values in which the function $f(\tau)$ differs appreciably from zero is of the order of the relaxation time which characterizes these processes.

The above statements are true also of metals, the only difference being that the function $f(\tau) - 4\pi\sigma$, rather than $f(\tau)$ itself, tends to zero as $\tau \to \infty$. This difference arises because the passage of a steady conduction current, though it does not cause any actual change in the physical state of the metal, in our equations leads formally to the presence of an induction $\mathbf{D}$ such that $(1/c)\partial\mathbf{D}/\partial t = 4\pi\sigma\mathbf{E}/c$ or

$$\mathbf{D}(t) = \int_{-\infty}^{t} 4\pi\sigma\,\mathbf{E}(\tau)\mathrm{d}\tau = 4\pi\sigma \int_{0}^{\infty} \mathbf{E}(t-\tau)\,\mathrm{d}\tau.$$

We have defined the function $\varepsilon(\omega)$ by (77.5):

$$\varepsilon(\omega) = 1 + \int_{0}^{\infty} e^{i\omega\tau} f(\tau)\,\mathrm{d}\tau. \tag{82.1}$$

It is possible to derive some very general relations concerning this function by regarding ω as a complex variable ($\omega = \omega' + i\omega''$). These relations could be formulated immediately, since the dielectric susceptibility $[\varepsilon(\omega) - 1]/4\pi$ is one of the generalized susceptibilities already discussed in *SP* 1, §123. We shall nevertheless repeat here some of the arguments and results, both to assist the reader and to emphasize certain differences between dielectrics and metals. From the definition (82.1) and the above-mentioned properties of the function $f(\tau)$, it follows that $\varepsilon(\omega)$ is a one-valued function which nowhere becomes infinite (i.e. has no singularities) in the upper half-plane. For, when $\omega'' > 0$, the integrand in (82.1) includes the exponentially decreasing factor $e^{-\omega''\tau}$ and, since the function $f(\tau)$ is finite throughout the region of integration, the integral converges. The function $\varepsilon(\omega)$ has no singularity on the real axis ($\omega'' = 0$), except possibly at the origin (where, for metals, $\varepsilon(\omega)$ has a simple pole).

In the lower half-plane, the definition (82.1) is invalid, since the integral diverges. Hence the function $\varepsilon(\omega)$ can be defined in the lower half-plane only as the analytical continuation of formula (82.1) from the upper half-plane, and in general has singularities.

The function $\varepsilon(\omega)$ has a physical as well as a mathematical significance in the upper half-plane: it gives the relation between $\mathbf{D}$ and $\mathbf{E}$ for fields whose amplitude increases as $e^{\omega'' t}$. In the lower half-plane, this physical interpretation is not possible, if only because the presence of a field which is damped as $e^{-|\omega''|t}$ implies an infinite field for $t \to -\infty$.

It is useful to notice that the conclusion that $\varepsilon(\omega)$ is regular in the upper half-plane is,

† It was to ensure this that the term $\mathbf{E}(t)$ was separated in (77.3), since otherwise the function $f(\tau)$ would have a delta-function singularity at $\tau = 0$.

physically, a consequence of the causality principle. The integration in (77.3) is, on account of this principle, taken only over times previous to t, and the region of integration in formula (82.1) therefore extends from 0 to ∞ rather than from $-\infty$ to ∞.

It is evident also from the definition (82.1) that

$$\varepsilon(-\omega^*) = \varepsilon^*(\omega). \tag{82.2}$$

This generalizes the relation (77.7) for real ω. In particular, for purely imaginary ω we have

$$\varepsilon(i\omega'') = \varepsilon^*(i\omega'') \tag{82.3}$$

i.e. the function $\varepsilon(\omega)$ is real on the positive imaginary axis.†

It should be emphasized that the property (82.2) merely expresses the fact that the operator relation $\mathbf{D} = \hat{\varepsilon}\,\mathbf{E}$ must give real values of $\mathbf{D}$ for real $\mathbf{E}$. If the function $\mathbf{E}(t)$ is given by the real expression

$$\mathbf{E} = \mathbf{E}_0 e^{-i\omega t} + \mathbf{E}_0^* e^{i\omega^* t}, \tag{82.4}$$

then, applying the operator $\hat{\varepsilon}$ to each term, we have

$$\mathbf{D} = \varepsilon(\omega)\,\mathbf{E}_0\, e^{-i\omega t} + \varepsilon(-\omega^*)\,\mathbf{E}_0^*\, e^{i\omega^* t},$$

and the condition for this to be real is just (82.2).

According to the results of §80, the imaginary part of $\varepsilon(\omega)$ is positive for positive real $\omega = \omega'$, i.e. on the right-hand half of the real axis. Since, by (82.2), $\operatorname{im}\varepsilon(-\omega') = -\operatorname{im}\varepsilon(\omega')$, the imaginary part of $\varepsilon(\omega)$ is negative on the left-hand half of this axis. Thus

$$\operatorname{im}\varepsilon \gtrless 0 \text{ for } \omega = \omega' \gtrless 0. \tag{82.5}$$

At $\omega = 0$, im ε changes sign, passing through zero for dielectrics and through infinity for metals. This is the only point on the real axis for which im $\varepsilon(\omega)$ can vanish.

When ω tends to infinity in any manner in the upper half-plane, $\varepsilon(\omega)$ tends to unity. This has been shown in §78 for the case where ω tends to infinity along the real axis. The general result is seen from formula (82.1): if $\omega \to \infty$ in such a way that $\omega'' \to \infty$, the integral in (82.1) vanishes because of the factor $e^{-\omega''\tau}$ in the integrand, while if ω'' remains finite but $|\omega'| \to \infty$ the integral vanishes because of the oscillating factor $e^{i\omega'\tau}$.

The above properties of the function $\varepsilon(\omega)$ are sufficient to prove the following theorem: the function $\varepsilon(\omega)$ does not take real values at any finite point in the upper half-plane except on the imaginary axis, where it decreases monotonically from $\varepsilon_0 > 1$ (for dielectrics) or from $+\infty$ (for metals) at $\omega = i0$ to 1 at $\omega = i\infty$. Hence, in particular, it follows that the function $\varepsilon(\omega)$ has no zeros in the upper half-plane. We shall not repeat here the proof of these results given in *SP* 1, §123; it need only be remembered that the generalized susceptibility is $\varepsilon(\omega) - 1$, not $\varepsilon(\omega)$.

We shall also not repeat the derivation of the relations between the imaginary and real parts of $\varepsilon(\omega)$, but give only the final formulae, with the notation appropriately modified. We write the function $\varepsilon(\omega)$ of the real variable ω, as in §77, in the form

† This is not in general true for the negative imaginary axis. Here the function $\varepsilon(\omega)$ may have branch points, and a cut along the negative axis may be necessary in order to define it as an analytic function in the lower half-plane. Equation (82.2) then signifies only that $\varepsilon(\omega)$ has complex conjugate values on the two sides of the cut.

$\varepsilon(\omega) = \varepsilon'(\omega) + i\varepsilon''(\omega)$. If $\varepsilon(\omega)$ relates to a dielectric, the relations in question are

$$\varepsilon'(\omega) - 1 = \frac{1}{\pi} P \int_{-\infty}^{\infty} \frac{\varepsilon''(x)}{x - \omega} dx, \tag{82.6}$$

$$\varepsilon''(\omega) = -\frac{1}{\pi} P \int_{-\infty}^{\infty} \frac{\varepsilon'(x) - 1}{x - \omega} dx \tag{82.7}$$

(H. A. Kramers and R. de L. Kronig, 1927). It should be emphasized that the only important property of the function $\varepsilon(\omega)$ used in the proof is that it is regular in the upper half-plane. Hence we can say that Kramers and Kronig's formulae, like this property of $\varepsilon(\omega)$, are a direct consequence of the causality principle.

Using the fact that $\varepsilon''(x)$ is an odd function, we can rewrite (82.6) as

$$\varepsilon'(\omega) - 1 = \frac{2}{\pi} P \int_{0}^{\infty} \frac{x\varepsilon''(x)}{x^2 - \omega^2} dx. \tag{82.8}$$

If a metal is concerned, the function $\varepsilon(\omega)$ has a pole at the point $\omega = 0$, near which $\varepsilon = 4\pi i\sigma/\omega$ (77.9). This gives an additional term in (82.7) (cf. *SP* 1, (123.18)):

$$\varepsilon''(\omega) = -\frac{1}{\pi} P \int_{-\infty}^{\infty} \frac{\varepsilon'(x)}{x - \omega} dx + \frac{4\pi\sigma}{\omega}, \tag{82.9}$$

but (82.6) and (82.8) remain unchanged. A further remark is also necessary as regards metals. We have said at the end of §77 that there may be ranges of frequency for metals in which the function $\varepsilon(\omega)$ becomes physically meaningless on account of the spatial non-uniformity of the field. In the formulae given here, however, the integration must be taken over all frequencies. In such cases $\varepsilon(\omega)$ must be taken, in the frequency ranges concerned, as the function obtained by solving the formal problem of the behaviour of the body in a fictitious uniform periodic electric field (and not in the necessarily non-uniform field of the electromagnetic wave).

Formula (82.8) is of particular importance: it makes possible a calculation of the function $\varepsilon'(\omega)$ if the function $\varepsilon''(\omega)$ is known even approximately (for example, empirically) for a given body. It is important to note that, for any function $\varepsilon''(\omega)$ satisfying the physically necessary condition $\varepsilon'' > 0$ for $\omega > 0$, formula (82.8) gives a function $\varepsilon'(\omega)$ consistent with all physical requirements, i.e. one which is in principle possible (the sign and magnitude of ε' are subject to no general physical restrictions). This makes it possible to use formula (82.8) even when the function $\varepsilon''(\omega)$ is approximate. Formula (82.7), on the other hand, does not give a physically possible function $\varepsilon''(\omega)$ for an arbitrary choice of the function $\varepsilon'(\omega)$, since the condition that $\varepsilon''(\omega) > 0$ is not necessarily fulfilled.

In dispersion theory the expression for $\varepsilon'(\omega)$ is customarily written in the form

$$\varepsilon'(\omega) - 1 = -\frac{4\pi e^2}{m} P \int_{0}^{\infty} \frac{f(x)}{\omega^2 - x^2} dx, \tag{82.10}$$

where e and m are the charge and mass of the electron, and $f(\omega)\,d\omega$ is called the *oscillator strength* in the frequency range $d\omega$. According to (82.8), this quantity is related to $\varepsilon''(\omega)$ by

$$f(\omega) = \frac{m}{2\pi^2 e^2}\,\omega\,\varepsilon''(\omega). \tag{82.11}$$

For metals, $f(\omega)$ tends to a finite limit as $\omega \to 0$.

For sufficiently large ω, x^2 can be neglected in comparison with ω^2 in the integrand in (82.8). Then

$$\varepsilon'(\omega) - 1 = -\frac{2}{\pi\omega^2}\int_0^\infty x\,\varepsilon''(x)\,dx.$$

For the permittivity at high frequencies, on the other hand, formula (78.1) holds, and a comparison shows that we have the *sum rule*

$$\frac{m}{2\pi^2 e^2}\int_0^\infty \omega\,\varepsilon''(\omega)d\omega = \int_0^\infty f(\omega)\,d\omega = N, \tag{82.12}$$

where N is the total number of electrons per unit volume.

If $\varepsilon''(\omega)$ is regular at $\omega = 0$, we can take the limit $\omega \to 0$ in formula (82.8), obtaining

$$\varepsilon'(0) - 1 = \frac{2}{\pi}\int_0^\infty \frac{\varepsilon''(x)}{x}\,dx. \tag{82.13}$$

If the point $\omega = 0$ is a singularity of $\varepsilon''(\omega)$ (as in metals), the limit of the integral (82.8) as $\omega \to 0$ is not what is obtained by simply deleting the term in ω. To calculate the limit, we must first replace $\varepsilon''(x)$ in the integrand by $\varepsilon''(x) - 4\pi\sigma/x$; the value of the integral is unchanged, because

$$P\int_0^\infty \frac{dx}{x^2-\omega^2} \equiv 0.$$

For a dielectric, formula (82.13) can be rewritten as

$$\varepsilon_0 - 1 = \frac{4\pi e^2 N}{m}\,\overline{\omega^{-2}}, \tag{82.14}$$

where the bar denotes averaging with respect to the oscillator strength:

$$\overline{\omega^{-2}} = \frac{1}{N}\int_0^\infty \frac{f(\omega)}{\omega^2}\,d\omega.$$

The expression (82.14) may be useful in estimating ε_0.

The following formula† relates the values of $\varepsilon(\omega)$ on the upper half of the imaginary axis

† Derived in *SP* 1, §123.

to those of $\varepsilon''(\omega)$ on the real axis:

$$\varepsilon(i\omega)-1=\frac{2}{\pi}\int_0^\infty \frac{x\varepsilon''(x)}{x^2+\omega^2}\,dx. \tag{82.15}$$

Integrating this relation over all ω, we obtain

$$\int_0^\infty [\varepsilon(i\omega)-1]\,d\omega=\int_0^\infty \varepsilon''(\omega)\,d\omega. \tag{82.16}$$

All the above results are applicable, apart from slight changes, to the magnetic permeability $\mu(\omega)$. The differences are due principally to the fact that the function $\mu(\omega)$ ceases to be physically meaningful at relatively low frequencies. Hence, for example, Kramers and Kronig's formulae must be applied to $\mu(\omega)$ as follows. We consider not an infinite but a finite range of ω (from 0 to ω_1), which extends only to frequencies where μ is still meaningful but no longer variable, so that its imaginary part may be taken as zero; let the real quantity $\mu(\omega_1)$ be denoted by μ_1.† Then formula (82.8) must be written as

$$\mu'(\omega)-\mu_1=\frac{2}{\pi}P\int_0^{\omega_1} \frac{x\mu''(x)}{x^2-\omega^2}\,dx. \tag{82.17}$$

Unlike ε_0, the value μ_0 of $\mu(0)$ may be either less than or greater than unity. The variation of $\mu(\omega)$ along the imaginary axis is again a monotonic decrease, from μ_0 to $\mu_1 < \mu_0$.

Lastly, it may be noted that the analytical properties of $\varepsilon(\omega)$ derived in this section are also possessed by $\eta(\omega)\equiv 1/\varepsilon(\omega)$. For example, $\eta(\omega)$ is analytic in the upper half-plane, because $\varepsilon(\omega)$ is analytic and has no zeros in that half-plane. The same Kramers–Kronig relations (82.6) and (82.7) apply to $\eta(\omega)$ as to $\varepsilon(\omega)$.

§83. A plane monochromatic wave

Maxwell's equations (77.2) for a monochromatic field are

$$i\omega\mu(\omega)\mathbf{H}=c\,\mathbf{curl}\,\mathbf{E},\qquad i\omega\varepsilon(\omega)\mathbf{E}=-c\,\mathbf{curl}\,\mathbf{H}. \tag{83.1}$$

These equations as they stand are complete, since equations (77.1) follow from (83.1) and so do not require separate consideration. Assuming the medium homogeneous, and eliminating $\mathbf{H}$ from equations (83.1), we obtain the second-order equation

$$\triangle\mathbf{E}+\varepsilon\mu(\omega^2/c^2)\mathbf{E}=0; \tag{83.2}$$

elimination of $\mathbf{E}$ gives a similar equation for $\mathbf{H}$.

Let us consider a plane electromagnetic wave propagated in an infinite homogeneous medium. In a plane wave in a vacuum, the spatial dependence of the field is given by a factor $e^{i\mathbf{k}\cdot\mathbf{r}}$, with a real wave vector $\mathbf{k}$. In considering wave propagation in matter, however, it is in

† In fact, ω_1 must be such that $\omega_1\tau \gg 1$, where τ is the shortest relaxation time for ferromagnetic and paramagnetic processes in a magnetic material.

general necessary to take $\mathbf{k}$ complex: $\mathbf{k} = \mathbf{k}' + i\mathbf{k}''$, where the vectors $\mathbf{k}'$ and $\mathbf{k}''$ are real.

Taking $\mathbf{E}$ and $\mathbf{H}$ as proportional to $e^{i\mathbf{k}\cdot\mathbf{r}}$, and carrying out the differentiation with respect to the coordinates in equations (83.1), we obtain

$$\omega\mu\mathbf{H} = c\mathbf{k}\times\mathbf{E}, \qquad \omega\varepsilon\mathbf{E} = -c\mathbf{k}\times\mathbf{H}. \tag{83.3}$$

Eliminating $\mathbf{E}$ and $\mathbf{H}$ from these two equations, we obtain for the square of the wave vector

$$k^2 \equiv k'^2 - k''^2 + 2i\mathbf{k}'\cdot\mathbf{k}'' = \varepsilon\mu\omega^2/c^2. \tag{83.4}$$

We see that $\mathbf{k}$ can be real only if ε and μ are real and positive. Even then, however, $\mathbf{k}$ may still be complex if $\mathbf{k}'\cdot\mathbf{k}'' = 0$; we shall meet with such a case in discussing total reflection in §86.

It must be borne in mind that, in the general case of complex $\mathbf{k}$, the term "plane wave" is purely conventional. Putting $e^{i\mathbf{k}\cdot\mathbf{r}} = e^{i\mathbf{k}'\cdot\mathbf{r}}\, e^{-\mathbf{k}''\cdot\mathbf{r}}$, we see that the planes perpendicular to the vector $\mathbf{k}'$ are planes of constant phase. The planes of constant amplitude, however, are those perpendicular to $\mathbf{k}''$, the direction in which the wave is damped. The surfaces on which the field itself is constant are in general not planes at all. Such waves are called *inhomogeneous plane waves*, in contradistinction to ordinary homogeneous plane waves.

The general relation between the electric and magnetic field components is given by formulae (83.3). In particular, taking the scalar product of these formulae with $\mathbf{k}$, we obtain

$$\mathbf{k}\cdot\mathbf{E} = 0, \qquad \mathbf{k}\cdot\mathbf{H} = 0, \tag{83.5}$$

and, squaring either and using (83.4),

$$\mathbf{E}^2 = \mu\mathbf{H}^2/\varepsilon. \tag{83.6}$$

It must be remembered, however, that because all three vectors $\mathbf{k}$, $\mathbf{E}$ and $\mathbf{H}$ are complex these formulae do not in general have the same evident significance as when the vectors are real.

We shall not give the cumbersome relations valid in the general case, but consider only the most important particular cases. Especially simple results are obtained for a wave propagated without damping in a non-absorbing (transparent) homogeneous medium. The wave vector is real, and its magnitude is

$$k = \sqrt{(\varepsilon\mu)}\,\omega/c = n\omega/c, \tag{83.7}$$

where $n = \sqrt{(\varepsilon\mu)}$ is called the *refractive index* of the medium. The electric and magnetic fields are both in a plane perpendicular to the vector $\mathbf{k}$ (a pure transverse wave); they are mutually perpendicular, and are related by

$$\mathbf{H} = \sqrt{(\varepsilon/\mu)}\,\mathbf{1}\times\mathbf{E}, \tag{83.8}$$

where $\mathbf{1}$ is a unit vector in the direction of $\mathbf{k}$. Hence it follows that $\varepsilon\mathbf{E}\cdot\mathbf{E}^* = \mu\mathbf{H}\cdot\mathbf{H}^*$, but this does not mean (as it would in the absence of dispersion) that the electric and magnetic energies in the wave are equal, since these energies are given by different expressions (namely, the two terms in formula (80.11)). The total electromagnetic energy density in this case may be written

$$\overline{U} = \frac{1}{16\pi\mu\omega}\frac{\mathrm{d}}{\mathrm{d}\omega}(\omega^2\varepsilon\mu)\,\mathbf{E}\cdot\mathbf{E}^* = \frac{c}{8\pi}\sqrt{\frac{\varepsilon}{\mu}}\frac{\mathrm{d}k}{\mathrm{d}\omega}\,\mathbf{E}\cdot\mathbf{E}^*. \tag{83.9}$$

The velocity u with which the wave is propagated in the medium is given by the familiar expression for the group velocity:†

$$u = \frac{d\omega}{dk} = \frac{c}{d(n\omega)/d\omega}. \tag{83.10}$$

Here $u = \overline{S}/\overline{U}$, in accordance with its significance as the velocity of transfer of energy in the wave packet; $\overline{U}$ is the energy density given by formula (83.9), and

$$\overline{S} = \frac{c}{8\pi}\sqrt{\frac{\varepsilon}{\mu}}\,\mathbf{E}\cdot\mathbf{E}^* \tag{83.11}$$

is the mean value of the Poynting vector. In the absence of dispersion, when the refractive index is independent of frequency, the expression (83.10) becomes simply c/n; cf. (75.13).

Next, let us consider a more general case, the propagation of an electromagnetic wave in an absorbing medium, the wave vector having a definite direction (i.e. $\mathbf{k}'$ and $\mathbf{k}''$ being parallel). Then the wave is literally plane, since the surfaces of constant field in it are planes perpendicular to the direction of propagation (a *homogeneous plane wave*).

In this case we can introduce the complex "length" k of the wave vector, given by $\mathbf{k} = k\mathbf{l}$ ($\mathbf{l}$ being a unit vector in the direction of $\mathbf{k}'$ and $\mathbf{k}''$), and from (83.4) we have $k = \sqrt{(\varepsilon\mu)}\,\omega/c$. The complex quantity $\sqrt{(\varepsilon\mu)}$ is usually written in the form $n + i\kappa$, with real n and κ, so that

$$k = \sqrt{(\varepsilon\mu)}\,\omega/c = (n + i\kappa)\,\omega/c. \tag{83.12}$$

The quantity n is called the *refractive index* of the medium, and κ the *absorption coefficient*; the latter gives the rate of damping of the wave during its propagation. It should be emphasized, however, that the damping of the wave need not be due to true absorption: dissipation of energy occurs only when ε or μ is complex, but κ may be different from zero if ε and μ are real and negative.

We may express n and κ in terms of the real and imaginary parts of the permittivity (taking $\mu = 1$). From the equation

$$n^2 - \kappa^2 + 2in\kappa = \varepsilon = \varepsilon' + i\varepsilon''$$

we have $n^2 - \kappa^2 = \varepsilon'$, $2n\kappa = \varepsilon''$. Solving these equations for n and κ, we have‡

$$\left.\begin{aligned} n &= \sqrt{\{\tfrac{1}{2}[\varepsilon' + \sqrt{(\varepsilon'^2 + \varepsilon''^2)}]\}},\\ \kappa &= \sqrt{\{\tfrac{1}{2}[-\varepsilon' + \sqrt{(\varepsilon'^2 + \varepsilon''^2)}]\}}. \end{aligned}\right\} \tag{83.13}$$

In particular, for metals and in the frequency range where formula (77.9) is valid, the imaginary part of ε is large compared with its real part, and is related to the conductivity by $\varepsilon'' = 4\pi\sigma/\omega$; neglecting ε' in comparison with ε'', we find that n and κ are equal, in agreement with (59.4):

$$n = \kappa = \sqrt{(2\pi\sigma/\omega)}. \tag{83.14}$$

† When considerable absorption occurs, the group velocity cannot be used, since in an absorbing medium wave packets are not propagated but rapidly "ironed out".

‡ Since $\varepsilon'' > 0$, the signs of n and κ must be the same, in accordance with the fact that the wave is damped in the direction of propagation. The choice of positive signs in (83.13) corresponds to a wave propagated in the positive x-direction.

The relation between the fields **E** and **H** in this homogeneous plane wave is again given by formula (83.8), but ε and μ are now complex. The formula again shows that the two fields and the direction of propagation are mutually perpendicular. If $\mu = 1$, we write $\sqrt{\varepsilon} = \sqrt{(n^2+\kappa^2)} \exp[i \tan^{-1}(\kappa/n)]$, which shows that the magnetic field is $\sqrt{(n^2+\kappa^2)}$ times the electric field in magnitude and $\tan^{-1}(\kappa/n)$ from it in phase; in particular, when (83.14) holds, the phase difference is $\frac{1}{4}\pi$.

PROBLEM

At a given instant $t = 0$ an electromagnetic perturbation occurs in some region of space. The perturbation is not maintained by external agencies, and is therefore damped in time. Find the damping decrement.

SOLUTION. We expand the initial perturbation as a Fourier integral with respect to the coordinates, and consider a component having a (real) wave vector **k**. The time dependence of this component is given, for sufficiently large t, by a factor $e^{-i\omega t}$ with a complex frequency ω, which is to be determined; the damping decrement is $-\operatorname{im}\omega$.

From the equations $-\dot{\mathbf{H}}/c = \mathbf{curl}\,\mathbf{E} = i\mathbf{k}\times\mathbf{E}$, $\dot{\mathbf{D}}/c = \mathbf{curl}\,\mathbf{H} = i\mathbf{k}\times\mathbf{H}$ we have, eliminating **H**,

$$\ddot{\mathbf{D}}/c^2 = \mathbf{k}\times(\mathbf{k}\times\mathbf{E}). \tag{1}$$

We take the direction of **k** as the x-axis. The longitudinal part of the perturbation therefore satisfies $\ddot{D}_x = 0$, whence $D_x = 0$.

The relation between D_x and E_x is given by an integral operator:

$$E_x(t) = \hat{\varepsilon}^{-1} D_x = \int_{-\infty}^{t} F(t-\tau) D_x(\tau)\, d\tau. \tag{2}$$

Since we have $D_x(t) = 0$ for $t > 0$ (there are no field sources when $t > 0$), it follows that

$$E_x(t) = \int_{-\infty}^{0} F(t-\tau) D_x(\tau)\, d\tau. \tag{3}$$

Hence we see that, for large t, the time dependence of E_x is given essentially by that of the function $F(t)$.

For a monochromatic field, (3) gives

$$\frac{1}{\varepsilon(\omega)} = \int_0^{\infty} F(\tau) e^{i\omega\tau}\, d\tau,$$

or, conversely,

$$F(t) = \frac{1}{2\pi}\int_{-\infty}^{\infty} \frac{1}{\varepsilon(\omega)} e^{-i\omega t}\, d\omega.$$

To estimate this integral for large t, we displace the path of integration into the lower half-plane of ω, where the integrand decreases rapidly. The singularities of the function $1/\varepsilon(\omega)$, i.e. the zeros and branch points of $\varepsilon(\omega)$, must be excluded from the contour. The integral is then essentially proportional to $e^{-i\omega_0 t}$, where ω_0 is the singularity nearest the real axis. This gives the solution for the longitudinal part of the perturbation.

For the transverse components, we have from (1) $\ddot{D}_{y,z}/c^2 + k^2 E_{y,z} = 0$. A similar analysis gives the result that the required "frequency" ω_0 is in this case the zero or branch point of the function $\omega^2\varepsilon(\omega) - c^2k^2$ which lies nearest the real axis.

§84. Transparent media

Let us apply the general formulae derived in §82 to media which absorb only slightly in a given range of frequencies, i.e. assuming that for these frequencies the imaginary part of the permittivity may be neglected.

In such a case there is no need to take the principal value in formula (82.8), since the point $x = \omega$ does not in practice lie in the region of integration. The integral can then be differentiated in the usual way with respect to the parameter ω, giving

$$\frac{d\varepsilon}{d\omega} = \frac{4\omega}{\pi} \int_0^\infty \frac{x\varepsilon''(x)}{(\omega^2 - x^2)^2} dx.$$

Since the integrand is positive throughout the region of integration, we conclude that

$$d\varepsilon(\omega)/d\omega > 0, \tag{84.1}$$

i.e. if absorption is absent the permittivity is a monotonically increasing function of the frequency.

Similarly in the same frequency range we obtain another inequality,

$$\frac{d}{d\omega}[\omega^2(\varepsilon - 1)] = \frac{4\omega}{\pi} \int_0^\infty \frac{x^3 \varepsilon''(x)}{(x^2 - \omega^2)^2} dx > 0,$$

or

$$d\varepsilon/d\omega > 2(1 - \varepsilon)/\omega. \tag{84.2}$$

If $\varepsilon < 1$, this inequality is more stringent than (84.1).†

It may be noted that the inequalities (84.1) and (84.2) (together with the corresponding ones for $\mu(\omega)$) ensure that the inequality $u < c$ is satisfied by the velocity of propagation of waves. For example, if $\mu = 1$ we have $n = \sqrt{\varepsilon}$ and, replacing ε by n^2 in (84.1) and (84.2),

$$d(n\omega)/d\omega > n, \qquad d(n\omega)/d\omega > 1/n. \tag{84.3}$$

Thus we obtain two inequalities for the velocity u (83.10): $u < c/n$ and $u < cn$, whence $u < c$ whether $n < 1$ or $n > 1$. These inequalities also show that $u > 0$, i.e. the group velocity is in the same direction as the wave vector. This is quite natural, even if not logically necessary.

Let us suppose that the weak absorption extends over a wide range of frequencies, from ω_1 to ω_2 ($\gg \omega_1$), and consider frequencies ω such that $\omega_1 \ll \omega \ll \omega_2$. The region of integration in (82.8) divides into two parts, $x < \omega_1$ and $x > \omega_2$. In the former region we can neglect x in comparison with ω, and in the latter region ω in comparison with x, in the denominator of the integrand:

$$\varepsilon(\omega) = 1 + \frac{2}{\pi} \int_{\omega_2}^\infty \varepsilon''(x) \frac{dx}{x} - \frac{2}{\pi\omega^2} \int_0^{\omega_1} x\varepsilon''(x) dx, \tag{84.4}$$

i.e. the function $\varepsilon(\omega)$ in this range is of the form $a - b/\omega^2$, where a and b are positive constants. The constant b can be expressed in terms of the oscillator strength N_1 responsible for the absorption in the range from 0 to ω_1 (cf. (82.12)):

$$\varepsilon(\omega) = a - 4\pi N_1 e^2/m\omega^2. \tag{84.5}$$

From this expression it follows, in particular, that, when the region of weak absorption is sufficiently wide, the permittivity in general passes through zero. In this connection it

† By adding the inequalities (84.1) and (84.2), we find that $d(\omega\varepsilon)/d\omega > 1$, which is stronger than (80.13).

should be recalled that a literally transparent medium is one in which $\varepsilon(\omega)$ is not only real but also positive; if ε is negative, the wave is damped inside the medium, even though no true dissipation of energy occurs.

For the frequency at which $\varepsilon = 0$ the induction $\mathbf{D}$ is zero identically, and Maxwell's equations admit a variable electric field satisfying the single equation $\mathbf{curl}\,\mathbf{E} = 0$, with zero magnetic field. In other words, longitudinal electric waves can occur. To determine their velocity of propagation, we must take into account the dispersion of the permittivity not only in frequency, but also with respect to the wave vector; we shall return to this topic in §105.

Lastly, suppose that within the broad transparency region there is a narrow absorption line near some frequency ω_0. Let us consider a neighbourhood of this frequency for which

$$\gamma \ll |\omega - \omega_0| \ll \omega_0, \tag{84.6}$$

where γ is the line width. In this range, x may be replaced by ω_0 in the integrand in (82.6) everywhere apart from the rapidly varying function $\varepsilon''(\omega)$. Then

$$\varepsilon(\omega) \cong \varepsilon'(\omega) \cong \frac{1}{\pi(\omega_0 - \omega)} \int \varepsilon''(x)\,dx, \tag{84.7}$$

the integration being taken over the absorption line.

PROBLEM

A plane electromagnetic wave with a sharply defined forward front is incident normally on the boundary of a half-space ($x > 0$) occupied by a transparent medium with $\mu = 1$. Determine the structure of the front of the transmitted wave (A. Sommerfeld and L. Brillouin, 1914).

SOLUTION. Let the wave be incident on the boundary of the medium at time $t = 0$, so that at $x = 0$ the field (E or H) of the incident wave is $E = 0$ for $t < 0$, $E \propto e^{-i\omega_0 t}$ for $t > 0$. Expanding this field as a Fourier integral with respect to time, we reduce the problem to that of waves of various frequencies and infinite extent incident on the boundary. The amplitude of the Fourier component with frequency ω is proportional to

$$\int_0^\infty e^{i(\omega-\omega_0)\tau}\,d\tau.$$

When a wave of frequency ω is incident, the transmitted wave is of the form $a(\omega)e^{-i\omega t + i\omega n x/c}$, where the amplitude $a(\omega)$ is a slowly varying function of frequency. Hence the wave field in the medium in the present problem is

$$E \propto \int_{-\infty}^{\infty} d\omega\, a(\omega) e^{-i\omega t + i\omega n x/c} \int_0^\infty e^{i(\omega-\omega_0)\tau}\,d\tau.$$

In the region near the wave front, the important values of ω in this integral are those close to ω_0. Using a new variable $\xi = \omega - \omega_0$, we replace $a(\omega)$ by $a(\omega_0)$, and expand the exponent in powers of ξ. Omitting unimportant constants and phase factors, we have

$$E \propto \int_0^\infty \int_{-\infty}^{\infty} \exp\left\{ i\xi\left(\tau - t + \frac{x}{u}\right) - \tfrac{1}{2} i\xi^2 x \frac{u'}{u^2} \right\} d\xi\, d\tau,$$

where $u = u(\omega_0)$ is the velocity of propagation (83.10), and $u' = [du/d\omega]_{\omega=\omega_0}$. Effecting the integration over ξ, we easily bring E to the form

$$E \propto \int_w^\infty e^{\pm i\eta^2}\,d\eta, \qquad w = (x - ut)/\sqrt{(2x|u'|)},$$

the sign in the exponent depending on that of u'. The intensity distribution near the wave front is given by

$$I \propto \left| \int_{w}^{\infty} e^{i\eta^2}\, d\eta \right|^2 .$$

This expression is of the same form as that which gives the intensity distribution near the edge of the shadow in Fresnel diffraction (see *Fields*, §60). For $w > 0$ the intensity decreases monotonically with increasing w, but for $w < 0$ it oscillates with decreasing amplitude about a constant value to which it tends as $w \to -\infty$.

At large distances preceding the front here considered there are found "precursors" propagated with velocity c. These correspond to the high-frequency Fourier components, for which $\varepsilon \to 1$.

CHAPTER X

THE PROPAGATION OF ELECTROMAGNETIC WAVES

§85. Geometrical optics

THE condition for geometrical optics to be applicable is that the wavelength λ should be small in comparison with the characteristic dimension l of the problem (see *Fields*, §53). The relation between geometrical and wave optics is that, for $\lambda \ll l$, any quantity ϕ which describes the wave field (i.e. any component of **E** or **H**) is given by a formula of the type $\phi = ae^{i\psi}$, where the amplitude a is a slowly varying function of the coordinates and time, and the phase ψ is a large quantity which is "almost linear" in the coordinates and the time; it is called the *eikonal*, and is of great importance in geometrical optics. The time derivative of ψ gives the frequency of the wave:

$$\partial\psi/\partial t = -\omega, \tag{85.1}$$

and the space derivatives give the wave vector

$$\mathbf{grad}\,\psi = \mathbf{k}, \tag{85.2}$$

and consequently the direction of the ray through any point in space.

For a steady monochromatic wave, the frequency is a constant and the time dependence of the eikonal is given by a term $-\omega t$. We then introduce a function ψ_1 (also called the eikonal), such that

$$\psi = -\omega t + (\omega/c)\psi_1(x, y, z). \tag{85.3}$$

Then ψ_1 is a function of the coordinates only, and its gradient is

$$\mathbf{grad}\,\psi_1 = \mathbf{n}, \tag{85.4}$$

where $\mathbf{n}$ is a vector such that

$$\mathbf{k} = \omega\mathbf{n}/c. \tag{85.5}$$

The magnitude of $\mathbf{n}$ is equal to the refractive index n of the medium.† Hence the equation for the eikonal in ray propagation in a medium with refractive index $n(x, y, z)$ (a given function of the coordinates) is

$$|\mathbf{grad}\,\psi_1|^2 \equiv \left(\frac{\partial\psi_1}{\partial x}\right)^2 + \left(\frac{\partial\psi_1}{\partial y}\right)^2 + \left(\frac{\partial\psi_1}{\partial z}\right)^2 = n^2. \tag{85.6}$$

† Only transparent media are considered in geometrical optics.

The equation of ray propagation in a steady state can also be derived from *Fermat's principle*, according to which the integral

$$\psi_1 = \int_A^B \mathbf{n}\cdot d\mathbf{l} = \int_A^B n\,dl$$

along the path of the ray between two given points A and B has a value less than for any other path between A and B. Equating to zero the variation of this integral, we have

$$\delta\psi_1 = \int_A^B (\delta n dl + n\delta dl) = 0.$$

Let $\delta\mathbf{r}$ be a displacement of the ray path under the variation. Then $\delta n = \delta\mathbf{r}\cdot\mathbf{grad}\, n$, $\delta\, dl = \mathbf{1}\cdot d\delta\mathbf{r}$, where $\mathbf{1}$ is a unit vector tangential to the ray. Substituting in $\delta\psi_1$ and integrating by parts in the second term (using the fact that $\delta\mathbf{r} = 0$ at A and B), we have

$$\delta\psi_1 = \int_A^B \delta\mathbf{r}\cdot\mathbf{grad}\, n\, dl + \int_A^B n\mathbf{l}\cdot d\delta\mathbf{r}$$

$$= \int_A^B \left(\mathbf{grad}\, n - \frac{d(n\mathbf{l})}{dl}\right)\cdot\delta\mathbf{r}\, dl = 0.$$

Hence

$$d(n\mathbf{l})/dl = \mathbf{grad}\, n. \tag{85.7}$$

Expanding the derivative and putting $dn/dl = \mathbf{1}\cdot\mathbf{grad}\ n$, we obtain

$$\frac{d\mathbf{l}}{dl} = \frac{1}{n}[\mathbf{grad}\, n - \mathbf{1}(\mathbf{1}\cdot\mathbf{grad}\, n)]. \tag{85.8}$$

This is the equation giving the form of the rays.

We know from differential geometry that the derivative $d\mathbf{l}/dl$ along the ray is equal to $\mathbf{N}/R$, where $\mathbf{N}$ is the unit vector along the principal normal and R the radius of curvature. Taking the scalar product of both sides of (85.8) with $\mathbf{N}$, and using the fact that $\mathbf{N}$ and $\mathbf{1}$ are perpendicular, we have

$$\frac{1}{R} = \mathbf{N}\cdot\frac{\mathbf{grad}\, n}{n}; \tag{85.9}$$

the rays are therefore bent in the direction of increasing refractive index.

The velocity of propagation of rays in geometrical optics is in the direction of $\mathbf{1}$, and is given by the derivative

$$\mathbf{u} = \partial\omega/\partial\mathbf{k}. \tag{85.10}$$

This is also called the *group velocity*, the ratio ω/k being called the *phase velocity*. However, the latter is not the velocity of physical propagation of any quantity.

It is easy to derive also the equation which gives the rate of change of the radiation intensity along a ray. The intensity I is the magnitude of the (time) averaged Poynting vector. This vector, like the group velocity, is in the direction of $\mathbf{l}: \overline{\mathbf{S}} = I\mathbf{l}$. In a steady state, the mean field energy density is constant at any given point in space. The equation of conservation of energy is therefore $\operatorname{div} \overline{\mathbf{S}} = 0$, or

$$\operatorname{div}(I\mathbf{l}) = 0. \tag{85.11}$$

This is the required equation.

Finally, let us consider how the direction of polarization of linearly polarized radiation varies along a ray (S. M. Rytov, 1938). As we know from differential geometry, a curve in space (in this case, the ray) is characterized at every point by three mutually perpendicular unit vectors along the tangent ($\mathbf{l}$), the principal normal ($\mathbf{N}$) and the binormal ($\mathbf{b}$), which form the *natural trihedral*. Since the electromagnetic waves are transverse, the vectors $\mathbf{E}$ and $\mathbf{H}$ are always coplanar with $\mathbf{N}$ and $\mathbf{b}$.

Let the direction of $\mathbf{E}$ at some point on the ray be the same as that of $\mathbf{N}$, i.e. let $\mathbf{E}$ lie in the osculating plane (that of $\mathbf{N}$ and $\mathbf{l}$). The deviation of the curve from the osculating plane over a length dl is of the third order of smallness with respect to dl. We can therefore say that, over a length dl of the ray, the vector $\mathbf{E}$ remains in the original osculating plane. The osculating plane at the other end of dl is inclined to the original one at an angle $d\phi = dl/T$, where T is the radius of torsion. This is therefore the angle turned through by the vector $\mathbf{E}$ relative to $\mathbf{N}$ in the normal plane. Thus, over a distance dl along the ray, the direction of polarization rotates in the normal plane, its angle to the principal normal varying in accordance with the equation

$$d\phi/dl = 1/T. \tag{85.12}$$

In particular, when the torsion is zero, i.e. the ray is a plane curve, the direction of the vector $\mathbf{E}$ in the normal plane is constant, as is in any case evident from symmetry.

PROBLEMS

PROBLEM 1. Find the law of transformation for the velocity of light (the group velocity) in a medium when the frame of reference is changed.

SOLUTION. By the definition of the group velocity $\mathbf{u}$, $d\omega = \mathbf{u}\cdot d\mathbf{k}$ and $d\omega' = \mathbf{u}'\cdot d\mathbf{k}'$, the primed and unprimed quantities referring to the frames K' and K; K' moves with velocity $\mathbf{v}$ relative to K. According to the Lorentz transformation formulae for a wave four-vector,

$$k'_x = \gamma(k_x - v\omega/c^2), \quad k'_y = k_y, \quad k'_z = k_z, \quad \omega = \gamma(\omega' + vk'_x),$$

where $\gamma = 1/\sqrt{(1-v^2/c^2)}$; the x and x' axes are parallel to $\mathbf{v}$. The last of these expressions gives

$$d\omega = \gamma(d\omega' + v\,dk'_x) = \gamma(\mathbf{u}'\cdot d\mathbf{k}' + v\,dk'_x).$$

Substituting $d\mathbf{k}'$ in terms of $d\mathbf{k}$ and $d\omega$ from the other expressions, and collecting terms in $d\omega$, gives

$$\gamma(1 + vv'_x/c^2)\,d\omega = \gamma(u'_x + v)dk_x + u'_y dk_y + u'_z dk_z.$$

Comparison with $d\omega = \mathbf{u}\cdot d\mathbf{k}$ shows that the velocities $\mathbf{u}'$ and $\mathbf{v}$ give $\mathbf{u}$ by the usual relativistic formulae for the addition of velocities, as expected.

PROBLEM 2. Determine the velocity of propagation of light in a medium moving relative to the observer.

SOLUTION. Let ω and $\mathbf{k}$ be the frequency and wave vector of the light wave in a fixed frame of reference K, and ω', $\mathbf{k}'$ the corresponding quantities in a frame K' moving with the medium at velocity $\mathbf{v}$ relative to K. In the frame K' the medium is at rest, and ω' and k' are therefore related by

$$ck' = \omega' n(\omega'). \tag{1}$$

According to the Lorentz transformation formulae for a wave four-vector we have, as far as terms of the first order in $v/c, \omega' = \omega - \mathbf{k}\cdot\mathbf{v}, \mathbf{k}' = \mathbf{k} - \omega\mathbf{v}/c^2$, $k' = k - \omega\mathbf{v}\cdot\mathbf{l}/c^2$, where $\mathbf{l} = \mathbf{k}/k$. Substituting these expressions in (1) and expanding the function $n(\omega')$ in powers of $\mathbf{v}$, we obtain to the same accuracy†

$$k = n\frac{\omega}{c} + \frac{\mathbf{v}\cdot\mathbf{l}\omega}{c^2}\left(1 - n\frac{\mathrm{d}(n\omega)}{d\omega}\right). \tag{2}$$

The velocity of propagation (the group velocity) in a medium at rest is found by differentiating the relation $ck = \omega n(\omega)$:

$$\mathbf{u}_0 = \frac{c}{\mathrm{d}(n\omega)/\mathrm{d}\omega}\mathbf{l}. \tag{3}$$

In a moving medium, it is found by differentiating the relation (2), which we first rewrite as

$$k = \omega n/c + \mathbf{k}\cdot\mathbf{v}\left(\frac{1}{cn} - \frac{1}{u_0}\right).$$

Again as far as the first-order terms,

$$\mathbf{u} = \mathbf{u}_0 + \mathbf{l}(\mathbf{l}\cdot\mathbf{v})\left(\frac{u_0}{cn} - \frac{u_0^2}{c^2} - \frac{n\omega}{c}\frac{\mathrm{d}u_0}{\mathrm{d}\omega}\right) + \mathbf{v}\left(1 - \frac{u_0}{cn}\right). \tag{4}$$

For propagation in the direction of motion ($\mathbf{v}\parallel\mathbf{l}$), we therefore have‡

$$u = u_0 + v(1 - u_0^2/c^2) - (vn\omega/c)\,\mathrm{d}u_0/\mathrm{d}\omega. \tag{5}$$

The first two terms can be derived by simply applying the relativistic expression for the addition of velocities.

If $\mathbf{v}$ and $\mathbf{l}$ are mutually perpendicular, then

$$\mathbf{u} = \mathbf{u}_0 + \mathbf{v}(1 - u_0/cn). \tag{6}$$

The phase velocity of the wave is given by (2):

$$\frac{\omega}{k} = \frac{c}{n} + \mathbf{v}\cdot\mathbf{l}\left(1 - \frac{1}{n^2} + \frac{\omega}{n}\frac{\mathrm{d}n}{\mathrm{d}\omega}\right);$$

when $\mathbf{v} \perp \mathbf{l}$, there is no first-order effect.

§86. Reflection and refraction of electromagnetic waves

Let us consider the reflection and refraction of a plane monochromatic electromagnetic wave at a plane boundary between two homogeneous media. Medium 1, from which the wave is incident, is assumed transparent, but not medium 2 (for the present). Quantities pertaining to the incident and reflected waves will be distinguished by the suffixes 0 and 1 respectively, and those for the refracted wave by the suffix 2 (Fig. 46, p. 294). The direction of the normal from the boundary plane into medium 2 is taken as the z-axis.

Since there is complete homogeneity in the xy-plane, the dependence of the solution of the field equations on x and y must be the same in all space. The components k_x, k_y of the wave vector must therefore be the same for all three waves. Consequently, the directions of propagation of the three waves lie in one plane, which we take as the xz-plane.

From the equations

$$k_{0x} = k_{1x} = k_{2x} \tag{86.1}$$

† The second term in (2), and therefore all first-order effects, vanish identically when $n^2 = \varepsilon = 1 - \text{constant}/\omega^2$.

‡ This formula represents the *Fizeau effect*, first predicted by A. Fresnel (1818). The influence of dispersion on this effect was discussed by H. A. Lorentz (1895).

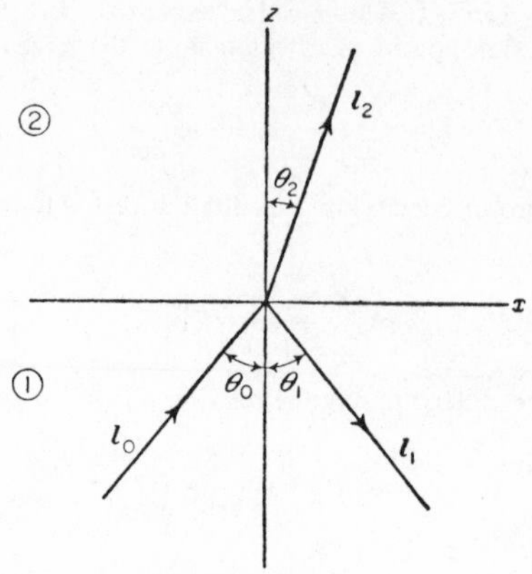

FIG. 46

we find

$$\left.\begin{aligned} k_{1z} &= -k_{0z} = -(\omega/c)\sqrt{\varepsilon_1}\cos\theta_0, \\ k_{2z} &= \sqrt{[(\omega/c)^2\varepsilon_2 - k_{0x}{}^2]} = (\omega/c)\sqrt{(\varepsilon_2 - \varepsilon_1\sin^2\theta_0)}; \end{aligned}\right\} \tag{86.2}$$

we take $\mu = 1$ in both media. The vector $\mathbf{k}_0$ is, by definition, real, and so is $\mathbf{k}_1$. The quantity k_{2z}, however, is complex in an absorbing medium, and the sign of the root must be taken so that im $k_{2z} > 0$, the refracted wave being damped towards the interior of medium 2.

If both media are transparent, equations (86.1) give the familiar laws of reflection and refraction:

$$\theta_1 = \theta_0, \qquad \frac{\sin\theta_2}{\sin\theta_0} = \sqrt{\frac{\varepsilon_1}{\varepsilon_2}} = \frac{n_1}{n_2}. \tag{86.3}$$

To determine the amplitudes of the reflected and refracted waves, we must use the boundary conditions at the surface of separation ($z = 0$), and we shall consider separately the two cases where the electric field $\mathbf{E}_0$ is in the plane of incidence and perpendicular to that plane; from the results we can obtain the solution for the general case, where $\mathbf{E}_0$ can be resolved into components in these two directions.

Let us first suppose that $\mathbf{E}_0$ is perpendicular to the plane of incidence. It is evident from symmetry that the same will be true of the fields $\mathbf{E}_1$ and $\mathbf{E}_2$ in the reflected and refracted waves. The vector $\mathbf{H}$ is in the xz-plane. The boundary conditions require† the continuity of $E_y = E$ and H_x; by (83.3) $H_x = -ck_z E_y/\omega$.

The field in medium 1 is the sum of the fields in the incident and reflected waves, so that we obtain the two equations $E_0 + E_1 = E_2$, $k_{0z}(E_0 - E_1) = k_{2z}E_2$. The exponential factors in E cancel because k_x (and therefore ω) is the same in all three waves. In what follows, $\mathbf{E}$ signifies the complex amplitude of a wave. The solution of the above equations gives

† The boundary conditions on the normal components of **B** and **D** give nothing new in the present problem, because the equations div $\mathbf{B} = \mathbf{0}$, div $\mathbf{D} = 0$ are consequences of equations (83.1).

Fresnel's formulae:

$$
\left.
\begin{aligned}
E_1 &= \frac{k_{0z}-k_{2z}}{k_{0z}+k_{2z}}E_0 = \frac{\sqrt{\varepsilon_1}\cos\theta_0-\sqrt{(\varepsilon_2-\varepsilon_1\sin^2\theta_0)}}{\sqrt{\varepsilon_1}\cos\theta_0+\sqrt{(\varepsilon_2-\varepsilon_1\sin^2\theta_0)}}E_0,\\
E_2 &= \frac{2k_{0z}}{k_{0z}+k_{2z}}E_0 = \frac{2\sqrt{\varepsilon_1}\cos\theta_0}{\sqrt{\varepsilon_1}\cos\theta_0+\sqrt{(\varepsilon_2-\varepsilon_1\sin^2\theta_0)}}E_0.
\end{aligned}
\right\} \tag{86.4}
$$

If both media are transparent, these formulae become, by (86.3),

$$
\left.
\begin{aligned}
E_1 &= \frac{\sin(\theta_2-\theta_0)}{\sin(\theta_2+\theta_0)}E_0,\\
E_2 &= \frac{2\cos\theta_0\sin\theta_2}{\sin(\theta_2+\theta_0)}E_0.
\end{aligned}
\right\} \tag{86.5}
$$

The case where **E** lies in the plane of incidence can be discussed similarly. Here it is more convenient to carry out the calculations for the magnetic field, which is perpendicular to the plane of incidence. A further two Fresnel's formulae are obtained:

$$
\left.
\begin{aligned}
H_1 &= \frac{\varepsilon_2 k_{0z}-\varepsilon_1 k_{2z}}{\varepsilon_2 k_{0z}+\varepsilon_1 k_{2z}}H_0 = \frac{\varepsilon_2\cos\theta_0-\sqrt{(\varepsilon_1\varepsilon_2-\varepsilon_1{}^2\sin^2\theta_0)}}{\varepsilon_2\cos\theta_0+\sqrt{(\varepsilon_1\varepsilon_2-\varepsilon_1{}^2\sin^2\theta_0)}}H_0,\\
H_2 &= \frac{2\varepsilon_2 k_{0z}}{\varepsilon_2 k_{0z}+\varepsilon_1 k_{2z}}H_0 = \frac{2\varepsilon_2\cos\theta_0}{\varepsilon_2\cos\theta_0+\sqrt{(\varepsilon_1\varepsilon_2-\varepsilon_1{}^2\sin^2\theta_0)}}H_0.
\end{aligned}
\right\} \tag{86.6}
$$

If both media are transparent, these formulae may be written

$$
\left.
\begin{aligned}
H_1 &= \frac{\tan(\theta_0-\theta_2)}{\tan(\theta_0+\theta_2)}H_0,\\
H_2 &= \frac{\sin 2\theta_0}{\sin(\theta_0+\theta_2)\cos(\theta_0-\theta_2)}H_0.
\end{aligned}
\right\} \tag{86.7}
$$

The *reflection coefficient* R is defined as the ratio of the (time) averaged energy flux reflected from the surface to the incident flux. Each of these fluxes is given by the averaged z-component of the Poynting vector (83.11) for the wave in question,

$$R = \frac{\sqrt{\varepsilon_1}\cos\theta_1|\mathbf{E}_1|^2}{\sqrt{\varepsilon_1}\cos\theta_0|\mathbf{E}_0|^2} = \frac{|\mathbf{E}_1|^2}{|\mathbf{E}_0|^2}.$$

For normal incidence ($\theta_0 = 0$) the two modes of polarization are equivalent, and the reflection coefficient is given by

$$R = \left|\frac{\sqrt{\varepsilon_1}-\sqrt{\varepsilon_2}}{\sqrt{\varepsilon_1}+\sqrt{\varepsilon_2}}\right|^2. \tag{86.8}$$

This formula is valid whether the reflecting medium is transparent or not. If we put

$\sqrt{\varepsilon_2} = n_2 + i\kappa_2$, and if medium 1 is a vacuum ($\varepsilon_1 = 1$), then

$$R = \frac{(n_2-1)^2+\kappa_2{}^2}{(n_2+1)^2+\kappa_2{}^2}. \tag{86.9}$$

The remaining discussion assumes that both media are transparent. The following general remark should be made first of all. The boundary between two different media is in reality not a geometrical surface but a thin transition layer. The validity of the formulae (86.1) does not rest on any assumptions concerning the nature of this layer. The derivation of Fresnel's formulae, on the other hand, is based on the use of the boundary conditions, and assumes that the thickness δ of the transition layer is small compared with the wavelength λ. The thickness δ is usually comparable with the distances between the atoms, which are always small compared with λ if the macroscopic description of the field is legitimate, and so the condition $\lambda \gg \delta$ is usually fulfilled. In the opposite limiting case the phenomenon of refraction is entirely different in character. For $\delta \gg \lambda$, geometrical optics is valid (λ being small compared with the dimensions of the inhomogeneities in the medium). In this case, therefore, the propagation of the wave can be regarded as the propagation of rays which undergo refraction in the transition layer but are not reflected from it. The reflection coefficient, therefore, would be zero.

Let us return now to Fresnel's formulae. In reflection from a transparent medium, the coefficients of proportionality between $\mathbf{E}_1$, $\mathbf{E}_2$ and $\mathbf{E}_0$ in these formulae are real.† This means that the wave phase either remains unchanged or changes by π, depending on the sign of the coefficients. In particular, the phase of the refracted wave is always the same as that of the incident wave. The reflection, on the other hand, may be accompanied by a change in phase.‡ For example, with normal incidence the phase of the wave is unchanged if $\varepsilon_1 > \varepsilon_2$, but if $\varepsilon_1 < \varepsilon_2$ the vectors $\mathbf{E}_1$ and $\mathbf{E}_0$ are in opposite directions, i.e. the wave phase changes by π.

The reflection coefficients for oblique incidence are, by (86.5) and (86.7),

$$R_\perp = \frac{\sin^2(\theta_2-\theta_0)}{\sin^2(\theta_2+\theta_0)}, \qquad R_\parallel = \frac{\tan^2(\theta_2-\theta_0)}{\tan^2(\theta_2+\theta_0)}. \tag{86.10}$$

Here, and in what follows, the suffixes $\perp$ and $\parallel$ refer to the cases where the field $\mathbf{E}$ is respectively perpendicular and parallel to the plane of incidence. The expressions (86.10) are unaltered when θ_2 and θ_0 are interchanged (but the phase of the reflected wave changes by π, as is seen from formulae (86.5) and (86.7)). The reflection coefficient for a wave incident from medium 1 at an angle θ_0 is therefore equal to that for a wave incident from medium 2 at an angle θ_2.

An interesting case is the reflection of light incident at an angle θ_0 such that $\theta_0+\theta_2 = \frac{1}{2}\pi$ (the reflected and refracted rays being thus perpendicular). Let this angle be θ_p; $\sin\theta_p = \sin(\frac{1}{2}\pi - \theta_2) = \cos\theta_2$, and the law of refraction (86.3) gives

$$\tan\theta_p = \sqrt{(\varepsilon_2/\varepsilon_1)}. \tag{86.11}$$

† We ignore for the moment the possibility of total reflection (see below).

‡ Reflection from an absorbing medium leads in general to the appearance of elliptical polarization. The explicit expressions for the amplitude and phase relations between the three waves are then extremely involved; they are given by J. A. Stratton, *Electromagnetic Theory*, Chapter IX, McGraw-Hill, New York, 1941.

For $\theta_0 = \theta_p$ we have $\tan(\theta_2 + \theta_0) = \infty$, and $R_{\parallel} = 0$. Hence, whatever the direction of polarization of light incident at this angle, the reflected light will be polarized so that the electric field is perpendicular to the plane of incidence. The reflected light is polarized in this way even when the incident light is natural: no component with any other polarization is reflected. The angle θ_p is called the *angle of total polarization* or the *Brewster angle.* It should be noticed that, whereas natural light can be totally polarized by reflection, this effect cannot be produced by refraction, whatever the angle of incidence.

The reflection and refraction of plane-polarized light always results in plane-polarized light, but the direction of polarization is in general not the same as in the incident light. Let γ_0 be the angle between the direction of $\mathbf{E}_0$ and the plane of incidence, and γ_1, γ_2 the corresponding angles for the reflected and refracted waves. Using formulae (86.5) and (86.7), we easily obtain the relations

$$\left.\begin{aligned} \tan\gamma_1 &= -\frac{\cos(\theta_0 - \theta_2)}{\cos(\theta_0 + \theta_2)}\tan\gamma_0, \\ \tan\gamma_2 &= \cos(\theta_0 - \theta_2)\tan\gamma_0. \end{aligned}\right\} \quad (86.12)$$

The angles γ_0, γ_1 and γ_2 are equal for all angles of incidence only in the obvious cases $\gamma_0 = 0$ and $\gamma_0 = \frac{1}{2}\pi$; they are also equal for normal incidence ($\theta_0 = \theta_2 = 0$) and for grazing incidence ($\theta_0 = \frac{1}{2}\pi$, in which case there is no refracted wave). In all other cases the formulae (86.12) give (by virtue of the inequalities $0 < \theta_0, \theta_2 < \frac{1}{2}\pi$ and, as we shall assume, $0 < \gamma_0 < \frac{1}{2}\pi$, $0 < \gamma_1, \gamma_2 < \pi$) the inequalities $\gamma_1 > \gamma_0$, $\gamma_2 < \gamma_0$. Thus the direction of $\mathbf{E}$ is turned away from the plane of incidence on reflection, but towards it on refraction.

A comparison of the two formulae (86.10) shows that, at all angles of incidence except $\theta_0 = 0$ or $\frac{1}{2}\pi$, $R_{\parallel} < R_{\perp}$. Hence, for example, when the incident light is natural the reflected light is partly polarized, and the predominant direction of the electric field is perpendicular to the plane of incidence. The refracted light is partly polarized, with the predominant direction of $\mathbf{E}$ lying in the plane of incidence.

The quantities $R_{\parallel}$ and $R_{\perp}$ depend quite differently on the angle of incidence. The coefficient $R_{\perp}$ increases monotonically with the angle θ_0 from the value (86.8) for $\theta_0 = 0$. The coefficient $R_{\parallel}$ takes the same value (86.8) for $\theta_0 = 0$, but as θ_0 increases $R_{\parallel}$ decreases to zero at $\theta_0 = \theta_p$ before monotonically increasing.

Here two distinct cases occur. If the reflection is from the "optically denser" medium, i.e. $\varepsilon_2 > \varepsilon_1$, then $R_{\parallel}$ and $R_{\perp}$ increase to the common value of unity at $\theta_0 = \frac{1}{2}\pi$ (grazing incidence). If, on the other hand, the reflecting medium is "optically less dense" ($\varepsilon_2 < \varepsilon_1$), both coefficients become equal to unity for $\theta_0 = \theta_r$, where

$$\sin\theta_r = \sqrt{(\varepsilon_2/\varepsilon_1)} = n_2/n_1; \quad (86.13)$$

θ_r is called the *angle of total reflection.* When $\theta_0 = \theta_r$ the angle of refraction $\theta_2 = \frac{1}{2}\pi$, i.e. the refracted wave is propagated along the surface separating the media.

Reflection from an optically less dense medium at angles $\theta_0 > \theta_r$ requires special consideration. In this case k_{2z} is purely imaginary (see (86.2)), i.e. the field is damped in medium 2. The damping of the wave without true absorption (i.e. dissipation of energy) signifies that the average energy flux from medium 1 into medium 2 is zero (by simple calculation it can easily be seen that the vector $\overline{\mathbf{S}}$ giving the average energy flux in medium 2 is in the x-direction). That is, all the energy incident on the boundary is reflected back into medium 1, so that the reflection coefficients are $R_{\perp} = R_{\parallel} = 1$. This phenomenon is called *total reflection.*† The equality of $R_\perp$ and $R_\parallel$ to unity can, of course, be obtained directly from Fresnel's formulae (86.4) and (86.6).

For $\theta_0 > \theta_r$ the proportionality coefficients between $\mathbf{E}_1$ and $\mathbf{E}_0$ become complex quantities, of the form $(a-ib)/(a+ib)$. The quantities $R_\perp$ and $R_\parallel$ are given by the squared moduli of these coefficients, which are equal to unity. The formulae give, besides the ratio of the magnitudes of the fields in the reflected and incident waves, the difference in their phases. For this purpose we write $E_{1\perp} = e^{-i\delta_\perp}E_{0\perp}$, $E_{1\parallel} = e^{-i\delta_\parallel}E_{0\parallel}$. Then‡

$$\left.\begin{aligned}\tan\tfrac{1}{2}\delta_\perp &= \sqrt{(\varepsilon_1\sin^2\theta_0-\varepsilon_2)}/\sqrt{\varepsilon_1}\cos\theta_0,\\ \tan\tfrac{1}{2}\delta_\parallel &= \sqrt{(\varepsilon_1{}^2\sin^2\theta_0-\varepsilon_1\varepsilon_2)}/\varepsilon_2\cos\theta_0.\end{aligned}\right\} \tag{86.14}$$

Thus total reflection involves a change in the wave phase which is in general different for the field components parallel and perpendicular to the plane of incidence. Hence, on reflection of a wave polarized in a plane inclined to the plane of incidence, the reflected wave will be elliptically polarized. The phase difference $\delta = \delta_\perp - \delta_\parallel$ is easily found to be such that

$$\tan\tfrac{1}{2}\delta = \frac{\cos\theta_0\sqrt{(\varepsilon_1\sin^2\theta_0-\varepsilon_2)}}{\sqrt{\varepsilon_1}\sin^2\theta_0}. \tag{86.15}$$

The difference is zero only for $\theta_0 = \theta_r$ or $\theta_0 = \frac{1}{2}\pi$.

PROBLEMS

PROBLEM 1. Find the manner in which the reflection coefficient approaches unity near the angle of total reflection.

SOLUTION. We put $\theta_0 = \theta_r - \delta$, where δ is a small quantity, and expand $\sin\theta_0$ and $\cos\theta_0$ in formulae (86.10) in powers of δ. The result is

$$R_\perp = 1 - 4\sqrt{(2\delta)}(n^2-1)^{-\frac{1}{4}},$$
$$R_\parallel = 1 - 4\sqrt{(2\delta)}n^2(n^2-1)^{-\frac{1}{4}},$$

where $n^2 = \varepsilon_1/\varepsilon_2 > 1$. The derivatives $dR/d\delta$ become infinite as $\delta^{-\frac{1}{2}}$ when $\delta \to 0$.

PROBLEM 2. Find the reflection coefficient for almost grazing incidence of light from a vacuum on the surface of a body for which ε is almost unity.

SOLUTION. Formulae (86.10) give the same reflection coefficient:

$$R_\perp \cong R_\parallel \cong [\phi_0 - \sqrt{(\phi_0{}^2+\varepsilon-1)}\,]^4/(\varepsilon-1)^2,$$

where $\phi_0 = \frac{1}{2}\pi - \theta_0$.

PROBLEM 3. Determine the reflection coefficient for a wave incident from a vacuum on a medium for which both ε and μ are different from unity.

SOLUTION. Calculations entirely analogous to those given above furnish the result

$$R_\perp = \left|\frac{\mu\cos\theta_0 - \sqrt{(\varepsilon\mu-\sin^2\theta_0)}}{\mu\cos\theta_0 + \sqrt{(\varepsilon\mu-\sin^2\theta_0)}}\right|^2,$$

$$R_\parallel = \left|\frac{\varepsilon\cos\theta_0 - \sqrt{(\varepsilon\mu-\sin^2\theta_0)}}{\varepsilon\cos\theta_0 + \sqrt{(\varepsilon\mu-\sin^2\theta_0)}}\right|^2.$$

† The reflection coefficient is always equal to unity in reflection from a medium with ε real and negative. In such a medium there is again no true absorption, but the wave cannot penetrate into it.

‡ If $(a-ib)/(a+ib) = e^{-i\delta}$, then $\tan\frac{1}{2}\delta = b/a$.

PROBLEM 4. A plane-parallel layer (region 2) lies between a vacuum (region 1) and an arbitrary medium (region 3). Light polarized parallel or perpendicular to the plane of incidence falls on the layer from the vacuum. Express the reflection coefficient R in terms of those for semi-infinite media of the substances in regions 2 and 3.

SOLUTION. We denote by A_0 and A_1 the amplitudes of the field (**E** or **H**, whichever is parallel to the layer) in the incident and reflected waves. The field in the layer consists of the refracted wave (amplitude A_2) and the wave reflected from region 3 (amplitude A_2'). The boundary condition between regions 1 and 2 gives

$$A_2' = a(A_1 - r_{12}A_0) \tag{1}$$

where a and r_{12} are constants. In reflection from a semi-infinite medium of the substance in region 2, A_2' is zero, and so from (1) we have $r_{12} = A_1/A_0$, i.e. r_{12} is the amplitude of reflection in that case. Another equation is obtained from (1) by interchanging A_1 and A_0 and replacing A_2' by A_2, which corresponds simply to a reversal of the z-component of the wave vector:

$$A_2 = a(A_0 - r_{12}A_1). \tag{2}$$

In region 3 there is only the transmitted wave, whose amplitude A_3 satisfies the conditions

$$A_2 e^{i\psi} = aA_3, \qquad A_2' e^{-i\psi} = -ar_{32}A_3 \tag{3}$$

analogous to (1), (2) with $A_1 = 0$. The exponential factors take account of the change in the wave phase over the thickness h of the layer, with

$$\psi = (\omega h/c)\sqrt{(\varepsilon_2 - \sin^2\theta_0)}. \tag{4}$$

Eliminating A_3 from equations (3), we obtain

$$A_2' e^{-i\psi} = r_{23}A_2 e^{i\psi}, \tag{5}$$

where $r_{23} = -r_{32}$.

From equations (1), (2) and (5) we find the amplitude of reflection from the layer:

$$r = \frac{A_1}{A_0} = \frac{r_{12}e^{-2i\psi} + r_{23}}{e^{-2i\psi} + r_{12}r_{23}}, \tag{6}$$

and the reflection coefficient $R = |r|^2$. The significance of r_{23} is found from the fact that, for $h = 0$, r must be the amplitude of reflection r_{13} from a semi-infinite medium of the substance in region 3. Hence

$$r_{23} = (r_{12} - r_{13})/(r_{12}r_{13} - 1). \tag{7}$$

Formulae (6) and (7) give the required solution. It should be emphasized that their derivation involves no assumptions concerning the properties of regions 2 and 3, which may be either transparent or not.

If regions 2 and 3 are transparent, then ψ, r_{12} and r_{13} are all real, and r_{23} is the amplitude of reflection at the boundary between semi-infinite media of the substances in regions 2 and 3. From (6) we have

$$R = \frac{(r_{12} + r_{23})^2 - 4r_{12}r_{23}\sin^2\psi}{(r_{12}r_{23} + 1)^2 - 4r_{12}r_{23}\sin^2\psi}. \tag{8}$$

As ψ varies, R varies between the limits $[(r_{12}+r_{23})/(r_{12}r_{23}+1)]^2$ and $[(r_{12}-r_{23})/(r_{12}r_{23}-1)]^2$. For normal incidence $r_{12} = (n_1 - n_2)/(n_1 + n_2)$, and r_{13} and r_{23} are given by similar relations. If $n_2{}^2 = n_1 n_3$, then $r_{12} = r_{23}$, and R may be zero for some value of the thickness of the layer. If region 3 is a vacuum, then $r_{13} = 0$, $r_{23} = -r_{12}$, and (6) gives

$$r = \frac{r_{12}(e^{-2i\psi} - 1)}{e^{-2i\psi} - r_{12}{}^2} = -\frac{\sinh i\psi}{\sinh[i\psi + \log(-r_{12})]}. \tag{9}$$

If also region 2 is transparent, we have

$$R = \frac{4R_{12}\sin^2\psi}{(1 - R_{12})^2 + 4R_{12}\sin^2\psi}.$$

The transmission coefficient D for the layer (between vacua) is $1 - R$ only if region 2 is transparent. Otherwise D must be calculated from equations (1)–(3), putting $r_{32} = r_{12}$. The amplitude of transmission d is

$$d = \frac{A_3}{A_0} = \frac{1 - r_{12}{}^2}{e^{-i\psi} - r_{12}{}^2 e^{i\psi}}, \tag{10}$$

and the transmission coefficient $D = |d|^2$.

PROBLEM 5. Determine the reflection and transmission coefficients for light incident normally on a slab with a very large complex permittivity ε.

SOLUTION. In this case $r_{12} = (1-\sqrt{\varepsilon})/(1+\sqrt{\varepsilon}) \cong -(1-2/\sqrt{\varepsilon})$, and formula (9) of Problem 4 gives $r = -[1-(2/\sqrt{\varepsilon})\coth i\psi]^{-1}$, $\psi = (\omega h/c)\sqrt{\varepsilon}$. If the slab is so thin that $\omega h/c \ll 1/\sqrt{|\varepsilon|}$, then we can put $r = -[1+2ic/\varepsilon\omega h]^{-1}$, and distinguish two cases:

$$\text{for } 1/|\varepsilon| \ll \omega h/c \ll 1/\sqrt{|\varepsilon|}, \qquad R = 1 - 4c\varepsilon''/\omega h|\varepsilon|^2,$$
$$\text{for } \omega h/c \ll 1/|\varepsilon|, \qquad R = \omega^2 h^2 |\varepsilon|^2/4c^2.$$

The transmission coefficient is, by formula (10),

$$\text{for } \omega h/c \sim 1/\sqrt{|\varepsilon|}, \qquad d = -2/\sqrt{\varepsilon}\sinh i\psi,$$
$$\text{for } \omega h/c \ll 1/\sqrt{|\varepsilon|}, \qquad d = (1 - i\varepsilon\omega h/2c)^{-1}.$$

Again two cases can be distinguished:

$$\text{for } 1/|\varepsilon| \ll \omega h/c \ll 1/\sqrt{|\varepsilon|}, \qquad D = 4c^2/\omega^2 h^2 |\varepsilon|^2,$$
$$\text{for } \omega h/c \ll 1/|\varepsilon|, \qquad D = 1 - \varepsilon''\omega h/c.$$

§87. The surface impedance of metals

The permittivity of metals is, in magnitude, large compared with unity at low frequencies (as $\omega \to 0$, it tends to infinity as $1/\omega$). The wavelength $\delta \sim c/\omega\sqrt{|\varepsilon|}$ in metals† is then small compared with the wavelength $\lambda \sim c/\omega$ in vacuum. If δ (but not necessarily λ) is also small compared with the radii of curvature of the metal surface, the problem of the reflection of arbitrary electromagnetic waves from the metal can be considerably simplified.

The smallness of δ implies that the derivatives of the field components inside the metal along the normal to the surface are large compared with the tangential derivatives. The field inside the metal near the surface can therefore be regarded as the field of a plane wave, and hence the fields $\mathbf{E}_t$ and $\mathbf{H}_t$ are related by

$$\mathbf{E}_t = \sqrt{(\mu/\varepsilon)}\mathbf{H}_t \times \mathbf{n}, \tag{87.1}$$

where $\mathbf{n}$ is a unit vector along the inward normal to the surface. Since $\mathbf{E}_t$ and $\mathbf{H}_t$ are continuous, their values outside the metal near the surface must be related in the same way. As M. A. Leontovich (1948) has pointed out, the equation (87.1) may be used as a boundary condition in determining the field outside the metal. Thus the problem of determining the external electromagnetic field can be solved without considering the field inside the metal.

The quantity $\sqrt{(\mu/\varepsilon)}$ is called the *surface impedance*‡ of the metal, and we denote it by $\zeta = \zeta' + i\zeta''$:

$$\zeta = \sqrt{(\mu/\varepsilon)}. \tag{87.2}$$

In the frequency range where ε can be expressed in terms of the ordinary conductivity of the metal, we have

$$\zeta = (1-i)\sqrt{(\omega\mu/8\pi\sigma)}; \tag{87.3}$$

this formula, with $\mu = 1$, has already been given in §59.

The (time) averaged energy flux through the surface of the metal is

$$\overline{\mathbf{S}} = (c/8\pi)\mathrm{re}(\mathbf{E}_t \times \mathbf{H}_t^*) = c\zeta'|\mathbf{H}_t|^2\mathbf{n}/8\pi. \tag{87.4}$$

---

† Large values of $\sqrt{\varepsilon(\omega)}$ are almost always complex. The electromagnetic field is damped inside the body, so that the wavelength in the body is also the depth of penetration of the field. If $\varepsilon(\omega)$ is expressed in terms of the conductivity by (77.9), the quantity δ is the same as the penetration depth used in §59.

‡ This name is also given to the quantity $c\zeta/4\pi$.

This is the energy which enters the metal and is dissipated therein. Hence we see, in particular, that

$$\zeta' > 0. \tag{87.5}$$

This inequality determines the sign of the root in (87.2).

As the frequency increases, the depth of penetration δ becomes of the same order as the mean free path l of the conduction electrons.† In this case the spatial non-uniformity of the field renders impossible a macroscopic description of it in terms of the permittivity ε. However, a boundary condition of the form

$$\mathbf{E}_t = \zeta \mathbf{H}_t \times \mathbf{n} \tag{87.6}$$

still holds at such frequencies. The field inside the metal near the surface can again be regarded as a plane wave, although it is no longer described by the usual macroscopic Maxwell's equations. In such a wave the fields **E** and **H** must be linearly related, and the only possible linear relation between the axial vector **H** and the polar vector **E** is (87.6). The coefficient ζ in this formula is the only quantity characterizing the metal which must be known in order to find the external electromagnetic field. The calculation, however, requires the use of kinetic theory (see *PK*, §86).

When the frequency increases further (usually into the infra-red region), the macroscopic description of the field again becomes possible, and ε is again meaningful. The reason is that, on absorbing a large quantum $\hbar\omega$, a conduction electron acquires a large amount of energy, and its mean free path is therefore reduced, the inequality $l \ll \delta$ being consequently again fulfilled. The impedance ζ is again inversely proportional to $\sqrt{\varepsilon}$.‡ In this frequency range the real part of $\varepsilon(\omega)$ is large and negative, and its imaginary part is small. The inequality $l \ll \delta$ is the condition for both ε' and ε'' to be macroscopically significant. The macroscopic significance of the large quantity ε' alone, however, can be ensured by the fulfilment of the less stringent condition $v/\omega \ll \delta$, where v is the velocity of the conduction electrons in the metal. If this condition holds, the spatial inhomogeneity of the field may be neglected in considering the motion of the electrons.§

The inequality $\zeta' > 0$ is always satisfied by the real part of the impedance. If formula (87.2) holds, we can also draw certain conclusions concerning the sign of ζ''. For example, if the dispersion of ε is more important than that of μ (i.e. if μ may be taken as real), the condition $\varepsilon'' > 0$ gives $\zeta'\zeta'' < 0$ and, since $\zeta' > 0$, $\zeta'' < 0$. This is the most usual case. If the dispersion of ζ is determined by that of μ, however, a similar argument shows that $\zeta'' > 0$.

The concept of impedance can also be applied to superconductors. A characteristic property of superconductors is that the penetration depth δ is small even in the static case ($\omega = 0$). At fairly low frequencies the magnetic field distribution can be taken to be the same as the static distribution. To determine the electric field we use the equation $\mathbf{curl}\,\mathbf{E} = i\omega\mathbf{H}/c$, taking the z-axis along the outward normal to the surface of the

† The mean free path depends considerably on the temperature of the metal. In practice, the temperatures considered are usually very low, in the helium range, and the phenomena under consideration occur in the range of very short radio waves.

‡ It should be borne in mind, however, that equation (87.6) can be used as a boundary condition only while $|\varepsilon|$ is large (i.e. ζ is small), and certainly does not hold at optical frequencies. We assume that $\mu \sim 1$, and so small ζ correspond to large $|\varepsilon|$. If $\mu \gg 1$, the inequality $\delta \ll \lambda$ must be fulfilled if the boundary condition (87.6) is valid, and therefore we must have $\sqrt{(\mu\varepsilon)} \gg 1$, so that $\zeta = \sqrt{(\mu/\varepsilon)}$ may not be small.

§ This situation is more fully discussed in *PK*, §87.

superconductor. Neglecting the tangential derivatives in comparison with the large z-derivatives, we have $\partial E_x/\partial z = i\omega H_y/c$, and similarly for E_y. Integrating with respect to z through the body gives

$$E_x(0) = \frac{i\omega}{c}\int_{-\infty}^{0} H_y\,dz,$$

$E_x(0)$ being the value of E_x for $z = 0$, i.e. at the surface of the body. We quantitatively define the penetration depth by the relation

$$\int_{-\infty}^{0} H_y dz = H_y(0)\delta. \qquad (87.7)$$

Then $E_x(0) = i\omega H_y(0)\delta/c$. Comparing this with the boundary condition (87.6), we find that the impedance of a superconductor (in the frequency range considered, which in practice extends to about the centimetre radio wave-band) is given by

$$\zeta = -i\omega\delta/c. \qquad (87.8)$$

This expression is the first term in an expansion of $\zeta(\omega)$ in powers of the frequency, and the expansion for superconductors thus begins with a term in ω. The next term, which is proportional to ω^2 and is real, is the first term in the expansion of ζ'.†

The impedance $\zeta(\omega)$, regarded as a function of the complex variable ω, has many properties analogous to those of the function $\varepsilon(\omega)$ (V. L. Ginzburg, 1954). The boundary condition, which for a monochromatic wave has the form (87.6), must in general be taken as the operator relation

$$\mathbf{E}_t = \hat{\zeta}\mathbf{H}_t \times \mathbf{n}, \qquad (87.9)$$

expressing the value of $\mathbf{E}_t$ at any instant in terms of the values of $\mathbf{H}_t$ at all previous instants (cf. §77). As in §82, it therefore follows that the function $\zeta(\omega)$ is regular in the upper half-plane of ω, including the real axis except for the point $\omega = 0$. The condition that $\mathbf{E}_t$ be real when $\mathbf{H}_t$ is real gives $\zeta(-\omega^*) = \zeta^*(\omega)$. Finally, since the energy dissipation is determined by the real part of $\zeta(\omega)$ (and not by the imaginary part as for $\varepsilon(\omega)$), it follows that $\zeta'(\omega)$ is positive, and does not vanish for any real ω except $\omega = 0$. Arguments similar to those given in §82 then lead to the conclusion that re $\zeta(\omega) > 0$ throughout the upper half-plane. Hence, in particular, $\zeta(\omega)$ has no zeros in the upper half-plane.

The regularity of $\zeta(\omega)$ in the upper half-plane again leads to Kramers and Kronig's formulae. A particularly important formula is

$$\zeta''(\omega) = -\frac{1}{\pi}P\int_{-\infty}^{\infty}\frac{\zeta'(x)-1}{x-\omega}\,dx.$$

Using the fact that $\zeta'(x)$ is even, we can also write

$$\zeta''(\omega) = -\frac{1}{\pi}P\int_{0}^{\infty}\frac{\zeta'(x)-1}{x-\omega}\,dx + \frac{1}{\pi}P\int_{0}^{\infty}\frac{\zeta'(x)-1}{x+\omega}\,dx$$

† The microscopic theory shows that the ω^2 term in the impedance also contains a factor varying logarithmically with ω; see *PK*, §§96, 97.

or

$$\zeta''(\omega) = -\frac{2\omega}{\pi} P \int_0^\infty \frac{\zeta'(x)}{x^2-\omega^2}\, dx. \tag{87.10}$$

The term -1 in the numerator of the integrand may be omitted, since the principal value of the integral of $1/(x^2-\omega^2)$ is zero.

The above statements concerning the function $\zeta(\omega)$ are equally applicable to the reciprocal function $1/\zeta(\omega)$; the operator $\hat{\zeta}^{-1}$ converts $\mathbf{E}_t$ into $\mathbf{H}_t \times \mathbf{n}$. In particular, (87.10) becomes

$$[\zeta^{-1}(\omega)]'' = -\frac{2\omega}{\pi} P \int_0^\infty \frac{[\zeta^{-1}(x)]'}{x^2-\omega^2}\, dx. \tag{87.11}$$

For small ζ this formula may be more useful than (87.10). In the form (87.11), however, it is not applicable to superconductors, for which ζ^{-1}, according to (87.8), has a pole at $\omega = 0$. A simple modification in the derivation, analogous to that which changes (82.7) into (82.9), gives

$$[\zeta^{-1}(\omega)]'' = -\frac{2\omega}{\pi} P \int_0^\infty \frac{[\zeta^{-1}(x)]'}{x^2-\omega^2}\, dx + \frac{c}{\omega\delta}. \tag{87.12}$$

To conclude this section we shall discuss, as an example of the use of the impedance, the reflection of a plane electromagnetic wave incident from a vacuum on the plane surface of a metal with surface impedance ζ. If the vector $\mathbf{E}$ is polarized perpendicular to the plane of incidence, the boundary condition (87.6) gives

$$E_0 + E_1 = \zeta(H_0 - H_1)\cos\theta_0 = \zeta(E_0 - E_1)\cos\theta_0,$$

the notation being the same as in §86. Hence, since ζ is small, we have $E_1/E_0 = -(1-2\zeta\cos\theta_0)$, and the reflection coefficient is

$$R_\perp = 1 - 4\zeta'\cos\theta_0. \tag{87.13}$$

If, on the other hand, $\mathbf{E}_0$ lies in the plane of incidence, the boundary condition in the form $\zeta\mathbf{H}_t = \mathbf{n}\times\mathbf{E}_t$ gives

$$\zeta(H_0 + H_1) = (E_0 - E_1)\cos\theta_0 = (H_0 - H_1)\cos\theta_0,$$

whence the reflection coefficient is

$$R_\parallel = \left|\frac{\cos\theta_0 - \zeta}{\cos\theta_0 + \zeta}\right|^2. \tag{87.14}$$

For angles of incidence not close to $\frac{1}{2}\pi$

$$R_\parallel = 1 - 4\zeta'\sec\theta_0. \tag{87.15}$$

If, on the other hand, $\phi_0 = \frac{1}{2}\pi - \theta_0 \ll 1$, then

$$R_\parallel = \left|\frac{\phi_0 - \zeta}{\phi_0 + \zeta}\right|^2. \tag{87.16}$$

This expression takes a minimum value $(|\zeta| - \zeta')/(|\zeta| + \zeta')$ for $\phi_0 = |\zeta|$.

Except for the special case (87.16), the reflection coefficient for a surface with small ζ is close to unity. A surface with $\zeta \to 0$ is perfectly conducting and also perfectly reflecting. The boundary condition at such a surface is simply $\mathbf{E}_t = 0$, analogously to that for the electrostatic field at the surface of a conductor. In a variable field, however, the fulfilment of this condition necessarily implies the fulfilment of a certain condition on the magnetic field: the equations $i\omega\mathbf{H}/c = \mathbf{curl}\,\mathbf{E}$ and $\mathbf{E}_t = 0$ on the surface imply that $H_n = 0$ there. Thus the normal component of the magnetic field must be zero on a perfectly conducting surface in a variable electromagnetic field. In this respect such a surface resembles the surface of a superconductor in a static magnetic field.

PROBLEM

Determine the intensity of thermal radiation (of a given frequency) from a plane surface of small impedance.

SOLUTION. According to Kirchhoff's law, the intensity $\mathrm{d}I$ of thermal radiation into an element of solid angle $\mathrm{d}o$ from an arbitrary surface is related to the intensity $\mathrm{d}I_0$ of radiation from the surface of a black body by $\mathrm{d}I = (1-R)\,\mathrm{d}I_0$, where R is the reflection coefficient for natural light incident on the surface concerned. Calculating $R = \frac{1}{2}(R_\perp + R_\parallel)$ from formulae (87.13) and (87.14) and using the isotropy of radiation from a black surface ($\mathrm{d}I_0 = I_0\,\mathrm{d}o/2\pi$), we have

$$I = 2I_0\zeta' \int_0^{\frac{1}{2}\pi} \left\{1 + \frac{1}{\cos^2\theta + 2\zeta'\cos\theta + \zeta'^2 + \zeta''^2}\right\} \cos\theta \sin\theta \, \mathrm{d}\theta.$$

Effecting the integration and omitting terms of higher order in ζ, we find

$$\frac{I}{I_0} = \zeta'\left[\log\frac{1}{\zeta'^2 + \zeta''^2} + 1 - \frac{2\zeta'}{\zeta''}\tan^{-1}\frac{\zeta''}{\zeta'}\right].$$

In particular, for a metal whose impedance is given by formula (87.3) ($\mu = 1$), we have

$$\frac{I}{I_0} = \sqrt{\frac{\omega}{8\pi\sigma}}\left[\log\frac{4\pi\sigma}{\omega} + 1 - \tfrac{1}{2}\pi\right].$$

§88. The propagation of waves in an inhomogeneous medium

Let us consider the propagation of electromagnetic waves in a medium which is electrically inhomogeneous but isotropic. In Maxwell's equations $\mathbf{curl}\,\mathbf{E} = i\omega\mathbf{H}/c$, $\mathbf{curl}\,\mathbf{H} = -i\varepsilon\omega\mathbf{E}/c$ (we put everywhere $\mu = 1$), ε is a function of the coordinates. Substituting for $\mathbf{H}$ from the first equation in the second, we obtain for $\mathbf{E}$ the equation

$$\triangle\mathbf{E} + (\varepsilon\omega^2/c^2)\mathbf{E} - \mathbf{grad}\,\mathrm{div}\,\mathbf{E} = 0. \tag{88.1}$$

Elimination of $\mathbf{E}$ gives for $\mathbf{H}$ the equation

$$\triangle\mathbf{H} + (\varepsilon\omega^2/c^2)\mathbf{H} + (1/\varepsilon)\,\mathbf{grad}\,\varepsilon \times \mathbf{curl}\,\mathbf{H} = 0. \tag{88.2}$$

These equations are considerably simplified in the one-dimensional case, where ε varies only in one direction in space. We take this direction as the z-axis, and consider a wave whose direction of propagation lies in the xz-plane. In such a wave all quantities are independent of y, and the uniformity of the medium in the x-direction means that the dependence on x can be taken as being through a factor $e^{i\kappa x}$, with κ a constant. For $\kappa = 0$ the field depends only on z, i.e. we have a wave which is said to pass *normally* through a layer of matter in which $\varepsilon = \varepsilon(z)$. If $\kappa \neq 0$, the wave is said to pass *obliquely*.

For $\kappa \neq 0$ two independent cases of polarization must be distinguished. In one, the vector **E** is perpendicular to the plane of propagation of the wave (i.e. it is in the y-direction), and the magnetic field **H** accordingly lies in that plane. Equation (88.1) becomes

$$\frac{\partial^2 E}{\partial z^2}+\left(\frac{\varepsilon\omega^2}{c^2}-\kappa^2\right)E=0. \qquad (88.3)$$

In the other case, the field **H** is in the y-direction, and **E** lies in the plane of propagation. Here it is more convenient to start from equation (88.2), which gives

$$\frac{\partial}{\partial z}\left(\frac{1}{\varepsilon}\frac{\partial H}{\partial z}\right)+\left(\frac{\omega^2}{c^2}-\frac{\kappa^2}{\varepsilon}\right)H=0. \qquad (88.4)$$

We shall call these two types of wave *E waves* and *H waves* respectively.

The equations can be solved in a general form in the important case where the conditions of propagation approximate to those of geometrical optics. In what follows we shall assume that the function $\varepsilon(z)$ is real.† In equation (88.3) the quantity $2\pi/\sqrt{f}$, where $f(z)=\varepsilon\omega^2/c^2-\kappa^2$, plays the part of a wavelength in the z-direction. The approximation of geometrical optics corresponds to the inequality

$$\frac{\mathrm{d}}{\mathrm{d}z}\frac{1}{\sqrt{f}}\ll 1, \qquad (88.5)$$

and the two independent solutions of equation (88.3) are of the form

$$\frac{\text{constant}}{f^{\frac{1}{4}}}e^{\pm i\int\sqrt{f}\mathrm{d}z}. \qquad (88.6)$$

The condition (88.5) is certainly not fulfilled near any *reflection point*, where $f=0$. Let $z=0$ be such a point, with $f>0$ for $z<0$ and $f<0$ for $z>0$. At sufficiently great distances on either side of $z=0$, the solution of equation (88.3) is of the form (88.6), but to establish the relation between the coefficients in the solutions for $z>0$ and $z<0$ it is necessary to examine the exact solution of equation (88.3) near $z=0$. In the neighbourhood of this point, $f(z)$ can be expanded as a power series in z: $f=-\alpha z$. The solution of the equation $\mathrm{d}^2E/\mathrm{d}z^2-\alpha zE=0$ which is finite for all z is

$$E=(A/\alpha^{1/6})\Phi(\alpha^{1/3}z), \qquad (88.7)$$

where

$$\Phi(\xi)=\frac{1}{\sqrt{\pi}}\int_0^\infty\cos\left(\tfrac{1}{3}u^3+u\xi\right)\mathrm{d}u$$

is the *Airy function*; we everywhere omit the factor $e^{-i\omega t+i\kappa x}$ in E.‡ The asymptotic form of

† Equation (88.3) bears a formal resemblance to Schrödinger's equation for one-dimensional motion of a particle in quantum mechanics, and the approximation of geometrical optics corresponds to the quasi-classical case. Here we shall give the final results; their derivation may be found in *QM*, Chapter VII.

‡ Here we make use of the Airy function as defined in other volumes of the *Course of Theoretical Physics*. It is now more usually defined as

$$\mathrm{Ai}\,\xi=\Phi(\xi)/\sqrt{\pi}.$$

the solution of equation (88.3) for large $|z|$ is

$$\left.\begin{aligned} E &= \frac{A}{f^{\frac{1}{4}}}\cos\left(\int_0^z \sqrt{f}\,\mathrm{d}z + \tfrac{1}{4}\pi\right) \quad \text{for } z<0,\\ E &= \frac{A}{2|f|^{\frac{1}{4}}}\exp\left(-\int_0^z \sqrt{|f|}\,\mathrm{d}z\right) \quad \text{for } z>0, \end{aligned}\right\} \tag{88.8}$$

with the same coefficient A as in (88.7). The first of these expressions represents the stationary wave obtained by superposing the wave incident in the positive z-direction and the wave reflected from the plane $z = 0$. The amplitudes of these waves are both equal to $\frac{1}{2}A/f^{\frac{1}{4}}$, i.e. the reflection coefficient is unity. Only an exponentially damped field penetrates into the region $z > 0$.

As the reflection point is approached, the wave amplitude increases, as is shown by the factor $f^{\frac{1}{4}}$ in the denominators in (88.8). To determine the field in the immediate vicinity of that point, the expression (88.7) must be used. This function decreases monotonically into the region $z > 0$ and oscillates in the region $z < 0$, the successive maxima of $|E|$ continually decreasing. The first and highest maximum is reached at $\alpha^{\frac{1}{3}}z = -1{\cdot}02$, and its value is $E = 0.949\,A\alpha^{-1/6}$.

So far we have spoken of solutions for E waves. It is easy to see that, in the approximation of geometrical optics, entirely similar formulae are valid for H waves. If we substitute in equation (88.4) $H = u\sqrt{\varepsilon}$, the derivatives of ε appear as products with u, but not with u'; neglecting therefore the terms containing these derivatives, which are small by (88.5), we obtain for the function $u(z)$ the equation

$$\frac{\mathrm{d}^2u}{\mathrm{d}z^2} + \left(\frac{\varepsilon\omega}{c^2} - \kappa^2\right)u = 0,$$

which is of the same form as (88.3). Hence the formulae for H differ from (88.6)–(88.8) only by a factor $\sqrt{\varepsilon}$.

A curious difference in the behaviour of the two types of wave occurs when an obliquely incident wave ($\kappa \neq 0$) is reflected from a layer in which $\varepsilon(z)$ passes through zero. The reflection takes place from the plane on which $f(z) = \varepsilon\omega^2/c^2 - \kappa^2 = 0$, i.e. the wave "does not reach" the plane where $\varepsilon = 0$. The E wave penetrates beyond the plane only as an exponentially damped field. When an H wave is reflected, however, there is superposed on a similar damped field a strong local field near the plane on which $\varepsilon = 0$ (K. Försterling, 1949).† This may be shown as follows.

Let $\varepsilon = 0$ at the point $z = 0$. Near this point, we write

$$\varepsilon = -az, \quad a > 0 \tag{88.9}$$

and equation (88.4) takes the form

$$\frac{\mathrm{d}^2H}{\mathrm{d}z^2} - \frac{1}{z}\frac{\mathrm{d}H}{\mathrm{d}z} - (a\omega^2 z/c^2 + \kappa^2)H = 0. \tag{88.10}$$

† This is a singularity of (88.4), and geometrical optics is therefore invalid near it, although $f(z)$ does not vanish and (88.5) may be valid.

According to the general theory of linear differential equations, one of the solutions of this equation, which we call $H_1(z)$, has no singularity at $z = 0$, and its expansion in powers of z begins with z^2:

$$H_1(z) = z^2 + \ldots .$$

The other independent solution has a logarithmic singularity, and its expansion is

$$H_2(z) = H_1(z)\log\kappa z + \frac{2}{\kappa^2} + \ldots .$$

The parameter a occurs only in the subsequent terms of the expansions. To find the field near $z = 0$, it is not necessary to discuss the choice of a linear combination of H_1 and H_2 that satisfies the conditions at infinity. We need only note that as $z \to 0$ this combination tends to a constant (H_0, say) and has a logarithmic singularity

$$H \cong H_0(1 + \tfrac{1}{2}\kappa^2 z^2 \log\kappa z);$$

here, the leading singularity term has been written together with the constant. The electric field is determined from $H_y = H$ by Maxwell's equations $E_x = -(ic/\varepsilon\omega)\partial H/\partial z$, $E_z = (ic/\varepsilon\omega)\partial H/\partial x$. The dependence of H on x is given by the factor $e^{i\kappa x}$, and the leading terms in E_x and E_z are therefore

$$E_x \cong H_0(i\kappa^2 c/a\omega)\log\kappa z, \qquad E_z \cong H_0(\kappa c/a\omega)(1/z). \tag{88.11}$$

These become infinite as $z \to 0$.

In reality, of course, the absorption which must be present in the medium, even though slight, means that the field attains large but not infinite values compared with the weak field in the adjoining region. It is noteworthy, however, that even an infinitesimal imaginary part of ε causes a finite dissipation of energy. Let $\varepsilon = -az + i\delta$ with $\delta \to +0$. Then the analytical continuation of the logarithm in (88.11) from the right-hand to the left-hand half of the z axis must be carried out from below in the complex z-plane, and for $z < 0$ we have

$$E_x = H_0(i\kappa^2 c/a\omega)(\log|\kappa z| - i\pi).$$

The (time) averaged energy flux along the z-axis:

$$\overline{S}_z = (c/8\pi)\,\mathrm{re}(E_x H_y{}^*)$$

(see (59.9a)) is zero for $z > 0$; for $z < 0$, the presence of a real part of E_x gives a non-zero energy flux towards the plane $z = 0$, where this energy is dissipated:†

$$\overline{S}_z = \kappa^2 c^2 H_0{}^2/8\omega a \tag{88.12}$$

(V. B. Gil'denburg, 1963).

† This result can also be derived from the expression (80.4) for the energy dissipated in unit volume:

$$Q = \frac{\omega\varepsilon''|\mathbf{E}|^2}{8\pi} \cong \frac{\kappa^2 c^2 H_0{}^2}{8\pi\omega}\lim_{\delta\to 0}\frac{\delta}{a^2 z^2 + \delta^2} = \kappa^2 c^2 H_0{}^2\delta(z)/8a\omega;$$

integration with respect to z gives (88.12).

PROBLEM

A surface H wave can be propagated along a plane boundary between two media whose permittivities ε_1 and $-|\varepsilon_2|$ are of opposite signs. The wave is damped in both media. Determine the relation between the frequency and the wave number.

SOLUTION. We take the boundary surface as the xy-plane, the wave being propagated in the x-direction and the field $\mathbf{H}$ being in the y-direction. Let the half-space $z > 0$ contain the medium with the positive permittivity ε_1, and the half-space $z < 0$ that with the negative permittivity ε_2. We seek the field in the wave damped as $z \to \pm\infty$ in the form

$$H_1 = H_0 e^{ikx-\kappa_1 z}, \qquad \kappa_1 = \sqrt{(k^2-\omega^2\varepsilon_1/c^2)} \qquad \text{for } z > 0,$$
$$H_2 = H_0 e^{ikx+\kappa_2 z}, \qquad \kappa_2 = \sqrt{(k^2+\omega^2|\varepsilon_2|/c^2)} \qquad \text{for } z < 0,$$

where k, κ_1 and κ_2 are real. The boundary condition that $H_y = H$ be continuous is already satisfied, and the continuity condition on E_x gives $(1/\varepsilon_1)\partial H_1/\partial z = (1/\varepsilon_2)\partial H_2/\partial z$ for $z = 0$, or $\kappa_1/\varepsilon_1 = \kappa_2/|\varepsilon_2|$. This equation can be satisfied if $\varepsilon_1 < |\varepsilon_2|$ (and if $\varepsilon_1\varepsilon_2 < 0$, as has been assumed). The relation between k and ω is

$$k^2 = \omega^2\varepsilon_1|\varepsilon_2|/c^2(|\varepsilon_2|-\varepsilon_1).$$

It is easily seen that surface E waves cannot be propagated.

§89. The reciprocity principle

The emission of monochromatic electromagnetic waves from a source consisting of a thin wire in an arbitrary medium is described by the equations

$$\mathbf{curl}\,\mathbf{E} = i\omega\mathbf{B}/c, \qquad \mathbf{curl}\,\mathbf{H} = -i\omega\mathbf{D}/c + 4\pi\mathbf{j}_{ex}/c, \tag{89.1}$$

where $\mathbf{j}_{ex}$ is the density of periodic currents flowing in the wire which are extraneous to the medium.

Let two different sources (of the same frequency) be placed in the medium; we denote by the suffixes 1 and 2 the fields due to these sources separately. The medium may be inhomogeneous and anisotropic. The only assumption which we shall make concerning the properties of the medium is that the linear relations $D_i = \varepsilon_{ik}E_k$, $B_i = \mu_{ik}H_k$ hold, the tensors ε_{ik} and μ_{ik} being symmetrical. Under these conditions it is possible to derive a relation between the fields of the two sources and the extraneous currents in them.

We take the scalar products of the two equations $\mathbf{curl}\,\mathbf{E}_1 = ik\mathbf{B}_1$, $\mathbf{curl}\,\mathbf{H}_1 = -ik\mathbf{D}_1 + 4\pi\mathbf{j}_{ex,1}/c$ with $\mathbf{H}_2$ and $\mathbf{E}_2$ respectively, and of the corresponding equations for $\mathbf{E}_2$ and $\mathbf{H}_2$ with $-\mathbf{H}_1$ and $-\mathbf{E}_1$. Adding all four together, we obtain

$$\begin{aligned}(\mathbf{H}_2\cdot\mathbf{curl}\,\mathbf{E}_1 - \mathbf{E}_1\cdot\mathbf{curl}\,\mathbf{H}_2) &+ (\mathbf{E}_2\cdot\mathbf{curl}\,\mathbf{H}_1 - \mathbf{H}_1\cdot\mathbf{curl}\,\mathbf{E}_2)\\ &= (i\omega/c)(\mathbf{B}_1\cdot\mathbf{H}_2 - \mathbf{H}_1\cdot\mathbf{B}_2) + (i\omega/c)(\mathbf{E}_1\cdot\mathbf{D}_2 - \mathbf{D}_1\cdot\mathbf{E}_2)\\ &\quad + (4\pi/c)(\mathbf{j}_{ex,1}\cdot\mathbf{E}_2 - \mathbf{j}_{ex,2}\cdot\mathbf{E}_1).\end{aligned}$$

But $\mathbf{B}_1\cdot\mathbf{H}_2 = \mu_{ik}H_{1k}H_{2i} = \mathbf{H}_1\cdot\mathbf{B}_2$, and $\mathbf{E}_1\cdot\mathbf{D}_2 = \mathbf{D}_1\cdot\mathbf{E}_2$, so that the first two terms on the right-hand side are zero. The left-hand side can be transformed by a formula of vector analysis, and the result is

$$\operatorname{div}[\mathbf{E}_1\times\mathbf{H}_2 - \mathbf{E}_2\times\mathbf{H}_1] = (4\pi/c)(\mathbf{j}_{ex,1}\cdot\mathbf{E}_2 - \mathbf{j}_{ex,2}\cdot\mathbf{E}_1).$$

We integrate this equation over all space; the integral on the left-hand side can be transformed into one over an infinitely remote surface, and is zero. Thus we have

$$\int \mathbf{j}_{ex,1}\cdot\mathbf{E}_2\,dV_1 = \int \mathbf{j}_{ex,2}\cdot\mathbf{E}_1\,dV_2. \tag{89.2}$$

The integrals are taken only over the volumes of sources 1 and 2 respectively, since the currents $\mathbf{j}_{ex,1}$ and $\mathbf{j}_{ex,2}$ are zero elsewhere. Since the wires are thin, the effect of each on the field of the other may be neglected, and therefore $\mathbf{E}_1$ and $\mathbf{E}_2$ in formula (89.2) are the fields due to each of the two sources at the position of the other. Formula (89.2) is the required relation; it is called the *reciprocity theorem.*

If the dimensions of the sources are small compared with the wavelength and with the distance between them, this formula can be simplified. The field of each source varies only slightly over the dimensions of the other, and in (89.2) we can take $\mathbf{E}_1$ and $\mathbf{E}_2$ outside the integrals and replace them by $\mathbf{E}_1(2)$ and $\mathbf{E}_2(1)$, 1 and 2 signifying the positions of the two sources:

$$\mathbf{E}_2(1)\cdot\int \mathbf{j}_{\mathrm{ex},1}\,\mathrm{d}V_1 = \mathbf{E}_1(2)\cdot\int \mathbf{j}_{\mathrm{ex},2}\,\mathrm{d}V_2.$$

The integral $\int \mathbf{j}_{\mathrm{ex}}\,\mathrm{d}V$ is just the time derivative of the total dipole moment $\mathscr{P}$ of the source. Since $\dot{\mathscr{P}} = -i\omega\mathscr{P}$, we have finally

$$\mathbf{E}_2(1)\cdot\mathscr{P}_1 = \mathbf{E}_1(2)\cdot\mathscr{P}_2. \tag{89.3}$$

This form of the reciprocity theorem applies, of course, only to dipole emission. If the dipole moment of the source is zero, or very small, the approximation made in going from the general formula (89.2) to (89.3) is inadequate; see Problem 1.

PROBLEMS

PROBLEM 1. Derive the reciprocity theorem for quadrupole emitters and for magnetic dipole emitters.

SOLUTION. If $\int \mathbf{j}_{\mathrm{ex}}\,\mathrm{d}V = 0$, the next terms in the expansion must be taken in the integrals (89.2):

$$\begin{aligned}\int \mathbf{j}_1\cdot\mathbf{E}_2\,\mathrm{d}V_1 &\cong \frac{\partial E_{2i}}{\partial x_k}\int x_k j_{1i}\,\mathrm{d}V_1\\ &= \frac{1}{4}\left(\frac{\partial E_{2i}}{\partial x_k}+\frac{\partial E_{2k}}{\partial x_i}\right)\int (x_k j_{1i}+x_i j_{1k})\,\mathrm{d}V_1\\ &\quad+\frac{1}{4}\left(\frac{\partial E_{2i}}{\partial x_k}-\frac{\partial E_{2k}}{\partial x_i}\right)\int (x_k j_{1i}-x_i j_{1k})\,\mathrm{d}V_1;\end{aligned}$$

we omit the suffix ex for brevity. The quadrupole moment tensor and the magnetic moment tensor are defined by

$$\dot{D}_{ik} = -i\omega D_{ik} = \int [3(x_i j_k + x_k j_i) - 2\delta_{ik}\mathbf{r}\cdot\mathbf{j}]\,\mathrm{d}V,$$

$$\mathscr{M} = \frac{1}{2c}\int \mathbf{r}\times\mathbf{j}\,\mathrm{d}V.$$

Using the equation $\mathbf{curl}\,\mathbf{E} = i\omega\mathbf{B}/c$ and assuming that ε = constant near the sources (and so $\operatorname{div}\mathbf{E} = 0$), we obtain

$$\int \mathbf{j}_1\cdot\mathbf{E}_2\,\mathrm{d}V = -\frac{i\omega}{12}\left(\frac{\partial E_{2i}}{\partial x_k}+\frac{\partial E_{2k}}{\partial x_i}\right)D_{1,ik} + i\omega\mathbf{B}_2(1)\cdot\mathscr{M}_1.$$

Hence we see that for quadrupole emitters the reciprocity theorem is

$$\left(\frac{\partial E_{2i}(1)}{\partial x_k}+\frac{\partial E_{2k}(1)}{\partial x_i}\right)D_{1,ik} = \left(\frac{\partial E_{1i}(2)}{\partial x_k}+\frac{\partial E_{ik}(2)}{\partial x_i}\right)D_{2,ik},$$

and for magnetic dipole emitters

$$\mathbf{B}_2(1)\cdot\mathscr{M}_1 = \mathbf{B}_1(2)\cdot\mathscr{M}_2.$$

PROBLEM 2. Determine the intensity of emission from a dipole source immersed in a homogeneous isotropic medium as a function of the permittivity ε and permeability μ of the medium.

SOLUTION. By substituting $\mathbf{E} = \sqrt{(\mu/\varepsilon)}\mathbf{E}'$, $\mathbf{H} = \mathbf{H}'$, $\omega = \omega'/\sqrt{(\varepsilon\mu)}$, equations (89.1) are brought to the form $\mathbf{curl}\,\mathbf{E}' = i\omega'\mathbf{H}'/c$, $\mathbf{curl}\,\mathbf{H}' = -i\omega'\mathbf{E}'/c + 4\pi\mathbf{j}_{\mathrm{ex}}/c$, which do not involve ε or μ. The solution of these equations for dipole emission gives a vector field potential in the wave region (see *Fields*, §67) $\mathbf{A}' = (1/cR_0)\int \mathbf{j}_{\mathrm{ex}}\,\mathrm{d}V$, where R_0 is the distance from the source; the phase factors are omitted, since they do not affect the calculation of the intensity. Hence we see that, for given $\mathbf{j}_{ex}$, we can put $\mathbf{A}' = \mathbf{A}_0$, where the suffix 0 signifies the value for the source field in a vacuum. The values of $\mathbf{H}'$ and $\mathbf{E}'$ are

$$\mathbf{H}' = i\mathbf{k}' \times \mathbf{A}' = i\sqrt{(\varepsilon\mu)}\mathbf{k} \times \mathbf{A}_0 = \sqrt{(\varepsilon\mu)}\mathbf{H}_0, \qquad \mathbf{E}' = \mathbf{H}'.$$

Hence $\mathbf{H} = \sqrt{(\varepsilon\mu)}\mathbf{H}_0$, $\mathbf{E} = \mu\mathbf{E}_0$ and $I = I_0\mu^{3/2}\varepsilon^{1/2}$. This is the required solution.

§90. Electromagnetic oscillations in hollow resonators

Let us consider the electric field in a hollow evacuated resonator with perfectly conducting walls. The equations of the monochromatic field in the vacuum are

$$\mathbf{curl}\,\mathbf{E} = i\omega\mathbf{H}/c, \qquad \mathbf{curl}\,\mathbf{H} = -i\omega\mathbf{E}/c. \tag{90.1}$$

The boundary conditions on the surface of a perfect conductor (i.e. one whose impedance $\zeta = 0$) are

$$\mathbf{E}_t = 0, \qquad H_n = 0. \tag{90.2}$$

To solve the problem, it suffices to consider either $\mathbf{E}$ or $\mathbf{H}$. For instance, eliminating $\mathbf{H}$ from equations (90.1), we find that $\mathbf{E}$ satisfies the wave equation

$$\triangle\mathbf{E} + (\omega^2/c^2)\mathbf{E} = 0, \tag{90.3}$$

together with the equation

$$\operatorname{div}\mathbf{E} = 0, \tag{90.4}$$

which does not follow from (90.1). Solving these equations with the boundary condition $\mathbf{E}_t = 0$, we find the field $\mathbf{E}$, and then $\mathbf{H}$ can be derived from the first of (90.1). The boundary condition $H_n = 0$ is automatically satisfied.

When the shape and size of the cavity are given, equations (90.3) and (90.4) have solutions only for certain values of ω, called the *eigenfrequencies* of the electromagnetic oscillations of the resonator concerned. For $\zeta = 0$ the electromagnetic field does not penetrate into the metal, and no loss occurs there. All the characteristic oscillations are therefore undamped, and all the eigenfrequencies are real. The latter are infinite in number, and the order of magnitude of the lowest eigenfrequency ω_1 is c/l, where l is the linear dimension of the cavity. This follows immediately from dimensional considerations, since l is the only dimensional parameter characterizing the problem if the shape of the resonator is given. The high eigenfrequencies ($\omega \gg c/l$) lie very close together, and the number of them per unit range of ω is $V\omega^2/2\pi^2c^3$, which depends on the volume V of the resonator but not on its shape (see *Fields*, §52).

The (time) averaged values of the electric and magnetic field energies in the resonator are respectively $\frac{1}{2}\int(|\mathbf{E}|^2/8\pi)\,\mathrm{d}V$ and $\frac{1}{2}\int(|\mathbf{H}|^2/8\pi)\,\mathrm{d}V$. We shall show that they are equal. Using the first equation (90.1), we write $\int\mathbf{H}\cdot\mathbf{H}^*\,\mathrm{d}V = (c^2/\omega^2)\int\mathbf{curl}\,\mathbf{E}\cdot\mathbf{curl}\,\mathbf{E}^*\,\mathrm{d}V$. The second integral can be integrated by parts:

$$\int\mathbf{curl}\,\mathbf{E}\cdot\mathbf{curl}\,\mathbf{E}^*\,\mathrm{d}V = \oint\mathbf{curl}\,\mathbf{E}^*\cdot\mathrm{d}\mathbf{f}\times\mathbf{E} + \int\mathbf{E}\cdot\mathbf{curl}\,\mathbf{curl}\,\mathbf{E}^*\,\mathrm{d}V.$$

Since $\mathbf{E}_t = 0$ on the boundary of the volume considered, the surface integral is zero, leaving

$$\begin{aligned}\int|\mathbf{H}|^2\mathrm{d}V &= \frac{c^2}{\omega^2}\int\mathbf{E}\cdot\mathbf{curl}\,\mathbf{curl}\,\mathbf{E}^*\,\mathrm{d}V\\ &= -(c^2/\omega^2)\int\mathbf{E}\cdot\triangle\mathbf{E}^*\,\mathrm{d}V\end{aligned}$$

or, by (90.3),

$$\int |\mathbf{H}|^2 dV = \int |\mathbf{E}|^2 dV. \tag{90.5}$$

This completes the proof.†

Undamped oscillations in a resonator are obtained if the impedance of its walls is assumed to be zero. Let us now ascertain the effect on the eigenfrequencies if the impedance of the walls is small but not zero.

The (time) averaged energy dissipated in the walls of the resonator in unit time can be calculated as the flux of energy into the walls from the electromagnetic field in the cavity. Using the boundary condition (87.6) on the surface of a body with impedance ζ, we write the normal component of the energy flux density as $\overline{S}_n = (c/8\pi)\,\mathrm{re}\,(\mathbf{E}_t \times \mathbf{H}_t^*) = c\zeta'|\mathbf{H}_t|^2/8\pi$, where ζ' is the real part of ζ. In this expression, which already contains the small factor ζ', we can as a first approximation take $\mathbf{H}$ to be the field obtained by solving the problem with $\zeta = 0$. The total energy dissipated is given by the integral

$$\frac{c}{8\pi}\oint \zeta' |\mathbf{H}|^2 df, \tag{90.6}$$

taken over the internal surface of the resonator. The field amplitude is damped in time with a decrement obtained by dividing (90.6) by twice the total energy of the field, namely $\frac{1}{2}\int(|\mathbf{E}|^2 + |\mathbf{H}|^2)\,dV/8\pi = \int |\mathbf{H}|^2\,dV/8\pi$.

The damping decrement is determined by the imaginary part $|\omega''|$ of the complex frequency $\omega = \omega' + i\omega''$.‡ Writing the formula in the complex form

$$\omega - \omega_0 = -\tfrac{1}{2}ic\frac{\oint \zeta |\mathbf{H}|^2 df}{\int |\mathbf{H}|^2 dV}, \tag{90.7}$$

ω and ω_0 being the frequencies with and without allowance for ζ, we can determine not only the damping decrement but also the change in the eigenfrequencies themselves. The latter is seen to be determined by the imaginary part of ζ. We have mentioned in §87 that usually $\zeta'' < 0$, and the eigenfrequencies are then reduced.

In actual calculations, it may be more convenient to transform the volume integral in the denominator of (90.7) into one over the surface. Since the vector $\mathbf{H}$ is tangential to the surface, we have identically

$$\oint(\mathbf{H}\cdot\mathbf{H}^*)(\mathbf{r}\cdot d\mathbf{f}) = \oint(\mathbf{H}\cdot\mathbf{H}^*)(\mathbf{r}\cdot d\mathbf{f}) - \oint(\mathbf{H}\cdot\mathbf{r})(\mathbf{H}^*\cdot d\mathbf{f}) - \oint(\mathbf{H}^*\cdot\mathbf{r})(\mathbf{H}\cdot d\mathbf{f}).$$

The integrals on the right are transformed by putting $d\mathbf{f} \to dV\,\mathbf{grad}$, and using (90.1) we obtain

$$\oint(\mathbf{H}\cdot\mathbf{H}^*)(\mathbf{r}\cdot d\mathbf{f}) = ik\int \mathbf{r}\cdot(\mathbf{H}\times\mathbf{E}^* - \mathbf{H}^*\times\mathbf{E})\,dV + \int \mathbf{H}\cdot\mathbf{H}^*\,dV.$$

Similarly, using the identity $\mathbf{r}\times(\mathbf{E}\times d\mathbf{f}) = \mathbf{E}(\mathbf{r}\cdot d\mathbf{f}) - (\mathbf{r}\cdot\mathbf{E})\,d\mathbf{f} = 0$ (by the boundary

---

† By $\mathbf{E}$ and $\mathbf{H}$ we always mean the fields corresponding to a particular eigenfrequency. It is not difficult to show that the fields corresponding to two different eigenfrequencies ω_a and ω_b are orthogonal:

$$\int \mathbf{E}_a\cdot\mathbf{E}_b^* dV = \int \mathbf{H}_a\cdot\mathbf{H}_b^* dV = 0.$$

‡ In radio engineering the *Q factor* or *quality* of the resonator, defined as the ratio $\omega'/2|\omega''|$, is generally used instead of the damping decrement.

condition $\mathbf{E}_t = 0$), we have

$$\oint(\mathbf{E}\cdot\mathbf{E}^*)(\mathbf{r}\cdot d\mathbf{f}) = -\oint(\mathbf{E}\cdot\mathbf{E}^*)(\mathbf{r}\cdot d\mathbf{f}) + \oint(\mathbf{E}\cdot\mathbf{r})(\mathbf{E}^*\cdot d\mathbf{f}) + \oint(\mathbf{E}^*\cdot\mathbf{r})(\mathbf{E}\cdot d\mathbf{f})$$
$$= ik\int\mathbf{r}\cdot(\mathbf{H}\times\mathbf{E}^* - \mathbf{H}^*\times\mathbf{E})\,dV - \int\mathbf{E}\cdot\mathbf{E}^*\,dV.$$

Subtracting and using (90.5), we obtain

$$\int|\mathbf{H}|^2 dV = \tfrac{1}{2}\oint(|\mathbf{H}|^2 - |\mathbf{E}|^2)\mathbf{r}\cdot d\mathbf{f}. \qquad (90.8)$$

The formulae for a resonator filled with a non-absorbing dielectric for which ε and μ differ from unity are obtained from those for an evacuated resonator by replacing

$$\omega,\quad \mathbf{E} \text{ and } \mathbf{H} \text{ by } \omega\sqrt{(\varepsilon\mu)},\quad \sqrt{\varepsilon}\mathbf{E} \text{ and } \sqrt{\mu}\mathbf{H}. \qquad (90.9)$$

This is seen from the fact that the transformation just given converts equations (90.1) into the correct Maxwell's equations for the medium:

$$\mathbf{curl}\,\mathbf{E} = i\omega\mu\mathbf{H}/c, \quad \mathbf{curl}\,\mathbf{H} = -i\omega\varepsilon\mathbf{E}/c.$$

In particular, the presence of the medium reduces each eigenfrequency by a factor $\sqrt{(\varepsilon\mu)}$.

PROBLEMS

PROBLEM 1. Determine the eigenfrequencies of a cuboidal resonator with perfectly conducting walls.

SOLUTION. We take the axes of x, y, z along three concurrent edges of the cuboid; let the lengths of these edges be a_1, a_2, a_3. The solutions of equations (90.3) and (90.4) which satisfy the boundary condition $\mathbf{E}_t = 0$ are

$$E_x = A_1 \cos k_x x \sin k_y y \sin k_z z \cdot e^{-i\omega t} \qquad (1)$$

and similarly for E_y, E_z, where

$$k_x = n_1\pi/a_1, \qquad k_y = n_2\pi/a_2, \qquad k_z = n_3\pi/a_3 \qquad (2)$$

(n_1, n_2, n_3 being positive integers). The constants A_1, A_2, A_3 are related by

$$A_1k_x + A_2k_y + A_3k_z = 0, \qquad (3)$$

and the eigenfrequencies are $\omega^2 = c^2(k_x^2 + k_y^2 + k_z^2)$.

The magnetic field is calculated from (1):

$$H_x = -(ic/\omega)(A_3k_y - A_2k_z)\sin k_x x \cos k_y y \cos k_z z \cdot e^{-i\omega t},$$

and similarly for H_y, H_z.

If two or all of the numbers n_1, n_2, n_3 are zero, $\mathbf{E} = 0$. Hence the lowest frequency corresponds to an oscillation in which one of these numbers is 0 and the other two are 1.

Since the relation (3) holds, the solution (1) (with given non-zero n_1, n_2, n_3) involves only two independent arbitrary constants, i.e. each eigenfrequency is doubly degenerate. The frequencies for which one of n_1, n_2, n_3 is zero are not degenerate.

PROBLEM 2. Determine the frequencies of electric dipole and magnetic dipole oscillations in a spherical resonator with radius a.

SOLUTION. In a stationary spherical electric dipole wave, the fields $\mathbf{E}$ and $\mathbf{H}$ are of the form

$$\mathbf{E} = e^{-i\omega t}\,\mathbf{curl}\,\mathbf{curl}\left(\frac{\sin kr}{r}\mathbf{b}\right), \qquad \mathbf{H} = -ike^{-i\omega t}\,\mathbf{curl}\left(\frac{\sin kr}{r}\mathbf{b}\right),$$

where $\mathbf{b}$ is a constant vector and $k = \omega/c$ (see *Fields*, §72). The boundary condition $\mathbf{n}\times\mathbf{E} = 0$ at $r = a$ gives $\cot ka = (ka)^{-1} - ka$. The smallest root of this equation is $ka = 2.74$. The frequency $\omega = 2.74\,c/a$ is the lowest eigenfrequency of a spherical resonator.

In a stationary spherical magnetic dipole wave, we have

$$\mathbf{E} = ike^{-i\omega t}\,\mathbf{curl}\left(\frac{\sin kr}{r}\mathbf{b}\right), \qquad \mathbf{H} = e^{-i\omega t}\,\mathbf{curl\,curl}\left(\frac{\sin kr}{r}\mathbf{b}\right).$$

The boundary condition on **E** gives the equation tan $ka = ka$, whose smallest root is $ka = 4.49$.

PROBLEM 3. A small sphere with electric and magnetic polarizabilities α_e and α_m is placed in a resonator. Determine the resulting shift in the resonator eigenfrequency.

SOLUTION. Let **E** and **H** be the fields in the resonator without the sphere. $\mathbf{E}_1$ and $\mathbf{H}_1$ those when the sphere is present. The fields **E** and **H** satisfy the equations (90.1); $\mathbf{E}_1$ and $\mathbf{H}_1$ satisfy

$$\mathbf{curl}\,\mathbf{E}_1 = i\omega_1\mathbf{H}_1/c, \quad \mathbf{curl}\,\mathbf{H}_1 = 4\pi\mathbf{j}_1/c - i\omega_1\mathbf{E}_1/c, \tag{1}$$

where $\mathbf{j}_1$ is the current density in the sphere. We multiply the first of these equations by $\mathbf{H}^*$, the second by $-\mathbf{E}^*$, take the complex conjugates of equations (90.1), and multiply them by $\mathbf{H}_1$ and $-\mathbf{E}_1$ respectively. Addition of all four equations then gives

$$\operatorname{div}[\mathbf{E}_1\times\mathbf{H}^* + \mathbf{E}^*\times\mathbf{H}_1] = i\delta\omega(\mathbf{H}_1\cdot\mathbf{H}^* + \mathbf{E}_1\cdot\mathbf{E}^*)/c - 4\pi\mathbf{j}_1\cdot\mathbf{E}^*/c,$$

where $\delta\omega = \omega_1 - \omega$ is the required frequency shift. This equation is next integrated over the volume of the resonator. The left-hand side is zero, by Gauss's theorem, since at the wall $\mathbf{E}_t = \mathbf{E}_{1t} = 0$. Since the sphere is small, the main contribution to the integral of the first term on the right arises at large distances from the sphere; at such distances, however, the field perturbation due to the sphere is small, and so we can put $\mathbf{E}_1 \cong \mathbf{E}$, $\mathbf{H}_1 \cong \mathbf{H}$. The integral of the second term is transformed as in §89 (and Problem 1 there), giving

$$\begin{aligned}\int \mathbf{j}_1\cdot\mathbf{E}^*\mathrm{d}V &= -i\omega(\mathscr{P}\cdot\mathbf{E}_0^* + \mathscr{M}\cdot\mathbf{H}_0^*)\\ &= -i\omega V_0(\alpha_e|\mathbf{E}_0|^2 + \alpha_m|\mathbf{H}_0|^2),\end{aligned}$$

where $\mathbf{E}_0 \equiv \mathbf{E}(\mathbf{r}_0)$, $\mathbf{H}_0 \equiv \mathbf{H}(\mathbf{r}_0)$, $\mathbf{r}_0$ is the position of the sphere, and V_0 is its volume; the sphere is supposed to be so small that the fields **E** and **H** do not vary appreciably across it.

Thus, with (90.5), we find the frequency shift

$$\frac{\delta\omega}{\omega} = -\frac{\alpha_e|\mathbf{E}_0|^2 + \alpha_m|\mathbf{H}_0|^2}{\int|\mathbf{H}|^2\,\mathrm{d}V/2\pi}V_0.$$

If the polarizabilities are complex, this formula gives both the frequency shift and the damping of the characteristic oscillations.

PROBLEM 4. A resonator is filled with a dispersionless transparent dielectric having permittivity ε_0. Determine the change in the eigenfrequency due to a small change $\delta\varepsilon(\mathbf{r})$ in the permittivity.

SOLUTION. The unperturbed field $\mathbf{E}_0$, $\mathbf{H}_0$ in the resonator satisfies the equations $\mathbf{curl}\,\mathbf{E}_0 = i\omega_0\mathbf{H}_0/c$, $\mathbf{curl}\,\mathbf{H}_0 = -i\omega_0\varepsilon_0\mathbf{E}_0/c$, and the perturbed field **E**, **H** satisfies $\mathbf{curl}\,\mathbf{E} = i(\omega_0+\delta\omega)\mathbf{H}/c$, $\mathbf{curl}\,\mathbf{H} = -i(\omega_0\varepsilon_0 + \omega_0\delta\varepsilon + \varepsilon_0\delta\omega)\mathbf{E}/c$, the term in $\delta\omega\delta\varepsilon$ being neglected. Using these four equations as in Problem 3, we find

$$\begin{aligned}\operatorname{div}[(\mathbf{E}\times\mathbf{H}_0^*) + (\mathbf{E}_0^*\times\mathbf{H})] &= i(\omega_0\delta\varepsilon + \varepsilon_0\delta\omega)\,\mathbf{E}\cdot\mathbf{E}_0^*/c + i\delta\omega\mathbf{H}\cdot\mathbf{H}_0^*/c\\ &\cong i(\omega_0\delta\varepsilon + \varepsilon_0\delta\omega)\mathbf{E}_0\cdot\mathbf{E}_0^*/c + i\delta\omega\mathbf{H}_0\cdot\mathbf{H}_0^*/c\end{aligned}$$

whence

$$\delta\omega/\omega_0 = -\int|\mathbf{E}_0|^2\,\delta\varepsilon\,\mathrm{d}V/2\varepsilon_0\int|\mathbf{E}_0|^2\,\mathrm{d}V.$$

In deriving the last expression, we have used the fact that for a dielectric-filled resonator the relation (90.5) becomes

$$\int|\mathbf{H}_0|^2\,\mathrm{d}V = \varepsilon_0\int|\mathbf{E}_0|^2\,\mathrm{d}V,$$

as is clear from (90.9).

§91. The propagation of electromagnetic waves in waveguides

A *waveguide* is a hollow pipe† of infinite length, i.e. a cavity infinite in one direction,

---

† The formulae below hold for an evacuated waveguide. Those for a waveguide filled with a non-absorbing dielectric are obtained by means of the transformation (90.9).

whereas the resonators discussed in §90 are of finite volume. The characteristic oscillations in a resonator are stationary waves, but those in a waveguide are "stationary" only in the transverse directions; waves travelling in the direction along the pipe can be propagated.

Let us consider a straight waveguide with any (simply-connected) cross-section uniform along its length. We shall first suppose that the walls of the waveguide are perfectly conducting, and take the z-axis along the waveguide. In a travelling wave propagated in the z-direction, all quantities depend on z through a factor $\exp(ik_z z)$, with k_z a constant.

The electromagnetic waves possible in such a waveguide can be divided into two types: in one, $H_z = 0$, and in the other $E_z = 0$ (Rayleigh, 1897).The former type, in which the magnetic field is purely transverse, are called *electric-type waves* or *E waves*. The latter, in which the electric field is purely transverse, are called *magnetic-type waves* or *H waves*.†

Let us first consider E waves. The x and y components of equations (90.1) give

$$\frac{\partial E_z}{\partial y} - ik_z E_y = i\frac{\omega}{c}H_x, \qquad -\frac{\partial E_z}{\partial x} + ik_z E_x = i\frac{\omega}{c}H_y,$$

$$ik_z H_y = i\frac{\omega}{c}E_x, \qquad ik_z H_x = -i\frac{\omega}{c}E_y.$$

Hence

$$\left.\begin{aligned} E_x &= \frac{ik_z}{\kappa^2}\frac{\partial E_z}{\partial x}, & E_y &= \frac{ik_z}{\kappa^2}\frac{\partial E_z}{\partial y}, \\ H_x &= -\frac{i\omega}{c\kappa^2}\frac{\partial E_z}{\partial y}, & H_y &= \frac{i\omega}{c\kappa^2}\frac{\partial E_z}{\partial x}, \end{aligned}\right\} \tag{91.1}$$

where $\kappa^2 = (\omega^2/c^2) - k_z^2$. Thus, in an E wave, all the transverse components of **E** and **H** can be expressed in terms of the longitudinal component of the electric field. This component must be determined by solving the wave equation, which takes the two-dimensional form

$$\triangle_2 E_z + \kappa^2 E_z = 0 \tag{91.2}$$

($\triangle_2$ being the two-dimensional Laplacian). The boundary conditions for this equation are that the tangential components of **E** should vanish on the walls of the waveguide, and can be satisfied by putting

$$E_z = 0 \text{ on the circumference of the cross-section.} \tag{91.3}$$

According to formulae (91.1), the two-dimensional vector whose components are E_x, E_y is proportional to the two-dimensional gradient of E_z. When the condition (91.3) holds, therefore, the tangential component of **E** in the xy-plane is also zero.

Similarly, in an H wave the transverse components of **E** and **H** can be expressed in terms of the longitudinal component of the magnetic field:

$$\left.\begin{aligned} H_x &= \frac{ik_z}{\kappa^2}\frac{\partial H_z}{\partial x}, & H_y &= \frac{ik_z}{\kappa^2}\frac{\partial H_z}{\partial y}, \\ E_x &= \frac{i\omega}{c\kappa^2}\frac{\partial H_z}{\partial y}, & E_y &= -\frac{i\omega}{c\kappa^2}\frac{\partial H_z}{\partial x}. \end{aligned}\right\} \tag{91.4}$$

† These types are also known as TM (transverse-magnetic) and TE (transverse-electric) waves.

The longitudinal field H_z is given by the solution of the equation

$$\triangle_2 H_z + \kappa^2 H_z = 0 \tag{91.5}$$

with the boundary condition

$$\partial H_z/\partial n = 0 \text{ on the circumference of the cross-section.} \tag{91.6}$$

According to formulae (91.4), this condition ensures that the normal component of **H** is zero.

Thus the problem of determining the electromagnetic field in a waveguide reduces to that of finding solutions of the two-dimensional wave equation $\triangle_2 f + \kappa^2 f = 0$, with the boundary condition $f = 0$ or $\partial f/\partial n = 0$ on the circumference of the cross-section. For a given cross-section, such solutions exist only for certain definite eigenvalues of the parameter κ^2.

For each eigenvalue κ^2 we have the relation

$$\omega^2 = c^2(k_z^2 + \kappa^2) \tag{91.7}$$

between the frequency ω and the wave number k_z of the wave. The velocity of propagation of the wave along the waveguide is given by the derivative

$$u_z = \frac{\partial\omega}{\partial k_z} = \frac{ck_z}{\sqrt{(k_z^2 + \kappa^2)}} = \frac{c^2 k_z}{\omega}. \tag{91.8}$$

For given κ, this varies from 0 to c when k_z varies from 0 to ∞.

The (time) averaged energy flux density along the waveguide is given by the z-component of the Poynting vector. A simple calculation, using formulae (91.1), gives for an E wave

$$\overline{S}_z = \frac{c}{8\pi}\mathrm{re}(\mathbf{E}\times\mathbf{H}^*)_z = \frac{\omega k_z}{8\pi\kappa^4}|\mathbf{grad}_2 E_z|^2.$$

The total energy flux q is obtained by integrating $\overline{S}_z$ over the cross-section of the waveguide. We have

$$\int |\mathbf{grad}_2 E_z|^2 \mathrm{d}f = \oint E_z^* \frac{\partial E_z}{\partial n}\mathrm{d}l - \int E_z^* \triangle_2 E_z \mathrm{d}f.$$

The first integral is taken along the circumference of the cross-section, and is zero on account of the boundary condition $E_z = 0$. In the second integral we replace $\triangle_2 E_z$ by $-\kappa^2 E_z$, and the result is

$$q = \frac{\omega k_z}{8\pi\kappa^2}\int |E_z|^2 \mathrm{d}f. \tag{91.9}$$

The expression obtained for an H wave is the same with H_z instead of E_z.

The electromagnetic energy density W (per unit length of the waveguide) may be calculated similarly. It is simpler, however, to derive W directly from q, since we must have $q = Wu_z$. From (91.8) and (91.9), therefore,

$$W = \frac{\omega^2}{8\pi\kappa^2 c^2}\int |E_z|^2 \mathrm{d}f. \tag{91.10}$$

It follows from (91.7) that, for each type of wave (for a given value of κ^2) there is a minimum possible frequency, namely $c\kappa$. At lower frequencies the propagation of waves of the type concerned is not possible. There is a smallest eigenvalue κ_{min}, which is not zero (see below). We therefore conclude that there is a frequency $\omega_{min} = c\kappa_{min}$ below which no waves can be propagated along the waveguide. The order of magnitude of ω_{min} is c/a, where a is the transverse dimension of the pipe.

This statement is valid, however, only for waveguides in which the cross-section is simply connected (as we have hitherto assumed). When the cross-section is multiply connected,† the situation is quite different. In such waveguides not only the E and H waves described above but also another type of wave, whose frequency is subject to no restriction, can be propagated. Such waves (called *principal waves*) are characterized by the fact that $k_z = \pm k$ (i.e. $\kappa = 0$); the velocity of propagation is equal to the velocity of light c. We shall derive the chief properties of such waves, and shall see why such waves cannot occur when the cross-section of the waveguide is simply connected.

All the field components in a principal wave satisfy the two-dimensional Laplace's equation, $\triangle_2 f = 0$. With the boundary condition $f = 0$, the only solution of this equation regular throughout the cross-section (whether or not multiply connected) is $f \equiv 0$. Hence we have $E_z = 0$ in a principal wave.

With the boundary condition $\partial f/\partial n = 0$, a regular solution is $f =$ constant. It is easy to see, however, that when f is H_z the constant must be zero (by a "constant", of course, we mean a quantity independent of x and y). For, integrating the equation

$$\operatorname{div} \mathbf{H} = \frac{\partial H_x}{\partial x} + \frac{\partial H_y}{\partial y} + \frac{i\omega}{c} H_z = 0$$

over the cross-section, we obtain $\oint H_n \mathrm{d}l + (i\omega/c) \int H_z \mathrm{d}f = 0$; since $H_n = 0$ on the circumference of the cross-section and H_z is constant over its area, it follows that $H_z = 0$.

Thus a principal wave is purely transverse. For $E_z = H_z = 0$, the x and y components of equations (90.1) give

$$H_x = -E_y, \qquad H_y = E_x, \tag{91.11}$$

i.e. the fields **E** and **H** are perpendicular and equal in magnitude. They are determined by the equations

$$\operatorname{div} \mathbf{E} = \frac{\partial E_x}{\partial x} + \frac{\partial E_y}{\partial y} = 0, \qquad (\mathbf{curl}\ \mathbf{E})_z = \frac{\partial E_y}{\partial x} - \frac{\partial E_x}{\partial y} = 0,$$

with the boundary condition $\mathbf{E}_t = 0$.

We see that the dependence of **E**, and therefore of **H**, on x and y is given by the solution of a two-dimensional electrostatic problem: $\mathbf{E} = -\mathbf{grad}_2 \phi$, where the potential ϕ satisfies the equation $\triangle_2 \phi = 0$ with the boundary condition $\phi =$ constant. In a simply-connected region, this boundary condition means that $\phi =$ constant (and so $\mathbf{E} = 0$) is the only solution regular throughout the region. This shows that waves of this type cannot be propagated along a waveguide whose cross-section is simply connected. In a multiply-connected region, on the other hand, the constant in the boundary condition need not be the same on the various separate parts of the boundary, and so Laplace's equation has solutions which are not trivial. The electric-field distribution over the cross-section of the waveguide is the same as the two-dimensional electrostatic field between the plates of a capacitor at a given potential difference.

† For example, the space between two pipes one inside the other, or the space outside two parallel pipes.

So far we have assumed the walls of the waveguide to be perfectly conducting † If the walls have a small but finite impedance, losses occur and the wave is therefore damped as it is propagated along the waveguide. The damping coefficient can be calculated in the same way as for the damping in time of electromagnetic oscillations in a resonator (§90).

The amount of energy dissipated in unit time per unit length of the walls of the waveguide is given by the integral $(c/8\pi)\zeta'\oint|\mathbf{H}|^2\,dl$, taken along the circumference of the cross-section; $\mathbf{H}$ is the magnetic field calculated on the assumption that $\zeta = 0$. Dividing this expression by twice the energy flux q along the waveguide, we obtain the required damping coefficient α. With this definition, α gives the rate of damping of the wave amplitude, which decreases along the waveguide as $e^{-\alpha z}$.

Expressing all quantities in terms of E_z or H_z by means of formulae (91.1) or (91.4), we obtain the following formulae for the absorption coefficients: for an E wave

$$\alpha = \frac{\omega\zeta'}{2\kappa^2 k_z c}\frac{\oint|\mathbf{grad}_2 E_z|^2 dl}{\int|E_z|^2 df} \tag{91.12}$$

and for an H wave

$$\alpha = \frac{c\kappa^2\zeta'}{2k_z\omega}\frac{\oint\{|H_z|^2 + (k_z^2/\kappa^4)|\mathbf{grad}_2 H_z|^2\}dl}{\int|H_z|^2 df}. \tag{91.13}$$

In an actual calculation it may be convenient to transform the surface integrals in the denominators into integrals along the circumference. The necessary formulae, whose derivation is similar to that of (90.8), are

$$\left.\begin{aligned}\int|E_z|^2 df &= \frac{1}{2\kappa^2}\oint(\mathbf{n}\cdot\mathbf{r})|\mathbf{grad}_2 E_z|^2 dl,\\ \int|H_z|^2 df &= \frac{1}{2\kappa^2}\oint(\mathbf{n}\cdot\mathbf{r})\{\kappa^2|H_z|^2 - |\mathbf{grad}_2 H_z|^2\}dl.\end{aligned}\right\} \tag{91.14}$$

When $k_z \to 0$, i.e. the frequency $\omega \to c\kappa$, the expressions (91.12) and (91.13) become infinite, but they are then no longer applicable, because their derivation presupposes that κ is small compared with k_z.

Formulae (91.12) and (91.13) are not valid for a principal wave (in a waveguide with a multiply-connected cross-section), in which E_z, H_z and κ are all zero. In this case all the field components can be expressed in terms of the scalar potential ϕ. Using the fact that the fields $\mathbf{H}$ and $\mathbf{E} = -\mathbf{grad}_2\phi$ in a principal wave are perpendicular and equal in magnitude, we obtain the absorption coefficient

$$\alpha = \frac{\zeta'\oint|\mathbf{grad}_2\phi|^2 dl}{2\int|\mathbf{grad}_2\phi|^2 df}. \tag{91.15}$$

The propagation of a principal wave along a waveguide can be relatively simply discussed when its absorption coefficient is not small (so that formula (91.15) is inapplicable) but the wavelength c/ω is large compared with the transverse dimension of the waveguide.

---

† In particular, this assumption is necessary for a rigorous separation of waves with $E_z = 0$ and those with $H_z = 0$.

As has been mentioned, the transverse electric field in a principal wave at any instant corresponds to the electrostatic field in a capacitor formed by the walls of the waveguide carrying equal and opposite charges. Let these charges be $\pm e(z)$ per unit length. They are related to the currents $\pm J(z)$ flowing on the walls by the equation of continuity $\partial e/\partial t = -\partial J/\partial z$, or, for a monochromatic field, $i\omega e = \partial J/\partial z$. Next, let C be the capacitance per unit length of the waveguide. The "potential difference" $\phi_2 - \phi_1$ between its walls is e/C; differentiating this with respect to z, we obtain the e.m.f. which maintains the current on the walls. (When absorption is present, the field is not purely transverse.) Equating the e.m.f. to ZJ, where Z is the impedance per unit length, we have

$$-\frac{\partial}{\partial z}\left(\frac{e}{C}\right) = ZJ$$

or

$$\frac{\partial}{\partial z}\left(\frac{1}{C}\frac{\partial J}{\partial z}\right) + i\omega ZJ = 0. \tag{91.16}$$

Substituting $Z = R - i\omega L/c^2$, where R and L are the resistance and self-inductance per unit length of the waveguide, we can return from the monochromatic current components to currents which are arbitrary functions of time. Assuming the capacitance C to be constant along the waveguide, we arrive at the *telegraph equation*:

$$\frac{1}{C}\frac{\partial^2 J}{\partial z^2} - R\frac{\partial J}{\partial t} - \frac{L}{c^2}\frac{\partial^2 J}{\partial t^2} = 0. \tag{91.17}$$

If there is no resistance ($R = 0$), this equation reduces, as it should, to the wave equation with a velocity of wave propagation $\sqrt{(c^2/LC)} = c$. The equation $LC = 1$ follows from the mathematical equivalence of the problems of determining $1/C$ and L for a given cross-section. The electric and magnetic fields between perfectly conducting surfaces are perpendicular and equal in magnitude (see (91.11)), and if this magnitude is given on the surfaces the charge density and the current density are respectively determined. Hence the coefficients of proportionality ($1/C$ and L) between the field energy and the squared charge and current respectively are the same.

PROBLEMS

PROBLEM 1. Find the values of κ for waves propagated in a waveguide whose cross-section is a rectangle with sides a and b. Find the damping coefficients.

SOLUTION. In E waves† $E_z = \text{constant} \times \sin k_x x \sin k_y y$, where $k_x = n_1\pi/a$, $k_y = n_2\pi/b$, with n_1 and n_2 positive integers $\geqslant 1$. In H waves $H_z = \text{constant} \times \cos k_x x \cos k_y y$, and one of n_1 and n_2 may be zero. In both types of wave $\kappa^2 = k_x{}^2 + k_y{}^2 = \pi^2(n_1{}^2/a^2 + n_2{}^2/b^2)$. The smallest value of κ corresponds to an H_{10} wave (the suffixes show the values of n_1 and n_2) and is $\kappa_{\min} = \pi/a$ (we assume that $a > b$).

The damping coefficients are calculated from formulae (91.12) and (91.13) and are: for E waves

$$\alpha = 2\zeta'\omega(k_x{}^2 b + k_y{}^2 a)/c\kappa^2 k_z ab,$$

for $H_{n_1 0}$ waves

$$\alpha = \frac{\zeta'\omega}{ck_z ab}\left(a + \frac{2\kappa^2}{k^2}b\right),$$

† We everywhere omit the factor $\exp(ik_z z - i\omega t)$.

and for $H_{n_1 n_2}$ waves (n_1, $n_2 \neq 0$)

$$\alpha = \frac{2c\kappa^2\zeta'}{\omega k_z ab}\left[a+b+\frac{k_z^{\,2}}{\kappa^4}(k_x^{\,2}a+k_y^{\,2}b)\right].$$

PROBLEM 2. The same as Problem 1, but for a waveguide whose cross-section is a circle with radius a.

SOLUTION. Solving the wave equation in polar coordinates r, ϕ, we have for E waves

$$E_z = \text{constant} \times J_n(\kappa r) \begin{matrix}\sin\\ \cos\end{matrix} n\phi$$

with the condition $J_n(\kappa a) = 0$, which gives the values of κ. In H waves the value of H_z is given by the same formula, but κ is determined by the condition $J_n'(\kappa a) = 0$. The smallest value of κ occurs for the H_1 wave, and is $\kappa_{\min} = 1.84/a$.

The damping coefficient is calculated from formulae (91.12)–(91.14). For E waves it is $\alpha = \omega\zeta'/cak_z$, and for H waves

$$\alpha = \frac{c\zeta'\kappa^2}{\omega k_z a}\left[1+\frac{n^2\omega^2}{c^2\kappa^2(a^2\kappa^2-n^2)}\right].$$

§92. The scattering of electromagnetic waves by small particles

Let us consider the scattering of electromagnetic waves by macroscopic particles whose dimensions are small compared with the wavelength $\lambda \sim c/\omega$ of the wave undergoing scattering (Rayleigh, 1871). When this condition holds, the electromagnetic field near the particle may be supposed uniform. Being in a uniform field periodic in time, each particle acquires definite electric and magnetic moments $\mathscr{P}$ and $\mathscr{M}$, whose dependence on time is given by factors of the form $e^{-i\omega t}$. The scattered wave can be described as being the result of radiation by these variable moments. At distances R from the particle which are large compared with λ (the wave region), the fields in the scattered wave are given by

$$\left.\begin{aligned}\mathbf{H}' &= \frac{\omega^2}{c^2 R}\{\mathbf{n}\times\mathscr{P}+\mathbf{n}\times(\mathscr{M}\times\mathbf{n})\},\\ \mathbf{E}' &= \mathbf{H}'\times\mathbf{n}\end{aligned}\right\} \qquad (92.1)$$

(see *Fields*, §71), where the unit vector $\mathbf{n}$ gives the direction of scattering, and the values of $\mathscr{P}$ and $\mathscr{M}$ must be taken at the time $t-R/c$. We denote the fields in the scattered wave by primed letters and those in the incident wave by unprimed letters. The (time) averaged intensity of radiation scattered into a solid angle do is $dI = \frac{1}{2}c|\mathbf{H}'|^2 R^2\, do/4\pi$; dividing by the energy flux density in the incident wave $c|\mathbf{H}|^2/8\pi = c|\mathbf{E}|^2/8\pi$, we obtain the *scattering cross-section.*

The calculation of $\mathscr{P}$ and $\mathscr{M}$ is particularly simple if the dimensions of the particle are small in comparison not only with λ but also with the "wavelength" δ corresponding to the frequency ω in the material of the particle. In this case we can calculate the polarizability of the particle from the formulae for an external uniform static field, the only difference being, of course, that the values taken for ε and μ are those corresponding to the given frequency ω, and not the static values. If, as usually happens, μ is close to unity, the magnetic dipole term in formula (92.1) may be omitted.

For a spherical particle with radius a we have (see (8.10))

$$\mathscr{P} = V\alpha\mathbf{E}, \qquad \alpha = 3(\varepsilon-1)/4\pi(\varepsilon+2), \qquad (92.2)$$

and the scattering cross-section is

$$d\sigma = (\omega/c)^4|\alpha|^2 V^2 \sin^2\theta\, do, \tag{92.3}$$

where θ is the angle between the scattering direction **n** and the direction of the electric field **E** in the linearly polarized incident wave. The total cross-section is

$$\sigma = 8\pi|\alpha|^2\omega^4 V^2/3c^4. \tag{92.4}$$

The frequency dependence of the cross-section is determined by the factor ω^4 and by the polarizability. At frequencies so low that α shows no dispersion, the scattering is proportional to ω^4. It may be noted also that the cross-section is proportional to the square of the volume of the particle.

If the incident wave is unpolarized (natural light) then the differential cross-section must be obtained by averaging (92.3) over all directions of the vector **E** in a plane perpendicular to the direction of propagation of the incident wave (i.e. perpendicular to its wave vector **k**). Denoting by ϑ and ϕ the polar angle and the azimuth of the direction of **n** relative to **k** (ϕ being measured from the plane of **k** and **E**), we have $\cos\theta = \sin\vartheta\cos\phi$ (Fig. 47), so that

$$d\sigma = (\omega/c)^4|\alpha|^2 V^2(1-\sin^2\vartheta\cos^2\phi)do. \tag{92.5}$$

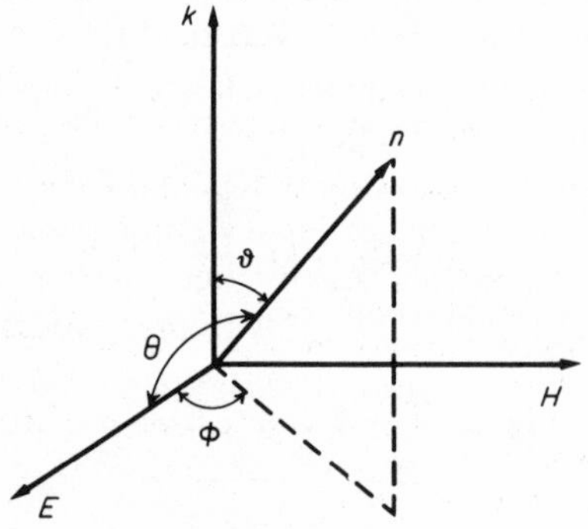

FIG. 47

On averaging over ϕ, we obtain the following formula for the cross-section for scattering of an unpolarized wave:

$$d\sigma = \tfrac{1}{2}(\omega/c)^4|\alpha|^2 V^2(1+\cos^2\vartheta)do, \tag{92.6}$$

where ϑ is the angle between the directions of incidence and scattering. The angular distribution (92.6) is symmetrical about the plane $\vartheta = \frac{1}{2}\pi$, i.e. the forward scattering and backward scattering are the same.

From formula (92.5) we can easily find the degree of depolarization of the scattered light. To do so, we notice that, for a given direction of **E**, **E**′ lies in the plane of **E** and **n**. The direction of the electric field **E**′ in the scattered wave therefore lies in the plane of **k** and **n** (the *plane of scattering*) or perpendicular to that plane, according as the azimuth ϕ of the vector **E**, measured from the plane of **k** and **n**, is 0 or $\frac{1}{2}\pi$. Let $I_\parallel$ and $I_\perp$ be the intensities of scattered radiation having these two polarizations. The degree of depolarization is defined as the ratio of the smaller to the larger of these quantities. By (92.5) we have

$$I_{\parallel}/I_{\perp} = \cos^2 \vartheta. \tag{92.7}$$

If the scattering particle has a large permittivity,

$$\delta \sim c/\omega\sqrt{|\varepsilon|} \ll \lambda.$$

The dimensions of the particle may then be small compared with λ but not small compared with δ. In the first approximation with respect to $1/\varepsilon$, the electric moment of the particle may be calculated as simply the moment of a conductor ($\varepsilon \to \infty$) in a uniform static external field. In calculating the magnetic moment, however, the induced currents in the particle are of importance, and the problem cannot be taken as static; instead, we must seek a solution of equation (83.2) (with $\mu = 1$):

$$\triangle \mathbf{H} + \varepsilon\omega^2 \mathbf{H}/c^2 = 0 \tag{92.8}$$

which becomes the field of the incident wave far from the particle. The magnetic and electric moments are of the same order of magnitude, and both terms in formula (92.1) must be retained. The angular distribution and amount of scattering are quite different from those discussed above (see Problem 2).

PROBLEMS

PROBLEM 1. Linearly polarized light is scattered by randomly oriented small particles whose electric polarizability tensor has three different principal values. Determine the depolarizing factor for the scattered light.

SOLUTION. Neglecting, as above, the magnetic moment, we have from (92.1)

$$\mathbf{E}' = (\omega^2/c^2 R)(\mathbf{n}\times\mathscr{P})\times\mathbf{n}.$$

The required depolarizing factor is given by the ratio of the principal values of the two-dimensional tensor $I_{\alpha\beta} = \langle E'_\alpha E'_\beta{}^* \rangle$, where the brackets denote an averaging over orientations of the scattering particle for a given direction of scattering **n**, and the suffixes α and β take two values in the plane perpendicular to **n** (see *Fields*, §50). It is more convenient, however, to average the three-dimensional tensor $\mathscr{P}_i \mathscr{P}_k{}^*$ and then project it on the plane perpendicular to **n**; these components of the tensor $\langle \mathscr{P}_i \mathscr{P}_k{}^* \rangle$ are proportional to the corresponding components $I_{\alpha\beta}$. Substituting $\mathscr{P}_i = \alpha_{ik} E_k$, we have

$$\langle \mathscr{P}_i \mathscr{P}_k{}^* \rangle = \langle \alpha_{il}\alpha_{km}{}^* \rangle E_l E_m{}^*.$$

In effecting the averaging we use the formula

$$\langle \alpha_{il}\alpha_{km}{}^* \rangle = A\delta_{il}\delta_{km} + B(\delta_{ik}\delta_{lm} + \delta_{im}\delta_{kl}).$$

This is the most general tensor of rank four which is symmetrical in *i*, *l* and *k*, *m* and contains only scalar constants. These constants are determined from two equations obtained by contracting the tensor, firstly with respect to *i*, *l* and *k*, *m*, secondly with respect to *i*, *k* and *l*, *m*. They are

$$A = \frac{1}{15}(2\alpha_{ii}\alpha_{kk}{}^* - \alpha_{ik}\alpha_{ik}{}^*),$$

$$B = \frac{1}{30}(3\alpha_{ik}\alpha_{ik}{}^* - \alpha_{ii}\alpha_{kk}{}^*).$$

In a linearly polarized wave, the field amplitude **E** (we omit the time factor $e^{-i\omega t}$) can always be defined so as to be real. Then we have

$$\langle \mathscr{P}_i \mathscr{P}_k{}^* \rangle = (A+B)E_i E_k + B\delta_{ik}E^2. \tag{1}$$

Let the *z*-axis be in the direction of **n**, and the *xz*-plane contain the directions of **n** and **E**; these axes are the principal axes of the tensor $I_{\alpha\beta}$. Taking the appropriate components of the tensor (1), we find the depolarizing factor $I_y/I_x = B/[(A+B)\sin^2\theta + B]$, where θ is the angle between **E** and **n**.

PROBLEM 2. Determine the cross-section for scattering by a sphere with radius a, for which ε is large; it is assumed that $\lambda \gg a \sim \delta$.

SOLUTION. The problem of calculating the magnetic moment acquired by a sphere with given ε (and $\mu = 1$) in a variable magnetic field **H** is the same as that solved in §59, Problem 1, except that k in the formulae derived there must be replaced by $\omega\sqrt{\varepsilon}/c$. Thus $\mathscr{M} = -a^3\gamma\mathbf{H}$, where

$$\gamma = \frac{1}{2}\left(1 + \frac{3}{ka}\cot ka - \frac{3}{(ka)^2}\right).$$

When $|ka| \ll 1$, $\gamma \cong -(ka)^2/30$; when $|ka| \gg 1$, $\cot ka \to -i$ and $\gamma \cong \frac{1}{2}(1 - 3i/ka)$.

The electric moment can be calculated, in the first approximation with respect to $1/\varepsilon$, as simply the moment of a conducting ($\varepsilon \to \infty$) sphere in a uniform static electric field: $\mathscr{P} = a^3\mathbf{E}$.

Taking into account the fact that **E** and **H** are perpendicular, we have after a simple calculation, using (92.1), the following formula for the scattering cross-section:

$$d\sigma = (a^6\omega^4/c^4)\{|\gamma|^2\cos^2\phi + \sin^2\phi - (\gamma+\gamma^*)\cos\vartheta + \cos^2\vartheta(\cos^2\phi + |\gamma|^2\sin^2\phi)\}\,do,$$

where ϕ and ϑ are the angles shown in Fig. 47. In scattering of unpolarized light we have

$$d\sigma = (a^6\omega^4/c^4)\{\tfrac{1}{2}[1+|\gamma|^2][1+\cos^2\vartheta] - (\gamma+\gamma^*)\cos\vartheta\}\,do,$$

and the degree of depolarization of the scattered light is $I_\parallel/I_\perp = |(\gamma - \cos\vartheta)/(1-\gamma\cos\vartheta)|^2$. The total scattering cross-section is $\sigma = 8\pi a^6\omega^4(1+|\gamma|^2)/3c^4$.

In the limit $ka \to \infty$ (i.e. when $\lambda \gg a \gg \delta$) we have $\gamma = \frac{1}{2}$, corresponding to scattering by a perfectly reflecting sphere into which neither the electric nor the magnetic field penetrates. The differential scattering cross-section is

$$d\sigma = \frac{5a^6\omega^4}{8c^4}(1 + \cos^2\vartheta - \tfrac{8}{5}\cos\vartheta)\,do.$$

Note that the angular distribution is highly asymmetrical about the plane $\vartheta = \frac{1}{2}\pi$: the scattering is mainly backward, the ratio of forward and backward scattered intensities being 1:9.

§93. The absorption of electromagnetic waves by small particles

The scattering of electromagnetic waves by particles is accompanied by absorption. The absorption cross-section is given by the ratio of the mean energy Q dissipated in a particle per unit time to the incident energy flux density. To calculate Q we can use the formula

$$Q = -\overline{\mathscr{P}\cdot\dot{\mathfrak{E}}} - \overline{\mathscr{M}\cdot\dot{\mathfrak{H}}}, \tag{93.1}$$

where $\mathscr{P}$ and $\mathscr{M}$ are the total electric and magnetic moments of the particle, and the external fields $\mathfrak{E}$ and $\mathfrak{H}$ are replaced by the fields **E** and **H** in the scattered wave; cf. (59.11).

Using the complex representation of quantities, we can write (see the last footnote to §59)

$$Q = -\tfrac{1}{2}\,\mathrm{re}\,(\mathscr{P}\cdot\dot{\mathbf{E}}^* + \mathscr{M}\cdot\dot{\mathbf{H}}^*) = \tfrac{1}{2}\omega V(\alpha_e'' + \alpha_m'')|\mathbf{E}|^2,$$

where α_e and α_m are the electric and magnetic polarizabilities of the particle. Dividing by the incident energy flux, we obtain

$$\sigma = 4\pi\omega V(\alpha_e'' + \alpha_m'')/c. \tag{93.2}$$

Let us apply this formula to absorption by a sphere with radius $a \ll \lambda$, assuming it non-magnetic ($\mu = 1$). The nature of the absorption depends considerably on the magnitude of the permittivity.

If ε is small, then we have both $a \ll \lambda$ and $a \ll \delta$. In this case the magnetic polarizability may be neglected in comparison with the electric polarizability. With the latter given by (92.2), we have

$$\sigma = 12\pi\omega a^3\varepsilon''/c\,[(\varepsilon'+2)^2 + (\varepsilon'')^2]. \tag{93.3}$$

If, on the other hand, $|\varepsilon| \gg 1$, the electric part of the absorption becomes small, and the magnetic absorption may be important even if we still have $\delta \gg a$. When this last condition holds (i.e. $|ka| \ll 1$), the magnetic polarizability is $\alpha_m = (ka)^2/40\pi = a^2\omega^2\varepsilon/40\pi c^2$ (see §92, Problem 2) and the absorption cross-section is

$$\sigma = \frac{12\pi\omega a^3\varepsilon''}{c}\left(\frac{1}{|\varepsilon|^2} + \frac{\omega^2 a^2}{90c^2}\right). \tag{93.4}$$

When ε increases further, the electric part of the absorption becomes small compared with the magnetic part. In the limit $\delta \ll a$ (i.e. $|ka| \gg 1$) we have $\alpha_m = -3/8\pi + 9i/8\pi ka = -3/8\pi + 9ic\zeta/8\pi\omega a$, where $\zeta = 1/\sqrt{\varepsilon}$ is the surface impedance of the sphere. Hence

$$\sigma = 6\pi a^2\zeta'. \tag{93.5}$$

It may be noticed that this formula could have been obtained more directly, without using the general expression for the magnetic polarizability $\alpha_m(\omega)$ of the sphere. When ζ is small, the energy dissipation Q can be calculated by integrating the mean Poynting vector (87.4) over the surface of the sphere, the distribution of the magnetic field over the surface being given by the solution (54.3) of the problem of a superconducting ($\zeta = 0$) sphere in a uniform magnetic field.

Knowing the absorption cross-section of the sphere, we can immediately determine the intensity of the thermal radiation emitted from the sphere. According to Kirchhoff's law (see *SP* 1, §63), the intensity dI in a frequency range $d\omega$ is given in terms of $\sigma(\omega)$ by

$$dI = 4\pi c\sigma(\omega)e_0(\omega)\, d\omega, \tag{93.6}$$

where $e_0(\omega) = \hbar\omega^3/4\pi^3c^3\,[\exp(\hbar\omega/T) - 1]$ is the spectral density of black-body radiation per unit volume and unit solid angle.

§94. Diffraction by a wedge

The ordinary approximate theory of diffraction (see *Fields*, §§59–61) is based on the assumption that the deviations from geometrical optics are small. It is thereby assumed, firstly, that all dimensions are large compared with the wavelength; this applies both to dimensions of bodies (screens) or apertures and to distances of bodies from the points of emission and observation of the light. Secondly, only small angles of diffraction are considered, i.e. the distribution of light is examined only in directions close to the edge of the geometrical shadow. Under these conditions, the actual optical properties of the substances involved are of no importance; all that matters is that they are opaque.

If these conditions are not fulfilled, the solution of the diffraction problem requires an exact solution of the wave equation, taking into account the appropriate boundary conditions on the surfaces of the bodies, which depend on their properties. The finding of such solutions offers exceptional mathematical difficulties, and has been effected for only a small number of problems. A simplifying assumption is usually made concerning the properties of the body at which the diffraction occurs, namely that it is perfectly conducting, and therefore perfectly reflecting.

The following remark may be made here. It might seem reasonable to solve the diffraction problem on the assumption that the surface of the body is "black", i.e. completely absorbs light incident on it. In reality, however, such an assumption concerning the body in stating the exact problem of diffraction would involve a contradiction. The reason is that, if the substance of the body is strongly absorbing, the coefficient of reflection is not small but, on the contrary, almost unity (see §87)). Hence a reflection coefficient close to zero implies a weakly absorbing substance and a thickness of the body which is large compared with the wavelength. In the exact theory of diffraction, parts of the surface of the body at a distance from its edge of the order of the wavelength are necessarily of importance; but the thickness of the body near its edge is always small, so that the assumption that it is "black" is certainly untenable.

Considerable theoretical interest attaches to the exact solution (A. Sommerfeld, 1894) of the problem of diffraction at the edge of a perfectly conducting wedge bounded by two intersecting semi-infinite planes. A complete exposition of the very complex mathematical theory, which involves the use of special devices, is beyond the scope of this book. Here we shall give, for reference, the final results.†

We take the edge of the wedge as the z-axis in a system of cylindrical polar coordinates r, ϕ, z. The front surface (OA in Fig. 48) corresponds to $\phi = 0$, and the rear surface (OB) to $\phi = \gamma$, where $2\pi - \gamma$ is the angle of the wedge, the region outside it being $0 \leqslant \phi \leqslant \gamma$. Let a plane monochromatic wave with unit amplitude be incident in the $r\phi$-plane on the front surface at an angle ϕ_0 to the surface; by symmetry, it is sufficient to consider angles $\phi_0 < \frac{1}{2}\gamma$. We shall distinguish two independent modes of polarization of the incident wave, and therefore of the diffracted wave also: the edge of the wedge (the z-axis) may be parallel to either **E** or **H**. The letter u will denote E_z and H_z respectively.

The electromagnetic field is then given in all space by the formula (the time factor $e^{-i\omega t}$ being everywhere omitted)

$$u(r, \phi) = v(r, \phi - \phi_0) \mp v(r, \phi + \phi_0), \tag{94.1}$$

where the upper and lower signs correspond to the polarizations with **E** and **H** respectively

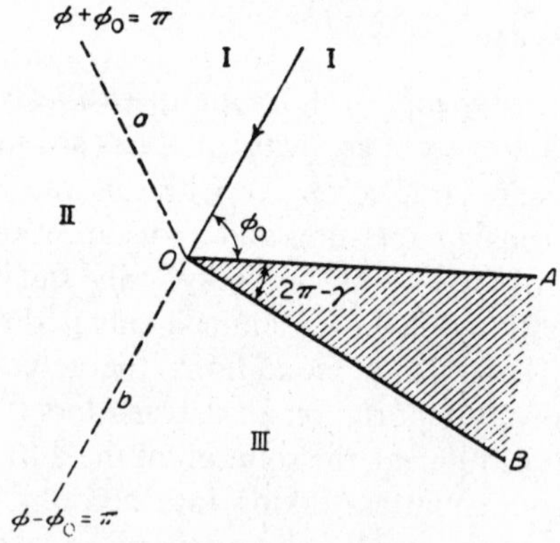

FIG. 48

† A detailed account of the calculations is given by A. Sommerfeld, *Optics*, Academic Press, New York, 1954; P. Frank and R. von Mises, *Differential and Integral Equations in Physics* (*Differential- und Integralgleichungen der Physik*), part 2, Chapter XX, 2nd ed., Vieweg, Brunswick, 1935. Another method of solution, due to M. I. Kontorovich and N. N. Lebedev, is given by G. A. Grinberg, *Selected Problems in the Mathematical Theory of Electric and Magnetic Phenomena* (*Izbrannye voprosy matematicheskoĭ teorii élektricheskikh i magnitnykh yavleniĭ*), Chapter XXII, Academy of Sciences, Moscow, 1948.

in the z-direction, and the function $v(r,\psi)$ is given by the complex integral

$$v(r,\psi) = \frac{1}{2\gamma}\int_C e^{-ikr\cos\zeta}\frac{d\zeta}{1-e^{-i\pi(\zeta+\psi)/\gamma}},\tag{94.2}$$

where $k = \omega/c$. The path of integration C in the ζ-plane consists of the two loops C_1 and C_2 shown in Fig. 49. The ends of these loops are at infinity in parts of the ζ-plane (shaded in Fig. 49) where im $\cos\zeta < 0$, and so the factor $e^{-ikr\cos\zeta}$ tends to zero at infinity. The integrand in (94.2) has poles on the real ζ-axis at the points $\zeta = -\psi + 2n\gamma$, where n is any integer. The integration may be taken along a path $D = D_1 + D_2$ (Fig. 49) instead of C, adding to the integral the residues of the integrand at the poles, if any, in the range $-\pi \leqslant \zeta \leqslant \pi$. We write v as

$$v(r,\psi) = v_0(r,\psi) + v_d(r,\psi),\tag{94.3}$$

where v_d is the integral (94.2) taken along the path D, and v_0 is the contribution from the residues at these poles. Each pole gives rise to a term $\exp[-ikr\cos(\psi - 2n\gamma)]$ in v_0, which represents either the incident wave or one of the waves reflected from the surface of the wedge in accordance with the laws of geometrical optics. The function v_d represents the diffractive distortion of the wave. The field at distances from the edge of the wedge large compared with the wavelength is of the greatest interest. When $kr \gg 1$, the asymptotic formula†

$$v_d(r,\psi) = \frac{\pi}{\gamma\sqrt{(2\pi kr)}}e^{i(kr+\frac{1}{4}\pi)}\frac{\sin(\pi^2/\gamma)}{\cos(\pi^2/\gamma) - \cos(\pi\psi/\gamma)}\tag{94.4}$$

holds, provided that the angle ψ satisfies the condition

$$[\cos(\pi^2/\gamma) - \cos(\pi\psi/\gamma)]^2 \gg 1/kr.\tag{94.5}$$

The dependence on r of the function v_d, and therefore of the field $u_d(r,\phi) = v_d(r,\phi-\phi_0) \mp v_d(r,\phi+\phi_0)$, is given by a factor $e^{ikr}/\sqrt{r}$, i.e. this field resembles a cylindrical wave emitted by the edge of the wedge.

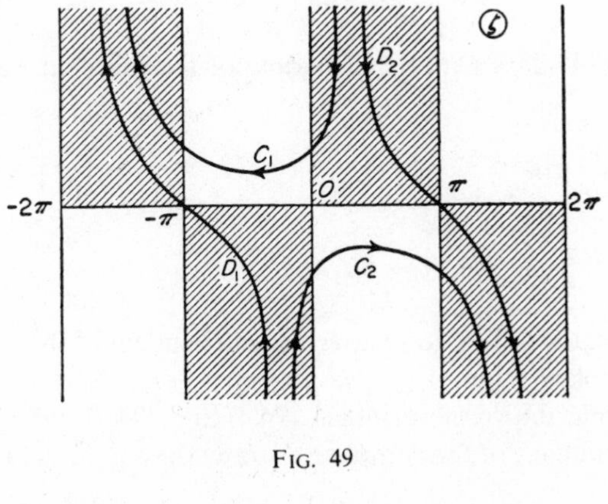

FIG. 49

† The next terms in this asymptotic expansion have been given by W. Pauli, *Physical Review* **54**, 924, 1938.

In the form given above, (94.1)–(94.5) are valid for any angles γ and ϕ_0. The more detailed discussion of these formulae will be effected on the assumption (for definiteness) that the angles ϕ_0 and γ are so related ($\gamma > \pi + \phi_0$) that, in geometrical optics, two boundaries are formed: the boundary *Ob* of the complete shadow (region III in Fig. 48), and the boundary *Oa* of the "shadow" of the wave reflected from the surface *OA*. In Fig. 48, $\phi_0 < \frac{1}{2}\pi$. If $\phi_0 > \frac{1}{2}\pi$, the boundary *Oa* lies to the right of the direction of the incident wave. If $\gamma < \pi + \phi_0$, there is no region of complete shadow, and reflection (single or multiple) takes place from both sides of the wedge.

In regions I, II, III the function $u_0(r, \phi) = v_0(r, \phi - \phi_0) \mp v_0(r, \phi + \phi_0)$ has the following forms:

$$\left.\begin{aligned} &\text{Region I: } u_0 = e^{-ikr\cos(\phi-\phi_0)} \mp e^{-ikr\cos(\phi+\phi_0)},\\ &\text{Region II: } u_0 = e^{-ikr\cos(\phi-\phi_0)},\\ &\text{Region III: } u_0 = 0. \end{aligned}\right\} \tag{94.6}$$

These expressions, which do not vanish as $kr \to \infty$, describe the incident (region II) or incident and reflected (region I) waves, undistorted by diffraction. The diffractive distortion of the field is given by formula (94.4), but the condition (94.5) ceases to hold when ψ approaches π and the difference $|\psi - \pi|$ is no longer large compared with $1/\sqrt{(kr)}$.

The values $\psi = \pi$ correspond to the geometrical boundaries of the shadow; for $\psi = \phi - \phi_0$ we have the boundary of the complete shadow, and for $\psi = \phi + \phi_0$ that of the shadow of the reflected wave. In the immediate neighbourhood of these values a different asymptotic expression must be used, which is valid if the inequality $|\psi - \pi| \ll 1$ holds. This condition, together with $kr \gg 1$, ensures the validity of the usual approximate theory of Fresnel diffraction. Accordingly we have near the boundary *Ob* of the complete shadow the asymptotic expression

$$\left.\begin{aligned} u(r,\phi) &= e^{-ikr\cos(\phi-\phi_0)} \frac{1-i}{\sqrt{(2\pi)}} \int_{-\infty}^{w} e^{i\eta^2}\, d\eta,\\ w &= -(\phi - \phi_0 - \pi)\sqrt{(\tfrac{1}{2}kr)}. \end{aligned}\right\} \tag{94.7}$$

Similarly, near the boundary *Oa* of the shadow of the reflected wave

$$\left.\begin{aligned} u(r,\phi) &= e^{-ikr\cos(\phi-\phi_0)} + e^{-ikr\cos(\phi+\phi_0)} \frac{1-i}{\sqrt{(2\pi)}} \int_{-\infty}^{w} e^{i\eta^2}\, d\eta,\\ w &= -(\phi + \phi_0 - \pi)\sqrt{(\tfrac{1}{2}kr)}. \end{aligned}\right\} \tag{94.8}$$

In this approximation the diffraction pattern is independent of the angle of the wedge and of the direction of polarization of the wave.

The ranges of applicability of formulae (94.4) and (94.7), (94.8) partly overlap. For example, near the boundary of the complete shadow the common range of applicability is given by

$$1 \gg |\phi - \phi_0 - \pi| \gg 1/\sqrt{(kr)},$$

and in this range

$$u(r,\phi) = u_0(r,\phi) + \frac{1}{\sqrt{(2\pi kr)}} e^{i(kr+\frac{1}{4}\pi)} \frac{1}{\phi-\phi_0-\pi}, \tag{94.9}$$

with u_0 given by (94.6). This expression can be obtained from (94.7) by using the asymptotic formulae for the Fresnel integral with large $|w|$:

$$\int_{-\infty}^{w} e^{i\eta^2} d\eta = (1+i)\sqrt{(\tfrac{1}{2}\pi)} + e^{iw^2}/2iw \quad \text{for } w > 0,$$

$$\int_{-\infty}^{w} e^{i\eta^2} d\eta = e^{iw^2}/2iw \qquad \text{for } w < 0.$$

§95. Diffraction by a plane screen

The exact formula (94.2) for diffraction by a wedge can be brought to a comparatively simple form in the particular case of diffraction by a half-plane ($\gamma = 2\pi$). The complex integral in (94.2) can be reduced to a Fresnel integral:

$$\left.\begin{aligned} v(r,\psi) &= \frac{1}{\sqrt{\pi}} e^{-i(kr\cos\psi+\frac{1}{4}\pi)} \int_{-\infty}^{w} e^{i\eta^2} d\eta, \\ w &= \sqrt{(2kr)}\cos\tfrac{1}{2}\psi. \end{aligned}\right\} \tag{95.1}$$

This formula holds for any values of r and ψ. For $kr \gg 1$ and angles $|\psi-\pi| \gg 1/\sqrt{(kr)}$ the asymptotic expression

$$v_d(r,\psi) = -e^{i(kr+\frac{1}{4}\pi)} \frac{1}{2\sqrt{(2\pi kr)}\cos\frac{1}{2}\psi} \tag{95.2}$$

(formula (94.4) with $\gamma = 2\pi$) holds.

Using formula (95.2), the solution of the problem of diffraction by a plane perfectly conducting screen of any shape can be obtained in closed form. The only assumptions are that the dimensions of the screen and the distance from it are large compared with the wavelength, and that the angles of diffraction are moderately large (this region overlaps the region of small angles in which the ordinary Fresnel-diffraction formulae are valid). The result is in the form of an integral along the edge of the screen, analogous to the expression of the diffraction field as an integral over a surface spanning the aperture in a screen, in the ordinary approximate theory. We shall not pause to give the calculations here.

In the exact theory of diffraction by plane perfectly conducting screens, there is a theorem (due to L. I. Mandel'shtam and M. A. Leontovich) in some ways analogous to Babinet's theorem in the approximate theory.

Let us consider a plane screen with an aperture of any shape, and take the plane of the screen as $z = 0$. Let an electromagnetic wave be incident from the side $z < 0$, and let $\mathbf{E}_0, \mathbf{H}_0$ be the total fields in the incident wave and the wave which would be reflected from the screen if there were no aperture. We assume the field continued beyond the screen ($z > 0$). Since $H_z = 0$, $\mathbf{E}_t = 0$ for $z = 0$ (by the boundary conditions at a perfectly conducting surface), the values of $\mathbf{E}_0$ and $\mathbf{H}_0$ for $z > 0$ and $z < 0$ are related by

$$\left.\begin{aligned} E_{0z}(x, y, z) &= E_{0z}(x, y, -z), \quad \mathbf{E}_{0t}(x, y, z) = -\mathbf{E}_{0t}(x, y, -z), \\ H_{0z}(x, y, z) &= -H_{0z}(x, y, -z), \mathbf{H}_{0t}(x, y, z) = \mathbf{H}_{0t}(x, y, -z). \end{aligned}\right\} \tag{95.3}$$

Next, let $\mathbf{E}'$ and $\mathbf{H}'$ be the fields which would occur if a flat plate corresponding to the aperture in size, shape and position, and having infinite magnetic permeability, were placed in the field $\mathbf{E}_0$, $\mathbf{H}_0$. Then the solution of the diffraction problem for the aperture in the screen is given by

$$\left.\begin{aligned} \mathbf{E} &= \tfrac{1}{2}(\mathbf{E}_0 + \mathbf{E}'), \quad \mathbf{H} = \tfrac{1}{2}(\mathbf{H}_0 + \mathbf{H}') \quad \text{for } z < 0, \\ \mathbf{E} &= \tfrac{1}{2}(\mathbf{E}_0 - \mathbf{E}'), \quad \mathbf{H} = \tfrac{1}{2}(\mathbf{H}_0 - \mathbf{H}') \quad \text{for } z > 0. \end{aligned}\right\} \tag{95.4}$$

To show this, we notice that the fields $\mathbf{E}'$, $\mathbf{H}'$ have the same symmetry (expressed by formulae (95.3)) as the fields $\mathbf{E}_0$, $\mathbf{H}_0$. They therefore satisfy on the plane $z = 0$ the conditions

$$\begin{aligned} \mathbf{E}'_t &= 0, \qquad H'_z = 0 \qquad \text{outside the aperture,} \\ \mathbf{E}'_{t1} &= -\mathbf{E}'_{t2}, \quad H'_{z1} = -H'_{z2} \quad \text{on the aperture,} \end{aligned}$$

the suffixes 1 and 2 corresponding to $z \to \pm 0$. They also satisfy the further conditions

$$E'_z = 0, \qquad \mathbf{H}'_t = 0 \text{ on the aperture,}$$

since the boundary conditions on the surface of a body with $\mu = \infty$ are obtained from those for a perfectly conducting body ($\varepsilon = \infty$) by interchanging **E** and **H**. Hence it is clear that the fields (95.4) satisfy the necessary conditions $\mathbf{E}_t = 0$, $H_z = 0$ on the surface of the screen ($z \to -0$) outside the aperture, and are continuous on the aperture. Finally, since $\mathbf{E}'$, $\mathbf{H}'$ tend to $\mathbf{E}_0$, $\mathbf{H}_0$ at infinity, the fields (95.4) tend to $\mathbf{E}_0$, $\mathbf{H}_0$ as $z \to -\infty$ and to zero as $z \to +\infty$. They therefore satisfy all the conditions of the problem. This proves the theorem.

Thus the problem of diffraction by an aperture in a screen with $\varepsilon = \infty$ is equivalent to a problem of diffraction by a complementary screen with $\mu = \infty$.

PROBLEMS

PROBLEM 1. A plane monochromatic wave is incident normally on a slit cut in a perfectly conducting plane screen, the width $2a$ of the slit being large compared with the wavelength. Determine the distribution of light intensity beyond the slit, at large distances from it and for large angles of diffraction.

SOLUTION. For $a \gg \lambda$, the diffraction field beyond the slit can be regarded as a superposition of fields arising from independent diffraction at each of the two edges of the slit and determined by means of the asymptotic formula (95.2).† When the distances $AP = r_1$ and $BP = r_2$ from the edges of the slit to the point of observation (Fig. 50) are large compared with a, we can put, in the factors e^{ikr_1} and e^{ikr_2}, $r_1 = r - a \sin\chi$, $r_2 = r + a \sin\chi$, and elsewhere $r_1 \cong r_2 \cong r$; the angles between the z-axis and AP, OP, BP can all be taken as the angle of diffraction χ.

The result is

$$u = \frac{e^{i(kr + \frac{1}{4}\pi)}}{\sqrt{(2\pi kr)}}\left\{\frac{\sin(ka\sin\chi)}{\sin\frac{1}{2}\chi} \pm i\frac{\cos(ka\sin\chi)}{\cos\frac{1}{2}\chi}\right\}.$$

† This assumption becomes invalid, however, for diffraction angles sufficiently close to $\frac{1}{2}\pi$ (when $\frac{1}{2}\pi - \chi \lesssim 1/\sqrt{(ka)}$).

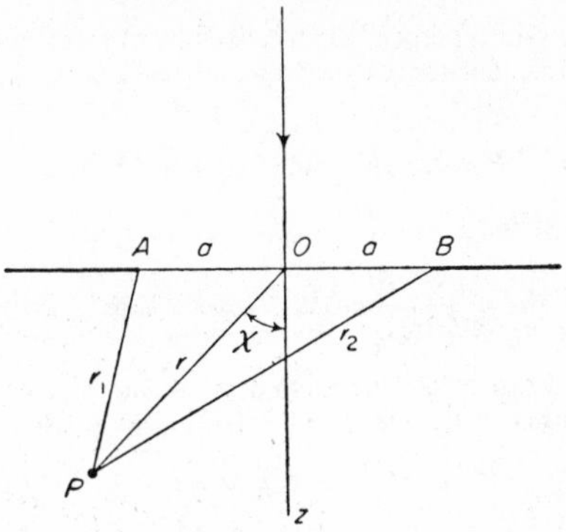

FIG. 50

Hence the intensity of light diffracted into an angle $d\chi$ is (relative to the total intensity of light incident on the slit)

$$dI = \frac{1}{4\pi ka}\left\{\left[\frac{\sin(ka\sin\chi)}{\sin\frac{1}{2}\chi}\right]^2 + \left[\frac{\cos(ka\sin\chi)}{\cos\frac{1}{2}\chi}\right]^2\right\} d\chi$$

$$= \frac{ka}{\pi}\left\{\left[\frac{\sin(ka\sin\chi)}{ka\sin\chi}\right]^2 \cos\chi + \frac{1}{[2ka\cos\frac{1}{2}\chi]^2}\right\} d\chi.$$

For small χ this expression becomes the formula for Fraunhofer diffraction by the slit:

$$dI = \frac{1}{\pi ka}\frac{\sin^2 ka\chi}{\chi^2} d\chi.$$

PROBLEM 2. A plane wave is incident on a perfectly conducting plane with a circular aperture whose radius a is small compared with the wavelength. Determine the intensity of diffracted light passing through the aperture (Rayleigh, 1897).

SOLUTION. As stated above, this problem is equivalent to that of diffraction by a circular disc with $\mu = \infty$, and, since $a \ll \lambda$, we have the case of scattering by a small particle. According to §92, in order to solve the latter problem it is necessary to determine the static electric and magnetic polarizabilities of the disc. The field $\mathbf{E}_0$ is perpendicular to the plane of the disc, and the boundary condition $E'_z = 0$ is formally identical with the condition in electrostatics at the boundary of a body with $\varepsilon = 0$. The field $\mathbf{H}_0$ is parallel to the disc, and the boundary condition $\mathbf{H}'_t = 0$ corresponds to a magnetostatic problem with $\mu = \infty$. Hence the electric and magnetic moments of the disc are (see §4, Problem 4, and §54, Problem) $\mathscr{P} = -2a^3\mathbf{E}_0/3\pi$, $\mathscr{M} = 4a^3\mathbf{H}_0/3\pi$. In going to the problem of diffraction by an aperture we must, in accordance with formulae (95.4), divide these expressions by 2 and then substitute them in the scattering formula (92.1).

Thus the intensity of radiation diffracted into a solid angle do is†

$$dI = \frac{c}{4\pi}\frac{\omega^4 a^6}{9\pi^2 c^4}\{\mathbf{n}\times\mathbf{E}_0 - 2\mathbf{n}\times(\mathbf{H}_0\times\mathbf{n})\}^2\, do$$

$$= \frac{c}{4\pi}\frac{\omega^4 a^6}{9\pi^2 c^4}\{(\mathbf{n}\times\mathbf{E}_0)^2 + 4(\mathbf{n}\times\mathbf{H}_0)^2 + 4\mathbf{n}\cdot\mathbf{H}_0\times\mathbf{E}_0\}\, do.$$

The total diffracted intensity is obtained by integration over a hemisphere, and is

$$I = \frac{\omega^4 a^6}{27\pi^2 c^3}(E_0^{\,2} + 4H_0^{\,2}).$$

The diffraction cross-section may be defined as the ratio of the intensity of diffracted radiation to the energy flux density in the incident wave $cE^2/4\pi$ (letters without suffixes refer to the incident wave). Two modes of polarization of the incident wave may be distinguished:

† The factor $e^{-i\omega t}$ is omitted; **E** and **H** are real.

(a) the vector **E** in the incident wave is perpendicular to the plane of incidence (the xz-plane), i.e. it is parallel to the plane of the screen (the xy-plane). The sum of the fields in the incident and reflected waves at the surface of the screen is

$$E_0 = 0, \quad H_{0x} = 2H \cos\alpha = 2E \cos\alpha$$

(α being the angle of incidence). Hence

$$d\sigma = \frac{16a^6\omega^4}{9\pi^2c^4}\cos^2\alpha\,(1-\sin^2\vartheta\cos^2\phi)\,do,$$

where ϑ is the angle between the direction of diffraction **n** and the normal to the screen (the z-axis), and ϕ is the azimuth of the vector **n** with respect to the plane of incidence. The total cross-section is

$$\sigma = (64\omega^4a^6/27\pi c^4)\cos^2\alpha.$$

(b) the vector **E** lies in the plane of incidence. Then $E_0 = E_{0z} = -2E\sin\alpha$, $H_0 = H_{0y} = 2H = 2E$. The differential cross-section is

$$d\sigma = \frac{16a^6\omega^4}{9\pi^2c^4}\{\cos^2\vartheta + \sin^2\vartheta\,(\cos^2\phi + \tfrac{1}{4}\sin^2\alpha) - \sin\vartheta\sin\alpha\cos\phi\}\,do,$$

and the total cross-section is $\sigma = (64a^6\omega^4/27\pi c^4)(1+\frac{1}{4}\sin^2\alpha)$.

For natural incident light $\sigma = (64a^6\omega^4/27\pi c^4)(1-\frac{3}{8}\sin^2\alpha)$.

CHAPTER XI

ELECTROMAGNETIC WAVES IN ANISOTROPIC MEDIA

§96. The permittivity of crystals

THE properties of an anisotropic medium with respect to electromagnetic waves are defined by the tensors $\varepsilon_{ik}(\omega)$ and $\mu_{ik}(\omega)$, which give the relation between the inductions and the fields:†

$$D_i = \varepsilon_{ik}(\omega)E_k, \quad B_i = \mu_{ik}(\omega)H_k. \tag{96.1}$$

In what follows we shall, for definiteness, consider the electric field and the tensor ε_{ik}; all the results obtained are valid for the tensor μ_{ik} also.

As $\omega \to 0$, the ε_{ik} tend to their static values, which have been shown in §13 to be symmetrical with respect to i and k. The proof was thermodynamical, and therefore holds only for states of thermodynamic equilibrium. In a variable field, a substance is of course not in equilibrium, and the proof in §13 is consequently invalid. To ascertain the properties of the tensor ε_{ik} we must use the generalized principle of the symmetry of the kinetic coefficients (see *SP* 1, §125).

The generalized susceptibilities $\alpha_{ab}(\omega)$ which appear in the formulation of this principle are defined in terms of the response of the system to a perturbation:

$$\hat{V} = -\hat{x}_a f_a(t)$$

(where the x_a are quantities describing the system) and are the coefficients in the linear relation between the Fourier components of the mean values $\overline{x_a}(t)$ and the generalized forces $f_a(t)$:

$$\overline{x_{a\omega}} = \alpha_{ab}(\omega) f_{b\omega}.$$

The change in the energy of the system with time under the perturbation is given by

$$\dot{\mathscr{U}} = -\dot{f}_a \overline{x_a}.$$

According to the symmetry principle,

$$\alpha_{ab}(\omega) = \alpha_{ba}(\omega),$$

if the system is not in an external magnetic field and has no magnetic structure; otherwise, $\alpha_{ba}(\omega)$ has to be taken for the "time-reversed" system.

It is easy to relate the components of the tensor $\varepsilon_{ik}(\omega)$ to the generalized susceptibilities. To do so, we note that the rate of change of the energy of a dielectric body in a variable

† It should be recalled that these quantities refer to the variable fields in the wave; the possible presence of a constant induction (in a pyroelectric or ferromagnetic crystal) is irrelevant to this discussion.

electric field is given by the integral

$$\int \frac{1}{4\pi} \mathbf{E} \cdot \frac{\partial \mathbf{D}}{\partial t} \mathrm{d}V. \tag{96.2}$$

A comparison with the above formulae shows that, if the components of the vector $\mathbf{E}$ at each point are taken as the quantities $\overline{x_a}$, the corresponding quantities f_a will be the components of $\mathbf{D}$. (The suffix a takes a continuous series of values, labelling both the components of the vectors and the points in the body.) The coefficients α_{ab} are then the components of the tensor $\varepsilon^{-1}{}_{ik}$. The symmetry properties of ε_{ik} are, of course, identical with those of its inverse. Since $\mathbf{E}$ and $\dot{\mathbf{D}}$ are multiplied in (96.2) only at the same point in the body, the interchange of the suffixes a and b is equivalent to simply interchanging the tensor suffixes. We thus conclude that the tensor ε_{ik} is symmetrical:†

$$\varepsilon_{ik}(\omega) = \varepsilon_{ki}(\omega). \tag{96.3}$$

It should be noted that the components of the polarizability tensor for the whole body, i.e. the coefficients in the equations $\mathscr{P}_i = V\alpha_{ik}\mathfrak{E}_k$, also come under the definition of generalized susceptibilities. For the rate of change of the energy of a body placed in a variable external field $\mathfrak{E}$ is

$$-\mathscr{P} \cdot \mathrm{d}\mathfrak{E}/\mathrm{d}t. \tag{96.4}$$

Hence we see that, if the x_a are the three components of the vector $\mathscr{P}$, then the corresponding f_a are those of the vector $\mathfrak{E}$, so that the coefficients α_{ab} are in this case $V\alpha_{ik}$.

Several of the formulae derived previously for an isotropic medium can be directly generalized to the anisotropic case. Repeating for the anisotropic case the derivation in §80, we find that the energy dissipation in a monochromatic electromagnetic field is

$$Q = \frac{i\omega}{16\pi}\left\{\left(\varepsilon_{ik}{}^* - \varepsilon_{ki}\right)E_i E_k{}^* + \left(\mu_{ik}{}^* - \mu_{ki}\right)H_i H_k{}^*\right\}, \tag{96.5}$$

which is analogous to (80.5). The condition that absorption be absent is $\varepsilon_{ik}{}^* = \varepsilon_{ki} = \varepsilon_{ik}$, i.e. the ε_{ik} must be real, as must the μ_{ik}.

When absorption is absent, the internal electromagnetic energy per unit volume can be defined as shown in §80. The formula for an anisotropic medium corresponding to (80.11) is

$$\bar{U} = \frac{1}{16\pi}\left\{\frac{\mathrm{d}}{\mathrm{d}\omega}(\omega\varepsilon_{ik})E_i E_k{}^* + \frac{\mathrm{d}}{\mathrm{d}\omega}(\omega\mu_{ik})H_i H_k{}^*\right\}. \tag{96.6}$$

In §87 we used the surface impedance ζ, in terms of which the boundary conditions at the surface of a metal can be formulated even if the permittivity is no longer meaningful. At the surface of an anisotropic body the boundary condition corresponding to (87.6) is

$$E_\alpha = \zeta_{\alpha\beta}(\mathbf{H} \times \mathbf{n})_\beta, \tag{96.7}$$

where $\zeta_{\alpha\beta}(\omega)$ is a two-dimensional tensor on the surface of the body. It should be borne in mind that the value of this tensor depends, in general, on the crystallographic direction of the surface concerned.

---

† The properties of this tensor in the presence of an external magnetic field will be discussed in §101.

The energy flux into the body is $(c/4\pi)\mathbf{E}\times\mathbf{H}\cdot\mathbf{n} = (c/4\pi)\mathbf{E}\cdot\mathbf{H}\times\mathbf{n} \equiv (c/4\pi)E_\alpha(\mathbf{H}\times\mathbf{n})_\alpha$. (Here **E** and **H** are real.) Hence we see that if, in applying the principle of the symmetry of the kinetic coefficients, we take the components E_α as the x_a, then the corresponding $\dot{f}_a$ will be $-(\mathbf{H}\times\mathbf{n})_\alpha$, i.e. f_a will be $-(i/\omega)(\mathbf{H}\times\mathbf{n})_\alpha$ (returning to the complex form). The coefficients α_{ab} are therefore the same, apart from a factor, as the components $\zeta_{\alpha\beta}$, and we conclude that

$$\zeta_{\alpha\beta} = \zeta_{\beta\alpha} \tag{96.8}$$

in the absence of an external magnetic field.

PROBLEM

Express the components of the tensor $\zeta_{\alpha\beta}$ in terms of those of $\eta_{\alpha\beta} \equiv \varepsilon^{-1}{}_{\alpha\beta}$, assuming that the latter exists and that the body is non-magnetic ($\mu_{ik} = \delta_{ik}$).

SOLUTION. In an anisotropic medium, the equation $\zeta^2 = 1/\varepsilon$ (87.2) becomes $\zeta_{\alpha\gamma}\zeta_{\gamma\beta} = \eta_{\alpha\beta}$. In components this gives

$$\zeta_{11}{}^2 + \zeta_{12}\zeta_{21} = \eta_{11}, \qquad \zeta_{22}{}^2 + \zeta_{12}\zeta_{21} = \eta_{22},$$
$$\zeta_{12}(\zeta_{11}+\zeta_{22}) = \eta_{12}, \qquad \zeta_{21}(\zeta_{11}+\zeta_{22}) = \eta_{21}.$$

The solution of these equations is

$$\zeta_{12} = \eta_{12}/\xi, \qquad \zeta_{21} = \eta_{21}/\xi,$$
$$\zeta_{11} = [\eta_{11} \pm \sqrt{(\eta_{11}\eta_{22}-\eta_{12}\eta_{21})}]/\xi, \quad \zeta_{22} = [\eta_{22} \pm \sqrt{(\eta_{11}\eta_{22}-\eta_{12}\eta_{21})}]/\xi,$$
$$\xi^2 = \eta_{11}+\eta_{22} \pm 2\sqrt{(\eta_{11}\eta_{22}-\eta_{12}\eta_{21})}.$$

The choice of signs is determined by the condition that the absorption of energy must be positive. We do not assume $\zeta_{12} = \zeta_{21}$, and thereby allow for the presence of an external magnetic field.

§97. A plane wave in an anisotropic medium

In studying the optics of anisotropic bodies (crystals) we shall take only the most important case, where the medium may be supposed non-magnetic and transparent in a given range of frequencies. Accordingly, the relation between the electric and magnetic fields and inductions is

$$D_i = \varepsilon_{ik}E_k, \qquad \mathbf{B} = \mathbf{H}. \tag{97.1}$$

The components of the dielectric tensor ε_{ik} are all real, and its principal values are positive.

Maxwell's equations for the field of a monochromatic wave with frequency ω are

$$i\omega\mathbf{H} = c\,\mathbf{curl}\,\mathbf{E}, \qquad i\omega\mathbf{D} = -c\,\mathbf{curl}\,\mathbf{H}. \tag{97.2}$$

In a plane wave propagated in a transparent medium all quantities are proportional to $e^{i\mathbf{k}\cdot\mathbf{r}}$, with a real wave vector **k**. Effecting the differentiation with respect to the coordinates, we obtain

$$\omega\mathbf{H}/c = \mathbf{k}\times\mathbf{E}, \qquad \omega\mathbf{D}/c = -\mathbf{k}\times\mathbf{H}. \tag{97.3}$$

Hence we see, first of all, that the three vectors **k**, **D**, **H** are mutually perpendicular. Moreover, **H** is perpendicular to **E**, and so the three vectors **D**, **E**, **k**, being all perpendicular to **H**, must be coplanar. Fig. 51 (p. 334) shows the relative position of all these vectors. With respect to the direction of the wave vector, **D** and **H** are transverse, but **E** is not. The diagram shows also the direction of the energy flux **S** in the wave. It is given by the vector product **E** × **H**, i.e. it is perpendicular to both **E** and **H**. The direction of **S** is not the same as

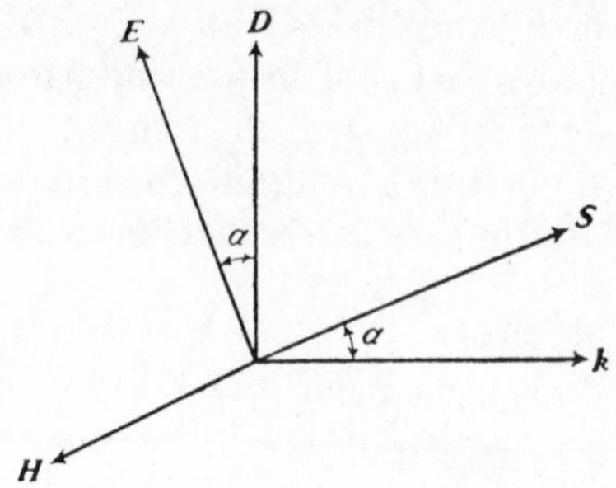

FIG. 51

that of **k**, unlike what happens for an isotropic medium. Clearly the vector **S** is coplanar with **E**, **D** and **k**, and the angle between **S** and **k** is equal to that between **E** and **D**.

We can define a vector **n** by

$$\mathbf{k} = \omega\mathbf{n}/c. \qquad (97.4)$$

The magnitude of this vector in an anisotropic medium depends on its direction, whereas in an isotropic medium $n = \sqrt{\varepsilon}$ depends only on the frequency.† Using (97.4), we can write the fundamental formulae (97.3) as

$$\mathbf{H} = \mathbf{n}\times\mathbf{E}, \qquad \mathbf{D} = -\mathbf{n}\times\mathbf{H}. \qquad (97.5)$$

The energy flux vector in a plane wave is

$$\mathbf{S} = c\mathbf{E}\times\mathbf{H}/4\pi = (c/4\pi)\{E^2\mathbf{n} - (\mathbf{E}\cdot\mathbf{n})\mathbf{E}\}; \qquad (97.6)$$

in this formula **E** and **H** are real.

So far we have not used the relation (97.1) which involves the constants ε_{ik} characterizing the material. This relation, together with equations (97.5), determines the function $\omega(\mathbf{k})$.

Substituting the first equation (97.5) in the second, we have

$$\mathbf{D} = \mathbf{n}\times(\mathbf{E}\times\mathbf{n}) = n^2\mathbf{E} - (\mathbf{n}\cdot\mathbf{E})\mathbf{n}. \qquad (97.7)$$

If we equate the components of this vector to $\varepsilon_{ik}E_k$ in accordance with (97.1), we obtain three linear homogeneous equations for the three components of **E**: $n^2 E_i - n_i n_k E_k = \varepsilon_{ik} E_k$ or

$$(n^2\delta_{ik} - n_i n_k - \varepsilon_{ik})E_k = 0 \qquad (97.8)$$

The compatibility condition for these equations is that the determinant of their coefficients should vanish:

$$\det|n^2\delta_{ik} - n_i n_k - \varepsilon_{ik}| = 0. \qquad (97.9)$$

In practice, this determinant is conveniently evaluated by taking as the axes of x, y, z the principal axes of the tensor ε_{ik} (called the *principal dielectric axes*). Let the principal values of the tensor be $\varepsilon^{(x)}$, $\varepsilon^{(y)}$, $\varepsilon^{(z)}$. Then a simple calculation gives

$$n^2(\varepsilon^{(x)}n_x^{\ 2} + \varepsilon^{(y)}n_y^{\ 2} + \varepsilon^{(z)}n_z^{\ 2}) - [n_x^{\ 2}\varepsilon^{(x)}(\varepsilon^{(y)} + \varepsilon^{(z)})$$
$$+ n_y^{\ 2}\varepsilon^{(y)}(\varepsilon^{(x)} + \varepsilon^{(z)}) + n_z^{\ 2}\varepsilon^{(z)}(\varepsilon^{(x)} + \varepsilon^{(y)})] + \varepsilon^{(x)}\varepsilon^{(y)}\varepsilon^{(z)} = 0. \qquad (97.10)$$

† The magnitude n is still called the *refractive index*, although it no longer bears the same simple relation to the law of refraction as in isotropic bodies.

The sixth-order terms cancel when the determinant is expanded; this is, of course, no accident and is due ultimately to the fact that the wave has two, not three, independent directions of polarization.

Equation (97.10), called *Fresnel's equation*, is one of the fundamental equations of crystal optics.† It determines implicitly the dispersion relation, i.e. the frequency as a function of the wave vector. (The principal values $\varepsilon^{(i)}$ are functions of frequency, and so are, in some cases (see §99), the directions of the principal axes of the tensor ε_{ik}.) For monochromatic waves, however, ω, and therefore all the $\varepsilon^{(i)}$, are usually given constants, and equation (97.10) then gives the magnitude of the wave vector as a function of its direction. When the direction of **n** is given, (97.10) is a quadratic equation, for n^2, with real coefficients. Hence two different magnitudes of the wave vector correspond, in general, to each direction of **n**.

Equation (97.10) (with constant coefficients $\varepsilon^{(i)}$) defines in the coordinates n_x, n_y, n_z the "wave-vector surface".‡ In general this is a surface of the fourth order, whose properties will be discussed in detail in the following sections. Here we shall mention some general properties of this surface.

We first introduce another quantity characterizing the propagation of light in an anisotropic medium. The direction of the light rays (in geometrical optics) is given by the group velocity vector $\partial\omega/\partial\mathbf{k}$. In an isotropic medium, the direction of this vector is always the same as that of the wave vector, but in an anisotropic medium the two do not in general coincide. The rays may be characterized by a vector **s**, whose direction is that of the group velocity, while its magnitude is given by

$$\mathbf{n}\cdot\mathbf{s} = 1. \tag{97.11}$$

We shall call **s** the *ray vector*. Its significance is as follows.

Let us consider a beam of rays (of a single frequency) propagated in all directions from some point. The value of the eikonal ψ (which is, apart from a factor ω/c, the wave phase; see §85) at any point is given by the integral $\int \mathbf{n}\cdot d\mathbf{l}$ taken along the ray. Using the vector **s** which determines the direction of the ray, we can put

$$\psi = \int \mathbf{n}\cdot d\mathbf{l} = \int (\mathbf{n}\cdot\mathbf{s}/s)\, dl = \int dl/s. \tag{97.12}$$

In a homogeneous medium, s is constant along the ray, so that $\psi = L/s$, where L is the length of the ray segment concerned. Hence we see that, if a segment equal (or proportional) to s is taken along each ray from the centre, the resulting surface is such that the phase of the rays is the same at every point. This is called the *ray surface*.

The wave-vector surface and the ray surface are in a certain dual relationship. Let the equation of the wave-vector surface be written $f(\omega, \mathbf{k}) = 0$. Then the group velocity vector is

$$\frac{\partial\omega}{\partial\mathbf{k}} = -\frac{\partial f/\partial\mathbf{k}}{\partial f/\partial\omega}, \tag{97.13}$$

i.e. is proportional to $\partial f/\partial\mathbf{k}$, or, what is the same thing (since the derivative is taken for constant ω), to $\partial f/\partial\mathbf{n}$. The ray vector, therefore, is also proportional to $\partial f/\partial\mathbf{n}$. But the vector

† The foundations of crystal optics were laid by A. J. Fresnel in the 1820s, on the basis of mechanical analogies, long before the development of the electromagnetic theory.

‡ A concept called the "surface of normals" or "surface of indices" has been used; it is obtained by taking a point at a distance $1/n$ (instead of n) in each direction, but is less convenient.

$\partial f/\partial \mathbf{n}$ is normal to the surface $f = 0$. Thus we conclude that the direction of the ray vector of a wave with given $\mathbf{n}$ is that of the normal at the corresponding point of the wave-vector surface.

It is easy to see that the reverse is also true: the normal to the ray surface gives the direction of the corresponding wave vectors. For the equation $\mathbf{s}\cdot\delta\mathbf{n} = 0$, where $\delta\mathbf{n}$ is an arbitrary infinitesimal change in $\mathbf{n}$ (for given ω), i.e. the vector of an infinitesimal displacement on the surface, expresses the fact that $\mathbf{s}$ is perpendicular to the wave-vector surface. Differentiating (again for given ω) the equation $\mathbf{n}\cdot\mathbf{s} = 1$, we obtain $\mathbf{n}\cdot\delta\mathbf{s}+\mathbf{s}\cdot\delta\mathbf{n} = 0$, and therefore $\mathbf{n}\cdot\delta\mathbf{s} = 0$, which proves the above statement.

This relation between the surfaces of $\mathbf{n}$ and $\mathbf{s}$ can be made more precise. Let $\mathbf{n}_0$ be the position vector of a point on the wave-vector surface, and $\mathbf{s}_0$ the corresponding ray vector. The equation (in coordinates n_x, n_y, n_z) of the tangent plane at this point is $\mathbf{s}_0\cdot(\mathbf{n}-\mathbf{n}_0) = 0$, which states that $\mathbf{s}_0$ is perpendicular to any vector $\mathbf{n}-\mathbf{n}_0$ in the plane. Since $\mathbf{s}_0$ and $\mathbf{n}_0$ are related by $\mathbf{s}_0\cdot\mathbf{n}_0 = 1$, we can write the equation as

$$\mathbf{s}_0\cdot\mathbf{n} = 1. \tag{97.14}$$

Hence it follows that $1/s_0$ is the length of the perpendicular from the origin to the tangent plane to the wave-vector surface at the point $\mathbf{n}_0$.

Conversely, the length of the perpendicular from the origin to the tangent plane to the ray surface at a point $\mathbf{s}_0$ is $1/n_0$.

To ascertain the location of the ray vector relative to the field vectors in the wave, we notice that the group velocity is always in the same direction as the (time) averaged energy flux vector. For let us consider a wave packet, occupying a small region of space. When the packet moves, the energy concentrated in it must move with it, and the direction of the energy flux is therefore the same as the direction of the velocity of the packet, i.e. the group velocity. It can be demonstrated from (97.5) that the group velocity is in the same direction as the Poynting vector. Differentiating (for given ω), we obtain

$$\delta\mathbf{D} = \delta\mathbf{H}\times\mathbf{n}+\mathbf{H}\times\delta\mathbf{n}, \qquad \delta\mathbf{H} = \mathbf{n}\times\delta\mathbf{E}+\delta\mathbf{n}\times\mathbf{E}. \tag{97.15}$$

We take the scalar product of the first equation with $\mathbf{E}$ and of the second with $\mathbf{H}$, obtaining

$$\mathbf{E}\cdot\delta\mathbf{D} = \mathbf{H}\cdot\delta\mathbf{H}+\mathbf{E}\times\mathbf{H}\cdot\delta\mathbf{n}, \qquad \mathbf{H}\cdot\delta\mathbf{H} = \mathbf{D}\cdot\delta\mathbf{E}+\mathbf{E}\times\mathbf{H}\cdot\delta\mathbf{n}.$$

But $\mathbf{D}\cdot\delta\mathbf{E} = \varepsilon_{ik}E_k\delta E_i = \mathbf{E}\cdot\delta\mathbf{D}$, and so, adding the two equations, we have

$$\mathbf{E}\times\mathbf{H}\cdot\delta\mathbf{n} = 0, \tag{97.16}$$

i.e. the vector $\mathbf{E}\times\mathbf{H}$ is normal to the wave-vector surface. This is the required result.†

Since the Poynting vector is perpendicular to $\mathbf{H}$ and $\mathbf{E}$, the same is true of $\mathbf{s}$:

$$\mathbf{s}\cdot\mathbf{H} = 0, \qquad \mathbf{s}\cdot\mathbf{E} = 0. \tag{97.17}$$

A direct calculation, using formulae (97.5), (97.11) and (97.17), gives

$$\mathbf{H} = \mathbf{s}\times\mathbf{D}, \qquad \mathbf{E} = -\mathbf{s}\times\mathbf{H}. \tag{97.18}$$

For example, $\mathbf{s}\times\mathbf{H} = \mathbf{s}\times(\mathbf{n}\times\mathbf{E}) = \mathbf{n}(\mathbf{s}\cdot\mathbf{E})-\mathbf{E}(\mathbf{n}\cdot\mathbf{s}) = -\mathbf{E}$.

† The result thus obtained relates to the instantaneous, as well as to the average, energy flux. In this proof, however, the symmetry of the tensor ε_{ik} is vital. The result is therefore not valid in the above form for media in which ε_{ik} is not symmetrical (*gyrotropic* media, §101). The statement is still valid, however, for the average value of the Poynting vector (§101, Problem 1).

If we compare formulae (97.18) and (97.5), we see that they differ by the interchange of

$$\mathbf{E} \text{ and } \mathbf{D}, \qquad \mathbf{n} \text{ and } \mathbf{s}, \qquad \varepsilon_{ik} \text{ and } \varepsilon^{-1}{}_{ik} \tag{97.19}$$

(the relation $\mathbf{n}\cdot\mathbf{s} = 1$ remaining valid, of course). The last of these pairs must be included in order that the relation (97.1) between **D** and **E** should remain valid. Thus the following useful rule may be formulated: an equation valid for one set of quantities can be converted into one valid for another set by means of the interchanges (97.19).

In particular, the application of this rule to (97.10) gives immediately an analogous equation for **s**:

$$s^2(\varepsilon^{(y)}\varepsilon^{(z)}s_x{}^2 + \varepsilon^{(x)}\varepsilon^{(z)}s_y{}^2 + \varepsilon^{(x)}\varepsilon^{(y)}s_z{}^2)$$
$$-[s_x{}^2(\varepsilon^{(y)}+\varepsilon^{(z)}) + s_y{}^2(\varepsilon^{(x)}+\varepsilon^{(z)}) + s_z{}^2(\varepsilon^{(x)}+\varepsilon^{(y)})] + 1 = 0. \tag{97.20}$$

This equation gives the form of the ray surface. Like the wave-vector surface, it is of the fourth order. When the direction of **s** is given, (97.20) is a quadratic equation for s^2, which in general has two different real roots. Thus two rays with different wave vectors can be propagated in any direction in the crystal.

Let us now consider the polarization of waves propagated in an anisotropic medium. Equations (97.8), from which Fresnel's equation has been derived, are unsuitable for this, because they involve the field **E**, whereas it is the induction **D** which is transverse (to the given **n**) in the wave. In order to take account immediately of the fact that **D** is transverse, we use for the time being a new coordinate system with one axis in the direction of the wave vector, and denote the two transverse axes by Greek suffixes, which take the values 1 and 2. The transverse components of equation (97.7) give $D_\alpha = n^2 E_\alpha$; substituting $E_\alpha = \varepsilon^{-1}{}_{\alpha\beta} D_\beta$, where $\varepsilon^{-1}{}_{\alpha\beta}$ is a component of the tensor inverse to $\varepsilon_{\alpha\beta}$, we have

$$\left(\frac{1}{n^2}\delta_{\alpha\beta} - \varepsilon^{-1}{}_{\alpha\beta}\right)D_\beta = 0. \tag{97.21}$$

The condition for the two equations ($\alpha = 1, 2$) in the two unknowns D_1, D_2 to be compatible is that their determinant should be zero:

$$\det|n^{-2}\delta_{\alpha\beta} - \varepsilon^{-1}{}_{\alpha\beta}| = 0. \tag{97.22}$$

This condition is, of course, the same as Fresnel's equation, which was written in the original coordinates x, y, z. We now see also, however, that the vectors **D** corresponding to the two values of n are along the principal axes of the symmetrical two-dimensional tensor of rank two $\varepsilon^{-1}{}_{\alpha\beta}$. According to general theorems it follows that these two vectors are perpendicular. Thus, in the two waves with the wave vector in the same direction, the electric induction vectors are linearly polarized in two perpendicular planes.

Equations (97.21) have a simple geometrical interpretation. Let us draw the tensor ellipsoid corresponding to the tensor $\varepsilon^{-1}{}_{ik}$, returning to the principal dielectric axes, i.e. the surface

$$\varepsilon^{-1}{}_{ik}x_i x_k = \frac{x^2}{\varepsilon^{(x)}} + \frac{y^2}{\varepsilon^{(y)}} + \frac{z^2}{\varepsilon^{(z)}} = 1 \tag{97.23}$$

(Fig. 52, p. 338). Let this ellipsoid be cut by a plane through its centre perpendicular to the given direction of **n**. The section is in general an ellipse; the lengths of its axes determine the values of n, and their directions determine the directions of the oscillations, i.e. the vectors **D**.

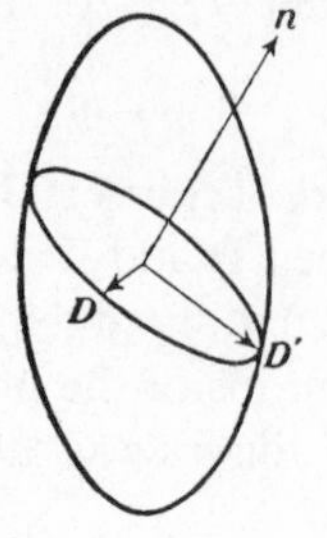

FIG. 52

From this construction (with, in general, $\varepsilon^{(x)}$, $\varepsilon^{(y)}$, $\varepsilon^{(z)}$ different) we see at once that, if the wave vector is in (say) the x-direction, the directions of polarization (**D**) will be the y and z directions. If the vector **n** lies in one of the coordinate planes, e.g. the xy-plane, one of the directions of polarization is also in that plane, and the other is in the z-direction.

The polarizations of two waves with the ray vector in the same direction have similar properties. Instead of the directions of the induction **D**, we must now consider those of the vector **E**, which is transverse to **s**, and equations (97.21) are replaced by the analogous equations

$$\left(\frac{1}{s^2}\delta_{\alpha\beta} - \varepsilon_{\alpha\beta}\right)E_\beta = 0. \tag{97.24}$$

The geometrical construction is here based on the tensor ellipsoid

$$\varepsilon_{ik}x_i x_k = \varepsilon^{(x)}x^2 + \varepsilon^{(y)}y^2 + \varepsilon^{(z)}z^2 = 1, \tag{97.25}$$

corresponding to the tensor ε_{ik} itself (called the *Fresnel ellipsoid*).

It should be emphasized that plane waves propagated in an anisotropic medium are linearly polarized in certain planes. In this respect the optical properties of anisotropic media are very different from those of isotropic media. A plane wave propagated in an isotropic medium is in general elliptically polarized, and is linearly polarized only in particular cases. This important difference arises because the case of complete isotropy of the medium is in a sense one of degeneracy, in which a single wave vector corresponds to two directions of polarization, whereas in an anisotropic medium there are in general two different wave vectors (in the same direction). The two linearly polarized waves propagated with the same value of **n** combine to form one elliptically polarized wave.

PROBLEM

Express the components of the ray vector **s** in terms of the components of **n** along the principal dielectric axes.

SOLUTION. Differentiating the left-hand side of the equation $f(\mathbf{n}) = 0$ (97.10) with respect to n_i and determining from the condition $\mathbf{n}\cdot\mathbf{s} = 1$ the proportionality coefficient between s_i and $\partial f/\partial n_i$, we obtain the following relations between the vectors **s** and **n**:

$$\frac{s_x}{n_x} = \frac{\varepsilon^{(x)}(\varepsilon^{(y)}+\varepsilon^{(z)}) - 2\varepsilon^{(x)}n_x^2 - (\varepsilon^{(x)}+\varepsilon^{(y)})n_y^2 - (\varepsilon^{(x)}+\varepsilon^{(z)})n_z^2}{2\varepsilon^{(x)}\varepsilon^{(y)}\varepsilon^{(z)} - n_x^2\varepsilon^{(x)}(\varepsilon^{(y)}+\varepsilon^{(z)}) - n_y^2\varepsilon^{(y)}(\varepsilon^{(x)}+\varepsilon^{(z)}) - n_z^2\varepsilon^{(z)}(\varepsilon^{(x)}+\varepsilon^{(y)})}$$

and similarly for s_y, s_z.

§98. Optical properties of uniaxial crystals

The optical properties of a crystal depend primarily on the symmetry of its dielectric tensor ε_{ik}. In this respect all crystals fall under three types: cubic, uniaxial and biaxial (see §13). In a crystal of the cubic system $\varepsilon_{ik} = \varepsilon\delta_{ik}$, i.e. the three principal values of the tensor are equal, and the directions of the principal axes are arbitrary. As regards their optical properties, therefore, cubic crystals are no different from isotropic bodies.

The uniaxial crystals include those of the rhombohedral, tetragonal and hexagonal systems. Here one of the principal axes of the tensor ε_{ik} coincides with the threefold, fourfold or sixfold axis of symmetry respectively; in optics, this axis is called the *optical axis* of the crystal, and in what follows we shall take it as the z-axis, denoting the corresponding principal value of ε_{ik} by $\varepsilon_{\parallel}$. The directions of the other two principal axes, in a plane perpendicular to the optical axis, are arbitrary, and the corresponding principal values, which we denote by $\varepsilon_{\perp}$, are equal.

If in Fresnel's equation (97.10) we put $\varepsilon^{(x)} = \varepsilon^{(y)} = \varepsilon_{\perp}$, $\varepsilon^{(z)} = \varepsilon_{\parallel}$, the left-hand side is a product of two quadratic factors:

$$(n^2 - \varepsilon_{\perp})[\varepsilon_{\parallel} n_z^{\,2} + \varepsilon_{\perp}(n_x^{\,2} + n_y^{\,2}) - \varepsilon_{\perp}\varepsilon_{\parallel}] = 0.$$

In other words, the quartic equation gives the two quadratic equations

$$n^2 = \varepsilon_{\perp}, \tag{98.1}$$

$$\frac{n_z^{\,2}}{\varepsilon_{\perp}} + \frac{n_x^{\,2} + n_y^{\,2}}{\varepsilon_{\parallel}} = 1. \tag{98.2}$$

Geometrically, this signifies that the wave-vector surface, which is in general of the fourth order, becomes two separate surfaces, a sphere and an ellipsoid. Fig. 53 shows a cross-section of these surfaces. Two cases are possible: if $\varepsilon_{\perp} > \varepsilon_{\parallel}$, the sphere lies outside the ellipsoid, but if $\varepsilon_{\perp} < \varepsilon_{\parallel}$ it lies inside. In the first case we speak of a *negative uniaxial crystal*, and in the second case of a *positive* one. The two surfaces touch at opposite poles on the n_z-axis. The direction of the optical axis therefore corresponds to only one value of the wave vector.

The ray surface is similar in form. By the rule (97.19), its equation is obtained from (98.1) and (98.2):

$$s^2 = 1/\varepsilon_{\perp}, \tag{98.3}$$

$$\varepsilon_{\perp} s_z^{\,2} + \varepsilon_{\parallel}(s_x^{\,2} + s_y^{\,2}) = 1. \tag{98.4}$$

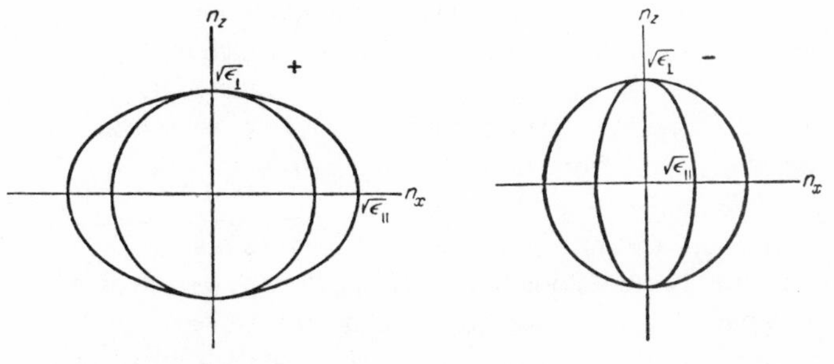

FIG. 53

In a positive crystal the ellipsoid lies within the sphere, and in a negative one outside.

Thus we see that two types of wave can be propagated in a uniaxial crystal. With respect to one type, called *ordinary waves*, the crystal behaves like an isotropic body with refractive index $n = \sqrt{\varepsilon_\perp}$. The magnitude of the wave vector is $\omega n/c$ whatever its direction, and the direction of the ray vector is that of **n**.

In waves of the second type, called *extraordinary waves*, the magnitude of the wave vector depends on the angle θ which it makes with the optical axis. By (98.2)

$$\frac{1}{n^2} = \frac{\sin^2\theta}{\varepsilon_\parallel} + \frac{\cos^2\theta}{\varepsilon_\perp}. \tag{98.5}$$

The ray vector in an extraordinary wave is not in the same direction as the wave vector, but is coplanar with that vector and the optical axis, their common plane being called the *principal section* for the given **n**. Let this be the zx-plane; the ratio of the derivatives of the left-hand side of (98.2) with respect to n_z and n_x gives the direction of the ray vector: $s_x/s_z = \varepsilon_\perp n_x/\varepsilon_\parallel n_z$. Thus the angle θ' between the ray vector and the optical axis and the angle θ satisfy the simple relation

$$\tan\theta' = (\varepsilon_\perp/\varepsilon_\parallel)\tan\theta. \tag{98.6}$$

The directions of **n** and **s** are the same only for waves propagated along or perpendicular to the optical axis.

The problem of the directions of polarization of the ordinary and extraordinary waves is very easily solved. It is sufficient to observe that the four vectors **E**, **D**, **s** and **n** are always coplanar. In the extraordinary wave **s** and **n** are not in the same direction, but lie in the same principal section. This wave is therefore polarized so that the vectors **E** and **D** lie in the same principal section as **s** and **n**. The vectors **D** in the ordinary and extraordinary waves with the same direction of **n** (or **E**, with the same direction of **s**) are perpendicular. Hence the polarization of the ordinary wave is such that **E** and **D** lie in a plane perpendicular to the principal section.

An exception is formed by waves propagated in the direction of the optical axis. In this direction there is no difference between the ordinary and the extraordinary wave, and so their polarizations combine to give a wave which is, in general, elliptically polarized.

The refraction of a plane wave incident on the surface of a crystal is different from refraction at a boundary between two isotropic media. The laws of refraction and reflection are again obtained from the continuity of the component $\mathbf{n}_t$ of the wave vector which is tangential to the plane of separation. The wave vectors of the refracted and reflected waves therefore lie in the plane of incidence. In a crystal, however, two different refracted waves are formed, a phenomenon known as *double refraction* or *birefringence*. They correspond to the two possible values of the normal component n_n which satisfy Fresnel's equation for a given tangential component $\mathbf{n}_t$. It should also be remembered that the observed direction of propagation of the rays is determined not by the wave vector but by the ray vector **s**, whose direction is different from that of **n** and in general does not lie in the plane of incidence.

In a uniaxial crystal, ordinary and extraordinary refracted waves are formed. The ordinary wave is entirely analogous to the refracted wave in isotropic bodies; in particular its ray vector (which is in the same direction as its wave vector) lies in the plane of incidence. The ray vector of the extraordinary wave in general does not lie in the plane of incidence.

PROBLEMS

PROBLEM 1. Find the direction of the extraordinary ray when light incident from a vacuum is refracted at a surface of a uniaxial crystal which is perpendicular to its optical axis.

SOLUTION. In this case the refracted ray lies in the plane of incidence, which we take as the xz-plane, with the z-axis normal to the surface. The x-component of the wave vector $n_x = \sin\vartheta$ (ϑ being the angle of incidence) is continuous; the component n_z for the refracted wave is found from (98.2):

$$n_z = \sqrt{\left(\varepsilon_\perp - \frac{\varepsilon_\perp}{\varepsilon_\parallel}\sin^2\vartheta\right)}.$$

The direction of the refracted ray is given by (98.6):

$$\tan\vartheta' = \frac{\varepsilon_\perp n_x}{\varepsilon_\parallel n_z} = \frac{\sqrt{\varepsilon_\perp}\sin\vartheta}{\sqrt{[\varepsilon_\parallel(\varepsilon_\parallel - \sin^2\vartheta)]}},$$

where ϑ' is the angle of refraction.

PROBLEM 2. Find the direction of the extraordinary ray when light is incident normally on a surface of a uniaxial crystal at any angle to the optical axis.

SOLUTION. The refracted ray lies in the xz-plane, which passes through the normal to the surface (the z-axis) and the optical axis. Let α be the angle between these axes. The ray vector $\mathbf{s}$, whose components are proportional to the derivatives of the left-hand side of equation (98.2) with respect to the corresponding components of $\mathbf{n}$, is proportional to

$$\frac{\mathbf{n}}{\varepsilon_\parallel} + (\mathbf{n}\cdot\mathbf{l})\mathbf{l}\left(\frac{1}{\varepsilon_\perp} - \frac{1}{\varepsilon_\parallel}\right),$$

where $\mathbf{l}$ is a unit vector in the direction of the optical axis. In the present case the wave vector $\mathbf{n}$ is in the z-direction, so that

$$s_x \propto \cos\alpha\sin\alpha\left(\frac{1}{\varepsilon_\perp} - \frac{1}{\varepsilon_\parallel}\right), \qquad s_z \propto \frac{\sin^2\alpha}{\varepsilon_\parallel} + \frac{\cos^2\alpha}{\varepsilon_\perp}.$$

Hence we find

$$\tan\vartheta' = \frac{s_x}{s_z} = \frac{(\varepsilon_\parallel - \varepsilon_\perp)\sin 2\alpha}{\varepsilon_\parallel + \varepsilon_\perp + (\varepsilon_\parallel - \varepsilon_\perp)\cos 2\alpha}.$$

§99. Biaxial crystals

In biaxial crystals the three principal values of the tensor ε_{ik} are all different. The crystals of the triclinic, monoclinic and orthorhombic systems are of this type. In those of the triclinic system, the position of the principal dielectric axes is unrelated to any specific crystallographic direction; in particular, it varies with frequency, as do all the components ε_{ik}. In crystals of the monoclinic system, one of the principal dielectric axes is crystallographically fixed; it coincides with the twofold axis of symmetry, or is perpendicular to the plane of symmetry. The position of the other two principal axes depends on the frequency. Finally, in crystals of the orthorhombic system, the position of all three principal axes is fixed: they must coincide with the three mutually perpendicular twofold axes of symmetry.

The study of the optical properties of biaxial crystals involves the consideration of Fresnel's equation in its general form. We shall assume for definiteness that

$$\varepsilon^{(x)} < \varepsilon^{(y)} < \varepsilon^{(z)}. \tag{99.1}$$

To ascertain the form of the fourth-order surface defined by equation (97.10), let us begin by finding its intersections with the coordinate planes. Putting $n_z = 0$ in equation

(97.10), we find that the left-hand side is the product of two factors:

$$(n^2 - \varepsilon^{(z)})(\varepsilon^{(x)} n_x{}^2 + \varepsilon^{(y)} n_y{}^2 - \varepsilon^{(x)}\varepsilon^{(y)}) = 0.$$

Hence we see that the section by the xy-plane consists of the circle

$$n^2 = \varepsilon^{(z)} \tag{99.2}$$

and the ellipse

$$\frac{n_x{}^2}{\varepsilon^{(y)}} + \frac{n_y{}^2}{\varepsilon^{(x)}} = 1, \tag{99.3}$$

and by the assumption (99.1) the ellipse lies inside the circle. Similarly we find that the sections by the yz and xz planes are also composed of an ellipse and a circle; in the yz-plane the ellipse lies outside the circle, and in the xz-plane they intersect. Thus the wave-vector surface intersects itself, and is as shown in Fig. 54, where one octant is drawn.

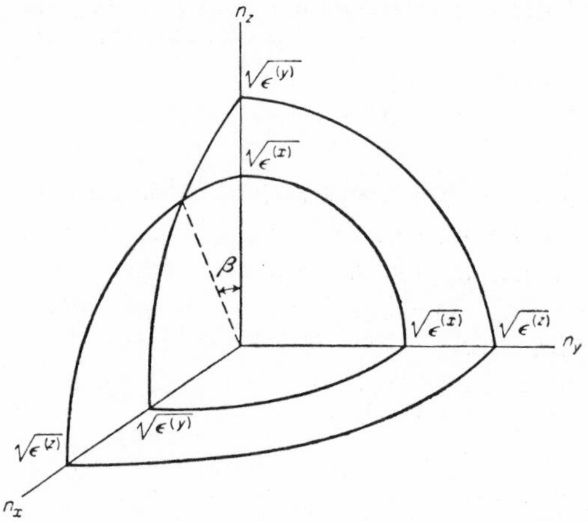

FIG. 54

This surface has four singular points of self-intersection, one in each quadrant of the xz-plane. The singular points of a surface whose equation is $f(n_x, n_y, n_z) = 0$ are given by the vanishing of all three first derivatives of the function f. Differentiating the left-hand side of (97.10), we obtain the equations

$$\left.\begin{aligned} n_x[\varepsilon^{(x)}(\varepsilon^{(y)} + \varepsilon^{(z)}) - \varepsilon^{(x)} n^2 - (\varepsilon^{(x)} n_x{}^2 + \varepsilon^{(y)} n_y{}^2 + \varepsilon^{(z)} n_z{}^2)] = 0, \\ n_y[\varepsilon^{(y)}(\varepsilon^{(x)} + \varepsilon^{(z)}) - \varepsilon^{(y)} n^2 - (\varepsilon^{(x)} n_x{}^2 + \varepsilon^{(y)} n_y{}^2 + \varepsilon^{(z)} n_z{}^2)] = 0, \\ n_z[\varepsilon^{(z)}(\varepsilon^{(x)} + \varepsilon^{(y)}) - \varepsilon^{(z)} n^2 - (\varepsilon^{(x)} n_x{}^2 + \varepsilon^{(y)} n_y{}^2 + \varepsilon^{(z)} n_z{}^2)] = 0; \end{aligned}\right\} \tag{99.4}$$

the equation (97.10) itself must, of course, be satisfied also. Since we know that the required directions of $\mathbf{n}$ lie in the xz-plane, we put $n_y = 0$, and the two remaining equations give immediately†

$$n_x^2 = \frac{\varepsilon^{(z)}(\varepsilon^{(y)}-\varepsilon^{(x)})}{\varepsilon^{(z)}-\varepsilon^{(x)}}, \qquad n_z^2 = \frac{\varepsilon^{(x)}(\varepsilon^{(z)}-\varepsilon^{(y)})}{\varepsilon^{(z)}-\varepsilon^{(x)}}. \tag{99.5}$$

The directions of these vectors **n** are inclined to the z-axis at an angle β such that

$$\frac{n_x}{n_z} = \pm\tan\beta = \pm\sqrt{\frac{\varepsilon^{(z)}(\varepsilon^{(y)}-\varepsilon^{(x)})}{\varepsilon^{(x)}(\varepsilon^{(z)}-\varepsilon^{(y)})}}. \tag{99.6}$$

This formula determines lines in two directions in the xz-plane, each of which passes through two opposite singular points and is at an angle β to the z-axis. These lines are called the *optical axes* or *binormals* of the crystal; one of them is shown dashed in Fig. 54. The directions of the optical axes are evidently the only ones for which the wave vector has only one magnitude.‡

The properties of the ray surface are entirely similar. To derive the corresponding formulae, it is sufficient to replace **n** by **s** and ε by $1/\varepsilon$. In particular, there are two *optical ray axes* or *biradials*, also lying in the xz-plane and at an angle γ to the z-axis, where

$$\tan\gamma = \sqrt{\frac{\varepsilon^{(y)}-\varepsilon^{(x)}}{\varepsilon^{(z)}-\varepsilon^{(y)}}} = \sqrt{\frac{\varepsilon^{(x)}}{\varepsilon^{(z)}}}\tan\beta. \tag{99.7}$$

Since $\varepsilon^{(x)} < \varepsilon^{(z)}$, $\gamma < \beta$.

The directions of corresponding vectors **n** and **s** are the same only for waves propagated along one of the coordinate axes (i.e. the principal dielectric axes). If **n** lies in one of the coordinate planes, **s** lies in that plane also. This rule, however, is subject to an important exception for wave vectors in the direction of the optical axes.

When the values of **n** given by (99.5) are substituted in the general formulae for **s** in terms of **n** (§97, Problem), these take the indeterminate form 0/0. The origin and meaning of this indeterminacy are quite evident from the following geometrical considerations. Near a singular point, the inner and outer parts of the wave-vector surface are cones with a common vertex. At the vertex, which is the singular point itself, the direction of the normal to the surface becomes indeterminate; and the direction of **s** as given by these formulae is just the direction of the normal. In fact the wave vector along the binormal corresponds to an infinity of ray vectors, whose directions occupy a certain conical surface, called the *cone of internal conical refraction*.§

To determine this cone of rays, we could investigate the directions of the normals near the singular point. It is more informative, however, to use a geometrical construction from the ray surface.

Fig. 55 (p. 344) shows one quadrant of the intersection of the ray surface with the xz-plane (continuous curves), and also the intersection of the wave-vector surface, on a different scale. The line OS is the biradial, and ON the binormal. Let $\mathbf{n}_N$ be the wave vector corresponding to the point N. It is easy to see that the singular point N on the wave-vector

† It is easy to see that the solution thus found is the only real solution of equations (99.4). If none of n_x, n_y, n_z is zero, the three equations (99.4) are inconsistent: they then involve only two unknowns, namely n^2 and $\varepsilon^{(x)}n_x^2+\varepsilon^{(y)}n_y^2+\varepsilon^{(z)}n_z^2$. If n_x or n_z is zero the solutions are imaginary.

‡ In the tensor ellipsoid (97.23) the binormals are the directions perpendicular to the circular sections of the ellipsoid. An ellipsoid has two such sections.

§ The phenomenon of conical refraction described below was predicted by W. R. Hamilton (1833).

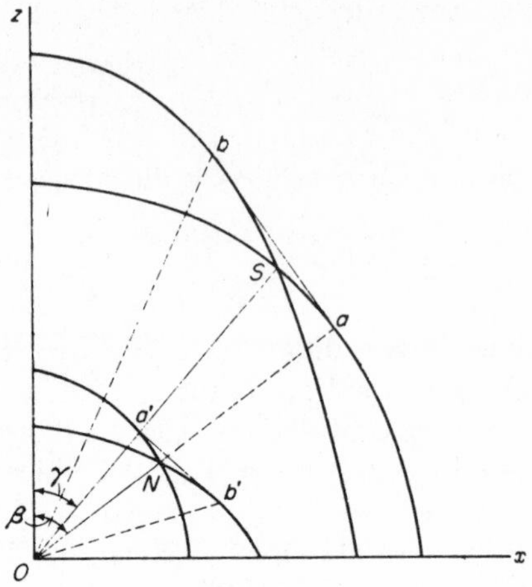

FIG. 55

surface corresponds to a singular tangent plane to the ray surface. This plane is perpendicular to ON, and touches the ray surface not at one point but along a curve, which is found to be a circle. In Fig. 55 the trace of this plane is shown by ab. This follows at once from the geometrical correspondence between the wave-vector surface and the ray surface (§97): if the tangent plane is drawn at any point $\mathbf{s}$ of the ray surface, then the perpendicular from the origin to this plane is in the same direction as the wave vector $\mathbf{n}$ corresponding to $\mathbf{s}$, and its length is $1/n$. In our case there must be an infinity of vectors $\mathbf{s}$ corresponding to the single value $\mathbf{n} = \mathbf{n}_N$; hence the points on the ray surface which represent these vectors $\mathbf{s}$ must lie in one tangent plane, which is perpendicular to $\mathbf{n}_N$. Thus in Fig. 55 the triangle Oab is the section of the cone of internal conical refraction by the xz-plane.

There is no especial difficulty in carrying out a quantitative calculation corresponding to this geometrical picture, but we shall not do so here, and give only the final formulae. The equations of the circle in which the cone of refraction cuts the ray surface are

$$\left.\begin{aligned} &(\varepsilon^{(z)}-\varepsilon^{(x)})s_y^{\,2}+\{s_x\sqrt{[\varepsilon^{(x)}(\varepsilon^{(z)}-\varepsilon^{(y)})]}-s_z\sqrt{[\varepsilon^{(z)}(\varepsilon^{(y)}-\varepsilon^{(x)})]}\}\times\\ &\qquad\times\left(s_x\sqrt{\frac{\varepsilon^{(z)}-\varepsilon^{(y)}}{\varepsilon^{(x)}}}-s_z\sqrt{\frac{\varepsilon^{(y)}-\varepsilon^{(x)}}{\varepsilon^{(z)}}}\right)=0,\\ &s_x\sqrt{[\varepsilon^{(z)}(\varepsilon^{(y)}-\varepsilon^{(x)})]}+s_z\sqrt{[\varepsilon^{(x)}(\varepsilon^{(z)}-\varepsilon^{(y)})]}=\sqrt{[\varepsilon^{(z)}-\varepsilon^{(x)}]}. \end{aligned}\right\} \quad (99.8)$$

The first of these equations is the equation of the cone of refraction if s_x, s_y, s_z are regarded as three independent variables. The second is the equation of the tangent plane to the ray surface. In particular, for $s_y = 0$ equation (99.8) gives the two equations

$$\frac{s_x}{s_z}=\sqrt{\frac{\varepsilon^{(z)}(\varepsilon^{(y)}-\varepsilon^{(x)})}{\varepsilon^{(x)}(\varepsilon^{(z)}-\varepsilon^{(y)})}},\qquad \frac{s_x}{s_z}=\sqrt{\frac{\varepsilon^{(x)}(\varepsilon^{(y)}-\varepsilon^{(x)})}{\varepsilon^{(z)}(\varepsilon^{(z)}-\varepsilon^{(y)})}},$$

which determine the directions of the extreme rays (respectively Oa and Ob in Fig. 55) in the section by the xz-plane. The former is along the binormal (cf. (99.6)), which is perpendicular to the tangent ab.

Similar results hold for the wave vectors corresponding to a given ray vector. The vector **s** along the biradial corresponds to an infinity of wave vectors, whose directions occupy the *cone of external conical refraction*. In Fig. 55 the triangle $Oa'b'$ is the section of this cone by the xz-plane. The corresponding formulae are again obtained by substituting **n** for **s** and $1/\varepsilon$ for ε in the formulae (99.8), and are

$$\left.\begin{aligned} &\varepsilon^{(y)}(\varepsilon^{(z)}-\varepsilon^{(x)})n_y^{\ 2}+[n_x\sqrt{(\varepsilon^{(z)}-\varepsilon^{(y)})}-n_z\sqrt{(\varepsilon^{(y)}-\varepsilon^{(x)})}\,]\times \\ &\qquad\times[n_x\varepsilon^{(x)}\sqrt{(\varepsilon^{(z)}-\varepsilon^{(y)})}-n_z\varepsilon^{(z)}\sqrt{(\varepsilon^{(y)}-\varepsilon^{(x)})}\,]=0, \\ &n_x\sqrt{(\varepsilon^{(y)}-\varepsilon^{(x)})}+n_z\sqrt{(\varepsilon^{(z)}-\varepsilon^{(y)})}=\sqrt{[\varepsilon^{(y)}(\varepsilon^{(z)}-\varepsilon^{(x)})]}. \end{aligned}\right\} \quad (99.9)$$

In observations of the internal conical refraction† we can use a flat plate cut perpendicular to the binormal (Fig. 56). The surface of the plate is covered by a diaphragm of small aperture, which selects a narrow beam from a plane light wave (i.e. one whose wave vector is in a definite direction) incident on the plate. The wave vector in the wave transmitted into the plate is in the direction of the binormal, and so the rays are on the cone of internal refraction. The wave vector in the wave leaving the other side of the plate is the same as in the incident wave, and so the rays are on a circular cylinder.

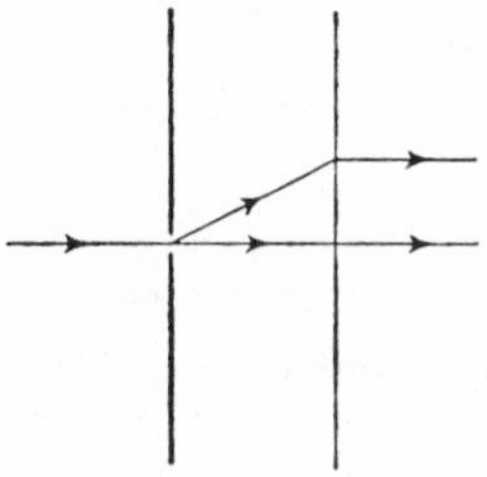

FIG. 56

To observe the external conical refraction, the plate must be cut perpendicular to the biradial, and both its surfaces must be covered by diaphragms having small apertures in exactly opposite positions. When the plate is illuminated by a convergent beam (i.e. one containing rays with all possible values of **n**), the diaphragms admit to the plate rays with **s** along the biradial, and therefore with directions of **n** occupying the surface of the cone of external conical refraction. The light leaving the second aperture is therefore on a conical surface, although this does not exactly coincide with the cone of external refraction, on account of the refraction on leaving the plate.

The laws of refraction at the surface of a biaxial crystal for an arbitrary direction of incidence are extremely complex, and we shall not pause to discuss them here,‡ but only mention that, unlike what happens for a uniaxial crystal, both refracted waves are "extraordinary" and the rays of neither lie in the plane of incidence.

As specified in §97, we are describing the optics of transparent crystals, but we may note here that there is a property of biaxial crystals that may occur when absorption is taken into account.

† We shall describe only the principle of the experiment.

‡ A detailed account of the calculations may be found in the article by G. Szivessy, *Handbuch der Physik*, vol. XX, Chapter 11, Springer, Berlin, 1928.

Let us consider a homogeneous plane wave propagated in the crystal, in which $\mathbf{n}$ is a complex vector but its real and imaginary parts are in the same direction: $\mathbf{n} = n\mathbf{v}$, where $\mathbf{v}$ is a real unit vector, $n = n(\omega)$ a complex quantity. For a given $\mathbf{v}$, the dispersion equation (97.21) can be expanded as

$$n^{-4} - n^{-2}(\eta_{11} + \eta_{22}) + \eta_{11}\eta_{22} - \eta_{12}{}^2 = 0,$$

where $\eta_{ik} \equiv \varepsilon^{-1}{}_{ik}$, and 1 and 2 are tensor suffixes in the plane perpendicular to $\mathbf{v}$. This equation quadratic in n^{-2} has a multiple root if

$$\eta_{22} - \eta_{11} = \pm 2i\eta_{12}; \tag{99.10}$$

then $n^{-2} = \frac{1}{2}(\eta_{11} + \eta_{22})$. When absorption is present, the tensor $\eta_{ik} = \eta_{ik}' + \eta_{ik}''$ is complex.

In biaxial crystals, the tensor ellipsoids of η_{ik}' and η_{ik}'' have three unequal axes; the ratios of the axes are different for the two tensors (and so are their directions, in triclinic and monoclinic crystals). Under these conditions, the two-dimensional tensors $\eta_{\alpha\beta}'$ and $\eta_{\alpha\beta}''$ cannot in general be simultaneously brought to diagonal form. The angle ϑ between the principal axes of the two tensors is a function of two independent variables, the angles which specify the direction of $\mathbf{v}$. For a given frequency ω, therefore, there can exist a one-parameter set of directions of $\mathbf{v}$ for which $\vartheta = \frac{1}{4}\pi$. With this value of ϑ, the imaginary part of the complex equation (99.10) is satisfied identically; the real part is

$$\eta_2' - \eta_1' = \mp(\eta_2'' - \eta_1''), \tag{99.11}$$

where the suffixes 1 and 2 denote the principal values of the tensors concerned.† For any choice of the x_1 and x_2 axes, equations (97.21) now give

$$D_2/D_1 = (\eta_{22} - \eta_{11})/2\eta_{12} = \pm i,$$

the two signs on the right corresponding to those in (99.10). Thus the conditions $\vartheta = \frac{1}{4}\pi$ and (99.11) together determine, for each value of ω, a particular direction of $\mathbf{v}$ in which only a wave with circular polarization of one sign, left or right according to the sign for which (99.10) is satisfied, can be propagated (W. Voigt, 1902). This direction in the crystal is called the *singular optical axis* or the *circular optical axis*.

In accordance with the general theory of linear differential equations, the second independent solution of the field equations then contains, not only the exponential factor $e^{in\mathbf{v}\cdot\mathbf{r}}$ (which includes the damping), but also the factor $a + b\mathbf{v}\cdot\mathbf{r}$ linear in the coordinates.‡ The polarization of this wave varies along the ray, but ultimately, as $\mathbf{v}\cdot\mathbf{r}$ increases, a circular polarization is established similar to that in the first wave, as is obvious if we note that in the limit concerned the substitution of the solution in the field equations involves differentiating only the exponential factor, and the difference between the two solutions then disappears.

We should again emphasize the difference between the singular axis and the case where the dispersion equation necessarily has a double root because of the symmetry of the crystal. For light propagated along the optical axis of a uniaxial crystal, the tensor $\eta_{\alpha\beta}$ has

† This is easily proved by taking the x_1 and x_2 axes along the principal axes of the tensor $\eta_{\alpha\beta}'$ and expressing the components of $\eta_{\alpha\beta}''$ in terms of its principal values.

‡ This solution has to be taken into account, for example, in problems of the reflection and refraction of light propagated along the singular axis.

the form $\eta\delta_{\alpha\beta}$, and the condition (99.10) is satisfied identically. Equations (97.21) then allow two independent solutions with different polarizations.

§100. Double refraction in an electric field

An isotropic body becomes optically anisotropic when placed in a static electric field. This anisotropy may be regarded as the result of a change in the permittivity due to the static field. Although this change is relatively very slight, it is important here because it leads to a qualitative change in the optical properties of bodies.

In this section we denote by **E** the static electric field in the body,† and expand the dielectric tensor ε_{ik} in powers of **E**. In an isotropic body in the zero-order approximation, we have $\varepsilon_{ik} = \varepsilon^{(0)}\delta_{ik}$. There can be no terms in ε_{ik} which are of the first order in the field, since in an isotropic body there is no constant vector with which a tensor of rank two linear in **E** could be constructed. The next terms in the expansion of ε_{ik} must therefore be quadratic in the field. From the components of the vector **E** we can form two symmetrical tensors of rank two, $E^2\delta_{ik}$ and E_iE_k. The former does not alter the symmetry of the tensor $\varepsilon^{(0)}\delta_{ik}$, and the addition of it amounts to a small correction in the scalar constant $\varepsilon^{(0)}$, which evidently does not result in optical anisotropy and is therefore of no interest. Thus we arrive at the following form of the dielectric tensor as a function of the field:

$$\varepsilon_{ik} = \varepsilon^{(0)}\delta_{ik} + \alpha E_iE_k, \tag{100.1}$$

where α is a scalar constant.

One of the principal axes of this tensor coincides with the direction of the electric field, and the corresponding principal value is

$$\varepsilon_{\parallel} = \varepsilon^{(0)} + \alpha E^2. \tag{100.2}$$

The other two principal values are both equal to

$$\varepsilon_{\perp} = \varepsilon^{(0)}, \tag{100.3}$$

and the position of the corresponding principal axes in a plane perpendicular to the field is arbitrary. Thus an isotropic body in an electric field behaves optically as a uniaxial crystal (the *Kerr effect*).

The change in optical symmetry in an electric field may occur in a crystal also (for example, an optically uniaxial crystal may become biaxial, and a cubic crystal may cease to be optically isotropic), and here the effect may be of the first order in the field. This linear effect corresponds to a dielectric tensor of the form

$$\varepsilon_{ik} = \varepsilon_{ik}{}^{(0)} + \alpha_{ikl}E_l, \tag{100.4}$$

where the coefficients α_{ikl} form a tensor of rank three symmetrical in the suffixes i and k. The symmetry of this tensor is the same as that of the piezoelectric tensor. The effect in question therefore occurs in the twenty crystal classes which admit piezoelectricity.

§101. Magnetic–optical effects

In the presence of a static magnetic field **H**,‡ the tensor $\varepsilon_{ik}(\omega;\, \mathbf{H})$ is no longer

† Not to be confused with the weak variable electric field of the wave.

‡ Not to be confused with the weak variable field of the electromagnetic wave.

symmetrical. The generalized principle of symmetry of the kinetic coefficients relates the components ε_{ik} and ε_{ki} in different fields:

$$\varepsilon_{ik}(\mathbf{H}) = \varepsilon_{ki}(-\mathbf{H}). \tag{101.1}$$

The condition that absorption be absent requires that the tensor should be Hermitian:

$$\varepsilon_{ik} = \varepsilon_{ki}^{*}, \tag{101.2}$$

as is seen from (96.5), but not that it should be real. Equation (101.2) implies only that the real and imaginary parts of ε_{ik} must be respectively symmetrical and antisymmetrical:

$$\varepsilon_{ik}' = \varepsilon_{ki}', \qquad \varepsilon_{ik}'' = -\varepsilon_{ki}''. \tag{101.3}$$

Using (101.1), we have

$$\left.\begin{aligned} \varepsilon_{ik}'(\mathbf{H}) &= \varepsilon_{ki}'(\mathbf{H}) = \varepsilon_{ik}'(-\mathbf{H}), \\ \varepsilon_{ik}''(\mathbf{H}) &= -\varepsilon_{ki}''(\mathbf{H}) = -\varepsilon_{ik}''(-\mathbf{H}), \end{aligned}\right\} \tag{101.4}$$

i.e. in a non-absorbing medium ε_{ik}' is an even function of **H**, and ε_{ik}'' an odd function.

The inverse tensor $\varepsilon^{-1}{}_{ik}$ evidently has the same symmetry properties, and is more convenient for use in the following calculations. To simplify the notation we shall write†

$$\varepsilon^{-1}{}_{ik} = \eta_{ik} = \eta_{ik}' + i\eta_{ik}'', \tag{101.5}$$

as already used above.

Any antisymmetrical tensor of rank two is equivalent (dual) to some axial vector; let the vector corresponding to the tensor η_{ik}'' be **G**. Using the antisymmetrical unit tensor e_{ikl}, we can write the relation between the components η_{ik}'' and G_i as

$$\eta_{ik}'' = e_{ikl}\, G_l, \tag{101.6}$$

or, in components, $\eta_{xy}'' = G_z$, $\eta_{zx}'' = G_y$, $\eta_{yz}'' = G_x$. The relation $E_i = \eta_{ik} D_k$ between the electric field and induction becomes

$$E_i = (\eta_{ik}' + ie_{ikl}\, G_l)\, D_k = \eta_{ik}'\, D_k + i\,(\mathbf{D} \times \mathbf{G})_i. \tag{101.7}$$

There is a similar linear relation

$$D_i = \varepsilon_{ik}'\, E_k + i\,(\mathbf{E} \times \mathbf{g})_i. \tag{101.8}$$

The connection between the coefficients in (101.7) and (101.8) is given by

$$\eta_{ik}' = \frac{1}{|\varepsilon|}\left\{\varepsilon'^{-1}{}_{ik}|\varepsilon'| - g_i g_k\right\}, \qquad G_i = -\frac{1}{|\varepsilon|}\varepsilon_{ik}'\, g_k, \tag{101.9}$$

where $|\varepsilon|$ and $|\varepsilon'|$ are the determinants of the tensors ε_{ik} and ε_{ik}'; cf. §22, Problem. A medium in which the relation between **E** and **D** is of this form is said to be *gyrotropic*. The vector **g** is called the *gyration vector*, and **G** the *optical activity vector*.

We shall give a general discussion of the nature of waves propagated in an arbitrary

† Of course, η_{ik}' and η_{ik}'' are not the tensors inverse to ε_{ik}' and ε_{ik}''.

gyrotropic medium, assumed anisotropic, with no restriction on the magnitude of the magnetic field.†

We take the direction of the wave vector as the z-axis. Then equations (97.21) become

$$\left(\eta_{\alpha\beta}-\frac{1}{n^2}\delta_{\alpha\beta}\right)D_\beta=\left(\eta_{\alpha\beta}'+i\eta_{\alpha\beta}''-\frac{1}{n^2}\delta_{\alpha\beta}\right)D_\beta=0, \qquad (101.10)$$

where the suffixes α, β take the values x, y. The directions of the x and y axes are taken along the principal axes of the two-dimensional tensor $\eta_{\alpha\beta}'$; and we denote the corresponding principal values of this tensor by $1/n_{01}^2$ and $1/n_{02}^2$. Then the equations become

$$\left.\begin{aligned}\left(\frac{1}{n_{01}^2}-\frac{1}{n^2}\right)D_x+iG_z D_y&=0,\\ -iG_z D_x+\left(\frac{1}{n_{02}^2}-\frac{1}{n^2}\right)D_y&=0.\end{aligned}\right\} \qquad (101.11)$$

The condition that the determinant of these equations vanishes gives an equation quadratic in n^2:

$$\left(\frac{1}{n^2}-\frac{1}{n_{01}^2}\right)\left(\frac{1}{n^2}-\frac{1}{n_{02}^2}\right)=G_z^2, \qquad (101.12)$$

whose roots give the two values of n for a given direction of $\mathbf{n}$:‡

$$\frac{1}{n^2}=\frac{1}{2}\left(\frac{1}{n_{01}^2}+\frac{1}{n_{02}^2}\right)\pm\sqrt{\left[\frac{1}{4}\left(\frac{1}{n_{01}^2}-\frac{1}{n_{02}^2}\right)^2+G_z^2\right]}. \qquad (101.13)$$

Substituting these values in equations (101.11), we find the corresponding ratios D_y/D_x:

$$\frac{D_y}{D_x}=\frac{i}{G_z}\left\{\frac{1}{2}\left(\frac{1}{n_{01}^2}-\frac{1}{n_{02}^2}\right)\mp\sqrt{\left[\frac{1}{4}\left(\frac{1}{n_{01}^2}-\frac{1}{n_{02}^2}\right)^2+G_z^2\right]}\right\}. \qquad (101.14)$$

The purely imaginary value of the ratio D_y/D_x signifies that the waves are elliptically polarized, and the principal axes of the ellipses are the x and y axes. The product of the two values of the ratio is easily seen to be unity. Thus, if in one wave $D_y = i\rho D_x$, where the real quantity ρ is the ratio of the axes of the polarization ellipse, then in the other wave $D_y = -iD_x/\rho$. This means that the polarization ellipses of the two waves have the same axis ratio, but are rotated 90° relative to each other, and the directions of rotation in them are opposite (Fig. 57, p. 350).

If the vectors $\mathbf{D}$ in the two waves are denoted by $\mathbf{D}_1$ and $\mathbf{D}_2$, these relations may be written $\mathbf{D}_1\cdot\mathbf{D}_2^* = D_{1x}D_{2x}^*+D_{1y}D_{2y}^*=0$. This is a general property of the eigenvectors on reduction to diagonal form of an Hermitian tensor (in this case, the tensor $\eta_{\alpha\beta}$).

† The medium is again assumed non-magnetic with respect to the variable field of the electromagnetic wave, i.e. $\mu_{ik}(\omega)=\delta_{ik}$. This, however, does not exclude a static field magnetizing the medium (i.e. the static permeability may differ from unity).

The properties derived for $\varepsilon_{ik}(\omega)$ are equally applicable to the tensor $\mu_{ik}(\omega)$ in a frequency range where the dispersion of the magnetic permeability is of importance.

‡ When there is no field, $\mathbf{G}=0$ and $n=n_{01}$ or n_{02}. It should be remembered, however, that when the field is present n_{01} and n_{02} in equation (101.12) are not in general the values of n for $\mathbf{H}=0$, since not only $\mathbf{G}$ but also the components η_{ik}' depend on the field.

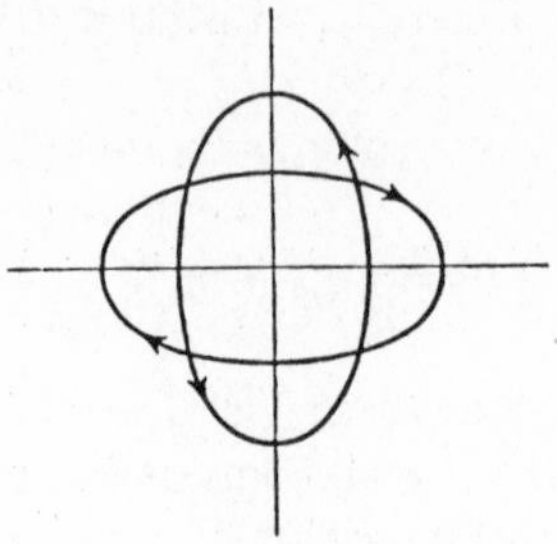

FIG. 57

The components G_i and η_{ik}' are functions of the magnetic field. If, as usually happens, the magnetic field is fairly weak, we can expand in powers of the field. The vector **G** is zero in the absence of the field, and so for a weak field we can put

$$G_i = f_{ik} H_k, \tag{101.15}$$

where f_{ik} is a tensor of rank two, in general not symmetrical. This dependence is in accordance with the general rule whereby, in a transparent medium, the components of the antisymmetrical tensor η_{ik}'' (and ε_{ik}'') must be odd functions of **H**. The symmetrical tensor components η_{ik}' are even functions of the magnetic field. The first correction terms (which do not appear in the absence of the field) in η_{ik}' are therefore quadratic in the field. When second-order quantities are neglected, formulae (101.9) reduce to the simpler form

$$\eta_{ik}' = \varepsilon'^{-1}{}_{ik}.$$

In the general case of an arbitrarily directed wave vector, the magnetic field has little effect on the propagation of light in the crystal, causing only a slight ellipticity of the oscillations, with an axis ratio of the polarization ellipse which is small (of the first order with respect to the field).

The directions of the optical axes (and neighbouring directions) form an exception. The two values of n are equal in the absence of the field. The roots of equation (101.12) then differ from these values by first-order quantities,† and the resulting effects are analogous to those in isotropic bodies, which we shall now consider.

The magnetic–optical effect in isotropic bodies (and in crystals of the cubic system) is of particular interest on account of its nature and its comparatively large magnitude.

Neglecting second-order quantities, we have $\eta_{ik}' = \varepsilon^{-1}\delta_{ik}$, where ε is the permittivity of the isotropic medium in the absence of the magnetic field. The relation between **D** and **E** is

$$\mathbf{E} = \frac{1}{\varepsilon}\mathbf{D} + i\mathbf{D}\times\mathbf{G}, \quad \mathbf{D} = \varepsilon\mathbf{E} + i\mathbf{E}\times\mathbf{g}; \tag{101.16}$$

in the same approximation, the vectors **g** and **G** are related by

$$\mathbf{G} = -\mathbf{g}/\varepsilon^2. \tag{101.17}$$

† It should be noticed that the two roots of (101.12) do not become equal. The geometrical significance of this is that the two parts of the wave-vector surface are separated.

The dependence of **g** (or **G**) on the external field reduces in an isotropic medium to simple proportionality:

$$\mathbf{g} = f\mathbf{H}, \tag{101.18}$$

in which the scalar constant f may be either positive or negative.

In equation (101.12) we now have $n_{01} = n_{02} \equiv n_0 = \sqrt{\varepsilon}$, the refractive index in the absence of the field. Hence $1/n^2 = \mp G_z + 1/n_0{}^2$ or, to the same accuracy,

$$n_{\mp}{}^2 = n_0{}^2 \pm n_0{}^4 G_z = n_0{}^2 \mp g_z. \tag{101.19}$$

Since the z-axis is in the direction of **n**, we can write this formula, to the same accuracy, in the vector form

$$\left(\mathbf{n} \pm \frac{1}{2n_0}\mathbf{g}\right)^2 = n_0{}^2. \tag{101.20}$$

Hence we see that the wave-vector surface in this case consists of two spheres with radius n_0, whose centres are at distances $\pm g/2n_0$ from the origin in the direction of **G**.

A different polarization of the wave corresponds to each of the two values of n: we have

$$D_x = \mp i D_y, \tag{101.21}$$

where the signs correspond to those in (101.19). The equality of the magnitudes of D_x and D_y, and their phase difference of $\mp\frac{1}{2}\pi$, signify a circular polarization of the wave, with the direction of rotation of the vector **D** respectively anticlockwise and clockwise looking along the wave vector (or, to use the customary expressions, with *right-hand* and *left-hand* polarization respectively).

The difference between the refractive indices in the left-hand and right-hand polarized waves has the result that two circularly polarized refracted waves are formed at the surface of a gyrotropic body. This phenomenon is called *double circular refraction*.

Let a linearly polarized plane wave be incident normally on a slab of thickness l. We take the direction of incidence as the z-axis, and that of the vector **E** (= **D**) in the incident wave as the x-axis. The linear oscillation can be represented as the sum of two circular oscillations with opposite directions of rotation, which are then propagated through the slab with different wave numbers $k_{\pm} = \omega n_{\pm}/c$. Arbitrarily taking the wave amplitude as unity, we have $D_x = \frac{1}{2}[\exp(ik_+ z) + \exp(ik_- z)]$, $D_y = \frac{1}{2}i[-\exp(ik_+ z) + \exp(ik_- z)]$, or, putting $k = \frac{1}{2}(k_+ + k_-)$ and $\kappa = \frac{1}{2}(k_+ - k_-)$,

$$D_x = \tfrac{1}{2}e^{ikz}(e^{i\kappa z} + e^{-i\kappa z}) = e^{ikz}\cos\kappa z,$$

$$D_y = \tfrac{1}{2}ie^{ikz}(-e^{i\kappa z} + e^{-i\kappa z}) = e^{ikz}\sin\kappa z.$$

When the wave leaves the slab we have

$$D_y/D_x = \tan\kappa l = \tan(l\omega g/2cn_0). \tag{101.22}$$

Since this ratio is real, the wave remains linearly polarized, but the direction of polarization is changed (the *Faraday effect*). The angle through which the plane of polarization is rotated is proportional to the path traversed by the wave; the angle per unit length in the direction of the wave vector is

$$(\omega g/2cn_0)\cos\theta, \tag{101.23}$$

where θ is the angle between **n** and **g**.

It should be noticed that, when the direction of the magnetic field is given, the direction of rotation of the plane of polarization (with respect to the direction of **n**) is reversed (left-hand becoming right-hand, and vice versa) when the sign of **n** is changed. If the ray traverses the same path twice in opposite directions, the total rotation of the plane of polarization is therefore double the value resulting from a single traversal.

For $\theta = \frac{1}{2}\pi$ (the wave vector perpendicular to the magnetic field), the effect linear in the field given by formulae (101.19) disappears, in accordance with the general rule stated above that only the component of **g** in the direction of **n** affects the propagation of light. For angles θ close to $\frac{1}{2}\pi$ we must therefore take account of the terms proportional to the square of the field, and in particular these terms must be included in the tensor η_{ik}'. By virtue of the axial symmetry about the direction of the field, two principal values of the symmetrical tensor η_{ik}' are equal, as for a uniaxial crystal. We shall take the x-axis in the direction of the field, and denote by $\eta_\parallel$ and $\eta_\perp$ the principal values of η_{ik}' in the directions parallel and perpendicular to the magnetic field. The difference $\eta_\parallel - \eta_\perp$ is proportional to H^2.

Let us consider the purely quadratic effect (called the *Cotton–Mouton effect*) which occurs when **n** and **g** are perpendicular. In equations (101.11) and (101.12) we have $G_z = 0$, and $1/n_{01}^2$, $1/n_{02}^2$ are respectively $\eta_\parallel, \eta_\perp$. Thus in one wave we have $1/n^2 = \eta_\parallel$, $D_y = 0$; this wave is linearly polarized, and the vector **D** is parallel to the x-axis. In the other wave $1/n^2 = \eta_\perp$, $D_x = 0$, i.e. **D** is parallel to the y-axis. Let linearly polarized light be incident normally on a slab in a magnetic field parallel to its surface. The two components in the slab (with vectors **D** in the xz and yz planes) are propagated with different values of n. Consequently the light leaving the slab is elliptically polarized.

Lastly, let us consider one other peculiar effect that occurs in a medium whose optical activity vector (101.15) is linear in the (static) magnetic field, namely the magnetization of a non-magnetic transparent medium by a variable electric field (L. P. Pitaevskiĭ, 1960).

We start from the general formula (31.6)

$$-\mathbf{B}/4\pi = \partial \tilde{U}/\partial \mathbf{H},$$

and take account of the contribution to $\tilde{U}$ from the variable electric field, which is given by (80.11). According to the theorem of small increments to thermodynamic quantities, the change $\delta\tilde{U}$ in this contribution when the permittivity changes by a small amount is (expressed in terms of the appropriate variables) the same as the change $\delta\tilde{F}$ in the free energy. For the latter we can use formula (14.1), with an obvious generalization to anisotropic media; the fact that this formula remains valid for a variable field (not a static field as in §14) in a dispersive transparent medium has been mentioned in §81.† We thus have

$$\begin{aligned} \delta\tilde{U} &= -\delta\varepsilon_{ik} E_i E_k^*/16\pi \\ &= \delta\eta_{ik} D_i D_k^*/16\pi; \end{aligned} \tag{101.24}$$

the extra factor $\frac{1}{2}$ takes account of the complex representation of **E**. The second equation (101.24) follows because the definition $\varepsilon_{il}\eta_{lk} = \delta_{ik}$ gives $\varepsilon_{il}\delta\eta_{lk} = -\eta_{lk}\delta\varepsilon_{il}$.‡

---

† The tilde above the symbol U refers here to the magnetic variables, not the electric ones. To simplify the notation, we omit the sign of time averaging from $\tilde{U}$.

‡ To derive (101.24) directly, it would be necessary to consider a dielectric-filled resonator instead of the oscillatory circuit discussed in §81. By calculating the change in frequency due to a small change in the permittivity (cf. §90, Problem 4) and using the adiabatic invariant theorem, we find the change in the resonator energy.

Now regarding the permittivity variation as the result of changing the static magnetic field, we write

$$-\frac{\mathbf{B}}{4\pi} = \frac{\partial \tilde{U}_0}{\partial \mathbf{H}} + \frac{\partial \eta_{ik}}{\partial \mathbf{H}} \frac{D_i D_k^*}{16\pi},$$

where $\tilde{U}_0$ refers to the medium in the absence of the electric field. If the medium itself is non-magnetic ($\mu = 1$), then $\partial \tilde{U}_0/\partial \mathbf{H} = -\mathbf{H}/4\pi$. The magnetization $\mathbf{M} = (\mathbf{B}-\mathbf{H})/4\pi$ is then

$$\mathbf{M} = -\frac{\partial \eta_{ik}}{\partial \mathbf{H}} \frac{D_i D_k^*}{16\pi}.$$

In the absence of the external magnetic field, the derivative $\partial \eta_{ik}/\partial \mathbf{H}$ is to be taken at $\mathbf{H} = 0$. With η_{ik} from (101.6), and (101.15), we obtain finally the following expression for the magnetization due to the variable electric field:

$$M_l = -(i/16\pi) e_{ikm} f_{ml} D_i D_k^*; \qquad (101.25)$$

this is quadratic in the electric field. If the medium is isotropic in the absence of the magnetic field, then $f_{ml} = f\delta_{ml}$ and

$$\mathbf{M} = -if\,\mathbf{D}\times\mathbf{D}^*/16\pi. \qquad (101.26)$$

For a linearly polarized field, the vector **D** can differ from a real quantity only by a phase factor; then **D** and **D*** are collinear, and (101.25) or (101.26) is zero. There is thus a magnetization only in the presence of a rotating electric field. This effect is in a sense the opposite of the rotation of the polarization plane in a magnetic field, and is expressed in terms of the same tensor f_{ik}; it is therefore called the *inverse Faraday effect*.

PROBLEMS

PROBLEM 1. Show by direct calculation that the direction of the (time) averaged Poynting vector in a wave propagated in a transparent gyrotropic medium is the same as that of the group velocity.

SOLUTION. According to (59.9a),

$$\overline{\mathbf{S}} = c \,\mathrm{re}\, \mathbf{E}^*\times\mathbf{H}/8\pi,$$

E and **H** being expressed in complex form. Proceeding as in the derivation of (97.16), we multiply equations (97.15) by **E*** and **H*** respectively:

$$\mathbf{E}^*\cdot\delta\mathbf{D} = \mathbf{H}^*\cdot\delta\mathbf{H} + (\mathbf{E}^*\times\mathbf{H})\cdot\delta\mathbf{n},$$

$$\mathbf{H}^*\cdot\delta\mathbf{H} = \mathbf{D}^*\cdot\delta\mathbf{E} + (\mathbf{E}\times\mathbf{H}^*)\cdot\delta\mathbf{n}.$$

Adding these and noting that $\mathbf{E}^*\cdot\delta\mathbf{D} = \mathbf{D}^*\cdot\delta\mathbf{E}$, since the tensor ε_{ik} is Hermitian, we find the required result:

$$\delta\mathbf{n}\cdot\mathrm{re}\,(\mathbf{E}^*\times\mathbf{H}) = 0.$$

PROBLEM 2. Determine the directions of the rays when a ray incident from a vacuum is refracted at the surface of an isotropic body in a magnetic field.

SOLUTION. The direction of the ray vector **s** is given by the normal to the wave-vector surface. Differentiating the left-hand side of equation (101.20) with respect to the components of the vector **n**, we find that **s** is proportional to $\mathbf{n} \pm \mathbf{g}/2n_0$. The square of the latter expression is n_0^2, and so the unit vector in the direction of the ray is given by

$$\frac{\mathbf{s}}{s} = \frac{1}{n_0}\left(\mathbf{n} \pm \frac{1}{2n_0}\mathbf{g}\right). \qquad (1)$$

Let the angle of incidence be θ. The refracted rays do not in general lie in the plane of incidence, and their directions are given by the angle θ' to the normal to the surface and the azimuth ϕ' measured from the plane of incidence. We take the latter as the xz-plane, with the z-axis perpendicular to the surface. The components n_x and n_y of the wave vector are unaltered by refraction. In the incident ray they are $n_x = \sin\theta$, $n_y = 0$. Substituting these values in (1), we find the x and y components of the unit vector $\mathbf{s}/s$, which give immediately the directions of the refracted rays:

$$\sin\theta' \cos\phi' = \frac{1}{n_0}\sin\theta \pm \frac{1}{2n_0^2} g_x,$$

$$\sin\theta' \sin\phi' = \pm \frac{1}{2n_0^2} g_y.$$

When the angle of incidence is not small, the azimuth ϕ' is small, and we can write

$$\phi' = \pm g_y/2n_0 \sin\theta,$$

$$\sin\theta' = \frac{\sin\theta}{n_0} \pm \frac{g_x}{2n_0^2}.$$

For normal incidence ($\theta = 0$) we take the xz-plane through the vector $\mathbf{g}$; then $\phi' = 0$, and $\theta' \cong \sin\theta' = \pm g_x/2n_0^2$. Although this formula does not involve g_z, it is not valid if $g_z = 0$, since the approximation linear in the field is inadequate when $\mathbf{n}$ and $\mathbf{g}$ are perpendicular.

PROBLEM 3. Determine the polarization of the reflected light when a linearly polarized wave is incident normally from a vacuum on the surface of a body rendered anisotropic by a magnetic field.

SOLUTION. For normal incidence the direction of the wave vector is unaltered by the passage of the wave into the medium. In all three waves (incident, refracted and reflected) the vectors $\mathbf{H}$ are therefore parallel to the surface (the xy-plane). The electric vector $\mathbf{E}$ in the incident and reflected waves is also parallel to the xy-plane; in the refracted wave $E_z \neq 0$, but the relation between the x and y components of $\mathbf{E}$ and $\mathbf{H}$ is the same as in an isotropic body ($H_x = -nE_y$, $H_y = nE_x$). If the polarization of the incident wave is the same as that of one of the two types of wave which can be propagated in the anisotropic (refracting) medium concerned, with the given direction of $\mathbf{n}$, then there is only one refracted wave, which has this polarization. The problem is then formally identical with that of reflection from an isotropic body, and the fields $\mathbf{E}_1$ and $\mathbf{E}_0$ in the reflected and incident waves are related by

$$\mathbf{E}_1 = (1-n)\,\mathbf{E}_0/(1+n), \tag{1}$$

where n is the refractive index corresponding to this polarization.

The linear polarization can be regarded as resulting from the superposition of two circular polarizations with opposite directions of rotation. If $\mathbf{E}_0$ in the incident wave is in the x-direction, we put $\mathbf{E}_0 = \mathbf{E}_0{}^+ + \mathbf{E}_0{}^-$, where $E_0{}^+{}_x = iE_0{}^+{}_y = \frac{1}{2}E_0$, $E_0{}^-{}_x = -iE_0{}^-{}_y = \frac{1}{2}E_0$. Using formula (1) for each wave, with $n_\pm$ given by (101.19), we obtain

$$E_{1x} = \tfrac{1}{2}E_0\left[\frac{1-n_+}{1+n_+} + \frac{1-n_-}{1+n_-}\right] \cong E_0\frac{1-n_0}{1+n_0},$$

$$E_{1y} = \tfrac{1}{2}iE_0\left[\frac{1-n_-}{1+n_-} - \frac{1-n_+}{1+n_+}\right] \cong iE_0\frac{g\cos\theta}{n_0(1+n_0)^2},$$

where θ is the angle between the direction of incidence and the vector $\mathbf{g}$. Hence we see that the reflected wave is elliptically polarized, the major axis of the ellipse being in the x-direction, and the ratio of the minor and major axes being $(g\cos\theta)/n_0(n_0^2 - 1)$.

PROBLEM 4. Determine the limiting form of the frequency dependence of the gyration vector at high frequencies.

SOLUTION. The calculations are similar to those in §78, except that the electron equation of motion must include the Lorentz force due to the static external magnetic field $\mathbf{H}$:

$$m\frac{d\mathbf{v}'}{dt} = e\,\mathbf{E}_0 e^{-i\omega t} + e\mathbf{v}' \times \mathbf{H}/c,$$

where $e = -|e|$ is the electron charge. If $\omega \gg |e|H/mc$, this equation can be solved by successive approximations. As far as terms of the first order in $\mathbf{H}$ we have

$$\mathbf{v}' = \frac{ie}{m\omega}\mathbf{E} - \frac{e^2}{m^2\omega^2 c}\mathbf{E}\times\mathbf{H},$$

and the induction is then

$$\mathbf{D} = \varepsilon(\omega)\,\mathbf{E} + if(\omega)\,\mathbf{E}\times\mathbf{H},$$

where $\varepsilon(\omega)$ is given by (78.1) and $f(\omega) = -4\pi Ne^3/cm^2\omega^3 = (|e|/2mc)\,d\varepsilon/d\omega$ (H. Becquerel, 1897).

§102. Mechanical–optical effects

Besides the electric–optical and magnetic–optical effects, there are other ways in which the optical symmetry of a medium can be changed by external agencies. These include, first of all, the effect of elastic deformations on the optical properties of solids. In particular, such deformations may render an isotropic solid body optically anisotropic. Such phenomena are described by the inclusion in $\varepsilon_{ik}(\omega)$ of additional terms proportional to the components of the strain tensor. The corresponding formulae are exactly the same as (16.1) and (16.6) for the static permittivity, except that the coefficients are now functions of frequency. In the deformation of an isotropic body, for example, we have

$$\varepsilon_{ik} = \varepsilon^{(0)}\delta_{ik} + a_1 u_{ik} + a_2 u_{ll}\delta_{ik}. \qquad (102.1)$$

The coefficients $a_1(\omega)$ and $a_2(\omega)$ are called *elastic–optical constants.*

Another case is the occurrence of optical anisotropy in a non-uniformly moving fluid. The corresponding general expression for the dielectric tensor is

$$\varepsilon_{ik} = \varepsilon^{(0)}{}_{ik} + \lambda_1\left(\frac{\partial v_k}{\partial x_i} + \frac{\partial v_i}{\partial x_k}\right) + \tfrac{1}{2}i\lambda_2\left(\frac{\partial v_k}{\partial x_i} - \frac{\partial v_i}{\partial x_k}\right), \qquad (102.2)$$

and represents the first terms in an expansion of ε_{ik} in powers of the derivatives of the velocity. The condition that absorption be absent (ε_{ik} is Hermitian) means that $\lambda_1(\omega)$ and $\lambda_2(\omega)$ must be real; $\varepsilon^{(0)}(\omega)$ is the permittivity of the fluid at rest. In an incompressible fluid $\partial v_l/\partial x_l \equiv \operatorname{div}\mathbf{v} = 0$, and the last two terms in (102.2) give zero on contraction.

To investigate the electromagnetic properties of the moving fluid, we have to combine the formulae (76.9)–(76.11) for the electrodynamics of moving dielectrics (with a velocity $\mathbf{v}$ that depends on the coordinates) with (102.2). Here, however, the terms which contain both the velocity and its derivatives are to be neglected, as being outside the accuracy of the formulae.

The second and third terms in (102.2) are respectively symmetrical and antisymmetrical in the suffixes i,k. For uniform rotation of the fluid we have $\mathbf{v} = \mathbf{\Omega}\times\mathbf{r}$, where $\mathbf{\Omega}$ is the angular velocity of rotation, and the symmetrical term is zero. The antisymmetrical term is $i\lambda_2 e_{ikl}\Omega_l$, so that the medium becomes gyrotropic, with gyration vector

$$\mathbf{g} = \lambda_2\mathbf{\Omega}. \qquad (102.3)$$

The quantity λ_2 contains contributions from two effects: the dispersion of the permittivity, and the influence of Coriolis forces on it.

In a frame of reference moving with a given element of the fluid, the amplitude $\mathbf{E}_0$ of a monochromatic wave (in the laboratory frame) rotates with angular velocity $-\mathbf{\Omega}$, i.e. becomes a function of time satisfying the equation

$$\partial\mathbf{E}_0/\partial t = -\mathbf{\Omega}\times\mathbf{E}_0.$$

In this sense the wave becomes quasi-monochromatic, and the relation between **D** and **E** is given by

$$\mathbf{D} = \varepsilon(\omega)\mathbf{E} + i\frac{d\varepsilon(\omega)}{d\omega}\frac{\partial \mathbf{E}_0}{\partial t}e^{-i\omega t}; \qquad (102.4)$$

the derivation is the same as that of (80.10), except that here $f(\omega) = \varepsilon(\omega)$. Substituting the value of $\partial \mathbf{E}_0/\partial t$ and comparing the result with the definition of the gyration vector $\mathbf{g}$ in (101.16), we find that the dispersion makes a contribution $d\varepsilon^{(0)}/d\omega$ to λ_2 (M. A. Player, 1976).

If we now put

$$\lambda_2 = \lambda_2^{(C)} + d\varepsilon^{(0)}/d\omega, \qquad (102.5)$$

then $\lambda_2^{(C)}$ is due only to the Coriolis forces, which are linear in $\mathbf{\Omega}$.

In a rotating frame of reference, the Hamiltonian of the system is

$$\hat{\mathscr{H}}' = \hat{\mathscr{H}} - \hat{\mathscr{M}}_{\text{mech}} \cdot \mathbf{\Omega},$$

where $\hat{\mathscr{H}}$ and $\hat{\mathscr{M}}_{\text{mech}}$ are the ordinary operators of the energy and angular momentum of the system (see *SP* 1, §34); the permittivity of a rotating medium must in principle be calculated from this Hamiltonian. The expression is, however, analogous to the Hamiltonian of a system in a magnetic field, written as far as the terms linear in $\mathbf{H}$:

$$\hat{\mathscr{H}} = \hat{\mathscr{H}}_0 - \mathscr{M} \cdot \mathbf{H},$$

where $\mathscr{M}$ is the magnetic moment operator (see *QM*, §113). The analogy becomes exact if the contribution to the permittivity in the frequency range considered arises only from the orbital motion of the electrons in the atoms. Then $\mathscr{M} = (e/2mc)\hat{\mathscr{M}}_{\text{mech}}$, where $e = -|e|$ is the electron charge, and the two Hamiltonians differ only in the replacement of $\mathbf{\Omega}$ by $e\mathbf{H}/2mc$. In this case, therefore, it is clear that

$$\lambda_2^{(C)}(\omega) = 2mcf(\omega)/e, \qquad (102.6)$$

where $f(\omega)$ is given by (101.18) (N. B. Baranova and B. Ya. Zel'dovich, 1978).†

The effects due to the coefficient λ_1 are significant in such systems as suspensions and colloidal solutions of anisotropically shaped particles. In this case the effect is due to the orienting of particles suspended in the fluid by the action of the velocity gradients. Since a uniform rotation cannot orient the particles, it follows that in this case $\lambda_2 \ll \lambda_1$ and the last term in (102.2) may be omitted. The effect given by the λ_1 term is called the *Maxwell effect*.

Lastly, it may be noted that the λ_1 term in (102.2) does not satisfy the generalized principle of symmetry of the kinetic coefficients, which would imply that we have $\varepsilon_{ik}(\omega; \mathbf{v}) = \varepsilon_{ki}(\omega; -\mathbf{v})$, since $\mathbf{v}$ changes sign under time reversal. This is not necessary, however. The reason is that the derivation of the principle presupposes that the processes described by the coefficients in question are the only cause of energy dissipation in the system. In the present case, however, besides the dissipation in the variable electromagnetic field in the wave, there is another cause of dissipation, which is unrelated to the field, namely the internal friction in the non-uniform fluid stream. As regards the theory of generalized

† Here it should be emphasized that a non-zero $\lambda_2^{(C)}$ may exist even in classical (not quantum) theory. The familiar proposition that, in the classical theory, the thermodynamic properties of bodies are independent of the Coriolis forces in uniform rotation (see *SP* 1, §34) applies only to the properties in statistical equilibrium. The permittivity when $\omega \neq 0$ relates to non-equilibrium kinetic properties.

susceptibilities, the λ_1 term describes the response of the system to a non-linear interaction, the contribution to the induction from the field **E** and the velocity gradients jointly.† A uniform rotation of the whole fluid does not involve any additional dissipation; the λ_2 term in (102.2), which is found even with this kind of rotation, therefore satisfies the symmetry principle: $\varepsilon_{ik}(\omega; \mathbf{\Omega}) = \varepsilon_{ki}(\omega; -\mathbf{\Omega})$.

PROBLEM

Determine the rotation of the plane of polarization for a wave propagated parallel to the axis of a rotating dielectric body.

SOLUTION. The problem amounts to determining the gyration vector, which is the sum of two parts: the contribution (102.3) from the change in the permittivity and its dispersion, and the "kinematic" part due to the presence of the velocity in the relations (76.10), (76.11); the latter part is to be calculated.

In Maxwell's equations

$$\mathbf{curl\,E} = i\omega\mathbf{B}/c, \quad \mathbf{curl\,H} = -i\omega\mathbf{D}/c, \quad \mathrm{div}\,\mathbf{B} = 0, \quad \mathrm{div}\,\mathbf{D} = 0, \tag{1}$$

we express **E** and **B** in terms of **D** and **H** according to (76.10) and (76.11) with $\mu = 1$, take the curl of the first equation, and use the remaining equations. The result is

$$\triangle\mathbf{D} + \frac{\varepsilon\omega^2}{c^2}\mathbf{D} + \frac{\varepsilon - 1}{c}\mathbf{curl\,curl}\,(\mathbf{v}\times\mathbf{H}) + \frac{i\omega(\varepsilon - 1)}{c^2}\mathbf{curl}\,(\mathbf{D}\times\mathbf{v}) = 0; \tag{2}$$

here we write ε in place of $\varepsilon^{(0)}$ as used in the text of §102. Since all formulae are valid only as far as the first order in **v**, the higher-order terms are omitted.

The last two terms in (2) give the required effect. We expand them and substitute $\mathbf{v} = \mathbf{\Omega}\times\mathbf{r}$, obtaining

$$\mathbf{curl}\,(\mathbf{v}\times\mathbf{H}) = -\mathbf{H}\times\mathbf{\Omega}, \quad \mathbf{curl}\,(\mathbf{D}\times\mathbf{v}) = \mathbf{D}\times\mathbf{\Omega}.$$

When these differentiations have been carried out, the coordinate dependence of all remaining quantities reduces to factors $e^{i\mathbf{k}\cdot\mathbf{r}}$, with $\mathbf{k}\parallel\mathbf{\Omega}$ according to the condition stated. Lastly, since in the zero-order approximation with respect to **v** we have $\mathbf{H} = c\mathbf{k}\times\mathbf{E}/\omega$, $k^2 = \varepsilon\omega^2/c^2$, equation (2) becomes

$$\triangle\mathbf{D} + \frac{\varepsilon\omega^2}{c^2}\mathbf{D} + 2i\omega\frac{\varepsilon - 1}{c^2}\mathbf{D}\times\mathbf{\Omega} = 0$$

or

$$\left(\frac{1}{{n_0}^2} - \frac{1}{n^2}\right)\mathbf{D} - \frac{2i(\varepsilon - 1)}{\omega\varepsilon^2}\mathbf{D}\times\mathbf{\Omega} = 0, \tag{3}$$

where ${n_0}^2 = \varepsilon$ and n is the refractive index in the rotating body. Comparison of (3) with (101.11) and (101.17) shows that the "kinematic" contribution to the gyration vector is $2(\varepsilon - 1)\mathbf{\Omega}/\omega$ (E. Fermi, 1923). The rotation of the plane of polarization of the wave is given by the total vector

$$\mathbf{g} = \left\{\frac{2(\varepsilon - 1)}{\omega} + \frac{d\varepsilon}{d\omega} + \lambda_2^{(C)}\right\}\mathbf{\Omega}.$$

In the high-frequency limit, when the atomic electrons in the substance may be regarded as free, ε is given by (78.1), the first two terms in the brackets cancelling.

† There exist, in principle, other such effects. For instance, in a conducting medium with no centre of symmetry, there can exist in ε_{ik} pseudotensor terms of the form $\delta_{ik}\mathbf{E}\cdot\mathbf{H}$ or $H_iE_k + H_kE_i$, where **E** and **H** are the static external fields (N. B. Baranova, Yu. V. Bogdanov and B. Ya. Zel'dovich, 1977). It is important to note that these terms, which formally violate the symmetry principle, can occur only in a conducting medium, where the static electric field causes additional dissipation.

CHAPTER XII

SPATIAL DISPERSION

§103. Spatial dispersion

So FAR, in discussing the dielectric properties of matter, we have assumed that the induction $\mathbf{D}(t, \mathbf{r})$ is determined by the electric field $\mathbf{E}(t', \mathbf{r})$ at the same point $\mathbf{r}$ in space, though (in the presence of dispersion) not only at the same time t but at all previous times $t' \leqslant t$. This assumption is not always valid. In general, the value of $\mathbf{D}(t, \mathbf{r})$ depends on those of $\mathbf{E}(t', \mathbf{r}')$ in some region of space about the point $\mathbf{r}$. The linear relation between $\mathbf{D}$ and $\mathbf{E}$ is then written in a form which generalizes (77.3):

$$D_i(t, \mathbf{r}) = E_i(t, \mathbf{r}) + \int_0^{\infty}\int f_{ik}(\tau; \mathbf{r}, \mathbf{r}')E_k(t-\tau, \mathbf{r}')\,dV'\,d\tau; \qquad (103.1)$$

here it has been written in the form applicable to an anisotropic medium. This *non-local* relation is a result of what is called *spatial dispersion*, in which connection the ordinary dispersion discussed in §77 is called *time dispersion* or *frequency dispersion.* For monochromatic field components, whose dependence on t is given by the factor $e^{-i\omega t}$, the relation becomes

$$D_i(\mathbf{r}) = E_i(\mathbf{r}) + \int f_{ik}(\omega; \mathbf{r}, \mathbf{r}')E_k(\mathbf{r}')\,dV'. \qquad (103.2)$$

It may be noted immediately that in the majority of cases spatial dispersion is much less important than time dispersion. The reason is that for ordinary dielectrics the kernel f_{ik} of the integral operator decreases considerably even at distances $|\mathbf{r}-\mathbf{r}'|$ that are large in comparison with the atomic dimensions a. In this chapter we shall consider, as hitherto, macroscopic fields averaged over physically infinitesimal volume elements. These fields must by definition vary only slightly over distances $\sim a$. As a first approximation we can then take $\mathbf{E}(\mathbf{r}') \cong \mathbf{E}(\mathbf{r})$ outside the integration over dV' in (103.1), and thus return to (77.3). In such cases the spatial dispersion manifests itself only in the form of small corrections. These, however, may lead (as we shall see) to qualitatively new physical phenomena and may therefore be significant.

A different situation may arise in conducting media (metals, electrolyte solutions, or plasmas): the motion of free current-carriers causes non-locality over distances which may be much greater than atomic dimensions. There may then be important spatial dispersion even in the macroscopic theory.†

† For an isotropic conducting medium, the conditions for spatial dispersion to be negligible may be different for the transverse and longitudinal permittivities. For the former, the characteristic distance r_0 over which the kernel of the integral in (103.2) is non-zero is the smaller of v/ω and l, where v is the mean velocity of the carriers and l their mean free path. For the longitudinal permittivity, r_0 is the smaller of v/ω and $(lv/\omega)^{\frac{1}{2}}$, the latter being the distance travelled by the carriers by diffusion along the field in a time $\sim 1/\omega$; the diffusion coefficient $D \sim lv$. The spatial dispersion is unimportant if $kr_0 \ll 1$.

Another manifestation of spatial dispersion is the *Doppler broadening* of absorption lines in a gas. If an atom at rest has at the frequency ω_0 an absorption line with negligible width, then for a moving atom this frequency is shifted by the Doppler effect through a distance $\mathbf{k}\cdot\mathbf{v}$, where $\mathbf{v}$ is the velocity of the atom ($v \ll c$). This causes the absorption spectrum of the gas as a whole to contain a line with width $\Delta\omega \sim kv_T$, where v_T is the mean thermal velocity of the atoms. In turn, this broadening means that the permittivity of the gas has an important spatial dispersion when $k \gtrsim |\omega-\omega_0|/v_T$.

The following comment is needed regarding the form in which (103.1) is written. No arguments based on symmetry (in space or time) can exclude the possible electric polarization of a dielectric in a non-uniform variable magnetic field. It may therefore be asked whether a magnetic-field term should be added to the right-hand side of (103.1) or (103.2). This is not necessary, however. The reason is that the fields **E** and **B** cannot be regarded as completely independent. They are related (in the monochromatic case) by the equation **curl E** $= i\omega\mathbf{B}/c$. By virtue of this equation, we can regard the dependence of **D** on **B** as a dependence on the spatial derivatives of **E**, i.e. as one manifestation of non-locality.

When spatial dispersion is taken into account, it is often appropriate, and does not affect the generality of the theory, to write Maxwell's equations in the form

$$\mathbf{curl\,E} = -\frac{1}{c}\frac{\partial\mathbf{B}}{\partial t}, \qquad \mathrm{div}\,\mathbf{B} = 0, \tag{103.3}$$

$$\mathbf{curl\,B} = \frac{1}{c}\frac{\partial\mathbf{D}}{\partial t}, \qquad \mathrm{div}\,\mathbf{D} = 0, \tag{103.4}$$

without using **H** in addition to the mean magnetic field $\overline{\mathbf{h}} = \mathbf{B}$. Instead, all terms which result from averaging the microscopic currents are included in the definition of **D**. The previous separation of the mean current into two parts as in (79.3) is in general not unique. In the absence of spatial dispersion, it is determined by the condition that **P** should be the electric polarization locally related to **E**. When there is no such relation, it is more convenient to put $\mathbf{M} = 0$, $\mathbf{B} = \mathbf{H}$, and

$$\overline{\rho\mathbf{v}} = \partial\mathbf{P}/\partial t, \tag{103.5}$$

corresponding to the form (103.3), (103.4) of Maxwell's equations.†

The components of the tensor $f_{ik}(\omega; \mathbf{r}, \mathbf{r}')$ (the kernel of the integral operator in (103.2)) satisfy the symmetry relations

$$f_{ik}(\omega; \mathbf{r}, \mathbf{r}') = f_{ki}(\omega; \mathbf{r}', \mathbf{r}). \tag{103.6}$$

This follows from the same arguments as were given in §96 for the tensor $\varepsilon_{ik}(\omega)$. The only difference is that the interchange of the suffixes a and b in the generalized susceptibilities α_{ab}, which implies that of the tensor suffixes i and k and also that of the points **r** and **r**′, now causes the interchange of the respective arguments in the functions $f_{ik}(\omega; \mathbf{r}, \mathbf{r}')$.‡

† We are, of course, referring to non-ferromagnetic bodies. Moreover, the following discussion relates only to infinite homogeneous media, and the form of the boundary conditions on the equations is not discussed.

The above treatment gives a somewhat different view of the assertion in §79 that the permeability μ has no meaning at optical frequencies. In that range, the effects due to the difference of μ from unity are in general indistinguishable from those of the spatial dispersion of the permittivity.

‡ As always in applications of the generalized principle of symmetry of the kinetic coefficients, if the body is in an external magnetic field or has a magnetic structure, the right-hand side of (103.6) is to be taken with the field reversed or the structure time-reversed.

We shall consider an infinite macroscopically homogeneous medium. Then the kernel of the integral operator in (103.1) or (103.2) depends only on the difference $\boldsymbol{\rho} = \mathbf{r} - \mathbf{r}'$. The functions **D** and **E** can then be usefully expanded as Fourier integrals with respect to coordinates as well as time, which reduces them to a set of plane waves whose dependence on **r** and t is given by a factor $\exp[i(\mathbf{k}\cdot\mathbf{r} - \omega t)]$. For such waves, the relation between **D** and **E** is

$$D_i = \varepsilon_{ik}(\omega, \mathbf{k}) E_k, \tag{103.7}$$

where

$$\varepsilon_{ik}(\omega, \mathbf{k}) = \delta_{ik} + \int_0^\infty \int f_{ik}(\tau, \boldsymbol{\rho}) e^{i(\omega\tau - \mathbf{k}\cdot\boldsymbol{\rho})} \mathrm{d}^3\rho \, \mathrm{d}\tau. \tag{103.8}$$

In this description, the spatial dispersion reduces to a dependence of the permittivity tensor on the wave vector.

The "wavelength" $1/k$ determines the distances over which the field varies considerably. We can therefore say that spatial dispersion expresses the dependence of the macroscopic properties of matter on the spatial inhomogeneity of the electromagnetic field, just as the frequency dispersion expresses their dependence on the time variation of the field. When $k \to 0$, the field becomes uniform, and accordingly $\varepsilon_{ik}(\omega, \mathbf{k})$ tends to the ordinary permittivity $\varepsilon_{ik}(\omega)$.†

From the definition (103.8),

$$\varepsilon_{ik}(-\omega, -\mathbf{k}) = \varepsilon_{ki}{}^*(\omega, \mathbf{k}), \tag{103.9}$$

a generalization of (77.7). The symmetry (103.6), expressed in terms of the functions $\varepsilon_{ik}(\omega, \mathbf{k})$, now gives

$$\varepsilon_{ik}(\omega, \mathbf{k}; \mathfrak{H}) = \varepsilon_{ki}(\omega, -\mathbf{k}; -\mathfrak{H}), \tag{103.10}$$

where the parameter $\mathfrak{H}$, denoting the external magnetic field (if any), is written explicitly. If the medium has a centre of symmetry, the components ε_{ik} are even functions of **k**; an axial vector is unchanged by inversion, and (103.10) thus becomes

$$\varepsilon_{ik}(\omega, \mathbf{k}; \mathfrak{H}) = \varepsilon_{ki}(\omega, \mathbf{k}; -\mathfrak{H}). \tag{103.11}$$

The spatial dispersion does not affect the derivation of equation (96.5) for the energy dissipation. The condition that absorption be absent is therefore again that the tensor $\varepsilon_{ik}(\omega, \mathbf{k})$ be Hermitian.

When spatial dispersion is present, the permittivity is a tensor, not a scalar, even in an isotropic medium: a distinctive direction is generated by the wave vector. If the medium not only is isotropic but also has a centre of symmetry, the tensor ε_{ik} must be constructed from the components of the vector **k** and the unit tensor δ_{ik} (in the absence of a centre of symmetry, there may also be a term containing the antisymmetric unit tensor e_{ikl}; see §104). The general form of such a tensor may be written as

$$\varepsilon_{ik}(\omega, \mathbf{k}) = \varepsilon_t(\omega, k)(\delta_{ik} - k_i k_k/k^2) + \varepsilon_l(\omega, k) k_i k_k/k^2, \tag{103.12}$$

where ε_t and ε_l depend only on the magnitude of the wave vector (and on ω). If **E** is parallel

† More precisely, the dependence on **k** disappears when $kr_0 \ll 1$, where r_0 is the size of the region in which $f_{ik}(\omega, \boldsymbol{\rho})$ is significantly different from zero.

to the wave vector, then $\mathbf{D} = \varepsilon_l \mathbf{E}$; if $\mathbf{E} \perp \mathbf{k}$, then $\mathbf{D} = \varepsilon_t \mathbf{E}$. The quantities ε_l and ε_t are accordingly called the *longitudinal* and *transverse permittivities*. When $\mathbf{k} \to 0$, the expression (103.12) should tend to $\varepsilon(\omega)\delta_{ik}$, which does not depend on the direction of $\mathbf{k}$; it is therefore clear that

$$\varepsilon_l(\omega, 0) = \varepsilon_t(\omega, 0) = \varepsilon(\omega). \tag{103.13}$$

The description of the electromagnetic properties of an isotropic medium by means of the permittivities ε_l and ε_t corresponds to Maxwell's equations written in the form (103.3) and (103.4). On the other hand, as $\mathbf{k} \to 0$ and the spatial dispersion disappears, we can revert to the description by means of ε and μ. There is consequently a certain relation between these quantities (see Problem 1).

The analogy between (103.8) and (77.5) enables us to apply to each component $\varepsilon_{ik}(\omega, \mathbf{k})$ as a function of the complex variable ω the results of the study of analytic properties in §§77 and 82. These are analytic functions with no singularity in the upper half-plane of ω, and they satisfy (for any fixed value of $\mathbf{k}$) the Kramers–Kronig dispersion relations. The same is true of the functions $\varepsilon_l(\omega, k)$ and $\varepsilon_t(\omega, k)$ in (103.12). Here it must be borne in mind that the function ε_l with $k \neq 0$ does not tend to infinity as $\omega \to 0$ even in a conducting medium, and therefore no subtraction is necessary here as it was in deriving (82.9); the fact that $\varepsilon(\omega)$ becomes infinite in a conductor as $\omega \to 0$ is due to the uniformity ($\mathbf{k} = 0$) of the static field.

The time average (as defined in §80) of the electromagnetic field energy density in a transparent medium with spatial dispersion is given by the previous formula (96.6). Since now $\mu \equiv 1$, we have

$$\bar{U} = \frac{1}{16\pi}\left[\frac{\partial(\omega\varepsilon_{ik})}{\partial\omega} E_i E_k{}^* + |\mathbf{B}|^2\right], \tag{103.14}$$

$\mathbf{E}$ and $\mathbf{B}$ being assumed written in complex form. In the energy flux density in such a medium, an additional term appears:

$$\bar{\mathbf{S}} = \frac{c}{8\pi}\,\mathrm{re}\,(\mathbf{E}^* \times \mathbf{B}) - \frac{\omega}{16\pi}\frac{\partial\varepsilon_{ik}}{\partial\mathbf{k}} E_i{}^* E_k. \tag{103.15}$$

This formula is obtained by generalizing the derivation of (80.11): here, we have to consider a wave spread over a small range, both in frequency and in wave-vector direction (see Problem 2).

PROBLEMS

PROBLEM 1. Find the relation between the functions $\varepsilon(\omega)$, $\mu(\omega)$ and the limiting values of $\varepsilon_l(\omega, k)$ and $\varepsilon_t(\omega, k)$ as $k \to 0$.

SOLUTION. We compare the expressions for the averaged microscopic current $\overline{\rho\mathbf{v}}$ in the forms (103.5) and (79.3). For a monochromatic field, we have in the first case

$$\overline{\rho v_i} = -i\omega[\varepsilon_{ik}(\omega, \mathbf{k}) - \delta_{ik}]E_k/4\pi,$$

and in the second case

$$\overline{\rho\mathbf{v}} = -i\omega[\varepsilon(\omega) - 1]\mathbf{E}/4\pi + ic[\mu(\omega) - 1]\mathbf{k} \times \mathbf{H}/4\pi.$$

Substituting in the first $\varepsilon_{ik}(\omega, \mathbf{k})$ from (103.12), and in the second $\mathbf{H} = c\mathbf{k} \times \mathbf{E}/\omega\mu$ in accordance with Maxwell's equation, and equating the two expressions (for $k \to 0$), we find

$$1 - \frac{1}{\mu(\omega)} = \frac{\omega^2}{c^2}\lim_{k \to 0}\frac{\varepsilon_t(\omega, k) - \varepsilon_l(\omega, k)}{k^2}$$

by comparing the terms in $\mathbf{k}(\mathbf{k}\cdot\mathbf{E})$. Together with (101.13), this gives the required relation.

PROBLEM 2. Derive the formula (103.15) for the (time) averaged energy flux density in a medium with spatial dispersion.

SOLUTION. We start, as in §80, from formula (80.2), putting there $\mathbf{H} = \mathbf{B}$ in accordance with the description of the field by equations (103.3) and (103.4), expressing all quantities in complex form and averaging over time:

$$-\operatorname{div}\frac{c}{8\pi}\operatorname{re}(\mathbf{E}\times\mathbf{H}^*) = \frac{1}{8\pi}\operatorname{re}\left(\mathbf{E}^*\cdot\frac{\partial\mathbf{D}}{\partial t}+\mathbf{B}^*\cdot\frac{\partial\mathbf{B}}{\partial t}\right). \tag{1}$$

Let us consider an almost monochromatic plane wave with

$$\mathbf{E} = \mathbf{E}_0(t, \mathbf{r})e^{i(\mathbf{k}_0\cdot\mathbf{r}-\omega_0 t)},$$

where $\mathbf{E}_0(t, \mathbf{r})$ is a function varying slowly in space and time. The derivative $\partial D_i/\partial t$ is written as $\hat{f}_{ik}E_k$, with the operator

$$\hat{f}_{ik} = \frac{\partial}{\partial t}\hat{\varepsilon}_{ik}; \tag{2}$$

for a strictly monochromatic wave,

$$\hat{f}_{ik}E_k = f_{ik}E_k = -i\omega\varepsilon_{ik}(\omega, \mathbf{k})E_k.$$

Expanding $\mathbf{E}_0(t, \mathbf{r})$ as a Fourier integral with respect to time and coordinates, we can express it as a superposition of components $\mathbf{E}_{0\alpha\mathbf{q}}e^{i(\mathbf{q}\cdot\mathbf{r}-\alpha t)}$, with $\alpha \ll \omega_0$ and $\mathbf{q} \ll \mathbf{k}_0$. We then proceed as in the derivation of (80.10). Applying the operator (2) to the function

$$\mathbf{E}_{\omega_0+\alpha,\mathbf{k}_0+\mathbf{q}} = \mathbf{E}_{0\alpha\mathbf{q}}\exp[i(\mathbf{k}_0+\mathbf{q})\cdot\mathbf{r}-i(\omega_0+\alpha)t]$$

gives

$$\begin{aligned}\hat{f}_{ik}\mathbf{E}_{\omega_0+\alpha,\mathbf{k}_0+\mathbf{q}} &= f_{ik}(\omega_0+\alpha, \mathbf{k}_0+\mathbf{q})\mathbf{E}_{\omega_0+\alpha,\mathbf{k}_0+\mathbf{q}}\\ &\cong\left[f_{ik}(\omega_0, \mathbf{k}_0)+\alpha\frac{\partial f_{ik}(\omega_0, \mathbf{k}_0)}{\partial\omega_0}+\mathbf{q}\cdot\frac{\partial f_{ik}(\omega_0, \mathbf{k}_0)}{\partial\mathbf{k}_0}\right]\mathbf{E}_{\omega_0+\alpha,\mathbf{k}_0+\mathbf{q}}.\end{aligned}$$

Now carrying out the inverse summation of the Fourier components, substituting $f_{ik}(\omega, \mathbf{k}) = -i\omega\varepsilon_{ik}(\omega, \mathbf{k})$ and omitting the suffix 0 from ω_0 and $\mathbf{k}_0$, we find

$$\frac{\partial D_i}{\partial t} = -i\omega\varepsilon_{ik}E_k+\left[\frac{\partial(\omega\varepsilon_{ik})}{\partial\omega}\frac{\partial E_{0k}}{\partial t}-\omega\frac{\partial\varepsilon_{ik}}{\partial\mathbf{k}}\cdot\mathbf{grad}\,E_{0k}\right]e^{i(\mathbf{k}\cdot\mathbf{r}-\omega t)}. \tag{3}$$

The second term in the square brackets is the difference between this expression and (80.10). Substitution of (3) in (1) brings the latter to the energy conservation equation:

$$\partial\bar{U}/\partial t = -\operatorname{div}\bar{\mathbf{S}},$$

with $\bar{U}$ and $\bar{\mathbf{S}}$ given by (103.14), (103.15).

§104. Natural optical activity

If the spatial dispersion is weak, the tensor $\varepsilon_{ik}(\omega, \mathbf{k})$ can be expanded in powers of $\mathbf{k}$. For ordinary solid or liquid dielectric media, the expansion is one in powers of a/λ, where a is the atomic dimension and λ the field wavelength.

As far as the first-order small terms, the expansion is

$$\varepsilon_{ik}(\omega, \mathbf{k}) = \varepsilon^{(0)}{}_{ik}(\omega)+i\gamma_{ikl}k_l, \tag{104.1}$$

where $\varepsilon^{(0)}{}_{ik} = \varepsilon_{ik}(\omega, 0)$, and γ_{ikl} is a tensor of rank three which depends on the frequency; as $\omega\to 0$, its components (which do not pertain to any expansion in powers of ω) tend to constants. The relation between $\mathbf{D}$ and $\mathbf{E}$ in a monochromatic field ($\propto e^{-i\omega t}$), corresponding to (104.1), in the coordinate representation, is

$$D_i = \varepsilon^{(0)}{}_{ik}E_k+\gamma_{ikl}\partial E_k/\partial x_l. \tag{104.2}$$

Applying to (104.1) the symmetry condition (103.10), we have, in the absence of an external field,

$$\gamma_{ikl}(\omega) = -\gamma_{kil}(\omega). \tag{104.3}$$

In the absence of dissipation, the Hermitian condition gives $\gamma_{ikl}^* = -\gamma_{kil}$; hence we find $\gamma_{ikl}^* = \gamma_{ikl}$ instead of (104.3). Thus the condition for absorption to be absent is that the tensor γ_{ikl} be real.

If we use the notation of §97 and express a plane wave vector as $\mathbf{k} = \omega\mathbf{n}/c$, the expression (104.1) becomes

$$\varepsilon_{ik} = \varepsilon^{(0)}{}_{ik} + i\omega\gamma_{ikl}n_l/c. \tag{104.4}$$

Instead of the antisymmetrical tensor of rank two $\gamma_{ikl}n_l$, we shall use the axial *gyration vector* $\mathbf{g}$, which is dual to it. This vector is given by

$$\omega\gamma_{ikl}n_l/c = e_{ikl}g_l, \tag{104.5}$$

i.e.

$$\varepsilon_{ik} = \varepsilon^{(0)}{}_{ik} + ie_{ikl}g_l, \tag{104.6}$$

which is formally the same as the expression used in §101. The only difference is that in §101 the vector $\mathbf{g}$ depended only on the properties of the medium (and on the applied magnetic field), whereas here the gyration vector depends also on the wave vector of the field. According to (104.5) the components of this vector are linear functions of the components of $\mathbf{n}$, i.e.

$$g_i = g_{ik}n_k. \tag{104.7}$$

Substituting (104.7) in (104.5), we find $\omega\gamma_{ikl}n_l/c = e_{ikm}g_{ml}n_l$, or, since $\mathbf{n}$ is arbitrary,

$$\omega\gamma_{ikl}/c = e_{ikm}g_{ml}, \tag{104.8}$$

which gives the relation between the components of the tensor γ_{ikl} of rank three and the pseudotensor g_{ik} of rank two.†

The particular crystallographic symmetry of the body places certain restrictions on the components of the tensor γ_{ikl} (or g_{ik}) and, in particular, may have the result that all the components are zero. For example, the tensor γ_{ikl} cannot exist in bodies having a centre of symmetry: when the sign of each coordinate is changed (inversion), all the components of a tensor of rank three (and of a pseudotensor of rank two) change sign, whereas by the symmetry of the body they must remain unchanged by this transformation.

Bodies in which the tensor g_{ik} is not zero are said to have *natural optical activity* or *natural gyrotropy*. Thus the existence of optical activity certainly implies that the body has no centre of symmetry.

Let us first consider the natural optical activity of isotropic bodies. If a liquid or gas consists of a substance having no stereoisomer, it is symmetrical not only with respect to any rotation but also with respect to reflection (inversion) about any point, and can have no optical activity. Such activity can occur only in fluids having two stereoisomeric forms, and the two forms must be present in different quantities. The fluid then has no centre of symmetry.

In an isotropic body, and in crystals of the cubic system, the pseudotensor g_{ik} reduces to

† In components $g_{xx} = \omega\gamma_{yzx}/c$, $g_{xy} = \omega\gamma_{yzy}/c$, $g_{yx} = \omega\gamma_{zxx}/c$, etc.

a pseudoscalar:

$$g_{ik} = f\delta_{ik}, \gamma_{ikl} = cfe_{ikl}/\omega. \tag{104.9}$$

A pseudoscalar is a quantity which changes sign on inversion of the coordinates. The two stereoisomers are formally converted into one another by the operation of inversion, and so their values of f are the same with opposite signs.

Thus, in an optically active isotropic body, the gyration vector $\mathbf{g} = f\mathbf{n}$, and the relation between the electric induction and field in the wave is given by

$$\mathbf{D} = \varepsilon^{(0)}\mathbf{E} + if\,\mathbf{E}\times\mathbf{n}. \tag{104.10}$$

Since $\mathbf{D}\cdot\mathbf{n} = 0$, it follows that $\mathbf{E}\cdot\mathbf{n} = 0$. That is, in such a wave not only the induction $\mathbf{D}$ (as in any medium) but also the field $\mathbf{E}$ is transverse to the direction of $\mathbf{n}$.

The change in the refractive index n when allowance is made for the natural optical activity is a small quantity. In determining this change we can therefore put $n = n_0 = \sqrt{\varepsilon^{(0)}}$ in the small term $\mathbf{E}\times\mathbf{g}$ in (104.10). Then the problem of calculating the difference $n - n_0$ is formally identical with that considered in §101 of the change in n due to the magnetic field, except that $\mathbf{g}$ has a different meaning and is always parallel to $\mathbf{n}$ (the z-axis in §101). By analogy with (101.19) we can therefore derive immediately the equation

$$n_\pm^{\ 2} = n_0^{\ 2} \pm g = n_0^{\ 2} \pm fn_0. \tag{104.11}$$

These two values correspond (cf. (101.21)) to the following ratios of the two components of $\mathbf{E}$ (or $\mathbf{D}$):

$$E_x = \pm iE_y, \tag{104.12}$$

i.e. to waves which are left-hand and right-hand circularly polarized. It may also be noted that the magnitude of $\mathbf{n}$ is independent of its direction, and therefore the direction of $\mathbf{n}$ is the same as that of the ray vector $\mathbf{s}$.

Thus we see that the optical properties of a naturally active isotropic body resemble those of an inactive body in a magnetic field: it exhibits double circular refraction, and when a linearly polarized wave is propagated in it the plane of polarization is rotated. The angle of rotation per unit path length of the ray is $\omega f/2c$.

The sign of the constant f, and therefore the direction of rotation, are opposite for the two stereoisomers, and we therefore speak of *dextrorotatory* and *laevorotatory* stereoisomers.

Unlike the rotation of the plane of polarization in a magnetic field, the magnitude and sign of the rotation in naturally active substances do not depend on the direction of propagation of the ray. Hence, if a linearly polarized ray traverses the same path in a naturally active medium twice in opposite directions, the plane of polarization is unchanged.

Let us now consider naturally active crystals. We shall not give here a systematic analysis of all possible cases of symmetry (see the Problem), but simply note that natural activity is impossible if a centre of symmetry is present, but possible if there is a plane of symmetry or a rotary-reflection axis. It should be emphasized that the conditions for the existence of natural activity in crystals are not the same as those which allow the existence of crystals in two mirror-image (*enantiomorphic*) forms; the latter conditions are more stringent, and require the absence of both a centre and a plane of symmetry. Thus a crystal can be optically active and yet be identical with its mirror image.

In a naturally active crystal (uniaxial or biaxial), when light is propagated with an arbitrary direction of the wave vector, we have essentially ordinary double refraction of linearly polarized waves; the allowance for the activity would amount essentially to replacing the strictly linear polarization by an elliptical polarization with an axis ratio of the first order of smallness.

The only exception is formed by the directions of the optical axes, along which, if the activity is neglected, the two roots of Fresnel's equation coincide. In these directions the phenomenon of natural activity of crystals is analogous to that of isotropic bodies: double circular refraction of the first order occurs, with a corresponding rotation of the plane of polarization of linearly polarized waves. These phenomena rapidly disappear as the wave vector deviates from the direction of the optical axis.

For a quantitative calculation of natural activity in crystals it is more convenient to use, not the expression giving **D** in terms of **E**, but the inverse, as in §101. As far as first-order quantities this is

$$E_i = \varepsilon^{(0)-1}{}_{ik} D_k + (\mathbf{D}\times\mathbf{G})_i, \tag{104.13}$$

where the vector **G** is related to the **g** previously used by $G_i = -\varepsilon^{(0)}{}_{ik} g_k / |\varepsilon^{(0)}|$; see (101.9). Owing to the formal correspondence between this expression and (101.7), the equations (101.11) and (101.12) are again valid. In these equations G_z is the component of **G** in the direction of **n**. If we write **G** in the form

$$G_i = G_{ik} n_k, \tag{104.14}$$

in analogy with (104.7), the component is proportional to

$$\mathbf{n}\cdot\mathbf{G} = G_{ik} n_i n_k. \tag{104.15}$$

This quadratic form determines the optical properties of a naturally active crystal. The tensor G_{ik} itself need not be symmetrical, but if it is separated into symmetrical and antisymmetrical parts the latter does not appear in the form (104.15). Thus we conclude that the tensor G_{ik} may be assumed symmetrical in discussing the optical properties of naturally active crystals.

PROBLEM

Find the restrictions imposed by crystal symmetry on the components of the tensor G_{ik}.

SOLUTION. Under any rotation, the pseudotensor G_{ik} behaves as a true tensor; in particular, the presence of an n-fold axis of symmetry with $n > 2$ results, as for a true symmetrical tensor of rank two, in complete isotropy in a plane perpendicular to the axis. The behaviour of the pseudotensor G_{ik} under reflection is determined by the fact that it is dual to a true tensor of rank three: under any reflection which changes the sign of a given component of a true tensor of rank two, the corresponding component of G_{ik} remains unchanged, and vice versa. For example, on reflection in the yz-plane the components G_{xx}, G_{yy}, G_{zz}, G_{yz} change sign, but G_{xy}, G_{xz} do not.

We give below the non-vanishing components of the tensor G_{ik} for all crystal classes which allow natural activity. The z-axis is taken along the threefold, fourfold, or sixfold axis of symmetry or (in the classes $\boldsymbol{C}_2$, $\boldsymbol{C}_{2v}$) along the only twofold axis of symmetry or (in the class $\boldsymbol{C}_s$) perpendicular to the plane of symmetry. When three mutually perpendicular axes of symmetry are present, they are the coordinate axes.

Class $\boldsymbol{C}_1$: all.
Class $\boldsymbol{C}_2$: G_{xx}, G_{yy}, G_{zz}, G_{xy}, the last of which may be made to vanish by a suitable choice of the x and y axes.
Class $\boldsymbol{C}_s$: G_{xz}, G_{yz}, one of which may be made to vanish by a suitable choice of the x and y axes.
Class $\boldsymbol{C}_{2v}$: G_{xy} (the xz and yz planes being planes of symmetry).
Class $\boldsymbol{D}_2$: G_{xx}, G_{yy}, G_{zz}.
Classes $\boldsymbol{C}_3$, $\boldsymbol{C}_4$, $\boldsymbol{C}_6$, $\boldsymbol{D}_3$, $\boldsymbol{D}_4$, $\boldsymbol{D}_6$: $G_{xx} = G_{yy}$, G_{zz}.
Class $\boldsymbol{S}_4$: $G_{xx} = -G_{yy}$, G_{xy}, one of which may be made to vanish by a suitable choice of the x and y axes.
Class $\boldsymbol{D}_{2d}$: G_{xy} (the x and y axes being in vertical planes of symmetry).
Classes $\boldsymbol{T}$, $\boldsymbol{O}$: $G_{xx} = G_{yy} = G_{zz}$.

It may be noted that, in uniaxial crystals of the classes $\boldsymbol{S}_4$ and $\boldsymbol{D}_{2d}$, the scalar (104.15) is zero if the vector $\mathbf{n}$ is in the z-direction, since $G_{zz} = 0$. This means that in these crystals there is no natural-activity effect in the direction of the optical axis.

In a biaxial crystal of the $\boldsymbol{C}_{2v}$ the optical axes are in one of the planes of symmetry. For vectors $\mathbf{n}$ lying in the xz or yz plane the scalar (104.15) is again identically zero, so that here also there is no effect in the direction of the optical axes. The only crystal class which allows rotation of the plane of polarization along the optical axis but not enantiomorphism is the monoclinic class $\boldsymbol{C}_s$.

§105. Spatial dispersion in optically inactive media

In crystals whose symmetry does not allow natural optical activity, the first terms (after the zero-order one) in the expansion of the permittivity $\varepsilon_{ik}(\omega, \mathbf{k})$ in powers of $\mathbf{k}$ are quadratic terms.

As usual in crystal optics, it is convenient in the subsequent analysis to write down this expansion for the inverse tensor $\eta_{ik} = \varepsilon^{-1}{}_{ik}$. We put

$$\eta_{ik} = \eta^{(0)}{}_{ik}(\omega) + \beta_{iklm}(\omega)k_l k_m. \tag{105.1}$$

The tensor β_{iklm} may be regarded as symmetrical in the second pair of suffixes, since it is multiplied by the symmetrical product $k_l k_m$. From (103.10) (with $\mathfrak{H} = 0$) it is also symmetrical in the first pair of suffixes:

$$\beta_{iklm} = \beta_{kilm} = \beta_{ikml}. \tag{105.2}$$

It need not, however, be symmetrical with respect to the interchange of both pairs. In the absence of absorption, since the tensor η_{ik} is Hermitian and symmetrical, the tensor β_{iklm} is real, and we shall assume this to be so.

In an isotropic medium, β_{iklm} must be expressed in terms of the unit tensor only, i.e. must have the form

$$\beta_{iklm} = \beta_1 \delta_{ik}\delta_{lm} + \tfrac{1}{2}\beta_2(\delta_{il}\delta_{km} + \delta_{kl}\delta_{im});$$

it contains only two independent components. In an isotropic body we also have $\eta^{(0)}{}_{ik} = \eta^{(0)}\delta_{ik}$, and so the tensor (105.1) becomes

$$\eta_{ik} = (\eta^{(0)} + \beta_1 k^2)\delta_{ik} + \beta_2 k_i k_k, \tag{105.3}$$

in accordance with the general expression (103.12) for the dielectric tensor in an isotropic medium with spatial dispersion. The wave propagation in the medium is governed by the equations (97.21). On substitution of (105.3) in these equations, however, the anisotropic term in β_2 disappears, since the vectors $\mathbf{D}$ and $\mathbf{k}$ are orthogonal in a plane wave, i.e. the medium remains optically isotropic, as it should.

Even in cubic crystals, however, the tensor β_{iklm} does not reduce to the unit tensor; depending on the crystal class, it has either three or four independent components. When spatial dispersion is neglected, cubic crystals are optically isotropic; the inclusion of the dispersion quadratic in $\mathbf{k}$ brings about a new property in them, namely optical anisotropy (H. A. Lorentz, 1878).

In a cubic crystal $\eta^{(0)}{}_{ik} = \delta_{ik}/\varepsilon^{(0)}$, and the expansion (105.1) becomes

$$\eta_{ik} = \frac{1}{\varepsilon^{(0)}}\delta_{ik} + \beta_{iklm}k_l k_m. \tag{105.4}$$

Substitution of this in equations (97.21) gives

$$\left[\left(\frac{1}{n^2}-\frac{1}{n_0{}^2}\right)\delta_{\alpha\beta}-\frac{\omega^2 n^2}{c^2}\beta_{\alpha\beta 33}\right]D_\beta=0, \tag{105.5}$$

where $n_0{}^2 = \varepsilon^{(0)}$ and the x_3-axis in Cartesian coordinates x_1, x_2, x_3 is parallell to the wave vector. From the sense of the expansion (105.4), the second term in the square brackets in these equations is to be regarded as a small correction (see §106 concerning the special case in which $1/n_0{}^2 = 0$). We can then replace n^2 in that term by $n_0{}^2$:

$$\left[\left(\frac{1}{n^2}-\frac{1}{n_0{}^2}\right)\delta_{\alpha\beta}-\frac{\omega^2 n_0{}^2}{c^2}\beta_{\alpha\beta 33}\right]D_\beta=0. \tag{105.6}$$

These equations have the same form as for waves in a non-cubic crystal when spatial dispersion is neglected. Their determinant is an expression quadratic in n^{-2}, whose vanishing determines the refractive indices of the two waves with the same direction of **k** but different polarizations. Thus the spatial dispersion in a cubic crystal removes the "polarization degeneracy": the velocities of the two waves become different, and direction-dependent.

The possibility of longitudinal electromagnetic waves in a transparent isotropic medium has been mentioned at the end of §84. The consistent formulation of the condition determining the relation between the frequency and the wave number (the dispersion relation) for these waves requires spatial dispersion to be taken into account; the condition is

$$\varepsilon_l(\omega, k) = 0. \tag{105.7}$$

For small k, the solution of this equation is

$$\omega(k) = \omega_{l0} + \tfrac{1}{2}\alpha k^2, \tag{105.8}$$

where α is a constant and ω_{l0} the frequency for which the permittivity $\varepsilon(\omega) = \varepsilon_l(\omega, 0)$. The wave propagation velocity.

$$\mathbf{u} = \partial\omega/\partial\mathbf{k} = \alpha\mathbf{k} \tag{105.9}$$

is proportional to the wave vector.

PROBLEM

Find the relations between the components of the tensor β_{iklm} in non-gyrotropic crystals of the cubic system.

SOLUTION. Natural gyrotropy does not occur in the crystal classes T_d, T_h, O_h.

In the classes T_d and T_h, we take the x, y and z axes along the three twofold axes of symmetry. The non-zero components of the tensor are

$$\beta_1 \equiv \beta_{xxxx} = \beta_{yyyy} = \beta_{zzzz}, \quad \beta_2 \equiv \beta_{xxzz} = \beta_{yyxx} = \beta_{zzyy},$$

$$\beta_3 \equiv \beta_{xyxy} = \beta_{yzyz} = \beta_{zxzx}, \quad \beta_4 \equiv \beta_{zzxx} = \beta_{xxyy} = \beta_{yyzz}.$$

In the class O_h, the three C_2 axes become C_4 axes, and so we also have $\beta_2 = \beta_4$.

§106. Spatial dispersion near an absorption line

In §§104 and 105, spatial dispersion effects have been regarded as small corrections, which they usually are. The situation is different, however, near a narrow absorption line in a crystal, where, by (84.7), $\varepsilon^{(0)}(\omega)$ is sharply increased. In this region, the inclusion of spatial dispersion causes a qualitative alteration in the picture.

The reason is that the addition to the permittivity of terms containing powers of k raises the order of the algebraic dispersion relation which determines the function $k(\omega)$. The formal solution therefore gives rise to additional roots. Far from the absorption line, these lie in the range of large k, i.e. outside the region where the theory is valid, and they must therefore be discarded. Near the absorption line, however, the permittivity varies considerably even for small k, and additional roots may arise which have a real physical significance, i.e. new transverse waves may occur.

For simplicity, we shall consider only isotropic media, and begin with the case where the medium is not gyrotropic, i.e. has no natural optical activity (S. I. Pekar, 1957; V. L. Ginzburg, 1958).

As already shown in §105, an isotropic medium remains optically isotropic even when spatial dispersion is taken into account. This means that the dispersion relation for transverse electromagnetic waves in such a medium is given by the usual equation $n^2 = \varepsilon$, in which ε is taken to be the transverse permittivity ε_t:

$$n^2 = \varepsilon_t(\omega, k). \tag{106.1}$$

It has been shown in §103 that $\varepsilon_t(\omega, k)$ and $\varepsilon_l(\omega, k)$ as functions of frequency satisfy the same Kramers–Kronig relations as $\varepsilon(\omega)$ does in the absence of spatial dispersion. We can therefore state that near the absorption line the function $\varepsilon_t(\omega, k)$ has the same form as in (84.7) but with constants that become functions of k; we write it as

$$\varepsilon_t(\omega, k) = \frac{A(k)}{\omega_t(k) - \omega}, \quad A(k) > 0.$$

If A is relatively small,† it may be meaningful to include not only the pole term but also the constant term (independent of ω) in ε_t; we denote this by $a(k)$, and assume that $a > 0$, i.e. that the medium is optically transparent far from the absorption line. The theoretical admissibility of including simultaneously the constant and pole terms in ε_t requires that they become comparable in the frequency range $|\omega - \omega_t| \ll \omega_t$, the only one where the pole expression is valid; we must thus have $A \ll a\omega_t$.

Since k is again assumed small, the functions can be expanded in powers of k. Here it is sufficient to replace $a(k)$ and $A(k)$ by the constants $a \equiv a(0) > 0$, $A \equiv A(0) > 0$, retaining the correction term only in the expansion of $\omega_t(k)$ in the small difference $\omega_t - \omega$:

$$\omega_t(k) \cong \omega_0 + \nu k^2.$$

Then the permittivity is

$$\varepsilon_t(\omega, k) = a + \frac{A}{\omega_0 + \nu k^2 - \omega}. \tag{106.2}$$

As a function of frequency, this is zero at a point in the range $\omega > \omega_t$. For $k = 0$, the zero is at $\omega = \omega_{l0} = \omega_0 + A/a$. Since the permittivities are the same when $k = 0$, ω_{l0} is the limiting frequency (as $k \to 0$) of the longitudinal wave; cf. (105.8). The zero of $\varepsilon_t(\omega, k)$ when $k \neq 0$ has no direct physical significance.

† For example, on account of some approximate selection rules which decrease the matrix elements determining the value of A.

The dispersion relation (106.1) now becomes

$$(n^2 - a)(\beta n^2 - \omega + \omega_0) = A, \tag{106.3}$$

with $\beta = v\omega^2/c^2 \cong v\omega_0{}^2/c^2$, which may be either positive or negative. The solutions of (106.3) may be regarded as resulting from the intersection of two branches of the spectrum: the ordinary light wave with $n^2 = a$, and the wave with $n^2 = (\omega - \omega_0)/\beta$ which arises from the pole of the permittivity. These branches are shown by the dot-and-dash lines in Fig. 58. Their "interaction", whose strength depends on A, moves the lines apart.†

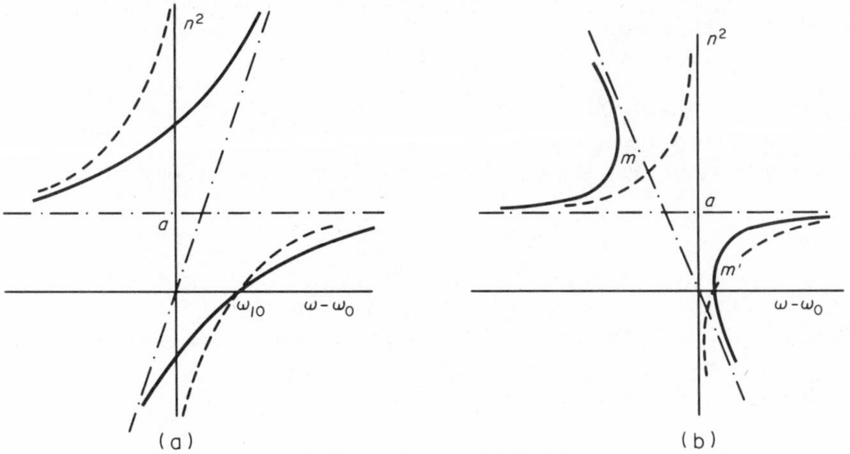

FIG. 58

The continuous curves in Fig. 58 show schematically the functions $n^2(\omega)$ given by the roots of equation (106.3); the dashed curves are the functions $n^2(\omega)$ when spatial dispersion is neglected ($\beta = 0$). Of course, only positive‡ roots n^2 correspond to waves propagated in the medium. When $\beta > 0$ (Fig. 58a), the upper continuous curve enters the region $\omega > \omega_0$, and this gives rise to an additional wave which would not exist without spatial dispersion; when $\omega > \omega_{l0}$, two different electromagnetic waves can be propagated in the medium. Figure 58b shows the function $n^2(\omega)$ when $\beta < 0$. To the left of the point m, whose coordinates are

$$n_m{}^2 = a + \sqrt{(A/|\beta|)}, \quad \omega_m - \omega_0 = -|\beta|\,[a + 2\sqrt{(A/|\beta|)}\,],$$

and at which $dn^2/d\omega = \infty$ and the two roots coincide, there are two positive roots, and two different waves can be propagated in the medium. The same is true of the region

† In the microscopic theory, the pole of the permittivity represents the existence of *excitons*, which are elementary Bose excitations in the dielectric; the sign of the constant β is the same as that of the effective exciton mass (see *SP* 2, §66). The corresponding branch of the wave spectrum is called the *exciton branch*. The part of the spectrum near the virtual intersection of the two branches is called the *polariton range*.

‡ For clarity, the diagram shows negative roots also. The purely imaginary values of n correspond to waves for which the medium is opaque (though there is no absorption in it); the wave may be said to be totally reflected from the medium.

between the point m', whose coordinates are

$$n_{m'}^2 = a - \sqrt{(A/|\beta|)}, \quad \omega_{m'} - \omega_0 = -|\beta| [a - 2\sqrt{(A/|\beta|)}],$$

and the point $\omega = \omega_{0l}$.

If A is not sufficiently small it is, strictly speaking, not consistent to retain the constant term in (106.2). On omitting this term (i.e. putting $a = 0$ in the above formulae) we get a picture that differs from Fig. 58 in that the horizontal asymptote of all the curves is the abscissa axis itself and not $n^2 = a$. The range $\omega \geqslant \omega_{l0}$ is not considered here.

Let us now examine the situation near an absorption line in a gyrotropic medium (V. L. Ginzburg, 1958). The permittivity $\varepsilon^{(0)}$, when spatial dispersion is neglected, may be expressed by the pole term

$$\varepsilon^0(\omega) = A/(\omega_0 - \omega). \tag{106.4}$$

We shall not now assume that A is particularly small, and the constant term is therefore omitted. For the relation between **E** and **D** a formula of the type (104.13) is to be used, expressed in terms of the inverse tensor $\eta_{ik} = \varepsilon^{-1}{}_{ik}$. In an isotropic medium

$$\mathbf{E} = \frac{1}{\varepsilon^{(0)}} \mathbf{D} + iF\mathbf{D} \times \mathbf{n}, \tag{106.5}$$

where the optical activity vector is written as $\mathbf{G} = F\mathbf{n}$. Near the absorption line, the components of the tensor η_{ik} simply pass through zero, and there is nothing to prevent its expansion in terms of the wave vector from converging.

The dispersion relation is

$$\left(\frac{1}{n^2} - \frac{1}{n_0{}^2}\right)^2 = F^2 n^2, \tag{106.6}$$

where $n_0{}^2 = \varepsilon^{(0)}$; cf. (101.12). Substitution of $\varepsilon^{(0)}$ from (106.4) gives

$$\left(\frac{1}{n^2} + \frac{\omega - \omega_0}{A}\right)^2 = F^2 n^2. \tag{106.7}$$

In Fig. 59 the continuous curves show schematically the dependence of the roots n^2 of this equation on $\omega - \omega_0$. One exists both for $\omega < \omega_0$ and for $\omega > \omega_0$, where there are no real values of n in the absence of spatial dispersion; this is $n_0{}^2(\omega)$, shown by the dashed curve in Fig. 59. The other two exist only when $\omega < \omega_m$, i.e. to the left of the point m at

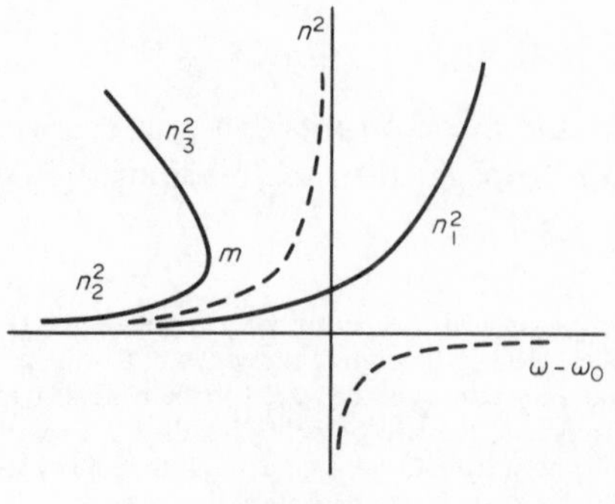

FIG. 59

which $\omega_0 - \omega_m = 3A(\frac{1}{2}F)^{\frac{2}{3}}$, $n_m^2 = (2/F)^{\frac{2}{3}}$. The curves $n_2^2(\omega)$ and $n_3^2(\omega)$ lie above $n_0^2(\omega)$; $n_1^2(\omega)$ lies below it. Hence (as is clear from equations (101.11), which determine the induction **D** in the wave) waves 2 and 3 have circular polarization of one sign, and wave 1 has circular polarization of the other sign.

Lastly, it should be emphasized that formulae (106.2) and (106.4) for the permittivity, and therefore the results based on them, apply only to frequencies sufficiently far from the centre of the line: $|\omega - \omega_0| \gg \gamma$, where γ is the line width. When $|\omega - \omega_0| \lesssim \gamma$, absorption (the imaginary part of the permittivity) has to be taken into account, and this may modify the picture substantially.

CHAPTER XIII

NON-LINEAR OPTICS

§107. Frequency transformation in non-linear media

THE theory of electromagnetic wave propagation in dielectric media described in the preceding chapters is based on the assumption of a linear relation between the electric field induction **D** and intensity **E**. This approximation is sufficiently accurate if (as is true in practice) **E** is much less than typical intra-atomic fields. Even then, however, the small non-linear corrections to **D**(**E**) cause qualitatively new effects and may therefore be important.

The most important feature of a non-linear medium is the generation in it of vibrations with new frequencies. For example, if a monochromatic wave with frequency ω_1 is incident on such a medium, then, as it is propagated in the medium, waves with frequencies $m\omega_1$ (m being an integer) are generated; if there is initially a set of monochromatic signals with frequencies ω_1 and ω_2, the combination frequencies $m\omega_1 + n\omega_2$ will arise in the course of time, and so on.

If the medium is non-dissipative, the frequency transformation process is subject to certain very general relationships, in addition to the obvious condition that the total energy of the vibrations at all frequencies must be conserved. Here it is assumed that the non-linearity is weak; the significance of this property will be made more precise later.

The origin and significance of the required relations are most clearly seen from the quantum standpoint, and this will be our initial basis. To simplify the discussion, we shall suppose that all frequencies in the system can be represented as linear combinations of two incommensurate fundamental frequencies ω_1 and ω_2:

$$\omega_{mn} = m\omega_1 + n\omega_2, \tag{107.1}$$

where m and n are positive or negative integers.

The total energy of the radiation in the medium can be expressed as the sum of the energies of all the quanta:

$$\mathscr{U} = \sum_{m,n} N_{mn}\hbar\omega_{mn},$$

where N_{mn} is the number of quanta with frequency ω_{mn}. The summation is taken over all m and n for which $\omega_{mn} > 0$ (since, of course, only positive frequencies are physically meaningful).

Frequency transformation processes cause a variation in the numbers N_{mn} with time, the total energy being conserved. Hence

$$\frac{d\mathscr{U}}{dt} = \hbar\omega_1 \sum_{m,n} m\frac{dN_{mn}}{dt} + \hbar\omega_2 \sum_{m,n} n\frac{dN_{mn}}{dt} = 0.$$

Since ω_1 and ω_2 are assumed incommensurate, and the numbers of quanta (and the changes in them) are integral, the two sums must separately be zero:†

$$\sum_{m,n} m \frac{dN_{mn}}{dt} = 0, \quad \sum_{m,n} n \frac{dN_{mn}}{dt} = 0. \tag{107.2}$$

Instead of the numbers of quanta, we will use the corresponding intensities, i.e. the total energies $\bar{\mathscr{U}}_{mn}$ in the radiation of the corresponding frequencies:

$$\bar{\mathscr{U}}_{mn} = \hbar\omega_{mn} N_{mn}. \tag{107.3}$$

The relations (107.2) then become

$$\sum_{m,n} \frac{m}{\omega_{mn}} \frac{d\bar{\mathscr{U}}_{mn}}{dt} = 0, \quad \sum_{m,n} \frac{n}{\omega_{mn}} \frac{d\bar{\mathscr{U}}_{mn}}{dt} = 0. \tag{107.4}$$

Here we must note a point that is especially clear in the classical picture of the oscillations. In referring to the time variation of $\bar{\mathscr{U}}_{mn}$, we mean only the systematic variation; that is, we are considering the energy averaged over time intervals much longer than the periods $1/\omega_1$, $1/\omega_2$, as indicated by the bar over $\bar{\mathscr{U}}_{mn}$ in (107.4). This is where the non-linear effects need to be weak: the characteristic time τ of the steady build-up which they cause in the oscillations excited must be much longer than the periods mentioned. Only then can it be meaningful to consider the time variation of quantities averaged over intervals Δt such that $1/\omega_1$, $1/\omega_2 \ll \Delta t \ll \tau$.

Equations (107.4) give the required relations, and constitute the *Manley–Rowe theorem* (J. M. Manley and H. E. Rowe, 1956).‡ It remains to give them a more definite form, avoiding the restriction $\omega_{mn} > 0$ in the summations. This is easily done by noting that, to each pair of numbers m, n for which $\omega_{mn} > 0$, there corresponds a pair $-m$, $-n$ for which the frequency is negative and $|\omega_{mn}|$ is the same. Defining

$$\bar{\mathscr{U}}_{-m,-n} = \bar{\mathscr{U}}_{mn} \tag{107.5}$$

and extending the summation to all integers from $-\infty$ to ∞, we double the sums, which are zero as before. Another simplification is then possible. In the first equation (107.4), we divide the sum over m into two, from 0 to ∞ and from $-\infty$ to 0, and in the second sum make the change $m, n \rightarrow -m, -n$; a similar change is made in the sum over n in the second equation. The result is

$$\sum_{m=0}^{\infty} \sum_{n=-\infty}^{\infty} \frac{m}{m\omega_1 + n\omega_2} \frac{d\bar{\mathscr{U}}_{mn}}{dt} = 0, \quad \sum_{n=0}^{\infty} \sum_{m=-\infty}^{\infty} \frac{n}{m\omega_1 + n\omega_2} \frac{d\bar{\mathscr{U}}_{mn}}{dt} = 0. \tag{107.6}$$

There is an obvious generalization to the case of a greater initial number of incommensurate frequencies.

The specific properties of the vibrating system may prohibit particular frequency transformation processes. The summations (107.6) are taken, in practice, only over allowed processes. For instance, in the simple case of a system that permits only the generation of the combination frequency $\omega_1 + \omega_2$, the numbers m and n take the values 0 and 1, and we find

$$-\frac{1}{\omega_1} \frac{d\bar{\mathscr{U}}_{10}}{dt} = -\frac{1}{\omega_2} \frac{d\bar{\mathscr{U}}_{01}}{dt} = \frac{1}{\omega_1 + \omega_2} \frac{d\bar{\mathscr{U}}_{11}}{dt}. \tag{107.7}$$

† It is, of course, conventional to write the rates of change of integers as derivatives.

‡ The quantum interpretation is due to M. T. Weiss (1957).

The meaning of these equations is obvious: the decrease in the numbers of quanta $\hbar\omega_1$ and $\hbar\omega_2$ is equal to the increase in the number of quanta $\hbar(\omega_1+\omega_2)$ generated.

§108. The non-linear permittivity

When the non-linearity is weak, the first correction to the linear dependence of **D** on **E** is quadratic in the field. In the presence of time dispersion† it can be represented in any anisotropic medium by

$$D_i^{(2)}(t) = \int_0^\infty\int f_{i,kl}(\tau_1, \tau_2)E_k(t-\tau_1)E_l(t-\tau_2)\,d\tau_1\,d\tau_2, \tag{108.1}$$

an expression analogous to (77.3). Of course, the existence of such a term places certain limitations on the permissible symmetry of the medium; in particular, it does not occur if the latter is invariant under inversion.

Although we shall be considering as typical a term of the form (108.1) quadratic in **E**, it must be noted that in the quadratic approximation the induction **D** may also contain terms bilinear in the components of **E** and **H**, or quadratic in **H**; these are usually less important, and they will not be discussed here. We shall also ignore the non-linear dependence of the magnetic induction **B** on **H**, because it is analogous to the dependence **D**(**E**).

We define the quantity

$$\varepsilon_{i,kl}(\omega_1, \omega_2) = \int_0^\infty\int e^{i(\omega_1\tau_1+\omega_2\tau_2)}f_{i,kl}(\tau_1, \tau_2)\,d\tau_1\,d\tau_2, \tag{108.2}$$

which may naturally be called the second-order *non-linear permittivity* by analogy with the linear permittivity defined by an expression of the form (77.5). It differs only by a factor from $\chi_{i,kl} = \varepsilon_{i,kl}/4\pi$, called the *non-linear susceptibility*. Because of the symmetry of E_k and E_l in (108.1), the tensor $f_{i,kl}$ is symmetrical in the suffixes k and l if the arguments are simultaneously transposed: $f_{i,kl}(\tau_1, \tau_2) = f_{i,lk}(\tau_2, \tau_1)$. The tensor $\varepsilon_{i,kl}$ therefore has the same type of symmetry:

$$\varepsilon_{i,kl}(\omega_1, \omega_2) = \varepsilon_{i,lk}(\omega_2, \omega_1). \tag{108.3}$$

In particular, when $\omega_1 = \omega_2$ the tensor is symmetrical in the last two suffixes:

$$\varepsilon_{i,kl}(\omega, \omega) = \varepsilon_{i,lk}(\omega, \omega). \tag{108.4}$$

Moreover, since the functions $f_{i,kl}$ are real, as follows from the definition (108.1) with real **E** and **D**, we have

$$\varepsilon_{i,kl}(-\omega_1, -\omega_2) = \varepsilon_{i,kl}{}^*(\omega_1, \omega_2). \tag{108.5}$$

The permittivity (108.2) arises naturally in the consideration of monochromatic fields or superpositions of these. In non-linear expressions such fields must of course be put in a real form. For example, if **E** is a monochromatic field with frequency ω, we must write $\mathbf{E}(t) = \mathrm{re}\{\mathbf{E}_0 e^{-i\omega t}\}$, and substitution in (108.1) gives

$$D_i^{(2)}(t) = \tfrac{1}{2}\mathrm{re}\{\varepsilon_{i,kl}(\omega, \omega)e^{-2i\omega t}E_{0i}E_{0k} + \varepsilon_{i,kl}(\omega, -\omega)E_{0i}E_{0k}{}^*\}. \tag{108.6}$$

† Spatial dispersion is neglected throughout this chapter.

This includes vibrations with twice the frequency (as well as a constant term corresponding to the frequency $\omega - \omega = 0$). In the general case, $\varepsilon_{i,kl}(\omega_1, \omega_2)$ describes the contribution to the induction that is proportional to $\exp(-i\omega_3 t)$ with $\omega_3 = \omega_1 + \omega_2$.

We shall discuss below only non-dissipative media, and refer to them as transparent, although they are not literally such (for waves of a given frequency), because of the possible transfer of energy to other frequencies. We shall also suppose that the medium has no magnetic structure.

First of all, we shall show that under these conditions the non-linear permittivity is real. This could be seen directly by expressing the components of the non-linear susceptibility tensor in terms of the matrix elements of the electric dipole interaction between the medium and the field, which acts as a small perturbation; the second-order susceptibility appears in third-order perturbation theory.† However, the reason why the result of such a calculation is real can be understood without actually doing it. The complete set of wave functions with respect to which the matrix elements are calculated may be chosen as real (for a medium having no magnetic structure, and therefore invariant under time reversal). The operator of the field interaction with the electric dipole moment of the medium is also real. Thus imaginary terms could occur only as a result of passing round poles due to the energy denominators in perturbation theory. However, the absence of dissipation in the medium signifies that none of the field frequencies coincides with a difference of energy levels in the system (or that the residues at the poles are zero because of certain selection rules); it is therefore unnecessary to pass round poles.

The transparency of the medium also gives rise to particular symmetry relations for the tensor $\varepsilon_{i,kl}$. These also could be deduced from specific expressions obtained in perturbation theory. Here again, however, the required result can be derived in a simpler manner. To do so, we assume that the field in the medium is the sum of three almost monochromatic fields with incommensurate frequencies $\omega_1, \omega_2, \omega_3$:

$$\left.\begin{aligned} \mathbf{E} &= \mathbf{E}_1 + \mathbf{E}_2 + \mathbf{E}_3 \\ &= \mathrm{re}\{\mathbf{E}_{01} e^{-i\omega_1 t} + \mathbf{E}_{02} e^{-i\omega_2 t} + \mathbf{E}_{03} e^{-i\omega_3 t}\}, \end{aligned}\right\} \tag{108.7}$$

and $\omega_3 = \omega_1 + \omega_2$. We shall suppose that the fields with frequencies $\omega_1, \omega_2, \omega_3$ are created by external sources which are afterwards switched off; the slight non-linearity of the medium causes their amplitudes $\mathbf{E}_{01}$ etc. to be slowly varying functions of time.

This slowness allows Maxwell's equations to be written down for the field of each fundamental frequency separately. In turn, these equations give in the usual way the energy conservation equation

$$\frac{1}{4\pi}(\overline{\mathbf{E}_1 \cdot \dot{\mathbf{D}}_1} + \overline{\mathbf{H}_1 \cdot \dot{\mathbf{H}}_1}) + \operatorname{div}\frac{c}{4\pi}(\overline{\mathbf{E}_1 \times \mathbf{H}_1}) = 0$$

and similarly for suffixes 2 and 3; $\mathbf{D}_1$ denotes the part of the induction $\mathbf{D}$ which contains the factors $e^{\pm i\omega_1 t}$, and the bar denotes averaging over time (which will be needed later). On integration over the whole volume of the field, the divergence term vanishes, leaving

$$\frac{1}{4\pi}\int(\overline{\mathbf{E}_1 \cdot \dot{\mathbf{D}}_1} + \overline{\mathbf{H}_1 \cdot \dot{\mathbf{H}}_1})\,\mathrm{d}V = 0.$$

† These calculations are analogous to, though considerably more laborious than, that of the linear generalized susceptibility in second-order perturbation theory (see *SP* 1, §126).

If we now separate explicitly in $\mathbf{D}_1$ the terms linear and non-linear in $\mathbf{E}$:

$$\mathbf{D}_1 = \mathbf{D}_1^{(1)} + \mathbf{D}_1^{(2)},$$

then the former together with $\mathbf{H}_1 \cdot \dot{\mathbf{H}}_1$ give $d\overline{\mathscr{U}}_1/dt$, the time variation of the field energy at frequency ω_1. This variation is consequently determined by

$$\frac{d\overline{\mathscr{U}}_1}{dt} = -\frac{1}{4\pi}\int \overline{\mathrm{re}\,\{\mathbf{E}_{01}e^{-i\omega_1 t}\}\cdot\dot{\mathbf{D}}_1^{(2)}}\,dV. \tag{108.8}$$

Here the derivative $\partial \mathbf{D}_1^{(2)}/\partial t$ is to be expressed in terms of the field by (108.1) and (108.2). The time averaging reduces to zero all terms except those in which the exponential factors cancel. Repetition of the calculations for the other frequencies gives finally

$$\left.\begin{aligned}
\frac{d\overline{\mathscr{U}}_1}{dt} &= -\frac{i\omega_1}{16\pi}\int \varepsilon_{k,il}(-\omega_3,\omega_2)E_{03i}{}^*E_{01k}E_{02l}\,dV + \text{c.c.},\\
\frac{d\overline{\mathscr{U}}_2}{dt} &= -\frac{i\omega_2}{16\pi}\int \varepsilon_{l,ki}(\omega_1,-\omega_2)E_{03i}{}^*E_{01k}E_{02l}\,dV + \text{c.c.},\\
\frac{d\overline{\mathscr{U}}_3}{dt} &= \frac{i\omega_3}{16\pi}\int \varepsilon_{i,kl}(\omega_1,\omega_2)E_{03i}{}^*E_{01k}E_{02l}\,dV + \text{c.c.},
\end{aligned}\right\} \tag{108.9}$$

where c.c. denotes the complex conjugate. The property (108.3) has been used in these calculations.

The total rate of change of the energy at all three frequencies is

$$\frac{d\overline{\mathscr{U}}}{dt} = \frac{d\overline{\mathscr{U}}_1}{dt} + \frac{d\overline{\mathscr{U}}_2}{dt} + \frac{d\overline{\mathscr{U}}_3}{dt}. \tag{108.10}$$

In a non-linear medium this sum need not in general vanish exactly, because of the possible transfer of energy to other combination frequencies such as $\omega_1 - \omega_2$ or $\omega_3 + \omega_2$. However, the fields at the frequencies $\omega_1, \omega_2, \omega_3$ arising from the external sources do not depend on the degree of non-linearity, and need not be small, unlike the fields at other frequencies, which occur only because the medium is non-linear. The contribution of the latter to the energy balance may therefore be neglected, and the condition that the sum (108.10) be zero imposed. Moreover, since such a medium is a non-linear system with only three frequencies, we can apply to it the Manley–Rowe theorem in the simple form (107.7). In the notation used here, the relations are

$$\frac{1}{\omega_1}\frac{d\overline{\mathscr{U}}_1}{dt} + \frac{1}{\omega_3}\frac{d\overline{\mathscr{U}}_3}{dt} = 0, \quad \frac{1}{\omega_2}\frac{d\overline{\mathscr{U}}_2}{dt} + \frac{1}{\omega_3}\frac{d\overline{\mathscr{U}}_3}{dt} = 0.$$

Substitution of (108.9) shows that the non-linear permittivity satisfies the following important symmetry relations†:

$$\varepsilon_{i,kl}(\omega_1,\omega_2) = \varepsilon_{k,il}(-\omega_3,\omega_2) = \varepsilon_{l,ki}(\omega_1,-\omega_3) \tag{108.11}$$

(J. A. Armstrong, N. Bloembergen, J. Ducuing, and P. S. Pershan, 1962). The symmetry

† It is important to note that the complex amplitudes are arbitrary; the complex conjugate terms in (108.9) and (108.10) are therefore independent and can be equated separately.

expressed by these equations becomes more obvious if the tensor components are given a third argument so that the sum of all three is zero:

$$\begin{aligned}\varepsilon_{i,kl}(-\omega_3;\omega_1,\omega_2) &= \varepsilon_{k,il}(\omega_1;-\omega_3,\omega_2)\\ &= \varepsilon_{l,ki}(\omega_2;\omega_1,-\omega_3)\\ &= \varepsilon_{i,lk}(-\omega_3;\omega_2,\omega_1);\end{aligned}$$

the last equation is (108.3). If we agree to relate the three successive argument frequencies to the three successive tensor indices, we can permute the latter in any manner, provided that the arguments are similarly permuted.

The requirement that dissipation be absent would by itself lead only to a weaker condition, that the sum (108.10) be zero:

$$\omega_3\varepsilon_{i,kl}(\omega_1,\omega_2)-\omega_1\varepsilon_{k,il}(-\omega_3,\omega_2)-\omega_2\varepsilon_{l,ki}(\omega_1,-\omega_3)=0. \qquad (108.12)$$

The above derivation cannot be applied directly when $\omega_1=\omega_2\equiv\omega$, since the Manley–Rowe relations then reduce to just the conservation of the total energy. The equation

$$\varepsilon_{i,kl}(\omega,\omega)=\varepsilon_{k,il}(-2\omega,\omega)=\varepsilon_{l,ki}(\omega,-2\omega) \qquad (108.13)$$

can, however, be derived simply on grounds of continuity by taking the limit of (108.11).

If both frequencies ω_1 and ω_2 tend to zero, the tensor $\varepsilon_{i,kl}$ becomes completely symmetrical. This symmetry expresses just the fact that in the static case the induction **D** can be found by differentiating the free energy with respect to **E**: $D_i=-4\pi\partial\tilde{F}/\partial E_i$, and so $\partial D_i/\partial E_k=\partial D_k/\partial E_i$. This shows that $\varepsilon_{i,kl}$ is symmetrical in the suffixes i and k, and therefore in all three suffixes.

If only one of the frequencies is zero, the relations (108.11) give

$$\varepsilon_{i,kl}(\omega,0)=\varepsilon_{l,ki}(\omega,-\omega), \qquad (108.14)$$

and

$$\varepsilon_{i,kl}(\omega,0)=\varepsilon_{k,il}(-\omega,0)=\varepsilon_{k,il}(\omega,0); \qquad (108.15)$$

by (108.5), the real functions $\varepsilon_{k,il}(\omega,0)$ are even functions of ω. The tensor $\varepsilon_{i,kl}(\omega,0)$ describes a linear electric–optical effect, the change in the permittivity of the crystal in a static electric field, and it is therefore the same as the tensor α_{ikl} defined in (100.4):

$$\varepsilon_{i,kl}(\omega,0)=\alpha_{ikl}(\omega);$$

from (108.15), it is symmetrical in the suffixes i, k, as it should be. The tensor $\varepsilon_{l,ki}(\omega,-\omega)$ describes another effect, the presence in the medium of a static dielectric polarization proportional to the square of the applied weak periodic field; compare the second term in (108.6). The equation (108.14) thus states the relation between these two effects.

By similar arguments using the non-linear susceptibility, which is a "cross" between electric and magnetic quantities, we could recover the relation between the magnetic–optical Faraday effect and the magnetization of a medium by a rotating electric field, as given by (101.15) and (101.25).

As already mentioned, for media invariant under spatial inversion, there is no second-order non-linearity. In such cases, the non-linear effects begin with the cubic terms in the expansion of **D(E)**. The corresponding third-order non-linear permittivity is a tensor of rank four, a function of three independent frequencies, $\varepsilon_{i,klm}(\omega_1,\omega_2,\omega_3)$. Its symmetry properties are exactly analogous to those of the second-order permittivity tensor: if we define a fourth frequency $\omega_4 = \omega_1 + \omega_2 + \omega_3$ and write it as $\varepsilon_{i,klm}(-\omega_4; \omega_1, \omega_2, \omega_3)$, then the suffixes can be permuted in any manner if the four arguments are permuted similarly.

The third-order non-linearity may be important even when there is quadratic non-linearity, owing to the specificity of the effects it causes.

§109. Self-focusing

In this section we shall discuss optical effects arising from the non-linear variation of the field at the primary wave frequency. That is, we consider the non-linear contribution to **D** at the frequency ω of the monochromatic field **E**. In the quadratic terms there is no such contribution, as they contain only the frequencies 2ω and zero. The first non-zero effect comes from the cubic non-linearity and occurs in terms having the form EEE^* (frequency $\omega + \omega - \omega = \omega$).

In the rest of this section, the medium (liquid or gas) will be assumed isotropic. The relevant third-order terms in the induction then have the general form

$$\mathbf{D}^{(3)} = \alpha(\omega)|\mathbf{E}|^2\mathbf{E} + \beta(\omega)\mathbf{E}^2\mathbf{E}^*; \tag{109.1}$$

they involve two independent coefficients, which in a transparent medium are real even functions of the frequency. This number of independent coefficients is in agreement with the symmetry properties of the tensor $\varepsilon_{i,klm}(-\omega; -\omega, \omega, \omega)$. With these values of the arguments, the tensor is symmetrical in the pairs of suffixes i, k and l, m; in an isotropic medium, such a tensor has two independent components. In the low-frequency limit, as noted in §108, the tensor must be symmetrical in all suffixes, and in an isotropic medium therefore proportional to the combination $\delta_{ik}\delta_{lm} + \delta_{il}\delta_{km} + \delta_{im}\delta_{kl}$. This means that

$$\alpha(0) = 2\beta(0). \tag{109.2}$$

The expression (109.1) can be simplified in the case of a linearly polarized field **E**. With this polarization, the complex vector **E** reduces to a real vector multiplied by a common phase factor; the expressions $|\mathbf{E}|^2\mathbf{E}$ and $\mathbf{E}^2\mathbf{E}^*$ then become

$$\mathbf{D}^{(3)} = (\alpha + \beta)|\mathbf{E}|^2\mathbf{E}. \tag{109.3}$$

A similar simplification occurs when the field **E** is circularly polarized; then $\mathbf{E}^2 = 0$, and (109.1) reduces to

$$\mathbf{D}^{(3)} = \alpha|\mathbf{E}|^2\mathbf{E}. \tag{109.4}$$

In either case, the induction is polarized in the same way as **E**. In the general case of elliptical polarization, however, the directions and ratios of the principal axes of the ellipses are not the same for **E** and $\mathbf{D}^{(3)}$.

The relation $\mathbf{D} = \mathbf{D}^{(1)} + \mathbf{D}^{(3)} = \varepsilon\mathbf{E} + \mathbf{D}^{(3)}$, where $\varepsilon(\omega)$ is the ordinary linear permittivity, must be substituted in Maxwell's equations, which are to be written (eliminating the magnetic field **H**) as

$$\mathbf{curl}\ \mathbf{curl}\ \mathbf{E} + \frac{1}{c^2}\frac{\partial^2 \mathbf{D}}{\partial t^2} = 0, \tag{109.5}$$

$$\operatorname{div} \mathbf{D} = 0. \tag{109.6}$$

It is important that these non-linear equations allow an exact solution as a monochromatic plane wave,

$$\mathbf{E} = \mathbf{E}_0 e^{i(\mathbf{k}\cdot\mathbf{r}-\omega t)} \tag{109.7}$$

with linear or circular polarization: for such waves $|\mathbf{E}|^2 = |\mathbf{E}_0|^2$, so that formulae (109.3) and (109.4) are of the same type as in the linear case with the permittivity depending on the field amplitude; we can therefore take the real part after solving the equations. The relation between **D** and **E** in these cases will be written as

$$\mathbf{D} = \left(\varepsilon + \frac{2c^2}{\omega^2}\eta|\mathbf{E}_0|^2\right)\mathbf{E}, \tag{109.8}$$

with the notation (which will later be convenient) $\eta = \omega^2(\alpha+\beta)/2c^2$ for a linearly polarized wave or $\eta = \omega^2\alpha/2c^2$ for a circularly polarized one.

Substitution of (109.8) in (109.6) gives div $\mathbf{E} = 0$, the field remaining transverse as in the linear theory. When this is taken into account, the substitution of (109.7) in (109.5) yields the dispersion relation

$$k^2 = \omega^2\varepsilon/c^2 + 2\eta|\mathbf{E}_0|^2. \tag{109.9}$$

The phase velocity ω/k now depends on the wave amplitude as well as on the frequency. If $\eta > 0$, the phase velocity decreases with increasing amplitude, and we have a *focusing medium* (the significance of this term will be explained later). If $\eta < 0$, the phase velocity increases with the amplitude, and this is a *defocusing medium*.

The use of the non-linear relation (109.1) presupposes, of course, that there is only slight non-linearity: the higher-order terms must be small in comparison with the terms in $\mathbf{D}^{(3)}$. Qualitatively new effects may arise by the "build-up" of non-linearity effects over long intervals of time and over large distances. A natural formulation of the problem here is to consider an almost monochromatic wave,

$$\mathbf{E} = \mathbf{E}_0(t, \mathbf{r})e^{i(k_0 x-\omega t)}, \tag{109.10}$$

where $\mathbf{E}_0(t, \mathbf{r})$ is a slowly varying function of time and coordinates, which has only a slight relative change over times $\sim 1/\omega$ and distances $\sim 1/k_0$. The wave vectors that occur in the Fourier expansion of this field are distributed over a small range of values about the vector $\mathbf{k}_0$, which is in the x direction and whose magnitude we regard as related to ω by

$$k_0{}^2 = \omega^2\varepsilon(\omega)/c^2, \tag{109.11}$$

as in the linear theory. We now derive an equation for $\mathbf{E}_0(t, \mathbf{r})$.

First of all, we note that in (109.5) **grad** div **E** is negligible in the term **curl curl E** = **grad** div $\mathbf{E} - \triangle\mathbf{E}$: by (109.6), the divergence of the field is

$$\operatorname{div}\mathbf{E} \cong -\eta(2c^2/\varepsilon\omega^2)\mathbf{E}_0\cdot\mathbf{grad}|\mathbf{E}_0|^2;$$

it is therefore non-zero only through the derivatives of the slowly varying function $\mathbf{E}_0$, and it is additionally small because the non-linear terms are small; such quantities are negligible. We thus have

$$\mathbf{curl}\,\mathbf{curl}\,\mathbf{E} \cong -\triangle\mathbf{E} \cong \left[k_0{}^2\mathbf{E}_0 - 2ik_0\frac{\partial\mathbf{E}_0}{\partial x} - \triangle_\perp\mathbf{E}_0\right]e^{i(\mathbf{k}_0\cdot\mathbf{r}-\omega t)}$$

where $\triangle_\perp = \partial^2/\partial y^2 + \partial^2/\partial z^2$; the term in the second derivative $\partial^2\mathbf{E}_0/\partial x^2$ is omitted, since it does not contain the large factor k_0. The transverse derivatives, however, may be large in comparison with the longitudinal ones.

The calculation of $\partial^2\mathbf{D}^{(1)}/\partial t^2$ is similar† to the derivation of (80.10) and gives

$$\frac{1}{c^2}\frac{\partial^2\mathbf{D}^{(1)}}{\partial t^2} \cong -\frac{1}{c^2}\left[\omega^2\varepsilon(\omega)\mathbf{E}_0 + i\frac{\partial(\omega^2\varepsilon)}{\partial\omega}\frac{\partial\mathbf{E}_0}{\partial t}\right]e^{i(\mathbf{k}_0\cdot\mathbf{r}-\omega t)}$$
$$= -\left(k_0{}^2\mathbf{E}_0 + 2i\frac{k_0}{u}\frac{\partial\mathbf{E}_0}{\partial t}\right)e^{i(\mathbf{k}_0\cdot\mathbf{r}-\omega t)},$$

with the group velocity u defined by

$$\frac{1}{u} = \frac{\mathrm{d}k_0}{\mathrm{d}\omega} = \frac{1}{c}\frac{\partial(\omega\sqrt{\varepsilon})}{\partial\omega}. \tag{109.12}$$

In the derivative of $\mathbf{D}^{(3)}$, it is sufficient to retain the term

$$\frac{1}{c^2}\frac{\partial^2\mathbf{D}^{(3)}}{\partial t^2} = -2\eta(\omega)|\mathbf{E}_0|^2\mathbf{E},$$

neglecting the term in the small derivative $\partial\mathbf{E}_0/\partial t$.

Substituting these expressions in (109.5), we have finally

$$ik_0\left(\frac{\partial}{\partial x} + \frac{1}{u}\frac{\partial}{\partial t}\right)\mathbf{E}_0 = -\tfrac{1}{2}\triangle_\perp\mathbf{E}_0 - \eta(\omega)|\mathbf{E}_0|^2\mathbf{E}_0. \tag{109.13}$$

The combination of derivatives on the left expresses the fact that the amplitude perturbations are transmitted in the direction of propagation of the wave, at the group velocity.

With this equation, we can examine the stability of an infinite plane wave described by the exact solution (109.7), (109.8) (V. I. Bespalov and V. I. Talanov, 1966). We shall see that the wave is unstable in a focusing medium.‡

According to (109.9), in the exact solution (109.7) $k \cong k_0 + \eta E_0{}^2/k_0$, with k_0 from (109.11); in a linearly polarized wave, the amplitude $\mathbf{E}_0$ may be defined as a real vector. Hence, if we write the wave (109.7) in the form (109.10), we must put in the latter $\mathbf{E}_0(x) = \mathbf{E}_0\exp(ix\eta E_0{}^2/k_0)$. This acts as the amplitude of the unperturbed wave. We shall consider the steady-state problem of the spatial evolution of perturbations along the direction of propagation of the wave. Accordingly, the amplitude of a wave subject to a small perturbation is written

$$\mathbf{E}_0(\mathbf{r}) = \{\mathbf{E}_0 + \delta\mathbf{E}(\mathbf{r})\}\exp(ix\eta E_0{}^2/k_0). \tag{109.14}$$

We shall assume that $\delta\mathbf{E}$ is parallel to $\mathbf{E}_0$.

Substitution of (109.14) in (109.13) gives

$$ik_0\partial\delta E/\partial x = -\tfrac{1}{2}\triangle_\perp\delta E - \eta E_0{}^2(\delta E + \delta E^*). \tag{109.15}$$

We put

$$\delta E = Ae^{i(\mathbf{q}\cdot\mathbf{r}+\gamma x)} + B^*e^{-i(\mathbf{q}\cdot\mathbf{r}+\gamma x)} \tag{109.16}$$

where $\mathbf{q}$ is a vector in the yz-plane. Substituting this expression in (109.15) and collecting

† The only difference is that here $\hat{f} = \partial^2\hat{\varepsilon}/\partial t^2$, $f(\omega) = -\omega^2\varepsilon(\omega)$.

‡ This was earlier noted by R. V. Khokhlov (1965).

the groups of terms in $\exp\{\pm i(\mathbf{q}\cdot\mathbf{r}+\gamma x)\}$, we obtain the two equations

$$(\tfrac{1}{2}q^2-\eta E_0{}^2+k_0\gamma)A-\eta E_0{}^2B=0,$$
$$-\eta E_0{}^2A+(\tfrac{1}{2}q^2-\eta E_0{}^2-k_0\gamma)B=0.$$

The condition that their determinant be zero gives

$$\gamma=\pm(q/2k_0)\sqrt{(q^2-4\eta E_0{}^2)}.$$

When $\eta>0$ and

$$q^2<4\eta E_0{}^2, \qquad (109.17)$$

γ is imaginary, so that δE in (109.16) contains an exponentially increasing term, and the wave is therefore unstable. The maximum instability growth rate is of the order of the non-linear correction to the wave vector.

One manifestation of this instability is the *self-focusing* of a light beam of finite width propagated in a focusing medium. The phenomenon occurs because, if the field amplitude decreases from the axis to the periphery of the beam, the permittivity of the medium (which depends on the amplitude) also decreases in that direction (if $\eta>0$), and the medium behaves as a focusing lens (G. A. Askar'yan, 1962). The behaviour of the beam is determined by the interplay of two opposing tendencies, namely this focusing and the broadening due to diffraction.

We shall show, first of all, that these tendencies may cancel out, in the sense that (if $\eta>0$) equation (109.13) has a solution in the form of a steady beam that undergoes no broadening. This *self-channelling* is a specifically non-linear effect. In the linear theory, any beam with a finite cross-section is broadened by diffraction. We shall take only the one-dimensional case where the field $\mathbf{E}$ depends only on one transverse coordinate y and is polarized in the z-direction and propagated in the x-direction.† An analytical solution can then be obtained (V. I. Talanov, 1965). We here ignore the fact that a beam of infinite width (in the z-direction) is certainly unstable, because perturbations with small q_z can occur in it, which are unstable by (109.17).

We write

$$E_{0z}=F(y)e^{i\kappa x} \qquad (109.18)$$

with a small quantity κ which acts as a correction to the wave number k_0. The function $F(y)$ is real. Substitution in (109.13) gives for it the equation

$$\tfrac{1}{2}\mathrm{d}^2F/\mathrm{d}y^2=k_0\kappa F-\eta F^3. \qquad (109.19)$$

The first integral is

$$\tfrac{1}{2}(\mathrm{d}F/\mathrm{d}y)^2-k_0\kappa F^2+\tfrac{1}{2}\eta F^4=\text{constant}.$$

We are interested in the solution for which F and $\mathrm{d}F/\mathrm{d}y$ tend to zero as $|y|\to\infty$. Accordingly, we take the constant to be zero, and a straightforward integration then gives

$$F=(2k_0\kappa/\eta)^{1/2}\,\mathrm{sech}\,[(2k_0\kappa)^{1/2}y], \qquad (109.20)$$

y being measured from the centre of the beam. The width of the beam in the y-direction is

$$\delta\sim(k_0\kappa)^{-1/2}\sim1/\sqrt{\eta}F(0).$$

† Under these conditions, the term **grad** div **E**, neglected previously as being small, is identically zero.

Since the energy flux along the beam is $W \sim E^2(0)\delta$, δ is proportional to $1/W$ and the beam becomes narrower as the power carried by it increases.

A self-channelled beam of this kind is a special case where the focusing properties of the medium are exactly balanced by diffraction. Other beams may either diverge or converge. First, let us derive a qualitative criterion for self-focusing of an actual beam with a finite cross-section (R. Y. Chiao, E. M. Garmire, and C. H. Townes, 1964). This can be done immediately by using the instability condition (109.17). In a beam with characteristic radius R, perturbations can occur whose wavelengths transverse to the beam axis are less than R, i.e. for which $q \gtrsim 1/R$. The condition (109.17) gives the upper limit of q values which lead to instability. The beam will therefore be unstable with respect to focusing if

$$E_0{}^2 R^2 \eta \gtrsim 1. \tag{109.21}$$

The power carried by the beam is the product $E_0{}^2 R^2$. The critical value of this power beyond which self-focusing begins is independent of the cross-sectional area of the beam.

It is also possible to establish an exact (not just order-of-magnitude) sufficient condition for self-focusing of a beam (S. N. Vlasov, V. A. Petrishchev, and V. I. Talanov, 1971).

For a steady linearly polarized light beam, but with no initial assumptions about the dependence on x, the equation for the function $E_0(x, \boldsymbol{\rho})$ is

$$ik_0 \partial E_0/\partial x = -\tfrac{1}{2}\triangle_\perp E_0 - \eta |E_0{}^2| E_0, \tag{109.22}$$

$\boldsymbol{\rho}$ being the two-dimensional position vector in the yz-plane, in which the differential operators $\triangle_\perp$ and $\mathbf{grad}_\perp$ act. It is easy to verify that this equation gives

$$\partial |E_0|^2/\partial x + \mathrm{div}_\perp \mathbf{j} = 0, \tag{109.23}$$

where

$$\mathbf{j} = (i/2k_0)(E_0\, \mathbf{grad}_\perp E_0{}^* - E_0{}^*\, \mathbf{grad}_\perp E_0).$$

Hence, in turn, it follows that the integral

$$N = \int_{-\infty}^{\infty} |E_0|^2 \, d^2\rho \tag{109.24}$$

is "conserved" (i.e. independent of x), as is

$$\mathscr{E} = \frac{1}{2k^2}\int_{-\infty}^{\infty} \{|\mathbf{grad}_\perp E_0|^2 - \eta |E_0|^4\} \, d^2\rho, \tag{109.25}$$

a result which is derived immediately by differentiating with respect to x and using (109.22). We assume, of course, that E_0 decreases sufficiently rapidly as $\rho \to \infty$, so that both integrals, as well as (109.26) below, converge.†

We will show that the behaviour of the beam is determined by the sign of the integral $\mathscr{E}$. When $\mathscr{E} > 0$ the beam is divergent on average; when $\mathscr{E} < 0$ it is focused. The proof is based

† As regards the derivatives present, equation (109.22) is similar to the two-dimensional Schrödinger's equation, with x in place of the time. In this analogy, N and $\mathscr{E}$ act as the particle number and the energy. The non-linearity of this equation does not affect the derivation of these conservation laws.

on a simple equation that can be derived for the mean beam radius R defined by

$$R^2(x) = \frac{1}{N} \int_{-\infty}^{\infty} \rho^2 |E_0|^2 \, d^2\rho. \qquad (109.26)$$

To derive this, we write, using (109.23),

$$\frac{d}{dx} \int |E_0|^2 \rho^2 d^2\rho = - \int \mathrm{div}_\perp \mathbf{j} \cdot \rho^2 d^2\rho = 2 \int \mathbf{j} \cdot \boldsymbol{\rho} \, d^2\rho.$$

Differentiating again with respect to x, we substitute $\partial E_0/\partial x$ from (109.22) and integrate twice by parts, obtaining $N d^2 R^2/dx^2 = 4\,\mathscr{E}$. Hence

$$R^2(x) = 2\mathscr{E}(x - x_0)^2/N + R_0{}^2, \qquad (109.27)$$

where x_0 and R_0 are constants. We see that, when $\mathscr{E} < 0$, complete focusing of the beam is attained at a finite distance in the direction of propagation, its radius R becoming zero.†

This result, derived with the approximate equation (109.13), cannot have a literal physical significance near the focus itself, where the assumptions made in obtaining that equation are invalid. We need only mention that, when the field energy density increases without limit in exact focusing, there is no justification for retaining only the lowest (the third) power of the non-linearity. However, the possibility of self-focusing of the beam, to such an extent that the non-linearity is no longer small, is important. It should be emphasized that the condition derived is sufficient but not necessary. A beam with $\mathscr{E} < 0$ certainly undergoes complete focusing, but the divergence on average of a beam with $\mathscr{E} \geqslant 0$ is not inconsistent with the focusing of some internal part of it.

§110. Second-harmonic generation

In §107, only some general relations have been considered which relate to frequency transformation processes characteristic of non-linear optics. We shall now describe the quantitative theory of a typical process of this type, namely second-harmonic generation, i.e. the excitation of an electromagnetic field with frequency 2ω by one with frequency ω (R. V. Khokhlov, 1960; J. A. Armstrong, N. Bloembergen, J. Ducuing and P. S. Pershan, 1962).

Second-harmonic generation is a non-linear second-order effect. It resides in the non-linear susceptibility tensor

$$\varepsilon_{i,kl}(-2\omega; \omega, \omega) \qquad (110.1)$$

and therefore does not occur in media which allow spatial inversion. The tensor (110.1) is symmetrical in the suffixes k and l; its symmetry properties in various crystals are the same as for the piezoelectric tensor (§17). We shall assume that there is no absorption in the medium, so that the $\varepsilon_{i,kl}$ are real quantities.

The problem of second-harmonic generation may be stated as follows. Let a monochromatic plane wave of frequency ω be incident on the plane surface of a crystal. As

† If the distribution of $|E_0|^2$ over the beam cross-section is unchanged along the beam, then R^2 = constant and $\mathscr{E} = 0$. The converse is not necessarily true, however: there may exist solutions with $\mathscr{E} = 0$, so that R = constant, but with the distribution depending on x.

well as the reflected and two (in a doubly refracting crystal) refracted waves with the same frequency, a reflected wave and refracted waves with frequency 2ω are also formed. The waves with this frequency in the crystal constitute the solution of equations (109.5) and (109.6), in which the non-linear term $\mathbf{D}^{(2)}$ in the induction is to be expressed in terms of the fundamental wave field. The amplitudes of all these waves are expressed in terms of the incident wave amplitude by means of the boundary conditions, which will not be discussed here. The amplitudes of the waves with frequency 2ω are, of course, small in proportion to the non-linear susceptibility.†

The refracted waves are propagated into the crystal as if it were infinite. The non-linearity effects build up during this process, and the harmonic intensities may become large, with energy transferred to them from the fundamental frequency. This is the process to be considered here. The conditions at the crystal surface then act only as "initial" conditions which specify a small but non-zero amplitude of the second-harmonic field. The same conditions specify, for a given direction of the incident wave, the wave vectors $\mathbf{k}_1$ and $\mathbf{k}_2$ of the first and second harmonics in the crystal.

It will be seen later that an effective energy transfer occurs only if the *synchronism condition* holds for the fundamental and the harmonic:‡

$$\mathbf{k}_2 \cong 2\mathbf{k}_1. \tag{110.2}$$

It should be emphasized that the presence of dispersion is of fundamental importance in formulating the problem of the generation of just one second harmonic. When dispersion is absent, the condition (110.2) is necessarily satisfied in refraction, as well as similar conditions for the higher harmonics also ($\mathbf{k}_3 \cong 3\mathbf{k}_1$, and so on). When dispersion is present, this is not so, and it can be assumed that if the synchronism condition is satisfied for the second harmonic it is not satisfied for the others. It should be emphasized that the condition (110.2) itself can in practice be satisfied only if the fundamental wave and the harmonic have different types of polarization and therefore different dispersion relations.

The field in the medium may be written as a superposition of two waves:

$$\mathbf{E} = \mathbf{E}_1 + \mathbf{E}_2 = \mathrm{re}\,[\mathbf{e}_1 E_{10} e^{i(\mathbf{k}_1\cdot\mathbf{r} - \omega t)} + \mathbf{e}_2 E_{20} e^{i(\mathbf{k}_2\cdot\mathbf{r} - 2\omega t)}], \tag{110.3}$$

where, according to the condition (110.2),

$$\mathbf{k}_2 = 2\mathbf{k}_1 + \mathbf{q} \tag{110.4}$$

with $q \ll k_1$. The wave amplitudes are written as products $\mathbf{E}_0 = \mathbf{e}E_0$, with $\mathbf{e}$ a unit polarization vector ($\mathbf{e}\cdot\mathbf{e}^* = 1$). In the linear approximation these amplitudes would be constant; when the non-linearity is taken into account, they are functions of the coordinates which vary only slowly (i.e. only slightly over distances $\sim 1/k_1$).

The equations for the amplitudes of the two waves are found by substituting (110.3) in Maxwell's equations (109.5), (109.6), and collecting terms with the same time dependence. We shall not go through these straightforward but laborious calculations in detail, but give only some comments on fundamental points.

† The calculation of the reflection and refraction conditions at the boundary of a non-linear medium is discussed for several particular cases by N. Bloembergen and P. S. Pershan, *Physical Review* **128**, 606, 1962.

‡ The nature of this condition is particularly clear from the quantum standpoint, the second-harmonic generation being regarded as a "fusion" of two photons into one. The equation $\hbar\mathbf{k}_2 = 2\hbar\mathbf{k}_1$ expresses the fact that momentum is conserved in this process.

We shall seek solutions describing the steady generation of the second harmonic in the crystal by a travelling fundamental wave. Such solutions are independent of time. With exact synchronism ($\mathbf{q} = 0$) the equations for the amplitudes would not explicitly contain the coordinates at all; with inexact synchronism, the coordinates appear only as $\mathbf{q}\cdot\mathbf{r}$ in the factor $e^{-i\mathbf{q}\cdot\mathbf{r}}$. Taking the direction of $\mathbf{q}$ as the z-axis, we can therefore look for solutions that depend only on z. With the above formulation (a wave ω incident on the surface of the crystal), the problem is homogeneous in planes parallel to the surface. The z axis is therefore perpendicular to the surface; the vector $\mathbf{q}$ necessarily is so also, according to the boundary conditions.

In the linear approximation, waves in an anisotropic (non-gyrotropic) medium are linearly polarized (see §97); for these $\mathbf{e}$ can be defined as a real vector, and such will be understood here by $\mathbf{e}_1$ and $\mathbf{e}_2$ for the two waves. If the amplitude $\mathbf{E}_0$ of each wave is resolved along the directions of $\mathbf{e}$, $\mathbf{k}$ and $\mathbf{e}\times\mathbf{k}$, then the components in the two latter directions are small because the derivatives $\mathrm{d}E_0/\mathrm{d}z$ (which appear as a result of the non-linearity) are small. The components along $\mathbf{e}$ are approximately equal to the magnitudes E_0. Equations for them are obtained by multiplying equation (109.5) by $\mathbf{e}_1$ and by $\mathbf{e}_2$. Since waves with E_0 = constant are exact solutions of Maxwell's equations in the linear approximation, the linear terms in these equations that do not involve derivatives with respect to z cancel out. The terms involving the components of $\mathbf{E}_0$ in the directions of $\mathbf{k}$ and $\mathbf{e}\times\mathbf{k}$, which might be of the same order as those containing the derivatives $\mathrm{d}E_0/\mathrm{d}z$, are found to disappear when the multiplication is carried out; this occurs because the induction $\mathbf{D}$ is orthogonal to both $\mathbf{k}$ and $\mathbf{e}\times\mathbf{k}$ (see (97.3)).

Since the amplitudes are assumed to vary only slowly with the coordinates, we can neglect the second derivatives of $\mathbf{E}_0$ with respect to z. Hence, for example, the expression

$$e_{2i}\left\{(\mathbf{curl\ curl})_{ik} - \frac{4\omega^2}{c^2}\varepsilon_{ik}(2\omega)\right\} e_{2k}E_{20}e^{i\mathbf{k}_2\cdot\mathbf{r}},$$

in the equation for $\mathbf{E}_2$ multiplied by $\mathbf{e}_2$, becomes approximately

$$2i\mathbf{e}_2\cdot[\mathbf{k}_2\times(\mathbf{l}\times\mathbf{e}_2)]\mathrm{d}E_{20}/\mathrm{d}z,$$

where $\mathbf{l}$ is a unit vector in the z-direction, and similarly for $\mathbf{E}_1$.

The final equations obtained by these procedures are

$$\left.\begin{aligned} \alpha_2\mathrm{d}E_{20}/\mathrm{d}z &= -i\eta e^{-iqz}E_{10}{}^2, \\ \alpha_1\mathrm{d}E_{10}{}^*/\mathrm{d}z &= i\eta e^{-iqz}E_{10}E_{20}{}^*, \end{aligned}\right\} \qquad (110.5)$$

with the notation†

$$\eta = (\omega^2/2c^2)\varepsilon_{i,kl}(\omega,\omega)e_{2i}e_{1k}e_{1l}, \qquad (110.6)$$

$$\left.\begin{aligned} \alpha_1 &= \mathbf{l}\cdot[\mathbf{e}_1\times(\mathbf{k}_1\times\mathbf{e}_1)], \\ \alpha_2 &= \tfrac{1}{2}\mathbf{l}\cdot[\mathbf{e}_2\times(\mathbf{k}_2\times\mathbf{e}_2)] \cong \mathbf{l}\cdot[\mathbf{e}_2\times(\mathbf{k}_1\times\mathbf{e}_2)]. \end{aligned}\right\} \qquad (110.7)$$

Multiplying the first equation (110.5) by $E_{20}{}^*$ and the second by E_{10} and adding, we obtain a first integral:

$$\alpha_1|E_{10}|^2 + \alpha_2|E_{20}|^2 = \text{constant} \equiv P. \qquad (110.8)$$

† The first equation (110.5) is derived from the terms in $e^{-2i\omega t}$, and the second from those in $e^{i\omega t}$. The relations (108.13) have been used on the right-hand sides.

This expresses the fact that the total energy flux in the two waves in the z-direction is constant.†

It is now convenient to change from complex quantities to real ones, namely the absolute values and phases of E_{10} and E_{20}. To make the equations as simple as possible, we define new unknowns ρ_1, ρ_2, ϕ_1, ϕ_2 as dimensionless quantities by

$$E_{10} = \sqrt{(P/\alpha_1)}\rho_1 e^{i\phi_1}, \quad E_{20} = \sqrt{(P/\alpha_2)}\rho_2 e^{i\phi_2}. \tag{110.9}$$

The equations (110.5) are invariant under the transformation $\phi_1 \to \phi_1 + c, \phi_2 \to \phi_2 + 2c$. It is therefore possible to separate the equations for the functions ρ_1, ρ_2 and the invariant combination $2\phi_1 - \phi_2$, to obtain a closed set of equations

$$d\rho_1/d\zeta = -\rho_1\rho_2 \sin\theta, \quad d\rho_2/d\zeta = \rho_1{}^2 \sin\theta, \tag{110.10}$$

$$d\theta/d\zeta = -s - (2\rho_2 - \rho_1{}^2/\rho_2)\cos\theta, \tag{110.11}$$

where

$$\theta = 2\phi_1 - \phi_2 - s\zeta, \tag{110.12}$$

the dimensionless variable

$$\zeta = z\eta\sqrt{(P/\alpha_1{}^2\alpha_2)}, \tag{110.13}$$

and the dimensionless parameter

$$s = (q/n)\sqrt{(\alpha_1{}^2\alpha_2/P)}. \tag{110.14}$$

The first integral (110.8) then becomes

$$\rho_1{}^2 + \rho_2{}^2 = 1. \tag{110.15}$$

Let us consider the case of exact synchronism, where $q = 0$ and therefore $s = 0$. Then equations (110.10) and (110.11) have another first integral

$$\rho_1{}^2\rho_2 \cos\theta = \text{constant} \equiv \delta, \tag{110.16}$$

where $\delta^2 \leqslant 4/27$, as is easily seen from equations (110.15) and (110.16) and the condition $|\cos\theta| \leqslant 1$. With these two integrals, the solution of equations (110.10) is reduced to quadratures, in the form of the elliptic integral

$$\zeta = \pm\tfrac{1}{2}\int\limits_{\rho_2{}^2(0)}^{\rho_2{}^2(\zeta)} \frac{du}{[u(1-u)^2 - \delta^2]^{1/2}}; \tag{110.17}$$

the choice of sign depends on the sign of the initial value of $\sin\theta$ for $\zeta = 0$. The cubic equation

$$u(1-u)^2 - \delta^2 = 0 \tag{110.18}$$

† The time-averaged energy flux density in the $\mathbf{E}_1$ wave is

$$\begin{aligned}\overline{S}_{1z} &= c\,\mathrm{re}\,\mathbf{l}\cdot[\mathbf{E}_{10}{}^* \times \mathbf{H}_{10}]/8\pi \\ &= c^2\mathbf{l}\cdot[\mathbf{E}_{10}{}^* \times (\mathbf{k}_1 \times \mathbf{E}_{10})]/8\pi\omega \\ &= c^2\alpha_1 |E_{10}|^2/8\pi\omega,\end{aligned}$$

and similarly for the $\mathbf{E}_2$ wave.

has (when $\delta^2 \leqslant 4/27$) three real positive roots, two of them less than unity, which we denote by $\rho_a{}^2$ and $\rho_b{}^2$ with $\rho_a{}^2 \leqslant \rho_b{}^2$.† The function $\rho_2{}^2(\zeta)$ defined by (110.17) varies periodically between these values, the period being

$$\int\limits_{\rho_a{}^2}^{\rho_b{}^2} \frac{du}{[u(1-u)^2-\delta^2]^{1/2}}. \qquad (110.19)$$

The function $\rho_1{}^2(\zeta) = 1-\rho_2{}^2(\zeta)$ varies similarly, one being a minimum when the other is a maximum.

The value $\rho_a{}^2$ is to be identified with the second harmonic $\rho_2{}^2(0)$ given by the boundary conditions at the crystal surface ($z = 0$). We see that, in the z-direction into the crystal, there is a periodic transfer of energy from the fundamental to the second harmonic and back. When $\rho_2(0)$ decreases, the period of this process becomes longer, tending logarithmically to infinity as $\rho_2(0) \to 0$. The limiting value $\rho_2(0) = \rho_a = 0$ corresponds to the solution

$$\rho_1 = \operatorname{sech}\zeta, \quad \rho_2 = \tanh\zeta, \qquad (110.20)$$

which is obtained from (110.17) for $\delta = 0$ by elementary integration; in this solution, the second-harmonic amplitude increases monotonically, and as $\zeta \to \infty$ all the energy is asymptotically transferred from the fundamental to the harmonic.

Let us now consider the opposite case where the amplitude ρ_2 remains everywhere small in comparison with ρ_1. We shall see that this corresponds to a considerable loss of synchronism of the waves.

When $\rho_2 \ll \rho_1$, we can as a first approximation regard ρ_1 as a constant $\rho_1(0)$, and write the equations for ρ_2 and θ as

$$d\rho_2/d\zeta = \rho_1{}^2(0)\sin\theta, \quad d\theta/d\zeta = -s+[\rho_1{}^2(0)/\rho_2]\cos\theta.$$

The solution of these equations that is zero at the initial point $\zeta = 0$ is

$$\rho_2(\zeta) = (2/s)\rho_1{}^2(0)\sin\tfrac{1}{2}s\zeta, \quad \theta = \tfrac{1}{2}\pi-\tfrac{1}{2}s\zeta. \qquad (110.21)$$

These give the variation of the field over one interval $0 \leqslant \zeta \leqslant 2\pi/s$ (i.e. $0 \leqslant x \leqslant 2\pi/q$), after which the process is repeated periodically.‡ The condition $\rho_2 \ll \rho_1$ signifies that we must have $\rho_1{}^2(0)/s \ll \rho_1(0)$, i.e. $s \gg \rho_1(0)$, or

$$qz_0 \gg 1, \quad z_0 \sim 1/\eta\rho_1(0)\sqrt{P}.$$

This is the condition for a comparatively large loss of synchronism. The value of q determines in all cases which effect limits the generation of the harmonic (i.e. the increase in the amplitude ρ_2)—whether it is the linear effect of loss of synchronism when $qz_0 \gg 1$, or the non-linear effects when $qz_0 \ll 1$.§

Hitherto in this section, we have everywhere referred to second-harmonic generation from the fundamental frequency. The equations also describe the converse effect, however,

† Between these roots, the polynomial on the left of (110.18) has a maximum at $u = \frac{1}{3}$, equal to $4/27-\delta^2$; when $\delta^2 = 4/27$, this is zero, two real roots coincide, and they disappear for larger values of δ.

‡ In each successive period, π is to be added to the constant term in the phase variable θ. At the point where $\rho_2 = 0$, the phase ϕ_2 has no meaning and the phase difference θ may have a discontinuity.

§ The whole of this discussion is based on the assumption that $q \ll k_1$. This condition has been fully utilized in deriving equations (110.5), which do not involve the small parameter q/k_1. In dealing with the case $qz_0 \gg 1$, of course, we assume that it is compatible with the condition $q \ll k_1$.

namely the *parametric amplification* of a weak signal with frequency ω in the field of strong radiation with frequency 2ω. Here we shall consider only this one process, but it is the simplest case of a more general phenomenon, the amplification of signals with different frequencies ω_2 and $\omega_1-\omega_2$ in the field of a strong wave with frequency ω_1 (S. A. Akhmanov and R. V. Khokhlov, 1962; R. H. Kingston, 1962).

We should first of all emphasize the following difference between this process and second-harmonic generation. The latter can begin from zero intensity of the harmonic. The amplification of the fundamental, however, depends on a non-zero initial intensity of it, however small: if $\rho_1(0) = 0$ at the starting-point, then it will remain so, for equations (110.10) show that the derivatives of ρ_1 and ρ_2 of all orders vanish with ρ_1.

Let us again consider the case of exact synchronism, and in addition let the initial value of the phase variable be $\theta(0) = -\frac{1}{2}\pi$; with exact synchronism, this value will be maintained. Then $\delta = 0$ since $\cos\theta = 0$, even though the initial values of ρ_1 and ρ_2 are not zero. The solution of equations (110.10) in this case is

$$\rho_1 = \operatorname{sech}(\zeta-\zeta_0), \quad \rho_2 = -\tanh(\zeta-\zeta_0), \tag{110.22}$$

where $\zeta_0 > 0$ is a constant. When ζ_0 is large, the initial value $\rho_1(0) = \operatorname{sech}\zeta_0$ is small. We see that the fundamental frequency waves are amplified in the z-direction into the crystal, at the expense of the harmonic intensity. The latter decreases to zero at $\zeta = \zeta_0$ and then increases again until asymptotically the whole intensity is concentrated in the harmonic.†

§111. Strong electromagnetic waves

The formulation of the problem in §110 in terms of the generation of just one particular harmonic was dependent on the presence of dispersion. Let us now consider the opposite case, where dispersion may be regarded as absent throughout the relevant range of frequencies, so that the induction $\mathbf{D}(t)$ at every point is determined by the value of $\mathbf{E}(t)$ at the same instant.‡ We shall regard the medium as isotropic; **D** and **E** are then parallel. In the present section, the non-linearity is not assumed small, and the function $D(E)$ is therefore arbitrary.

Neglecting absorption and dispersion is of fundamental significance, in the sense that the field equations then involve no parameters having the dimension of frequency or (equivalently) of length. This makes it possible to construct an exact solution which generalizes the usual one-dimensional plane wave of the linear approximation (A. V. Gaponov and G. I. Freĭdman, 1959).§

Let the wave be propagated in the x-direction, the electric field be in the y-direction, and the magnetic field be in the z-direction; E_y and H_z will be denoted simply by E and H. Maxwell's equations

$$\mathbf{curl\ H} = \frac{1}{c}\frac{\partial \mathbf{D}}{\partial t}, \quad \mathbf{curl\ E} = -\frac{1}{c}\frac{\partial \mathbf{H}}{\partial t}$$

† When $\zeta > \zeta_0$, the phase variable is to be given the value $\theta = \frac{1}{2}\pi$, and the sign of the hyperbolic tangent in $\rho_2(\zeta)$ changed.

‡ For uniformity of exposition in this chapter, we shall continue to refer to a non-linear relation between **D** and **E**, assuming the medium non-magnetic. Actually, the phenomena in question usually relate to media with a non-linear dependence of **B** on **H**.

§ This solution is analogous to the simple waves in one-dimensional flow of a compressible fluid (see *FM*, §94).

become

$$-\frac{\partial H}{\partial x} = \frac{1}{c}\frac{\partial D}{\partial t} = \frac{\varepsilon}{c}\frac{\partial E}{\partial t}, \quad -\frac{\partial E}{\partial x} = \frac{1}{c}\frac{\partial H}{\partial t}, \tag{111.1}$$

where by definition

$$\varepsilon(E) = \mathrm{d}D/\mathrm{d}E; \tag{111.2}$$

As $E \to 0$, $\varepsilon(E)$ tends to the ordinary permittivity ε_0.

We shall seek a solution in which the functions $E(t, x)$ and $H(t, x)$ may be expressed as functions of each other: $H = H(E)$. Equations (111.1) may then be written as

$$\frac{\varepsilon}{c}\frac{\partial E}{\partial t} + \frac{\mathrm{d}H}{\mathrm{d}E}\frac{\partial E}{\partial x} = 0, \quad \frac{1}{c}\frac{\mathrm{d}H}{\mathrm{d}E}\frac{\partial E}{\partial t} + \frac{\partial E}{\partial x} = 0. \tag{111.3}$$

In order for these two equations to be satisfied by non-zero values of the unknowns $\partial E/\partial t$ and $\partial E/\partial x$, their determinant must be zero. From this condition, $(\mathrm{d}H/\mathrm{d}E)^2 = \varepsilon(E)$, and so

$$H = \pm \int_0^E \sqrt{\varepsilon(E)}\,\mathrm{d}E. \tag{111.4}$$

Now substituting $\mathrm{d}H/\mathrm{d}E$ from (111.4) in one of the equations (111.3), we have

$$-\frac{(\partial E/\partial t)_x}{(\partial E/\partial x)_t} = \left(\frac{\partial x}{\partial t}\right)_E = \pm c/\sqrt{\varepsilon}.$$

Hence $x \mp ct/\sqrt{\varepsilon}$ may be any function of E. Denoting the inverse function by f, we have

$$E = f(x \mp ct/\sqrt{\varepsilon(E)}); \tag{111.5}$$

the two signs here correspond to the two directions of propagation of the wave. When the function f has been chosen, (111.5) determines implicitly the function $E(t, x)$. In weak fields, when we can take $\varepsilon = \varepsilon_0$, (111.5) becomes an ordinary plane wave with phase velocity $c/\sqrt{\varepsilon_0}$. The solution found exists only for $\varepsilon > 0$, in accordance with the stability condition (18.8).†

As the wave is propagated, the profile which it has at the initial instant becomes distorted, since different parts of it move at different velocities. Usually, $\varepsilon(E)$ decreases with increasing E, and tends to saturation. Then the points on the profile where E is greater will travel at higher speeds, so that the slope of the forward front becomes steeper; Fig. 60 shows the profile at successive times. At some instant, the profile turns over and would then have to become no longer one-valued. In reality, at this instant an *electromagnetic shock*

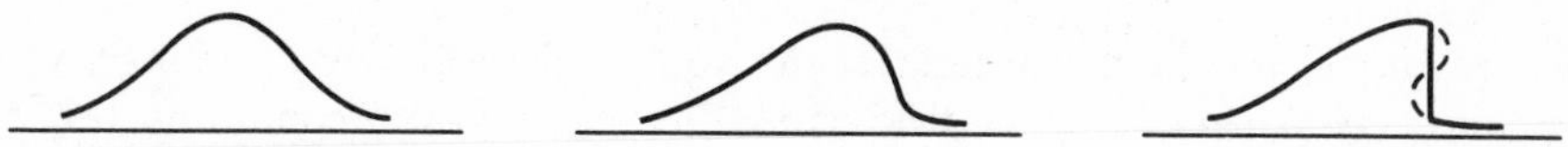

FIG. 60

† The derivation given here assumes that the density, temperature, etc. of the medium are not affected by the field oscillations. This is justified both by the smallness of the striction effects and by the largeness of the wave propagation velocity relative to that of sound.

wave (a discontinuity of E and H) is formed. The boundary conditions at the discontinuity have the same form (76.13) as on any moving surface. For a transverse plane wave they are

$$\left.\begin{aligned} H_2 - H_1 &= v(D_2 - D_1)/c, \\ E_2 - E_1 &= v(H_2 - H_1)/c, \end{aligned}\right\} \tag{111.6}$$

where the suffixes 1 and 2 denote the values of quantities before and behind the front respectively. Multiplication of these two equations gives for the shock wave velocity

$$v^2 = c^2(E_2 - E_1)/(D_2 - D_1). \tag{111.7}$$

In a shock wave there is dissipation of energy. Let Q be the rate of dissipation per unit area of the discontinuity surface. To calculate Q, we write the law of energy conservation for a cylindrical volume element in the medium with one end on each side of the discontinuity:

$$c(E_2H_2 - E_1H_1)/4\pi = v(U_2 - U_1) + Q. \tag{111.8}$$

The left-hand side is the difference of the energy fluxes through the two ends; the right-hand side is the sum of the rate of change of the internal energy through the movement of the boundary of regions 1 and 2, and the energy dissipated at the boundary. The difference of the internal energies is (for constant density and temperature)

$$U_2 - U_1 = \frac{1}{4\pi}\int_{D_1}^{D_2} E\,dD + \frac{1}{8\pi}(H_2{}^2 - H_1{}^2).$$

Using also equations (111.6) and (111.7), we can bring (111.8) to the form

$$Q = \frac{v}{4\pi}\left\{\tfrac{1}{2}(D_2 - D_1)(E_2 + E_1) - \int_{D_1}^{D_2} E\,dD\right\}.$$

If the shock wave is weak, i.e. the discontinuities in it are small, in calculating Q we can express the relation between D and E as an expansion

$$D(E) = D_1 + \varepsilon(E_1)(E - E_1) + \tfrac{1}{2}\varepsilon'(E_1)(E - E_1)^2,$$

where $\varepsilon'(E) = d^2D/dE^2$. A straightforward calculation gives

$$Q = -v\varepsilon'(E_1)(E_2 - E_1)^3/48\pi. \tag{111.9}$$

The energy dissipation in a weak electromagnetic shock wave is therefore of the third order in the field discontinuity in the shock. Since $Q > 0$, when $\varepsilon' < 0$ we must have $E_2 > E_1$, as shown in Fig. 60.

The presence of the shock wave makes the solution derived above no longer valid: the expressions (111.4), (111.5) for the field contradict the boundary conditions (111.6). It is important to note, however, that the wave remains approximately (up to and including second-order quantities) a simple wave so long as the shock wave may be regarded as weak.† To this accuracy, the expression for the velocity of the discontinuity can be written

† The situation here is exactly analogous to the formation of ordinary hydrodynamic shock waves in a strong sound wave (see *FM*, §95), and the arguments will not be repeated in full.

as

$$v = c\left[\tfrac{1}{2}\varepsilon(E_1 + E_2)\right]^{-1/2}. \qquad (111.10)$$

In the same approximation, the position of the discontinuity on the wave profile is determined by the equality of the two areas between the vertical line and the dashed curve in Fig. 60.

PROBLEM

A plane wave $E = f_i(t - x/c)$ is incident normally from a vacuum on to the surface of a medium. Determine the reflected wave (L. J. F. Broer, 1963).

SOLUTION. The field in the vacuum (the half-space $x < 0$) consists of the incident wave and the reflected wave (suffix r):

$$E = f_i(t - x/c) + f_r(t + x/c),$$

$$H = f_i(t - x/c) - f_r(t + x/c);$$

in the vacuum, the field equations are linear and the two solutions may be added. In the medium ($x > 0$) there is only the transmitted wave, in which

$$E = f_t[t - (x/c)\sqrt{\varepsilon(E)}], \quad H = \int_0^E \sqrt{\varepsilon}\, dE.$$

The continuity condition on the electric field at $x = 0$ gives $f_t(t) = f_i(t) + f_r(t)$. The continuity of H at $x = 0$ then gives

$$f_i(t) - f_r(t) = \int_0^{f_i(t)+f_r(t)} \sqrt{\varepsilon(E)}\, dE,$$

and this implicitly determines the function f_r.

§112. Stimulated Raman scattering

The third-order non-linear effects include the influence of radiation with some frequency ω_1 (the *pump wave*) on the propagation of a wave with a different frequency ω_2 in the same medium. Such effects are contained in the non-linear permittivity

$$\varepsilon_{i,klm}(\omega_2, \omega_1, -\omega_1) \qquad (112.1)$$

which contributes to the induction with frequency ω_2.†

In an isotropic medium, the induction $\mathbf{D}_2$ at the frequency ω_2, including the contribution mentioned, is given by

$$\mathbf{D}_2 = \varepsilon_2 \mathbf{E}_2 + \alpha_2(\mathbf{E}_1 \cdot \mathbf{E}_1^*)\mathbf{E}_2 + \beta_2 \mathbf{E}_1(\mathbf{E}_1^* \cdot \mathbf{E}_2) + \gamma_2 \mathbf{E}_1^*(\mathbf{E}_1 \cdot \mathbf{E}_2), \qquad (112.2)$$

where

$$\mathbf{E}_1 = \mathbf{E}_{10} e^{i(\mathbf{k}_1 \cdot \mathbf{r} - \omega_1 t)}, \quad \mathbf{E}_2 = \mathbf{E}_{20} e^{i(\mathbf{k}_2 \cdot \mathbf{r} - \omega_2 t)}. \qquad (112.3)$$

---

† The condition that the third-order permittivity $\varepsilon_{i,klm}(\omega_1, \omega_2, \omega_3)$ be real requires that not only the frequencies $\omega_1, \omega_2, \omega_3$ and their sum ω_4 but also their sums in pairs should not coincide with differences of energy levels in the system; for the susceptibility (112.1), these sums are $\omega_1 + \omega_2$ and $|\omega_1 - \omega_2|$. This may be seen by tracing the origin of the energy denominators in the expression mentioned in §108 for the non-linear susceptibility in terms of the matrix elements of the interaction between the medium and the field; this expression has been given by J. A. Armstrong, N. Bloembergen, J. Ducuing and P. S. Pershan, *Physical Review* **127**, 1918, 1962.

In the first term in (112.2), $\varepsilon_2 = \varepsilon_2(\omega_2)$ is the ordinary linear permittivity; in the other terms, α_2, β_2 and γ_2 are three independent components of the tensor (112.1), their number being obvious from the way in which (112.2) is constructed from the three vectors $\mathbf{E}$, $\mathbf{E}_1{}^*$ and $\mathbf{E}_2$. We see that the non-linear action of the field $\mathbf{E}_1$ on the field with frequency ω_2 can be described by means of an anisotropic permittivity

$$\varepsilon_{2ik} = (\varepsilon_2 + \alpha_2 \mathbf{E}_1 \cdot \mathbf{E}_1{}^*)\delta_{ik} + \beta_2 E_{1i} E_{1k}{}^* + \gamma_2 E_{1i}{}^* E_{1k}. \tag{112.4}$$

In a non-dissipative medium, the coefficients α_2, β_2 and γ_2 are, like ε_2, real, and the tensor (112.4) is Hermitian. When $\omega_1 \to 0$ and accordingly $\mathbf{E}_1$ is real, it includes as a particular case double refraction in a static electric field, described by (100.1). When $\omega_1 \neq 0$, (112.4) also describes the gyrotropy of the medium induced by the field $\mathbf{E}_1$. Comparison of (112.2) with (101.8) gives the gyration vector

$$\mathbf{g} = \tfrac{1}{2}i(\beta_2 - \gamma_2)\mathbf{E}_1{}^* \times \mathbf{E}_1. \tag{112.5}$$

This is zero if the field $\mathbf{E}_1$ is linearly polarized.

A greater variety of phenomena can occur if the non-linear interactions of the field with the medium are accompanied by dissipation. In that case, the coefficients α_2, β_2 and γ_2 are complex; the linear permittivity will again be supposed real. It is found that such dissipation can cause either attenuation or amplification of the field $\mathbf{E}_2$. In the latter case we have *stimulated Raman scattering*.†

Since the linear permittivities $\varepsilon(\omega_1)$ and $\varepsilon(\omega_2)$ are real, there is no dissipation in the medium at the frequencies ω_1 and ω_2: the quanta $\hbar\omega_1$ and $\hbar\omega_2$ themselves are not absorbed. Let the difference $\omega_1 - \omega_2$, but not the sum $\omega_1 + \omega_2$, lie in the frequency range where the medium is capable of absorption. The dissipation takes place only by the conversion of quanta of higher energy into quanta of lower energy, the excess released being transferred to the medium. When $\omega_1 > \omega_2$, therefore, the pump wave amplifies the wave with the lower frequency ω_2. The time-averaged energy gained by the field $\mathbf{E}_2$ per unit time and volume through the weak non-linear effects is just minus the expression (96.5):

$$\begin{aligned}\frac{d\overline{U}_2}{dt} &= -\frac{i\omega_2}{8\pi}(\varepsilon_{2ik}{}^* - \varepsilon_{2ki})E_{2i}E_{2k}{}^* \\ &= -\frac{\omega_2}{4\pi}\left\{\alpha_2''|\mathbf{E}_1|^2|\mathbf{E}_2|^2 + \beta_2''|\mathbf{E}_1\cdot\mathbf{E}_2|^2 + \gamma_2''|\mathbf{E}_1\cdot\mathbf{E}_2{}^*|^2\right\};\end{aligned} \tag{112.6}$$

cf. the derivation of equations (108.9). A similar expression gives the change in energy of the field $\mathbf{E}_1$:

$$\frac{d\overline{U}_1}{dt} = -\frac{\omega_1}{4\pi}\left\{\alpha_1''|\mathbf{E}_1|^2|\mathbf{E}_2|^2 + \beta_1''|\mathbf{E}_1\cdot\mathbf{E}_2|^2 + \gamma_1''|\mathbf{E}_1\cdot\mathbf{E}_2{}^*|^2\right\}, \tag{112.7}$$

where α_1, β_1 and γ_1 are independent components of the permittivity tensor $\varepsilon_{i,klm}(\omega_1, \omega_2, -\omega_2)$, which describes the effect of a field with frequency ω_2 on one with frequency ω_1.

† From the quantum microscopic standpoint, this refers to the emission of a photon $\hbar\omega_2$ when a photon $\hbar\omega_1$ is incident on atoms in a field of photons $\hbar\omega_2$. The energy $\hbar(\omega_1 - \omega_2)$ is transferred to the medium, with the formation of an elementary excitation of the medium such as a phonon or an exciton. The literature of the subject uses specific names for various types of scattering process. In our purely phenomenological description, we use the term mentioned, as a conventional general name.

Arguments similar to those used in §107 to derive the Manley–Rowe theorem show that

$$\frac{1}{\omega_1}\frac{d\bar{U}_1}{dt} = -\frac{1}{\omega_2}\frac{d\bar{U}_2}{dt}: \qquad (112.8)$$

for each quantum $\hbar\omega_2$ created, one quantum $\hbar\omega_1$ disappears. Hence it follows that

$$\alpha_1'' = -\alpha_2'', \quad \beta_1'' = -\beta_2'', \quad \gamma_1'' = -\gamma_2''. \qquad (112.9)$$

The energy dissipated is given by the decrease in the total energy of the two fields:

$$Q = -\frac{d\bar{U}_1}{dt} - \frac{d\bar{U}_2}{dt} = \frac{\omega_1-\omega_2}{\omega_2}\frac{d\bar{U}_2}{dt}. \qquad (112.10)$$

When $\omega_1 > \omega_2$, the condition $Q > 0$ shows that $d\bar{U}_2/dt > 0$; that is, the wave with the lower frequency is amplified, as stated above. The condition that the expression (112.6) be positive is given by the inequalities†

$$\alpha_2'' < 0, \alpha_2''+\beta_2'' < 0, \alpha_2''+\gamma_2'' = 0, \alpha_2''+\beta_2''+\gamma_2'' < 0. \qquad (112.11)$$

This effect is independent of the phase relations between the fields, because the pump wave field appears in the equations as expressions bilinear in $\mathbf{E}_1$ and $\mathbf{E}_1^*$, from which the phase factors vanish. The ultimate result is that field synchronism is not necessary for amplification of the field ω_2, in contrast to the harmonic generation and parametric signal amplification effects discussed in §110.

It is possible to relate the characteristics of the stimulated Raman scattering to those of the ordinary (spontaneous) scattering to be discussed in Chapter XV. The relevant calculations are given in §118, Problem.

The above relations are valid, as already mentioned, if energy is absorbed by the medium only at the difference frequency $\omega_1-\omega_2$. There is a different situation if the sum frequency $\omega_1+\omega_2$, not the difference, lies in the absorption range. In this case, for each quantum $\hbar\omega_2$ absorbed, a quantum $\hbar\omega_1$ is also absorbed, and an energy $\hbar(\omega_1+\omega_2)$ is transferred to the medium (*two-photon absorption*). In this case, of course, the waves of both frequencies are attenuated.

† This may be seen by considering the values of (112.6) for various polarizations of the fields $\mathbf{E}_1 = \mathbf{e}_1 E_1$ and $\mathbf{E}_2 = \mathbf{e}_2 E_2$: linear polarizations in parallel or perpendicular directions, circular polarizations with the same or opposite signs. In the first two cases $\mathbf{e}_1$ and $\mathbf{e}_2$ are real, with $\mathbf{e}_1\cdot\mathbf{e}_2 = 1$ or 0; in the second two they are complex, with either $\mathbf{e}_1\cdot\mathbf{e}_2 = 0$, $\mathbf{e}_1\cdot\mathbf{e}_2^* = 1$ or $\mathbf{e}_1\cdot\mathbf{e}_2 = 1$, $\mathbf{e}_1\cdot\mathbf{e}_2^* = 0$.

CHAPTER XIV

THE PASSAGE OF FAST PARTICLES THROUGH MATTER

§113. Ionization losses by fast particles in matter: the non-relativistic case

A FAST charged particle, in passing through matter, ionizes the atoms and thereby loses energy.† In gases, the ionization losses can be regarded as being due to collisions between the fast particle and individual atoms. In a solid or liquid medium, however, several atoms may interact simultaneously with the particle. The effect of this on the energy loss by the particle can be macroscopically regarded as resulting from the dielectric polarization of the medium by the charge. Let us first consider this effect for non-relativistic velocities of the particle. We shall see that the polarization of the medium then has only a slight effect on the losses. The derivation of this result is of interest because the method can be extended to other cases.

Let us first of all ascertain the conditions under which the phenomenon can be macroscopically considered. The spectral resolution of the field produced at a distance r from the path of a particle moving with velocity v consists chiefly of terms whose frequency is of the order v/r (the reciprocal of the "collision time"). The ionization of an atom can be effected by field components of frequency $\omega \gtrsim \omega_0$, where ω_0 is some mean frequency corresponding to the motion of the majority of the electrons in the atom. The particle therefore interacts simultaneously with many atoms if v/ω_0 is large compared with the interatomic distances. In solids and liquids these distances are of the same order of magnitude as the dimension a of the atoms themselves. Thus we obtain the condition $v \gg a\omega_0$, i.e. the velocity of the ionizing particle must be large compared with the velocities of the atomic electrons (or at least of the majority of them).‡

Let us now determine the field produced by a charged particle moving through matter. In the non-relativistic case it is sufficient to consider only the electric field, defined by the scalar potential ϕ. This potential satisfies Poisson's equation

$$\hat{\varepsilon}\triangle\phi = -4\pi e\delta(\mathbf{r}-\mathbf{v}t), \tag{113.1}$$

in which the permittivity is written as an operator, and the expression $e\delta(\mathbf{r}-\mathbf{v}t)$ on the right-hand side is the density due to a point charge e moving with constant velocity $\mathbf{v}$.§

† We speak, as is customary, of "ionization losses", but these are, of course, understood to include losses due to the excitation of atoms to discrete energy levels.

‡ The corresponding condition for the energy E of the particle is $E \gg MI/m$, where M is the mass of the particle, m that of the electron, and I some mean ionization energy for the majority of the electrons in the atom.

§ We assume that the particle moves in a straight line, and thereby neglect scattering, as is always permissible in problems of this type.

If the charge on the particle is ze, then all the formulae pertaining to energy loss in this and the following sections should be multiplied by z^2.

We expand ϕ as a Fourier space integral:

$$\phi = \int_{-\infty}^{\infty} \phi_{\mathbf{k}} \exp(i\mathbf{k}\cdot\mathbf{r}) \frac{d^3 k}{(2\pi)^3} \qquad (113.2)$$

Taking the Laplacian of this equation, we find that the Fourier component of $\triangle\phi$ is $(\triangle\phi)_{\mathbf{k}} = -k^2\phi_{\mathbf{k}}$.

Taking the Fourier component of equation (113.1) gives

$$\begin{aligned}\hat{\varepsilon}(\triangle\phi)_{\mathbf{k}} &= -\int 4\pi e\delta(\mathbf{r}-\mathbf{v}t)\exp(-i\mathbf{k}\cdot\mathbf{r})\,dV \\ &= -4\pi e\exp(-it\mathbf{v}\cdot\mathbf{k}).\end{aligned}$$

Thus $\hat{\varepsilon}\phi_{\mathbf{k}} = (4\pi e/k^2)\exp(-it\mathbf{v}\cdot\mathbf{k})$, and $\phi_{\mathbf{k}}$ therefore depends on time through a factor $\exp(-it\mathbf{v}\cdot\mathbf{k})$. The operator $\hat{\varepsilon}$ acting on a function $\exp(-i\omega t)$ multiplies it by $\varepsilon(\omega)$. Hence

$$\phi_{\mathbf{k}} = \frac{4\pi e}{k^2\varepsilon(\mathbf{k}\cdot\mathbf{v})}\exp(-it\mathbf{v}\cdot\mathbf{k}).$$

The Fourier components of the field and of the potential are related by $\mathbf{E}_{\mathbf{k}}\exp(i\mathbf{k}\cdot\mathbf{r}) = -\mathbf{grad}\,[\phi_{\mathbf{k}}\exp(i\mathbf{k}\cdot\mathbf{r})] = -i\mathbf{k}\phi_{\mathbf{k}}\exp(i\mathbf{k}\cdot\mathbf{r})$. Thus

$$\mathbf{E}_{\mathbf{k}} = -i\mathbf{k}\phi_{\mathbf{k}} = -\frac{4\pi ie\mathbf{k}}{k^2\varepsilon(\mathbf{k}\cdot\mathbf{v})}\exp(-it\mathbf{v}\cdot\mathbf{k}). \qquad (113.3)$$

The total field strength is obtained by inverting the Fourier transform:

$$\mathbf{E} = \int_{-\infty}^{\infty} \mathbf{E}_{\mathbf{k}}\exp(i\mathbf{k}\cdot\mathbf{r})\frac{d^3k}{(2\pi)^3}. \qquad (113.4)$$

The energy loss by the moving particle is just the work done by the force $e\mathbf{E}$ exerted on the particle by the field which it produces. Taking the value of the field at the point occupied by the particle, namely $\mathbf{r} = \mathbf{v}t$, we obtain in the integrand in (113.4) a factor $\exp(it\mathbf{v}\cdot\mathbf{k})$ which cancels with the factor $\exp(-it\mathbf{v}\cdot\mathbf{k})$ in the expression (113.3) for $\mathbf{E}_{\mathbf{k}}$. Hence the force $\mathbf{F}$ is

$$\mathbf{F} = -4\pi ie^2\int_{-\infty}^{\infty}\frac{\mathbf{k}}{k^2\varepsilon(\mathbf{k}\cdot\mathbf{v})}\frac{d^3k}{(2\pi)^3}$$

It is evident that the direction of the force $\mathbf{F}$ is opposite to that of the velocity $\mathbf{v}$; let the latter be the x-direction. Putting $k_x v = \omega$, $q = \sqrt{(k_y{}^2 + k_z{}^2)}$ and replacing $dk_y dk_z$ by $2\pi q\,dq$, we can write the magnitude of $\mathbf{F}$ as

$$F = \frac{ie^2}{\pi}\int_{-\infty}^{\infty}\int_0^{q_0}\frac{q\omega\,dq\,d\omega}{\varepsilon(\omega)(q^2v^2+\omega^2)}. \qquad (113.5)$$

The choice of q_0 is discussed below.

The following remark should be made concerning the integration with respect to ω in formula (113.5). As $\omega \to \infty$ the function $\varepsilon(\omega) \to 1$, and the integral is logarithmically divergent. This happens because we ought to have subtracted from the field $\mathbf{E}$ the field which would be present if the particle were moving in a vacuum (i.e. if $\varepsilon = 1$); this field evidently does not affect the energy lost by the particle in matter.

If this subtraction were effected, $1/\varepsilon$ in the integrand of (113.5) would become $1/\varepsilon - 1$, and the integral would converge. The same result can be obtained by taking the integration from $-\Omega$ to $+\Omega$ and then letting Ω tend to infinity. Since the function $\varepsilon'(\omega)$ is even, the real part of the integrand is an odd function of the frequency, and gives zero. The integral of the imaginary part of the integrand converges.

In what follows we shall sometimes find it convenient to use the notation

$$1/\varepsilon(\omega) = \eta(\omega) = \eta' + i\eta'', \tag{113.6}$$

with $\eta'(\omega)$ and $\eta''(\omega)$ respectively even and odd functions, and $\eta'' = -\varepsilon''/|\varepsilon|^2 < 0$. Formula (113.5) can be rewritten in the explicitly real form

$$F = \frac{2e^2}{\pi}\int_0^\infty\int_0^{q_0} \frac{q\omega|\eta''(\omega)|}{(q^2v^2+\omega^2)}\,dq\,d\omega. \tag{113.7}$$

The energy loss per unit path length is the work done by the force over that distance, which is just F; it is called the *stopping power* of the substance with respect to the particle.

According to the general rules of quantum mechanics, the Fourier component of the field whose wave vector is $\mathbf{k}$ transmits to the δ-electron released in ionization a momentum $\hbar\mathbf{k}$. For sufficiently large q ($\gg \omega_0/v$) we have $k^2 = q^2 + \omega^2/v^2 \approx q^2$, so that the momentum transferred is approximately $\hbar\mathbf{q}$. A given value of $\mathbf{q}$ corresponds to collisions with impact parameter $\sim 1/q$. Hence the condition for the macroscopic treatment to be valid is $1/q \gg a$. Accordingly, we take as the upper limit of integration a value q_0 such that $\omega_0/v \ll q_0 \ll 1/a$. The quantity $F(q_0)$ is the energy loss of a fast particle with transfer of momentum not exceeding $\hbar q_0$ to the atomic electron.

Integrating with respect to q in (113.7), we obtain

$$F(q_0) = \frac{2e^2}{\pi v^2}\int_0^\infty \omega|\eta''(\omega)|\log\frac{q_0 v}{\omega}\,d\omega. \tag{113.8}$$

This formula cannot be further transformed in a general manner, but it can be written in a more convenient form as follows. We first calculate the integral

$$\int_0^\infty \omega\eta''(\omega)\,d\omega = -\tfrac{1}{2}i\int_{-\infty}^\infty (\omega/\varepsilon)\,d\omega.$$

To do so, we notice that, if the integration is taken in the complex ω-plane along a contour consisting of the real axis and a very large semicircle σ in the upper half-plane, the integral is zero, since the integrand has no poles in the upper half-plane. For large values of ω, the function $\varepsilon(\omega)$ is given by formula (78.1):

$$\varepsilon(\omega) = 1 - \frac{4\pi e^2 N}{m\omega^2}. \tag{113.9}$$

The integration along the large semicircle σ can be carried out by using this formula, and the result is†

$$-\int_0^\infty \omega\eta''(\omega)\,d\omega = -\frac{2\pi i Ne^2}{m}\int_\sigma \frac{d\omega}{\omega} = 2\pi^2 Ne^2/m. \tag{113.10}$$

We define a mean frequency of the motion of the atomic electrons by

$$\log\bar{\omega} = \frac{\int_0^\infty \omega\eta''(\omega)\log\omega\,d\omega}{\int_0^\infty \omega\eta''(\omega)\,d\omega}$$

$$= \frac{m}{2\pi^2 Ne^2}\int_0^\infty \omega|\eta''(\omega)|\log\omega\,d\omega. \tag{113.11}$$

Then formula (113.8) can be written

$$F(q_0) = (4\pi Ne^4/mv^2)\log(q_0 v/\bar{\omega}). \tag{113.12}$$

The following remark should be made here. It might seem from the form of (113.7) or (113.11) that the main contribution to the ionization losses (113.12) comes from frequencies at which there is considerable absorption. This is not so; these formulae may contain a considerable contribution from ranges in which ε'' is small. The reason is that in such ranges the function $\varepsilon(\omega) \approx \varepsilon'(\omega)$ may pass through zero. It is seen from formula (113.5) that the zeros of $\varepsilon(\omega)$ are poles of the integrand. In reality, of course, $\varepsilon''(\omega)$ is not exactly zero, and so the zeros of $\varepsilon(\omega)$ are not on the real axis but just below it. Hence, when the expression used for $\varepsilon(\omega)$ is real and passes through zero, the contour must be indented upwards at the pole of the integrand, and so a contribution to the integral occurs. For example, if the function $\varepsilon(\omega)$ is given by (84.5), the contribution to the energy loss (113.12) from the poles $\pm\omega_1$ (where $\varepsilon(\omega_1) = 0$) is easily seen, by direct calculation from (113.7), to be $(4\pi Ne^4 mv^2 a^2)\log(q_0 v/\omega_1)$.

In order to find the energy loss $F(q_1)$ with transfer of momentum not exceeding some value $\hbar q_1 > \hbar q_0$, we must "join" formula (113.12) to that given by the quantum theory of collisions, corresponding to energy loss by collisions with single atoms. This can be done by using the fact that the ranges of applicability of the two formulae overlap. As we know from the theory of collisions, the energy loss with transfer of momentum in a range of $\hbar dq$ is

$$dF = (4\pi Ne^4/mv^2)\,dq/q, \tag{113.13}$$

and this formula is applicable (in the non-relativistic case) for any value of $q \gg \omega_0/v$ which is compatible with the laws of conservation of momentum and energy, provided that the energy transferred is small compared with the initial energy of the fast particle.‡ The

† This is the same as (82.12), as it should be, since, as $|\omega| \to \infty$, $|\varepsilon| \to 1$ and $\eta'' \to -\varepsilon''$.

‡ See *QM*, §149. The "effective retardation" used there differs from F by a factor $N_a = N/Z$, the number density of atoms.

Formula (113.13) applies to collisions with free electrons. Its range of applicability as hitherto determined ($q \gg \omega_0/v$), however, extends to values of q for which the atomic electrons cannot be regarded as free. The condition for this is $q \gg \omega_0/v_0$, where v_0 is the order of magnitude of the velocity of the majority of the atomic electrons; the energy $\hbar^2 q^2/2m$ of the δ-electron is then large compared with atomic energies.

energy loss with all values of q between q_0 and q_1 is accordingly $(4\pi Ne^4/mv^2)\log(q_1/q_0)$. When this quantity is added to formula (113.12), q_0 is replaced by q_1, so that

$$F(q_1) = (4\pi Ne^4/mv^2)\log(q_1 v/\bar{\omega}). \tag{113.14}$$

If a momentum $\hbar q_1$ large compared with the atomic momenta is given to an atomic electron, its energy is $E_1 = \hbar^2 q_1{}^2/2m$. Thus we can write

$$F(E_1) = (2\pi Ne^4/mv^2)\log(2mv^2 E_1/\hbar^2\bar{\omega}^2). \tag{113.15}$$

Formulae (113.14) and (113.15) give the energy loss of a fast particle by ionization with a transfer of energy not exceeding a value E_1 that is small compared with the original energy of the particle. It must be emphasized that with this condition the formulae are equally valid for fast electrons and fast heavy particles. Formula (113.15) differs from the formula derived from a microscopic discussion, neglecting interactions between atoms (*QM*, (149.14)) only by the definition of the "ionization energy" I, which is here represented by $\hbar\bar{\omega}$. The mean (with respect to the electrons) ionization energy of an atom is usually almost independent of its interaction with other atoms, being determined mainly by the electrons of the inner shells, which are almost unaffected by that interaction. Moreover, this quantity appears here only in a logarithm, and so the exact definition of it has even less effect on the magnitude of the energy loss.

In a collision between a heavy particle and an electron, even the maximum transferable momentum $\hbar q_{\max}$ is small compared with the momentum Mv of the particle. The change in the energy of the heavy particle is therefore $\mathbf{v}\cdot\hbar\mathbf{q}$; equating this to the energy of the electron gives $\hbar^2 q^2/2m = \hbar\mathbf{q}\cdot\mathbf{v} \leqslant \hbar qv$, whence $\hbar q_{\max} = 2mv$, and $E_{1,\max} = 2mv^2$. Substituting for E_1 in (113.15), we obtain as the total ionization energy loss by the fast particle

$$F = \frac{4\pi Ne^4}{mv^2}\log\frac{2mv^2}{\hbar\bar{\omega}}. \tag{113.16}$$

This differs from the usual expression (*QM*, (150.10)) only in the definition of the ionization energy $\hbar\bar{\omega}$.

We can see how $\hbar\bar{\omega}$ defined by (113.11) becomes, in a rarefied medium, the mean ionization energy of a single atom given by *QM*, (149.11). To do so, we note that in a rarefied gas, which for simplicity we suppose to consist of uniform atoms, the permittivity is $\varepsilon = 1 + 4\pi N_a\alpha(\omega)$, where N_a is the number of atoms per unit volume, $\alpha(\omega)$ the polarizability of one atom; here $|\varepsilon - 1| \ll 1$. The imaginary part of $\eta = 1/\varepsilon$ is $|\eta''| \cong 4\pi N_a\alpha''(\omega)$. The polarizability of the atom is given by *QED*, (85.13); separating the imaginary part by means of *QED* (75.19), we have when $\omega > 0$

$$|\eta''| = \tfrac{4}{3}\pi N_a \sum_n |\mathbf{d}_{0n}|^2 \delta(E_n - E_0 - \hbar\omega),$$

where E_0 and E_n are the energies of the ground state and excited states of the atom. Substitution of this expression in (113.11) and carrying out the integration, with $N = N_a Z$, gives the definition in *QM*, (149.11).

§114. Ionization losses by fast particles in matter: the relativistic case

At velocities comparable with that of light, the effect of the polarization of the medium on its stopping power with respect to a fast particle may become very important even in gases.†

To derive the appropriate formulae, we use a method analogous to that used in §113, but it is now necessary to begin from the complete Maxwell's equations. When extraneous charges are present with volume density ρ_{ex}, and extraneous currents wih density $\mathbf{j}_{ex}$, these equations are‡

$$\operatorname{div} \mathbf{H} = 0, \qquad \mathbf{curl}\, \mathbf{E} = -\frac{1}{c}\frac{\partial \mathbf{H}}{\partial t}, \tag{114.1}$$

$$\operatorname{div} \hat{\varepsilon}\mathbf{E} = 4\pi\rho_{ex}, \qquad \mathbf{curl}\, \mathbf{H} = \frac{1}{c}\frac{\partial \hat{\varepsilon}\mathbf{E}}{\partial t} + \frac{4\pi}{c}\mathbf{j}_{ex}. \tag{114.2}$$

In the present case the extraneous charge and current distribution are given by

$$\rho_{ex} = e\delta(\mathbf{r}-\mathbf{v}t), \qquad \mathbf{j}_{ex} = e\mathbf{v}\,\delta(\mathbf{r}-\mathbf{v}t). \tag{114.3}$$

We introduce scalar and vector potentials, with the usual definitions:

$$\mathbf{H} = \mathbf{curl}\, \mathbf{A}, \qquad \mathbf{E} = -\frac{1}{c}\frac{\partial \mathbf{A}}{\partial t} - \mathbf{grad}\, \phi, \tag{114.4}$$

so that equations (114.1) are satisfied identically. The additional condition

$$\operatorname{div} \mathbf{A} + \frac{1}{c}\frac{\partial \hat{\varepsilon}\phi}{\partial t} = 0 \tag{114.5}$$

is imposed on the potentials $\mathbf{A}$ and ϕ; this is a generalization of the usual *Lorentz condition* in the theory of radiation. Then, substituting (114.4) in (114.2), we obtain the following equations for the potentials:

$$\left.\begin{aligned} \triangle \mathbf{A} - \frac{\hat{\varepsilon}}{c^2}\frac{\partial^2 \mathbf{A}}{\partial t^2} &= -\frac{4\pi}{c} e\mathbf{v}\,\delta(\mathbf{r}-\mathbf{v}t), \\ \hat{\varepsilon}\left(\triangle\phi - \frac{\hat{\varepsilon}}{c^2}\frac{\partial^2\phi}{\partial t^2}\right) &= -4\pi e\,\delta(\mathbf{r}-\mathbf{v}t). \end{aligned}\right\} \tag{114.6}$$

We expand $\mathbf{A}$ and ϕ as Fourier space integrals. Taking the Fourier components of equations (114.6), we have

$$k^2\mathbf{A}_{\mathbf{k}} + \frac{\hat{\varepsilon}}{c^2}\frac{\partial^2 \mathbf{A}_{\mathbf{k}}}{\partial t^2} = \frac{4\pi e\mathbf{v}}{c}\exp(-it\mathbf{v}\cdot\mathbf{k}),$$

$$\hat{\varepsilon}\left(k^2\phi_{\mathbf{k}} + \frac{\hat{\varepsilon}}{c^2}\frac{\partial^2\phi_{\mathbf{k}}}{\partial t^2}\right) = 4\pi e\exp(-it\mathbf{v}\cdot\mathbf{k}).$$

† This effect was pointed out by E. Fermi (1940), who performed the calculation for the particular case of a gas whose atoms are regarded as harmonic oscillators. The general derivation given here is due to L. Landau.

‡ We put $\mu(\omega) \equiv 1$, since matter does not exhibit magnetic properties at the frequencies important as regards ionization losses.

Hence we see that $\mathbf{A}_\mathbf{k}$ and $\phi_\mathbf{k}$ depend on time through a factor $\exp(-it\mathbf{v}\cdot\mathbf{k})$. We again put $\omega = \mathbf{k}\cdot\mathbf{v} = k_x v$, and obtain

$$\left.\begin{aligned}\mathbf{A}_\mathbf{k} &= \frac{4\pi e}{c}\frac{\mathbf{v}}{k^2-\omega^2\varepsilon(\omega)/c^2}e^{-i\omega t},\\ \phi_\mathbf{k} &= \frac{4\pi e}{\varepsilon(\omega)}\frac{1}{k^2-\omega^2\varepsilon(\omega)/c^2}e^{-i\omega t}.\end{aligned}\right\} \tag{114.7}$$

The Fourier component of the electric field is

$$\mathbf{E}_\mathbf{k} = i\omega\mathbf{A}_\mathbf{k}/c - i\mathbf{k}\phi_\mathbf{k}. \tag{114.8}$$

From these formulae the force $\mathbf{F} = e\mathbf{E}$ acting on the particle is found in the same way as in §113.† Using the same notation, we now have

$$F = \frac{ie^2}{\pi}\int_{-\infty}^{\infty}\int_0^{q_0}\frac{\left(\dfrac{1}{v^2}-\dfrac{\varepsilon}{c^2}\right)\omega q\,\mathrm{d}q\,\mathrm{d}\omega}{\varepsilon\left[q^2+\omega^2\left(\dfrac{1}{v^2}-\dfrac{\varepsilon}{c^2}\right)\right]}. \tag{114.9}$$

As $c\to\infty$ this formula tends, of course, to (113.5).

Let us first carry out the integration with respect to frequency. In order to effect an integration in the complex ω-plane, we first ascertain the poles of the integrand in the upper half-plane. The function $\varepsilon(\omega)$ has no singularity and no zero in this half-plane, and so the required poles can only be the zeros of the expression

$$\omega^2\left(\frac{\varepsilon}{c^2}-\frac{1}{v^2}\right)-q^2.$$

We shall show that, for any value of the positive real quantity q^2, this expression vanishes for only one value of ω.

To prove this,‡ we use a theorem in the theory of functions of a complex variable: the integral

$$\frac{1}{2\pi i}\int_C\frac{\mathrm{d}f(\omega)}{\mathrm{d}\omega}\frac{\mathrm{d}\omega}{f(\omega)-a}, \tag{114.10}$$

taken along a closed contour C, is equal to the difference between the numbers of zeros and poles of $f(\omega)-a$ in the region bounded by C. Let

$$f(\omega) = \omega^2\left(\frac{\varepsilon(\omega)}{c^2}-\frac{1}{v^2}\right),$$

$a = q^2$ be a positive real number, and C be a contour consisting of the real axis and a very large semicircle (Fig. 61). The function $f(\omega)$ has no pole in the upper half-plane

† The magnetic force $e\mathbf{v}\times\mathbf{H}/c$ is seen by symmetry to be zero, and in any case is perpendicular to the velocity of the particle and so does no work on it.

‡ The following argument is analogous to the proof (*SP* 1, §123) that $\varepsilon(\omega)$ has no zero in the upper half-plane.

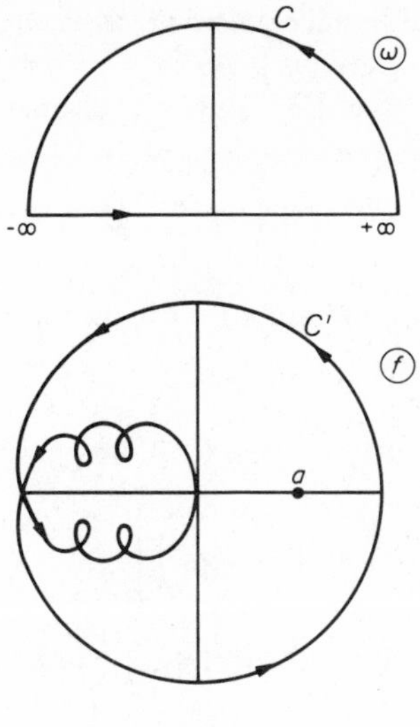

FIG. 61

or on the real axis†; the integral (114.10) therefore gives the number of zeros of the function $f(\omega) - a$ in the upper half-plane. To calculate its value, we write it as

$$\frac{1}{2\pi i} \int_{C'} \frac{\mathrm{d}f}{f-a}, \tag{114.11}$$

the integration being taken along a contour C' in the plane of the complex variable f which maps the contour C from the ω-plane. For $\omega = 0, f = 0$. For positive real ω we have $\operatorname{im} f > 0$, and for negative real ω, $\operatorname{im} f < 0$. At infinity $f \to -\omega^2 [(1/v^2) - (1/c^2)]$, and therefore f goes round a large circle when ω goes round the large semicircle. Hence we see that the path of integration C' in the f-plane is of the kind shown schematically in Fig. 61. When a is real and positive, as in Fig. 61, in going round C' the argument of the complex number changes by 2π, and the integral (114.11) is equal to unity. This completes the proof.‡

Furthermore, it is easy to see that this single root of the equation $f(\omega) - q^2 = 0$ lies on the imaginary ω-axis: for purely imaginary ω the function $f(\omega)$, like $\varepsilon(\omega)$, is real and takes all values from 0 to ∞, including q^2.

Let us now return to the integral with respect to ω in (114.9):

$$\int_{-\infty}^{\infty} \frac{\left(\dfrac{1}{\varepsilon v^2} - \dfrac{1}{c^2}\right) \omega \, \mathrm{d}\omega}{q^2 - \omega^2 \left(\dfrac{\varepsilon}{c^2} - \dfrac{1}{v^2}\right)}.$$

---

† For metals $\varepsilon(\omega)$ has a pole at $\omega = 0$, but $\omega^2 \varepsilon$ always tends to zero with ω.

‡ If a is negative the argument of $f - a$ changes by 4π on going round C', so that the integral (114.11) is equal to 2, i.e. the function $f(\omega) + |a|$ has two zeros in the upper half-plane.

This can be written as the difference between the integral along the contour C and that along the large semicircle. The latter is $\int \mathrm{d}\,\omega/\omega = i\pi$, and the former is $2\pi i$ times the residue of the integrand at its only pole. Let $\omega(q)$ be the function defined by the equation

$$\omega^2\left(\frac{\varepsilon}{c^2}-\frac{1}{v^2}\right)=q^2. \tag{114.12}$$

Then, since the residue of an expression $f(z)/\phi(z)$ at a pole $z=z_0$ is $f(z_0)/\phi'(z_0)$, the integral along C is

$$2\pi i\frac{\omega\left(\dfrac{1}{\varepsilon v^2}-\dfrac{1}{c^2}\right)}{-\dfrac{\mathrm{d}}{\mathrm{d}\omega}\left[\omega^2\left(\dfrac{\varepsilon}{c^2}-\dfrac{1}{v^2}\right)\right]}=2\pi i\frac{\omega\left(\dfrac{1}{\varepsilon v^2}-\dfrac{1}{c^2}\right)}{-\,\mathrm{d}q^2/\mathrm{d}\omega}.$$

Collecting these expressions and substituting in (114.9), we have

$$F=e^2\int\limits_0^{q_0}\left[\frac{\omega\left(\dfrac{1}{\varepsilon v^2}-\dfrac{1}{c^2}\right)}{q\,\mathrm{d}q/\mathrm{d}\omega}+1\right]q\,\mathrm{d}q$$

or, replacing the integration with respect to q in the first term by one with respect to ω,

$$\begin{aligned}F&=e^2\int\limits_{\omega(0)}^{\omega(q_0)}\left[\frac{1}{v^2\varepsilon(\omega)}-\frac{1}{c^2}\right]\omega\,\mathrm{d}\omega+\tfrac{1}{2}e^2{q_0}^2\\&=\frac{e^2}{v^2}\int\limits_{\omega(0)}^{\omega(q_0)}\left[\frac{1}{\varepsilon(\omega)}-1\right]\omega\,\mathrm{d}\omega+\tfrac{1}{2}e^2{q_0}^2\\&\quad+\tfrac{1}{2}e^2\left(\frac{1}{v^2}-\frac{1}{c^2}\right)[\omega^2(q_0)-\omega^2(0)].\end{aligned} \tag{114.13}$$

Large values of q correspond to large absolute values ω of the root of equation (114.12). Using therefore the expression (113.9) for $\varepsilon(\omega)$, we find

$$\omega^2(q_0)=-v^2\gamma^2\left({q_0}^2+\frac{4\pi Ne^2}{mc^2}\right),$$

where we have put $\gamma=1/\sqrt{[1-v^2/c^2]}$. Substitution in (114.13) gives

$$F=\frac{e^2}{v^2}\int\limits_{\omega(0)}^{ivq_0\gamma}\left[\frac{1}{\varepsilon(\omega)}-1\right]\omega\,\mathrm{d}\omega-\frac{2\pi Ne^4}{mc^2}-\frac{e^2}{2v^2\gamma^2}\omega^2(0); \tag{114.14}$$

in the integral, only the leading term $ivq_0\gamma$ need be retained in $\omega(q_0)$.

The integration in (114.14) is over purely imaginary values of ω. We use the real variable $\omega''\equiv\omega/i$, with the lower limit $\xi\equiv\omega(0)/i$, and again put $1/\varepsilon=\eta$ (113.6). The required integral is

$$-\int_{\xi}^{vq_0\gamma}[\eta(i\omega'')-1]\omega''\,d\omega''.$$

The values of the function $\eta(\omega)$ on the imaginary axis can be expressed in terms of its imaginary part on the real axis:

$$\eta(i\omega'')-1=\frac{2}{\pi}\int_0^{\infty}\frac{x\eta''(x)}{x^2+\omega''^2}\,dx$$

(cf. (82.15)). Hence the integral is (if we neglect x in comparison with $vq_0\gamma$)

$$\frac{2}{\pi}\int_0^{\infty}\int_{\xi}^{vq_0\gamma}\frac{x|\eta''(x)|\omega''\,d\omega''\,dx}{x^2+\omega''^2}=\frac{1}{\pi}\int_0^{\infty}x|\eta''(x)|\log\frac{v^2{q_0}^2\gamma^2}{x^2+\xi^2}\,dx.$$

We substitute this result in (114.14), and for simplicity put

$$\log\Omega\equiv\tfrac{1}{2}\overline{\log(\omega^2+\xi^2)},\tag{114.15}$$

where the bar denotes an averaging with weight $\omega|\eta''(\omega)|$, as in (113.11). Then

$$F(q_0)=\frac{4\pi Ne^4}{mv^2}\log\frac{vq_0\gamma}{\Omega}-\frac{2\pi Ne^4}{mc^2}+\frac{e^2}{2v^2\gamma^2}\xi^2.\tag{114.16}$$

Two cases must be considered in the further examination of this formula. Let us first suppose that the medium is a dielectric, and that the velocity of the particle satisfies the condition

$$v^2<c^2/\varepsilon_0,\tag{114.17}$$

where $\varepsilon_0=\varepsilon(0)$ is the electrostatic value of the permittivity. On the imaginary axis the function $\varepsilon(\omega)$ decreases monotonically from $\varepsilon_0>1$ for $\omega=0$ to 1 for $\omega=i\infty$. The expression on the left-hand side of equation (114.12) therefore increases monotonically from 0 to ∞, and for $q=0$ (114.12) gives $\omega=0$. Thus we must put $\xi=0$ in (114.16); then Ω becomes the mean atomic frequency $\bar{\omega}$ (113.11), and

$$F(q_0)=\frac{4\pi Ne^4}{mv^2}\left[\log\frac{vq_0\gamma}{\bar{\omega}}-\frac{v^2}{2c^2}\right].\tag{114.18}$$

For $v\ll c$ this formula becomes (113.12), as it should.

The value of q_0 is such that $q_0\ll 1/a$, where a is the order of magnitude of the interatomic distances (in solids and liquids equal to the dimension of the atoms). In order to extend the formula to higher values of the transferred momentum and energy, it must be "joined" to the formulae of the ordinary theory of collisions, as in §113, but the joining must now be carried out in two stages, First, using formula (113.13), we enter the range of q corresponding to energy transfers large compared with atomic energies but not yet relativistic. Formula (114.18) is unchanged in form, but may now involve the δ-electron energy $\hbar^2{q_1}^2/2m$. Calling this E_1, we have

$$F(E_1)=\frac{2\pi Ne^4}{mv^2}\left[\log\frac{2mv^2E_1\gamma^2}{\hbar^2\bar{\omega}^2}-\frac{v^2}{c^2}\right].\tag{114.19}$$

We can now go on to the relativistic values of E_1 by using a formula of relativistic collision theory, according to which the stopping power with energy transfer between E' and $E'+\mathrm{d}E'$ is

$$(2\pi Ne^4/mv^2)\,\mathrm{d}E'/E' \tag{114.20}$$

if E' is small compared with the maximum transfer $E_{1,\max}$ compatible with the laws of conservation of momentum and energy for a collision between the fast particle concerned and a free electron.† Since the integration of (114.20) gives a term in log E', it is clear that formula (114.19) is unchanged in form, and it is therefore valid for all $E_1 \ll E_{1,\max}$.

In the retardation of a fast heavy particle (with mass $M \gg m$ and energy E which, though relativistic, is such that $E \ll M^2c^2/m$), the maximum energy transfer to an electron is $E_{1,\max} \cong 2mv^2\gamma^2$ and is still small in comparison with E (see *QED*, (82.23)). For such particles the differential expression for the energy lost to free electrons is

$$\frac{2\pi Ne^4}{mv^2}\left(\frac{1}{E'}-\frac{1}{2mc^2\gamma^2}\right)\mathrm{d}E'$$

for all E'; see *QED*, (82.24). The energy loss additional to (114.19), with energy transfer from E_1 to $E_{1,\max}$ (with $E_1 \ll E_{1,\max}$) is then

$$\frac{2\pi Ne^4}{mv^2}\left(\log\frac{E_{1,\max}}{E_1}-\frac{E_{1,\max}}{2mc^2\gamma^2}\right)=\frac{2\pi Ne^4}{mv^2}\left(\log\frac{2mv^2\gamma^2}{E_1}-\frac{v^2}{c^2}\right). \tag{114.21}$$

Adding this to (114.19), we find the total stopping power with respect to the heavy particle:

$$F=\frac{4\pi Ne^4}{mv^2}\left(\log\frac{2mv^2\gamma^2}{\hbar\bar{\omega}}-\frac{v^2}{c^2}\right). \tag{114.22}$$

Formula (114.22) differs from that of the usual theory only in that the "ionization energy" is $\hbar\bar{\omega}$; cf. *QED*, (82.26).

Let us now turn to the second case, namely that where

$$v^2 > c^2/\varepsilon_0, \tag{114.23}$$

which, in particular, always holds for metals, where $\varepsilon(0)=\infty$. The expression $\omega^2(\varepsilon/c^2-1/v^2)$ on the left-hand side of equation (114.12) then has two zeros on the imaginary ω-axis, one at $\omega = 0$ and the other at $\omega = i\xi$, where ξ is defined by

$$\varepsilon(i\xi) = c^2/v^2. \tag{114.24}$$

In the range from 0 to $i\xi$ the expression $\omega^2(\varepsilon/c^2-1/v^2)$ is negative, and for $|\omega| > \xi$ it takes all positive values from 0 to ∞. As $q \to 0$, therefore, the root of equation (114.12) in this case tends to ξ, which is the value to be substituted in (114.15) and (114.16).

Two limiting cases may be considered. If ξ is small compared with the atomic frequencies ω_0, then the last term in (114.16) may be neglected, and $\Omega \cong \bar{\omega}$. Thus we return to formula (114.18). The opposite limiting case, where $\xi \gg \omega_0$, is of particular interest. Since, for large ξ, the function $\varepsilon(i\xi)$ tends to 1, it is evident from (114.24) that this case corresponds to ultra-relativistic velocities of the particle. Using formula (113.9) for $\varepsilon(\omega)$, we can write from equation (114.24)

$$\xi^2 = 4\pi Ne^2v^2\gamma^2/mc^2 \cong 4\pi Ne^2\gamma^2/m.$$

† See *QED*, (81.15) and (82.24). The stopping power F is obtained on multiplying these expressions for the cross-section by the energy loss $m\Delta$ and by N.

As the velocity of the particle increases, the condition $\xi \gg \omega_0$ is ultimately fulfilled in any medium, i.e. whatever the electron density N (even in a gas). The velocity required is, however, the greater, the smaller N, i.e. the more rarefied the medium.

From (114.15) we then have simply $\Omega \cong \xi$. Putting also $v \cong c$, we find that the last two terms in (114.16) cancel, leaving

$$F(q_0) = (2\pi Ne^4/mc^2) \log (mc^2 q_0{}^2/4\pi Ne^2).$$

Extending this formula, in the same manner as above, to large values of the momentum and energy transfer, we find the following expression for the energy loss of an ultra-relativistic particle with an energy transfer not exceeding E_1 ($\ll E_{1,\max}$):

$$F(E_1) = (2\pi Ne^4/mc^2) \log (m^2c^2E_1/2\pi Ne^2 \hbar^2). \tag{114.25}$$

This result is considerably different from that obtained in the ordinary theory, which neglects the polarization of the medium. According to that theory (see *QED*, §82), in the ultra-relativistic range the stopping power $F(E_1)$ continues to increase (though only logarithmically) with the energy of the particle:†

$$F(E_1) = \frac{2\pi Ne^4}{mc^2} \log \left(\frac{2mc^2 \gamma^2 E_1}{I^2} - 2 \right).$$

The polarization of the medium results in a screening of the charge, and the increase in the losses is thereby finally stopped; it tends to the constant value (independent of γ) given by formula (114.25).

For heavy particles a formula can also be derived for the total stopping power with any energy transfer up to $E_{1,\max}$ (if the latter is small compared with the energy of the particle itself). Again using the expression (114.21), in which we can now put $v = c$, we find

$$F = \frac{2\pi Ne^4}{mc^2} \left[\log \frac{m^3 c^4 \gamma^2}{\pi Ne^2 \hbar^2} - 1 \right]. \tag{114.26}$$

We see that the total stopping power continues to increase with the velocity of the particle, owing to close collisions with a large energy transfer, for which the polarization of the medium has no screening effect. This increase, however, is rather slower than that given by the theory when the polarization is neglected. According to that theory,

$$F = \frac{4\pi Ne^4}{mc^2} \left[\log \frac{2mc^2 \gamma^2}{I} - 1 \right];$$

see *QED*, (82.28). The coefficient of the $\log \gamma$ term here is twice that in (114.26).

It may also be noted that the presence of the electron density N in the argument of the logarithm in formulae (114.25) and (114.26) results in the following property of energy losses of ultra-relativistic particles: when such a particle passes through layers of different substances containing the same number of electrons per unit surface area, the losses are smaller in media with larger N.

† This formula is obtained by adding *QED* (82.20) and (82.25), with E_1 for $m\Delta_{\max}$ in the latter. For a small energy transfer E_1, the formulae apply to both fast electrons and fast heavy particles.

§115. Cherenkov radiation

A charged particle moving in a transparent medium emits, in certain circumstances, an unusual type of radiation, first observed by P. A. Cherenkov and S. I. Vavilov, and theoretically interpreted by I. E. Tamm and I. M. Frank (1937). It must be emphasized that this radiation is entirely unrelated to the bremsstrahlung which is almost always emitted by a rapidly moving electron. The latter radiation is emitted by the moving electron itself when it collides with atoms. The Cherenkov effect, however, involves radiation emitted by the medium under the action of the field of the particle moving in it. The distinction between the two types of radiation appears with particular clarity when the particle has a very large mass: the bremsstrahlung disappears, but the Cherenkov radiation is unaffected.

The wave number and frequency of an electromagnetic wave propagated in a transparent medium are related by $k = n\omega/c$, where $n = \sqrt{\varepsilon}$ is the refractive index, which is real. We again suppose the medium isotropic and non-magnetic. We have seen that the frequency of the Fourier component of the field of a particle moving uniformly in the x-direction in a medium is related to the x-component of the wave vector by $\omega = k_x v$. If this component is a freely propagated wave, these two relations must be consistent. Since $k > k_x$, it follows that we must have

$$v > c/n\,(\omega). \tag{115.1}$$

Thus radiation of frequency ω occurs if the velocity of the particle exceeds the phase velocity of waves of that frequency in the medium concerned.†

Let θ be the angle between the direction of motion of the particle and the direction of emission. We have $k_x = k\cos\theta = (n\omega/c)\cos\theta$ and, since $k_x = \omega/v$, we find that

$$\cos\theta = c/nv. \tag{115.2}$$

Thus a definite value of the angle θ corresponds to radiation of a given frequency. That is, the radiation of each frequency is emitted forwards, and is distributed over the surface of a cone with vertical angle 2θ, where θ is given by (115.2). The distributions of the radiation in angle and in frequency are thus related in a definite manner.

The emission of electromagnetic waves, if it occurs, involves a loss of energy by the moving particle. This loss forms part, though a small part, of the total losses calculated in §114. (The bremsstrahlung is not included therein.) In this sense the term "ionization losses" is not quite accurate. We shall now find the corresponding part of the total losses, and thus determine the intensity of the Cherenkov radiation.

According to (114.9), the energy loss in the frequency interval $d\omega$ is

$$dF = -d\omega \frac{ie^2}{\pi} \sum \omega \left(\frac{1}{c^2} - \frac{1}{\varepsilon v^2}\right) \int \frac{q\,dq}{q^2 - \omega^2\left(\dfrac{\varepsilon}{c^2} - \dfrac{1}{v^2}\right)},$$

where the summation is over terms with $\omega = \pm|\omega|$. We introduce as a new variable

$$\xi = q^2 - \omega^2\left(\frac{\varepsilon}{c^2} - \frac{1}{v^2}\right).$$

† The problem of radiation from an electron moving uniformly in a vacuum at a velocity $v > c$ was discussed by A. Sommerfeld (1904) before the theory of relativity became known.

Then

$$dF = -d\omega \frac{ie^2}{2\pi} \sum \omega \left(\frac{1}{c^2} - \frac{1}{\varepsilon v^2} \right) \int \frac{d\xi}{\xi}.$$

In integrating along the real ξ-axis we must pass round the singular point $\xi = 0$ (for which $q^2 + k_x{}^2 = k^2$) in some manner, which is determined by the fact that, although we suppose $\varepsilon(\omega)$ real (the medium being transparent), it actually has a small imaginary part, which is positive for $\omega > 0$ and negative for $\omega < 0$. Accordingly, ξ has a small negative or positive imaginary part, and the path of integration ought to pass below or above the real axis respectively. This means that, when the path of integration is displaced to the real axis, we must pass below or above the singular point respectively. This gives a contribution to dF, and the real parts cancel in the sum. Indenting the path of integration with infinitesimal semicircles, we find

$$\sum \omega \int d\xi/\xi = \omega \left\{ \int_{\cup} d\xi/\xi - \int_{\cap} d\xi/\xi \right\} = 2i\pi\omega.$$

Thus the final formula is

$$dF = \frac{e^2}{c^2} \left(1 - \frac{c^2}{v^2 n^2} \right) \omega \, d\omega, \tag{115.3}$$

which gives the intensity of the radiation in a frequency interval dω. According to (115.2), this radiation is emitted in an angle interval.

$$d\theta = \frac{c}{vn^2 \sin\theta} \frac{dn}{d\omega} d\omega. \tag{115.4}$$

The total intensity of the radiation is obtained by integrating (115.3) over all frequencies for which the medium is transparent.

It is easy to determine the polarization of the Cherenkov radiation. As we see from (114.7) the vector potential of the radiation field is parallel to the velocity **v**. The magnetic field $\mathbf{H_k} = i\mathbf{k} \times \mathbf{A_k}$ is therefore perpendicular to the plane containing **v** and the ray direction **k**. The electric field (in the "wave region") is perpendicular to the magnetic field, and therefore lies in that plane.

PROBLEM

Find the cone of Cherenkov radiation wave vectors for a particle moving uniformly in a uniaxial non-magnetic crystal: (a) along the optical axis, (b) at right angles to the optical axis (V. L. Ginzburg, 1940).

SOLUTION. (a) When a charge moves in a uniaxial crystal, the Cherenkov radiation is in general on two cones corresponding to the ordinary and extraordinary waves. In motion along the optical axis, however, the ordinary wave is not emitted, even though a condition such as (115.1) may be satisfied: this wave always has linear polarization with the vector **E** perpendicular to the principal cross-section (that is, the plane through the optical axis—which we take as the z-axis—and the direction of any given **k**), and the emission of such a wave in the case concerned is evidently impossible, since the work $e\mathbf{E}\cdot\mathbf{v} = 0$ and the particle does not lose energy. The extraordinary radiation cone is found by substituting in (98.5) the value of n from (115.2), which is valid even if the medium is not isotropic; in the present case, the angle θ between **k** and **v** is the same as the angle between **k** and the optical axis. The result is

$$\tan^2\theta = (\varepsilon_\parallel/\varepsilon_\perp)(v^2\varepsilon_\perp/c^2 - 1),$$

and we must have $v > c/\sqrt{\varepsilon_\perp}$. This is a circular cone on which the intensity distribution is uniform over the generators (as is in any case obvious from symmetry). The vertical angle 2ϑ of the ray vector cone is related to θ by $\tan\vartheta = (\varepsilon_\perp/\varepsilon_\parallel)\tan\theta$.

(b) In this case, there are two Cherenkov cones. We take the direction of **v** as the x-axis, and the optical axis as the z-axis; θ is the angle between **k** and the x-axis, and ϕ the azimuth of **k** measured from the xy-plane (Fig. 62). The cone angle for the ordinary waves is given by

$$\cos\theta = c/v\sqrt{\varepsilon_\perp},$$

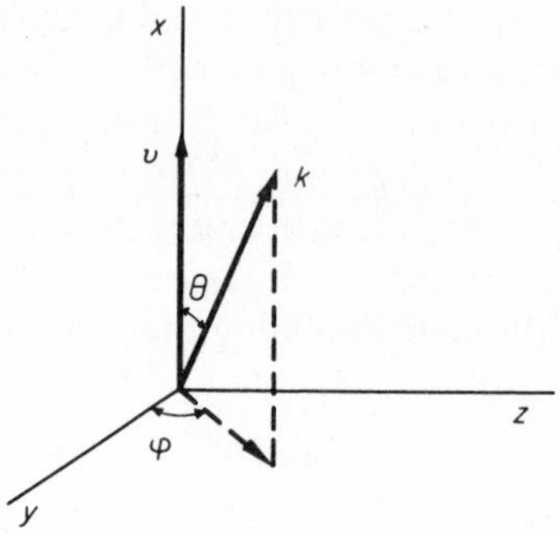

FIG. 62

and we must have $v > c/\sqrt{\varepsilon_\perp}$. This is a circular cone, but the radiation intensity depends on the azimuth†; in particular, there is no radiation in the xz-plane ($\phi = \frac{1}{2}\pi$), since $\mathbf{v}\cdot\mathbf{E} = 0$. The extraordinary wave cone is not circular; its vertical angle depends on ϕ:

$$\cos^2\theta = \frac{(\varepsilon_\parallel - \varepsilon_\perp)\sin^2\phi + \varepsilon_\perp}{(\varepsilon_\parallel - \varepsilon_\perp)\sin^2\phi + \varepsilon_\perp\varepsilon_\parallel v^2/c^2},$$

and we must have $v > c/\sqrt{\varepsilon_\parallel}$. The extraordinary radiation is polarized with the vector **D** in the principal cross-section, perpendicular to **k**. If **k** lies in the xy-plane ($\phi = 0$), then **D** and also **E** are in the z-direction; here $\mathbf{v}\cdot\mathbf{E} = 0$, so that the intensity of the extraordinary radiation is zero in the xy-plane.

§116. Transition radiation

Cherenkov radiation has the property that it occurs with uniform motion of a charged particle (whereas a charge moving uniformly in a vacuum does not radiate). A different class of phenomena that are similar in this respect is represented by *transition radiation*, which occurs with uniform motion of a charged particle in a spatially inhomogeneous medium, for example when passing from one medium to another (V. L. Ginzburg and I. M. Frank, 1945). It differs in principle from the Cherenkov radiation, in that it occurs for any velocity of the particle, not necessarily exceeding the phase velocity of light in the medium. Like the Cherenkov radiation, it is unrelated to the bremsstrahlung which also occurs when charged particles are incident on a surface separating two media, and the distinction is particularly clear in the limit of a particle with infinite mass, for which the bremsstrahlung is zero but the transition radiation is not.

Let us consider the transition radiation when a charged particle passes (with constant velocity **v**) across the boundary between a vacuum and a dielectric (non-magnetic) medium with complex permittivity ε. The motion takes place along the x-axis, at right angles to the interface plane ($x = 0$; Fig. 63).

† The determination of the intensity distribution would need a calculation of the stopping power, similar to that given in §114 for an isotropic medium. These calculations and certain other topics relating to Cherenkov radiation are described in the review articles by B. M. Bolotovskiĭ, *Uspekhi fizicheskikh nauk* **62**, 201, 1957; *Soviet Physics Uspekhi* **4**, 781, 1962.

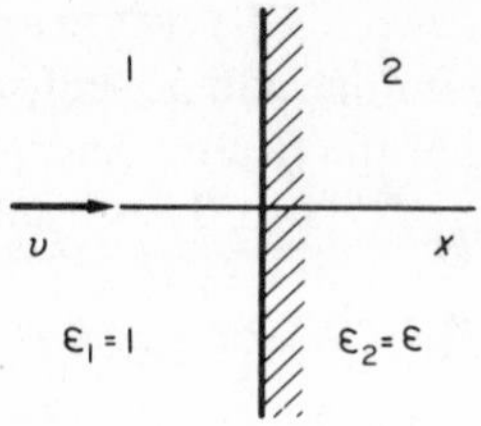

FIG. 63

The electromagnetic field is determined by the equations (114.1)–(114.3) or by the equivalent ones (114.6). All quantities in these equations can be expanded as Fourier integrals with respect to the time and the coordinates y and z, for which the medium is homogeneous:

$$\mathbf{E} = \int \mathbf{E}_{\omega\mathbf{q}}(x) e^{i(\mathbf{q}\cdot\mathbf{r} - \omega t)} d\omega\, d^2q/(2\pi)^3, \tag{116.1}$$

etc. where $\mathbf{q}$ is the wave vector in the yz-plane. In each of the two half-spaces, we seek the field as the sum of a particular solution of the inhomogeneous equations (114.6) (the field of the charge, denoted by the index (e)), and the general solution of the homogeneous equations (114.6) with zero on the right (the free radiation field, denoted by the index (r)). The former is given by formulae analogous to (114.7):

$$\left.\begin{aligned} \phi^{(e)}_{\omega\mathbf{q}} &= \frac{4\pi e}{\varepsilon v}\left[q^2 + \frac{\omega^2}{v^2} - \frac{\varepsilon\omega^2}{c^2}\right]^{-1} e^{i\omega x/v}, \\ \mathbf{A}^{(e)}_{\omega\mathbf{q}} &= \varepsilon\,\mathbf{v}\,\phi^{(e)}_{\omega\mathbf{q}}/c. \end{aligned}\right\} \tag{116.2}$$

The electric field is therefore

$$\mathbf{E}^{(e)}_{\omega\mathbf{q}} = i\left[\omega\mathbf{v}\left(\frac{\varepsilon}{c^2} - \frac{1}{v^2}\right) - \mathbf{q}\right]\phi^{(e)}_{\omega\mathbf{q}}; \tag{116.3}$$

the expression for $\mathbf{H}^{(e)}_{\omega\mathbf{q}}$ will not be needed, and will not be written out.

The second part of the solution, corresponding to the free radiation field, may be put in the form of the electric field immediately. The longitudinal component of $\mathbf{E}^{(r)}_{\omega\mathbf{q}}$ is $(\mathbf{E}^{(r)}_{\omega\mathbf{q}})_x = iae^{\pm ik_x x}$ with a coefficient a as yet unknown; then

$$k_x{}^2 + q^2 = \varepsilon\omega^2/c^2. \tag{116.4}$$

The transverse component, which must obviously be along $\mathbf{q}$, the only distinctive direction in this plane, is then determined by the equation $\operatorname{div}\mathbf{D} = 0$, i.e. $\varepsilon(\mathbf{q} \pm k_x\mathbf{n})\cdot\mathbf{E}_{\omega\mathbf{q}} = 0$ (where $\mathbf{n}$ is a unit vector along the x-axis):

$$\mathbf{E}^{(r)}_{\omega\mathbf{q}} = ia\left[\mathbf{n} \mp \frac{\mathbf{q}}{q^2}\sqrt{\left(\frac{\varepsilon\omega^2}{c^2} - q^2\right)}\right]\exp\left[\pm ix\sqrt{\left(\frac{\varepsilon\omega^2}{c^2} - q^2\right)}\right]. \tag{116.5}$$

The plus and minus signs in the exponent refer to the half-spaces $x > 0$ and $x < 0$ respectively: the waves are propagated away from the interface.†

† In equations (116.2)–(116.5), ε is to be taken as $\varepsilon_1 = 1$ in $x < 0$ and as $\varepsilon = \varepsilon_2$ in $x > 0$. In the formulae below, ε everywhere means ε_2.

The constants a_1 and a_2 in the two half-spaces are determined from the conditions of continuity at the interface for the normal component $\varepsilon\mathbf{n}\cdot\mathbf{E}_{\omega\mathbf{q}}$ of the electric induction and the tangential component $\mathbf{q}\cdot\mathbf{E}_{\omega\mathbf{q}}$ of the electric field; the continuity of the magnetic field provides no further information, of course. The result for the coefficient in the vacuum $x<0$ is

$$a_1 = \frac{4\pi e\beta\kappa^2(\varepsilon-1)\,[1-\beta^2+\beta\sqrt{(\varepsilon-\kappa^2)}]}{\omega(1-\beta^2+\beta^2\kappa^2)\,[1+\beta\sqrt{(\varepsilon-\kappa^2)}]\,[\sqrt{(\varepsilon-\kappa^2)}+\varepsilon\sqrt{(1-\kappa^2)}]}, \tag{116.6}$$

where $\beta = v/c$, $\kappa = qc/\omega$.

Let us now calculate the total energy $\mathscr{U}_1$ radiated by the particle into the vacuum, i.e. backwards relative to its motion. A simple way of doing this is to consider the emitted wave-train for long times t, when it is already at a great distance to the left; the radiation field is then separated from the intrinsic field of the charge. The energy $\mathscr{U}_1$ is found by integrating the radiation field energy density over all space. If the origin is moved along the axis into the region of the wave-train, the integration with respect to x can be taken from $-\infty$ to ∞, because the field is attenuated in both directions.

In the wave region, the electric and magnetic energy densities are equal. Hence

$$\mathscr{U}_1 = \frac{1}{4\pi}\int \mathrm{d}y\,\mathrm{d}z \int_{-\infty}^{\infty} \mathrm{d}x\cdot\mathbf{E}_1{}^2.$$

Substituting $\mathbf{E}_1$ as the expansion (116.1), we write the square of the integral as a double integral:

$$\mathbf{E}^2(t,\mathbf{r}) = \int \mathbf{E}_{\omega\mathbf{q}}(x)\cdot\mathbf{E}_{\omega'\mathbf{q}'}{}^*(x)\exp\{i[\mathbf{r}\cdot(\mathbf{q}-\mathbf{q}')-t(\omega-\omega')]\}\frac{\mathrm{d}\omega\,\mathrm{d}\omega'\,\mathrm{d}^2q\,\mathrm{d}^2q'}{(2\pi)^6}.$$

Integration over $\mathrm{d}y\,\mathrm{d}z$ gives the delta function $(2\pi)^2\delta(\mathbf{q}-\mathbf{q}')$, which is then eliminated by the integration over d^2q'. Thus

$$\mathscr{U}_1 = \frac{1}{4\pi}\int_{-\infty}^{\infty}\mathrm{d}x\int \mathbf{E}_{\omega\mathbf{q}}(x)\cdot\mathbf{E}_{\omega'\mathbf{q}}{}^*(x)\,e^{-it(\omega-\omega')}\frac{\mathrm{d}\omega\,\mathrm{d}\omega'\,\mathrm{d}^2q}{(2\pi)^4}. \tag{116.7}$$

Substituting (116.5) (with $\varepsilon = 1$, $a = a_1$) for $\mathbf{E}_{\omega\mathbf{q}}$, and carrying out the integration with respect to x, we find

$$\mathscr{U}_1 = \tfrac{1}{2}\int |a_1|^2\frac{\omega^2}{q^2c^2}\,\delta\left[\sqrt{\left(\frac{\omega^2}{c^2}-q^2\right)}-\sqrt{\left(\frac{\omega'^2}{c^2}-q^2\right)}\right]\frac{\mathrm{d}^2q\,\mathrm{d}\omega\,\mathrm{d}\omega'}{(2\pi)^4};$$

here we have also used the fact that, because of the delta function, $\omega=\omega'$; for the same reason, the phase factors which arise in a_1 when the origin is moved in (116.5) disappear in the product $a_1(\omega,\mathbf{q})a_1{}^*(\omega',\mathbf{q})$. The integration with respect to ω and ω' is taken from $-\infty$ to ∞; eliminating the delta function by the second of these leaves

$$\mathscr{U}_1 = \int_0^{\infty}\int |a_1(\omega,\mathbf{q})|^2\frac{\omega^2}{cq^2}\sqrt{\left(1-\frac{c^2q^2}{\omega^2}\right)}\frac{\mathrm{d}^2q\,\mathrm{d}\omega}{(2\pi)^4}; \tag{116.8}$$

since the integrand is an even function of ω, the integral with respect to ω has been written as twice the integral from 0 to ∞.

The integration over d^2q is to be taken over the range $q^2 < \omega^2/c^2$ in which k_x is real, so that the field (116.5) actually represents the wave propagated.† We use the angle θ between the radiation wave vector $\mathbf{k} = (k_x, \mathbf{q})$ and the direction of the vector $-\mathbf{v}$, so that $\theta = 0$ corresponds to radiation in the direction exactly opposite to the motion of the particle. Then $q = (\omega/c)\sin\theta$, and a change from integration over d^2q to one over $2\pi q\, dq = (2\pi\omega^2/c^2)\sin\theta\cos\theta d\theta$ gives

$$\mathscr{U}_1 = \frac{1}{c(2\pi)^3}\int_0^\infty\int_0^{\frac{1}{2}\pi} |a_1|^2\,\omega^2\frac{\cos^2\theta}{\sin\theta}\,d\theta\,d\omega$$

$$\equiv \int_0^\infty\int_0^{\frac{1}{2}\pi} \mathscr{U}_1(\omega,\theta)\cdot 2\pi\sin\theta\,d\theta\,d\omega.$$

The function $\mathscr{U}_1(\omega,\theta)$ gives the distribution of the radiation in frequency and angle. With a_1 from (116.6), we have finally

$$\mathscr{U}_1(\omega,\theta) = \frac{e^2\beta^2\sin^2\theta\cos^2\theta}{\pi^2c(1-\beta^2\cos^2\theta)^2}\left|\frac{(\varepsilon-1)[1-\beta^2+\beta\sqrt{(\varepsilon-\sin^2\theta)}]}{[1+\beta\sqrt{(\varepsilon-\sin^2\theta)}][\varepsilon\cos\theta+\sqrt{(\varepsilon-\sin^2\theta)}]}\right|^2 \tag{116.9}$$

(V. L. Ginzburg and I. M. Frank, 1945). The transition radiation is linearly polarized, and the electric vector is seen from (116.5) to be coplanar with **k** and **v**. For non-relativistic velocities, the radiation intensity is proportional to v^2, i.e. to the energy of the particle.

At an interface with an ideal conductor ($\varepsilon = \infty$), formula (116.9) becomes

$$\mathscr{U}_1(\omega,\theta) = \frac{e^2v^2}{\pi^2c^3}\frac{\sin^2\theta}{(1-\beta^2\cos^2\theta)^2}.$$

In the ultra-relativistic case ($\beta \approx 1$), the radiation has a maximum at small angles $\theta \sim \sqrt{(1-\beta^2)}$: formula (116.9) then becomes

$$\mathscr{U}_1(\omega,\theta) = \frac{e^2}{\pi^2c}\left|\frac{\sqrt{\varepsilon}-1}{\sqrt{\varepsilon}+1}\right|^2\frac{\theta^2}{[\theta^2+(1-\beta^2)]^2}.$$

This gives as the frequency distribution of the total radiation, with logarithmic accuracy,

$$\mathscr{U}_1(\omega) \cong \int_0^{\sim 1}\mathscr{U}_1(\omega,\theta)\cdot 2\pi\theta\,d\theta$$

$$\cong \frac{e^2}{\pi c}\left|\frac{\sqrt{\varepsilon}-1}{\sqrt{\varepsilon}+1}\right|^2\log\frac{1}{1-\beta^2} \tag{116.10}$$

† When $q^2 > \omega^2/c^2$, the exponential factors in (116.5) must be written as $\exp[\pm x\sqrt{(q^2-\omega^2/c^2)}]$, corresponding to surface waves attenuated away from the interface. Such waves are necessarily present with transition radiation; they will not be discussed here.

Using, for sufficiently high frequencies, the limiting expression (78.1) for the permittivity:

$$\varepsilon = 1 - \omega_0{}^2/\omega^2, \quad \omega_0{}^2 = 4\pi N e^2/m, \tag{116.11}$$

we find that for $\omega \gg \omega_0$ the intensity decreases as $1/\omega^4$. Thus the frequencies $\omega \lesssim \omega_0$ do in fact give the main contribution to the transition radiation (in the backward direction).

Lastly, let us consider the transition radiation when a particle goes from a medium into a vacuum. This problem differs from the previous one only by a change in the sign of the velocity **v**. The differential intensity of radiation into the vacuum when a charged particle leaves the medium is therefore given by an expression derived from (116.9) by putting $-v$ for v; θ is now the angle between the directions of **k** and **v** (and so $\theta = 0$ corresponds to radiation in exactly the direction of motion of the particle):†

$$\mathscr{U}_1(\omega, \theta) = \frac{e^2\beta^2 \sin^2\theta \cos^2\theta}{\pi^2 c(1-\beta^2\cos^2\theta)^2}\left|\frac{(\varepsilon - 1)[1-\beta^2-\beta\sqrt{(\varepsilon-\sin^2\theta)}]}{[1-\beta\sqrt{(\varepsilon-\sin^2\theta)}][\varepsilon\cos\theta+\sqrt{(\varepsilon-\sin^2\theta)}]}\right|^2. \tag{116.12}$$

In the ultra-relativistic case, the radiation has a maximum intensity at small angles $\theta \sim \sqrt{(1-\beta^2)}$. Formula (116.12) then becomes

$$\mathscr{U}_1(\omega, \theta) = \frac{e^2\theta^2}{\pi^2 c}\frac{|\sqrt{\varepsilon}-1|^2}{(1-\beta^2+\theta^2)^2|1-\beta\sqrt{(\varepsilon-\theta^2)}|^2}. \tag{116.13}$$

If ε is not too close to unity, the last factor in the denominator can be replaced by $|1-\sqrt{\varepsilon}|^2$, and the frequency distribution is, with logarithmic accuracy,

$$\mathscr{U}_1(\omega) = \frac{e^2}{\pi c}\log\frac{1}{1-\beta^2}.$$

For high frequencies, we again use (116.11) and put, in (116.13),

$$1-\beta\sqrt{(\varepsilon-\theta^2)} \cong \tfrac{1}{2}\left(1-\beta^2+\frac{\omega_0{}^2}{\omega^2}+\theta^2\right), \quad \sqrt{\varepsilon}-1 \cong -\omega_0{}^2/2\omega^2.$$

The integration over angles gives

$$\mathscr{U}_1(\omega) = \frac{e^2}{\pi c}\log\frac{\omega_0{}^2}{\omega^2(1-\beta^2)} \quad \text{for} \quad \omega_0 \ll \omega \ll \omega_0/\sqrt{(1-\beta^2)},$$

$$\mathscr{U}_1(\omega) = \frac{e^2}{6\pi c}\left(\frac{\omega_0}{\omega}\right)^4\frac{1}{(1-\beta^2)^2} \quad \text{for} \quad \omega \gg \omega_0/\sqrt{(1-\beta^2)}.$$

These expressions show that the main contribution to the forward radiation comes from high frequencies, $\omega \sim \omega_0/\sqrt{(1-\beta^2)}$ (G. M. Garibyan, 1959). The radiation energy integrated over all frequencies is then proportional to the particle energy:‡

$$\mathscr{U}_1 = e^2\omega_0/3c\sqrt{(1-\beta^2)}. \tag{116.14}$$

† For a transparent medium (ε may be taken as real), we consider only velocities $v < c/\sqrt{\varepsilon}$. Otherwise, there would be the question of separating the contribution of the Cherenkov radiation emitted forwards by the particle in the medium and passing through the interface into the vacuum. This process corresponds to the pole of the intensity (116.12) where $\beta\sqrt{(\varepsilon-\sin^2\theta)} = 1$. The angle θ given by this equation is just the angle of emergence of the Cherenkov cone ray in the medium after refraction at the interface.

‡ A more detailed account of topics related to transition radiation is to be found in the review articles by F. G. Bass and V. M. Yakovenko, *Soviet Physics Uspekhi* **8**, 420, 1965; V. L. Ginzburg and V. N. Tsytovich, *Physics Reports* **49**, 1, 1979.

CHAPTER XV

SCATTERING OF ELECTROMAGNETIC WAVES

§117. The general theory of scattering in isotropic media

In the theory of propagation of electromagnetic waves in transparent media discussed in previous chapters, a phenomenon has been neglected which, though not prominent, is of fundamental importance: *scattering*. Scattering results in the appearance of *scattered waves* of small intensity, whose frequencies and directions are not those of the main wave.

Scattering is ultimately due to the change in the motion of the charges in the medium under the influence of the field of the incident wave, resulting in the emission of the scattered waves. The microscopic mechanism of scattering must be investigated by quantum methods, but this investigation is not needed in developing the macroscopic theory described below. We shall therefore give only some brief remarks on the nature of the processes which cause the change in the wave frequency on scattering.

The basic scattering process consists in the absorption of the original quantum $\hbar\omega$ by the scattering system and the simultaneous emission by that system of another quantum $\hbar\omega'$. The frequency ω' of the scattered quantum may be either less or greater than ω; these two cases are called respectively *Stokes scattering* and *anti-Stokes scattering*. In the former case the system absorbs an amount of energy $\hbar(\omega-\omega')$; in the latter case it emits $\hbar(\omega'-\omega)$ and makes a transition to a state of lower energy. In the simple case of a gas, for example, scattering takes place at individual molecules, and the change in frequency may be due either to a transition of the molecule to another energy level or to a change in the kinetic energy of its motion.

Another kind of process occurs when the primary quantum $\hbar\omega$ remains unchanged but causes the scattering system to emit two quanta: one of energy $\hbar\omega$, with the same frequency and direction, and a "scattered" quantum $\hbar\omega'$. The energy $\hbar(\omega+\omega')$ is obtained from the scattering system. Processes of this type, however, are, under ordinary conditions, very rare in comparison with those of the first type.†

Proceeding now to consider the macroscopic theory of scattering, we must first make precise the meaning of the averaging processes performed in that theory. The averaging of quantities in macroscopic electrodynamics can be regarded as comprising two operations. If, for clarity, we take the classical view, then we can distinguish the averaging over a physically infinitesimal volume with a given position of all the particles in it, and the averaging of the result with respect to the motion of the particles. In the theory of scattering, however, this procedure is impossible, because the averaging with respect to the motion of the particles annuls the very phenomenon which is to be discussed. Thus, for example, the field and induction of the scattered wave which appear in the theory of

† We shall see in §118 that this *stimulated emission* is unimportant at all temperatures $T \ll \hbar(\omega+\omega')$. It may become significant for radio waves.

scattering must be taken to be those resulting from the first averaging only. It should be noted that in the quantum treatment one can of course refer to averaging over a volume only for an operator of a physical quantity, not for the quantity itself. The second stage of the averaging consists in determining the expectation value of the operator by means of the quantum probabilities. Hence the electromagnetic quantities mentioned below should, strictly speaking, be understood as quantum operators. This, however, does not affect the final results of the theory given in the present section, and to simplify the formulae all quantities will be treated as classical.

The monochromatic components of the fields in the scattered wave, taken in this sense, will be denoted in this section by $\mathbf{E}'$, $\mathbf{H}'$, $\mathbf{D}'$ and $\mathbf{B}'$. The fields in the incident wave will be denoted by the unprimed letters $\mathbf{E}$, $\mathbf{H}$. In the present chapter we always suppose the incident wave to be monochromatic with frequency ω.

In the propagation of the scattered wave we have the relation $\mathbf{D}' = \varepsilon(\omega')\mathbf{E}'$ between the electric induction and field (the scattering medium being assumed isotropic), but this relation does not reveal the phenomenon of scattering, i.e. the formation of the scattered wave from the incident wave. To describe this, additional small terms must be included in the expression for $\mathbf{D}'$. In the first approximation, these terms must be linear in the field of the incident wave. The most general form of the relation is then

$$D'_i = \varepsilon' E'_i + \alpha_{ik} E_k + \beta_{ik} E_k^*. \tag{117.1}$$

Here ε' denotes $\varepsilon(\omega')$; α_{ik} and β_{ik} are tensors which characterize the scattering properties of the medium. In general they are not symmetrical, and their components are functions both of the frequency ω' of the scattered wave and of the primary frequency ω. The fact that α and β are tensors does not, of course, contradict the assumed isotropy of the medium. Only the fully averaged properties of the medium are isotropic; the local deviations from the average properties, which include the additional terms in (117.1), need not be isotropic.

The last term in (117.1) pertains to the part of the scattering which results from processes of stimulated emission. All the terms on the right-hand side of equation (117.1) must correspond to the same frequency ω' as $\mathbf{D}'$ on the left-hand side. Since $\mathbf{E}^*$ has the frequency $-\omega$, the frequency of the quantities β_{ik} must be $\omega+\omega'$ to make the frequency of the products $\beta_{ik}E_k^*$ equal to ω'. But $\omega+\omega'$ is the frequency which characterizes processes of stimulated emission. Because this effect is small, as mentioned above, we can neglect the corresponding term in (117.1), and in what follows we shall write

$$D'_i = \varepsilon' E'_i + \alpha_{ik} E_k. \tag{117.2}$$

Similar formulae give the relation between $\mathbf{B}'$ and $\mathbf{H}'$. We shall, however, neglect the magnetic properties of the medium, which are usually of no importance as regards the scattering of light, and therefore put $\mathbf{B}' = \mathbf{H}'$.

Maxwell's equations for the field in the scattered wave are $\mathbf{curl}\,\mathbf{E}' = i\omega'\mathbf{H}'/c$, $\mathbf{curl}\,\mathbf{H}' = -i\omega'\mathbf{D}'/c$. Eliminating $\mathbf{H}'$ from these equations, we find $\mathbf{curl}\,\mathbf{curl}\,\mathbf{E}' = \omega'^2\mathbf{D}'/c^2$. Substituting from (117.2) $\mathbf{E}' = \mathbf{D}'/\varepsilon' - \boldsymbol{\alpha}\cdot\mathbf{E}/\varepsilon'$, where $\boldsymbol{\alpha}\cdot\mathbf{E}$ denotes the vector whose components are $\alpha_{ik}E_k$, and using the equation $\operatorname{div}\mathbf{D}' = 0$, we obtain for $\mathbf{D}'$ the equation

$$\triangle\mathbf{D}' + k'^2\mathbf{D}' = -\mathbf{curl}\,\mathbf{curl}\,(\alpha\cdot\mathbf{E}), \tag{117.3}$$

where $k' = \omega\sqrt{\varepsilon'}/c$ is the wave number of the scattered wave.

For an exact formulation of the conditions under which equation (117.3) is to be solved, we divide the scattering medium into small regions (whose dimensions are still large compared with molecular distances). On account of the molecular nature of the scattering processes, their correlation at different points in the medium (assumed non-crystalline) extends in general only to molecular distances.† Hence the scattered light from the various regions is non-coherent. We can therefore treat scattering from one region as if the light were not scattered at all in the remainder of the medium. In this way we calculate the field of the scattered wave at a large distance from the scattering region. Using a well-known approximation for the retarded potentials at a large distance from the source (see *Fields*, §66), we can immediately derive the required solution of equation (117.3):

$$\mathbf{D}' = \frac{1}{4\pi}\,\mathbf{curl}\,\mathbf{curl}\frac{\exp(ik'R_0)}{R_0}\int \boldsymbol{\alpha}\cdot\mathbf{E}\exp(-i\mathbf{k}'\cdot\mathbf{r})\,dV. \tag{117.4}$$

Here $\mathbf{R}_0$ is the radius vector from some point within the scattering volume (the integration being over that volume) to the point where the field is to be calculated; the vector $\mathbf{k}'$ is in the direction of $\mathbf{R}_0$. The integral in (117.4) is independent of the coordinates of the point considered; retaining in the differentiation, as usual, only terms in $1/R_0$, we obtain

$$\mathbf{D}' = -\frac{\exp(ik'R_0)}{4\pi R_0}\mathbf{k}'\times[\mathbf{k}'\times\int \boldsymbol{\alpha}\cdot\mathbf{E}\exp(-i\mathbf{k}'\cdot\mathbf{r})\,dV].$$

Since, at the point considered, the medium is regarded as not scattering, the relation between $\mathbf{D}'$ and $\mathbf{E}'$ there is given by $\mathbf{D}' = \varepsilon'\mathbf{E}'$ simply. In the field of the incident wave $\mathbf{E}$ we separate a factor periodic in space, putting

$$\mathbf{E} = \mathbf{E}_0 e^{i\mathbf{k}\cdot\mathbf{r}} = E_0\,\mathbf{e}e^{i\mathbf{k}\cdot\mathbf{r}}; \tag{117.5}$$

in the second equation, the complex amplitude $\mathbf{E}_0$ is written as $E_0\mathbf{e}$, where E_0 is a real quantity ($E_0{}^2 = |\mathbf{E}_0|^2$), and $\mathbf{e}$ is a complex unit vector ($\mathbf{e}\cdot\mathbf{e}^* = 1$) which defines the polarization of the wave. With the notation

$$\mathbf{G} = \int (\boldsymbol{\alpha}\cdot\mathbf{e})e^{-i\mathbf{q}\cdot\mathbf{r}}\,dV, \quad \mathbf{q} = \mathbf{k}-\mathbf{k}', \tag{117.6}$$

we then have

$$\begin{aligned}\mathbf{E}' &= -e^{ik'R_0}\frac{E_0}{4\pi R_0\varepsilon'}\mathbf{k}'\times(\mathbf{k}'\times\mathbf{G})\\ &= e^{ik'R_0}\frac{E_0 k'^2}{4\pi R_0\varepsilon'}\mathbf{G}_\perp.\end{aligned} \tag{117.7}$$

The vector $\mathbf{E}'$ is perpendicular to the direction $\mathbf{k}'$ of the scattered wave, and is given by the component $\mathbf{G}_\perp$ perpendicular to $\mathbf{k}'$.

Having thus determined the non-averaged field in the scattered wave, we can now investigate the intensity and polarization of the scattered light. To do so, we form the tensor

$$I_{ik} = \langle E'_i E'^*_k\rangle, \tag{117.8}$$

where the brackets denote the final averaging over the motion of the particles, which so far has not been carried out. The averaging of a quadratic expression gives, of course, a result

† Exceptions may occur for particular cases of scattering, which will be discussed in §120. In such cases the dimensions of the scattering regions must be supposed large in comparison with the wavelength of the light.

which is not zero. Since $\mathbf{E}'$ is perpendicular to $\mathbf{k}'$, the tensor I_{ik} has non-zero components only in the plane perpendicular to $\mathbf{k}'$. These components form a two-dimensional tensor $I_{\alpha\beta}$ in that plane (Greek suffixes take two values). The tensor $I_{\alpha\beta}$ is, by definition, Hermitian: $I_{\alpha\beta} = I_{\beta\alpha}^*$. It can be diagonalized, and the ratio of its two principal values gives the degree of depolarization, while their sum is proportional to the total intensity.†

The products $E'_i E'_k{}^*$ involve products of integrals G_i, which must also be averaged. Writing the product as a double integral, we have

$$\langle G_i G_k^* \rangle = e_l e_m^* \int\int \langle \alpha_{il}^{(1)} \alpha_{km}^{(2)*} \rangle \exp[-i\mathbf{q}\cdot(\mathbf{r}_1 - \mathbf{r}_2)] \, dV_1 \, dV_2. \qquad (117.9)$$

The indices (1) and (2) mean that the values of α are taken at two different points in space.

In averaging the integrand it must be remembered that the correlation between the values of α at different points in the body extends in general only over molecular distances. After averaging, therefore, the integrand will be appreciably different from zero only for $|\mathbf{r}_2 - \mathbf{r}_1| \sim a$, where a is of the order of molecular distances. The exponent is $\sim a/\lambda$, where λ is the wavelength of the scattered wave; but $a/\lambda \ll 1$ if the macroscopic theory is applicable, and so we can replace the exponential factor by unity.‡

Next, the integration with respect to the coordinates $\mathbf{r}_1$ and $\mathbf{r}_2$ can be replaced by one with respect to $\frac{1}{2}(\mathbf{r}_1 + \mathbf{r}_2)$ and $\mathbf{r} = \mathbf{r}_1 - \mathbf{r}_2$. Since the integrand depends, after averaging, on $\mathbf{r}$ only, we have

$$\langle G_i G_k^* \rangle = V e_l e_m^* \int \langle \alpha_{il}^{(1)} \alpha_{km}^{(2)*} \rangle \, dV, \qquad (117.10)$$

where V is the volume of the scattering region. It is evident *a priori* that the scattering must be proportional to V. It should be noted that the direction of the wave vector $\mathbf{k}$ in the incident wave appears neither in (117.10) nor, consequently, in the following formulae.

The integrals in (117.10) form a tensor of rank four, which depends only on the properties of the scattering medium. Since the medium is isotropic, this tensor must be expressible in terms of the unit tensor δ_{ik} (and scalar constants). Before giving the appropriate expression, we should note that the tensor α_{ik}, like any tensor of rank two, can in the general case be represented as the sum of three independent parts:

$$\alpha_{ik} = \alpha\delta_{ik} + s_{ik} + a_{ik}, \qquad (117.11)$$

where α is a scalar, s_{ik} is an irreducible (i.e. with zero trace) symmetrical tensor, and a_{ik} is an antisymmetrical tensor:

$$\alpha = \tfrac{1}{3}\alpha_{ii}, \; s_{ik} = \tfrac{1}{2}(\alpha_{ik} + \alpha_{ki} - \tfrac{2}{3}\alpha_{ll}\delta_{ik}), \; a_{ik} = \tfrac{1}{2}(\alpha_{ik} - \alpha_{ki}). \qquad (117.12)$$

On averaging the product $\alpha_{il}^{(1)}\alpha_{km}^{(2)*}$, a non-zero result can occur only for products of components of each of the three parts of α_{ik} separately; clearly, the unit tensor cannot be used to form an expression whose symmetry properties correspond to those of cross-products. It follows that the tensor of rank four can be written as

$$\int \langle \alpha_{il}^{(1)} \alpha_{km}^{(2)*} \rangle \, dV = G_0 \delta_{il}\delta_{km} + \tfrac{1}{10} G_s (\delta_{ik}\delta_{lm} + \delta_{im}\delta_{kl} - \tfrac{2}{3}\delta_{il}\delta_{km}) + \tfrac{1}{6} G_a (\delta_{ik}\delta_{lm} - \delta_{im}\delta_{kl}), \qquad (117.13)$$

† See *Fields*, §50. The diagonalization of an Hermitian tensor means putting it in the form $I_{ik} = \lambda_1 n_{1i} n_{1k}^* + \lambda_2 n_{2i} n_{2k}^*$, where $\mathbf{n}_1, \mathbf{n}_2$ are, in general, perpendicular complex unit vectors: $\mathbf{n}_1 \cdot \mathbf{n}_1^* = 1$, $\mathbf{n}_2 \cdot \mathbf{n}_2^* = 1$, $\mathbf{n}_1 \cdot \mathbf{n}_2^* = 0$. The principal values λ_1, λ_2 of an Hermitian tensor are real.

‡ This procedure requires further discussion in the case of Rayleigh scattering (§120).

where the symmetry of the three terms corresponds to the products of the scalar, symmetrical and antisymmetrical parts of the tensors $\alpha_{il}^{(1)}$ and $\alpha_{km}^{(2)}$. Contracting this expression with respect to various pairs of suffixes, we obtain three equations, from which the significance of the coefficients in (117.13) becomes evident:

$$\left.\begin{aligned} G_0 &= \int \langle \alpha^{(1)} \alpha^{(2)*} \rangle \mathrm{d}V, \\ G_s &= \int \langle s_{ik}^{(1)} s_{ik}^{(2)*} \rangle \mathrm{d}V, \\ G_a &= \int \langle a_{ik}^{(1)} a_{ik}^{(2)*} \rangle \mathrm{d}V. \end{aligned}\right\} \tag{117.14}$$

These quantities are real and positive.†

The tensor I_{ik} thus becomes

$$I_{ik} = \text{constant} \times [G_0 e_i e_k^* + \tfrac{1}{10} G_s(\delta_{ik} + e_i^* e_k - \tfrac{2}{3} e_i e_k^*) + \tfrac{1}{6} G_a(\delta_{ik} - e_i^* e_k)], \tag{117.15}$$

the constant being independent of the direction of scattering and of the incident wave polarization. This tensor is, of course, not yet transverse to $\mathbf{k}'$. The required tensor $I_{\alpha\beta}$ is obtained by projecting the tensor (117.15) on a plane perpendicular to $\mathbf{k}'$; to do this, it is sufficient to take a coordinate system with one axis in the direction of $\mathbf{k}'$ and find the components of the tensor along the other two axes.

In the general case, scattering can be regarded as a superposition of three independent processes of *scalar*, *symmetric* and *antisymmetric scattering*, corresponding to the three terms in (117.15).‡

If this separation is not important, it may be convenient to put (117.15) in another form, combining similar terms:

$$I_{ik} = \text{constant} \times \{\tfrac{1}{2}(a+c) e_i e_k^* + \tfrac{1}{2}(a-c) e_i^* e_k + b\delta_{ik}\}, \tag{117.16}$$

where

$$a = G_0 + \tfrac{1}{30} G_s - \tfrac{1}{6} G_a, \quad b = \tfrac{1}{10} G_s + \tfrac{1}{6} G_a, \quad c = G_0 - \tfrac{1}{6} G_s + \tfrac{1}{6} G_a. \tag{117.17}$$

Formula (117.15) or (117.16) determines the angular distribution and polarization properties of the scattered light. In particular, on projecting this tensor on some polarization vector $\mathbf{e}'$ (which defines the direction of $\mathbf{E}'$), we obtain the intensity of the scattered light component polarized in a particular way, which could be distinguished by means of a suitable analyser:

$$\begin{aligned} I_{ik} e'^*_i e'_k = \text{constant} \times \{ & G_0 |\mathbf{e}\cdot\mathbf{e}'^*|^2 + \tfrac{1}{10} G_s(1 + |\mathbf{e}\cdot\mathbf{e}'|^2 \\ & - \tfrac{2}{3}|\mathbf{e}\cdot\mathbf{e}'^*|^2) + \tfrac{1}{6} G_a(1 - |\mathbf{e}\cdot\mathbf{e}'|^2)\} \end{aligned} \tag{117.18}$$

† The fact that the tensor (117.13) is real, and therefore so are the coefficients in it, is evident from the automatic symmetry of this tensor with respect to interchange of the pairs of suffixes *il* and *km*, this interchange being equivalent to taking the complex conjugate (since the points 1 and 2 are equivalent). The coefficients are positive because they can be expressed as the square of a modulus (or as the sum of such squares) by a transformation opposite to that used in going from (117.9) to (117.10). For example,

$$G_0 = \frac{1}{V} \langle |\int \alpha^{(1)} \mathrm{d}V|^2 \rangle.$$

‡ Formula (117.15) and the subsequent deductions made from it differ only in the definition of G_0, G_s and G_a from those in the Placzek quantum theory of scattering by separate freely oriented molecules (*QED*, §60).

or

$$I_{ik}e'^{*}_{i}e'_{k} = \text{constant} \times \{\tfrac{1}{2}(a+c)|\mathbf{e}\cdot\mathbf{e}'^{*}|^2 + \tfrac{1}{2}(a-c)|\mathbf{e}\cdot\mathbf{e}'|^2 + b\}. \tag{117.19}$$

Let us consider the scattering of a linearly polarized wave. This corresponds to a real vector **e**; see *Fields*, §§48, 50. The components of the scattered light tensor $I_{\alpha\beta}$ are therefore all real also. This means that the scattered light is partially polarized, and can be divided into two independent (non-coherent) waves, each of which is linearly polarized. Since there is only one distinctive direction in the plane perpendicular to $\mathbf{k}'$, given by the projection of the vector **e** on that plane, it is evident that one of these waves is polarized with the vector $\mathbf{e}'$ in the plane of **e** and $\mathbf{k}'$, with intensity I_1, say, and the other is polarized perpendicular to that plane, with intensity I_2, say.†

When **e** is real, the expression (117.16) reduces to

$$I_{ik} = \text{constant} \times (ae_ie_k + b\delta_{ik}). \tag{117.20}$$

This contains only two independent constants, not three. Accordingly,

$$I_{ik}e'_ie'_k = \text{constant} \times [a(\mathbf{e}\cdot\mathbf{e}')^2 + b]$$

and, taking $\mathbf{e}'$ in the two directions mentioned above, we find the angular distributions of the two non-coherent components of the scattered light:

$$I_1 = \text{constant} \times (a\sin^2\theta + b), \quad I_2 = \text{constant} \times b, \tag{117.21}$$

where θ is the angle between **e** and the direction of scattering $\mathbf{k}'$. The second distribution is isotropic.

When natural light traverses the medium, the scattered light is partially polarized. The corresponding tensor I_{ik} is obtained from (117.16) by averaging over all directions of **e** in the plane perpendicular to **k**. This averaging is effected by means of the formula

$$\overline{e_ie_k^*} = \tfrac{1}{2}(\delta_{ik} - n_in_k) \tag{117.22}$$

(where $\mathbf{n} = \mathbf{k}/k$), which represents a tensor of rank two depending only on the direction of **n**, giving unity on contraction, and satisfying the condition

$$\overline{n_ie_ie_k^*} = \overline{(\mathbf{n}\cdot\mathbf{e})e_k^*} = 0.$$

In the scattering of natural light, therefore,

$$I_{ik} = \text{constant} \times \{\tfrac{1}{2}a(\delta_{ik} - n_in_k) + b\delta_{ik}\}. \tag{117.23}$$

It is evident from symmetry that the two non-coherent components of the scattered light are linearly polarized, with the vector $\mathbf{e}'$ in the plane of **k** and $\mathbf{k}'$ (the scattering plane) and perpendicular to it; let the intensities of these components be $I_\parallel$ and $I_\perp$ respectively. The formula

$$I_{ik}e'_ie'_k = \text{constant} \times \{\tfrac{1}{2}a[1 - (\mathbf{n}\cdot\mathbf{e}')^2] + b\}$$

gives

$$I_\parallel = \text{constant} \times (\tfrac{1}{2}a\cos^2\vartheta + b), \quad I_\perp = \text{constant} \times (\tfrac{1}{2}a + b), \tag{117.24}$$

where ϑ is the scattering angle (between **k** and $\mathbf{k}'$).

---

† We must again emphasize the necessity of distinguishing the polarization state of the scattered light as such from the polarization $\mathbf{e}'$ that is registered by a detector.

We shall also give expressions for the angular distribution and polarization properties of each of the three types of scattering separately. These are found from (117.21) and (117.24) by simply substituting for a and b the corresponding terms in (117.17).

In scalar scattering of linearly polarized light, the scattered light too is completely polarized, and its angular intensity distribution is given by

$$I = \tfrac{3}{2}\sin^2\theta; \tag{117.25}$$

here and henceforward, the expressions for I are normalized so as to give unity on averaging over directions. In scattering of natural light, however, the angular distribution of the intensity and the degree of depolarization (the ratio of the smaller and larger of $I_\parallel$ and $I_\perp$) of the scattered light are given by†

$$I = I_\perp + I_\parallel = \tfrac{3}{4}(1+\cos^2\vartheta),\quad I_\parallel/I_\perp = \cos^2\vartheta. \tag{117.26}$$

For symmetric scattering of polarized light, we have

$$I = I_1 + I_2 = \tfrac{3}{20}(6+\sin^2\theta),\quad I_2/I_1 = 3/(3+\sin^2\theta), \tag{117.27}$$

and for scattering of natural light

$$I = \tfrac{3}{40}(13+\cos^2\vartheta),\quad I_\parallel/I_\perp = \tfrac{1}{7}(6+\cos^2\vartheta). \tag{117.28}$$

Lastly, for antisymmetric scattering of polarized light,

$$I = \tfrac{3}{4}(1+\cos^2\theta),\quad I_1/I_2 = \cos^2\theta, \tag{117.29}$$

and for scattering of natural light

$$I = \tfrac{3}{8}(2+\sin^2\vartheta),\quad I_\perp/I_\parallel = 1/(1+\sin^2\vartheta).$$

§118. The principle of detailed balancing applied to scattering

The general principle of detailed balancing in quantum mechanics (see *QM*, §144) can be used to obtain a relation between the intensities in various scattering processes.

Let $\mathrm{d}w_{21}$ be the probability that a quantum $\hbar\omega_1$ is scattered (on a path of unit length) and gives rise to a quantum $\hbar\omega_2$ in the solid angle element $\mathrm{d}o_2$;‡ let $\mathrm{d}w_{12}$ be the probability of the converse process, in which a quantum $\hbar\omega_2$ yields a quantum $\hbar\omega_1$ in the solid angle element $\mathrm{d}o_1$. According to the principle of detailed balancing we have $\mathrm{d}w_{21}/k_2{}^2\mathrm{d}o_2 = \mathrm{d}w_{12}/k_1{}^2\mathrm{d}o_1$, where k_1 and k_2 are the wave numbers of the two quanta. Substituting $k_1{}^2 = \varepsilon_1\omega_1{}^2/c^2$, $k_2{}^2 = \varepsilon_2\omega_2{}^2/c^2$ (where $\varepsilon_1 = \varepsilon(\omega_1)$, $\varepsilon_2 = \varepsilon(\omega_2)$), we obtain

$$\varepsilon_1\omega_1{}^2\mathrm{d}w_{21}/\mathrm{d}o_2 = \varepsilon_2\omega_2{}^2\mathrm{d}w_{12}/\mathrm{d}o_1. \tag{118.1}$$

Here it is assumed that the initial and final states of the scattering system correspond to discrete energy levels E_1 and E_2, related by $E_1 + \hbar\omega_1 = E_2 + \hbar\omega_2$. This statement of the

---

† The passage from the formulae for $I(\theta)$ to those for $I(\vartheta)$ corresponds to averaging in accordance with

$$\overline{\sin^2\theta} = \tfrac{1}{2}(1+\cos^2\vartheta);$$

compare the derivation of (92.6) from (92.3).

‡ The two quanta also have definite polarizations, but for brevity these are not shown. The order of suffixes in $\mathrm{d}w_{21}$ corresponds to that which is usual in quantum mechanics, with the initial state on the right and the final state on the left.

problem is not quite true to reality, since the energy levels of a macroscopic body are extremely closely spaced and can be regarded as quasi-continuous.

Instead of the scattering probability dw_{21} with an exactly determined frequency change, we must therefore use the probability of scattering into a frequency range $d\omega_2$, i.e. of the body's entering a state whose energy lies in a range $dE_2 = \hbar d\omega_2$. Denoting this probability (again per unit path length) by dh_{21}, we have $dh_{21} = dw_{21}d\Gamma_2 = dw_{21}(d\Gamma_2/dE_2)\hbar\, d\omega_2$, where $d\Gamma_2$ is the number of quantum states of the body in the energy range dE_2. Instead of (118.1), we therefore have

$$\frac{d\Gamma_1}{dE_1}\varepsilon_1\omega_1^2\frac{dh_{21}}{do_2 d\omega_2} = \frac{d\Gamma_2}{dE_2}\varepsilon_2\omega_2^2\frac{dh_{12}}{do_1\, d\omega_1}.$$

According to a well-known relation between the statistical weight of a macroscopic state of a body and its entropy $\mathscr{S}$, the derivative $d\Gamma/dE$ is essentially $\exp \mathscr{S}$, so that $(d\Gamma_1/dE_1):(d\Gamma_2/dE_2) = \exp(\mathscr{S}_1 - \mathscr{S}_2)$. Since the relative change in the energy of the body resulting from the scattering of one quantum is negligible, the change in entropy is also small, and can be taken as $\mathscr{S}_1 - \mathscr{S}_2 = (d\mathscr{S}/dE)(E_1 - E_2) = \hbar(\omega_2 - \omega_1)/T$. Using this result, we can write the final expression of the principle of detailed balancing for scattering in the form

$$e^{-\hbar\omega_1/T}\varepsilon_1\omega_1^2\frac{dh_{21}}{do_2\, d\omega_2} = e^{-\hbar\omega_2/T}\varepsilon_2\omega_2^2\frac{dh_{12}}{do_1\, d\omega_1}. \tag{118.2}$$

The quantity dh_{21}, whose dimensions are cm^{-1}, is called the *differential extinction coefficient* for scattering of light. It can also be defined as follows: dh_{21} is the ratio of the number of quanta scattered in the direction do_2 and the frequency range $d\omega_2$ per unit time and volume to the incident photon flux density. By integrating dh_{21} over all directions and frequencies of the scattered light, we btain the *total extinction coefficient*, which represents the damping decrement of the photon flux density as the light passes through the scattering medium.

Let $\omega_2 < \omega_1$. The relation (118.2) connects the intensities (extinction coefficients) of Stokes ($1 \to 2$) and anti-Stokes ($2 \to 1$) scattering. We see that the latter is in general less than the former by approximately the factor $e^{-\hbar(\omega_1-\omega_2)/T}$. This is a very general result, and corresponds to the fact that the transfer of energy from the body to the electromagnetic field reduces the probability of the process by a factor $e^{-\Delta E/T}$, where ΔE is the energy transferred. In particular, the stimulated emission, in which the body gives up an energy $\hbar(\omega_1+\omega_2)$ in each scattering process, is therefore usually very weak. The probability of such a process, when $\hbar(\omega_1+\omega_2) \gg T$, contains the small factor $e^{-\hbar(\omega_1+\omega_2)/T}$.

The general relation (118.2) is much simplified in the important case of scattering with a relatively small change in frequency. We shall denote ω_1 by ω simply, and the small difference $\omega_2 - \omega_1$ by Ω ($|\Omega| \ll \omega$), and put for brevity

$$dh_{21}/do_2\, d\omega_2 = I(\omega, \Omega). \tag{118.3}$$

In the non-exponential factors $\varepsilon\omega^2$ in (118.2) we can neglect the difference Ω; these factors then cancel, leaving

$$I(\omega, \Omega)e^{-\hbar\omega/T} = I(\omega+\Omega, -\Omega)e^{-\hbar(\omega+\Omega)/T}.$$

In the first argument of the function $I(\omega+\Omega, -\Omega)$, which gives the initial frequency of the light, we can neglect Ω, i.e. refer the scattered intensity to the slightly different frequency of the incident light. Then

$$I(\omega, \Omega) = I(\omega, -\Omega)e^{-\hbar\Omega/T}. \tag{118.4}$$

In this approximation I on each side of the equation refers to the same frequency of the incident light. In other words, the relation (118.4) gives a simple relation between Stokes and anti-Stokes scattering of the same light with the same magnitude of the frequency change Ω.

PROBLEM

Determine the relation between the intensities of stimulated Raman scattering (§112) and ordinary (*spontaneous*) scattering.

SOLUTION. The probability of stimulated scattering is found from that of spontaneous scattering by multiplying by $N_{\mathbf{k}_2}$, the number of photons in the quantum state with wave vector $\mathbf{k}_2$. In order to relate this number to the field $\mathbf{E}_2$ of the scattered wave, we must regard the latter as almost monochromatic, and equate the expressions for the field energy (per unit volume) in terms of the number of quanta and in terms of the field:

$$\int \hbar\omega_2 N_{\mathbf{k}_2}\frac{d^3k_2}{(2\pi)^3} = \frac{c\sqrt{\varepsilon_2}}{8\pi}\frac{dk_2}{d\omega_2}|\mathbf{E}_2|^2, \tag{1}$$

the right-hand side being written in accordance with (83.9). On the left-hand side, the factors which vary only slightly over a narrow frequency range can be taken outside the integral after substituting

$$d^3k_2 = k_2^2(dk_2/d\omega_2)do_2 d\omega_2 = (\varepsilon^2\omega_2^2/c^2)(dk_2/d\omega_2)do_2 d\omega_2.$$

Then

$$\int \hbar\omega_2 N_{\mathbf{k}_2}\frac{d^3k_2}{(2\pi)^3} = \frac{\hbar\omega_2^3\sqrt{\varepsilon_2}}{8\pi^3c^2}\frac{dk_2}{d\omega_2}\int N_{\mathbf{k}_2}d\omega_2 do_2,$$

and (1) gives

$$\int N_{\mathbf{k}_2}d\omega_2 do_2 = |\mathbf{E}_2|^2\frac{\pi^2c^3}{\hbar\omega_2^3\sqrt{\varepsilon_2}}.$$

The incident photon flux density is, from (83.11),

$$\frac{1}{\hbar\omega_1}\bar{S}_1 = \frac{c\sqrt{\varepsilon_1}}{8\pi\hbar\omega_1}|\mathbf{E}_1|^2.$$

Thus the energy transferred to the field $\mathbf{E}_2$ by stimulated scattering of photons $\hbar\omega_1$ is, with the notation (118.3),

$$\begin{aligned}\frac{d\bar{U}_2}{dt} &= \hbar\omega_2\frac{\bar{S}_1}{\hbar\omega_1}I(\omega_1, \Omega)\int N_{\mathbf{k}_2}d\omega_2 do_2\\ &= \frac{\pi c^4}{8\hbar\omega_1\omega_2^2}\sqrt{\frac{\varepsilon_1}{\varepsilon_2}}|\mathbf{E}_1|^2\,|\mathbf{E}_2|^2 I(\omega_1, \Omega)\end{aligned} \tag{2}$$

with $\Omega = \omega_2 - \omega_1$. The field $\mathbf{E}_2$ also loses energy by the stimulated scattering of photons $\hbar\omega_2$ with their conversion into photons $\hbar\omega_1$. The energy thus gained by the field $\mathbf{E}_1$ is expressed by formula (2) with the suffixes 1 and 2 interchanged. The energy lost by the field $\mathbf{E}_2$ is then found by multiplying by $-\omega_2/\omega_1$; see (112.8). It is

$$-\frac{\pi c^4\sqrt{\varepsilon_2}}{8\hbar\omega_1^3\sqrt{\varepsilon_1}}|\mathbf{E}_1|^2\,|\mathbf{E}_2|^2 I(\omega_2, -\Omega). \tag{3}$$

Adding the two expressions, and expressing $I(\omega_2, -\Omega)$ in terms of $I(\omega_1, \Omega)$ by means of (118.2), we find the total energy change of the field with frequency ω_2:

$$\frac{d\bar{U}_2}{dt} = \frac{\pi c^4}{8\hbar\omega_1\omega_2^2}\sqrt{\frac{\varepsilon_1}{\varepsilon_2}}\left(1 - e^{-\hbar\Omega/T}\right)I(\omega_1, \Omega)\,|\mathbf{E}_1|^2\,|\mathbf{E}_2|^2. \tag{4}$$

The growth rate (when $\omega_2 < \omega_1$) of the scattered radiation intensity at frequency ω_2, per unit path, is

$q_2 = (1/\overline{S}_2)\mathrm{d}\overline{U}_2/\mathrm{d}t$, with dimensions 1/cm. The final form of the relation is then

$$q_2 = \frac{(2\pi)^3}{k_2{}^2\hbar\omega_1}\left(1-e^{-\hbar\Omega/T}\right)\overline{S}_1\frac{\mathrm{d}h_{21}}{\mathrm{d}o_2\mathrm{d}\omega_2}, \tag{5}$$

where $k_2 = \omega_2\sqrt{\varepsilon_2}/c$, and $\overline{S}_1$ is the energy flux density of the incident light at frequency ω_1.

§119. Scattering with small change of frequency

The theory given in §117 is entirely general, and is applicable to all cases of scattering in an isotropic medium, whatever the mechanism of scattering. Such a general discussion, of course, cannot proceed very far, and a further investigation of the phenomenon of scattering requires some restrictive assumptions.

Usually, the scattering of light involves only a relatively small change in frequency, $\Omega = \omega' - \omega$. The calculations given below pertain to this case. Besides the condition $|\Omega| \ll \omega$, we shall suppose that the relative change in the refractive index of the medium over the frequency range Ω is small. This condition means that the frequency ω must not lie close to a range or line in which the scattering medium is also absorbing.

If ω is in the optical range, the microscopic mechanism of scattering with small Ω may involve various kinds of motion of atoms and molecules (i.e. of nuclei, as opposed to the purely electronic motions which give rise to optical transitions), including intramolecular vibrations of atoms, rotations or vibrations of molecules, etc.†

Let $q = q(t)$ denote the set of coordinates describing the motion which causes the scattering. (For simplicity, we shall first give a classical discussion.) Since this motion is relatively slow, the macroscopic description of scattering can be regarded from a different standpoint by introducing the permittivity tensor $\varepsilon_{ik}(q)$, whose components at any instant depend only on the values of the coordinates q at that instant as parameters. This property follows from the assumed slowness of the relative change in ε. The permittivity thus defined pertains to the field averaged with respect to the electron motion for a given position of the nuclei. When the averaging of the field (including that with respect to the motion of the nuclei) is carried out, the permittivity reduces to the scalar $\varepsilon(\omega)$. Let the deviation of ε_{ik} from this value be $\delta\varepsilon_{ik}$:

$$\varepsilon_{ik}(q) = \varepsilon\delta_{ik} + \delta\varepsilon_{ik}(q). \tag{119.1}$$

The tensor ε_{ik} gives the relation between the field and the induction as functions of time. It should be emphasized that the incident wave is still assumed to be monochromatic with frequency ω, but the field $\mathbf{E}'$ in the scattered wave is now regarded as a function of time, not resolved into monochromatic components. The total field consists of the field $\mathbf{E}$ in the incident wave and the field $\mathbf{E}'$ in the scattered wave. Thus $D_i + D'_i = \varepsilon_{ik}(E_k + E'_k)$. Cancelling $D_i = \varepsilon E_i$ and omitting the second-order term $\delta\varepsilon_{ik}E'_k$, we obtain

$$D'_i = \varepsilon E'_i + \delta\varepsilon_{ik}(q)E_k. \tag{119.2}$$

The relation (119.2) is of the same form as (117.2). There is a difference, however, in that with this approach it is clear that the tensor $\alpha_{ik} = \delta\varepsilon_{ik}$ is symmetrical. This follows at once from the general theorem concerning the symmetry of the permittivity tensor. Furthermore, since this tensor is real for a transparent medium, the tensor $\delta\varepsilon_{ik}$ is also real.

† Here it is assumed that the relevant values of Ω are much smaller than the electronic transition frequencies. This condition may be violated in a gas consisting of molecules which have a degenerate electronic ground state.

Since the tensor α_{ik} has no antisymmetrical part, there is no antisymmetric scattering (§117) with small change in frequency.

Let us calculate the total scattered intensity with all frequency changes $\Omega \equiv \omega' - \omega \ll \omega$. This can easily be done as follows. In equation (117.3) for the field in the scattered wave we can replace k' by $k = \omega\sqrt{\varepsilon}/c$ (and take the value $\alpha_{ik} \equiv \delta\varepsilon_{ik}$ for $\omega' = \omega$); this equation does not then involve ω', i.e. it is the same for every component of the spectral resolution of the field. The equation is therefore valid for the unresolved field in the scattered wave, which we shall denote by the same letter $\mathbf{E}'$. Using the solution (117.7), we obtain

$$\langle|\mathbf{E}'|^2\rangle = \frac{E_0{}^2k^4}{16\pi^2\varepsilon^2R_0{}^2}\langle|\mathbf{G}|^2\rangle\sin^2\theta = \frac{E_0{}^2\omega^4}{16\pi^2R_0{}^2c^4}\langle|\mathbf{G}|^2\rangle\sin^2\theta,$$

where θ is the angle between $\mathbf{k}$ and $\mathbf{G}$, and the brackets denote, as in §117, the final average with respect to the motion of the particles.

We define the extinction coefficient h as the ratio of the total intensity of light scattered in all directions per unit volume of the scattering medium to the incident flux density:†

$$h = \frac{1}{V|\mathbf{E}|^2}\int\langle|\mathbf{E}'|^2\rangle R_0{}^2\mathrm{d}o', \tag{119.3}$$

in which we have put $\varepsilon(\omega') \cong \varepsilon(\omega)$.

For the reason given in §117, in calculating the mean value $\langle|\mathbf{G}|^2\rangle$ the exponential factor in the integrand of $\mathbf{G}$ is replaced by unity, so that

$$\langle|\mathbf{G}|^2\rangle = Ve_le_m{}^*\int\langle\delta\varepsilon_{il}{}^{(1)}\delta\varepsilon_{im}{}^{(2)}\rangle\,\mathrm{d}V;$$

cf. (117.9), (117.10). The expression in the angle brackets is a tensor of rank two, and, since the medium is isotropic, it gives on averaging

$$\langle\delta\varepsilon_{il}{}^{(1)}\delta\varepsilon_{im}{}^{(2)}\rangle = \tfrac{1}{3}\delta_{lm}\langle\delta\varepsilon_{ik}{}^{(1)}\delta\varepsilon_{ik}{}^{(2)}\rangle.$$

The mean square

$$\langle|\mathbf{G}|^2\rangle = \tfrac{1}{3}V\int\langle\delta\varepsilon_{ik}{}^{(1)}\delta\varepsilon_{ik}{}^{(2)}\rangle\,\mathrm{d}V$$

is then independent of the direction of scattering, and the result of the integration in (119.3) is

$$h = \frac{\omega^4}{18\pi c^4}\int\langle\delta\varepsilon_{ik}{}^{(1)}\delta\varepsilon_{ik}{}^{(2)}\rangle\,\mathrm{d}V. \tag{119.4}$$

The integrand here is the correlation function of the permittivity fluctuations at different points $\mathbf{r}_1$ and $\mathbf{r}_2$ in the medium at the same time; the integration is taken with respect to the difference of coordinates $\mathbf{r} = \mathbf{r}_1 - \mathbf{r}_2$. If we return to integration over $\mathrm{d}V_1$ and $\mathrm{d}V_2$, formula (119.4) becomes

$$h = \frac{\omega^4}{18\pi c^4}V\langle\delta\varepsilon_{ik}{}^2\rangle_V, \tag{119.5}$$

where $\langle\ldots\rangle_V$ denotes the mean square fluctuation in the volume V. It may be noted that the total extinction coefficient is independent of the polarization of the incident light.

---

† This definition differs by a factor ω'/ω from the definition (in terms of the number of scattered quanta) given in §118. In the present case this factor may be taken as unity, and the two definitions are equivalent.

The formulae derived above enable us to consider scattering macroscopically as occurring at fluctuational inhomogeneities in the medium. In this treatment, the angular distribution and spectral composition of the scattered light are determined by the space–time characteristics of the fluctuations, i.e. by the correlation function†

$$\langle \delta\varepsilon_{il}(t_1, \mathbf{r}_1)\delta\varepsilon_{km}(t_2, \mathbf{r}_2)\rangle$$

between the fluctuations at different points at different instants.

In order to see this, we expand the time-dependent quantities $\delta\varepsilon_{ik}$ as a Fourier integral:

$$\delta\varepsilon_{ik}(t) = \int_{-\infty}^{\infty} \delta\varepsilon_{ik\Omega} e^{-i\Omega t}\frac{d\Omega}{2\pi}, \quad \delta\varepsilon_{ik\Omega} = \int_{-\infty}^{\infty} \delta\varepsilon_{ik}(t)e^{i\Omega t}dt.$$

Each component $\delta\varepsilon_{ik\Omega}e^{-i\Omega t}$ then acts as α_{ik} in the relation (117.2) between the monochromatic components of **E** and **D**′, with $\omega' = \omega + \Omega$. Next, representing the expressions quadratic in the field as double integrals, similarly to (117.9), we easily find the differential extinction coefficient with respect to frequency and direction:

$$dh = \frac{\omega^4}{16\pi^2 c^4}\left\{e'^*_i e'_k e_l e_m^* (\delta\varepsilon_{il}\delta_{km})_{\Omega\mathbf{q}}\right\}\cdot do' \frac{d\Omega}{2\pi}, \qquad (119.6)$$

where we have used, in accordance with the notation of *SP* 2, Chapters VIII and IX, the space–time Fourier expansion component of the correlation function:

$$(\delta\varepsilon_{il}\delta\varepsilon_{km})_{\Omega\mathbf{q}} = \int\int_{-\infty}^{\infty} \langle \delta\varepsilon_{il}(t_1, \mathbf{r}_1)\delta\varepsilon_{km}(t_2, \mathbf{r}_2)\rangle e^{i(\Omega t - \mathbf{q}\cdot\mathbf{r})} dt\, dV \qquad (119.7)$$

($t = t_1 - t_2, \mathbf{r} = \mathbf{r}_1 - \mathbf{r}_2$). Formula (119.6) relates to the scattered light component with polarization **e**′ recorded by an analyser. The term "spectral distribution" refers to the strong dependence on Ω within the scattering line; the slowly varying factor ω'^4 is replaced by ω^4. The factor $e^{-i\mathbf{q}\cdot\mathbf{r}}$ is retained in the integrand (119.6). Its replacement by unity in the expression for the spectral distribution may be inadmissible even if it can be done for the scattering integrated over frequencies (see §120).

Hitherto, the discussion has been presented in terms of classical mechanics. In the quantum description, the coordinates q, and therefore the $\delta\varepsilon_{ik}$, are replaced by the corresponding quantum-mechanical operators in the Heisenberg representation. It can be shown (see the end of this section) that formula (119.6) then remains valid if $(\delta\varepsilon_{il}\delta\varepsilon_{km})_{\Omega\mathbf{q}}$ is taken as

$$(\delta\varepsilon_{il}\delta\varepsilon_{km})_{\Omega\mathbf{q}} = \int\int_{-\infty}^{\infty} \langle \delta\hat{\varepsilon}_{km}(t_2, \mathbf{r}_2)\delta\hat{\varepsilon}_{il}(t_1, \mathbf{r}_1)\rangle e^{i(\Omega t - \mathbf{q}\cdot\mathbf{r})} dt\, dV. \qquad (119.8)$$

The angle brackets now denote complete (both quantum and statistical) averaging over the state of the medium. Since the operators $\delta\hat{\varepsilon}_{ik}$ at different times and positions do not

† This function depends, of course, only on the difference $t = t_1 - t_2$. The second stage of averaging (denoted by the angle brackets), as applied to the product of fluctuations shown, may be regarded as an averaging over the "initial" instant t_2 for a given t.

commute, the order of the operators in (119.8) is significant. This non-commutativity is the reason for the dependence of the scattering intensity on the sign of Ω in (118.4); in the classical limit, there is no such dependence. The quantum formula necessarily satisfies the relation stated.

The extinction coefficient integrated over frequencies is obtained by integration with respect to Ω; in view of the rapid convergence outside the absorption line, the integration may be extended from $-\infty$ to ∞ (Ω is the difference $\omega'-\omega$, and its positive and negative values are therefore physically distinct). The integral

$$\int_{-\infty}^{\infty} e^{i\Omega t}\, d\Omega/2\pi = \delta(t),$$

and the delta function is then eliminated by integrating with respect to t, so that the time-difference correlation function becomes a single-time function. The differential extinction coefficient with respect to directions is

$$dh = \frac{\omega^4}{16\pi^2 c^4}\{e'_i{}^* e'_k e_l e_m{}^* (\delta\varepsilon_{il}\delta\varepsilon_{km})_{\mathbf{q}}\}\, do', \tag{119.9}$$

where

$$(\delta\varepsilon_{il}\delta\varepsilon_{km})_{\mathbf{q}} = \int \langle \delta\varepsilon_{il}{}^{(1)}\delta\varepsilon_{km}{}^{(2)}\rangle e^{-i\mathbf{q}\cdot\mathbf{r}}\, dV \tag{119.10}$$

is the Fourier component of the single-time correlation function. When $\mathbf{q}=0$, the angular distribution depends only on the polarization factors, and we return to the formulae derived previously.

The retention of the factor $e^{-i\mathbf{q}\cdot\mathbf{r}}$ (not equal to unity) in the integral (119.10) considerably complicates the angular distribution and the polarization properties of the scattered light.† In particular, it is not correct to divide the scattering into two parts (scalar and symmetric) given by the first two terms in (117.18). An essential point here is that the tensor $(\delta\varepsilon_{il}\delta\varepsilon_{km})_{\mathbf{q}}$ need not be a true tensor as it must when $\mathbf{q}=0$; it may contain pseudotensor components (see Problem 2). These terms are allowable if the isotropic medium consists of molecules with right–left asymmetry and is therefore not invariant under inversion.

Finally, let us discuss briefly the quantum derivation of (119.6) and (119.8).

The perturbation operator in the Hamiltonian of the system comprising the medium and the field is the integral

$$\hat{V} = -\int \delta\hat{\varepsilon}_{ik}\frac{\hat{E}_i\hat{E}_k}{8\pi}\, dV, \tag{119.11}$$

where $\hat{\mathbf{E}}$ is the quantized electromagnetic field operator. After averaging over the stationary states of the system and statistical averaging over the Gibbs distribution, this operator gives the change $\delta\mathscr{F}$ in the free energy when there is a slow change in the permittivity; see (101.24).

The operator $\hat{\mathbf{E}}$ is expressed in terms of the annihilation and creation operators for photons in the state ω, $\mathbf{k}$, $\mathbf{e}$:

$$\hat{\mathbf{E}} = i\sum_{\mathbf{k},\mathbf{e}}\left(\frac{2\pi\hbar\omega u}{c\sqrt{\varepsilon}}\right)^{1/2}\{\hat{c}_{\mathbf{ke}}\mathbf{e}e^{i(\mathbf{k}\cdot\mathbf{r}-\omega t)} + \hat{c}_{\mathbf{ke}}{}^{+}\mathbf{e}^* e^{-i(\mathbf{k}\cdot\mathbf{r}-\omega t)}\}, \tag{119.12}$$

† In §§121–123 we shall meet with cases where it is not permissible to put $\mathbf{q}=0$ in (119.10) even for the scattering integrated over frequencies.

where $u = d\omega/dk$ (the normalization volume is taken as unity). This expression differs from the one for the field in a vacuum (see *QED*, §2) by the factor $(u/\sqrt{\varepsilon})^{1/2}$ in the normalization coefficients, which originates from the factor $\sqrt{\varepsilon}/u$ in the energy density (83.9) for a plane electromagnetic wave in the medium.†

The probability of a transition in which a photon **k** is absorbed, a photon **k**′ is emitted, and the medium goes from a specified initial state (suffix n) to any final state (f) is (see *QM*, (40.5))

$$dw = \sum_f \left| \frac{1}{\hbar} \int_{-\infty}^{\infty} \langle \mathbf{k}'f | V | \mathbf{k}n \rangle dt \right|^2 \frac{d^3k'}{(2\pi)^3}, \tag{119.13}$$

with the matrix element

$$\langle \mathbf{k}'f | V | \mathbf{k}n \rangle = \frac{2\pi\hbar\omega u}{c\sqrt{\varepsilon}} \int (\delta\varepsilon_{ik})_{fn} \frac{e'^*_i e_k}{4\pi} e^{i(\Omega t - \mathbf{q}\cdot\mathbf{r})} dV,$$

$$\Omega = \omega' - \omega, \quad \mathbf{q} = \mathbf{k}' - \mathbf{k}.$$

The integral in (119.13) is written as a double integral over $dV_1\, dV_2\, dt_1\, dt_2$, and we use the fact that

$$\sum_f (\delta\varepsilon_{il}{}^{(1)})_{fn} (\delta\varepsilon_{km}{}^{(2)})_{fn}{}^* = (\delta\varepsilon_{km}{}^{(2)} \delta\varepsilon_{il}{}^{(1)})_{nn}.$$

The integrand depends only on the differences $\mathbf{r}_1 - \mathbf{r}_2$ and $t_1 - t_2$. The probability dw therefore contains as a factor the total observation time t. The extinction coefficient sought is defined as $dh = dw/tu$. The final statistical averaging over states of the medium gives the required result.

PROBLEMS

PROBLEM 1. Find the general form of the polarization dependence for scattering in an isotropic medium, taking into account the momentum **q** transferred to the medium (B. Ya. Zel'dovich, 1972).

SOLUTION. The problem amounts to that of finding all the independent rank-four tensor combinations having the symmetry of the tensor $(\delta\varepsilon_{il}\delta_{km})_{\Omega\mathbf{q}}$ which can be formed from the unit tensor δ_{ik}, the antisymmetric unit tensor e_{ikl} and the components of the vector $\mathbf{v} = \mathbf{q}/q$; they must be symmetrical in each pair of suffixes il and km, and invariant under interchange of these two pairs (which is equivalent to interchanging the points $\mathbf{r}_1$ and $\mathbf{r}_2$, and therefore to changing the sign of **r**) with a simultaneous change in the sign of $\mathbf{v}$. These conditions are satisfied by the combinations

(1) $\delta_{il}\delta_{km}$,

(2) $\delta_{ik}\delta_{lm} + \delta_{lk}\delta_{im}$,

(3) $\delta_{il}v_k v_m + \delta_{km}v_i v_l$,

(4) $\delta_{ik}v_l v_m + \dots$,

(5) $v_i v_k v_l v_m$,

(6) $v_p e_{pik}\delta_{lm} + \dots$,

(7) $v_p e_{pik} v_l v_m + \dots$;

† The normalization coefficient in (119.12) is obtained from the condition that the eigenvalues of the field energy density operator should be $\Sigma\, (N_{\mathbf{k}e} + \frac{1}{2})\, \hbar\omega$, where $N_{\mathbf{k}e}$ are the photon quantum state occupation numbers. The photon energy in the medium is $h\omega$, and the momentum is $\hbar\mathbf{k}$ ($k = \omega\sqrt{\varepsilon}/c$); these are in the exponents in (119.12). To avoid misunderstanding, it should be emphasized that the momentum $\hbar\mathbf{k}$ includes not only the contribution from the field as such but also the momentum acquired by the medium in the photon emission process.

three terms obtained by symmetrization have been omitted from (4), (6) and (7). These combinations together correspond to an angular distribution

$$f_1|\mathbf{e}\cdot\mathbf{e}'^*|^2+f_2\{1+|\mathbf{e}\cdot\mathbf{e}'|^2\}+f_3\{(\mathbf{e}\cdot\mathbf{e}'^*)(\boldsymbol{\nu}\cdot\mathbf{e}^*)(\boldsymbol{\nu}\cdot\mathbf{e}')+\text{c.c.}\}+f_4\{|\boldsymbol{\nu}\cdot\mathbf{e}|^2+|\boldsymbol{\nu}\cdot\mathbf{e}'|^2+$$

$$[(\mathbf{e}\cdot\mathbf{e}')(\boldsymbol{\nu}\cdot\mathbf{e}^*)(\boldsymbol{\nu}\cdot\mathbf{e}'^*)+\text{c.c.}]\}+f_5|(\boldsymbol{\nu}\cdot\mathbf{e})(\boldsymbol{\nu}\cdot\mathbf{e}')|^2+if_6\boldsymbol{\nu}\cdot\{\mathbf{e}\times\mathbf{e}^*+\mathbf{e}'^*\times\mathbf{e}'+$$

$$[\mathbf{e}\times\mathbf{e}'(\mathbf{e}^*\cdot\mathbf{e}'^*)-\text{c.c.}]\}+if_7\boldsymbol{\nu}\cdot\{\mathbf{e}'^*\times\mathbf{e}'(\boldsymbol{\nu}\cdot\mathbf{e})(\boldsymbol{\nu}\cdot\mathbf{e}^*)+\mathbf{e}\times\mathbf{e}^*(\boldsymbol{\nu}\cdot\mathbf{e}')(\boldsymbol{\nu}\cdot\mathbf{e}'^*)+[\mathbf{e}\times\mathbf{e}'(\boldsymbol{\nu}\cdot\mathbf{e}^*)(\boldsymbol{\nu}\cdot\mathbf{e}'^*)-\text{c.c.}]\}$$

in which $f_1, \ldots, f_7$ are real functions of Ω and q. In a medium which allows inversion, only the first five terms appear, and the first two are equivalent to the first two in (117.18). In a medium having no centre of symmetry, the last two terms also occur, but they are zero if the polarizations $\mathbf{e}$ and $\mathbf{e}'$ are both linear. If the difference between the frequencies ω and ω' is neglected in $\mathbf{q}$, then $\boldsymbol{\nu} = (\mathbf{n}'-\mathbf{n})/|\mathbf{n}'-\mathbf{n}|$, where $\mathbf{n} = \mathbf{k}/k$, $\mathbf{n}' = \mathbf{k}'/k$. In this approximation, the term in f_7 is identically zero, as may be shown by a somewhat lengthy calculation in which each of the vectors $\mathbf{e}$ and $\mathbf{e}'$ is resolved into two components in the scattering plane and perpendicular to it.

PROBLEM 2. Determine the radiation from the motion of a fast particle with a speed below that of light in a light-scattering medium (S. P. Kapitza, 1960).

SOLUTION. The radiation in this case may be regarded as due to the scattering of the particle field by fluctuations in the permittivity of the medium. The energy emitted per unit time from unit volume when the field scattered is monochromatic may be written as $W = h\bar{S} = hc\sqrt{\varepsilon}\cdot|\mathbf{E}|^2/8\pi$, where $\bar{S}$ is given by (83.11) and h is the light extinction coefficient. In this form, the expression is valid for a field $\mathbf{E}$ of any origin.

The field of the moving particle has a continuous spectrum of frequencies. Hence, in order to obtain the radiation in the frequency range $d\omega$ (from unit volume over the whole time of passage), we must replace $|\mathbf{E}|^2$ by $2|\mathbf{E}_\omega(\mathbf{r})|^2\, d\omega/2\pi$ (see *Fields*, §66), where $\mathbf{E}_\omega$ is a time Fourier component of the field. Integration over the volume gives the frequency distribution of the total radiation:

$$dW_\omega = d\omega\frac{hc\sqrt{\varepsilon}}{8\pi^2}\int|\mathbf{E}_\omega(\mathbf{r})|^2\, dV.$$

For this expression to be valid, it is necessary that the field should change only slightly over atomic distances, or more precisely over the correlation distance of the permittivity fluctuations in the medium. Moreover, in order to neglect the frequency shift in scattering, the velocities of the molecules in the medium must be much less than the particle velocity v.

The field of the moving particle is given by (114.7) and (114.8). We have

$$\mathbf{E} = \int \mathbf{E}_\mathbf{k} e^{i\mathbf{k}\cdot\mathbf{r}}\, d^3k/(2\pi)^3.$$

The field frequency in the particle motion is $\omega = vk_x$. Hence $d^3k = dk_x d^2q = v^{-1}\, d\omega\, d^2q$, so that

$$\mathbf{E}_\omega(\mathbf{r})e^{-i\omega t} = \frac{1}{v}\int \mathbf{E}_\mathbf{k} e^{i\mathbf{k}\cdot\mathbf{r}} d^2q/(2\pi)^2,$$

with $\mathbf{k}\cdot\mathbf{r} = \omega x/v + \mathbf{q}\cdot\mathbf{r}$. From this,

$$\int|\mathbf{E}_\omega|^2\, dV = \frac{1}{v^2}\int \mathbf{E}_\mathbf{k}\cdot\mathbf{E}_{\mathbf{k}'}\, e^{i(\mathbf{q}-\mathbf{q}')\cdot\mathbf{r}}\frac{d^2q\, d^2q'}{(2\pi)^4}\, dV.$$

The integral over dx gives just the particle path length l, and that over $dy\,dz$ gives the delta function $(2\pi)^2\delta(\mathbf{q}-\mathbf{q}')$. Thus

$$\int|\mathbf{E}_\omega|^2\, dV = \frac{l}{v^2}\int|\mathbf{E}_\mathbf{k}|^2\frac{d^2q}{(2\pi)^2}.$$

In this integral, the important values of q are fairly large, such that

$$\omega\left(\frac{1}{v^2}-\frac{\varepsilon}{c^2}\right)^{\frac{1}{2}} \ll q \ll 1/a,$$

where a denotes the atomic dimensions: in this range, the expression for $\mathbf{E}_\mathbf{k}$ reduces to

$$\mathbf{E}_\mathbf{k} \cong -i\mathbf{k}\phi_\mathbf{k} \cong -i\mathbf{q}(4\pi e/\varepsilon q^2)e^{-i\omega t},$$

and the integral is logarithmically divergent. With logarithmic accuracy, the integral is to be cut off at limits corresponding to those of the range mentioned. This gives the following final expression for the frequency distribution of the radiation intensity per unit path:

$$dF = dW/l = \frac{he^2c}{\pi v^2 \varepsilon^{3/2}} \log \frac{vc}{a\omega(c^2 - \varepsilon v^2)^{1/2}} \tag{1}$$

with the extinction coefficient given by (119.5).

This radiation is analogous to transition radiation in being independent of the mass of the particle. For comparable velocities (though these are on opposite sides of the limit $c/\sqrt{\varepsilon}$), the intensity is much less than that of the Cherenkov radiation. For example, in gases, with $v \sim c$, a comparison of (1) and (115.3) shows that as a rough estimate

$$dF/dF_{\text{Ch}} \sim \alpha^2/\lambda^3 d^3 \sim a^6/\lambda^3 d^3,$$

where d is the distance between molecules and $\lambda \sim \omega/c$; for h, we have used (120.4) with $n - 1 \sim N\alpha$, and $\alpha \sim a^3$ is the polarizability of the molecule. An estimate for a liquid is obtained by putting $d \sim a$, which gives $dF/dF_{\text{Ch}} \sim (a/\lambda)^3$.

§120. Rayleigh scattering in gases and liquids

Two types of scattering can be distinguished, depending on the change in frequency of the light: (1) *Raman scattering*, which is the *Raman–Landsberg–Mandel'shtam effect* and results in the appearance in the scattered light of lines whose frequency differs from that of the incident light, (2) *Rayleigh scattering*, in which the frequency is essentially unchanged.

Raman scattering in gases results from a change, due to the incident light, in the vibrational, rotational or electronic state of the molecule. Rayleigh scattering, on the other hand, does not involve a change in the internal state of the molecule. In the limiting case of a rarefied gas, when the mean free path l of the molecules is large compared with the wavelength λ of the light, scattering takes place independently at each molecule, and can be discussed microscopically, using quantum mechanics.

Here we shall discuss the opposite limiting case, where $l \ll \lambda$,† and the Rayleigh scattering in gases can be divided into two parts. One part is due to irregularities in the orientation of the molecules (called *fluctuations of anisotropy*). The other part is scattering by fluctuations in the gas density. The orientation of the molecules is entirely changed by a few collisions, i.e. after a time of the order of the mean free time τ. Hence the scattering by fluctuations of anisotropy results in the appearance of a relatively broad line with its peak at $\omega' = \omega$ and width $\sim \hbar/\tau$. The scattering by fluctuations of density gives a much sharper line superposed on the other. As we shall see below, fluctuations of density in volumes $\sim \lambda^3$ are of importance in the scattering of light with wavelength λ. Since these volumes are large, the fluctuations in them occur comparatively slowly, and so the scattered line is narrow. In what follows we shall regard this sharp line as being undisplaced.

The scattering by density fluctuations is scalar scattering; since the density ρ is a scalar, so is the change $\delta\varepsilon$ in the permittivity resulting from a change in ρ. The change in the permittivity in fluctuations of anisotropy, on the other hand, is described by a symmetrical tensor $\delta\varepsilon_{ik}$ with zero trace. The latter property follows from the fact that the effect must vanish on averaging over all directions. Thus the scattering by anisotropy fluctuations is symmetric scattering.

In liquids the situation is less simple. Raman scattering can arise only from a change in the vibrational or electronic state of the molecule; rotational Raman lines do not occur for

† More precisely, the necessary condition is $l \ll \lambda \sin \frac{1}{2}\vartheta$, where ϑ is the scattering angle. This is because the expression (119.7) involves the frequency only in the expression q (120.5).

scattering in liquids. The reason is that, because of the strong interaction between molecules in a liquid, they cannot rotate freely so as to acquire discrete rotational energy levels. The rotation of the molecules, therefore, like any motion in which their relative position changes, contributes in a liquid only to the relatively broad scattering line at $\omega' = \omega$, which in this case may be regarded as entirely the effect of Rayleigh scattering. The relaxation time of such motions depends on the viscosity of the liquid.

The possibility of separating from the total Rayleigh scattering in a liquid a part due to thermodynamic fluctuations (of density or temperature) depends on the magnitudes of the various relaxation times. It is necessary that the relaxation times of all processes of establishment of equilibrium in the liquid should be small in comparison with the times characterizing the fluctuations concerned. In such cases a narrow undisplaced line is observed against a less sharp background called the *wing* of the Rayleigh line. The undisplaced line is due to scalar scattering. The wing, however, does not in general correspond in liquids, as it does in gases, to purely symmetric scattering with no scalar part.

The angular distribution in the undisplaced line is given by the general expressions (117.25) and (117.26) for scalar scattering. It is therefore sufficient to calculate the total extinction coefficient. Substitution of $\delta\varepsilon_{ik} = \delta\varepsilon\delta_{ik}$ in (119.5) gives

$$h = \frac{\omega^4}{6\pi c^4} V \langle \delta\varepsilon^2 \rangle_V. \tag{120.1}$$

If $\delta\rho$ and δT are the changes in density and temperature, then

$$\delta\varepsilon = (\partial\varepsilon/\partial\rho)_T \delta\rho + (\partial\varepsilon/\partial T)_\rho \delta T.$$

According to the known results (see *SP* 1, §112), the fluctuations of density and temperature are statistically independent ($\langle \delta\rho\delta T \rangle = 0$), and their mean squares are

$$\langle (\delta T)^2 \rangle_V = T^2/\rho c_v V, \quad \langle (\delta\rho)^2 \rangle_V = (\rho T/V)(\partial\rho/\partial P)_T,$$

where c_v is the specific heat per unit mass. Thus we have

$$h = \frac{\omega^4}{6\pi c^4}\left[\rho T\left(\frac{\partial\rho}{\partial P}\right)_T\left(\frac{\partial\varepsilon}{\partial\rho}\right)_T^2 + \frac{T^2}{\rho c_v}\left(\frac{\partial\varepsilon}{\partial T}\right)_\rho^2\right], \tag{120.2}$$

a formula first derived by A. Einstein (1910).

This can be expressed in terms of other thermodynamic derivatives. Taking as the independent variables another pair of statistically independent quantities, the pressure P and the entropy s per unit mass, we write

$$\delta\varepsilon = (\partial\varepsilon/\partial P)_s \delta P + (\partial\varepsilon/\partial s)_P \delta s$$

and use the familiar expressions for the fluctuations:

$$\langle (\delta s)^2 \rangle_V = c_p/\rho V, \quad \langle (\delta P)^2 \rangle_V = \rho T u^2/V,$$

where u is the adiabatic velocity of sound in the medium: $u^2 = (\partial P/\partial\rho)_s$. With the further substitutions

$$\left(\frac{\partial\varepsilon}{\partial s}\right)_P = \left(\frac{\partial\varepsilon}{\partial T}\right)_P\left(\frac{\partial T}{\partial s}\right)_P = \frac{T}{c_p}\left(\frac{\partial\varepsilon}{\partial T}\right)_P, \quad \left(\frac{\partial\varepsilon}{\partial P}\right)_s = \left(\frac{\partial\varepsilon}{\partial\rho}\right)_s\frac{1}{u^2},$$

we obtain Einstein's formula in the form

$$h = \frac{\omega^4}{6\pi c^4}\left[\frac{T^2}{\rho c_p}\left(\frac{\partial \varepsilon}{\partial T}\right)_P^2 + \frac{\rho T}{u^2}\left(\frac{\partial \varepsilon}{\partial \rho}\right)_s^2\right]. \tag{120.3}$$

For gases formula (120.3) becomes much simpler. The permittivity of a gas (at optical frequencies) is almost independent of temperature, and hence the first term in the brackets can be neglected. The density dependence is that $\varepsilon - 1$ is proportional to ρ, and hence

$$\rho(\partial \varepsilon/\partial \rho)_T \cong \varepsilon - 1 \cong 2(n-1),$$

where $n = \sqrt{\varepsilon}$ is the refractive index. Since, from the equation of state of a perfect gas, $(1/\rho)(\partial \rho/\partial P)_T = 1/NT$, where N is the number of particles in unit volume, we find that

$$h = 2\omega^4 (n-1)^2/3\pi c^4 N. \tag{120.4}$$

This formula was first derived by Rayleigh (1881).

Let us now examine the fine structure of the undisplaced line. This requires a consideration of the time variation of the fluctuations. In this respect, thermodynamic fluctuations fall into two classes. Adiabatic fluctuations of pressure in a fluid are propagated as undamped waves with the velocity of sound u; we here neglect the absorption of sound, since it causes only a broadening of the line (see below). Fluctuations of entropy at constant pressure, however, are not propagated relative to the fluid, and are damped only gradually as a result of thermal conduction.

Because acoustic perturbations are propagated as waves, the time variation of the pressure fluctuations is correlated even over distances much greater than those between molecules. This fact was not important in calculating the total intensity (integrated over frequencies) of the scattering line, which is determined by the correlation between the fluctuations at different points at the same time, a correlation which extends only over short distances. The frequency distribution of the scattering intensity, however, is determined by the fluctuation correlation function for different times, and the presence of long-range correlation makes it necessary to retain the factor $e^{-i\mathbf{q}\cdot\mathbf{r}}$ in (119.7).

In an undamped sound wave, the frequency Ω and the wave vector $\mathbf{q}$ are related by $\Omega^2 = u^2\mathbf{q}^2$. Accordingly, the frequency resolution of the correlation function for pressure fluctuations (and therefore for the corresponding permittivity fluctuations) consists of two sharp lines at frequencies $\Omega = \pm qu$. The magnitude of the vector $\mathbf{q} = \mathbf{k}' - \mathbf{k}$ corresponding to light scattering is related to the scattering angle ϑ (between $\mathbf{k}$ and $\mathbf{k}'$) by

$$q = |\mathbf{k}' - \mathbf{k}| \cong (2n\omega/c)\sin\tfrac{1}{2}\vartheta; \tag{120.5}$$

since $\Omega = \omega' - \omega$ is small, we have put here $\omega' \cong \omega$. Denoting the corresponding value of Ω by Ω_0, we therefore have

$$\Omega_0 = \pm(2n\omega u/c)\sin\tfrac{1}{2}\vartheta. \tag{120.6}$$

Thus the scattering by pressure fluctuations results in the appearance of a doublet (called the *Mandel'shtam–Brillouin doublet*) in which the distance $2|\Omega_0|$ between the components depends on the angle of scattering (L. I. Mandel'shtam, 1918; L. Brillouin, 1922).†

† As an illustration, for the typical values $u = 1.5 \times 10^5$ cm/sec, $n = 1.5$, scattered light wavelength $\lambda \cong 5 \times 10^{-5}$ cm, scattering angle $\vartheta = 90°$, the width of the doublet is $\Omega_0/2\pi c = 0.05$ cm^{-1}. The Rayleigh line wing width, however, may reach $200 - 150$ cm^{-1}.

The fluctuations of entropy have zero frequency, as stated above, and so scattering by them gives a further central line with $\Omega = 0$ (L. Landau and G. Placzek, 1933).†

Let us determine the intensity distribution of the undisplaced scattering between the doublet and the central line. By the intensity of the doublet we mean the sum of those of its components, i.e. twice that of either one separately.‡ The total extinction coefficient given by (120.2) or (120.3) is $h = h_d + h_{cl}$.

Since the doublet lines are due to scattering by adiabatic pressure fluctuations, their intensity is given by the second term in (120.3), which arises from these fluctuations. The adiabatic derivative $(\partial\varepsilon/\partial\rho)_s$ can be related to the isothermal derivative by changing to the variables ρ and T:

$$\left(\frac{\partial\varepsilon}{\partial\rho}\right)_s = \left(\frac{\partial\varepsilon}{\partial\rho}\right)_T + \frac{T}{c_v\rho^2}\left(\frac{\partial P}{\partial T}\right)_\rho\left(\frac{\partial\varepsilon}{\partial T}\right)_\rho.$$

If the temperature dependence of ε at constant density is neglected, then $(\partial\varepsilon/\partial\rho)_s = (\partial\varepsilon/\partial\rho)_T$. To the same accuracy, in the total derivative written in the form (120.2) we may neglect the second term; see Problem 1 for a calculation in which these approximations are not made. Lastly, using a known thermodynamic formula for the ratio of the adiabatic and isothermal compressibilities (see *SP* 1, (16.14))

$$\left(\frac{\partial\rho}{\partial P}\right)_s = \frac{c_v}{c_p}\left(\frac{\partial\rho}{\partial P}\right)_T, \tag{120.7}$$

we obtain the *Landau–Placzek formula* for the doublet part of the total undisplaced line intensity:

$$h_d/h = c_v/c_p. \tag{120.8}$$

To determine the shape of the lines, it is necessary to consider the different-times correlation function and take account of the dissipative processes which cause the damping of the fluctuations. For the pressure fluctuations, these are viscosity and thermal conduction. The Fourier components of the correlation function for adiabatic pressure fluctuations are

$$(\delta P^2)_{\Omega\mathbf{q}} = \frac{\rho T u^3 \gamma}{(\Omega \mp qu)^2 + u^2\gamma^2}, \tag{120.9}$$

where

$$\gamma = \frac{q^2}{2\rho u}\left[\tfrac{4}{3}\eta + \zeta + \kappa\left(\frac{1}{c_v} - \frac{1}{c_p}\right)\right]; \tag{120.10}$$

see *SP* 2, §89. The quantity γ is the sound absorption coefficient per unit length; η and ζ are the viscosity coefficients and κ the thermal conductivity of the medium (see *FM*, §77). The intensity distribution in the line (in each doublet component), for a given direction of scattering, is proportional to (120.9). Normalization to unity gives

$$dI = \frac{\Gamma}{2\pi\left[(\Omega - \Omega_0)^2 + \frac{1}{4}\Gamma^2\right]}d\Omega, \tag{120.11}$$

† In superfluid liquid helium (the isotope ^{4}He), entropy perturbations are propagated as weakly damped vibrations called second sound waves, whose velocity u_2 is, however, much less than that of ordinary sound. The central scattering line in superfluid helium is therefore split into a narrow doublet, whose width is given by the same formula (120.6) with u_2 in place of u (V. L. Ginzburg, 1943).

‡ The difference between the intensities of the two components is usually, according to (118.4), quite unimportant, since $\hbar\Omega_0 \ll T$.

where $\Gamma = 2u\gamma$. This is called the *dispersion form* of the line, and Γ is the line width. Taking q from (120.5), we find

$$\Gamma = \frac{2\omega^2 n^2}{\rho c^2}(1-\cos\vartheta)\left[\tfrac{4}{3}\eta+\zeta+\kappa\left(\frac{1}{c_v}-\frac{1}{c_p}\right)\right]. \tag{120.12}$$

Isobaric entropy fluctuations decay only by thermal conduction. Their correlation function is

$$(\delta s^2)_{\Omega\mathbf{q}} = \frac{2c_p}{\rho}\frac{\chi q^2}{\Omega^2+\chi^2 q^4}, \tag{120.13}$$

where $\chi = \kappa/\rho c_p$ is the thermometric conductivity. The shape of the central line is given by a similar dispersion formula (120.11) but with $\Omega_0 = 0$, and the line width is

$$\Gamma = 2\chi q^2 = (4\chi n^2\omega^2/c^2)(1-\cos\vartheta). \tag{120.14}$$

As already mentioned at the beginning of this section, the above theory is applicable to scattering in a liquid if all the relaxation times in it are small compared with the times characterizing the fluctuations. It should be borne in mind that, in any liquid, there are relaxation times of various orders of magnitude. The most rapid relaxation process, apparently, is the decay of elastic stresses in the liquid. The corresponding *Maxwellian relaxation time* is $\tau_M \sim \eta/G$, where G is the modulus of rigidity. The reorientation of the molecules, i.e. the decay of the anisotropy fluctuations, takes place less rapidly. The corresponding *Debye relaxation time* is $\tau_D \sim \eta a^3/T$, where a is the dimension of the molecule; the difference between τ_M and τ_D is particularly large in liquids with large molecules. Finally, various other slow relaxation processes leading to the dispersion of sound are also possible (e.g. chemical reactions, slow transfer of energy to vibrational degrees of freedom of the molecule). The important processes as regards scattering are those for which $1/\tau$ is comparable with the frequency of the sound disturbances which cause the scattering. We shall merely mention that, when the viscosity of the liquid is sufficiently high, and so $\tau_M \gg 1/qu$, the liquid behaves as an amorphous solid with respect to the scattering of light.

PROBLEMS

PROBLEM 1. Find an exact formula for the ratio of intensities of the central line and the doublet in the undisplaced scattering line (I. L. Fabelinskiĭ, 1956).

SOLUTION. As noted in the text, the second term in (120.3) gives the doublet intensity. The first term arises from isobaric entropy fluctuations, and therefore gives the intensity of the central line. Thus

$$\frac{h_{cl}}{h_d} = \frac{T}{c_p\rho^2}\left(\frac{\partial\varepsilon}{\partial T}\right)_P^2 \Big/ \left(\frac{\partial\varepsilon}{\partial\rho}\right)_s^2\left(\frac{\partial\rho}{\partial P}\right)_s.$$

From the thermodynamic relations (120.7) and

$$\left(\frac{\partial\rho}{\partial P}\right)_s = \left(\frac{\partial\rho}{\partial T}\right)_T - \frac{T}{c_p\rho^2}\left(\frac{\partial\rho}{\partial T}\right)_P^2$$

(see *SP* 1, (16.15)), we find

$$\left(\frac{\partial\rho}{\partial P}\right)_s = \frac{c_v}{c_p - c_v}\frac{T}{c_p\rho^2}\left(\frac{\partial\rho}{\partial T}\right)_P^2.$$

The final result is

$$\frac{h_{cl}}{h_d} = \left(\frac{c_p}{c_v} - 1\right)\left[\left(\frac{\partial \varepsilon}{\partial T}\right)_P \Big/ \left(\frac{\partial \varepsilon}{\partial \rho}\right)_s \left(\frac{\partial \rho}{\partial T}\right)_P\right]^2.$$

In the Landau–Placzek approximation, the expression in the square brackets is unity.

PROBLEM 2. Light is scattered in a gas whose molecules are linear, with polarizabilities $\alpha_\parallel$ and $\alpha_\perp$ along and across the axis respectively. Determine the intensity resulting from the various types of scattering.

SOLUTION. The total intensity of scattered light (for given vibrational and electronic states of the molecules) includes the Rayleigh scattering and the rotational part of the Raman scattering. Since the scattering takes place independently at the individual molecules of the gas, the total extinction coefficient is most simply obtained from formula (92.4), by multiplying by the number of particles per unit volume N and replacing $|\alpha V|^2$ by $\frac{1}{3}\alpha_{ik}{}^2 = \frac{1}{3}(\alpha_\parallel{}^2 + 2\alpha_\perp{}^2)$:

$$h = \frac{8\pi\omega^4 N}{9c^4}(\alpha_\parallel{}^2 + 2\alpha_\perp{}^2); \tag{1}$$

the polarizabilities as defined here and in §92 differ by a factor V.

The undisplaced Rayleigh line is due to the scalar part of the polarizability, i.e. it is the same as if the polarizability tensor of the molecule were $\frac{1}{3}\alpha_{ll}\delta_{ik}$. The same formula, (92.4), therefore gives

$$h_{\text{undisp}} = \frac{8\pi\omega^4 N}{27c^4}(\alpha_\parallel + 2\alpha_\perp)^2. \tag{2}$$

The difference $h_{\text{total}} - h_{\text{undisp}}$ includes the background (scattering by anisotropy fluctuations) and the rotational Raman scattering. In order to separate the former, we must first average the polarizability tensor of the molecule with respect to rotation about some particular axis (perpendicular to the axis of the molecule). The polarizability along the axis of rotation averaged in this way is evidently $\alpha_\perp$, and that along any direction in a plane perpendicular to the axis of rotation is $\frac{1}{2}(\alpha_\perp + \alpha_\parallel)$. In other words, a molecule rotating about a given axis is to be regarded as a particle for which the principal values of the polarizability tensor are $\alpha_\perp, \frac{1}{2}(\alpha_\perp + \alpha_\parallel), \frac{1}{2}(\alpha_\perp + \alpha_\parallel)$. Using these, we calculate the symmetrical tensor $\alpha_{ik} - \frac{1}{3}\alpha_{ll}\delta_{ik}$, whose trace is zero, and then a procedure similar to the derivation of formulae (1) and (2) gives

$$h_{\text{backg}} = \frac{8\pi\omega^4 N}{9c^4}\frac{(\alpha_\perp - \alpha_\parallel)^2}{6}. \tag{3}$$

Finally, the intensity of the rotational Raman scattering is obtained by subtracting (2) and (3) from (1):

$$h_R = \frac{8\pi\omega^4 N}{9c^4}\frac{(\alpha_\perp - \alpha_\parallel)^2}{2}.$$

§121. Critical opalescence

The isothermal compressibility $(\partial\rho/\partial P)_T$ increases without limit as the critical point is approached. The expression (120.2) for the total intensity due to Rayleigh scattering therefore increases also. This indicates a marked increase in scattering near the critical point, called *critical opalescence*.† The formula (120.2) itself is, however, inapplicable, because near the critical point the single-time correlation between the density fluctuations (and therefore the permittivity fluctuations) at different points in space extends to a distance of the order of the correlation radius r_c, which increases without limit as the critical point is approached (see *SP* 1, §§152, 153). Here, therefore, we cannot in general replace the factor $e^{-i\mathbf{q}\cdot\mathbf{r}}$ in (119.9) by unity, even when calculating the total scattered intensity (and not just when calculating its spectral fine structure).

† The idea that this phenomenon is due to an increase in the density fluctuations was put forward by M. Smoluchowski (1908). In relation to the van der Waals theory of the critical point (see *SP* 1, §152) it was discussed by L. S. Ornstein and F. Zernike (1914).

This replacement is valid for all scattering angles only if

$$kr_c \ll 1, \tag{121.1}$$

where $k = n\omega/c$ is the wave number of the scattered light. In that case, we can again use (120.2) for the total extinction coefficient, and it is sufficient to retain just the increasing first term (only the density fluctuations increase, not the temperature fluctuations). As the critical point is approached from any direction in the ρT-plane except along the critical isotherm $T = T_c$, the compressibility increases according to†

$$(\partial\rho/\partial P)_T \propto |T - T_c|^{-\gamma}, \quad \gamma \cong 1.26.$$

Since there is no reason for $(\partial\varepsilon/\partial\rho)_T$ to become zero or infinite, the scattering intensity behaves similarly:

$$h \propto |T - T_c|^{-\gamma}. \tag{121.2}$$

This increase involves an increased intensity of only the central component of the Rayleigh line, not of all the fine-structure components. From (120.8), $h_d \equiv hc_v/c_p$. The factor c_p in the denominator compensates the factor $(\partial\rho/\partial P)_T$ in h, since they increase in the same manner. The doublet intensity therefore increases only as c_v, i.e. much more slowly‡:

$$h_d \propto |T - T_c|^{-\alpha}, \quad \alpha \cong 0.1. \tag{121.3}$$

Sufficiently close to the critical point, the inequality (121.1) for a given k is certainly violated, and it is no longer permissible to replace the factor $e^{-i\mathbf{q}\cdot\mathbf{r}}$ by unity; this occurs later for small scattering angles.§ Let dh be the differential extinction coefficient for scattering into a solid-angle element do', i.e. for a given value of $\mathbf{q} = \mathbf{k}' - \mathbf{k}$. Considering, for definiteness, the scattering of unpolarized light, and using the result that the angular dependence related to the polarization state of the light in the case of scattering by scalar fluctuations is given by the factor I (117.26), we have

$$dh = \frac{\omega^4}{6\pi c^4}(\delta\varepsilon^2)_{\mathbf{q}} \cdot \tfrac{3}{4}(1 + \cos^2\vartheta)\frac{do'}{4\pi}, \tag{121.4}$$

with

$$(\partial\varepsilon^2)_{\mathbf{q}} = (\partial\varepsilon/\partial\rho)_T{}^2(\delta\rho^2)_{\mathbf{q}}.$$

In the immediate neighbourhood of the critical point, where

$$kr_c \gg 1, \tag{121.5}$$

for not too small scattering angles $qr_c \gg 1$ will also hold. In this angle range, the important distances in the integral $(\delta\rho^2)_{\mathbf{q}}$ are $r \sim 1/q \gg r_c$, where the density fluctuation correlation function follows a power law‖:

$$\langle \delta\rho^{(1)}\delta\rho^{(2)} \rangle \propto r^{-(1+\zeta)}, \quad \zeta \cong 0.04.$$

---

† All the results used in this section concerning the variation of thermodynamic quantities near the critical point are given in *SP* 1, §153. The notation for the critical indices is the same as there.

‡ All these arguments presuppose, of course, that the inequality (121.1) is compatible with the assumption that the substance is already in the "fluctuation region" of the neighbourhood of the critical point.

§ It is, however, meaningful to consider scattering itself only down to angles of the order of the diffraction angle $\sim \lambda/L$, where L is the linear size of the body.

‖ See *SP* 1, (148.7). Because of the formal equivalence between critical-point problems and problems of second-order phase transitions (with a one-dimensional order parameter), this formula applies near a critical point also.

The Fourier component of the correlation function is then

$$(\delta\rho^2)_{\mathbf{q}} \propto q^{-2+\zeta} \tag{121.6}$$

with a coefficient that is independent of the temperature. Thus we arrive at the following angle and frequency dependence of the extinction coefficient in the range under consideration:

$$dh \propto \omega^{2-\zeta} \frac{1+\cos^2\vartheta}{(1-\cos\vartheta)^{1-\zeta/2}} do'. \tag{121.7}$$

For a fixed scattering angle in the range $qr_c \gg 1$, the intensity ceases to increase as the critical point is approached. At the critical point itself, (121.7) is valid for all angles.

§122. Scattering in liquid crystals

In liquid crystals there is strong scattering of light, similar in several respects to critical opalescence. Here we shall discuss the phenomenon only in nematic-type liquid crystals (P. G. de Gennes, 1968).

Such crystals have already been mentioned at the end of §17, where an expression was given for their permittivity tensor:

$$\varepsilon_{ik} = \varepsilon_0(\omega)\delta_{ik} + \varepsilon_a(\omega)d_i d_k; \tag{122.1}$$

ε_0 and ε_a are functions of the frequency.

The permittivity fluctuations are due firstly to fluctuations in the direction of the director **d**, and are therefore anisotropy fluctuations. A simultaneous equal rotation of this direction at all points leaves the energy of the body unchanged; the long-wavelength fluctuations thus involve only small energy losses, and are therefore large. Large fluctuations of the permittivity lead in turn to strong scattering of light.

The fluctuating quantity **d** may be expressed as

$$\mathbf{d} = \mathbf{d}_0 + \boldsymbol{\nu}, \tag{122.2}$$

where $\mathbf{d}_0$ is its constant mean value and $\boldsymbol{\nu}$ a small change in the fluctuation; since $\mathbf{d}^2 = \mathbf{d}_0{}^2 = 1$, and $\boldsymbol{\nu}$ is small, we have

$$\boldsymbol{\nu}\cdot\mathbf{d}_0 = 0. \tag{122.3}$$

The change in the permittivity in the fluctuation is given in terms of $\boldsymbol{\nu}$ by

$$\delta\varepsilon_{ik} = \varepsilon_a(d_{0i}\nu_k + d_{0k}\nu_i). \tag{122.4}$$

The influence of the density and temperature fluctuations on $\delta\varepsilon_{ik}$ is negligible.

Using (122.4), we express the single-time correlation function of the permittivity fluctuations in terms of that of the director fluctuations:

$$(\delta\varepsilon_{il}\delta\varepsilon_{km})_{\mathbf{q}} = \varepsilon_a{}^2\{d_{0i}d_{0k}(\nu_l\nu_m)_{\mathbf{q}} + d_{0i}d_{0m}(\nu_k\nu_l)_{\mathbf{q}} + d_{0k}d_{0l}(\nu_i\nu_m)_{\mathbf{q}} + d_{0l}d_{0m}(\nu_i\nu_k)_{\mathbf{q}}\}. \tag{122.5}$$

The function $(\nu_i\nu_k)_{\mathbf{q}}$ has been derived in *SP* 1, §141. From (122.3), this tensor has non-zero components only in the plane perpendicular to $\mathbf{d}_0$. Denoting the tensor suffixes in this plane by α and β, we have

$$\left.\begin{aligned} (\nu_\alpha \nu_\beta)_{\mathbf{q}} &= a(\delta_{\alpha\beta} - q_\alpha q_\beta/q_\perp^2) + b q_\alpha q_\beta/q_\perp^2, \\ a = T/(a_2 q_\perp^2 + a_3 q_\parallel^2), &\quad b = T/(a_1 q_\perp^2 + a_3 q_\parallel^2), \end{aligned}\right\} \tag{122.6}$$

where a_1, a_2, a_3 are positive moduli giving the dependence of the free energy of a nematic crystal on the derivatives of the director, $q_\parallel$ and $q_\perp$ the components of $\mathbf{q}$ along $\mathbf{d}_0$ and in the plane perpendicular to it.

If we assume that the anisotropy of the permittivity is relatively small, i.e. $|\varepsilon_a| \ll \varepsilon_0$, we can use for the differential extinction coefficient (with respect to directions) formula (119.9), derived for isotropic media (but without putting $\mathbf{q} = 0$, of course). We shall not give here the resulting lengthy expression for the angular distribution and the polarization dependences.

A characteristic feature of this scattering is that its intensity increases at low scattering angles according to

$$dh \propto do'/q^2. \tag{122.7}$$

This large increase, like that given by (121.7) at the critical point, is due to the slow (power-law) decrease of the fluctuation correlation function with increasing distance. The angle integral diverges logarithmically at the lower limit. It has to be cut off at values of $q \sim 1/L$, corresponding to diffraction by the body as a whole (cf. the fourth footnote to §121), and the total scattering intensity thus varies logarithmically with the size L of the body.

It may be noted, finally, that an external magnetic field restricts the increase of the correlation function $(\nu_i \nu_k)_{\mathbf{q}}$ for small $\mathbf{q}$ (see *SP* 1, §141), and thus suppresses scattering, making the liquid crystal transparent.

§123. Scattering in amorphous solids

Rayleigh scattering in amorphous solids† differs considerably from that in fluids. In an isotropic solid there are two velocities of propagation of sound, u_l (longitudinal) and u_t (transverse). The fine structure of the Rayleigh line therefore includes not one but two Mandel'shtam–Brillouin doublets. They are due to scattering by transverse and longitudinal "sound waves", and their distances from the centre of the line are respectively $\Omega_l = \pm u_l q$ and $\Omega_t = \pm u_t q$. Since $u_l > u_t$, it follows that $|\Omega_l| > |\Omega_t|$. The central component of the line is again due to scattering by fluctuations which are not propagated relative to the medium. In this case the main fluctuations of the latter type are those of structure. In an amorphous body, where the atoms are not arranged in an ordered manner, these fluctuations are comparatively large and vary only slowly with time (on account of the extreme slowness of the diffusion processes in a solid). Scattering by these fluctuations leads to a strong line whose width is almost zero. As regards polarization and angular distribution, this line results from a superposition of scalar and symmetric scattering.

Next, let us consider the doublet components of the Rayleigh line in amorphous solids. In a solid, the influence of any deformation (in this case, fluctuations) extends to considerable distances. Hence even the single-time fluctuations at different points in the body are correlated over distances large compared with $1/q$. We thus have again a situation

† The considerably more complex theory of scattering in solid crystals will not be discussed in this book.

where, even in calculating the total intensity (and the polarization) of the scattered light, we cannot put $q = 0$ in the fluctuation correlation function.

The field in the scattered wave is

$$\mathbf{E}' = -\frac{\omega^2 \exp(ikR_0)}{4\pi R_0 c^2} E_0 \mathbf{n}' \times (\mathbf{n}' \times \mathbf{G}), \tag{123.1}$$

where

$$G_i = \int \delta\varepsilon_{ik} \exp(-i\mathbf{q}\cdot\mathbf{r}) dV . e_k, \tag{123.2}$$

and $\mathbf{n}'$ is a unit vector in the direction of scattering. The change in the permittivity resulting from the deformation of an isotropic body is

$$\delta\varepsilon_{ik} = a_1 u_{ik} + a_2 u_{ll} \delta_{ik}, \tag{123.3}$$

where u_{ik} is the strain tensor (see (102.1)). Since the integral (123.2) isolates from $\delta\varepsilon_{ik}$ the Fourier space component with wave vector $\mathbf{q}$, u_{ik} in (123.3) must be taken as the deformation in a sound wave with this wave vector. We therefore write the displacement vector as

$$\mathbf{u} = \mathrm{re}\{\mathbf{u}_0 \exp(i\mathbf{q}\cdot\mathbf{r})\} = \tfrac{1}{2}[\mathbf{u}_0 \exp(i\mathbf{q}\cdot\mathbf{r}) + \mathbf{u}_0{}^* \exp(-i\mathbf{q}\cdot\mathbf{r})], \tag{123.4}$$

whence the strain tensor is

$$u_{ik} = \frac{1}{2}\left(\frac{\partial u_i}{\partial x_k} + \frac{\partial u_k}{\partial x_i}\right)$$

$$= \mathrm{re}\{\tfrac{1}{2} i (u_{0i} q_k + u_{0k} q_i) \exp(i\mathbf{q}\cdot\mathbf{r})\},$$

and the volume integral is

$$\int u_{ik} \exp(-i\mathbf{q}\cdot\mathbf{r}) dV = \tfrac{1}{4} i V (u_{0i} q_k + u_{0k} q_i). \tag{123.5}$$

Let us first consider scattering by transverse sound waves. Since in a transverse wave $\mathbf{u}$ is perpendicular to $\mathbf{q}$, and $u_{ll} = 0$, $\delta\varepsilon_{ik} = a_1 u_{ik}$. Using (123.5), we therefore have

$$\mathbf{G} = \tfrac{1}{4} i V a_1 \{\mathbf{u}_0(\mathbf{q}\cdot\mathbf{e}) + \mathbf{q}(\mathbf{u}_0\cdot\mathbf{e})\}. \tag{123.6}$$

A transverse sound wave can have two independent directions of polarization; the vector $\mathbf{u}$ may be in the plane of $\mathbf{k}$ and $\mathbf{k}'$, or perpendicular to that plane. Since $\mathbf{E}$ is perpendicular to $\mathbf{k}$, it is easy to see that in the first case the component of $\mathbf{G}$ in the plane perpendicular to $\mathbf{k}'$ is zero. Thus transverse sound waves polarized in the plane of $\mathbf{k}$ and $\mathbf{k}'$ do not scatter light.

If the vector $\mathbf{u}$ is perpendicular to the plane of $\mathbf{k}$ and $\mathbf{k}'$, a simple calculation, using (123.1) and (123.6), gives for the field in the scattered wave

$$\left.\begin{aligned} E'_\parallel &= \frac{\omega^2 E_0 \exp(ikR_0)}{4\pi R_0 c^2} . \tfrac{1}{4} a_1 i V q u_0 \cos\tfrac{1}{2}\vartheta . e_\perp, \\ E'_\perp &= \frac{\omega^2 E_0 \exp(ikR_0)}{4\pi R_0 c^2} . \tfrac{1}{4} a_1 i V q u_0 \cos\tfrac{1}{2}\vartheta . e_\parallel. \end{aligned}\right\} \tag{123.7}$$

Here ϑ is, as usual, the angle between $\mathbf{k}$ and $\mathbf{k}'$, and the suffixes $\parallel$ and $\perp$ denote components in the plane of scattering and perpendicular to that plane. The coefficient of proportionality in these two formulae involves the same fluctuation u_0. This means that no depolarization occurs on scattering: linearly polarized light remains so (though it is polarized in a different plane).

Since the coefficients in formulae (123.7) are exactly the same, the extinction coefficient dh does not depend on the state of polarization of the incident light, and is

$$dh = \left(\frac{q\omega^2 a_1}{16\pi c^2}\right)^2 V|u_0|^2 \cos^2\tfrac{1}{2}\vartheta\, do. \tag{123.8}$$

It remains to determine the mean square amplitude of the fluctuation u_0.

From the point of view of the general theory of thermodynamic fluctuations, the sound wave (123.4) may be regarded as a combination of two classical oscillators (waves propagated to the right and to the left), each having a mean kinetic energy $\frac{1}{2}T$. Since the frequency of the oscillations is here $\Omega = qu_t$, the mean kinetic energy is $\frac{1}{2}V\langle\rho\dot{\mathbf{u}}^2\rangle = \frac{1}{4}V\rho(u_t q)^2\langle|u_0|^2\rangle$. Equating this to $2.\frac{1}{2}T$, we have

$$\langle|u_0|^2\rangle = 4T/V\rho u_t^2 q^2. \tag{123.9}$$

Finally, substituting (123.9) in (123.8), we obtain

$$dh = \frac{a_1^2\omega^4 T}{64\pi^2 c^4 u_t^2\rho}\cos^2\tfrac{1}{2}\vartheta\, do. \tag{123.10}$$

The angular dependence of the scattering is totally different from that which occurs in fluids.

Let us now consider scattering by longitudinal sound waves. In these waves $\mathbf{u}$ is parallel to $\mathbf{q}$, and from (123.3) and (123.4) we find

$$\mathbf{G} = \tfrac{1}{2}iVu_0 q\left\{a_1\frac{\mathbf{q}(\mathbf{q}\cdot\mathbf{e})}{q^2} + a_2\mathbf{e}\right\}E_0.$$

A simple calculation gives for the field in the scattered wave

$$\left.\begin{aligned} E_\perp &= \frac{\omega^2\exp(ikR_0)}{4\pi R_0 c^2}\cdot\tfrac{1}{2}iVu_0 q a_2 E_0 e_\perp, \\ E_\parallel &= \frac{\omega^2\exp(ikR_0)}{4\pi R_0 c^2}\cdot\tfrac{1}{2}iVu_0 q[\tfrac{1}{2}a_1 + (\tfrac{1}{2}a_1 + a_2)\cos\vartheta]E_0 e_\parallel. \end{aligned}\right\} \tag{123.11}$$

In this case also there is no depolarization on scattering. The angular distribution and the extinction coefficient, however, depend on the state and direction of the polarization of the incident light. We shall not pause to write out the relevant formulae, which are somewhat cumbersome. The calculations are wholly similar to those given above, and the expression for $\langle|u_0|^2\rangle$ differs only in that u_t is replaced by u_l in (123.9).

CHAPTER XVI

DIFFRACTION OF X-RAYS IN CRYSTALS

§124. The general theory of X-ray diffraction

THE phenomenon of X-ray diffraction in crystals occupies a special place in the electrodynamics of matter, since the wavelengths concerned are comparable with the interatomic distances. For this reason the usual macroscopic approach to matter as a continuous medium is entirely invalid, and we must begin by considering scattering by individual charged particles, and essentially by electrons; the scattering by nuclei is unimportant, because of their greater mass.

The frequencies of the motion of electrons in the atom are of order $\omega_0 \sim v/a$, where v is their velocity and a the dimension of the atom. If $\lambda \sim a$, then, since $v \ll c$, these frequencies are small compared with the X-ray frequency $\omega \sim c/\lambda$. This makes it possible to write the equation of motion of an electron in the field of the electromagnetic wave as

$$m\dot{\mathbf{v}}' = e\mathbf{E}, \tag{124.1}$$

i.e. the electrons may be regarded as free (see §78).

From (124.1) we find the additional velocity acquired by the electron under the action of the wave field: $\mathbf{v}' = ie\mathbf{E}/m\omega$.

Let $n(\mathbf{r})$ be the number density of electrons in a crystal, averaged over the quantum-mechanical states of the electrons and over the statistical distribution of the thermal motion of the nuclei in the lattice. It should be emphasized that the usual macroscopic averaging over physically infinitesimal volume elements is *not* included, i.e. $n(\mathbf{r})$ is the actual quantum density of the electrons in the crystal lattice. The corresponding current density due to the wave field is:

$$\mathbf{j}' = en\mathbf{v}' = ie^2n\mathbf{E}/m\omega. \tag{124.2}$$

We substitute this current in the microscopic Maxwell's equations:

$$\mathbf{curl}\,\mathbf{E} = i\omega\mathbf{H}/c, \tag{124.3}$$

$$\mathbf{curl}\,\mathbf{H} = -i\omega\mathbf{E}/c + 4\pi\mathbf{j}'/c$$

$$= -\frac{i\omega}{c}\left(1 - \frac{4\pi e^2 n}{m\omega^2}\right)\mathbf{E}. \tag{124.4}$$

We thereby take account of its reciprocal effect on the field, i.e. scattering. It is, of course, assumed that this effect is small, i.e. that the inequality

$$4\pi e^2 n/m\omega^2 \ll 1 \tag{124.5}$$

holds. Putting $\mathbf{D} = \varepsilon\mathbf{E}$, where

$$\varepsilon = 1 - \frac{4\pi e^2 n}{m\omega^2}, \tag{124.6}$$

in accordance with the usual definition of the induction, we reduce equation (124.4) to the usual form $\mathbf{curl\,H} = -i\omega\mathbf{D}/c$. Thus, in this sense, the expression (124.6) for the permittivity (cf. (78.1)) can be used even for wavelengths $\lambda \sim a$, though it must of course be remembered that the symbols **E** and **D** no longer retain their previous meanings: they now pertain to the field which has not been averaged over physically infinitesimal volumes, and ε is accordingly a function of the coordinates.

In the scattering of X-rays by heavy atoms it may happen that the condition $\omega \gg \omega_0$ is fulfilled for the outer electron shells but not for the inner ones, where $\omega \lesssim \omega_0$ and so the inequality $\lambda \gg a$ holds. In this case the permittivity can still be regarded as the coefficient of proportionality between **D** and **E**, but the formula corresponding to (124.6) gives only the contribution of the outer electrons. That of the inner electrons must in principle be calculated by averaging over the volume of their shells. Thus, if we put $\mathbf{D} = \varepsilon\mathbf{E}$ with ε a function of the coordinates, all possible cases are allowed for. In what follows we shall, for definiteness, use the expression (124.6).

In effecting the averaging of the electron density in (124.2) to obtain $n(\mathbf{r})$ independent of time, we exclude a possible change of frequency on scattering. That is, we consider only strictly coherent scattering, with no change in frequency.

Eliminating **H** from the two equations (124.3) and (124.4), we obtain $\mathbf{curl\,curl\,E} = \omega^2\mathbf{D}/c^2$. Here we substitute $\mathbf{E} = \mathbf{D} + 4\pi e^2 n\mathbf{E}/m\omega^2$ and expand the expression **curl curl E**, using the fact that $\operatorname{div}\mathbf{D} = 0$, as follows from (124.4). Then

$$\triangle\mathbf{D} + \omega^2\mathbf{D}/c^2 = \mathbf{curl\,curl}\,(4\pi e^2 n\mathbf{E}/m\omega^2). \qquad (124.7)$$

On the right-hand side of this equation, which already contains the small quantity $4\pi e^2 n/m\omega^2$, **E** must be taken as the given field of the incident wave. Let us find the solution of equation (124.7) in the region outside the scattering crystal and at large distances from it. Since this equation is of the same form as equation (117.3), the required solution is obtained immediately by analogy with (117.4):†

$$\mathbf{E} = \frac{e^2}{m\omega^2}\frac{\exp(ikR_0)}{R_0}\mathbf{k}'\times(\mathbf{k}'\times\mathbf{E}_0)\int n\exp(-i\mathbf{q}\cdot\mathbf{r})\,dV. \qquad (124.8)$$

Here R_0 is the distance from the origin, which is within the crystal, to the point considered; $\mathbf{q} = \mathbf{k}' - \mathbf{k}$; $k = k' = \omega/c$; $\mathbf{E}_0$ is the amplitude of the incident wave. We put **E** instead of **D** on the left-hand side because the two are equal in the vacuum outside the crystal.

To characterize the intensity of X-ray diffraction we use an effective cross-section σ, defined as the ratio of the intensity diffracted into a solid angle do' to the energy flux density in the incident wave. By (124.8) we have

$$d\sigma = \left(\frac{e^2}{mc^2}\right)^2 \sin^2\theta \left|\int n\exp(-i\mathbf{q}\cdot\mathbf{r})\,dV\right|^2 do', \qquad (124.9)$$

where θ is the angle between $\mathbf{E}_0$ and $\mathbf{k}'$. If the incident radiation is "natural" (not polarized), the factor $\sin^2\theta$ in this formula becomes $\frac{1}{2}(1+\cos^2\vartheta)$, where ϑ is the angle between **k** and

† In solving equation (117.3) it was not possible to consider the field outside the body, since the boundary conditions on the surface would have had to be taken into account (the quantity ε' on the left-hand side being different inside and outside the body). The left-hand side of equation (124.7), however, is the same in all space.

$\mathbf{k}'$ (see the last footnote to §117):

$$d\sigma = \frac{1}{2}\left(\frac{e^2}{mc^2}\right)^2 (1+\cos^2\vartheta)\,|\int n \exp(-i\mathbf{q}\cdot\mathbf{r})dV|^2 do'. \qquad (124.10)$$

In what follows we shall, for definiteness, consider this particular case.

We see that the intensity of radiation diffracted in a given direction is essentially proportional to the squared modulus of the integral

$$\int n \exp(-i\mathbf{q}\cdot\mathbf{r})dV, \qquad (124.11)$$

i.e. the Fourier space component of the electron density. As $\mathbf{q} \to 0$ this integral becomes simply the electron density $\bar{n}$ averaged over a lattice cell. If n is replaced by $\bar{n}$ in equations (124.3) and (124.4), we obtain the usual macroscopic Maxwell's equations, with permittivity $\varepsilon(\omega) = 1 - 4\pi e^2\bar{n}/m\omega^2$. According to these equations, when X-rays pass through a crystal they are refracted according to the ordinary laws of refraction, with refractive index $\sqrt{\varepsilon}$. Thus diffraction through small angles amounts to ordinary refraction, which is of no interest here. In what follows we shall always assume that $\mathbf{q}$ is appreciably different from zero.

The electron density, like any function of position in a crystal lattice, can be expanded as a Fourier series:

$$n = \sum_{\mathbf{b}} n_{\mathbf{b}} \exp(i\mathbf{b}\cdot\mathbf{r}), \qquad (124.12)$$

where the summation is taken over all periods $\mathbf{b}$ of the reciprocal lattice (see *SP* 1, §133). When (124.12) is substituted in (124.11) and the result is integrated over the volume of the crystal, we obtain practically zero except for values of $\mathbf{q}$ close to some $\mathbf{b}$. Between these values the intensity is negligible. We can therefore consider each diffraction maximum separately, putting

$$n = n_{\mathbf{b}} \exp(i\mathbf{b}\cdot\mathbf{r})$$

with the appropriate value of $\mathbf{b}$. Substitution in (124.10) gives

$$d\sigma = \frac{1}{2}\left(\frac{e^2}{mc^2}\right)^2 (1+\cos^2\vartheta)|n_{\mathbf{b}}|^2\,|\int \exp[-i(\mathbf{k}'-\mathbf{k}-\mathbf{b})\cdot\mathbf{r}]dV|^2 do'. \qquad (124.13)$$

The strongest maxima occur in directions for which the equation

$$\mathbf{k}' - \mathbf{k} = \mathbf{b} \qquad (124.14)$$

(*Laue's equation*) is exactly satisfied, and are called *principal maxima*. For given $\mathbf{b}$, however, a principal maximum does not occur for an arbitrary direction and frequency of the incident radiation. If the equation (124.14) is written as $\mathbf{k}' = \mathbf{k} + \mathbf{b}$ and squared, and we use the fact that $k^2 = k'^2$, we have

$$2\mathbf{b}\cdot\mathbf{k} = -b^2. \qquad (124.15)$$

This equation determines the values of the wave vector $\mathbf{k}$ for which principal maxima occur with the given value of $\mathbf{b}$. Geometrically, equation (124.15) represents a plane in $\mathbf{k}$-space perpendicular to the vector $\mathbf{b}$ at a distance $\frac{1}{2}b$ from the origin. In particular, we see that $k \geqslant \frac{1}{2}b$.

Since $|\mathbf{k}' - \mathbf{k}| = 2k \sin\frac{1}{2}\vartheta$, it follows from (124.14) that

$$2k \sin\tfrac{1}{2}\vartheta = b \qquad (124.16)$$

(*Bragg and Vul'f's equation*), which determines the angle of diffraction at the principal maximum.

Any vector **b** of the reciprocal lattice determines a family of crystal planes represented by the equations

$$\mathbf{r}\cdot\mathbf{b} = 2\pi \times \text{an integer}.$$

These planes are perpendicular to **b**, and the vectors **k** and **k**′ corresponding to the condition (124.14) make equal angles of incidence and reflection with the planes (Fig. 64). For this reason, diffraction at a principal maximum is sometimes spoken of as reflection from the corresponding crystal planes.

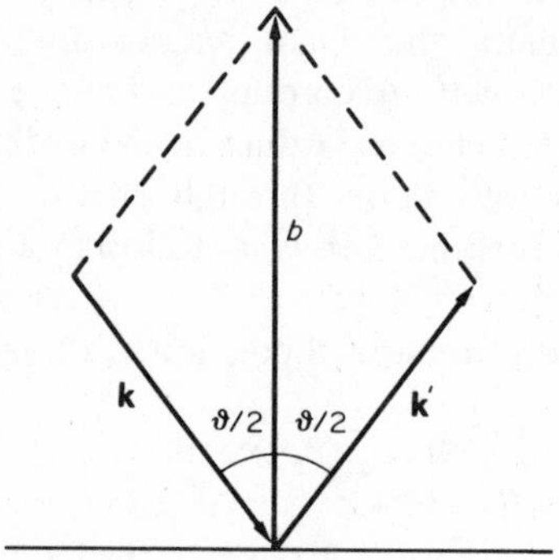

FIG. 64

The total intensity of the diffraction spot near a maximum is obtained by integrating (124.13) over a solid angle about the direction of **k**′. Let us determine the total intensity near a principal maximum. We denote by $\mathbf{k}'_0$ the value of **k**′ corresponding to Laue's equation for a given **k**: $\mathbf{k}'_0 = \mathbf{k}+\mathbf{b}$, and put also $\boldsymbol{\varkappa} = \mathbf{k}' - \mathbf{k}'_0$. Near the maximum, $\boldsymbol{\varkappa}$ is small; since **k**′ and $\mathbf{k}'_0$ differ only in direction, $\boldsymbol{\varkappa}$ is perpendicular to $\mathbf{k}'_0$. The solid angle element can therefore be written

$$do' = d\kappa_x d\kappa_y/k'^2 = d\kappa_x d\kappa_y/k^2, \tag{124.17}$$

where the z-axis is taken in the direction of $\mathbf{k}'_0$. Thus

$$\sigma = \frac{1}{2k^2}\left(\frac{e^2}{mc^2}\right)^2 (1+\cos^2\vartheta)|n_{\mathbf{b}}|^2 \iint d\kappa_x d\kappa_y \left|\int \exp(-i\boldsymbol{\varkappa}\cdot\mathbf{r})dV\right|^2.$$

In the volume integral we can effect the integration with respect to z, since $\exp(-i\boldsymbol{\varkappa}\cdot\mathbf{r})$ is independent of z: $\int \exp(-i\boldsymbol{\varkappa}\cdot\mathbf{r})dV = \int Z\exp(-i\boldsymbol{\varkappa}\cdot\mathbf{r})df$, where $df = dx\,dy$ and $Z = Z(x, y)$ is the length of the body in the direction of $\mathbf{k}'_0$. Finally, using a well-known formula in the theory of Fourier integrals:

$$\frac{1}{(2\pi)^2}\int |\phi_\kappa|^2 d\kappa_x d\kappa_y = \int \phi^2 dxdy, \tag{124.18}$$

where

$$\phi_\kappa = \int \phi(x, y)\exp(-i\boldsymbol{\varkappa}\cdot\mathbf{r})dxdy$$

are the two-dimensional Fourier components, we obtain

$$\sigma = \frac{2\pi^2}{k^2}\left(\frac{e^2}{mc^2}\right)^2 (1+\cos^2\vartheta)|n_{\mathbf{b}}|^2 \int Z^2 \mathrm{d}f$$

$$= \frac{8\pi^2}{b^2}\left(\frac{e^2}{mc^2}\right)^2 \sin^2\tfrac{1}{2}\vartheta(1+\cos^2\vartheta)|n_{\mathbf{b}}|^2 \int Z^2 \mathrm{d}f. \qquad (124.19)$$

The integral is of the order of L^4, where L is the linear dimension of the body. Thus the total diffraction cross-section, and therefore the total intensity of the spot, are proportional to $V^{4/3}$, where V is the volume of the body. The maximum intensity, however, follows a different law. For $\mathbf{k}' - \mathbf{k} = \mathbf{b}$, the integral in (124.13) is just V, and so $d\sigma$ is proportional to V^2:

$$\left(\frac{\mathrm{d}\sigma}{\mathrm{d}o'}\right)_{\max} = \frac{1}{2}\left(\frac{e^2}{mc^2}\right)^2 (1+\cos^2\vartheta)|n_{\mathbf{b}}|^2 V^2. \qquad (124.20)$$

The sharpness of the maximum is shown by the fact that the maximum intensity is proportional to a higher power of V than the total intensity. The width of the peak is evidently proportional to $V^{4/3}/V^2 = V^{-2/3}$.

The theory given above is valid only if the diffraction effect is small. We now see that this requirement imposes a certain condition on the size of the crystal: σ must be small compared with the geometrical cross-section of the body ($\sim L^2$), whence

$$\frac{e^2}{mc^2}\frac{L}{k}|n_{\mathbf{b}}| \ll 1. \qquad (124.21)$$

If this condition is not satisfied, then perturbation theory as used in the derivation of (124.8) is not valid.†

PROBLEMS

PROBLEM 1. Determine the intensity distribution in the diffraction spot round a principal maximum in diffraction by a crystal in the form of a cuboid with sides L_x, L_y, L_z.

SOLUTION. As above, we use the vector $\boldsymbol{\varkappa} = \mathbf{k}' - \mathbf{k}'_0$, and take the axes of coordinates parallel to the sides of the cuboid, with the origin at its centre.

The integral $\int \exp(-i\boldsymbol{\varkappa}\cdot\mathbf{r})\mathrm{d}V$ becomes a product of three integrals of the form

$$\int_{-\frac{1}{2}L}^{\frac{1}{2}L} \exp(-i\kappa x)\,\mathrm{d}x = \frac{2}{\kappa}\sin\tfrac{1}{2}\kappa L.$$

Thus

$$\mathrm{d}\sigma = 32\left(\frac{e^2}{mc^2}\right)^2 (1+\cos^2\vartheta)|n_{\mathbf{b}}|^2 \frac{1}{\kappa_x^2\kappa_y^2\kappa_z^2}\sin^2\tfrac{1}{2}\kappa_x L_x \sin^2\tfrac{1}{2}\kappa_y L_y \sin^2\tfrac{1}{2}\kappa_z L_z\,\mathrm{d}o'.$$

The components of the vector $\boldsymbol{\varkappa}$ are not independent, being related by the condition $\boldsymbol{\varkappa}\cdot\mathbf{k}'_0 = 0$.

PROBLEM 2. The same as Problem 1, but for diffraction by a spherical crystal with radius a.

† The *dynamical theory* of scattering, which is not subject to the restriction (124.21), is given by Z. G. Pinsker, *Dynamical Scattering of X-Rays in Crystals*, Springer, Berlin, 1978. The theory given above is called the *kinematic theory*.

SOLUTION. We again put $\varkappa = \mathbf{k}' - \mathbf{k}'_0$, and take the z-axis in the direction of $\varkappa$, with the origin at the centre of the sphere. Then

$$\int \exp(-i\kappa z)\,\mathrm{d}V = \int_{-a}^{a} \pi(a^2 - z^2)\exp(-i\kappa z)\,\mathrm{d}z$$

$$= \frac{4\pi}{\kappa^3}(\sin\kappa a - \kappa a\cos\kappa a).$$

Thus

$$\mathrm{d}\sigma = 8\pi^2\left(\frac{e^2}{mc^2}\right)^2(1+\cos^2\vartheta)|n_{\mathbf{b}}|^2\frac{1}{\kappa^6}(\sin\kappa a - \kappa a\cos\kappa a)^2\,\mathrm{d}o'.$$

PROBLEM 3. Determine the total intensity of the diffraction spot round a subsidiary maximum.

SOLUTION. In this case the wave vector $\mathbf{k}$ of the incident wave does not satisfy the condition (124.15). As shown above, (124.15) is the equation of a plane perpendicular to the vector $\mathbf{b}$. Let the small displacement of the end-point of the vector $\mathbf{k}$ from this plane be $\eta\mathbf{b}$, where $\eta \ll 1$. That is, we put $\mathbf{k} = \mathbf{k}_0 + \eta\mathbf{b}$, where $\mathbf{k}_0$ satisfies equation (124.15) (Fig. 65).

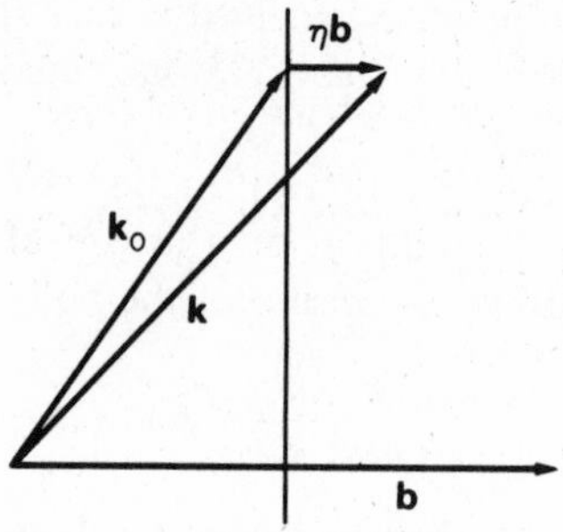

FIG. 65

The maximum intensity in the spot occurs for a direction of $\mathbf{k}'$ for which the difference $\mathbf{k}' - (\mathbf{k}+\mathbf{b})$ has its least magnitude (so that the integral in (124.13) has its maximum value). The magnitude of the difference of two vectors, one of which is arbitrary in direction, has its least value when they are parallel. Hence, since $k' = k$, we have

$$|\mathbf{k}' - \mathbf{k} - \mathbf{b}|_{\min} = k - |\mathbf{k}+\mathbf{b}|$$

$$= \frac{k^2 - (\mathbf{k}+\mathbf{b})^2}{k + |\mathbf{k}+\mathbf{b}|}.$$

Since $\mathbf{k}$ is close to $\mathbf{k}_0$ and we are considering the region near the maximum, $\mathbf{k}' \cong \mathbf{k}+\mathbf{b}$ and the denominator can be replaced by $2k$. In the numerator, we expand the squared parenthesis and obtain

$$-2\mathbf{k}\cdot\mathbf{b} - \mathbf{b}^2 = [-2\mathbf{k}_0\cdot\mathbf{b} - \mathbf{b}^2] - 2\eta\mathbf{b}^2 = -2\eta b^2.$$

Thus $|\mathbf{k}' - \mathbf{k} - \mathbf{b}|_{\min} \cong -\eta b^2/k$.
Next, we put

$$\mathbf{k}' = (\mathbf{k}+\mathbf{b})\left(1 - \frac{\eta b^2}{k^2}\right) + \varkappa,$$

and take the z-axis in the direction of $\mathbf{k}+\mathbf{b}$. This reduces the problem to the calculation of the integral (cf. the derivation of formula (124.19))

$$\int\int \mathrm{d}\kappa_x \mathrm{d}\kappa_y |\int \exp\{i\eta b^2 z/k - i\varkappa\cdot\mathbf{r}\}\,\mathrm{d}V|^2$$

$$= \iint \mathrm{d}\kappa_x\,\mathrm{d}\kappa_y\left|\int \exp(-i\varkappa\cdot\mathbf{r})\frac{\sin(\frac{1}{2}\eta b^2 Z/k)}{\frac{1}{2}\eta b^2/k}\,\mathrm{d}f\right|^2.$$

Finally, using formula (124.18), we obtain

$$\sigma = \frac{2\pi^2}{k^2}\left(\frac{e^2}{mc^2}\right)^2 (1+\cos^2\vartheta)\,|n_{\mathbf{b}}|^2 \int \frac{\sin^2(\frac{1}{2}\eta b^2 Z/k)}{(\frac{1}{2}\eta b^2/k)^2}\,\mathrm{d}f.$$

As $\eta \to 0$ this formula becomes (124.19). If $\frac{1}{2}\eta b^2 Z/k \gg 1$ (which is compatible with $\eta \ll 1$), the squared sine can be replaced by its mean value $\frac{1}{2}$, and we have

$$\sigma = 4\pi^2\left(\frac{e^2}{mc^2}\right)^2 \frac{1+\cos^2\vartheta}{\eta^2 b^4}|n_{\mathbf{b}}|^2 S,$$

where S is the area of the "shadow", i.e. the projection of the body on the xy-plane.

§125. The integral intensity

The formulae derived in §124 give the diffracted intensity when a strictly plane monochromatic wave is incident on a crystal. Let us now consider some cases where these conditions are not fulfilled.

First, let the incident wave be plane but not monochromatic,† its spectral resolution including waves with wave vectors $\mathbf{k}$ whose directions are the same but whose magnitudes $k = \omega/c$ are not. Let $\rho(k)$ be the frequency distribution of the incident radiation intensity, normalized by the condition $\int \rho(k)\,\mathrm{d}k = 1$.

The total intensity of the diffraction spot is determined by the cross-section, which is obtained by multiplying the expression (124.13) by $\rho(k)$ and integrating with respect to o' and k:

$$\sigma = \frac{1}{2}\left(\frac{e^2}{mc^2}\right)^2 |n_{\mathbf{b}}|^2 \int\int \left|\int \exp\left[-i(\mathbf{k}'-\mathbf{k}-\mathbf{b})\cdot\mathbf{r}\right]\mathrm{d}V\right|^2 (1+\cos^2\vartheta)\rho(k)\,\mathrm{d}o'\,\mathrm{d}k. \qquad (125.1)$$

We put temporarily $\mathbf{K} = \mathbf{k}' - \mathbf{k} - \mathbf{b}$ and write the squared modulus as a double integral:

$$\left|\int \exp(-i\mathbf{K}\cdot\mathbf{r})\,\mathrm{d}V\right|^2 = \int\int \exp\left[i\mathbf{K}\cdot(\mathbf{r}_2-\mathbf{r}_1)\right]\mathrm{d}V_1\,\mathrm{d}V_2.$$

Using instead of $\mathbf{r}_1$ and $\mathbf{r}_2$ the variables $\frac{1}{2}(\mathbf{r}_1+\mathbf{r}_2)$ and $\mathbf{r} = \mathbf{r}_2 - \mathbf{r}_1$ and integrating with respect to the first gives $\left|\int \exp(i\mathbf{K}\cdot\mathbf{r})\,\mathrm{d}V\right|^2 = V\int \exp(i\mathbf{K}\cdot\mathbf{r})\,\mathrm{d}V$. In the remaining integral we can effect the integration over all space,‡ and the result is

$$\left|\int \exp(i\mathbf{K}\cdot\mathbf{r})\,\mathrm{d}V\right|^2 = (2\pi)^3 V\delta(\mathbf{K}). \qquad (125.2)$$

Substituting this result in (125.1), we obtain

$$\sigma = 4\pi^3\left(\frac{e^2}{mc^2}\right)^2 |n_{\mathbf{b}}|^2 V(1+\cos^2\vartheta_0)\int\int \delta(\mathbf{k}'-\mathbf{k}-\mathbf{b})\rho(k)\,\mathrm{d}o'\,\mathrm{d}k; \qquad (125.3)$$

on account of the presence of the delta function, the factor $1+\cos^2\vartheta$ in the integrand can be replaced by its value at $\vartheta = \vartheta_0$, where ϑ_0 is the angle between the $\mathbf{k}$ and $\mathbf{k}'$ which satisfy Laue's conditions (denoted by $\mathbf{k}_0$ and $\mathbf{k}'_0 = \mathbf{k}_0 + \mathbf{b}$).

The integration with respect to o' can be carried out by noticing that it is equivalent to an integration with respect to

$$\mathrm{d}^3k' = k'^2\,\mathrm{d}k'\,\mathrm{d}o' = \tfrac{1}{2}k'\,\mathrm{d}(k'^2)\,\mathrm{d}o',$$

if an additional factor $(2/k)\delta(k'^2 - k^2)$ is included in the integrand. Thus the integral in

† Corresponding to *Laue's method* in the X-ray analysis of crystals.

‡ This is possible because we require only the total intensity of the diffraction spot, and not its width.

(125.3) becomes

$$\iint \frac{2}{k}\delta(\mathbf{k}'-\mathbf{k}-\mathbf{b})\delta(k'^2-k^2)\rho(k)\,\mathrm{d}^3k'\,\mathrm{d}k.$$

Effecting the integration with respect to d^3k' by means of the first delta function, we can replace k'^2 by $(\mathbf{k}+\mathbf{b})^2$ in the second delta function, and the result is

$$\sigma = (2\pi)^3\left(\frac{e^2}{mc^2}\right)^2 |n_{\mathbf{b}}|^2 V(1+\cos^2\vartheta_0)\int \frac{1}{k}\delta(2\mathbf{b}\cdot\mathbf{k}+b^2)\rho(k)\mathrm{d}k. \qquad (125.4)$$

Finally, we have to carry out the integration over k (the direction $\mathbf{n} = \mathbf{k}/k$ being given). The argument of the delta function is zero for $k = k_0$, and the integral is $\frac{1}{2}\rho(k_0)/k_0|\mathbf{b}\cdot\mathbf{n}| = \frac{1}{2}\rho(k_0)/|\mathbf{b}\cdot\mathbf{k}_0| = \rho(k_0)/b^2$. Thus

$$\sigma = (2\pi)^3\left(\frac{e^2}{mc^2}\right)^2 |n_{\mathbf{b}}|^2 V(1+\cos^2\vartheta_0)\rho(k_0)/b^2. \qquad (125.5)$$

Let us now consider another case, where the incident wave is monochromatic but its components have varying directions of $\mathbf{k}$ which differ by rotation about some axis;† let $\mathbf{1}$ be a unit vector along that axis, and ψ the angle of rotation about it. Let $\rho(\psi)$ be the angular distribution of the incident radiation intensity, normalized by the condition

$$\int_0^{2\pi} \rho(\psi)\mathrm{d}\psi = 1.$$

The calculations leading to formula (125.4) are valid in this case also, except that the integration with $\rho(k)\,\mathrm{d}k$ must be replaced by one with $\rho(\psi)\,\mathrm{d}\psi$:

$$\sigma = (2\pi)^3\left(\frac{e^2}{mc^2}\right)^2 |n_{\mathbf{b}}|^2 V(1+\cos^2\vartheta_0)\int \frac{1}{k}\delta(2\mathbf{b}\cdot\mathbf{k}+b^2)\rho(\psi)\mathrm{d}\psi. \qquad (125.6)$$

We again denote by $\mathbf{k}_0$ the value of $\mathbf{k}$ for which the argument of the delta function is zero, and measure ψ from the plane of $\mathbf{1}$ and $\mathbf{k}_0$. For small ψ, $\mathbf{k} = \mathbf{k}_0 + (\mathbf{1}\times\mathbf{k}_0)\psi$. Then the integral in (125.6) becomes

$$\begin{aligned}\int \frac{1}{k}\delta[2\mathbf{b}\cdot(\mathbf{1}\times\mathbf{k}_0)\psi]\rho(\psi)\mathrm{d}\psi &= \tfrac{1}{2}\rho(0)/k|\mathbf{b}\cdot(\mathbf{1}\times\mathbf{k}_0)] \\ &= \tfrac{1}{2}\rho(0)/k^2|\mathbf{b}\cdot(\mathbf{1}\times\mathbf{n}_0)| \\ &= 2\rho(0)\sin^2(\tfrac{1}{2}\vartheta_0)/b^2|\mathbf{b}\cdot(\mathbf{1}\times\mathbf{n}_0)|.\end{aligned}$$

Thus

$$\sigma = \frac{2(2\pi)^3}{b^2}\left(\frac{e^2}{mc^2}\right)^2 \sin^2\tfrac{1}{2}\vartheta_0(1+\cos^2\vartheta_0)|n_{\mathbf{b}}|^2 V\frac{\rho(0)}{|\mathbf{b}\cdot(\mathbf{1}\times\mathbf{n}_0)|}. \qquad (125.7)$$

Finally, let us consider the diffraction of a plane monochromatic wave from a body consisting of crystallites arranged at random.‡

† Corresponding to *Bragg's method* (the *rotation method*) in X-ray analysis. The rotation referred to is that of the crystal about $\mathbf{1}$, not that of the direction of $\mathbf{k}$.

‡ Corresponding to *Debye and Scherrer's method* (the *powder method*) in X-ray analysis.

Let $\mathbf{k}'_0$ and $\mathbf{b}_0$ be values of $\mathbf{k}'$ and $\mathbf{b}$ such that Laue's condition $\mathbf{k}'_0 = \mathbf{k}+\mathbf{b}_0$ is satisfied. The directions of $\mathbf{k}'_0$ and $\mathbf{b}_0$ are not uniquely determined, since Laue's condition is, of course, still fulfilled when the triangle $\mathbf{k}$, $\mathbf{b}_0$, $\mathbf{k}'_0$ is rotated about the direction of $\mathbf{k}$. Thus the principal maximum corresponds to directions of $\mathbf{k}'$ occupying a conical surface with vertical angle $2\vartheta_0$. Instead of a diffraction spot we now have a ring.

The required total cross-section is determined by a formula which differs from (125.4) only in that the integration with $\rho(k)\,dk$ is replaced by an averaging over the directions of $\mathbf{b}$:

$$\sigma = (2\pi)^3 V\left(\frac{e^2}{mc^2}\right)^2 |n_\mathbf{b}|^2(1+\cos^2\vartheta_0)\int\frac{1}{k}\delta(2\mathbf{b}\cdot\mathbf{k}+b^2)\frac{do_\mathbf{b}}{4\pi}, \tag{125.8}$$

where $do_\mathbf{b}$ is an element of solid angle about the direction of $\mathbf{b}$. Denoting by α the angle between $\mathbf{k}$ and $\mathbf{b}$, we can write the integral in (125.8) as

$$\int\frac{1}{k}\delta(2bk\cos\alpha+b^2)\frac{2\pi d\cos\alpha}{4\pi} = \frac{1}{4bk^2} = \frac{1}{b^3}\sin^2\tfrac{1}{2}\vartheta_0.$$

Thus

$$\sigma = (2\pi)^3\left(\frac{e^2}{mc^2}\right)^2 |n_\mathbf{b}|^2\frac{V}{b^3}(1+\cos^2\vartheta_0)\sin^2\tfrac{1}{2}\vartheta_0. \tag{125.9}$$

Each of the three cases considered in this section corresponds to a particular method of averaging the diffraction pattern. The dependence of the total averaged diffraction intensity on the volume of the body reduces, as we should expect, to a simple proportionality. In the pattern which is not averaged, the intensity and its distribution over the spot depend more markedly on the volume.

§126. Diffuse thermal scattering of X-rays

In §§124 and 125 we have taken $n(x, y, z)$ to be the time-averaged electron density in the crystal: various density oscillations were thereby excluded, and consequently so was the corresponding (non-coherent) scattering of X-rays. One cause of non-coherent scattering is the thermal fluctuations of density. This scattering is diffusely distributed in all directions, but it is characterized by a relatively high intensity near directions corresponding to the sharp lines of the structural scattering described in the preceding sections. Here we shall discuss these maxima of the thermal scattering (W. H. Zachariasen, 1940).

The thermal oscillations of the crystal lattice can be represented as combinations of sound waves. As we shall see, the maxima of the thermal scattering arise from wavelengths large compared with the lattice constant. The change in the electron density due to such a wave can be regarded, at any point, as due to a simple displacement of the lattice by an amount equal to the local value of the displacement vector $\mathbf{u}$ in the wave. Thus the change in the density (*not* averaged with respect to time) when a given sound wave passes can be expressed in terms of the mean density by

$$\delta n = n(\mathbf{r}-\mathbf{u})-n(\mathbf{r}) \cong -\mathbf{u}\cdot\partial n/\partial\mathbf{r}.$$

In considering diffuse scattering near a given line, we must replace n by $n_\mathbf{b}\exp(i\mathbf{b}\cdot\mathbf{r})$ with the appropriate $\mathbf{b}$, so that

$$\delta n = -i\mathbf{b}\cdot\mathbf{u}n_\mathbf{b}\exp(i\mathbf{b}\cdot\mathbf{r}). \tag{126.1}$$

The scattering by density fluctuations is, of course, not coherent with that by the mean density, and the two therefore do not interfere. Hence the cross-section for diffuse scattering can be obtained from (124.10), substituting δn for n and then carrying out the statistical averaging over fluctuations:

$$d\sigma = \frac{1}{2}\left(\frac{e^2}{mc^2}\right)^2 |n_{\mathbf{b}}|^2(1+\cos^2\vartheta)\langle|\int \mathbf{b}\cdot\mathbf{u}\exp(-i\mathbf{K}\cdot\mathbf{r})dV|^2\rangle do', \qquad (126.2)$$

where $\mathbf{K} = \mathbf{k}' - \mathbf{k} - \mathbf{b}$. The scattered intensity is large for directions where $K \ll b$.

The integral $\int \mathbf{u}\exp(-i\mathbf{K}\cdot\mathbf{r})\,dV$ gives the Fourier space component of $\mathbf{u}$ whose wave vector is $\mathbf{K}$, and we can therefore take $\mathbf{u}$ to be simply the displacement vector in a sound wave having this wave vector. The inequality $K \ll b$ thus implies that the wavelength of the scattering sound wave is large compared with the dimension of the crystal lattice cell. Thus we can put

$$\mathbf{u} = \tfrac{1}{2}[\mathbf{u}_0\exp(i\mathbf{K}\cdot\mathbf{r}) + \mathbf{u}_0{}^*\exp(-i\mathbf{K}\cdot\mathbf{r})], \qquad (126.3)$$

so that $\int(\mathbf{b}\cdot\mathbf{u})\exp(-i\mathbf{K}\cdot\mathbf{r})\,dV = \tfrac{1}{2}V\mathbf{b}\cdot\mathbf{u}_0$ and the cross-section is

$$d\sigma = \frac{1}{8}\left(\frac{e^2}{mc^2}\right)^2 |n_{\mathbf{b}}|^2(1+\cos^2\vartheta)b_i b_k\langle u_{0i}u_{0k}\rangle V^2\, do'. \qquad (126.4)$$

The products of the components of $\mathbf{u}_0$ are averaged as in §123 for a sound wave in an isotropic medium. The elastic energy per unit volume of a deformed crystal is $\tfrac{1}{2}\lambda_{iklm}u_{ik}u_{lm}$, where u_{ik} is the strain tensor and λ_{iklm} the elastic modulus tensor (see *TE*, §10). Hence the mean elastic energy of the whole crystal is $\tfrac{1}{2}V\lambda_{iklm}\langle u_{ik}u_{lm}\rangle$. We substitute

$$u_{ik} = \frac{1}{2}\left(\frac{\partial u_i}{\partial x_k} + \frac{\partial u_k}{\partial x_i}\right)$$
$$= \tfrac{1}{2}\mathrm{re}\{(iK_k u_{0i} + iK_i u_{0k})\exp(i\mathbf{K}\cdot\mathbf{r})\}.$$

The terms containing $\exp(\pm 2i\mathbf{K}\cdot\mathbf{r})$ give zero on averaging. Using also the symmetry of the tensor λ_{iklm} with respect to interchange of i, k, or l, m, or i, k and l, m, we obtain $\tfrac{1}{4}V\lambda_{iklm}K_k K_m\langle u_{0i}u_{0l}{}^*\rangle$ or $\tfrac{1}{4}Vg_{ik}\langle u_{0i}u_{0k}{}^*\rangle$, where

$$g_{ik} = \lambda_{ilkm}K_l K_m. \qquad (126.5)$$

According to the general theory of thermodynamic fluctuations, we can at once write down the required mean values:†

$$\langle u_{0i}u_{0k}{}^*\rangle = (4T/V)g^{-1}{}_{ik}, \qquad (126.6)$$

where $g^{-1}{}_{ik}$ is the tensor inverse to g_{ik}, and the scattering cross-section is

$$d\sigma = \frac{1}{2}\left(\frac{e^2}{mc^2}\right)^2 TV|n_{\mathbf{b}}|^2(1+\cos^2\vartheta)b_i b_k g^{-1}{}_{ik}\, do'. \qquad (126.7)$$

Thus the diffusely scattered intensity is, as we should expect, proportional to the volume of the crystal. A characteristic feature of this scattering is the way in which its intensity is distributed over the area of the spot. Apart from the factor $1+\cos^2\vartheta$, which is almost

† See *SP* 1, §111. If the probability distribution for fluctuating quantities $x_1, x_2, \ldots$ is of the form $\exp(-\tfrac{1}{2}\lambda_{ik}x_i x_k)$, then $\langle x_i x_k\rangle = \lambda^{-1}{}_{ik}$. A factor 2 in (126.6) appears because each of the complex u_{0i} involves two independent quantities.

constant for a given spot, the intensity is given by the expression $g^{-1}{}_{ik}b_ib_k$. This expression is the product of $1/K^2$ and a fairly involved function of the direction of the vector **K** with respect to the crystal axes. For scattering near a principal maximum the diffusely scattered intensity is itself a maximum where $\mathbf{K} = 0$ (the expression (126.7) becomes infinite for $\mathbf{K} = 0$ and is, of course, invalid). If the condition (124.15) $2\mathbf{b}\cdot\mathbf{k} = -b^2$ is not satisfied, however, **K** cannot be zero, and the maximum of the diffusely scattered intensity lies at some **K** different from zero, which in general does not coincide with the maximum of the structural scattering. In either case the diffuse scattering forms a background whose intensity falls off essentially as $1/K^2$, that is, considerably more slowly than the intensity in the sharp structural-scattering line superposed upon it.

§127. The temperature dependence of the diffraction cross-section

Let us now calculate the temperature dependence of the cross-section for coherent scattering of X-rays. This amounts to determining the temperature dependence of the microscopic electron density in the crystal averaged with allowance for the thermal motion of the atoms.

We shall assume the atoms to be so heavy that the majority of their electrons are localized in non-overlapping shells which are only slightly deformed by the lattice vibrations. We shall also assume that the lattice consists of atoms of only one kind, with one atom in each unit cell; it must be emphasized that this assumption has no fundamental significance, and is made only to simplify the writing of the formulae.

The exact (not averaged) microscopic electron density may then be expressed as

$$n(\mathbf{r}) = \sum_{\mathbf{n}} F(\mathbf{r}-\mathbf{r}_{\mathbf{n}}) = \sum_{\mathbf{n}} F(\mathbf{r}-\mathbf{r}_{\mathbf{n}0}-\mathbf{u}_{\mathbf{n}}), \tag{127.1}$$

where $F(\mathbf{r})$ is the electron density in a single atom (the *atomic form factor*); the summation is taken over all the atoms in the lattice, and $\mathbf{r}_{\mathbf{n}}$ are the position vectors of the nuclei, labelled by the vector suffix **n** (with integral components). Denoting the equilibrium position vectors of the nuclei (at the lattice sites) by $\mathbf{r}_{\mathbf{n}0}$, and the displacement vectors of the atoms from these positions by $\mathbf{u}_{\mathbf{n}}$, we have $\mathbf{r}_{\mathbf{n}} = \mathbf{r}_{\mathbf{n}0} + \mathbf{u}_{\mathbf{n}}$, as used in the second equation (127.1).

Expanding the density (127.1) as a Fourier series (124.12) in the volume V of the lattice, we can write the expansion coefficients as

$$n_{\mathbf{b}} = \frac{1}{V}\sum_{\mathbf{n}} \exp[-i(\mathbf{r}_{\mathbf{n}0}+\mathbf{u}_{\mathbf{n}})\cdot\mathbf{b}]F_{\mathbf{b}},$$

where

$$F_{\mathbf{b}} = \int F(\mathbf{r})\exp(-i\mathbf{b}\cdot\mathbf{r})\,dV \tag{127.2}$$

are the Fourier components of the atomic form factor. All the products $\mathbf{r}_{\mathbf{n}0}\cdot\mathbf{b}$ are integral multiples of 2π. Hence all the factors $\exp(-i\mathbf{r}_{\mathbf{n}0}\cdot\mathbf{b}) = 1$, and so

$$n_{\mathbf{b}} = (F_{\mathbf{b}}/V)\sum_{\mathbf{n}} \exp(-i\mathbf{u}_{\mathbf{n}}\cdot\mathbf{b}). \tag{127.3}$$

This expression can now be averaged over the motion of the atoms. It is evident that the mean values of the terms in the sum do not depend on **n**. Hence

$$\langle n_{\mathbf{b}}\rangle = (F_{\mathbf{b}}/v)\langle\exp(-i\mathbf{b}\cdot\mathbf{u})\rangle, \tag{127.4}$$

where $\mathbf{u}$ is the displacement vector of any atom, and $v = V/N$ is the unit cell volume (N being the number of cells in the volume V). The averaging in (127.4) is to be taken as a complete statistical averaging, i.e. over the wave functions of the stationary states and then over the Gibbs distribution.

To carry out this averaging, we have to regard $\mathbf{u}$ as a quantum-mechanical operator

$$\hat{\mathbf{u}} = \sum_{\mathbf{k},\alpha}\left(\frac{\hbar}{2MN\omega_\alpha(\mathbf{k})}\right)^{1/2}\{\hat{c}_{\mathbf{k}\alpha}\mathbf{e}_{\mathbf{k}\alpha}\exp(i\mathbf{k}\cdot\mathbf{r}_\mathbf{n})+\hat{c}_{\mathbf{k}\alpha}{}^{+}\mathbf{e}_{\mathbf{k}\alpha}{}^{*}\exp(-i\mathbf{k}\cdot\mathbf{r}_\mathbf{n})\}; \tag{127.5}$$

see *SP* 1, §72. The summation is taken over all values of the phonon wave vector $\mathbf{k}$ (in the volume V) and over the independent phonon polarizations labelled by $\alpha = 1, 2, 3$; $\omega_\alpha(\mathbf{k})$ are the phonon frequencies, $\mathbf{e}_{\mathbf{k}\alpha}$ the phonon polarization vectors, and M the atomic mass; $\hat{c}_{\mathbf{k}\alpha}$ and $\hat{c}_{\mathbf{k}\alpha}{}^{+}$ are the annihilation and creation operators for phonons in the states $\mathbf{k}$, α.

For operators having the form (127.5), Wick's theorem holds: the average product of any even number of operators is equal to the sum of all possible products of pair means (the average product of an odd number of operators is zero).† An essential result here is

$$\langle\exp\hat{L}\rangle = \exp(\tfrac{1}{2}\langle\hat{L}^2\rangle), \tag{127.6}$$

which is valid for any operator $\hat{L}$ satisfying Wick's theorem; its correctness is easily seen by expanding $\exp\hat{L}$ in series and averaging each term in the expansion.‡

Applying (127.6) to (127.4), we obtain

$$\langle n_\mathbf{b}\rangle = (F_\mathbf{b}/v)\exp\{-\tfrac{1}{2}\langle(\mathbf{b}\cdot\mathbf{u})^2\rangle\}.$$

The diffraction cross-section is proportional to the square of this quantity, and its temperature dependence is therefore given by the separate factor

$$D = \exp\{-\langle(\mathbf{b}\cdot\mathbf{u})^2\rangle\}, \tag{127.7}$$

called the *Debye–Waller factor* (P. Debye, 1912; I. Waller, 1925).

It remains to calculate the mean square $\langle(\mathbf{b}\cdot\mathbf{u})^2\rangle$. Of all the products of pairs of operators $\hat{c}_{\mathbf{k}\alpha}$, $\hat{c}_{\mathbf{k}\alpha}{}^{+}$, the only non-zero mean values are

$$\langle\hat{c}_{\mathbf{k}\alpha}{}^{+}\hat{c}_{\mathbf{k}\alpha}\rangle = N_{\mathbf{k}\alpha},\quad \langle\hat{c}_{\mathbf{k}\alpha}\hat{c}_{\mathbf{k}\alpha}{}^{+}\rangle = N_{\mathbf{k}\alpha}+1,$$

where $N_{\mathbf{k}\alpha}$ are the mean phonon state occupation numbers in equilibrium. Hence

$$\langle(\mathbf{b}\cdot\mathbf{u})^2\rangle = \sum_{\mathbf{k},\alpha}\frac{\hbar}{2MN\omega_\alpha(\mathbf{k})}|\mathbf{e}_{\mathbf{k}\alpha}\cdot\mathbf{b}|^2\langle\hat{c}_{\mathbf{k}\alpha}\hat{c}_{\mathbf{k}\alpha}{}^{+}+\hat{c}_{\mathbf{k}\alpha}{}^{+}\hat{c}_{\mathbf{k}\alpha}\rangle$$

$$= \sum_{\mathbf{k},\alpha}\frac{\hbar}{MN\omega_\alpha(\mathbf{k})}|\mathbf{b}\cdot\mathbf{e}_{\mathbf{k}\alpha}|^2(N_{\mathbf{k}\alpha}+\tfrac{1}{2}).$$

The numbers $N_{\mathbf{k}\alpha}$ are given by the Bose distribution

$$N_{\mathbf{k}\alpha} = [\exp\{h\omega_\alpha(\mathbf{k})/T\}-1]^{-1}. \tag{127.8}$$

---

† The theorem is proved in *SP* 2, §13, for the "macroscopic limit" ($N\to\infty$), corresponding to the application in statistical physics.

‡ The product of an even number $2n$ of factors $\hat{L}$ can be resolved into pairs in $(2n-1)(2n-3)\ldots 1$ ways (having chosen one of the $2n$ factors, we can "pair" it with any of the other $2n-1$ factors; then choosing one of the remaining $2n-2$ operators, we can pair it in $2n-3$ ways and so on). Hence

$$\langle\hat{L}^{2n}\rangle = [(2n-1)(2n-3)\ldots 1]\langle\hat{L}^2\rangle^n.$$

On substitution in (127.7), the term involving the zero-point vibrations gives a factor independent of temperature, which may be omitted (or rather, included in the definition of $F_{\mathbf{b}}$). Finally, changing from summation over $\mathbf{k}$ to integration over $V d^3k/(2\pi)^3$, we have

$$D = \exp\left\{ -\frac{\hbar v}{M} \sum_{\alpha} \int \frac{|\mathbf{b}\cdot\mathbf{e}_{\mathbf{k}\alpha}|^2}{\omega_\alpha(\mathbf{k})} N_{\mathbf{k}\alpha} \frac{d^3k}{(2\pi)^3} \right\}. \tag{127.9}$$

As $T \to 0$, $N_{\mathbf{k}\alpha}$ tends to zero, and accordingly D tends to unity; as the temperature increases, D becomes smaller. The effect of the temperature amounts to a general lowering of the scattered line intensity, with no change in the shape or width of the line.

The mean thermal displacement of the atoms from the lattice sites is usually small, even at high temperatures, in comparison with the lattice constant. The exponent in D, for scattering lines with only slight changes in $\mathbf{b}$, is then much less than unity, and the temperature effect is a small correction. The decrease in intensity becomes considerable, however, for scattering lines which correspond to large $\mathbf{b}$.

For temperatures much higher than the Debye temperature, $\langle \mathbf{u}^2 \rangle \propto T$, and the exponent in D is also proportional to T. At low temperatures, the thermal phonons belong mainly to the acoustic branches of the spectrum, for which $\omega \propto k$; in this case, the integration over ω in (127.9) can be extended to infinity, and the exponent is then proportional to T^2.

At low temperatures, when only the long-wavelength phonons are important, the validity of formulae (127.7)–(127.9) does not depend on the possibility of expressing the density $n(\mathbf{r})$ as the sum over atoms (127.1). In long-wavelength vibrations, large regions of the lattice move as a whole with their respective electron density values, and only on this does the derivation of the formulae in question depend.

APPENDIX

CURVILINEAR COORDINATES

WE give below, for reference, certain formulae relating to vector operations in curvilinear coordinates, both general and particular.

In an arbitrary system of orthogonal curvilinear coordinates u_1, u_2, u_3, the squared element of length is $\mathrm{d}l^2 = h_1{}^2\mathrm{d}u_1{}^2 + h_2{}^2\mathrm{d}u_2{}^2 + h_3{}^2\mathrm{d}u_3{}^2$, where the h_i are functions of the coordinates. The element of volume is

$$\mathrm{d}V = h_1 h_2 h_3 \,\mathrm{d}u_1\,\mathrm{d}u_2\,\mathrm{d}u_3.$$

The various vector operations can be expressed in terms of the functions h_i as follows. For vector operations on a scalar:

$$(\mathbf{grad}\, f)_i = \frac{1}{h_i}\frac{\partial f}{\partial u_i},$$

$$\triangle f = \frac{1}{h_1 h_2 h_3}\sum \frac{\partial}{\partial u_1}\left(\frac{h_2 h_3}{h_1}\frac{\partial f}{\partial u_1}\right),$$

where the summation is over cyclic interchanges of the suffixes 1, 2, 3. For vector operations on a vector:

$$\operatorname{div} \mathbf{A} = \frac{1}{h_1 h_2 h_3}\sum \frac{\partial}{\partial u_1}(h_2 h_3 A_1),$$

$$(\mathbf{curl}\, \mathbf{A})_1 = \frac{1}{h_2 h_3}\left[\frac{\partial}{\partial u_2}(h_3 A_3) - \frac{\partial}{\partial u_3}(h_2 A_2)\right].$$

The remaining components of **curl A** are obtained by cyclic interchanges of the suffixes.

Cylindrical coordinates r, ϕ, z.

Element of length: $\mathrm{d}l^2 = \mathrm{d}r^2 + r^2\,\mathrm{d}\phi^2 + \mathrm{d}z^2$;

$$h_r = 1, \qquad h_\phi = r, \qquad h_z = 1.$$

Vector operations:

$$\triangle f = \frac{1}{r}\frac{\partial}{\partial r}\left(r\frac{\partial f}{\partial r}\right) + \frac{1}{r^2}\frac{\partial^2 f}{\partial \phi^2} + \frac{\partial^2 f}{\partial z^2},$$

$$\operatorname{div} \mathbf{A} = \frac{1}{r}\frac{\partial}{\partial r}(r A_r) + \frac{1}{r}\frac{\partial A_\phi}{\partial \phi} + \frac{\partial A_z}{\partial z},$$

$$(\mathbf{curl}\, \mathbf{A})_r = \frac{1}{r}\frac{\partial A_z}{\partial \phi} - \frac{\partial A_\phi}{\partial z},$$

$$(\textbf{curl A})_\phi = \frac{\partial A_r}{\partial z} - \frac{\partial A_z}{\partial r},$$

$$(\textbf{curl A})_z = \frac{1}{r}\frac{\partial}{\partial r}(r A_\phi) - \frac{1}{r}\frac{\partial A_r}{\partial \phi},$$

$$(\triangle \mathbf{A})_r = \triangle A_r - \frac{A_r}{r^2} - \frac{2}{r^2}\frac{\partial A_\phi}{\partial \phi},$$

$$(\triangle \mathbf{A})_\phi = \triangle A_\phi - \frac{A_\phi}{r^2} + \frac{2}{r^2}\frac{\partial A_r}{\partial \phi},$$

$$(\triangle \mathbf{A})_z = \triangle A_z.$$

In the expressions for the components of $\triangle \mathbf{A}$, $\triangle A_i$ signifies the result of the operator $\triangle$ acting on A_i regarded as a scalar.

Spherical coordinates r, θ, ϕ.

Element of length: $\mathrm{d}l^2 = \mathrm{d}r^2 + r^2\mathrm{d}\theta^2 + r^2 \sin^2\theta\, \mathrm{d}\phi^2$;

$$h_r = 1, \qquad h_\theta = r, \qquad h_\phi = r \sin\theta.$$

Vector operations:

$$\triangle f = \frac{1}{r^2}\frac{\partial}{\partial r}\left(r^2\frac{\partial f}{\partial r}\right) + \frac{1}{r^2 \sin\theta}\frac{\partial}{\partial \theta}\left(\sin\theta\frac{\partial f}{\partial \theta}\right) + \frac{1}{r^2 \sin^2\theta}\frac{\partial^2 f}{\partial \phi^2},$$

$$\operatorname{div}\mathbf{A} = \frac{1}{r^2}\frac{\partial}{\partial r}(r^2 A_r) + \frac{1}{r\sin\theta}\frac{\partial}{\partial\theta}(A_\theta \sin\theta) + \frac{1}{r\sin\theta}\frac{\partial A_\phi}{\partial\phi}$$

$$(\textbf{curl A})_r = \frac{1}{r\sin\theta}\left[\frac{\partial}{\partial\theta}(A_\phi \sin\theta) - \frac{\partial A_\theta}{\partial\phi}\right],$$

$$(\textbf{curl A})_\theta = \frac{1}{r\sin\theta}\frac{\partial A_r}{\partial\phi} - \frac{1}{r}\frac{\partial}{\partial r}(r A_\phi),$$

$$(\textbf{curl A})_\phi = \frac{1}{r}\left[\frac{\partial}{\partial r}(rA_\theta) - \frac{\partial A_r}{\partial\theta}\right],$$

$$(\triangle \mathbf{A})_r = \triangle A_r - \frac{2}{r^2}\left[A_r + \frac{1}{\sin\theta}\frac{\partial}{\partial\theta}(A_\theta \sin\theta) + \frac{1}{\sin\theta}\frac{\partial A_\phi}{\partial\phi}\right],$$

$$(\triangle \mathbf{A})_\theta = \triangle A_\theta + \frac{2}{r^2}\left[\frac{\partial A_r}{\partial\theta} - \frac{A_\theta}{2\sin^2\theta} - \frac{\cos\theta}{\sin^2\theta}\frac{\partial A_\phi}{\partial\phi}\right],$$

$$(\triangle \mathbf{A})_\phi = \triangle A_\phi + \frac{2}{r^2\sin\theta}\left[\frac{\partial A_r}{\partial\phi} + \cot\theta\frac{\partial A_\theta}{\partial\phi} - \frac{A_\phi}{2\sin\theta}\right].$$

INDEX